D0086946

DUXBURY

EXPERIMENTAL DESIGNS USING ANOVA

Barbara G. Tabachnick
California State University, Northridge

Linda S. Fidell
California State University, Northridge

Australia • Brazil • Canada • Mexico • Singapore • Spain
United Kingdom • United States

Experimental Designs Using ANOVA

Barbara Tabachnick and Linda Fiddell

Acquisitions Editor: *Carolyn Crockett*
Assistant Editor: *Ann Day*
Technology Project Manager: *Fiona Chong*
Marketing Manager: *Stacy Best*
Marketing Assistant: *Brian Smith*
Project Manager, Editorial Production: *Belinda Krohmer*
Creative Director: *Rob Hugel*
Art Director: *Lee Friedman*
Print Buyer: *Rebecca Cross*
Permissions Editor: *Kiely Sisk*
Production Service: *Interactive Composition Corporation*
Copy Editor: *Tara Joffe*
Cover Designer: *Irene Morris*
Cover Image: *Getty Images*
Cover Printer: *Phoenix Color Corp*
Compositor: *Interactive Composition Corporation*
Printer: *RR Donnelley Crawfordsville*

Printed in the United States
1 2 3 4 5 6 7 09 08 07 06

Library of Congress Control Number: 2006920045
ISBN 0534405142

Thomson Higher Education
10 Davis Drive
Belmont, CA 94002-3098
USA

Contents

4 One-Way Randomized-Groups Analysis of Variance, Fixed-Effects Designs 98

Preface

Experimental Designs Using ANOVA is a textbook or reference book for people who have a background in basic descriptive and inferential statistics. Although we review basic descriptive statistics in Chapter 2 and the logic of basic ANOVA in Chapter 3, the reviews are slanted toward screening data and understanding the assumptions of analyses. In most universities, this book is likely to be appropriate for senior or first-year graduate study or for those who are taking part in a year-long research design and statistics course. An earlier version of this book was entitled *Computer-Assisted Research Design and Analysis,* emphasizing the use of statistical software both for analyzing data and designing the more complex experiments.

One distinctive feature of this book is the inclusion of the regression approach to ANOVA alongside the traditional approach. As we included more and more of the regression approach in our courses, we developed appreciation for the clarity and flexibility of this approach to ANOVA. The approach easily clarifies issues associated with unequal *n* and interaction contrasts, among others. It also greatly facilitates understanding of fractional factorials and other screening designs, which are the topics of Chapter 10. Although screening designs are not routinely included in a social science curriculum, they should be. They offer a powerful approach to pilot studies, as well as an organized method of research for areas where few cases are available.

With knowledge of how to set up a data set for the regression approach and with the availability of regression software, you will learn to appreciate the full flexibility of ANOVA conducted through regression. For the less complex designs, a traditional ANOVA approach to the technique is first taken. Then, the analysis is repeated through methods associated with the general linear model. For the more complicated designs, the easier of the two approaches is demonstrated. Although we believe the regression approach is helpful, others may not agree; it is described in clearly defined subsections that could be omitted if desired.

Experimental Designs Using ANOVA includes details on how to perform both simple and complicated analyses by hand—through traditional means, through regression, and through the most recent (at the time of writing) versions of SPSS and SAS, with occasional demonstration of SYSTAT when it offers unique solutions to problems of design or analysis. Syntax and output for both basic and complicated analyses are shown, together with interpretive comments about the output. We also show syntax for SYSTAT for many of these analyses.

Chapters 4–7 review basic and complicated analyses for randomized-groups, repeated-measures, and mixed designs.[1] We then include illustrations of analyses for basic and complicated screening, Latin-square, ANCOVA, and nesting and random-effects designs.

Because more material is included than some instructors may want to cover in depth in one semester, instructors should choose the analyses most appropriate for their disciplines. However, because all chapters follow a similar outline, students can continue study on their own or readily find material they need when they encounter an unusual problem later on. We have used this book in manuscript form three times prior to publication. In the first 10 weeks of class, we cover material through Chapter 7 (mixed randomized-groups, repeated-measures designs). Students found the review of factorial designs at the end of Chapter 7 especially helpful. The remaining five weeks are used to cover the basics of ANCOVA, Latin-square, screening, and random-effects designs and to familiarize students with some of the complications of those designs. Thus, the first eight chapters comprise a fast-paced quarter course.

Chapters 4–11 follow the same common format as our more advanced text, *Using Multivariate Statistics*: general purpose and description, kinds of research questions that can be answered by the analysis, and then limitations and major assumptions of the analysis, together with recommendations if there is a violation of an assumption. The next section has an unrealistically small and often silly data set that is analyzed by traditional methods, through regression, and through the two software packages. These examples reflect our hobbies (skiing, belly dancing, and reading—primarily mystery and science fiction) and are meant to be immediately accessible to people from a variety of backgrounds. Except for being perhaps slightly amusing, they are intended to "get out of the way" so you can concentrate on the design and the statistical analysis. Or, if you prefer, you can substitute you own variable names and labels.

The next section of each chapter discusses varieties of the basic design, if there are any, followed by a section that covers issues particular to the analysis. The next section illustrates a complete analysis of a real-life data set (the data sets come from a variety of research areas). Although we are social scientists, we have consulted in

[1] We decided to use the terms that are used more widely in statistics rather than the between-subjects and within-subjects terms used in the social sciences, which are confusing in some settings.

many research areas and have learned the necessity of quickly "getting up to speed" in a new area. The complete examples are eclectic and are described more fully in Appendix B. The data sets for the complete examples are on the accompanying CD. Those who find some of the examples less than compelling are urged to substitute a data set from their own research area and to follow the steps of analysis with it. The last section of each chapter compares several programs available for the analysis. Problem Sets are included at the end of each chapter, and answers to most of them are on the Student Suite CD that is bound in this book.

Once students have completed this course, they should be prepared to take a course in multivariate statistics, with *Using Multivariate Statistics* (5th ed., Allyn and Bacon, 2007), our other book, as text. Both books use the same outline for all technique chapters, making the transition from univariate to multivariate statistics easier.

We are aware of controversy surrounding use of tests of statistical significance instead of focusing on practical significance (effect size). We have chosen to continue the practice of using statistical significance levels to focus interpretation but have also emphasized measures of effect size, as well as confidence limits around effect size, for most analyses. The CD contains free software for effect sizes and confidence limits around them.

The book reflects our practical approach, emphasizing computer assistance in design and analysis of research, in addition to conceptual understanding fostered by presentation and interpretation of fundamental equations.

Students from two classes survived manuscript versions of the first incarnation of this book and made numerous very helpful comments. Much that is right about this book is due to their very careful reading of it.

We'd like to thank all of those anonymous reviewers who helped us sort out our thoughts so that we could try to make this book as clear as possible. John Jamieson of Lakehead University in Canada has provided helpful information about ANOVA for this book (and for our other book); Dave DeYoung of the California Department of Motor Vehicles has, as always, been generous with helpful suggestions. The reviews of the first edition by Dennis Doverspike of the University of Akron and Barry McLaughlin of University of California, Santa Cruz, were especially helpful.

Dianna L. Newman, University at Albany SUNY

Joseph J. Benz, University of Nebraska at Kearney

Dawn M. McBride, Illinois State University

Mei Wang, University of Chicago

Barbara Tabachnick
Linda Fidell

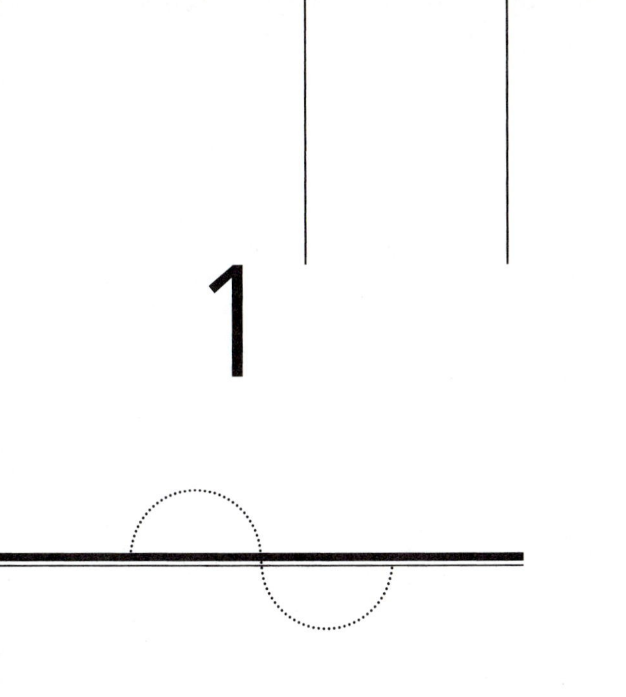

1 Introduction

All behavior varies. This is true of the behavior of people, assembly lines, stock markets, fabrics, equipment, bacteria, and so on. Cases that are treated one way behave differently from those treated another. Cases that are ostensibly treated the same also behave differently from each other and over time. Sometimes differences in behavior are associated with known differences in treatment, and other times the reason for the variability is unknown. Identification of the various factors that contribute to the variability in behavior is the goal of the set of statistical procedures described in this book.

For instance, you might be interested in factors that contribute to success in a statistics course. Such factors might include the textbook used in the course, the professor's teaching style, the student's cognitive style, the number of assignments, the timing of assignments, the clarity of assignments, the number of prior statistics courses, the grades earned in those courses, the class size, the color of the classroom walls, the method of study, the timing of study, the student's age, and so on. Among the many factors that might be effective, some have important effects and others have no effect or minimal effect. The statistical methods described here help you decide which factors have important effects and which are largely irrelevant.

Many different types of analysis of variance (ANOVA) techniques are covered here: one-way and factorial randomized-groups ANOVA, repeated-measures ANOVA, incomplete and partial factorial ANOVA, ANOVA with a covariate, and ANOVA with randomly chosen levels of treatment. Each analysis is appropriate in a different situation, depending on the design of the study. Guidelines for choosing the analysis to match your design are presented at the end of this chapter (see Section 1.5 and Table 1.5).

We assume that you have had one or two courses in descriptive and inferential statistics where you covered central tendency and variance (standard deviation), the logic of the statistical hypothesis test, some fundamental terms in research design, and perhaps the *t* test, simple correlation and regression, and chi-square. Although we revisit these topics briefly in Chapter 2, we recommend reviewing your own earlier text if our review is not sufficient. We also assume some exposure to research design in your own field of study.

1.1 THE NATURE OF RESEARCH

1.1.1 IVs and DVs

The elements of a research design are often identified as independent variables (IVs) and dependent variables (DVs). The IVs are the factors that may or may not be effective in changing scores (values) on the DV. In an experimental design, the researcher manipulates the IVs, whereas in a nonexperimental design, the researcher chooses the levels of the IVs for study. The DV is the score produced by the case. From the example above, potential IVs are the textbook used in the course, the professor's teaching style, the student's cognitive style, the number of assignments, the number of prior statistics courses, the class size, the student's age, and so on. The DV is success in the course, perhaps measured by points cumulated over examinations. The terminology comes from the nature of experiments where the outcome (the DV) is thought to be dependent on the IV.

IVs have levels. Each IV has at least two levels but may have many more. For instance, textbook has two levels (this book or another). Class size has three levels (15, 30, or 45 students). Timing of study has two levels (distributed or cramming). Classroom wall color has five levels (yellow, institutional green, blue, lavender, or pink). The researcher then studies differences in success associated with different levels of various IVs.

1.1.2 What Is an Experiment?

Although the statistical procedures are indifferent to the data sources, the strength of the conclusions that can be drawn from a study differ depending on whether the study is experimental or nonexperimental. Experimental studies have at least three special characteristics: random assignment of cases to levels of IVs, manipulation of those levels, and control of extraneous variables. Levels of the IV are manipulated by deciding what the levels are, how they are implemented, and how and when cases are assigned and exposed to them.

To randomly assign cases to levels of treatment, the researcher uses a random process to determine which cases experience which level of an IV. Random assignment is used for at least two reasons. The first is to remove researcher bias from the study—to keep the researcher from assigning (however unconsciously) bigger, stronger, smarter cases to the level of treatment he or she hopes is more effective. The second reason is to let random processes *average out* differences among cases. The researcher's fondest hope is that all differences among cases are spread evenly among the levels of treatment *before* treatment is delivered.

For example, it is assumed that students come into statistics courses with preexisting differences in motivation, study habits, competing demands on time, preferences for instructors, and so on. If a flip of a coin (or some other random process) is used to assign students to classes with different textbooks, then the differences in motivation, study habits, and so on should be randomly and approximately evenly spread through the classes. The classes, in short, should be roughly equivalent at the start of the study.

Flipping a coin, however, is usually not the major method for random assignment, because it is unlikely to result in equal sample sizes in each group. Instead a variety of alternative random assignment strategies have been devised that differ depending on the experimental design chosen. Examples include use of a random number table, generation of random numbers through software, poker chips with colors that represent groups, and so on. Randomization for the various designs in this book is discussed in Chapters 4–11 in sections labeled "Allocation of Cases." Randomization is often discussed in research design and introductory statistics or texts, such as Keppel, Saufley, and Tokunaga (1992).

Manipulation of the levels of treatment is undertaken so that the researcher controls both the precise nature of each level of treatment and the timing of its delivery. The researcher hopes to demonstrate that different treatments are accompanied by different average responses and that the differences in average response are produced by manipulating which group gets what type of treatment. In the example, the researcher first randomly assigns students to classroom and then randomly assigns textbook to classroom. Because the students in the classrooms should be relatively equal to begin with, differences in success are attributed to textbook, provided that all other relevant variables have been controlled. In a nonexperimental study, the researcher merely looks at scores for classes using different textbooks, without any assurance of the equivalence of classes or any control over how the text is used and other features of the course.

Control of extraneous variables ensures that the only thing that changes during the study is level of treatment. One way to control extraneous variables is to hold them constant for the duration of the study. For example, the same instructor teaches both classes, in the same room, at the same time of day, with the same assignments, and so on (i.e., the same class but with different textbooks). A second way to deal with extraneous variables is to spread them randomly through the levels of treatment (much like differences among cases are controlled) in the hopes that their effects will average out. If, for example, several classes are assigned to different classrooms at different times of day and the two textbooks are randomly assigned to them, the effects of room and time should be neutralized in the study. A third way to deal with extraneous variables is to take control of them and counterbalance their effects. For example, for four classes, two are scheduled for morning and two for afternoon, with each textbook used for one morning and one afternoon class. Finally, one can turn an extraneous variable into another IV. For example, time of day is turned into a second IV and studied in conjunction with textbook. These possibilities are discussed at length as they arise in the context of various research designs.

One class of extraneous variables can occur when researchers or cases are aware of the nature of the experiment and their part in it. Researchers or cases may be biased toward or against one or more of the IV levels. It is sometimes possible to control these extraneous variables by making the cases "blind" to the level of treatment to which they are exposed (e.g., by making placebos and experimental drugs look the same) or by double blinding both the cases and those who administer the treatments.

When cases have been randomly assigned so differences among them are not associated with treatment, when the researcher has delivered levels of treatment under precisely controlled conditions at known times, and when all extraneous variables have been controlled, differences in scores are attributed to differences in the level of treatment, because there is no other explanation for systematic changes in the DV.

If a study has several IVs, only one of them needs to be manipulated for the study to be considered an experiment. However, studies without any manipulated IVs are considered observational, or correlational or nonexperimental. Studies in which some, but not all, of the requirements for an experiment are met are called *quasi-experiments*. For example, the researcher might have two classes available for study, with one textbook randomly assigned to one class and the other textbook to the other class and all other features of instruction equivalent. This is still a quasi-experiment rather than an experiment, because in this case, it is not possible to randomly assign individual cases within one or the other of the two classes. Some other types of studies that are not experiments include causal-comparative, pseudo-experimental, organismic, ex post facto, surveys, case studies, and archival research.

Even if a study is properly experimental in all respects, inference of a causal relationship between DV and IV is hazardous until the study is successfully replicated one or more times. Researchers only begin to believe that there is a casual relationship between two variables when a study is repeated and similar results are found. After that, the argument for a causal relationship is strengthened by asking the research question using a variety of research strategies and achieving similar results each time.

1.1.3 What If It Isn't an Experiment?

When either random assignment of cases, manipulation of the levels of at least one IV, or control of extraneous variables is not used, the study is not an experiment and the researcher cannot attribute changes in DV scores to differences in the level of an IV. The researcher is only able to say that differences in the DV are related to or associated with differences in levels of the IV.

The inference is limited, in part, because of the third variable problem. When extraneous third variables have not been properly controlled (including potential differences among groups of cases), there is always the chance that some of them have contributed to apparent treatment effects. And worse, if some changes in extraneous variables happen to coincide with changes in the level of treatment (that is, if the levels of the IV and the values of the extraneous variable are confounded), it is not possible to determine which is responsible for change in the DV. The variability in the DV may appear to be coming from different levels of treatment when it is actually coming from changes in an extraneous variable. For example, if two different instructors are used along with two different textbooks, differences in success of students could be due to either textbook or instructor, and it is impossible to determine which has been effective.

Inference is also limited by the directionality problem that occurs when the researcher does not manipulate the timing of delivery of the levels of the IV. In this case, it isn't always clear which came first: Did the IV produce changes in the DV, or did the DV produce changes in the IV? The researcher knows that IV and DV are changing together but not which occurred first. For example, if the researcher doesn't first assign the textbook and then observe differences in success, he or she will not be able to determine whether differences in textbook led to differences in success or differences in success led to more effective use of textbook, regardless of the textbook used.

The statistics, however, are blissfully unaware of all these subtleties and simply assume that the researcher has done everything right. Statistics crunch the numbers and produce estimates of probability, regardless of the source of the numbers. It is up to the researcher to be aware of the limitations of the design and to reach conclusions justified by it.

1.1.4 Relationship between Design and Analysis

There is almost a one-to-one correspondence between the design of the study and the statistics used to analyze it. In fact, one step in planning a study is to plan the statistics, because you want to be sure that the planned study can be analyzed. One of us encountered a data set for a doctoral dissertation recently, with data already collected by the graduate student over several years, without thought as to how the data were to be analyzed and for which there was no analysis.

There are several considerations in the choice of statistics. One consideration is scale on which the DV is measured. Some analyses are appropriate when the DV is continuous (interval or ratio), other analyses when it is discrete (nominal), and yet others when it is ordinal. Another consideration is whether cases provide a score to just one level of treatment or scores to several levels of treatment. A third consideration is relevant when there is more than one IV—it is important to determine whether all possible *combinations* of all levels of all IVs are presented to cases. A fourth consideration is the potential presence of other variables, called *covariates,* whose effects need to be neutralized in the analysis. A fifth consideration is whether specific levels of treatment were deliberately chosen or whether levels of treatment were randomly selected. These and other considerations are part of the decision about the appropriate analysis, as discussed throughout the book.

1.1.5 Monitoring Processes to Aid in Experimental Design

How do you decide what is important to study? Which IVs should be included and at what levels? What is the appropriate DV? In basic research, theory guides not only the selection of IVs and their levels but also the selection of the DV. In applied research, you often know the DV you are interested in but are less certain about which IVs influence that DV.

For example, in industry, you know you want to improve the quality of a product, and you may have some guesses about factors to improve quality. One way to pinpoint potential IVs is to monitor the production process, recording quality over a period of time while also keeping track of things that change over time, such as temperature of the production facility, maintenance schedule of the machines that fabricate the product, characteristics of the operators who produce each product, changes in materials used for the product, and so on. The same principle pertains in other applied research settings. For example, what is happening when a class of students is more successful than usual? What is happening when symptoms of a disease change over time? What is happening as pilots respond to in-flight interruptions? Some of the quality control charts introduced in Chapter 2 are useful for monitoring many types of processes at once in an effort to uncover the IVs that are important to include in a more formal study.

It also is sometimes useful to monitor processes continuously during the course of a study to help refine the design or to call an end to the experiment, as in the case of an obviously successful drug trial, for example.

1.2 TYPES OF RESEARCH DESIGNS

One of the basic considerations in choosing a design and its associated analysis is to determine how cases are allocated to treatments. Does a case participate in just one level of treatment (or one combination of levels of treatments), or does the case participate in several levels of treatment (and/or their combinations)? Stated differently, does each case provide a single DV score, or does it provide several DV scores, each in response to a different level of treatment? Other considerations are the possible features of the cases themselves (e.g., age) that are almost certainly going to affect outcome. Such features should be treated as nonexperimental IVs in the design and analysis of research.

1.2.1 Randomized-Groups Designs

Randomized-groups designs (see Chapters 4 and 5) have different cases in every level (or combinations of levels) of the IV(s), and each case provides a single

TABLE 1.1 Assignment of cases in a one-way randomized-groups design

Treatment		
a_1	a_2	a_3
s_1	s_4	s_7
s_2	s_5	s_8
s_3	s_6	s_9

DV.[1] The terminology comes from experimental methodology, in which cases are randomly assigned to levels of treatment so that the group of cases in each level is different—hence *randomized groups.* In the social sciences, this is called a between-subjects design because differences due to level of treatment are tested between different groups of subjects. In other contexts, cases are screws, trainees, fabrics, college undergraduates, or whatever; whatever the context, each case is measured only once. Table 1.1 shows the allocation of cases to three levels of treatment (designated a_1, a_2, and a_3) in a randomized-groups design. Note that the nine different cases in this example are lowercase *s* designated with a subscript number.

In the example, the cases are students randomly assigned to classes that teach the same subject but that use three different textbooks (or classes of three different sizes, or whatever). The researcher is looking for differences in success associated with differences in level of treatment.

This design is necessary when, for instance, cases are used up in the process of measurement. It is also necessary when either the treatment or the measurement process produces relatively permanent changes in the cases. With this design, each case is naïve with respect to treatment because it only participates in one level of treatment; thus, this design is also desirable when lack of naïveté produces its own changes in behavior. For example, having cases read more than one résumé in an experiment designed to study stereotyping may reveal too much about the purposes of the study. However, the design requires the availability (willingness?) of more cases than some of the other designs require.

1.2.2 Repeated-Measures Designs

In repeated-measures designs, each case participates in two or more levels of treatment and is measured once at each level (see Chapter 6). Thus, there will be as many DV scores for each case as there are levels (or combinations of levels) of the IV(s). In the social sciences, this is called a within-subjects design, because differences

[1] There is a single DV, unless the design is multivariate where several different DVs are measured at one time for each case. This complicates the analysis, however, as described in Chapter 7 of Tabachnick and Fidell (2007).

TABLE 1.2 Assignment of cases in a one-way repeated-measures design

Treatment (or time period)		
a_1	a_2	a_3
s_1	s_1	s_1
s_2	s_2	s_2
s_3	s_3	s_3

due to treatments are measured within the same set of subjects. Table 1.2 shows the allocation of cases to three levels of treatment (designated a_1, a_2, and a_3) in a repeated-measures design. Note that the three different cases in this design are represented by subscripts. In the example, students take three tests during the semester, providing one score for each test. The DV (success) is measured three times on the same measurement scale.

This design is appropriate when the desire is to monitor changes in the DV that occur over time, with repeated exposure to treatment, or with repeated measurement. It is also appropriate when the naïveté of the cases is not important or is controlled through use of a Latin square (Chapter 9) or some other procedure. The design has additional advantages when the cases have individual differences that the researcher wants to take into account. The design is efficient in its use of cases: It uses far fewer cases than the randomized-groups design (in this example, three instead of nine), and once set up, the measurement often goes more quickly than when many different cases have to be set up. However, the design has a restrictive assumption that makes its use problematic in many situations. Chapter 6 also describes a variation of the design in which different cases are matched and then treated as if they were the same case during analysis.

1.2.3 One-Way and Factorial Designs

Many research studies have a single IV with two or more levels. In this book, a single IV is identified by a capital letter A. The number of levels of the IV is designated with a lowercase letter, a, and each separate level is designated with a subscript number, such as a_1, a_2, and a_3. Designs with only one IV are called one-way designs. Tables 1.1 and 1.2 show allocation of cases in a one-way randomized-groups design and a one-way repeated-measures design, respectively.

The design is two-way when a second IV is added, three-way with a third IV, and so on (see Chapters 5 and 6). Each additional IV is designated with an additional capital letter, B, then C, and so on, with their respective number of levels designated b, c, and so on, and each level of each IV designated as b_1 and b_2 (if $b = 2$) and c_1, c_2, and c_3 (if $c = 3$). Second and third IVs often are in factorial arrangement with A. In a factorial arrangement, all *combinations* of all levels of all IVs are included. For instance, if there are two IVs with $a = 3$ and $b = 2$, cases are tested under the following combinations of levels: a_1b_1, a_1b_2, a_2b_1, a_2b_2, a_3b_1, and a_3b_2. Sometimes it is undesirable or inconvenient to present all possible combinations of levels of all IVs, and the design is not factorial, but rather incomplete, unbalanced, or partially factorial, as discussed in Chapters 9 and 10.

1.2.4 Blocking Designs

Often a second or third IV is a blocking variable introduced as a form of control for an extraneous variable. A blocking variable is often a characteristic of the cases that the researcher turns into a nonexperimental IV. The analysis then provides information about differences associated with the manipulated IV, differences produced by the blocking IV, and differences produced by the combination of the two IVs. For example, the researcher might block on age of student and then study the effects of class size on success in a statistics class. When the cases are people, they are often blocked on variables such as gender, age, income, educational level, or skill.

A blocking variable can also be a feature of the testing situation, perhaps not of major interest itself but with potential impact on the DV. For example, the classroom in which the statistics course is taught is not of substantive interest but is, nonetheless, a potential factor in the design and could be turned into a blocking variable.

1.2.5 Crossing and Nesting

To understand the factors and their error terms in research design and analysis, you have to come to grips with the concepts of crossing and nesting. Crossing occurs when all levels of one variable completely cross (occur in combination with) all levels of another variable. When the IVs are completely crossed, the design is *factorial.* Consider the two-way design of Table 1.3(a), in which there are two IVs, A and B, in factorial arrangement. Each IV has two levels, and the levels of A completely cross the levels of B (and vice versa); that is, all combinations of effects occur in the design, a_1b_1, a_1b_2, a_2b_1, and a_2b_2. For example, suppose A is textbook (this one and another) and B is class size (large and small). In a factorial design, each textbook occurs in combination with (is used in) both large classes and small ones.

A third IV, C, is fully crossed with A and B in Table 1.3(b). In this design, C crosses levels of A, levels of B, and combinations of levels of A and B. Essentially, the AB design of Table 1.3(a) is repeated at each level of C. If C is classroom, then all combinations of textbook, class size, and classroom appear in the design—this textbook with the large class in the classroom with pink walls, the other textbook with the large class in the classroom with green walls, and so on—until all combinations of the three variables are present.

The third IV, C, is added to the factorial arrangement of A and B in a different way in Table 1.3(c). This design has *different* levels of C at each *combination* of levels of A and B. For instance, c_1 and c_2 are in the a_1b_1 combination, c_5 and c_6 are in the a_2b_1 combination, and so forth. The eight levels of C are said to *nest* in each AB

TABLE 1.3 Crossing and nesting in arrangements of IVs

(a) Two-way factorial design

	Treatment *A*	
Treatment *B*	a_1	a_2
b_1	a_1b_1	a_2b_1
b_2	a_1b_2	a_2b_2

(b) Three-way factorial design

		Treatment *A*	
Treatment *C*	**Treatment *B***	a_1	a_2
c_1	b_1	$a_1b_1c_1$	$a_2b_1c_1$
	b_2	$a_1b_2c_1$	$a_2b_2c_1$
c_2	b_1	$a_1b_1c_2$	$a_2b_1c_2$
	b_2	$a_1b_2c_2$	$a_2b_2c_2$

(c) Treatment *C* nested in *AB* combinations

	Treatment *A*	
Treatment *B*	a_1	a_2
b_1	c_1: $a_1b_1c_1$	c_5: $a_2b_1c_5$
	c_2: $a_1b_1c_2$	c_6: $a_2b_1c_6$
b_2	c_3: $a_1b_2c_3$	c_7: $a_2b_2c_7$
	c_4: $a_1b_2c_4$	c_8: $a_2b_2c_8$

combination. The backslash notation is used to designate nesting; in this example, the notation is C/AB (C nested in AB combinations). For example, suppose C is still classrooms and there are eight classrooms in which statistics classes are taught. Classrooms c_1 and c_2 are randomly assigned to receive this textbook and a large class, c_3 and c_4 receive this textbook and a small class, c_5 and c_6 receive the other textbook and a large class, while c_7 and c_8 receive the other textbook and a small class.

In Table 1.3(b) and (c), levels of the third IV C either cross AB combinations or nest within them. Most commonly, it is cases that cross the levels of the other IV (Table 1.2) or nest within the levels of the other IV (Table 1.1). In the repeated-measures design, cases cross levels of A ($A \times S$), while in the randomized-groups design, cases nest in levels of A (S/A).

Becoming familiar with this terminology is important because you need to understand the pattern of crossing and nesting in your design in order to pick the right analysis. Ask yourself whether each IV crosses over all levels of all other IVs. In more complicated designs (see Chapters 9 and 10), some IVs cross completely, and others do not. In other words, not all combinations of levels may appear in the design; you may, for instance, have the $a_1b_1c_1$ combination but not the $a_1b_1c_2$ combination. Then ask yourself whether *cases* cross all levels of all IVs (as in the repeated-measures design of Table 1.2) or whether cases are nested in levels of IVs (as in the randomized-groups design of Table 1.1). Later, you will discover in designs that are said to be mixed, cases may nest in levels of some IVs while they cross levels of other IVs.[2]

1.3 TYPES OF TREATMENTS

The levels of an IV sometimes differ from each other in type and sometimes in amount. Some choices about the appropriate analysis depend on whether the levels differ in type (a qualitative IV) or in amount (a quantitative IV).

1.3.1 Qualitative IVs

Qualitative IVs have levels that differ in type. For instance, textbook is a qualitative IV whose levels differ in type (this text versus another text), or the IV is type of personnel training procedure (manual-based or video-based). Frequently used blocking IVs are sex (male and female) or gender (men and women) whose levels also (happily) differ in type.

1.3.2 Quantitative IVs

Quantitative IVs have levels that differ in amount. For example, number of assignments (5, 10, or 15) is a quantitative IV, or the IV is amount of personnel training (4 hours or 8 hours). Frequently used quantitative blocking IVs are age (20–29, 30–39, and 40–49) or years of education (8, 12, or 16) or income. Note that in these examples, the levels are spaced the same distance apart. This is not a requirement, but it does simplify some of the computations.

When an IV is quantitative, a special analysis called a *trend analysis* is available that looks for a linear (straight-line) relationship between the DV and the levels of the IV or a quadratic (curved) relationship between the DV and the levels of the IV or both.[3] This fascinating topic is discussed in detail in Chapters 4–6.

A study is likely to have some IVs that are qualitative, some that are quantitative, and some that are hard to fathom. For instance, what do you call an IV that has two levels, one with none of something and one with lots of something? Do the levels differ in quality or quantity? Since there are only two levels, you really don't have to decide, because the analysis is the same and, in any event, addition of a third level would clarify the situation. In this case, most researchers would call the "none of something" level the control condition and the "lots of something" level the treatment, and let it go at that.

[2] This is one use of the term *mixed.* You will discover that *mixed* is also used to designate designs with fixed and random effects, described in Chapter 11. Most confusing!

[3] There are also much more complicated relationships, depending on the number of levels of the IV.

1.4 OUTCOME MEASURES

There are also various types of DVs: continuous, ordinal, and discrete. Determining which type of DV you have is not a trivial matter—it is one of the fundamental considerations when choosing the appropriate analysis.

1.4.1 Types of Outcome Measures

Outcomes are most usefully classified as continuous (interval or ratio), ordinal, or discrete (nominal).[4] Discrete outcomes with only two categories are called *dichotomous.* Continuous outcomes assess the *amount* of the DV; there are numerous possible values and the size of the DV depends on the sensitivity of the measuring device. As the measuring device becomes more sensitive, so does the precision with which the DV is assessed. Examples of continuous DVs are time to complete a task, amount of fabric used in various manufacturing processes, score on an exam, and variables such as attitudes.

Ordinal DVs reveal the rank order of the outcomes vis-à-vis each other, but they do not indicate the extent to which one outcome exceeds another. For instance, if the quality of several different paint jobs is rank ordered, the researcher knows which is the best but not by how much. The best paint job might be a lot better than the next best one, or it might be just a little bit better; there is no way to know. Ordinal DVs are a bit dangerous, because fewer statistics are available for their analysis.

Discrete DVs are like qualitative IVs in that they differ in type. DV responses are classified into categories. With discrete DVs, there are usually only a few possible values, and the values are mutually exclusive (there are no values between the values, so to speak) and often exhaustive (every DV can be classified). For instance, employees are classified as properly trained or not; eggs are divided into small, medium, large, extra-large, or jumbo; respondents answer either "yes" or "no"; or manufactured parts either pass or don't pass quality control.

In the eggs example, notice that the DV is continuous if the eggs are weighed but discrete (and a little ordinal) if the eggs are classified into categories. This brings up the point that although a continuous DV can be changed to an ordinal or discrete one, the reverse is not true because there isn't enough information to change a discrete DV to a continuous one.

Often the issue of type of DV is simply decided by the researcher. The decision is basically a matter of how much faith the researcher has in the measuring process and whether a difference makes a meaningful difference. For instance, although eggs could be weighed to small fractions of a kilogram, does it make any sense to do so? What is the meaningful level of measurement for this particular DV? There is lively debate as to the extent to which noncontinuous DVs are appropriately analyzed by statistical procedures designed for continuous outcomes, such as ANOVA. Ordinal

[4] The term *categorical* is often used to refer to either ordinal or nominal DVs.

data are sometimes treated as if continuous when the number of categories is large (say seven or more) and the data meet other assumptions of the analysis. An example of a DV with ambiguous measurement is one measured on a Likert-type scale in which consumers rate their attitudes toward a product as "strongly like," "moderately like," "mildly like," "neither like nor dislike," "mildly dislike," "moderately dislike," or "strongly dislike." Occasionally, even dichotomous DVs are treated as if continuous when the two categories of the DV occur fairly equally in each level of the IV. For the most part, the designs and analyses in this book are limited to outcomes that either are continuous or can be treated as continuous.

1.4.2 Number of Outcome Measures

A single DV is measured in the designs discussed in this book. The same DV can be measured once or more than once, but only one DV is assessed. For instance, the instructor might measure success in the statistics course with a single test at the end or measure success as the average of scores achieved on several tests. In either situation, success is inferred from only one DV—test score. Designs with a single DV are often referred to as *univariate.*

Many designs exist, however, in which several different DVs are measured, either all at once or all several times. For instance, the researcher could be interested in both test scores and attitude toward statistics. Scores and attitude are both DVs, and both could be measured only once for each case at the end of the course or several times for each case during the course. Designs with several different DVs are called *multivariate* and are described by Tabachnick and Fidell (2007).

1.5 OVERVIEW OF RESEARCH DESIGNS

1.5.1 Continuous Outcomes

In the previous example, the scores on tests are a continuous DV. The most common analysis for a continuous DV is ANOVA,[5] which subdivides IVs into levels and then asks if the levels have reliably different means. Do the treatments at some levels increase means while others lower means? Details of the analysis are presented in Chapter 3, along with an explanation for why an analysis of *variance* makes inferences about *means*.

There are about half a dozen different kinds of ANOVA and a near infinite number of ways to combine them. The flexibility of the analytic procedure is astonishing. For this reason, it is used in almost all branches of science. Once you understand the great variety of research designs that can be analyzed through ANOVA, you can tailor your design to precisely meet your research needs.

[5] ANOVA was also the stage name used by one of us in our somewhat less than reputable belly-dancing days.

1.5.1.1 Randomized-Groups ANOVA

The one-way randomized-groups design is very popular. This design has one IV (A) with two or more levels and with different cases in every level of A. A one-way randomized-groups design with three levels of A is illustrated in Table 1.1. If A represents different class sizes (say 15, 30, or 45 students), the researcher asks if the average DV score is different for classes with different numbers of students. One-way randomized-groups ANOVA is discussed in Chapter 4.

If there are two or more IVs each with two or more levels deliberately chosen by the researcher, if all combinations of levels are present, and if each combination of levels has different cases, the design is a factorial randomized-groups design, as illustrated in Table 1.3(a). If the researcher in the example is interested in the possible impact of textbook in addition to class size, all combinations of the two variables are used and each student generates a single DV score. In this case, the researcher asks if success is different in classes of different size, if success is different with different textbooks, and if some combinations of textbook and class size work particularly well (say, this textbook with a small class but another textbook with a large class). Factorial randomized-groups ANOVA is the topic of Chapter 5.

1.5.1.2 Repeated-Measures ANOVA

Another general class of ANOVA is used with repeated-measures designs. These designs are characterized by having the same outcome measured more than one time for each case, as illustrated in Table 1.2. If several tests are given during the course and each score is used, a one-way repeated-measures ANOVA is appropriate. The researcher asks if there are reliable mean differences in test scores as the semester progresses. One-way and factorial repeated-measures ANOVAs are discussed in Chapter 6.

The mixed design combines features of the randomized-groups design with features of the repeated-measures design. There are different cases in levels of the randomized-groups IV, but cases are measured repeatedly at the levels of the repeated-measures IV, as seen in Table 1.4. For example, suppose one class is assigned this textbook while another is assigned the other textbook, but three tests are given in the semester and each test score is used separately. The design is randomized groups with respect to textbook but repeated measures with respect to test scores. ANOVA for mixed designs is discussed in Chapter 7.

TABLE 1.4 Mixed design where A is a randomized-groups IV and B is a repeated-measures IV

	Treatment B		
Treatment A	b_1: Test 1	b_2: Test 2	b_3: Test 3
a_1: This text	s_1	s_1	s_1
	s_2	s_2	s_2
a_2: Another text	s_3	s_3	s_3
	s_4	s_4	s_4

When a very large number of DV scores (say 50 or more) are measured on an ongoing basis, time series analysis (as discussed in Tabachnick and Fidell, 2007), rather than repeated-measures ANOVA, is appropriate. The researcher studies changes that occur over time before and after introduction of a treatment. For instance, water consumption for a municipality is measured monthly for several years before and after a campaign is mounted to conserve water. Or records of number of accidents per week are kept for a year or more and compared with the number in the weeks after the start of a safety campaign.

1.5.1.3 Analysis of Covariance (ANCOVA)

Another variety of ANOVA is analysis of covariance, known as ANCOVA. The term *covariance* comes from use of a covariate (CV) in the analysis. A CV is a continuous variable that is related to the DV but, at least in experimental work, independent of the IV(s). In experimental work, the CV often is the same as the DV but is measured *before* levels of treatment are applied. The goal of the analysis is to examine mean differences on the DV associated with levels of the IV after DV scores are adjusted for pretreatment differences on the CV. For example, a pretest of knowledge of statistics is measured for all students *before* the levels of treatment are applied; the pretest scores serve as the CV. These pretest scores reflect differences in motivation, past instruction, number of previous courses, time since previous course, and so on, that students bring into the experiment. Scores on the DV are adjusted for differences on the CV, and then adjusted mean differences associated with levels of treatment are analyzed. The researcher asks if there are mean differences due to treatment after cases are equalized for preexisting differences on the CV. ANCOVA is the topic of Chapter 8.

1.5.1.4 Latin-Square ANOVA

Sometimes the researcher wants to control for nuisance variables in the randomized-groups design or for the order in which levels of the repeated measure are presented to cases in the repeated-measures design. A Latin-square design is used if a fully crossed (full factorial) design is not desirable or feasible. The Latin square is only part of the full factorial arrangement of levels of IVs. Latin-square analyses provide tests of mean differences among the levels of each IV but not among their interactions. Latin-square analyses are described in Chapter 9.

1.5.1.5 ANOVA for Screening and Other Incomplete Designs

Another fairly common situation occurs in the early stages of a research program in which the researcher is not yet sure which IVs are important or in which all combinations of treatments are not feasible. Screening and other incomplete designs are appropriate in these situations because numerous IVs are included but are not fully crossed. In other words, there is no attempt to implement the IVs in a full factorial design with all their combinations. The effects of the levels of each IV are assessed but not the interactions among IVs—or at least not all of them. For instance, in the early stages of research, the researcher may want to examine potential effects of the student's cognitive style, number of assignments, timing of assignments, color of walls in classroom, and so on, in addition to textbook and class size. A fully factorial arrangement of all these IVs, with at least two levels each, is nearly overwhelming. However, use of a screening or other incomplete design, as described in Chapter 10, makes the project manageable.

1.5.1.6 Random-Effects ANOVA

A final variety of ANOVA has to do with the way levels of the IV are selected. In most research designs, the researcher deliberately selects the levels of the IV to reflect its plausible range of values. This is a fixed-effects IV. Results generalize to the levels actually included in the study. Another option, however, is to randomly select the levels of the IV. The researcher lists the population of all possible levels and then randomly selects a few of them. This is a random-effects IV. For instance, the researcher lists all possible statistics instructors at a university and then randomly selects two of them to teach the courses. Results generalize to the population of all possible levels. Random-effects ANOVA, including the nested design of Table 1.3(c), are described in Chapter 11.

1.5.2 Rank Order Outcomes

There are far fewer analytical techniques for ordinal DVs. Although there are nonparametric counterparts to one-way randomized-groups and repeated-measures ANOVA, they are not covered in this book. For instance, the rank sums test (when $a = 2$) and the Kruskall-Wallis test (when $a > 2$) are available for one-way randomized-groups designs, while the sign test (when $a = 2$) and the Friedman test (when $a > 2$) are available for one-way repeated-measures designs. These analyses are described in Siegel and Castellan (1988) and are available through most statistical packages. ANOVA is sometimes applied to ordinal outcomes, as discussed in Section 1.4.1, when the research design is one for which no nonparametric technique is available and the distribution of outcomes is (or can be transformed to be) relatively normal.

1.5.3 Discrete Outcomes

Analysis of a discrete DV is accomplished through simple chi-square analysis if there is only one discrete IV or through log-linear models or logistic regression if there are several IVs. Chi-square analysis is described in Chapter 2 (and in most elementary statistics textbooks). For example, suppose the researcher categorizes the students' scores into either pass or fail. Chi-square is used to determine whether there is a relationship between success and textbook.

1.5.3.1 Log-Linear Models

When there is one discrete DV and several discrete IVs, log-linear models (variously called logit analysis or multiway frequency analysis) are used. These models are extensions of chi-square analysis that yield information about frequencies on the DV associated with each IV separately and then with combinations of IVs. For instance, suppose the researcher includes textbook, classroom wall color, and student's cognitive style as IVs and then classifies the test score for each case as either pass or fail. The researcher would then ask if success is associated with each IV separately and all combinations of them. Log-linear models are described in Chapter 16 of Tabachnick and Fidell (2007).

1.5.3.2 Logistic Regression

Logistic regression is used when there is one discrete DV and one or more IVs, which may be discrete, continuous, or a combination of discrete and continuous. For instance, if the DV is classified as pass or fail, while IVs are textbook and student's cognitive style (both discrete), along with student's age (continuous), logistic regression is used for the analysis. Logistic regression is described in Chapter 10 of Tabachnick and Fidell (2007).

1.5.4 Time as an Outcome

The DV in some applications is elapsed time until something happens. For instance, the DV might be the length of time that students study statistics when they are given two different types of classes for the first course, or length of time patients survive following two types of chemotherapy, or length of time parts made of different alloys survive in a testing chamber. When the DV is length of time until something happens, survival (or failure[6]) analysis is appropriate, as described in Chapter 11 of Tabachnick and Fidell (2007).

1.6 OVERVIEW OF ISSUES ENCOUNTERED IN MOST DESIGNS

1.6.1 Statistical Inference

This section is intended as a brief review of a complicated topic rather than a complete exposition of it. If you mastered the intricacies of statistical inference in a previous course, or if you believe that statistical inference is misguided (see Section 1.6.5), you can skip this section.

Statistical inference is used to make rational decisions under conditions of uncertainty. Inferences (decisions) are made about populations based on data from samples that contain incomplete information. Different samples taken from the same population probably differ from one another and from the population. Therefore, inferences regarding the population are always a little risky.

[6] Is the cup half full or half empty?

FIGURE 1.1 Overlapping distributions representing the null (H_0) and alternative (H_a) hypotheses. A Type I error (with probability α) occurs if the result comes from the H_0 distribution but the decision is made to reject H_0. A Type II error (with probability β) occurs if the result comes from the H_a distribution but the decision is made to retain H_0.

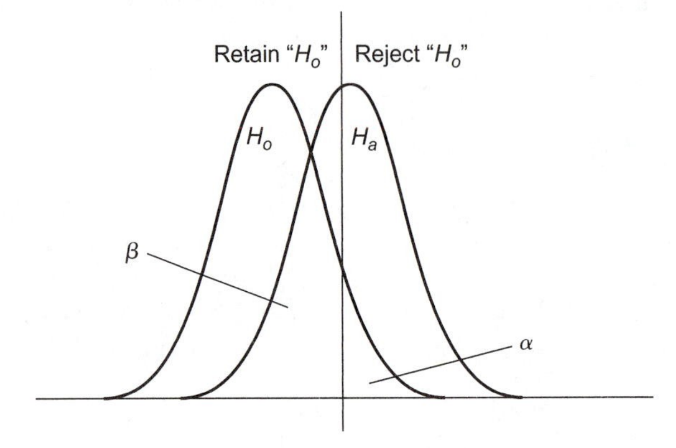

The traditional solution to this problem is statistical decision theory. Two hypothetical states of reality are set up, each representing a probability distribution of means that come from alternative hypotheses about the true nature of the population. One distribution represents the hypothesis that treatment makes no difference, affectionately known as the null hypothesis, H_0. The other distribution represents the hypothesis that treatment makes a difference, known as the alternative hypothesis, H_a. Given sample results, a best guess is made as to which distribution the sample was taken from, a tricky proposition because, as you recall from your introductory statistics course, the distributions overlap (Figure 1.1). Formalized statistical rules are used to define *best.* Section 3.2.3.3 discusses the application of statistical inference to ANOVA.

H_0 and H_a represent alternative realities, only one of which is true. When the researcher is forced to decide whether to retain or reject H_0, four things can happen. If the null hypothesis is true, a correct decision is made if the researcher retains H_0, and an error (Type I error) is made if the researcher rejects it. If the probability of making the wrong decision is α, the probability of making the right decision is $1 - \alpha$. If, on the other hand, H_a is true, the probability of making the right decision by rejecting H_0 is $1 - \beta$, and the probability of making the wrong decision is β (a Type II error). This information is summarized in a *confusion matrix* (aptly named, according to beginning statistics students) showing the probabilities of each of these four outcomes:

	Reality	
Statistical Decision	H_0	H_a
Retain H_0	$1 - \alpha$	β (Type II error)
Reject H_0	α (Type I error)	$1 - \beta$
	1.00	1.00

1.6.2 Power

The lower right cell of the confusion matrix represents the most desirable outcome and the power of the research. Power is the probability of rejecting H_0 when H_a is true; it is the probability of deciding that there is an IV–DV relationship when there really is such a relationship. The researcher usually believes that H_a is true—that there really is a relationship—and fervently hopes the data lead to rejection of H_0.

Many of the choices in designing research are made with an eye toward increasing power, because research with low power is often not worth the effort. One way to increase power is to increase the probability of making a Type I error. Given the choice between an α error rate of .05 or .01, selecting the .05 level increases power. However, if Type I error rates are increased too much, they will reach unacceptable levels.

A second strategy is to apply stronger treatments, which has the effect of moving the distributions farther apart so they don't overlap so much. For example, distributions will be farther apart if differences are sought between supercomputer and PC performance than if differences are sought between PC and Macintosh performance.

A third strategy is to reduce the variability among scores in each treatment by exerting greater experimental control. Experimental control can be achieved by, among other things, selecting a homogeneous sample of cases prior to randomly assigning them to levels of treatment, delivering treatments consistently, and controlling extraneous variables. This has the effect of reducing the overlap among the distributions by making them narrower.

A fourth strategy to increase power is to increase sample size. Most of the equations that test for significance have sample size lurking in the denominator. As sample size becomes larger, the denominator becomes smaller, and the value of the significance test becomes larger. Larger samples are more likely to produce evidence for a relationship than are smaller samples. However, when samples are very large, there is the danger of *too much* power. The null hypothesis is probably never exactly true, and any treatment is likely to have some effect, however trivial. For that reason, a *minimal meaningful difference* should guide the selection of sample size: The sample size should be large enough to reveal a meaningful difference but not so large as to reveal any difference whatsoever. Part of the impetus to report effect size along with significance level is due to the likelihood of finding a significant result with a trivial difference when sample sizes are very large (see Section 1.6.3).

Researchers usually think about these trade-offs when they consider the size of the sample they need in the initial planning stages. The factors to be considered are the anticipated size of the effect (e.g., the expected mean difference), the estimated variability in the scores, the desired alpha level (usually .05), and the desired power (usually .80). Once these four factors are determined, the sample size can be calculated, often using readily available software (see Section 4.5.2). If sample size is not calculated ahead of time, the researcher may not have a powerful enough study to produce a significant difference, even with a true H_a. More detailed discussions of power are available in Cohen (1965, 1988), Rossi (1990), or Sedlmeier and Gigerenzer (1989).

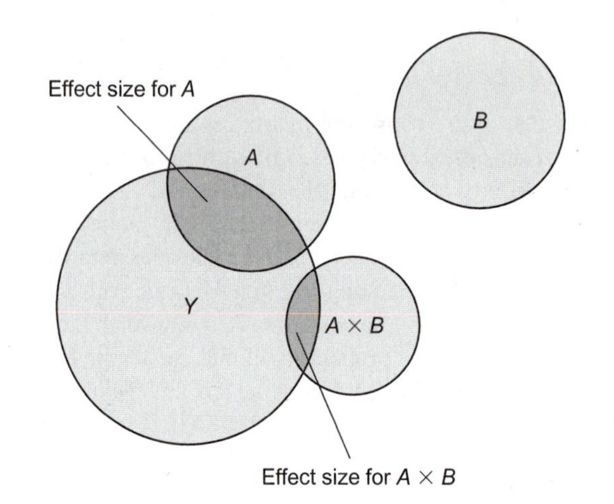

FIGURE 1.2 Conceptual diagram of effects sizes for A, B, and $A \times B$

1.6.3 Effect Size

Statistical significance tests assess the *reliability* of the association between the IV(s) and DV. For example, is the null hypothesis likely to be rejected if the study is repeated? Measures of effect size assess *how much* association there is by assessing the degree of relationship between the IV(s) and DV. A researcher must assess the degree of relationship to avoid publicizing trivial results as though they had practical utility or spending money to change to a new procedure that actually makes very little difference.

Most measures of effect size assess the proportion of variance in the DV associated with levels of an IV—that is, they assess the amount of total variance in the DV that is predictable from knowledge of the levels of the IV. If circles represent the total variances of the DV and the IV, measures of effect size assess the degree of overlap of the circles, as seen in Figure 1.2. In this example, the first IV, A, is rather strongly related to the DV, Y; the second IV, B, is unrelated to Y; and the interaction of $A \times B$ is modestly related to Y.

1.6.4 Parameter Estimates

What are the likely population means for cases exposed to the various levels of the IV(s)? If the entire population is measured instead of only a sample of cases, what means are expected? The population means are the unknown parameters; the observed DV scores provide estimates of them. In other words, sample means are used as parameter estimates. In fact, statistics are called *parameters* when they describe populations. Because we never know the parameters exactly without measuring the entire population, we use the sample statistics to create a range of values within which the population values are likely to fall. These are called *confidence intervals.*

The ability to estimate parameters for a population from a sample depends on how representative the sample is of the population. If you take a random sample from the population (and then, perhaps, randomly assign cases to levels of treatment), sample statistics provide estimates of the population values. If the sample is one of "convenience," however, estimating population values from it is hazardous.

1.6.5 To Test or Not to Test?

There is a recurring controversy about the value of using statistical inference (Harlow, Mulaik, and Steiger, 1997). This has grown out of a history of overemphasis on simply deciding whether a null hypothesis can be rejected and an underemphasis on considerations of power, effect size, and parameter estimates. Historically, a research project was considered for publication only if the statistical test was significant (the null hypothesis was rejected), without regard for the degree of association between the IV and the DV. The result was a lot of journal articles about trivial treatment effects and file drawers full of unpublished studies rejected for publication because of Type II errors. To ameliorate these effects, some have recommended abandoning the inferential test altogether.

Our position is that significance testing is still useful for a number of purposes but needs to be accompanied by consideration of effect size, parameter estimates, and power. In large, complex studies, inference provides a convenient guide for determining which results deserve further attention. However, complex designs provide numerous significance tests, some likely to be spurious if all are tested at liberal α levels, such as .05. By considering effect size and power along with the significance test, one minimizes the likelihood of interpreting very small and perhaps unreliable treatment differences. By reporting effect sizes and confidence intervals (or including information that allows the reader to compute confidence intervals), one minimizes publication of reliable, but trivial, results.

A new development to assist decision making is estimation of a confidence interval around an effect size. This interval contains information about both effect size and the significance test, because an effect size interval that includes zero is also a nonsignificant result. Software for estimating effect sizes and their confidence intervals is demonstrated in several of the technique chapters.

1.7 GENERALIZING RESULTS

The trick with statistics is using the values from a sample of cases to make decisions about a population of cases. The sample of cases itself is often of no or very limited interest; it is merely a convenient mechanism for estimating values for the population. There is usually no need for inferential statistics if you measure the whole population or if you have no desire to generalize beyond your sample.

Ideally, the researcher carefully defines the population of interest and then randomly samples from it.[7] The sample is said to be representative if it has the same distributions of relevant characteristics as are found in the population. A great deal

[7] Strategies for random sampling are discussed in detail in such texts as Rea and Parker (1997).

of work and interest has gone into strategies for gathering random samples in nonexperimental, observational, or correlational research. Indeed, the usefulness and quality of the work is often determined by the representativeness of the sample. Less attention has been paid to this issue in experimental studies, although it probably deserves more focus.

The concept of sampling has somewhat different connotations in nonexperimental and experimental research. In nonexperimental research, you investigate relationships among variables in a predefined population from which you have randomly drawn cases. In experimental research, you often use convenient cases and then attempt to *create* different populations by treating subgroups of cases differently. The objective in experiments is to ensure that all cases come from the same population before you treat them differently. Instead of random sampling, you randomly assign cases to treatment groups (levels of the IV) to ensure that, before treatment, the subsamples are homogeneous. Statistical tests provide evidence as to whether, after treatment, the samples are still homogeneous.

In experiments, *strict* statistical generalization about treatment effectiveness is limited to the population from which you randomly assigned cases to treatments (for example, the subjects who signed up for your study). In practice, *logical* generalization usually goes beyond that to the type of cases who participated in the study. For example, if a particular group of freshmen and sophomores in introductory psychology courses at your university signed up for your study, you often generalize to freshmen and sophomores in universities like yours.

You can also randomly select the levels of the IV, as described in the random effects designs of Chapter 11. Generalization about the effectiveness of treatment is to the population of levels when the levels of the IV are randomly selected from a population of levels.

1.8 COMPUTER ASSISTANCE

Nobody does statistics by hand any more, except in initial learning stages in which application of pencil to paper and finger to calculator has a salutatory effect on understanding (and on developing appreciation for the computer). Instead, the work of statistical analysis—and, increasingly, experimental design—is done with one or more software packages. Throughout this book, we demonstrate syntax and output from the two most popular statistical packages, as well as from several additional packages, for designing experiments, calculating sample size, and the like.

1.8.1 Programs for Statistical Analysis

Throughout this book, syntax and output are presented from SPSS 12.0 and SAS 9.1 (manuals for these programs are available on disk). These are the most recent releases at the time of writing. However, newer releases of some of programs may be available in a year or two. In addition, your facility may be running an older version of one or more of these programs. Pay some attention to which release you are running because the syntax and output may look different, because errors in older releases are often corrected in newer ones, and because new descriptive and inferential statistics may have become available.

You are fortunate if you have access to both SPSS and SAS. Programs within the packages do not completely overlap, and some problems are better handled through one package than the other. For example, doing several versions of the same basic analysis on the same set of data is particularly easy with SPSS, as are graphs, whereas SAS has extensive front-end data-handling capabilities. We illustrate both packages because different facilities have different programs. You may start out using SPSS, for instance, and then move to a facility that has SAS. It can take several days to familiarize yourself with a new package, and it helps to have some examples of it side-by-side with examples from the package you already know.

Most routine analyses are available through menu systems in both packages. On-disk help and manuals are available, and there is additional documentation in various newsletters, pocket guides, technical reports, and other materials that are printed as the programs are continuously corrected, revised, and upgraded. At this point, hard copy documentation for both packages is quite extensive but usually not readily available for the most recent releases. The documentation accompanying the packages on disk is usually adequate, especially with the availability of the index tab in help menus to locate specific topics. Without considerable experience, however, you may find it very difficult to produce the analysis you need. When the menu system is incomplete, you have to use syntax to get the analysis: Let the menu system take you as far as it will and then modify it through syntax if you need to. For this reason, we present the syntax that produced each output, although for many examples, we used the menu system to write the syntax.

From time to time in this book, you will also encounter information about SYSTAT (version 11). Although this package is widely used, it is not as popular as SPSS and SAS for most of the analyses shown in this book. For example, along with syntax from SPSS and SAS, the tables scattered throughout the chapters in this book also show the syntax for the same problem using SYSTAT. In addition, SYSTAT is also included in the "Comparison of Programs" section at the ends of Chapters 4–11. You may discover that one or the other has just the right capability for one of your applications. We do not, however, discuss or illustrate use of SYSTAT with examples in the body of most chapters.

One of the problems with big, flexible statistical packages is that they usually produce something, but not necessarily the analysis you want. Because the output is visually appealing, it is tempting to believe that it is also the desired analysis. That may be a rash assumption. To compare the results, start by running a small data set in which the result is known and which has the same design as your problem. When the results are identical, you know you have the right syntax, and you know how to read the results in the program you are running. You are now ready to run your larger problem.

Another problem is that the packages present results in different order, and often with different jargon, so you have to learn how to read the output. Such learning is facilitated by annotated output in the manuals and online help that provides a verbal description of various statistics. We attempt to foster this learning by presenting and describing output from both programs. Although you may only want to consider output from one of the packages right now, the other is available when you need it.

Chapters 4–11 offer explanations and illustrations from a variety of analyses within each package and compare the features of the programs. Our hope is that once you understand the techniques, you will be able to generalize to other programs within the packages, as well as to packages (such as STATA and MINITAB) not reviewed here. However, be prepared to detest and find vastly inferior the second and all subsequent packages you encounter. In our experience, people form unwarranted loyalties to the first package they use, whichever it is.

1.8.2 Programs for Designing Experiments

Computer programs are available to assign cases to treatments in randomized-groups designs, as well as other, more complicated designs that involve different orders of treatments for different cases and treatments that are not fully crossed. SAS and SYSTAT have built-in programs for experimental design. NCSS (Number Cruncher Statistical Systems; Hintze, 2001) also has extensive capabilities for designing experiments; both NCSS and SYSTAT are reviewed, along with the other programs for complicated designs, in Chapters 9 and 10.

Another set of programs helps the data analyst estimate sample sizes needed to run studies with desired power. These programs include SAS GLMPOWER, NCSS PASS, and PC-SIZE, all of which are described in Section 4.5.2. SPSS has a stand-alone program, SamplePower, for estimation of sample sizes.

A final type of program useful to the researcher (and the teacher) is one that generates mock data for a research design. Complicated designs are often best understood by doing the associated statistical analyses. However, it is sometimes too late to figure out how to analyze data after they are collected. When in doubt about the mechanics of an analysis, a good strategy is to generate a small sample of simulated data and run the proposed analyses on them. The small, hypothetical sample data sets of Chapters 4–11 were generated, for the most part, through DATASIM (Bradley, 1988), a program that generates data sets for a variety of ANOVA designs.

1.9 ORGANIZATION OF THE BOOK

Chapter 2 gives an overview of data entry and basic descriptive statistics for screening data, including several newer graphical methods. It also briefly summarizes bivariate correlation and regression, as well as chi-square analysis. Chapter 3 provides a review of the rationale behind one-way randomized-groups ANOVA, including the basic assumptions of analysis, and a demonstration of analysis through regression. For those with an adequate statistical background, Chapters 2 and 3 can be skimmed or skipped altogether. However, even if you understand the statistics, it may be helpful to glance through these chapters to familiarize yourself with the notational system we use and our way of discussing statistics.

Chapters 4–11 cover specific research designs and associated statistical techniques. The chapters follow a common format, and although they proceed from simple to more complex analyses, they are intended to be taken up in any order, or omitted, at your discretion. Once you are familiar with the format, it should be relatively easy to find the answers to questions about an analysis in a particular chapter.

Each chapter begins with an overview of the design and analysis, as well as the major research questions that can be asked using them. Then conceptual and practical limitations and assumptions of analysis are summarized, including recommendations for proceeding after violation of an assumption. Next, a tiny, inadequate, and usually silly data set is analyzed by hand to illustrate the computations and by software to illustrate the corresponding syntax and output for both major packages. These examples are intended to be readily grasped by people from diverse backgrounds and to be perhaps entertaining, if not compelling from a research perspective. A section then follows describing variations of the technique, if any. The next section discusses special issues and complexities that arise with the particular analysis. Then there is a guided tour through one or two real-world data sets for which the analysis is appropriate. These data sets come from a variety of research areas that those who become proficient in statistical analyses may encounter if they serve as consultants. This tour section concludes with an example of results in American Psychological Association format,[8] describing the outcome of the statistical analysis appropriate for submission to a professional journal. The tour concludes with both a tabular and a verbal comparison of the features of SPSS and SAS, as well as SYSTAT (version 11). Problem sets are, of course, available at the end of each chapter for the edification of the student.

1.10 CHOOSING A DESIGN: SOME GUIDELINES

Table 1.5 is a flowchart to assist you in choosing the appropriate type of analysis for your data set. To use the chart, you must first identify the type of DV you have: Is it continuous, discrete, or ordinal?[9] The next step is to identify the number of IVs; the number of levels of each IV; the arrangement of IVs (factorial or incomplete); the type of each IV (randomized groups, repeated measures, mixed, or random effects); and then, if you have a continuous DV, whether you have a covariate in your design. For the inexperienced researcher, these decisions are not always easy to make. After a while, however, this will become old hat, so that a quick glance at Table 1.5 will verify your plan of analysis.

[8] You may need to adjust the write-up of the results to the practices of your own journal.

[9] This book deals only with analyses for continuous DVs, or those that can be treated as if continuous.

TABLE 1.5 Guidelines for selection of statistical analysis

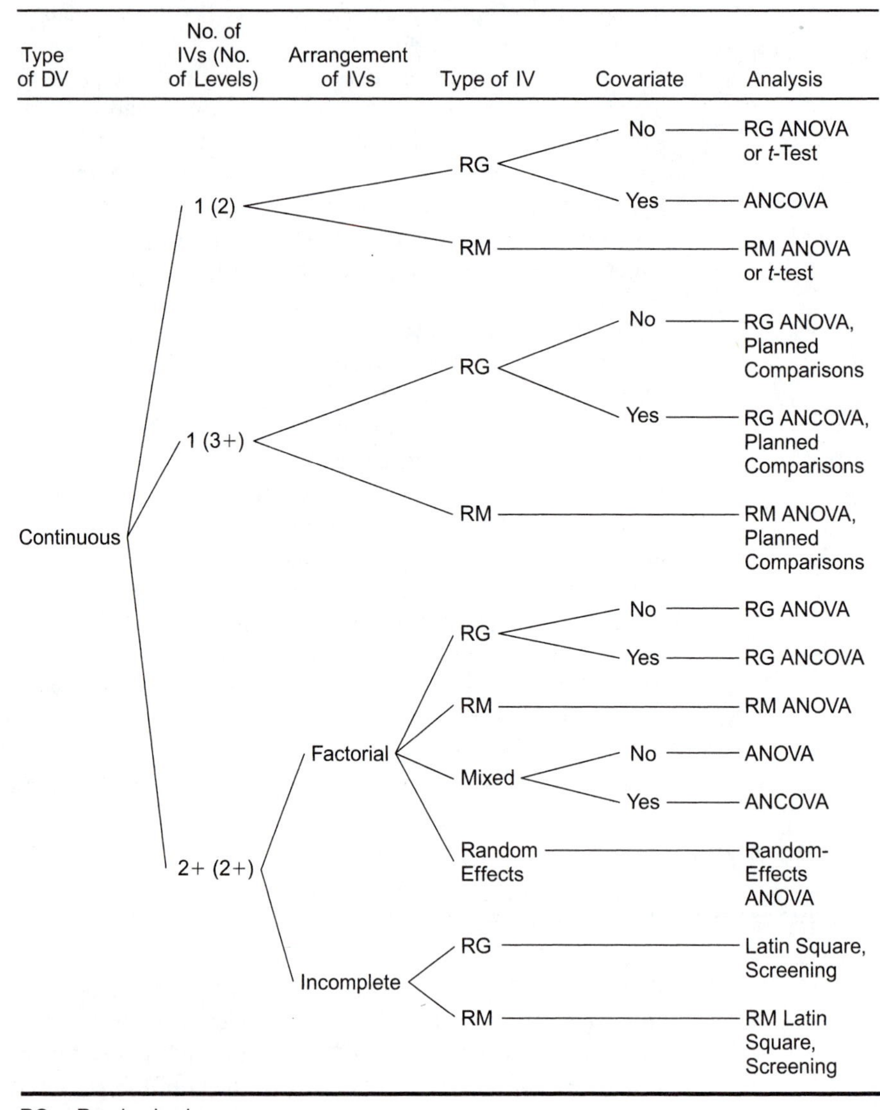

Type of DV	No. of IVs (No. of Levels)	Arrangement of IVs	Type of IV	Covariate	Analysis
Continuous	1 (2)		RG	No	RG ANOVA or *t*-Test
				Yes	ANCOVA
			RM		RM ANOVA or *t*-test
	1 (3+)		RG	No	RG ANOVA, Planned Comparisons
				Yes	RG ANCOVA, Planned Comparisons
			RM		RM ANOVA, Planned Comparisons
	2+ (2+)	Factorial	RG	No	RG ANOVA
				Yes	RG ANCOVA
			RM		RM ANOVA
			Mixed	No	ANOVA
				Yes	ANCOVA
			Random Effects		Random-Effects ANOVA
		Incomplete	RG		Latin Square, Screening
			RM		RM Latin Square, Screening

RG = Randomized-groups
RM = Repeated-measures

TABLE 1.5 Continued

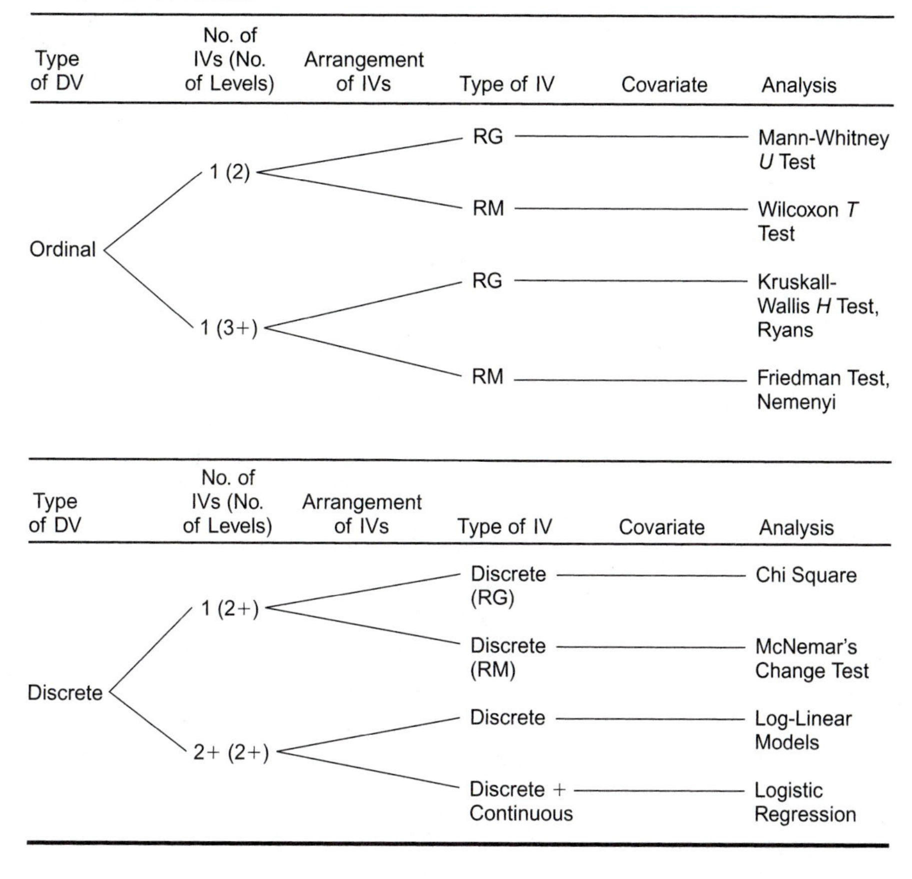

Type of DV	No. of IVs (No. of Levels)	Arrangement of IVs	Type of IV	Covariate	Analysis
Ordinal	1 (2)		RG		Mann-Whitney *U* Test
			RM		Wilcoxon *T* Test
	1 (3+)		RG		Kruskall-Wallis *H* Test, Ryans
			RM		Friedman Test, Nemenyi

Type of DV	No. of IVs (No. of Levels)	Arrangement of IVs	Type of IV	Covariate	Analysis
Discrete	1 (2+)		Discrete (RG)		Chi Square
			Discrete (RM)		McNemar's Change Test
	2+ (2+)		Discrete		Log-Linear Models
			Discrete + Continuous		Logistic Regression

1.11 PROBLEM SETS

1.1 An experiment is designed to determine how speaker size affects loudness. The researcher measures loudness of 30 speakers randomly chosen from the factory floor—10 speakers have 8 in. diameter, 10 have 10 in. diameter, and 10 have 12 in. diameter.

a. Identify the IV and its levels.

b. Identify whether the IV is formed by randomized groups or repeated measures.

c. Identify whether the IV is quantitative or qualitative.

d. Identify the DV. What type of outcome is it?

e. What kind of design is this?

f. What type of statistical error would you be making if speaker size really affected loudness but your experiment failed to uncover that difference?

g. How would you change speaker size to make this experiment more powerful?

1.2 Dr. Optimist believes he has discovered a "smarts" pill that will increase IQ by 15 points. He first measures the IQ of 90 college juniors and then randomly assigns them to one of two groups: 45 of them take the smarts pill and 45 take a placebo. He then measures the IQ of all the students at three intervals: 1 month after starting, 6 months after starting, and 1 year after starting.

a. Identify all IVs and their levels.
b. Identify whether each IV is formed by randomized groups or repeated measures.
c. Identify whether each IV is quantitative or qualitative.
d. Identify the DV. What type of outcome is it?
e. What kind of design is this?
f. What type of statistical error would Dr. Optimist have made if he reported that his pill worked when it really was no better than a placebo?

1.3 A manufacturer wonders whether adding polyester to a pantsuit makes it last longer. She randomly assigns half of her female and half of her male employees to a group that wears uniforms with polyester added and the remaining employees to a group that wears uniforms without polyester. She then observes the uniforms to see how long it takes for them to wear out.

a. Identify all IVs and their levels.
b. Identify whether each IV is formed by randomized groups or repeated measures.
c. Identify whether each IV is quantitative or qualitative.
d. Identify whether each IV is manipulated.
e. What kind of design is this?

1.4 Create a potential experiment with three IVs, each having three levels, and a continuous DV to be analyzed as a factorial randomized-groups ANOVA.

a. Identify all IVs and their levels.
b. Identify whether each IV is quantitative or qualitative.
c. Identify whether each IV is manipulated.

1.5 Suppose a researcher at a hospital wants to identify the combination of disinfectant and temperature most effective with a new strain of bacteria. Each disinfectant is used at all temperatures. Several different wards in the hospital are identified, and two are randomly assigned to use each combination of disinfectant and temperature. The cases are various pieces of equipment in use in various wards. Identify the pattern of crossing and nesting in this design.

Extending the Concepts

1.6 Expand on Table 1.3 to show a four-way factorial design and then a design in which Treatment D is nested within ABC combinations.

1.7 Give yourself a pat on the back for surviving your first assignment.

2

Organizing, Describing, and Screening Data

2.1 ORGANIZING DATA FOR A COMPUTER

Both statistical packages reviewed in this chapter use data organized in a similar fashion; a data set organized for SPSS is also appropriate for SAS (or SYSTAT or NCSS). The data set can be entered into a file through a spreadsheet, a word processor (and saved in ASCII format), or a database, and then read by the statistical package, or it can be entered directly into the specialized data-entry program of the statistical package. The packages also have conversion facilities that allow them, with varying levels of difficulty, to import data created by some of the other packages.[1]

The statistical packages have a great deal of flexibility for manipulating variables once they are entered into a file. You can convert continuous variables to discrete variables; change numeric values attached to levels of discrete variables; rescale values of continuous variables using logarithmic or other transformations; add together, average, or otherwise manipulate values from two or more variables; and

[1]SPSS requires data to be converted to its own format. SAS command mode is capable of using ASCII files directly. SPSS can read and write SAS files, but the converse is not true.

so on. In other words, the numbers entered into the data set are potentially just the starting point for the values that are analyzed.

The packages also have some built-in flexibility for handling missing values, which occur when, for one reason or another, some data for a case are unavailable. The default option for the programs is to delete a case with data missing on one or more of the variables currently in use. In randomized-groups designs, deletion of the case is appropriate. Deletion of a case, however, often results in unequal numbers of cases in various levels of the IV(s). Problems occasionally associated with unequal n (unbalanced or nonorthogonal designs) are discussed in Chapter 5. In repeated-measures designs, one or more of the DV measures may be missing for a case when other DVs are available. Missing values may then be imputed (guessed at) by various means, instead of dropping the case. This problem and the various remedies for it are discussed in Chapter 6.

A note of caution is appropriate here: Statistical packages are rather simple-minded in that they don't know or care where the numbers come from. You provide the numbers, and they cheerfully and quickly crank out the results of the statistical method you have chosen. Nevertheless, the methods implemented in statistical packages often have underlying assumptions that, if seriously violated, may make the output uninterpretable. Computers do the crunching; you have to do the thinking.

2.1.1 Discrete, Continuous, and Ordinal Data

In a data file, data for each case are entered in a different row. The data in the row can be any mix of discrete, continuous, or ordinal variables, entered in any sequence, but all cases have to use the same order. The order itself doesn't matter, because variables are named and then manipulated by their names. For purposes of proofreading, however, it is usually easier to enter the data in the order in which they are collected.[2]

Each row of data typically is for a different case,[3] and each column of data is for a different variable. The values that are entered into a column depend on the type of variable. If the variable is a discrete/categorical/nominal/enumerated/group membership variable, the researcher assigns an arbitrary number to each category and inserts the appropriate number for the case in the column. Most programs permit you to assign names rather than arbitrary numbers for these kinds of variables, but you may later find that assignment of names has reduced your flexibility. For example, if a discrete variable represents group membership (experimental versus control) or the outcome of quality assessment (pass or fail), the researcher arbitrarily assigns, say, 1 to experimental group/pass and 2 to control group/fail; thus, the column for the variable contains 1s and 2s, each appropriate for a particular case. If the variable is to be treated as continuous, the entries in the column are the actual values (scores) that

[2] Large data sets cannot really be proofread. Instead you have to rely on the *reasonableness* of the descriptive statistics for the distributions of each variable to reveal errors in data entry.

[3] Sometimes there is so much data for each case that you may want to wrap the data for each case into more than one row so that you can view or print out all the values. In any event, each case must begin with a new row.

each case receives on the continuous variable. If the variable is ordinal/rank order, the entry is the rank order assigned to each case or by each case.

By the way, we strongly recommend that you avail yourself of the programs' built-in capacity to attach labels to the values of discrete variables. For instance, if you are assessing whether a case passes or fails quality control, you might use either 1 to represent "pass" or 1 to represent "fail." When you create the data set, you are sure that you will never forget your choice. However, three months and a couple of data sets later, you are likely to be less sure of your decision. Interpretive labels attached to the values during creation of the data set and printed automatically on the output go a long way toward removing this source of confusion. Another option is to create a code book for the data set and be meticulous with its location.

If a datum is missing for a case, remember to leave the appropriate column empty or insert a missing-values code. A missing-values code is a default code used by the program or a code chosen by the researcher to indicate missing values. Most current programs use a period (.) as a default missing-values code. If you don't acknowledge the missing value somehow, the numbers after the missing value may be shifted into the wrong columns.

2.1.2 Randomized-Groups Designs

The minimum randomized-groups design has one discrete variable and one continuous variable. The discrete variable represents levels of the IV and the continuous variable the DV. The data set has a minimum of two columns. The column for the IV contains numbers that represent the level of the IV experienced by each case. The column for the DV contains the scores for the cases. It is common practice to use a third column to represent case number. The data organization for a one-way randomized-groups design with six cases and three levels of A appears in Table 2.1(a).

A randomized-groups design with two IVs, A and B, requires a minimum of three columns: one to code level of A, one to code level of B, and one for the scores on the DV. The data organization for a two-way factorial design with eight cases, two levels of A, and two levels of B appears in Table 2.1(b). An additional column is required to code the levels for each additional IV, whether or not it is in factorial arrangement with the other IVs. If a covariate (CV; see Chapter 8) is also measured, an additional column is required for its values. The data organization for a two-way factorial randomized-groups design with a DV and a CV is shown in Table 2.1(c).

Much of the time, data sets contain more variables than are currently being analyzed. It is common in the social sciences, for instance, to have demographic information for each case, in addition to its status on IVs and the DV. Each additional variable requires an additional column. This information may never be analyzed or may be useful only with certain outcomes of the major analysis. We have usually found it easier to enter all information when the data set is created than to go back and add it later. In other words, it is usually better to put in all the information you have and never use some of it than to go back and (painfully) add it later. You can always save a subset of the variables needed for a particular analysis into another file with a unique name.

TABLE 2.1 Organization of data for randomized-groups designs

(a) One-way randomized-groups design with three levels of *A*

Case	*A*	DV
1	1	14.23
2	1	17.70
3	2	16.94
4	2	17.05
5	3	21.55
6	3	19.46

(b) Two-way factorial randomized-groups design with two levels of *A* and two levels of *B*

Case	*A*	*B*	DV
1	1	1	6
2	1	1	10
3	1	2	22
4	1	2	17
5	2	1	4
6	2	1	11
7	2	2	19
8	2	2	15

(c) Two-way factorial randomized-groups design with two levels of *A*, two levels of *B*, a DV, and a CV

Case	*A*	*B*	DV	CV
1	1	1	6	122
2	1	1	10	146
3	1	2	22	108
4	1	2	17	112
5	2	1	4	107
6	2	1	11	134
7	2	2	19	93
8	2	2	15	72

2.1.3 Repeated-Measures Designs

Each case provides a score at each level of an IV, or each combination of levels of IVs, in a repeated-measures design. Table 2.2(a) shows the layout of data for a one-way repeated-measures design with five cases and three levels of *A*. Table 2.2(b) shows the layout of data for a two-way repeated-measures design with five cases, three levels of *A*, and two levels of *B*. Table 2.2(c) shows the layout of data for a mixed design, where *A* is a randomized-groups IV and *B* is a repeated-measures IV. The layout of data for a mixed design is a combination of the layout for randomized groups and for repeated measures. As designs become larger, more columns are needed, but the organizational principles remain the same.

The scores for each case at each level of a repeated-measures IV, as opposed to the codes for levels of the IV, are entered in the data set.[4] The name of the repeated-measures

[4] Sometimes it is useful to enter repeated-measures data with IV codes (and case codes) in randomized-groups format (cf. Chapter 9).

TABLE 2.2 Organization of data for repeated-measures and mixed designs

(a) One-way repeated-measures design with three levels of *A*

Case	DV at a_1	DV at a_2	DV at a_3
1	0.0047	0.0042	0.0032
2	0.0083	0.0081	0.0078
3	0.0063	0.0059	0.0055
4	0.0075	0.0075	0.0070
5	0.0051	0.0047	0.0043

(b) Two-way factorial repeated-measures design with three levels of *A* and two levels of *B*

Case	DV at a_1b_1	DV at a_1b_2	DV at a_2b_1	DV at a_2b_2	DV at a_3b_1	DV at a_3b_2
1	138	140	155	139	157	155
2	126	126	131	128	130	133
3	140	139	144	138	141	140
4	120	118	119	122	123	130
5	149	144	145	139	148	145

(c) Two-way factorial mixed design with two levels of *A* (randomized groups) and two levels of *B* (repeated measures)

Case	*A*	DV at b_1	DV at b_2
1	1	0.11	0.17
2	1	0.14	0.20
3	2	0.23	0.24
4	2	0.22	0.20

IV, its number of levels, and the labels for the levels are defined "on the fly," so to speak, through facilities provided by the software. If there is more than one repeated-measures IV, it is very helpful to use some naming convention like the one suggested in Table 2.2(b), in which the DV corresponding to particular combinations of levels of *A* and *B* are very clearly indicated.

2.2 DESCRIBING DATA FROM ONE VARIABLE

This section and the one that follows are intended as reminders of material already mastered in a first course on descriptive and inferential statistics. If you are encountering this material for the first time, you may wish to study it in greater detail in a more elementary book such as Spatz (1997).

There are two ways to summarize the data from a single variable—statistical or graphical. Statistical methods involve presenting certain summary values that represent an entire distribution of numbers with just a few statistics. A measure of central tendency is usually presented, along with a measure of dispersion. The measure of central tendency is the single number that best represents the entire distribution; it is a single point on a scale (score line) and is the value that one would guess if asked to make a prediction about the score of a new case. The measure of dispersion indicates the variability in the numbers and the extent to which the scores in the distribution differ from central tendency. The dispersion measure is a distance along a scale. If the measure of dispersion is small, the value of central tendency is a good prediction, because all cases have scores that are close to each other. If the measure of dispersion is large, the prediction is not as good, because cases have scores that are very different and are widely spread along the scale.

The second method, and the one gaining popularity, is graphic. The researcher seeks to present information about a distribution in a visually compelling, immediately comprehensible display. Different graphical methods are useful for displaying different kinds of data, and new methods are constantly being developed. Modern computers allow some truly stunning graphical methods that convey the information about a single variable or the relationships among two or more variables, sometimes in motion. Some of these new methods are summarized by Cleveland (1994), who also provides numerous helpful hints for improving the usefulness of all graphical displays. Tukey (1977) shows many of these methods, which are useful in exploratory data analysis. Hoaglin, Mosteller, and Tukey (1991) demonstrate graphical displays that are especially useful for exploratory ANOVA. Tufte (1983) is also helpful.

The following discussion briefly summarizes statistical and graphical methods useful for different types of variables. Output from SPSS or SAS is used to illustrate some of these methods. However, the choice of package for type of data is more or less random because they both provide similar information. The data set (DESCRPT.* on the data disk that accompanies this text[5]) is randomly generated with 50 cases and five variables (unimaginatively labeled Var_*A*, Var_*B*, Var_*C*, Var_*D*, and Var_*E*); the first four are continuous and the last discrete. The distributions for the continuous variables have been generated to illustrate certain properties of data, such as normality, skewness, and outliers.

2.2.1 Discrete Variables

Recall that discrete variables have only a few levels or values (categories). IVs in ANOVA are always discrete (categorical) variables. (DVs also may be discrete, but usually are not in ANOVA.) For example, a discrete IV might be type of drug with three levels: placebo, stimulant, and tranquilizer. (A discrete DV might be improvement with two values: not improved and improved.)

2.2.1.1 Central Tendency and Dispersion for Discrete Variables

Because numbers are often arbitrarily assigned to name categories of discrete variables, there are very few statistics that are useful for summarizing their distributions. The mode is used as the measure of central tendency, where the mode is the category

[5] The README.TXT file on the disk describes the naming conventions for the data sets.

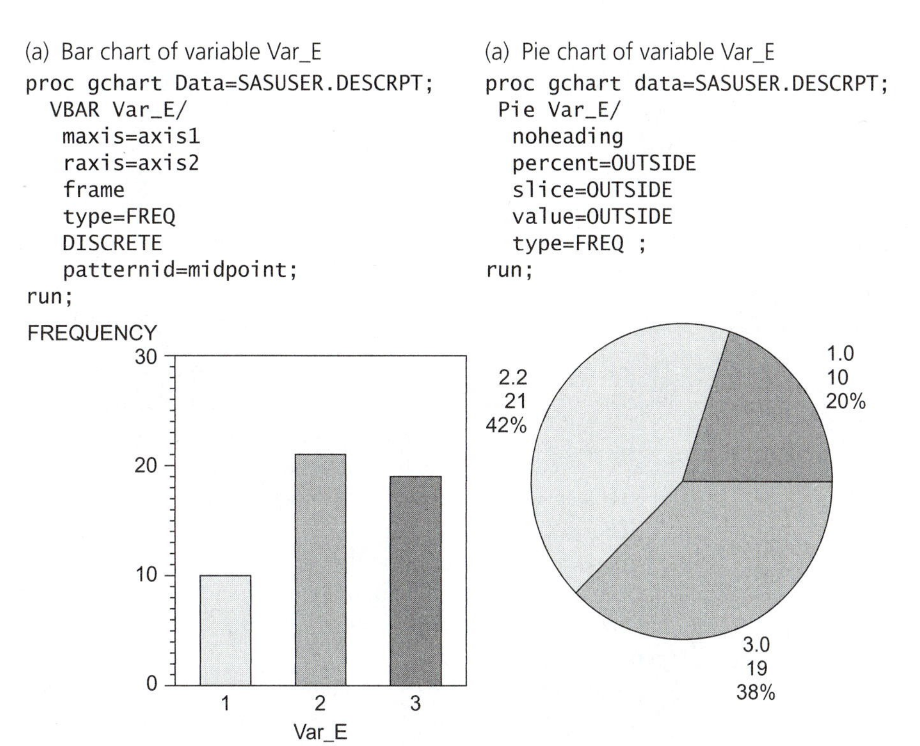

FIGURE 2.1 Syntax, bar chart, and pie chart from SAS GCHART for discrete Var_*E*

with the greatest number of cases. Sometimes there are two such categories, in which case the distribution is said to be bimodal. It is also often useful to report the percentage (and sometimes number) of cases in each category to give your reader an idea of the dispersion of cases into categories.

2.2.1.2 Graphical Methods for Discrete Variables

Frequency histograms, bar charts, and pie charts are often used to display data from a discrete variable. The length of the bar or the size of the slice that represents each category gives a clear, rapid, visual summary of the popularity of the various categories and the dominance of the modal category, as shown in the charts produced by SAS GCHART (accessible through ASSIST), shown in Figure 2.1.

It is apparent from either the bar chart or the pie chart that the middle category (2) is the most popular, with 21 of the cases (42% of 50 cases); this is the category that one would predict for a new case. Category 3 has 19 cases (38%), and category 1 has 10 cases (20%). Category 2, then, is not a great deal more popular than category 3, but both are more popular than category 1.

2.2.2 Ordinal Variables

Rank-order/ordinal data are obtained when the researcher assesses the relative positions of cases in a distribution of cases (e.g., most talented, least efficient), when the researcher has others rank order several items (e.g., most important to me, etc.), or when the researcher has assessed numerical scores for cases but does not trust them. In the last situation, the researcher believes that the case with the highest score has the most (or least) of something but is not comfortable analyzing the numerical scores themselves, so the researcher treats the data as ordinal. Numbers reveal which case is in what position, but there is no assurance that the distance between the first and second cases is the same as, for instance, the distance between the second and third cases or between any other adjacent pair.

2.2.2.1 Central Tendency and Dispersion for Ordinal Variables

The median (the value of the middle case) is used as the measure of central tendency, because the size of adjacent differences is not stable. When there is an odd number of cases, the median is the value associated with the middle case; when there is an even number of cases, the median is the average of the values for the two middle cases. The median is the value at the 50th percentile—50% of the cases have values above the median and 50% below.

Dispersion is assessed in two ways: by range and by quartiles. The range is the highest score minus the lowest score. The quartiles are the values that cut off the upper and lower 25% of the cases. In other words, the lower quartile is the 25th percentile, the median is the 50th percentile, and the upper quartile is the 75th percentile.

The data set DESCRPT.* contains four continuous variables and one discrete variable. Suppose that the researcher has doubts about the measurement of Var_*A* and feels more justified considering it ordinal than continuous—that is, the researcher is willing to take advantage of the order property of the numbers but is uncomfortable about treating the scale as continuous. Appropriate summary statistics are as follows: median for central tendency and range and interquartile range (75th percentile minus the 25th percentile)[6] for dispersion, shown in Table 2.3 using SPSS FREQUENCIES. The /NTILES=4 instruction requests quartiles. Also requested are relevant statistics and a bar chart based on frequencies (i.e., a frequency histogram). The /ORDER ANALYSIS instruction is produced by default using the SPSS menu system.

The range is equal to the maximum score minus the minimum score $(72 - 32 = 40)$. The interquartile range is the value associated with the 75th percentile minus the value associated with the 25th percentile $(58 - 43.75 = 14.25)$, and the semi-interquartile range is half the interquartile range $(14.25/2 = 7.125)$.

2.2.2.2 Frequency Histogram for Ordinal Variables

A frequency histogram or bar chart is appropriate if a graphical display of an ordinal variable created from an originally continuous variable is desired. Figure 2.2 is a frequency histogram produced by SPSS FREQUENCIES. From this display, one can judge the overall shape of the distribution and, with the information in Table 2.3, visualize the position of the median and the width of the interquartile range.

[6] The semi-interquartile range that is sometimes reported is one-half the interquartile range.

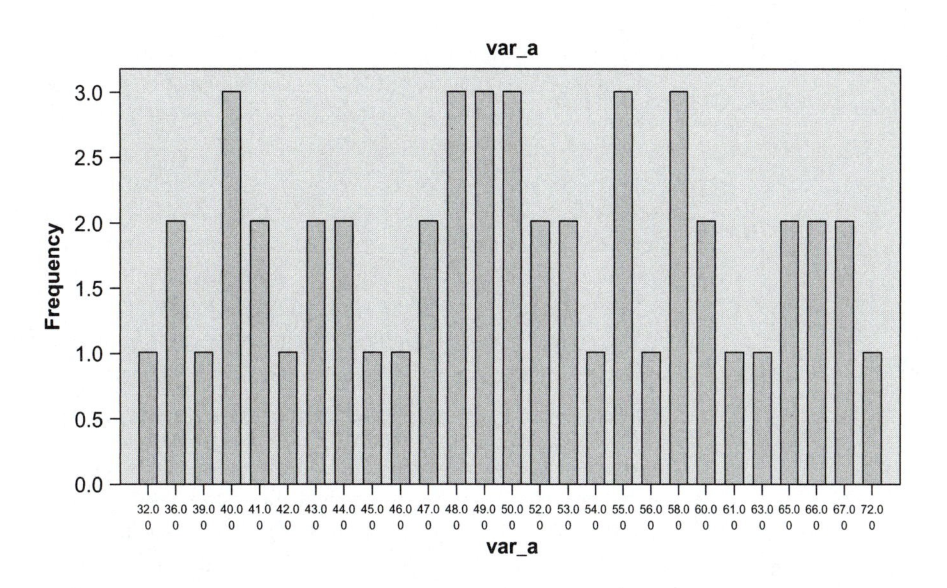

FIGURE 2.2 Frequency histogram of Var_*A* as produced by SPSS FREQUENCIES (syntax is in Table 2.3)

TABLE 2.3 SPSS syntax and descriptive statistics for Var_*A* treated as ordinal

```
FREQUENCIES
  VARIABLES=var_a
  /NTILES= 4
  /STATISTICS=RANGE MINIMUM MAXIMUM MEDIA
  /BARCHART FREQ
  /ORDER ANALYSIS.
```

Statistics

var_a

N	Valid	50
	Missing	0
Median		50.0000
Range		40.00
Minimum		32.00
Maximum		72.00
Percentiles	25	43.7500
	50	50.0000
	75	58.0000

2.2.3 Continuous Variables

Numerous statistics are available for continuous variables. The mean measures central tendency; standard deviation and variance measure dispersion. Skewness and kurtosis describe the shape of the distribution so that normality may be assessed.

Measures of central tendency and dispersion for discrete and ordinal data also are interesting for continuous variables. Histograms, bar charts, stem-and-leaf charts, box plots, and normal probability plots are just a few of the types of graphs available for visual inspection of continuous distributions.

The data set DESCRPT.* contains four continuous variables built to illustrate different features of distributions. All the variables center around 50 and could potentially range between 0 and 100. Var_*A* is fairly normally distributed with a somewhat wide dispersion. Var_*B* is similar, but with narrow dispersion. Var_*C* is positively skewed, and Var_*D* contains an outlier (a case with an outrageous value). These properties of the variables emerge as the distributions are investigated.

2.2.3.1 Central Tendency and Dispersion for Continuous Variables

The mean is the appropriate measure of central tendency for a continuous variable with a distribution that is symmetrical above and below its central value. It is the sum of the scores divided by the number of scores. The mean is also something like a center of gravity for a distribution, because it is the value around which the sum of the deviations is zero. In other words, if you subtract the mean from each score and sum the resulting deviations, the sum is always zero. (These relationships are seen later for one numerical example.)

This property of the mean creates havoc with otherwise common-sense measures that might be developed to assess dispersion. A first thought about an appropriate measure of dispersion might be something like the sum of the deviations of the scores from the mean. However, because the sum of the deviations is always zero, this is unhelpful. For this reason, the deviations are *squared* before being summed to compute standard deviation. A problem then arises with the sum of squared deviations as a potential measure of dispersion; variables from large data sets are likely to have larger dispersion than variables from smaller data sets simply because more squared deviations are added together. For this reason, the squared deviations are *averaged* when computing standard deviation. Finally, squaring the deviations changes the scale from the one in which the variable was originally measured. For this reason, when computing standard deviation, the *square root* of the average squared deviation is taken:

$$s_N = \sqrt{\frac{\sum (Y - \bar{Y})^2}{N}} \tag{2.1}$$

The standard deviation of a sample of raw scores is the *square root* of the *average of the squared deviations* of the scores about their mean, where N is the number of scores.

Equation 2.1 for standard deviation is appropriate when the researcher wants a summary of the dispersion of a particular sample of numbers. Frequently, however, the researcher is more interested in estimating the dispersion in a population from the dispersion in a sample. In that case, Equation 2.1 yields a value that is, on average, slightly too small—on average, the sample standard deviation underestimates the standard deviation in a population because the values in the sample are a bit too close to their own mean. Thus, Equation 2.1 provides a *biased estimate* of standard deviation when applied to a population. The correction for this underestimation is simple—decrease N by 1 in the denominator. Equation 2.2 provides an *unbiased estimate* and is used to estimate the dispersion in a population, σ (sigma), from scores in a sample:

$$s_{N-1} = \sqrt{\frac{\sum(Y - \bar{Y})^2}{N - 1}} \tag{2.2}$$

The standard deviation used to estimate a population dispersion value from a sample uses $N - 1$ in the denominator, instead of N.

If the square root is not taken of the average squared deviation, the result is the variance:

$$s^2_{N-1} = \frac{\sum(Y - \bar{Y})^2}{N - 1} \tag{2.3}$$

The variance is a numerical assessment of dispersion in squared units of the scale in which the variable is measured. This makes variance harder to understand, although you do develop a feel for it after awhile. A conceptual relationship between variance and diagrams is a helpful and powerful way to understand relationships among variables (Figure 2.3), as introduced in Chapter 1 and used in later chapters.

In a conceptual diagram, the total variance of a variable is illustrated as a circle. The extent of a relationship between two variables is represented by the degree of overlap between their two circles. If the relationship is strong, the overlap is high; if the relationship is low or nonexistent, there is small or no overlap. In Figure 2.3, the circle labeled X represents the total variance in X, the circle labeled Y represents the total variance in Y (which, as illustrated by the bigger circle, has more variance than X), and the overlapping area represents a moderate degree of relationship between X and Y.

There are two equations for standard deviation and variance; one is easier to understand, and the other is easier to compute and less subject to rounding error. The easier-to-understand standard deviation equation is shown in Equation 2.2.[7] Should you find yourself on a desert island with your calculator but no computer, the

FIGURE 2.3 Variance and degree of relationship between variables

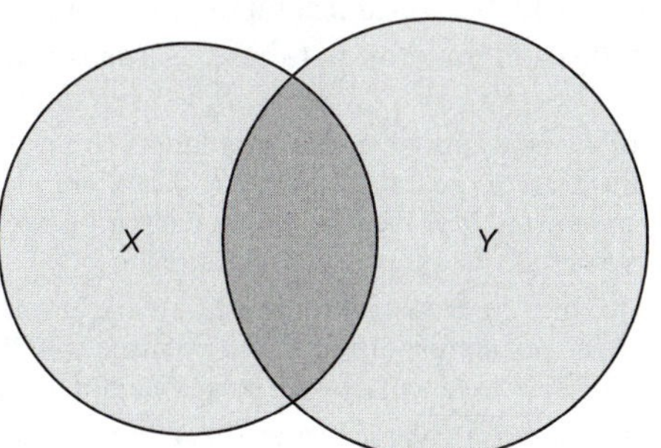

[7] Believe it or not.

easier-to-calculate raw score version is as follows:

$$s_{N-1} = \sqrt{\frac{\sum Y^2 - \frac{\left(\sum Y\right)^2}{N}}{N-1}} \qquad \textbf{(2.4)}$$

2.2.3.2 Parameter Estimates, Confidence Intervals, and Sampling Distributions

The desire to estimate central tendency in a population from the mean of a sample leads to parameter estimation and the computation of confidence intervals. The mean in the population is the parameter to be estimated in ANOVA.[8] Confidence limits are boundaries placed around the sample mean to create an interval of values, indicating that populations are likely to have means μ (mu) that could have produced the sample mean.[9] A common size (and the default for many programs) is the 95% confidence interval; the interval based on the sample mean contains the true population mean in 95% of all samples (Mulaik, Raju, and Harshman, 1997). If, instead, 99% confidence limits are used, the confidence interval is wider, and the interval based on the sample mean contains the true population mean in 99% of all samples. Confidence intervals with other limits are also possible, depending on the particular application. Confidence intervals are computed using the following equation:

$$\overline{Y} - t(s) \leq \mu \leq \overline{Y} + t(s) \qquad \textbf{(2.5)}$$

A confidence interval for a population mean, μ, is the sample mean plus or minus the t value[10] at the required probability level times the standard error of the mean (i.e., the standard deviation of the sampling distribution the sample means, $s_{\overline{Y}}$).

The standard error is a special kind of standard deviation that comes from a sampling distribution in which statistics, not raw scores, are distributed.

The standard error of the mean comes from a distribution of sample means. A sampling distribution is more theoretical than actual because it is a distribution of all the values of the *statistics* (means, in this case);[11] it is computed for all samples of a given size drawn from a population. For instance, you draw a first sample of, say, 37 cases from a population of 4000 cases and compute the mean. That mean becomes one entry in the sampling distribution. You put these 37 cases back into the population,

[8] The observed values of statistics estimated for populations are called *parameter estimates;* the population values themselves are called *parameters* and are usually depicted by Greek letters. Population parameters are hypothesized, estimated, or known; they are not calculated unless you happen to have all the data from the population.

[9] Confidence intervals also may be estimated for the difference between two means. These are the bounds within which the population difference is expected to fall. If the interval includes zero, the null hypothesis of no difference between means cannot be rejected.

[10] Tables of probabilities associated with t values are in all introductory statistics texts, such as Spatz (1997).

[11] The sample size, N, is used in the equation even when the known population standard deviation (σ) replaces its estimated value (S_{N-1}). The standard error of the mean then is labeled $\sigma_{\overline{Y}}$.

draw a different sample of 37 cases, compute their mean, and keep going until you have drawn and computed the mean for some huge number of different samples of size 37 from this population of 4000 cases.

The standard error of the mean is the standard deviation of the distribution of means. Fortunately, you do not have to draw these samples; the simple equation for using a sample standard deviation to estimate a standard error for a sampling distribution of means is

$$s_{\overline{Y}} = \frac{S_{N-1}}{\sqrt{N}} \tag{2.6}$$

The standard error for the sampling distribution of means (the standard error of the mean) is the standard deviation used to estimate dispersion in a population divided by the square root of sample size.

One pleasant feature of sampling distributions is that their shapes are often known. For instance, the Central Limit Theorem tells us that a sampling distribution of means has an approximately normal distribution if samples of about size 30 or larger are drawn from the population of raw scores. The distribution of raw scores does not have to be normal, or anywhere near normal, for the sampling distribution of means to be normal. Therefore, as long as one is dealing with the sampling distribution, and not the raw scores, known properties of a normal curve (or whichever curve is appropriate) can be used. This advantage of sampling distributions allows one to go back and forth between statistical values and probabilities in statistical inference.

The sample of numbers in Table 2.4 is used to illustrate computations of standard deviation, variance, and a 95% confidence interval. The mean is

$$\overline{Y} = \frac{\sum Y}{N} = \frac{43}{10} = 4.3$$

TABLE 2.4 Data set for illustration of mean, standard deviation, variance, and standard error

Y	$(Y - \overline{Y})$	$(Y - \overline{Y})^2$
4	$4 - 4.3 = -0.3$	0.09
1	$1 - 4.3 = -3.3$	10.89
7	$7 - 4.3 = 2.7$	7.29
3	$3 - 4.3 = -1.3$	1.69
2	$2 - 4.3 = -2.3$	5.29
9	$9 - 4.3 = 4.7$	22.09
7	$7 - 4.3 = 2.7$	7.29
3	$3 - 4.3 = -1.3$	1.69
6	$6 - 4.3 = 1.7$	2.89
1	$1 - 4.3 = -3.3$	10.89
$\sum Y = 43$	$\sum(Y - \overline{Y}) = 0$	$\sum(Y - \overline{Y})^2 = 70.1$

Following Equation 2.1, the standard deviation of the sample is

$$s_N = \sqrt{\frac{70.1}{10}} = 2.65$$

The estimated standard deviation for the population (Equation 2.2) is

$$s_{N-1} = \sqrt{\frac{70.1}{9}} = 2.79$$

The estimated population variance (Equation 2.3) is

$$s^2_{N-1} = (2.79)^2 = 7.78$$

The standard error of the mean (standard deviation for the sampling distribution of means [Equation 2.6]) is

$$s_{\overline{Y}} = \frac{2.79}{\sqrt{10}} = 0.88$$

Therefore, the 95% confidence interval (with $t = 2.26$ when there are 9 degrees of freedom[12]) for the population mean (Equation 2.5) is

$$4.3 - (2.26)(.88) \leq \mu \leq 4.3 + (2.26)(.88)$$

$$2.31 \leq \mu \leq 6.29$$

Thus, we are 95% confident that the population mean, μ, is in the interval between 2.31 and 6.29.

Fun though all of this is, most people rely on computers to provide simple descriptive statistics. Table 2.5 contains appropriate descriptive statistics for the continuous variables through the SAS Interactive Data Analysis menu system. Menu selections are listed because, unlike other SAS procedures, Interactive Data Analysis does not produce syntax in the SAS log.

It takes a bit of experience to learn to read output, but it is well worth the effort. Numerous descriptive statistics for four variables, each with 50 cases, are accurately summarized in SAS in a fraction of a second. The same information could be produced by hand, but it would take about half a day and would be of dubious accuracy. Summarized for each variable is N, the sum of the weights (Sum Wgts), the Mean, the Sum of the values of the variables, the standard deviation (Std Dev), Variance, Skewness, Kurtosis, the uncorrected sum of squares (USS), the corrected sum of squares (CSS), the coefficient of variation (CV), the standard error of the mean (Std Mean), and the 95% Confidence Intervals for the mean, standard deviation, and variance.

Notice in Table 2.5 that standard deviation (Std Dev) and the confidence interval for the mean are quite a bit smaller for Var_*B* than they are for Var_*A*, indicating that the values in Var_*B* are more like each other, although all variables center at about 50. The CV value reported for each variable is the coefficient of variation, or the standard deviation divided by the mean. If the mean is very large and the dispersion very small, the CV is quite small, signaling a potential problem with the accuracy with which older computers render the numbers. Std Mean is the standard error of the mean (standard

[12] Degrees of freedom (df) are discussed more thoroughly in Chapter 3 and elsewhere.

TABLE 2.5 Menu selections and output from SAS INTERACTIVE for Var_*A*, Var_*B*, Var_*C*, and Var_*D*

1. Open SAS Interactive Data Analysis with the appropriate data set (here, SASUSER.DESCRPT).
2. Choose Analyze and then Distribution(Y).
3. Select Y variables: Var_A, Var_B, Var_C, and Var_D.
4. In the Output dialog box, select Moments and Basic Confidence Intervals.

Var_A

Moments			
N	50.0000	Sum Wgts	50.0000
Mean	51.1600	Sum	2558.0000
Std Dev	9.4747	Variance	89.7698
Skewness	0.2202	Kurtosis	−0.6498
USS	135266.000	CSS	4398.7200
CV	18.5197	Std Mean	1.3399

95% Confidence Intervals			
Parameter	Estimate	LCL	UCL
Mean	51.1600	48.4673	53.8527
Std Dev	9.4747	7.9145	11.8067
Variance	89.7698	62.6398	139.3989

Var_B

Moments			
N	50.0000	Sum Wgts	50.0000
Mean	49.7200	Sum	2486.0000
Std Dev	1.9171	Variance	3.6751
Skewness	0.0373	Kurtosis	−0.6842
USS	123784.000	CSS	180.0800
CV	3.8557	Std Mean	0.2711

95% Confidence Intervals			
Parameter	Estimate	LCL	UCL
Mean	49.7200	49.1752	50.2648
Std Dev	1.9171	1.6014	2.3889
Variance	3.6751	2.5644	5.7069

Var_C

Moments			
N	50.0000	Sum Wgts	50.0000
Mean	49.8000	Sum	2490.0000
Std Dev	11.6759	Variance	136.3265
Skewness	1.3582	Kurtosis	1.8361
USS	130682.000	CSS	6680.0000
CV	23.4456	Std Mean	1.6512

Var_D

Moments			
N	50.0000	Sum Wgts	50.0000
Mean	51.1200	Sum	2556.0000
Std Dev	8.1407	Variance	66.2710
Skewness	−0.2307	Kurtosis	−0.0263
USS	133910.000	CSS	3247.2800
CV	15.9247	Std Mean	1.1513

95% Confidence Intervals			
Parameter	Estimate	LCL	UCL
Mean	51.1200	48.8064	53.4336
Std Dev	8.1407	6.8002	10.1444
Variance	66.2710	46.2428	102.9088

deviation of the sampling distribution of $\overline{Y}$). USS and CSS are sums of squared deviations around the mean, uncorrected for the mean and corrected, respectively. These values are helpful when calculating ANOVA by hand, as described in Chapter 3. Notice that all variables have Sum Wgts = 50, indicating that each case has been allocated equal weight (by default). Descriptions of skewness and kurtosis values follow.

2.2.3.3 Normality

Normal distributions are symmetrical about the mean, with a defined shape and height. Mean, median, and mode are the same, and the percentages of cases between the mean and various standard deviation units from the mean are known. For this reason, you can rescale a normally distributed continuous variable to a z score scale (with mean 0 and standard deviation 1) and look up the probability that corresponds to a particular range of raw scores. The legitimacy of using the z score scale and its associated probabilities depends on the normality of the distribution of the continuous variable. The following equation is used for conversion of a raw score to a z score:[13]

$$z = \frac{Y - \overline{Y}}{s_{N-1}} \tag{2.7}$$

There are both statistical and graphical methods of assessing the normality of the distribution of a continuous variable. Skewness and kurtosis are descriptive statistics; skewness assesses the symmetry of the distribution and kurtosis its peakedness. Distributions are not normal if they are not symmetrical or if they are too peaked or too flat. A positively skewed distribution has a pileup of cases at small values and a few cases with large values that lengthen the tail of the distribution at high values. A negatively skewed distribution has a pileup of cases at large values and a few cases with small values that lengthen the tail of the distribution at small values. A leptokurtic distribution (one with positive kurtosis) is too peaked; a platykurtic distribution (one with negative kurtosis) is too flat (think "flatty/platy"). Both have a different percentage of cases between standard deviation units than does the normal distribution, so z score conversions, inferential tests, and the like with non-normal distributions, are often misleading. A normal distribution is called *mesokurtic.* Figure 2.4 shows a normal curve and several curves that depart from normality.

In a normal distribution, skewness and kurtosis are zero. The standard error of skewness is

$$s_s = \sqrt{\frac{6}{N}} \tag{2.8}$$

For the fictitious data set DESCRPT.*, with $N = 50$, the standard error of skewness is

$$s_s = \sqrt{\frac{6}{N}} = 0.346$$

The standard error of kurtosis is

$$s_k = \sqrt{\frac{24}{N}} \tag{2.9}$$

[13] The sample standard deviation (s_N) is appropriate for some uses of z scores if there is to be no generalization of them beyond the sample tested. For example, classroom grades are sometimes based on sums (or averages) of z scores of exams over the course of a semester.

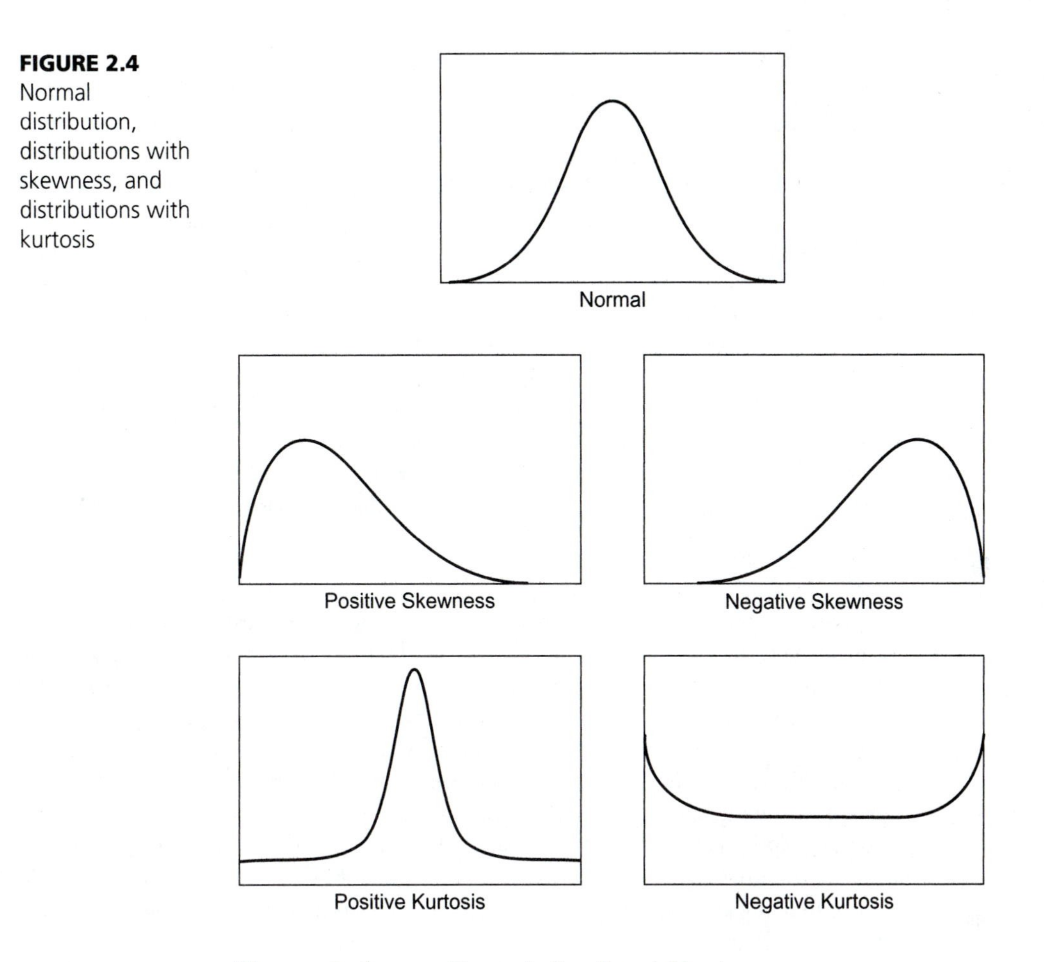

FIGURE 2.4 Normal distribution, distributions with skewness, and distributions with kurtosis

The standard error of kurtosis for all variables is

$$s_k = \sqrt{\frac{24}{50}} = 0.693$$

To determine whether a distribution departs from normality, you can use either of these standard errors by dividing the skewness or kurtosis value for the distribution by its respective standard error and looking up the result as a z score from a standard normal table of values. For skewness,

$$z_s = \frac{s - 0}{s_s} \tag{2.10}$$

and for kurtosis,

$$z_k = \frac{k - 0}{s_k} \tag{2.11}$$

Table 2.5 shows skewness and kurtosis values for the continuous variables in the data set. After calculating the standard errors using Equations 2.8 and 2.9, the z score for

Var_*A* for skewness is 0.64 (0.220/0.346), and the z score for kurtosis is −0.94. For Var_*C*, which was generated to have skewness, the z score for skewness is 3.92, and the z score for kurtosis is 2.65.[14] Conventional but conservative (.01 or .001) two-tailed alpha levels (along with inspection of the shape of the distribution) are used to evaluate the significance of skewness and kurtosis with small to moderate samples.[15] By these criteria, Var_*A* is normal, but Var_*C* has statistically significant positive skewness. If the sample is large, however, it is better to inspect the shape of the distribution instead of using formal inference, because the equations for standard error of both skewness and kurtosis contain N, and normality is likely to be rejected with large samples even when the deviation is slight.

Some advanced analyses require evaluation of normality by testing skewness and kurtosis. If the z scores for skewness and/or kurtosis are too discrepant, requirements for normality are often satisfied by transforming the DV (discussed in Section 3.5.7). The problem with transformation is that results are interpreted in terms of the transformed DV, not the original scale of the DV, an often awkward state of affairs.

2.2.3.4 Outliers

Outliers are deviant cases that may have undue impact on the results of analysis. They can increase means or lower means and, by doing so, create artificial significance (increase Type I error rate; see Section 1.6.1) or cover up real significance (increase Type II error rate). They almost always increase measures of dispersion, making it less likely to find significance. Their inclusion in a data set, in short, makes the outcome of analysis unpredictable and not generalizable, unless to a population that happens to include the same sort of outlier.

An outlier is a score that is unusually far from the mean and apparently disconnected from the rest of the scores. The z score distribution is used to assess the distance of a raw score from its mean. Equation 2.7 is used to convert a raw score to a z score. If the sample is fairly large, say, 100 or more, a case with an absolute z value of 3.3 or greater is probably an outlier, because the probability of sampling a score of this size in random sampling from the population of interest is .001 or less (two-tailed). If the sample is smaller, an absolute value of $z = 2.58$ ($p < .01$, two-tailed) is appropriate. However, visual inspection of the distribution is needed before firmly concluding that a case is an outlier, as seen in what follows.

If you discover a case that might be an outlier, the first step is to check the score for the case to make sure it is accurately entered into the data file. If the datum is accurately entered, next check that missing values have been correctly specified. If not, the computer may be reading the missing-values code as a raw score. If the datum is accurately entered and missing values are specified, next decide whether the

[14] Var_*D* also has a z score indicative of skewness, but it is due to the presence of an outlier, as becomes clear in the next section.

[15] There also are formal statistical tests for the significance of the departure of a distribution from normality, such as Shapiro and Wilks' W statistic, but they are very sensitive and often signal departures from normality that do not really matter.

case is properly part of the population from which you intended to sample. If the case is not part of the population, it is deleted with no loss of generalizability of results to your intended population.[16] The description of outliers is a description of the kinds of cases to which the results do not apply; investigation of the conditions associated with production of the outlying score can reveal the effects of unintended changes in your research program (i.e., shifts in delivery of treatment).

If you decide that the outlier is properly sampled from your population of interest, it remains in the analysis, but steps are taken to reduce its impact—either the distribution is transformed or outlying scores are changed. The first option for reducing impact is to transform *all* the scores to change the shape of the entire distribution to more nearly normal (cf. Chapter 3). In this case, the outlier is considered to have come from a non-normal distribution with tails that are too heavy, so that too many cases fall at extreme values. The same transformations—square root, logarithmic, or inverse—used to produce normality (Section 2.2.3.3) are thus appropriate to reduce the impact of an outlier. A case that is an outlier in the untransformed distribution is still on the tail of the transformed distribution, but its impact is reduced.

The second option is to change the score for just the outlying case so that it is deviant, but not as deviant as it was. For instance, the outlying case can be assigned a raw score that is one unit larger (or smaller) than the next most extreme score in the distribution. This is an attractive alternative to reduce the impact of an outlier if measurement is rather arbitrary anyway.

Researchers are often reluctant to deal with outliers. They feel that the sample should be analyzed "as is." However, not dealing with outliers is also dealing with them, just in another way. The outliers remain in the analysis but change the population to which one can generalize to a population that contains a similar outlier. Not dealing with outliers is a bit like burying one's head in the sand and hoping that the problem will simply disappear.

Any deletions, changes of scores, or transformations are reported together with the rationale. You should always be skeptical as you read a paper that makes no mention of outliers; the authors either dealt with them but failed to mention it or did not consider the problem, which raises questions about the findings.

In the example data set, Var_*D* was created with an outlying score. Features of SPSS EXPLORE that are appropriate for identifying outliers are shown in Table 2.6. The PLOT instruction requests a BOXPLOT; the STATISTICS instruction requests EXTREME values (to identify outliers), as well as DESCRIPTIVES. The remaining instructions are default values generated by the SPSS menu system.[17]

The output segment labeled "Descriptives" contains most of the important descriptive statistics for a continuous variable, and that labeled "Extreme Values" shows information relevant for identifying outliers. The cases with the highest and lowest five values are listed along with the values. For Var_*A*, the highest and lowest

[16] Deletion of cases in ANOVA produces unequal sample sizes in cells representing levels of the IV(s). This, in turn, raises other issues (see, e.g., Sections 4.5.4 and 6.5.6).

[17] If you want box plots from two or more groups side by side, use the box-and-whisker plots available through quality control menus.

TABLE 2.6 Identification of outliers through SPSS EXPLORE (syntax and output)

```
EXAMINE
  VARIABLES=var_a var_d
  /PLOT BOXPLOT
  /COMPARE GROUP
  /STATISTICS DESCRIPTIVES EXTREME
  /CINTERVAL 95
  /MISSING LISTWISE
  /NOTOTAL.
```

Explore

Descriptives

			Statistic	Std. Error
var_a	Mean		51.1600	1.33992
	95% Confidence	Lower Bound	48.4673	
	Interval for Mean	Upper Bound	53.8527	
	5% Trimmed Mean		51.1000	
	Median		50.0000	
	Variance		89.770	
	Std. Deviation		9.47469	
	Minimum		32.00	
	Maximum		72.00	
	Range		40.00	
	Interquartile Range		14.25	
	Skewness		.220	.337
	Kurtosis		-.650	.662
var_d	Mean		50.5200	1.37269
	95% Confidence	Lower Bound	47.7615	
	Interval for Mean	Upper Bound	53.2785	
	5% Trimmed Mean		51.1111	
	Median		50.0000	
	Variance		94.214	
	Std. Deviation		9.70638	
	Minimum		13.00	
	Maximum		69.00	
	Range		56.00	
	Interquartile Range		10.50	
	Skewness		-1.218	.337
	Kurtosis		3.498	.662

TABLE 2.6 Continued

Extreme Values

			Case Number	Value
var_a	Highest	1	20	72.00
		2	2	67.00
		3	16	67.00
		4	36	66.00
		5	39	66.00
	Lowest	1	45	32.00
		2	22	36.00
		3	8	36.00
		4	1	39.00
		5	30	40.00[a]
var_d	Highest	1	48	69.00
		2	40	64.00
		3	5	63.00
		4	14	62.00
		5	38	62.00[b]
	Lowest	1	12	13.00
		2	34	32.00
		3	2	33.00
		4	46	35.00
		5	32	38.00

a. Only a partial list of cases with the value 40.00 are shown in the table of lower extremes.

b. Only a partial list of cases with the value 62.00 are shown in the table of upper extremes.

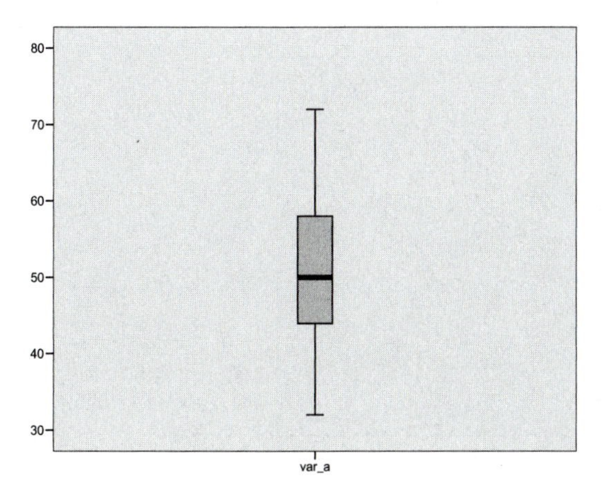

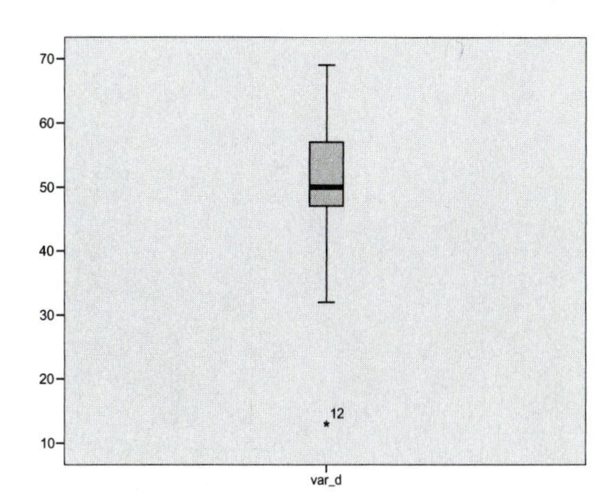

values do not differ much from the values near them, but for Var_*D*, the lowest value of 13 differs considerably from the next higher value of 32, which, in turn, does not differ much from the next higher value of 34. The z score associated with the value of 13 is extreme: at $(13 - 50.52)/9.71 = 3.86$, according to Equation 2.7.

Further evidence that the score is an outlier comes from the box plot for Var_*D*, where case 12 is identified as below the interval containing the rest of the values. In the box plot for Var_*A*, there is no case with a value outside the interval. The interval in the box plot is based on the interquartile range—the range that is between the 75th and 25th percentile and that contains 50% of the cases. These borders are called *hinges.*

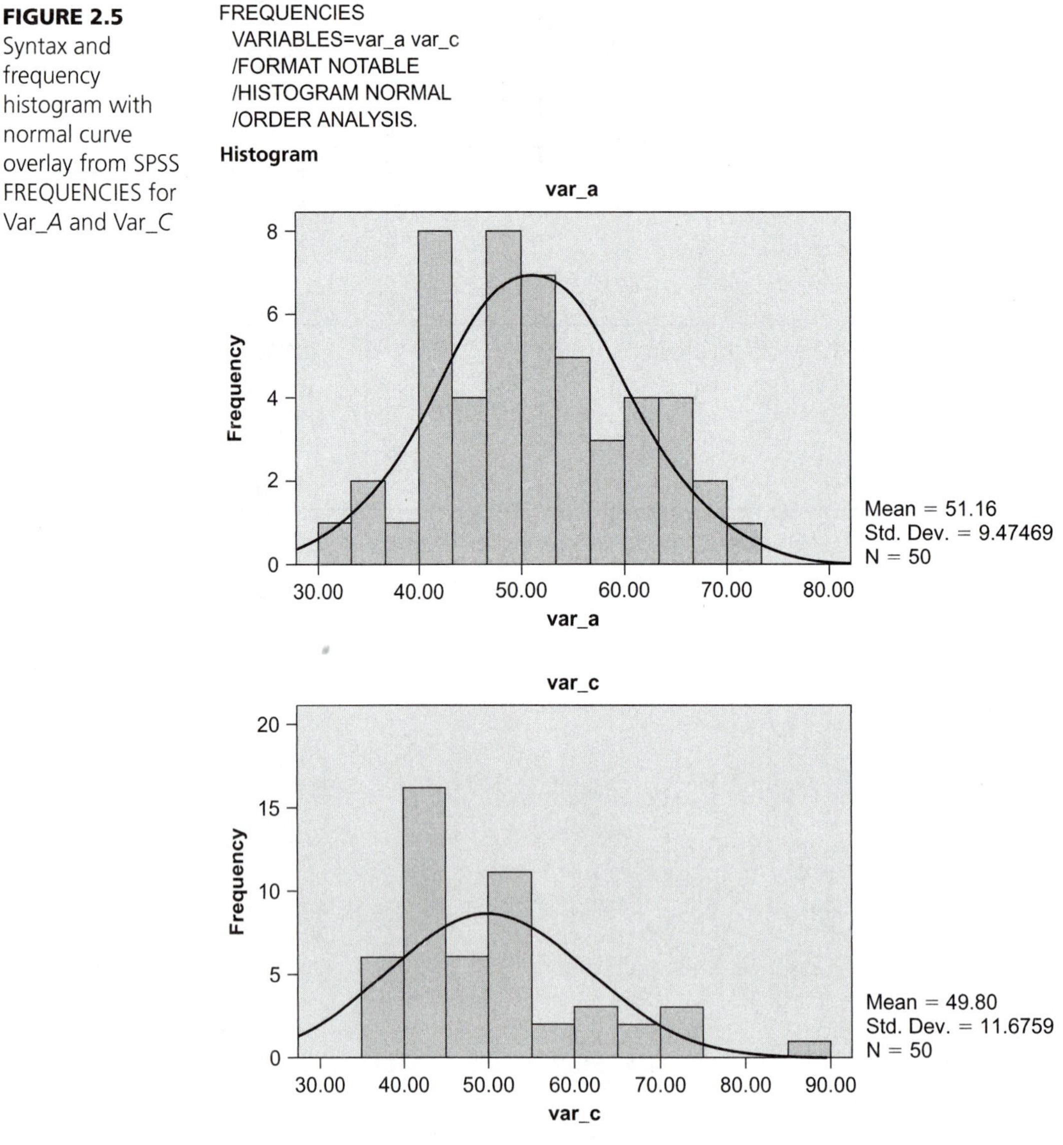

FIGURE 2.5 Syntax and frequency histogram with normal curve overlay from SPSS FREQUENCIES for Var_*A* and Var_*C*

The median is the line through the box. If the median is off center, there is some skewness to the data. The lines outside the box are 1.5 times the interquartile range from their own hinges; that is, the top line is 1.5(75th percentile–25th percentile) above the 75th percentile, and the lower line is 1.5(75th percentile–25th percentile) below the 25th percentile. These two lines are called the *upper inner fences* and *lower inner fences,* respectively.[18] Any score, such as that for case 12, that is above or below the inner fences is likely to be an outlier.

If this were a real data set, with the scale of measurement of Var_*D* somewhat arbitrary, the researcher might decide (after checking the accuracy of data entry and missing-values codes) to convert the 13 to a value around 30.

2.2.3.5 Graphical Methods for Continuous Variables

Along with examining the statistics, you also want to look at graphical representations of the distributions of continuous variables. Graphical methods include, among others, frequency histograms, stem-and-leaf charts, and normal probability plots. SPSS FREQUENCIES produced the frequency histograms in Figure 2.5 for Var_*A*, which is normally distributed, and for Var_*C*, which is not. The normal curve overlay is selected, along with the frequency histogram, to assist in judgment of normality. The positive skewness of Var_*C* is readily apparent from the graph.

Stem-and-leaf plots are also available for assessing the overall shape of distributions, but they take a little getting used to. They actually display every number in the distribution by using the largest digit (thousands, hundreds, tens, or whatever) to identify the rows and then the smaller digit(s) as the entries in the rows. Figure 2.6 shows a stem-and-leaf plot for Var_*A* generated by SPSS EXPLORE.

FIGURE 2.6 Syntax and stem-and-leaf plot produced by SPSS EXPLORE for Var_*A*

```
EXAMINE
  VARIABLES=var_a
  /PLOT STEMLEAF
  /COMPARE GROUP
  /STATISTICS NONE
  /CINTERVAL 95
  /MISSING LISTWISE
  /NOTOTAL.
```

```
var_a Stem-and-Leaf Plot

 Frequency    Stem &  Leaf

     1.00        3 .  2
     3.00        3 .  669
    10.00        4 .  0001123344
    10.00        4 .  5677888999
     8.00        5 .  00022334
     7.00        5 .  5556888
     4.00        6 .  0013
     6.00        6 .  556677
     1.00        7 .  2

Stem width: 10
Each leaf:     1 case(s)
```

[18] There can be outer fences as well (which are three times the interquartile range from their respective hinges), but only if there are very extreme data points.

The rows (stems) in Figure 2.6 are in tens (as identified near the end of the table), and the elements in the rows (leaves) are in single units. The first column gives the frequency count for values in that range. Thus, there is one number in the first row (32) and three numbers in the next row (36, 36, and 39).

Normal probability plots also require some explanation. In these plots, the scores are first sorted and then ranked. Then an expected normal value is computed

FIGURE 2.7
Syntax and normal probability plots of Var_*A* and Var_*C* produced by SPSS EXPLORE

```
EXAMINE
  VARIABLES=var_a var_c
  /PLOT NPPLOT
  /COMPARE GROUP
  /STATISTICS NONE
  /CINTERVAL 95
  /MISSING LISTWISE
  /NOTOTAL.
```

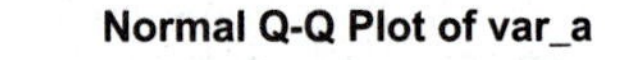

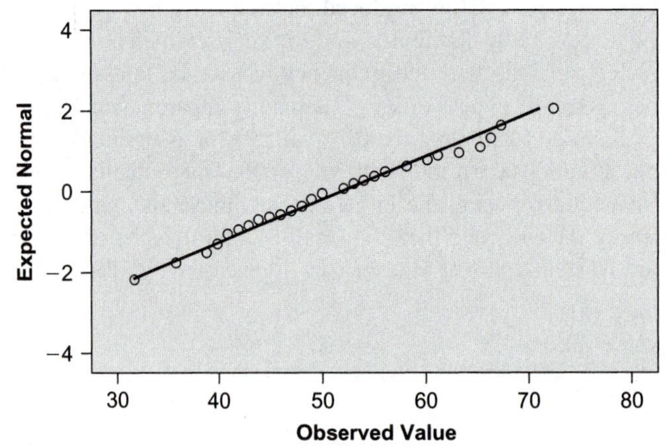

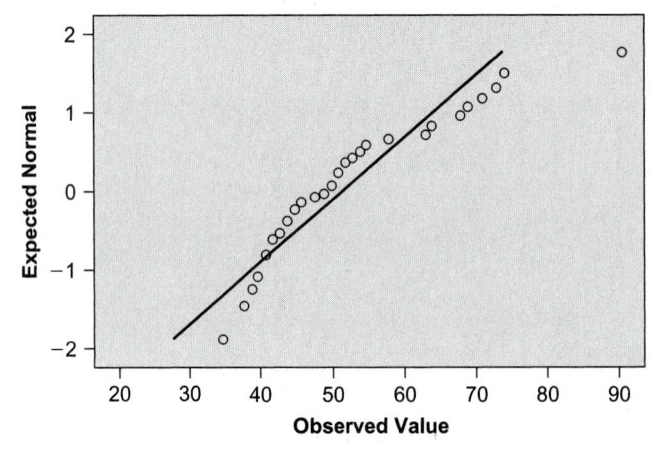

and plotted against the actual normal value for each case. The expected normal value is the z score that a case with that rank holds in a normal distribution; the actual normal value is the z score it has in the actual distribution. If the actual distribution is normal, the two z scores are similar, and the points fall along the diagonal running from lower left to upper right, with some minor deviations due to random processes. Deviations from normality shift the points away from the diagonal. Figure 2.7 contains normal probability plots (requested as NPPLOT) for Var_*A* and Var_*C* produced by SPSS EXPLORE.

As shown in Figure 2.7, the data points fall very close to the diagonal in the plot for Var_*A* but some distance from it in the plot for Var_*C*. The low values and the high values of Var_*C* have z scores that are a bit too low, whereas the z scores for the middle values are a bit too high. The data point far from the others in the upper right part of the plot for Var_*C* looks suspiciously like an outlier and, if the data set were real, deserves further investigation.

Usually it is preferable to examine graphical distributions of variables before or along with an examination of the statistics. In any case, you often find that you have to use both methods to fully understand the variables you are analyzing.

2.3 DESCRIBING RELATIONSHIPS BETWEEN TWO VARIABLES

Often, the researcher is interested in the relationship between two variables, as well as the distributions of both of them. Both statistical methods and graphical methods of examining relationships depend on the types of both variables. If both variables are discrete, chi-square analysis and bar charts are used. If both are continuous, correlation and regression, along with scattergrams, are used. If one is discrete and the other continuous, ANOVA and line graphs are used.

All of these two-variable analyses contain both descriptive and inferential components. Inferential tests for chi-square and correlation are discussed in this chapter, along with descriptive components. Remaining inferential analyses are in subsequent chapters.

2.3.1 Both Variables Discrete

Chi-square (χ^2) analysis is used to examine the relationship between two discrete variables. For instance, χ^2 is the appropriate analysis if one wants to examine a potential relationship between three different machines for producing a part and the number of parts that pass (and fail) quality control.

The null hypothesis (cf. Section 1.6.1) in χ^2 analysis generates expected frequencies against which observed frequencies are tested. If the observed frequencies are similar to the expected frequencies, the value of χ^2 is small, and the null hypothesis is retained; if they are sufficiently different, the value of χ^2 is large, and the null hypothesis is rejected. The relationship between the size of χ^2 and the difference in observed and expected frequencies can be readily seen from the computational equation for χ^2:

$$\chi^2 = \sum_{ij} \frac{(fo - Fe)^2}{Fe} \qquad \textbf{(2.12)}$$

where *fo* represents observed frequencies and *Fe* represents the expected frequencies in each cell. Summation is over all the cells in a two-way table. Expected frequencies are estimated during the analysis.

Usually the expected frequencies for a cell are generated from its row sum and its column sum.

$$\text{Cell } Fe = \frac{(\text{row sum})(\text{column sum})}{N} \qquad \textbf{(2.13)}$$

When this procedure is used to generate the expected frequencies, the null hypothesis tested is that the variable on the row (say, pass/fail quality control) is independent of the variable on the column (type of machinery). If the fit of expected-to-observed frequencies is good (so that χ^2 is small), then one concludes that the two variables are probably independent; a poor fit leads to a large χ^2, rejection of the null hypothesis, and the conclusion that the two variables are probably related.

The degrees of freedom (df) for the analysis are

$$(R - 1)(C - 1) \qquad \textbf{(2.14)}$$

or the number of rows minus 1 times the number of columns minus 1. You probably remember the concept of df from your previous course, where it was discussed as the number of numeric elements that are free to take on any value before the last ones can be found by subtraction. Or you may have encountered the loss of df as the number of parameters that are estimated.

Table 2.7 contains a fictitious data set appropriate for χ^2 analysis that relates three pieces of machinery to the number of items that either pass or fail inspection. Table 2.7(a) *describes* the relationship between machines and outcomes—whether they pass or fail. Columns include the three machines for producing parts, and rows include the information for each machine. Table 2.7(b) contains information for the *inferential* component of the relationship. Columns include observed and expected frequencies for each cell (each combination of type of machine and outcome). The final column contains the χ^2 component for each cell, and ends with the sum over the six cells. The calculated χ^2 value (sum of individual cell values) in Table 2.7(b) is 7.18. In this case, there are 2 df—(2 – 1)(3 – 1)—so that the critical χ^2 value for rejecting the null hypothesis is 5.99 at $\alpha \leq .05$ (cf. Table A.2 in Appendix A). Because 7.18 exceeds 5.99, the null hypothesis of no relationship between the variables is rejected in favor of the conclusion that the different machines produce different quality control pass rates.

Bar charts are often used to graphically display the relationship between two discrete variables. Figure 2.8 contains the display for the data in Table 2.7(a), produced by

TABLE 2.7 Data set for illustration of chi-square analysis

(a) Observed outcomes for each type of machine

	Type of Machine			
Outcome	a_1	a_2	a_3	
Pass (b_1)	73	62	28	163
Fail (b_2)	27	18	22	67
	100	80	50	230

(b) χ^2 analysis of outcomes for each type of machine

Cell	*fo*	*Fe*	$(fo - Fe)^2/Fe$
a_1b_1	73	$163(100)/230 = 70.9$	$(73 - 70.9)2/70.9 = .06$
a_1b_2	27	$67(100)/230 = 29.1$	$(27 - 29.1)2/29.1 = .15$
a_2b_1	62	$163(80)/230 = 56.7$	$(62 - 56.7)2/56.7 = .50$
a_2b_2	18	$67(80)/230 = 23.3$	$(18 - 23.2)2/23.2 = 1.17$
a_3b_1	28	$163(50)/230 = 35.4$	$(28 - 35.4)2/35.4 = 1.55$
a_3b_2	22	$67(50)/230 = 14.6$	$(22 - 14.6)2/14.6 = 3.75$
			$\chi^2 = 7.18$

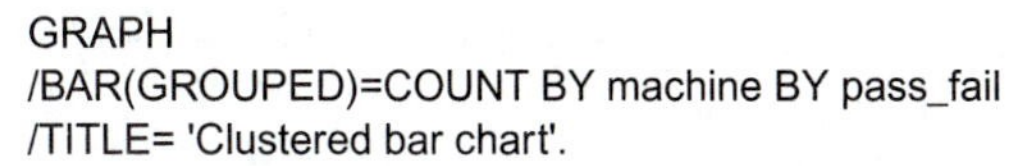

```
GRAPH
/BAR(GROUPED)=COUNT BY machine BY pass_fail
/TITLE= 'Clustered bar chart'.
```

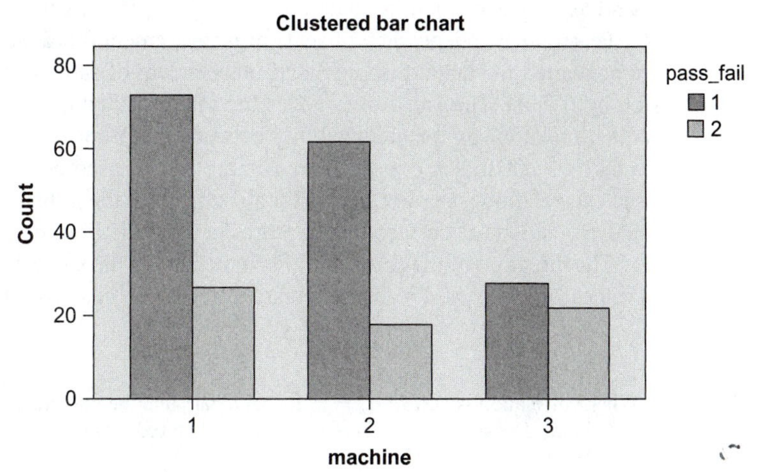

FIGURE 2.8 Syntax and SPSS GRAPHS clustered bar chart for two discrete variables

SPSS Graphs using the data in CHISQ.SAV. One concludes from this display and from rejection of the null hypothesis that the third machine produces a less favorable pass/fail ratio than do the other two machines. The statistics and charts of Section 2.2.1, derived separately for each of the discrete variables, also might be interesting.

2.3.2 Both Variables Continuous

Correlation and regression are used when both variables are continuous. Correlation is used to measure the size and direction of the relationship between the two variables; regression is used to predict one variable from the other. Correlation and regression are first cousins both logically (if there is no relationship, prediction is also not possible) and computationally, as indicated in what follows. Correlation and regression are preceded by separate descriptive statistics and/or graphs for each of the continuous variables (Section 2.2.3).

2.3.2.1 Correlation

The Pearson product-moment correlation coefficient, r, is a frequently used measure of association between two continuous variables, X and Y. The most easily understood equation for Pearson r is

$$r = \frac{\sum z_X z_Y}{N - 1} \qquad \textbf{(2.15)}$$

where Pearson r is the average cross product of standardized X and Y variable scores—

$$z_Y = \frac{Y - \overline{Y}}{S_{N-1}} \quad \text{and} \quad z_X = \frac{X - \overline{X}}{S_{N-1}}$$

—and S_{N-1} is as defined in Equation 2.2.

Pearson r is independent of scale of measurement (because both X and Y scores are converted to standard scores) and independent of sample size (because of division by $N - 1$). The value of r ranges between $+1.00$ and -1.00, where .00 represents no relationship or predictability between the X and Y variables. An r value of $+1.00$ or -1.00 indicates perfect predictability of one score when the other is known. When correlation is perfect, scores for all subjects in the X distribution have the same relative positions as corresponding scores in the Y distribution.[19]

The direction of the relationship is indicated by the sign of r. When r is positive, Y values increase as X values increase. When r is negative, Y values decrease as

[19] When correlation is perfect, $z_X = z_Y$ for each pair, and the numerator of Equation 2.15 is, in effect, $\Sigma z_X z_X$. Because $\Sigma {z_X}^2 = N - 1$, Equation 2.15 reduces to $(N - 1)/(N - 1)$, or 1.00.

X values increase. The absolute value of r indicates the size of the relationship, whether positive or negative. Thus, an r value of .65 indicates a relationship of the same size as an r value of $-.65$. However, the direction of the relationship is opposite in the two cases. There is also a raw score form of Equation 2.15, helpful on that desert island:

$$r = \frac{\sum XY - ((\Sigma X)(\Sigma Y))/N}{\sqrt{(\Sigma X^2 - (\Sigma X)^2/N)(\Sigma Y^2 - (\Sigma Y)^2/N)}} \tag{2.16}$$

Compare the numerator of Equation 2.16 with the numerator of Equation 2.4. Notice that they are the same except that a Y has been inserted into the numerator in place of one of the Xs (or an X for one of the Ys). With this replacement, the numerator becomes (the sum of squares for) covariance instead of (the sum of squares for) variance. Correlation, then, really is asking about the substitutability of one variable for the other. When correlation is perfect, you can replace X with Y (or the reverse) with no loss of information.

Table 2.8 demonstrates calculation of values used in correlation and regression with fictitious continuous variables labeled X and Y. Using Equation 2.16, Pearson r for the data set in Table 2.8 is

$$r = \frac{225 - \dfrac{(49)(58)}{10}}{\sqrt{\left(313 - \dfrac{49^2}{10}\right)\left(392 - \dfrac{58^2}{10}\right)}} = -.93$$

TABLE 2.8 Data set for illustration of correlation and regression

Case	X	X^2	Y	Y^2	XY
1	3	9	8	64	24
2	7	49	3	9	21
3	9	81	2	4	18
4	2	4	7	49	14
5	4	16	8	64	32
6	6	36	5	25	30
7	1	1	8	64	8
8	8	64	4	16	32
9	7	49	4	16	28
10	2	4	9	81	18
	$\Sigma X = 49$	$\Sigma X^2 = 313$	$\Sigma Y = 58$	$\Sigma Y^2 = 392$	$\Sigma XY = 225$

The null hypothesis in correlation is that there is no relationship between the two variables in the population: $\rho = 0$, where ρ (rho) is the population correlation. This hypothesis most easily is tested by referring to tables of r values that differ reliably from zero with probability levels .05 and .01, given the degrees of freedom for the test (the number of pairs of scores minus 2). The null hypothesis is rejected if the computed value of r exceeds the tabled value. These tables are available in introductory statistics texts. For example, the critical value at $\alpha = .05$ for the 8 df of Table 2.8 (10 pairs of scores minus 2) is .63 (absolute). We can reject the null hypothesis of no relationship between X and Y because this is exceeded by the calculated value of .93 (absolute).

Squared correlation (r^2), also called effect size or coefficient of determination, is used to interpret the degree of relationship if it is found to be statistically significant. In a conceptual diagram (Figure 2.3), the variance of each variable is represented as a circle, and the relationship between the two variables is represented by the overlapping area between them. If $r = 0$, then the circles do not overlap. If $r = .32$, then $r^2 = .10$, and about 10% of the area in the circles overlaps. Absolute values of correlations smaller than about .30, then, are usually considered minimal, even if they are found to be statistically significant, because less than 10% of the variance in each variable can be attributed to its linear association with $\overline{Y}$ and vice versa. When $|r| = .71$, $r^2 = .50$, and about half the area in the circles is overlapping. Depending on the research area, such a correlation might be considered small, moderate, or very strong.

The data in Table 2.8 show a strong, negative relationship. If $r = -.93$, then $r^2 = .86$ for an 86% overlap between the variance of X and the variance of Y. The direction of the correlation is negative, meaning that scores on Y decrease as scores on X increase.

2.3.2.2 Regression

Whereas correlation assesses the size and direction of the linear relationship between two variables, regression is used to predict a score on one variable from a score on the other. In bivariate (two-variable) simple linear regression, a straight line relationship between the two variables is found. The best-fitting straight line goes through the means of X and Y and minimizes the sum of the squared vertical distances between the data points and the line.

To find the best-fitting straight line for predicting Y from X, solve the following equations:

$$Y' = bX + a \tag{2.17}$$

where Y is the predicted score, b is the slope of the line (change in Y divided by change in X), X is the value for which Y is to be predicted, and a is the value of Y when X is 0.00.

The differences between the predicted and the obtained values of Y for each case represent errors of prediction, called *residuals.* The best-fitting straight line is the line that minimizes the squared errors of prediction, or residuals. For this reason, this type of regression is called *least squares regression.*

To solve Equation 2.17, both b and a are found:

$$b = \frac{\Sigma XY - ((\Sigma X)(\Sigma Y))/N}{\Sigma X^2 - (\Sigma X)^2/N} \tag{2.18}$$

The bivariate regression coefficient (b) is the ratio of the covariance of the variables to the variance of the one from which predictions are made.

Note the differences and similarities between Equation 2.16 (for correlation) and Equation 2.18 (for the regression coefficient). Both have covariance as the numerator but differ in the denominator. In correlation, the variances of both are used in the denominator. In regression, the variance of the predictor variable serves as the denominator: If Y is predicted from X, variance of X is the denominator, but if X is predicted from Y, variance of Y is the denominator. The value of the intercept, a, is also calculated to complete the solution:

$$a = \overline{Y} - b\overline{X} \tag{2.19}$$

The intercept is the mean of the observed values of the predicted variable minus the product of the regression coefficient times the mean of the predictor variable.

The slope for regression for predicting Y from X in the data of Table 2.8, using Equation 2.18, is

$$b = \frac{225 - ((49)(58))/10}{313 - 49^2/10} = -.81$$

And the intercept, from Equation 2.19, is

$$a = 5.8 - (-.81)4.9 = 9.8$$

Equation 2.17 is used to predict Y from X as follows:

$$Y' = (-.81)X + 9.8$$

If a new case has an X score of 3, the predicted Y score for the case is

$$Y' = (-.81)3 + 9.8 = 7.4$$

This result is consistent with the negative signs of r and b, which means that scores on Y go down as scores on X go up.

The null hypothesis in regression is that the regression coefficient in the population (ρ) is equal to zero—Y cannot be predicted from X. One way to test the null hypothesis is to calculate r from the same data. The test for r is also the test for b. In the example, then, we can reject the hypothesis of no prediction because we rejected the hypothesis of no relationship.

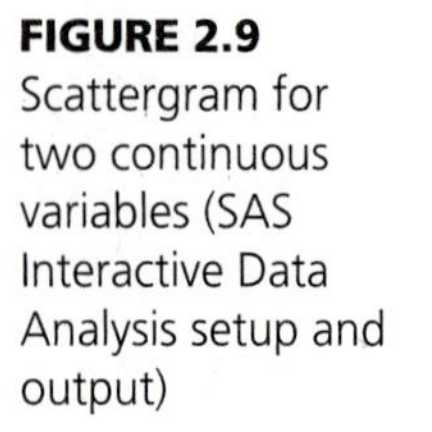

FIGURE 2.9 Scattergram for two continuous variables (SAS Interactive Data Analysis setup and output)

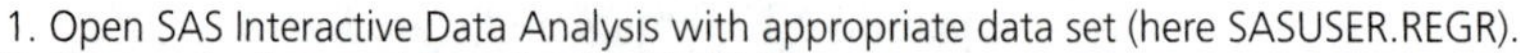

1. Open SAS Interactive Data Analysis with appropriate data set (here SASUSER.REGR).
2. Choose Analyze and then Scatter Plot(X Y).
3. Select x and y variables

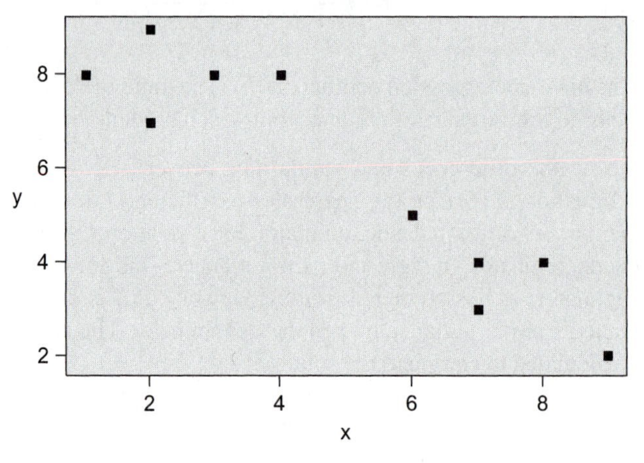

The relationship between the two variables is usually visualized using a scattergram, as shown in Figure 2.9, which was produced by SAS Interactive Data Analysis. Data are in REGR.sas7bdat.

It is clear from the graph that Y scores go down as X scores go up, consistent with the negative correlation and the negative regression coefficient. The line representing the slope ends at the lower right corner if $b = -1.00$ and $a = 0$.

Although regression is typically used with two continuous variables, it can also be used when one of the variables is dichotomous. In this case, one is said to be taking the regression approach to ANOVA. (This application is demonstrated in Chapter 3.) This approach sometimes offers greater conceptual clarity than the normal ANOVA approach and is used when helpful throughout this book.

2.3.3 One Discrete and One Continuous Variable

The relationship between a discrete and a continuous variable is usually studied through ANOVA when the discrete variable is the IV and the continuous variable is the DV. The many varieties and complexities of ANOVA are the subject of the rest of this book. This section concentrates on graphical displays. The files on disk named DESGROUP.* contain 48 cases: three levels of an IV with 16 cases in each level and a single DV.

SPSS GRAPHS (Interactive) together with SPSS TABLES (Basic Tables) can be cajoled into providing helpful information for outlier identification, as shown in Table 2.9. Histograms with a normal curve overlay are shown for each level of the IV by activating the PANEL option. A potential outlier appears as a data point far from its

TABLE 2.9 Relationship between a discrete IV and a continuous DV (SPSS GRAPHS and TABLES syntax and output)

(a) No outliers

```
|GRAPH/VIEWNAME='Histogram'/X1 = VAR(dv) TYPE = SCALE/Y = $count
/COORDINATE = VERTICAL/PANEL VAR(iv) /X1LENGTH=3.0 /YLENGTH=3.0 /X2LENGTH=3.0
/CHARTLOOK='NONE'
/CATORDER VAR(IV) (ASCENDING VALUES OMITEMPTY)
/Histogram SHAPE = HISTOGRAM CURVE = ON X1INTERVAL AUTO X1START = 0.
EXE.
```

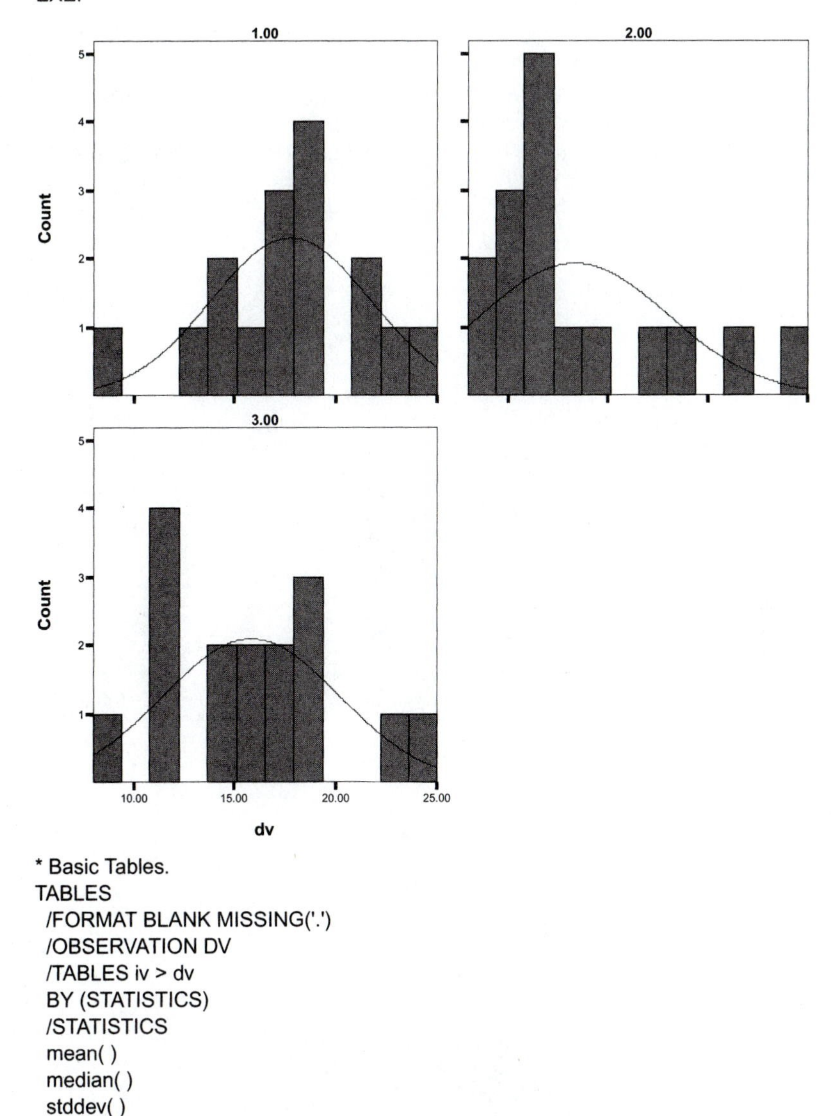


```
* Basic Tables.
TABLES
 /FORMAT BLANK MISSING('.')
 /OBSERVATION DV
 /TABLES iv > dv
 BY (STATISTICS)
 /STATISTICS
 mean( )
 median( )
 stddev( )
 minimum( )
 maximum( ).
```

TABLE 2.9 Continued

	Mean	Median	Std Deviation	Minimum	Maximum
1.00	17.75	17.50	3.92	9.00	25.00
2.00	13.31	12.00	4.67	8.00	24.00
3.00	15.81	16.00	4.31	8.00	24.00

(b) Probable outlier in Group 1 of the IV

```
|GRAPH /VIEWNAME='HISTOGRAM' /X1 = VAR(DV_out) TYPE = SCALE /Y = $count
/COORDINATE = VERTICAL
/PANEL VAR(iv) /X1LENGTH=3.0 /YLENGTH=3.0 /X2LENGTH=3.0 /CHARTLOOK='NONE'
/CATORDER VAR(IV) (ASCENDING VALUES OMITEMPTY)
/Histogram SHAPE = HISTOGRAM CURVE = ON X1INTERVAL AUTO X1START = 0.
EXE.
```

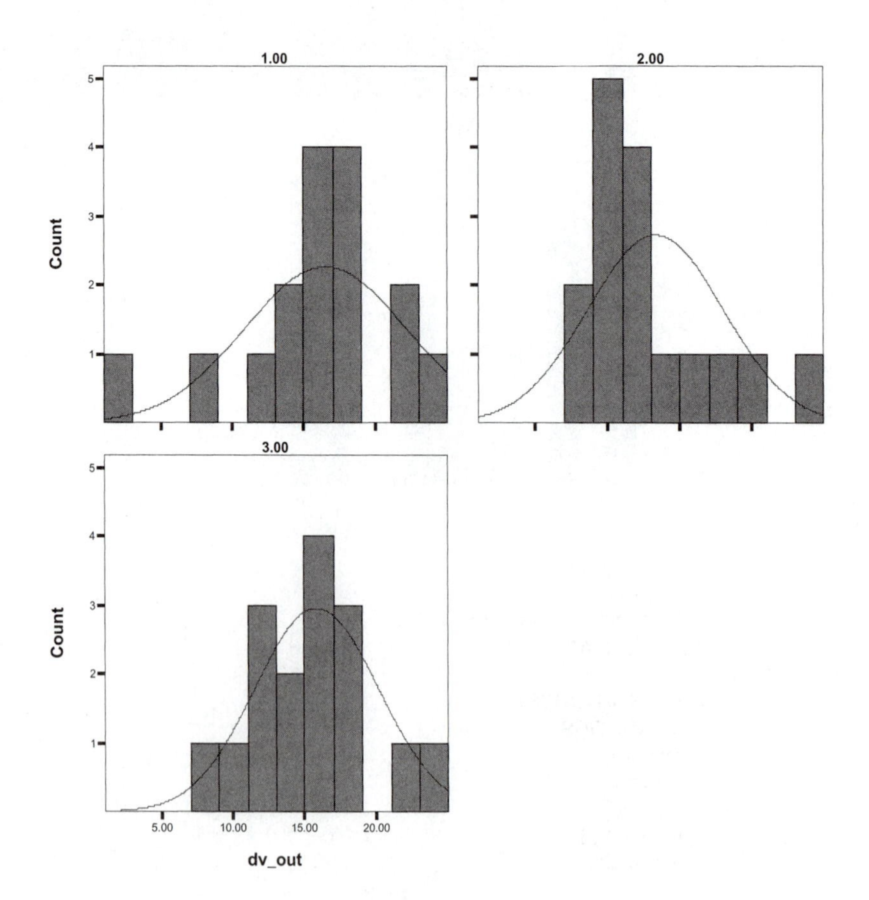

TABLE 2.9 Continued

```
* Basic Tables.
TABLES
  /FORMAT BLANK MISSING('.')
  /OBSERVATION dv_out
  /TABLES iv > dv_out
  BY (STATISTICS)
  /STATISTICS
  mean( )
  median( )
  stddev( )
  minimum( )
  maximum( ).
```

	Mean	Median	Std Deviation	Minimum	Maximum
1.00	16.50	17.00	5.63	1.00	25.00
2.00	13.31	12.00	4.67	8.00	24.00
3.00	15.81	16.00	4.31	8.00	24.00

own mean, disconnected from the rest of the data in its own level, and with a substantial z score. Syntax and output for two runs with hypothetical data in DESGROUP.SAV are shown in the table.

There are no outliers in any of the three levels of the IV in Table 2.9(a), but there is a probable outlier in the first level of the IV in Table 2.9(b), created by changing a value of 21 to 1 (in what might be a common typographical data entry error). This data point has a z score of $(16.50 - 1)/5.63 = 2.75$ and is apparently disconnected from the rest of the data in its group. The outlier pulls down the mean for the first group.

A line graph of the group means is useful when a significant difference between means is found in ANOVA, as shown in Figure 2.10, produced by SPSS GLM for the data used in Table 2.9(a). The graph is requested in the PLOT=PROFILE instruction, with the label of the IV in parentheses. The remaining syntax is produced by the SPSS menu system once the IV and DV are identified and is explained in subsequent chapters.

There is a single line on this graph because there is only one IV. If there were a second IV, separate lines (say, dotted and dashed) would be used to shown the mean values of one IV at each level of the second IV. It is clear from Figure 2.10 that the highest mean DV score from the sample is at the first level of the IV, the lowest mean at the second level, and an intermediate mean at the third level. Although the means are not the same, the graph does not reveal whether they are reliably different. For that, a formal statistical analysis is required. Also, note that the scale of Figure 2.10 may exaggerate differences among levels because the lowest value of the Y axis is not zero.

The graphical procedures just discussed for the relationship between a discrete IV and a continuous variable DV are appropriate for experimental data but probably less so for observational data. When observations are underway, either as an end in

FIGURE 2.10 Line graph of mean DV scores at each level of the IV (syntax and output from SPSS GLM)

```
UNIANOVA
  DV BY IV
  /METHOD = SSTYPE(3)
  /INTERCEPT = INCLUDE
  /PLOT = PROFILE( iv IV )
  /CRITERIA = ALPHA(.05)
  /DESIGN = IV.
```

Profile Plots

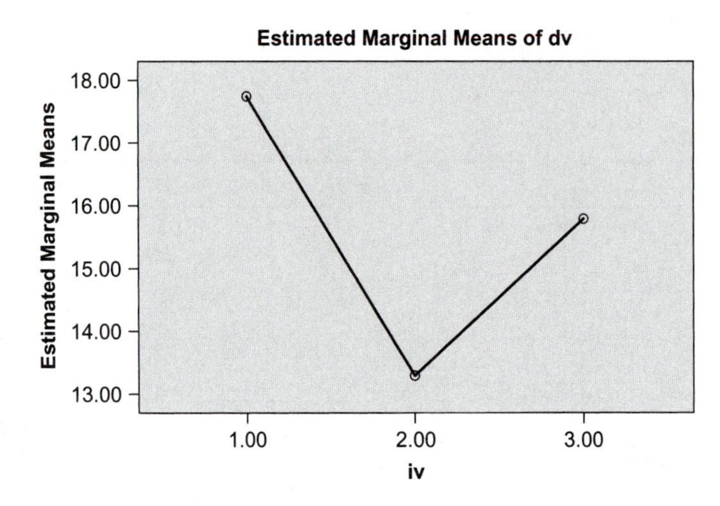

themselves or in preparation for conducting an experiment, some graphical procedures developed for quality control may be more helpful. One of these, the stratified control chart, is appropriate when the DV is continuous, and another, the attribute control chart, when the DV is discrete and you have counted the number of such discrete outcomes in a period of time. Both of these charts plot DV scores against other observed variables.

Stratification of control charts helps pinpoint factors associated with observations that are extreme by showing mean values as well as upper confidence limits (UCL) and lower confidence limits (LCL). In a typical example, diameters of manufactured parts are measured every day for 20 days. There are three different machines (A386, A455, and C334) on which the parts are manufactured and four different operators (DRJ, CMB, RMM, and MKS) using the machines. Data are in DIAMETER.SD7, provided by SAS as part of the SASUSER directory.

Figure 2.11 shows the syntax and stratified control chart for SAS SHEWART. Stratification by machines is indicated by putting it in parentheses after the IV: DAY (MACHINE). Operators are identified by different symbols on the chart. The syntax after the forward slash is produced by default through the SAS QC menu system.

Machines are indicated at the top of the chart and operators by the use of different symbols for points in the chart itself. The chart shows that the diameter is too large when RMM is operating the A455 machine. There is also some indication that

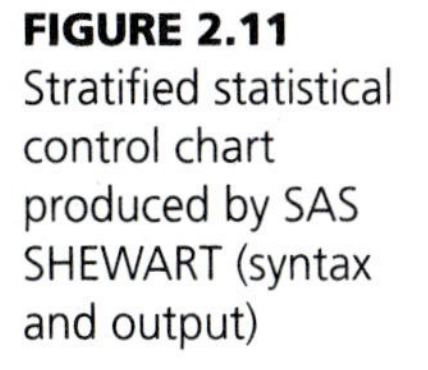

FIGURE 2.11 Stratified statistical control chart produced by SAS SHEWART (syntax and output)

```
symbol1 w=1 v=PLUS c=RED;
symbol2 v=STAR c=GREEN;
symbol3 v=CIRCLE c=BLUE;
symbol4 v=SQUARE c=CYAN;
proc shewhart data= SASUSER.DIAMETER;
  XRCHART DIAMETER * DAY (MACHINE) = OPERATOR/
    llimits=2 blockpos=1 blocklabtype=TRUNCATED cconnect=BLACK
    coutfill=RED tableall(noprint exceptions) ;
run;
```

RMM produces more variable diameters, particularly on machine A386. However, there was no random assignment of operators to machines and no observations of CMB on machine A455, so the data are strictly correlational.

In the social or biological sciences, a researcher might be interested in fluctuating anxiety scores or temperature or the like. Such scores are monitored over a period of time while also tracking the occurrence of other variables (e.g., operators) that are potentially related but not manipulated. The stratified control chart is used to assist in discovering the variables that are most closely associated with changes in the DV.

Attribute control charts are helpful when the outcome is not measured as continuous scores but instead are measured as a head-count of discrete values that occur in a period of time. Suppose, for a typical example, that you have operators who have two levels of training, expert and novice, and who are sampling 200 parts each hour,

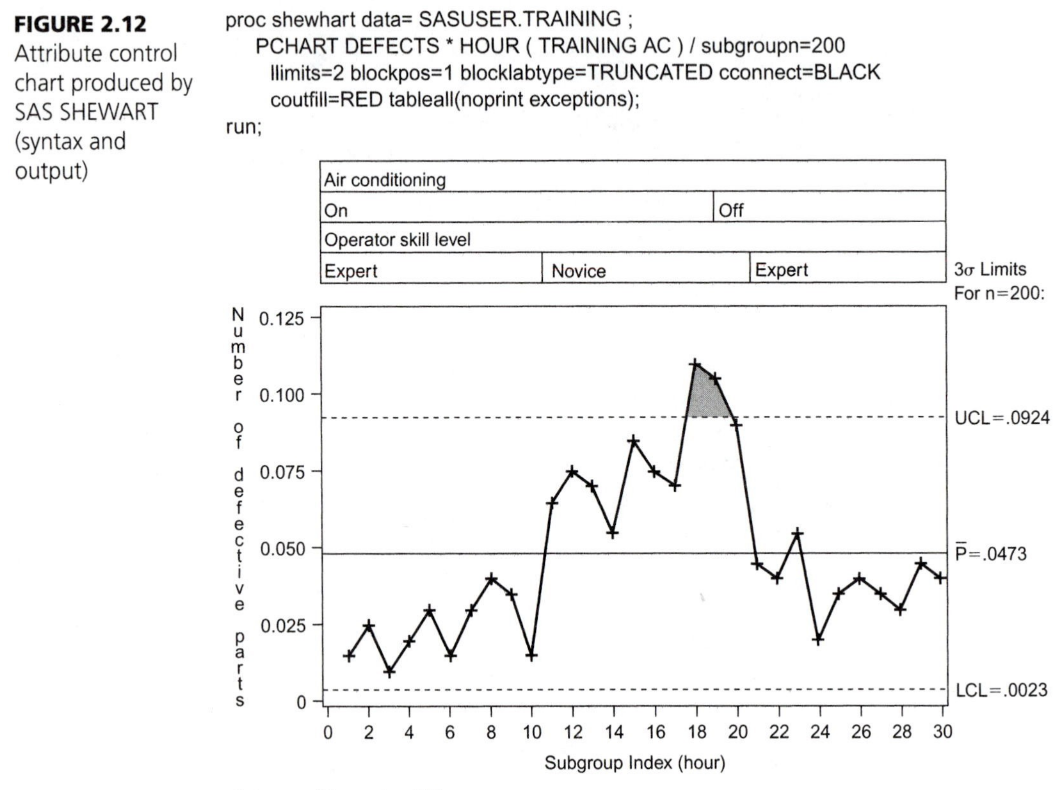

FIGURE 2.12 Attribute control chart produced by SAS SHEWART (syntax and output)

inspecting for defects (the DV) under two environmental conditions (air conditioning on and off). The data are in TRAINING.SD7, provided by SAS as part of the SASUSER directory.

Figure 2.12 shows syntax and an attribute chart through SAS SHEWART. HOUR is the subgrouping variable (the major IV). HOUR (TRAINING AC) indicates that TRAINING and AC are consecutive stratification variables (stratification is optional for attribute control analysis). The instructions show that there are 200 items produced each hour. The remaining syntax is produced by default through the SAS QC menus.

Figure 2.12 shows that the process goes out of control under a novice operator, beginning while the air conditioning is on. The process seems to be coming back into control as the air conditioning is turned off. Attribute control charts are also produced by NCSS.

In the social sciences, one might count the number of temper tantrums of a child each day and also record the values of several other variables potentially related to those tantrums. Use of the attribute control chart could help reveal which of the other variables are potentially related to the occurrence of more numerous tantrums.

2.4 PROBLEM SETS

2.1 Compute the sample standard deviation, the estimated population standard deviation, and the standard error for the continuous variable shown. Compute the 95% confidence interval for the population mean for the same data. Use a statistical software program to produce the same statistics.

45	45	23	23
42	43	10	40
33	16	25	42
27	31	43	32
35	44	42	37
28	18	29	19
14	78	38	50
32	29	25	26

2.2 Use appropriate graphical means to inspect the data in Problem 2.1. Use both statistical and graphical methods to determine whether the distribution is relatively normal.

2.3 Use statistical and graphical methods to inspect the distribution of the continuous variable shown for the presence of outliers. (Note that this distribution is not quite the same as that in Problem 2.1.)

45	45	23	23
42	43	10	40
33	16	25	42
27	31	43	32
35	44	42	37
28	18	29	19
14	38	38	50
32	29	25	26

2.4 Compute chi-square for the relationship between the two discrete variables shown. Is there a relationship? Evaluate the relationship using Table A.2 in Appendix A.

	Response to Question		
	Yes	No	Sum
Treatment group	34	19	53
Control group	26	27	53
Sum	60	46	106

2.5 Compute correlation and regression for the two continuous variables shown. Generate a scatter plot to inspect the relationship. Generate a conceptual diagram to represent the size of the relationship.

Plant	Height of Plant	Number of Seeds	Plant	Height of Plant	Number of Seeds
1	10	50	10	14	44
2	11	44	11	11	39
3	13	43	12	12	39
4	11	39	13	11	43
5	12	38	14	11	51
6	14	60	15	13	49
7	14	43	16	12	42
8	14	45	17	9	38
9	12	60			

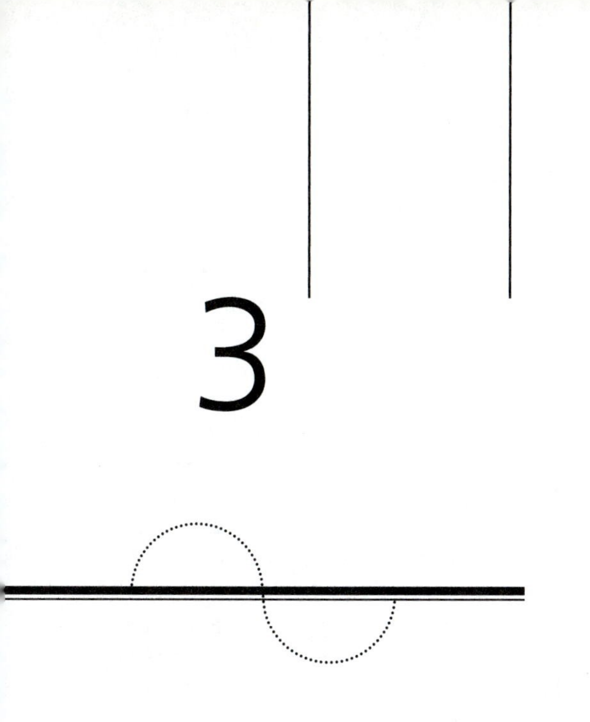

3

Basic ANOVA

Logic of Analysis and Tests of Assumptions

3.1 INTRODUCTION

Analysis of variance tests the statistical significance[1] of mean differences—central tendency—among different groups of scores. The different groups of scores may correspond to different levels of a single IV or to different combinations of levels of two or more IVs. The groups of scores may come from different cases or from the same cases measured repeatedly. If a difference between means is statistically significant, the difference is expected (with a certain probability) to reappear if the study is replicated. A nonsignificant difference implies that you cannot rule out the possibility that the mean differences that do exist in the sample data occurred by chance.

[1] We use the term *statistically significant* throughout the text to remind the reader that rejecting the null hypothesis does not mean that the finding is important but rather that the mean difference is large enough that a difference is likely to reappear in a replication of the study.

The simplest design for which ANOVA is appropriate has one IV with two levels and a single DV measured on a continuous scale. The IV consists of one control group of cases and one treatment group of cases, and the researcher asks if means on the DVs for each group are reliably different. If another treatment group is added to the design, the researcher inquires about the reliability of the differences among the three means, and so on.[2]

For example, in a drug-testing program in which a control group is given a placebo and a treatment group is given a new analgesic, the researcher asks if the average report of pain (the DV) is reliably different in the two groups. However, even if the new analgesic is completely worthless in terms of relieving pain, the control group and the treatment group are likely to have slightly different means in reported pain. Indeed, it would be surprising if the means were identical, given variability in pain assessments. A yardstick is needed to test the *reliability* of the difference between the two means. In ANOVA, that yardstick compares the difference between the means with an estimate of naturally occurring variability in pain to determine whether the difference is big enough to be statistically significant (where "big enough," as you recall, is statistically defined).

3.2 ANALYSIS OF VARIANCE

One of us had endless difficulty as a student understanding why an analysis that tested for differences in central tendency was called an analysis of variance. The following is an effort to spare you similar trouble.

3.2.1 The General Linear Model

ANOVA is part of the general linear model (GLM). Some software packages acknowledge this fact by using GLM as a general designation for their ANOVA programs. However, GLM is a good deal more general than ANOVA, and a thorough discussion of it is beyond the scope of this book.[3]

A simple form of GLM appropriate for ANOVA is

$$Y = \mu + \alpha + e \qquad \textbf{(3.1)}$$

This equation asserts that a score, Y, can be decomposed into three elements: the population mean (μ), the effects of an independent variable (α), and error (e)—that is,

[2] Levels of the IV (groups) to be investigated are chosen explicitly by the researcher in this *fixed-effects* design. *Random-effects* designs (Chapter 11) are those in which levels of the IV to be investigated are chosen randomly by the researcher from a larger set of levels.

[3] Chapter 17 of Tabachnick and Fidell (2007) treats you to a brief overview of GLM, including the multivariate general linear model (MGLM). You might want to wait until you are well into this book before you tackle it.

each score can be conceptualized as the grand mean modified by the potential effects of an IV, A, and error. ANOVA tests whether the effects of A are statistically significant. The idea is that different levels of A systematically change scores from the value of the grand mean—the treatments associated with some levels of A tend to raise scores, the treatments associated with other levels of A tend to lower them, and the treatments associated with yet other levels may leave the mean unaffected.

All of this is, of course, tempered by error. Error is composed of uncontrolled influences on scores. The nature of error differs from discipline to discipline and from study to study. In some settings, it is variation in the operation of equipment; in others, variation in the cases; in yet others, variation in the testing situation or the researcher; or all of these. The error term is the composite of all the effects you have not measured or controlled and do not know about (although you may well have suspicions). Like levels of an IV, the effect of some kinds of error is to raise scores, whereas the effect of other kinds of error is to lower scores. It should be clear from Equation 3.1 that a very large error term can swamp the effects of the IV, burying it in the noise, so to speak. For this reason, a great deal of effort is directed toward minimizing the error term.

However, the error term is a useful metric to judge whether the effects of the IV are statistically significant. Even if the IV is a complete bust, one would not expect the means of different groups of scores to be exactly equal. They should be a bit different due simply to the vagaries of repeatedly taking different samples of scores from the same population for each level of A. For this reason, a yardstick is needed against which to test the mean differences. That yardstick should be measured on the same scale as are the means and should reflect naturally occurring variability in scores, without the influence of an IV. In ANOVA, the error term, e, is used for this important function. The idea, then, is to test mean differences against naturally occurring variability in scores as represented by e.

3.2.2 Generation of a Data Set

The easiest way to understand ANOVA is to generate a tiny data set using the GLM and then see how ANOVA processes it. As a first step, let's set the grand mean, μ, to 5 for a data set with 10 cases, as shown in Table 3.1.[4] In this table, all 10 cases have a score of 5 at this point.

The next step is to add the effects of the IV. Suppose that the effect of the treatment at a_1 is to raise scores by 2 units and the effect of the treatment at a_2 is to lower scores by 2 units. At this point, as shown in Table 3.2, all the scores and the mean at a_1 are 7 and all the scores and the mean at a_2 are 3.

[4] This data set is too small for many research applications that use ANOVA. Small data sets are often analyzed using either the Wilcoxon Mann Whitney Test, with adjustment for small sample size, or another appropriate nonparametric test. We use a very small data set here for didactic purposes.

TABLE 3.1 Generation of a data set for ANOVA, $Y = \mu$

a_1		a_2	
Case	Score	Case	Score
s_1	5	s_6	5
s_2	5	s_7	5
s_3	5	s_8	5
s_4	5	s_9	5
s_5	5	s_{10}	5

TABLE 3.2 Generation of a data set for ANOVA, $Y = \mu + \alpha$

a_1		a_2	
Case	Score	Case	Score
s_1	$5 + 2 = 7$	s_6	$5 - 2 = 3$
s_2	$5 + 2 = 7$	s_7	$5 - 2 = 3$
s_3	$5 + 2 = 7$	s_8	$5 - 2 = 3$
s_4	$5 + 2 = 7$	s_9	$5 - 2 = 3$
s_5	$5 + 2 = 7$	s_{10}	$5 - 2 = 3$
	$\sum Y_{a_1} = 35$		$\sum Y_{a_2} = 15$
	$\sum Y^2_{a_1} = 245$		$\sum Y^2_{a_2} = 45$
	$\overline{Y}_{a_1} = 7$		$\overline{Y}_{a_2} = 3$

The changes produced by treatment are the deviations of the scores from μ. Over all of these cases, the sum of the squared deviations is

$$5(2)^2 + 5(-2)^2 = 40$$

This is the sum of the (squared) effects of treatment if all cases are influenced identically by the various levels of A and there is no error.

The third step is to complete the GLM with addition of error. Table 3.3 contains the complete GLM for the tiny data set with numbers representing error added to each score. Using the variance form of Equation 2.4, the variance for the a_1 group is

$$a_1\colon s^2_{N-1} = \frac{\Sigma Y^2 - (\Sigma Y)^2/N}{N - 1} = \frac{251 - 35^2/5}{4} = 1.5$$

and the variance for the a_2 group is

$$a_2\colon s^2_{N-1} = \frac{51 - 15^2/5}{4} = 1.5$$

TABLE 3.3 Generation of a data set for ANOVA, $Y = \mu + \alpha + e$

a_1		a_2		
Case	Score	Case	Score	Sum
s_1	$5+2+2=9$	s_6	$5-2+0=3$	
s_2	$5+2+0=7$	s_7	$5-2-2=1$	
s_3	$5+2-1=6$	s_8	$5-2+0=3$	
s_4	$5+2+0=7$	s_9	$5-2+1=4$	
s_5	$5+2-1=6$	s_{10}	$5-2+1=4$	
	$\sum Y_{a_1} = 35$		$\sum Y_{a_2} = 15$	$\sum Y = 50$
	$\sum Y_{a_1}^2 = 251$		$\sum Y_{a_2}^2 = 51$	$\sum Y^2 = 302$
	$\bar{Y}_{a_1} = 7$		$\bar{Y}_{a_2} = 3$	$\bar{Y} = 5$

The average of these two variances is also 1.5. Keep in mind that these numbers represent error variance—that is, they represent random variability in scores within each group where all cases are ostensibly treated the same and, therefore, are uncontaminated by effects of the IV.

The variance for this group of 10 numbers, *ignoring group membership,* is

$$s_{N-1}^2 = \frac{302 - 50^2/10}{9} = 5.78$$

3.2.3 The Basic Analysis

The simplest design for ANOVA has one IV with two levels and different cases at each level. For the traditional approach, DV scores are gathered and arranged in a table similar to Table 3.3, with a columns representing levels of the IV and n scores within each level.[5] Table 3.4 contains the same scores as Table 3.3, but in standard ANOVA format, without listing individual cases or GLM breakdown.[6] New notation in the table is described in subsequent sections.

3.2.3.1 Basic ANOVA: The Deviation Approach

For development of equations, each score is designated Y_{ij}, where $i = 1, 2, \ldots, n$ observations at each level of A and j represents group or level. GM represents the grand

[5] Throughout the book, n is used for number of scores within a single level, and N is used for total number of scores.

[6] It would be most helpful if cases *did* report the breakdown for their scores in terms of the GLM, but that would put a lot of statisticians out of work. Instead, we have to rely on ANOVA to discover the breakdown.

TABLE 3.4 Standard setup for ANOVA data set

A		
a_1	a_2	Sum
9	3	
7	1	
6	3	
7	4	
6	4	
$\Sigma Y_{a_1} = A_1 = 35$	$\Sigma Y_{a_2} = A_2 = 15$	$\Sigma Y = T = 50$
$\sum Y_{a_1}^2 = 251$	$\sum Y_{a_2}^2 = 51$	$\sum Y^2 = 302$
$\overline{Y}_{a_1} = 7$	$\overline{Y}_{a_2} = 3$	$\overline{Y} = \textbf{GM} = 5$

mean of all scores over all groups. The difference between each score and the grand mean (Y_{ij} – GM) is broken into two components: (1) the difference between the score and its own group mean and (2) the difference between that group mean and the grand mean.

$$(Y_{ij} - \textbf{GM}) = (Y_{ij} - \overline{Y}_j) + (\overline{Y}_j - \textbf{GM}) \tag{3.2}$$

This result is achieved by first subtracting and then adding information about group membership—the group means—to the equation. The first difference, between a score and its own group mean $(Y_{ij} - \overline{Y}_j)$, represents error (all the cases in each level have, after all, been treated identically, so deviations from the group mean are unexplained). The second difference, between a group mean and the grand mean $(\overline{Y}_j - \textbf{GM})$ represents the effects of the IV plus error (more later . . .). For the first case, then

$$(9 - 5) = (9 - 7) + (7 - 5)$$

The score for the first case can be conceptualized as either a difference between the score and the grand mean (9 – 5) or as the difference between the score and its own group mean (9 – 7) plus the difference between its own group mean and the grand mean (7 – 5).

Each term in Equation 3.2 is then squared and summed separately to produce the sum of squares for error and the sum of squares for treatment, respectively. The basic partition holds because, conveniently, the cross-product terms produced by squaring and summing cancel each other out. Across all scores, the partition is

$$\sum_i \sum_j (Y_{ij} - \textbf{GM})^2 = \sum_i \sum_j (Y_{ij} - \overline{Y}_j)^2 + \sum_n \sum_j (\overline{Y}_j - \textbf{GM})^2 \tag{3.3}$$

This is the deviation form of basic ANOVA, where each quantity is represented as a difference (a deviation) between two elements. Representing variance as a deviation is the easiest way to *understand* the partition, although equations presented in Section 3.2.3.2 are easier to use for computation.

Each of these terms is a *sum of squares* (SS), or a sum of squared differences between two individual elements. In other words, each term is a special case of the numerator of a variance equation, such as Equation 2.4. The term to the left of the equal sign is the total sum of squared differences between scores and the grand mean, SS_{total} or SS_T. This sum, when "averaged," is the total variance in the set of scores, ignoring group membership. The first term to the right of the equal sign is the sum of squared deviations between each score and its group mean, summed over all groups. This component is variously called SS within groups (SS_{wg}) or SS error (SS_{error}) or $SS_{S/A}$. The second term to the right of the equal sign is the sum of squared deviations between each group mean and the grand mean, variously called SS between groups (SS_{bg}) or SS_A. The partition of total variance of Equation 3.3 is also frequently symbolized (with reversed terms on the right side) as

$$SS_{total} = SS_{bg} + SS_{wg} \tag{3.4}$$

It is important to realize that the total variance in the set of scores is partitioned into two sources, one representing the effects of the IV and the other representing all other effects, error. In this sense, it is truly an analysis of variance. Because the effects of the IV are assessed by changes in the central tendencies of the groups, the inferences that come from ANOVA are about differences in central tendency.

But sums of squares are not yet variances. To become variances, they must be "averaged." The denominators for averaging SS in ANOVA are called degrees of freedom (df). The degrees of freedom partition the same way as do the sums of squares:

$$df_{total} = df_{bg} + df_{wg} \tag{3.5}$$

Total degrees of freedom is the total number of scores, N, minus 1, lost when the grand mean is estimated. N can also be represented as the number of groups, a, times the number of scores in each group, n (if the number of scores in each group is equal). Therefore,

$$df_{total} = N - 1 = an - 1 \tag{3.6}$$

Within-groups (error) degrees of freedom is the number of scores, N, minus the number of groups, a. One df is lost when the mean for each a group is estimated. Therefore,

$$df_{wg} = N - a = an - a = a(n - 1) \tag{3.7}$$

Between-groups (treatment) degrees of freedom is the a group means minus 1, lost when the grand mean is estimated:

$$df_{bg} = a - 1 \tag{3.8}$$

Verifying the equality proposed in Equation 3.5, we have

$$(N - 1) = N - a + a - 1$$

FIGURE 3.1 Partitions of (a) total sum of squares into sum of squares due to differences between the groups and differences among cases within the groups and (b) total degrees of freedom into degrees of freedom associated with the groups and associated with cases within the groups

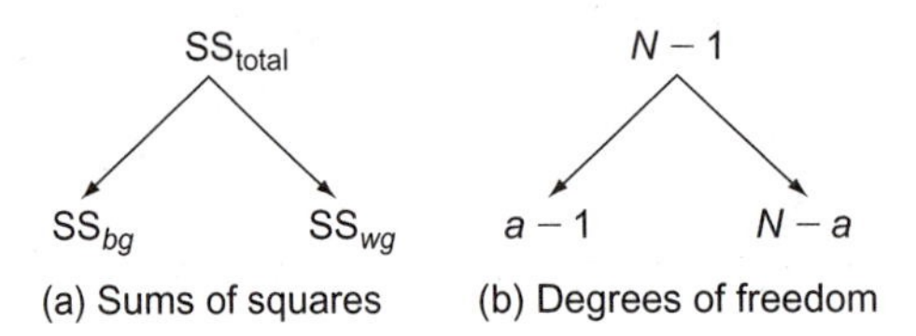

As in the partition of sums of squares, information about group membership is subtracted out of the equation and then added back in.

Degrees of freedom are best understood as the number of numbers that are free to vary before the last number gets "fixed." Imagine you are told that the mean of a group of numbers is 5, that there are four numbers, and that the first three are 2, 8, and 7. What is the fourth number? It has to be 3, because $5 = (2 + 8 + 5 + x)/4$. Or, if you know that the first three numbers are 111, −24, and 2, the fourth number has to be. . . .[7] The point is that this problem has 3 df because the first three numbers can be anything, but the fourth number is fixed because of the mean. In ANOVA, df_{total} is $N - 1$ because the numbers have to work out to the GM. The a means for each group also have to work out to the GM, so there are $a - 1$ df. Finally, there are two ways to think about df for the error term: (1) Each group has $n - 1$ df (because each group of numbers has to work out to the mean for its own group of size n), and because there are a such groups, there are $a(n - 1)$ df. Or, (2) there are N scores, and a df are lost as the mean of each group is estimated, leaving $N - a$ df. Figure 3.1 is a graphic representation of the partition.

For the simplest ANOVA design, the partition of Equation 3.4 is

$$\text{SS}_T = \text{SS}_A + \text{SS}_{S/A} \tag{3.9}$$

In this notation, the total sum of squares (SS_T) is partitioned into a sum of squares for treatment A, SS_A, and a sum of squares error, $\text{SS}_{S/A}$ (scores nested within A).[8]

Variance is an "averaged" sum of squares. The division of a sum of squares by df produces variance, which is called mean square (MS) in ANOVA. ANOVA produces three variances: one associated with total variability among scores (MS_T); one with variability between groups (MS_{bg} or MS_A); and one with variability within groups (MS_{wg} or $\text{MS}_{S/A}$). The equations for mean squares are:

$$\mathbf{MS}_A = \frac{\mathbf{SS}_A}{\mathbf{df}_A} \tag{3.10}$$

and

$$\mathbf{MS}_{S/A} = \frac{\mathbf{SS}_{S/A}}{\mathbf{df}_{S/A}} \tag{3.11}$$

[7] −69

[8] The order of terms on the right has been reversed from that in Equation 3.3.

MS_T is total variance in scores across all groups, s^2_{N-1}, and is generally not a useful quantity for ANOVA.

The F distribution is a sampling distribution of the ratio of two variances. In ANOVA, MS_A and $MS_{S/A}$ provide the variances for the F ratio to test the null hypothesis that $\mu_1 = \mu_2 = \cdots = \mu_a$.

$$F = \frac{MS_A}{MS_{S/A}} \qquad \text{df} = (a-1), (N-a) \tag{3.12}$$

If the null hypothesis is true and there are no treatment effects to make the numerator larger, the F ratio boils down to a ratio between two estimates of the same error. In this case, the likely value for F is a number around 1 (more later . . .).

Once F is obtained (computed), it is tested against critical F, looked up in a table (such as Table A.1),[9] with numerator df $= a - 1$ and denominator df $= N - a$ at desired alpha level. If obtained F is equal to or larger than critical F, the null hypothesis that the means are equal is rejected in favor of the hypothesis that there is a difference among the means in the a groups.

Anything that increases obtained F increases power, or the probability of rejecting a false null hypothesis. Equation 3.12 tells us that power is increased by increasing differences among means in the numerator (MS_A) or by decreasing the error variability in the denominator ($MS_{S/A}$).

3.2.3.2 Standard Computational Form

The deviation equations above are useful for understanding the partition of sums of squares in ANOVA but inconvenient for calculating sums of squares from a data set.[10] The more convenient equation for calculating a sum of squares is

$$SS_Y = \Sigma Y^2 - \frac{(\Sigma Y)^2}{N} \tag{3.13}$$

Actually, ΣY^2 is divided by 1, the number of "scores" in Y, and thus can be expressed as $\Sigma Y^2/1$.

All of the SS equations earlier in this section have a more convenient form, but first we need to develop a notational scheme and apply it to the data in Table 3.4.

Y is used to designate a single score (9, 7, . . ., 4 in the example).

A is used to designate the IV.

a is used to designate the number of levels of A (2 in this example).

a_1 and a_2 are used to designate specific levels of treatment A.

A_1 and A_2 are used to designate totals at each level of treatment (35 and 15, respectively).

[9] Actually, when the analysis is done by computer, the exact probability associated with the obtained F is given, so it is no longer necessary to look up the critical value in a table.

[10] For one thing, calculating means and then deviations from them is fraught with rounding error.

T is used to designate the grand total ($\Sigma Y = 35 + 15 = 9 + 7 + \cdots + 4 = 50$).
ΣY^2 is the total of all the squared scores ($9^2 + 7^2 + \cdots + 4^2 = 302$).
n is used to designate the number of scores in each group (5 in this example).[11]
N or an designates the total number of scores (10 in this example).

In this notational scheme, the equation for SS_A, sums of squares treatment, is

$$SS_A = \frac{\Sigma A^2}{n} - \frac{T^2}{an} \qquad \textbf{(3.14)}$$

Unlike ΣY^2 of Equation 3.13, ΣA^2 has n scores, rather than 1, in each value of A.
The equation for $SS_{S/A}$, sum of squares error, is

$$SS_{S/A} = \Sigma Y^2 - \frac{\Sigma A^2}{n} \qquad \textbf{(3.15)}$$

And the equation for SS_T is

$$SS_T = \Sigma Y^2 - \frac{T^2}{an} \qquad \textbf{(3.16)}$$

Each equation has two elements, the second subtracted from the first. Notice that these elements repeat over the three equations; although two appear in each equation, there are only three of them altogether. Notice also that in each element, the denominator is the number of scores in the values that were summed in the numerator before squaring. For instance, ΣY^2 is divided by 1 (because each Y that is squared is a single score), ΣA^2 is divided by n (because each A that is squared is the sum of n scores), and T^2 is divided by an (because the T that is squared is the sum of an scores). Table 3.5 shows sums of squares for the example.

Take a moment to compare SS_A with the value of 40 for sum of squared deviations calculated in Section 3.2.2 from Table 3.2. If all cases respond identically to treatment, SS_A can be interpreted as the SS due to treatment effects alone. It can also be interpreted as an answer to the question, "What is the SS if all cases are at the mean for their own group?"

Table 3.6 shows degrees of freedom for the example. Mercifully, the partitions of SS and the partitions of df both sum to their respective totals. If they do not sum

TABLE 3.5 Calculation of sums of squares for basic one-way ANOVA

SS_A	$\frac{35^2 + 15^2}{5} - \frac{50^2}{5} = 40.0$
$SS_{S/A}$	$302 - \frac{35^2 + 15^2}{5} = 12.0$
SS_T	$302 - \frac{50^2}{5} = 52.0$

[11] If there were unequal numbers of scores in each group, n would be used with a subscript, n_j.

TABLE 3.6 Calculation of degrees of freedom for basic ANOVA

df_A	$= a - 1$	$= 2 - 1$	$= 1$
$df_{S/A}$	$= N - a$	$= 10 - 2$	$= 8$
df_T	$= an - 1$	$= 10 - 1$	$= 9$

TABLE 3.7 Calculation of mean squares for basic ANOVA

$$MS_A = \frac{SS_A}{df_A} = \frac{40}{1} = 40$$

$$MS_{S/A} = \frac{SS_{S/A}}{df_{S/A}} = \frac{12}{8} = 1.5$$

TABLE 3.8 Source table for basic ANOVA

Source	SS	df	MS	F
A	40.0	1	40.0	26.67
S/A	12.0	8	1.5	
T	52.0	9		

to SS_T and df_T, there has been a computational error. Variance (mean square) for each source is obtained by dividing SS by df. Table 3.7 shows calculation of mean squares for treatment and error sources of variance. Finally, Table 3.8 is a source table that summarizes the results of Tables 3.5–3.7 and adds the F ratio of Equation 3.12.

Take a moment to compare these values with those in Section 3.2.2 calculated from the data in Table 3.3. The variances within each group are 1.5, as is their average. $MS_{S/A}$ is, then, the average within group variance. This fact becomes important later when we consider the homogeneity of variance assumption of ANOVA (see Section 3.4.3). Also, compute MS_T from Table 3.8 ($52/9 = 5.78$) and compare it with the total variance computed for all the scores from the data in Table 3.3; MS_T is the total variance for the scores, ignoring group membership.

3.2.3.3 Statistical Inference in ANOVA

Section 1.6 described the rationale for deciding whether the differences among sample means are likely to have occurred by chance. Recall that the denominator of the F ratio is an estimate of how much the means are likely to differ on the basis of random variation or error alone. The numerator of the F ratio is an estimate of the difference among means due to treatment plus error, whereas the denominator is differences in scores due to error. If there are no treatment effects (i.e., if the null hypothesis is true and $\mu_{a_1} = \mu_{a_2} = \cdots = \mu_{a_j}$), then differences among means equal differences among

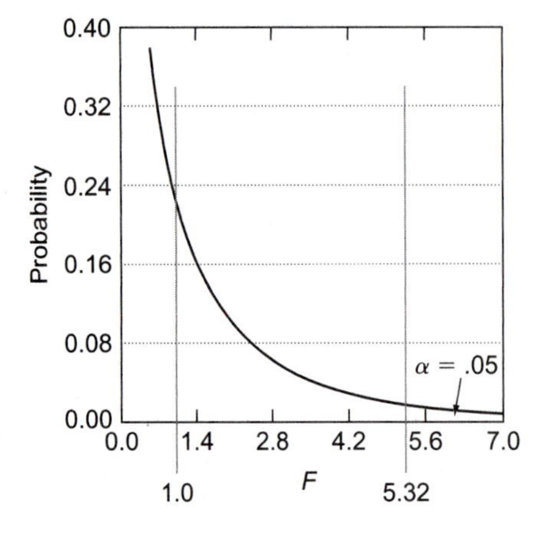

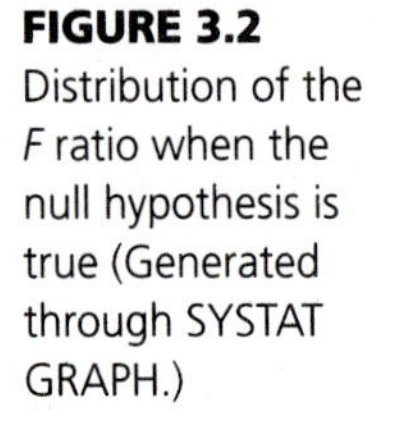
FIGURE 3.2
Distribution of the *F* ratio when the null hypothesis is true (Generated through SYSTAT GRAPH.)

scores and both the numerator and denominator represent error. Thus, the F ratio is likely to be around 1.0.[12] (It might be less than 1.0, or it might be a bit greater than 1.0, or it might be much greater than 1.0, although that would be improbable.) Figure 3.2 shows the shape of the F ratio when the null hypothesis is true. The vertical line on the left represents the mean of this distribution at $F = 1$.

Therefore, the notion that differences among means are due to random variation is unlikely to be true if the F ratio is large enough. But how large is large enough? The answer depends on degrees of freedom for both the numerator (df_A) and the denominator ($\text{df}_{S/A}$). Table A.1 shows the critical F ratio for combinations of numerator and denominator df and for various levels of Type I error (willingness to risk deciding erroneously that the mean differences are due to treatment rather than to random variation). If the calculated (obtained) F ratio of Equation 3.12 is equal to or greater than the critical F ratio from Table A.1, then the null hypothesis that means differ only by random variation is rejected, and we rejoice in our research success.

Say, for the example, that our willingness to risk a Type I error is 5%. (For the remainder of this book, we will assume $\alpha = .05$, unless otherwise stated.) With numerator df $= 1$ and denominator df $= 8$, the critical value of F in Table A.1 is 5.32. The calculated (obtained) F of 26.67 is larger than 5.32; therefore, the null hypothesis is rejected. The right vertical line in Figure 3.2 shows the distribution at the point where $F = 5.32$. Five percent of the distribution falls to the right of that point, which means that 5% of the time, an F ratio this large or larger could occur even when the null hypothesis is true. On the other hand, it is much more likely that an F ratio this large is the result of a *different* distribution—one in which the null hypothesis is *not* true and $F > 1.00$. Therefore, we conclude that $\mu_{a_1} \neq \mu_{a_2}$, or that the mean for the first group, 7, is different from (larger than) the mean of the second group, 3.

[12] However, it is unlikely to be exactly 1.0 because error is estimated differently in the numerator and denominator (more later . . .).

3.2.3.4 More Later . . .

It is time for a confession regarding the data generated from GLM in Table 3.3. Notice that within both a_1 and a_2, the sum of the numbers for error is 0. When the sum of the errors inside each group is 0, the means for each group are the same as the population values. However, nature is rarely so accommodating. Ordinarily, the error terms are spread around through the whole data set, summing to 0 over the whole data set, not within each group. When that is the case, the means for each group are shifted a bit, too, depending on exactly how the errors are distributed, as shown in Table 3.9.

The numbers for deviations due to error in Table 3.9 are the same deviations as those in Table 3.3, but now error is spread around randomly. The sum of the error effects at a_1 is -1, and the mean has changed from 7 to 6.8; similarly, the sum of the error effects at a_2 is $+1$, and the mean has changed from 3 to 3.2. Although this is alarming, recall that researchers almost always have to use sampled values to estimate population values, and the estimation is rarely exact. When estimating the population values (i.e., parameters) for the mean of each group, confidence intervals are created around the sample means so that the intervals contain the population means with a specified confidence (e.g., 95%).

This slippage of the means also explains why the F ratio is sometimes written as

$$F = \frac{\text{treatment} + \text{error}}{\text{error}} \tag{3.17}$$

When the group means are shifted by the distribution of errors, the numerator reflects both the effects of treatment and the effects of error. If there are no treatment effects, the F ratio should be about 1, but probably not exactly 1 due to the differences in the way error is estimated in the numerator and denominator. If there are treatment effects, the F ratio should exceed 1, perhaps by enough to exceed the critical value so we can have faith in the mean difference.

TABLE 3.9 A more typical data set for ANOVA, $Y = \mathbf{GM} + A + e$

a_1 Case	a_1 Score	a_2 Case	a_2 Score
s_1	$5 + 2 + 2 = 9$	s_6	$5 - 2 + 0 = 3$
s_2	$5 + 2 - 2 = 5$	s_7	$5 - 2 + 0 = 3$
s_3	$5 + 2 + 0 = 7$	s_8	$5 - 2 + 2 = 5$
s_4	$5 + 2 + 0 = 7$	s_9	$5 - 2 + 0 = 3$
s_5	$5 + 2 - 1 = 6$	s_{10}	$5 - 2 - 1 = 2$
$\Sigma Y_{a_1} = A_1 =$	34	$\Sigma Y_{a_2} = A_2 =$	16
$\sum Y_{a_1}^2 =$	240	$\sum Y_{a_2}^2 =$	56
$\overline{Y}_{a_1} =$	6.8	$\overline{Y}_{a_2} =$	3.2

3.3 THE REGRESSION APPROACH TO ANOVA

Analysis of variance is a special case of regression, and both are special cases of the GLM. All ANOVA problems can be solved through regression, but the reverse is not true: Not all regression problems can be solved using traditional ANOVA. Although continuous variables can be "backed down" to ordinal or discrete variables, there may be loss of information in doing so. Regression is most often used to study the relationship between two continuous variables, but it can also be used if one of the variables is discrete and represents group membership, which is exactly the setting for ANOVA.

Bivariate (two-variable) regression for two continuous variables is summarized in Section 2.3.2. Bivariate regression is sufficient to analyze the simple ANOVA problem of Section 3.2. As the ANOVA problem grows, however, multiple (many-variable) regression is needed for the solution, as shown in Chapter 4 and elsewhere.

Because both the traditional approach and the regression approach produce identical results, many of our students seem to think that we include the regression approach to ANOVA out of some sadistic impulse. However, about halfway through the course, they often realize that the regression approach is often conceptually cleaner than the traditional ANOVA approach. The design of the study is completely replicated in the columns necessary for the regression approach, and separate elements of the design are readily identified and available for analysis. As the design becomes more complicated, the clarity and specificity of the regression approach remain. In fact, screening designs and problems created by unequal n are easily understood through the regression approach but very difficult to fathom without it. Finally, as the design becomes more and more complicated—perhaps with numerous IVs, unequal n, and covariates, and perhaps even with continuous IVs—it is possible to retain complete control over the analysis by using the regression approach.

In the chapters that follow, we illustrate this flexibility and use whichever approach is easier. Even if you never choose to use the regression approach yourself, you will benefit in terms of increased flexibility and depth in your understanding of ANOVA.[13]

3.3.1 Bivariate Regression and Basic ANOVA

In a typical data set for bivariate regression, such as that in Table 2.8, the cases each have scores for two continuous variables, X and Y. In the regression approach to ANOVA, the Y column (vector) is usually written first, and the X column is then filled

[13] There is yet another approach to ANOVA, the cell means model, which requires matrix algebra for a solution. The decision was made to omit demonstration of the cell means model in this book, although it is sometimes used internally by software programs. Interested readers are referred to Kirk (1995) and Woodward, Bonett, and Brecht (1990).

TABLE 3.10 Columns for regression approach to ANOVA with two treatment levels

Level of *A*	Case	*Y*	*X*	*YX*
	s_1	9	1	9
	s_2	7	1	7
a_1	s_3	6	1	6
	s_4	7	1	7
	s_5	6	1	6
	s_6	3	–1	–3
	s_7	1	–1	–1
a_2	s_8	3	–1	–3
	s_9	4	–1	–4
	s_{10}	4	–1	–4
Sum		50	0	20
Squares summed		302	10	
N		10		
Mean		5		

with coding that represents group membership. There are several different types of coding (see Section 3.5.5), each with its own advantages and disadvantages. We use contrast coding here because of its many advantages in the regression approach to ANOVA. By using this coding for the example, every case in a_1 receives an X "score" (code) of 1, and every case in a_2 receives a score of -1.

The data set of Table 3.4 is organized in regression format in Table 3.10. The first column indicates the level of A (and is included for your convenience); the second column, the case; the third column, Y; the fourth column, X (contrast coded for group membership); and the fifth column, the product of columns Y and X. The number of X columns needed to separate the groups is equal to $\text{df}_A = a - 1$. Since $a = 2$, in this example, a single X column is required to separate a_1 from a_2. The sum of each column, the sum of the squared values in the column, the number of scores, and the relevant means are given at the bottom of the table.

The next step is to calculate the preliminary sum of squares (SS) and the sum of products (SP) for the problem. Computational equations and values for the example in Table 3.10 appear in Table 3.11. Notice the happy event that occurs in the equations for SS(X) and SP(YX): The second terms reduce to 0. This is because the codes used in the X column sum to 0. This feature simplifies the computations and is one of the advantages of contrast coding when there are equal sample sizes in each group.

Equation 3.1 describes the general linear model in terms suitable for the deviation (traditional) approach to basic ANOVA. The corresponding equation suitable for the regression approach to two-group ANOVA is

$$Y = a + bX + e \tag{3.18}$$

TABLE 3.11 Preliminary sums of squares and sum of products for regression approach to basic ANOVA

SS(*Y*)	$= \Sigma Y^2 - \dfrac{(\Sigma Y)^2}{N}$	$= 302 - \dfrac{50^2}{10}$	$= 52$
SS(*X*)	$= \Sigma X^2 - \dfrac{(\Sigma X)^2}{N}$	$= 10 - \dfrac{0^2}{10}$	$= 10$
SP(*YX*)	$= \Sigma XY - \dfrac{(\Sigma X)(\Sigma Y)}{N}$	$= 20 - \dfrac{(50)(0)}{10}$	$= 20$

TABLE 3.12 Slope and intercept estimates for regression approach to basic ANOVA

b	$= \dfrac{\Sigma XY - ((\Sigma X)(\Sigma Y))/N}{\Sigma X^2 - (\Sigma X)^2/N}$	$= \dfrac{20}{10}$	$= 2$
a	$= \bar{Y} - b\bar{X}$	$= 5 - 2(0)$	$= 5$
Y'	$= a + bX$	For a_1, $X = 1$	
		$= 5 + (2)(1) = 7$	$= 7$
		For a_2, $X = -1$	
		$= 5 + (2)(-1) = 3$	$= 3$

This equation asserts that a score, Y, can be decomposed into three elements: The intercept, a, corresponds to the mean, μ, of Equation 3.1 with the contrast coding described earlier. Note that a has nothing to do with the number of levels of the IV, which is also designated a, just to maximize confusion. The slope, b, when multiplied by the code (1 or -1) representing the level of the IV (X), corresponds to the main effect of the independent variable: α of Equation 3.1. The error term, e, is the same in both equations.

This equation may seem familiar; it is the bivariate regression Equation 2.17, with the addition of the error term and the change of predicted Y' to the actual value of the DV, Y. Indeed, error is defined as the difference between an observed score and the predicted score:

$$e = Y - Y' \qquad \textbf{(3.19)}$$

Thus, Equations 2.16–2.18 are used to find slope, intercept, and predicted score for two-level randomized-group ANOVA, as shown in Table 3.12.

With contrast coding of the X column (1 -1) and equal sample sizes, the intercept, a, is equal to the grand mean. When the contrast code for a case is inserted in the prediction equation (as shown in the lower right portion of the table), the predicted score for each case is the mean of that case's own treatment group. Thus, for cases in level a_1, with a contrast code of 1, Y' is 7, whereas for cases in level a_2, with a contrast code of -1, Y' is 3.

The preliminary SS and SP values in Table 3.11 are also used to compute a final sum of squares for regression, SS(reg.), used to test the null hypothesis that the slope,

TABLE 3.13 Final sums of squares for basic ANOVA through regression

SS(reg.)	$\frac{(SP(YX))^2}{SS(X)}$	$= \frac{20^2}{10}$	$= 40$
SS(total)	$= SS(Y)$		$= 52$
SS(resid.)	$= SS(total) - SS(reg.)$	$= 52 - 40$	$= 12$

TABLE 3.14 Calculation of degrees of freedom for basic ANOVA through regression

df(reg.)	= number of predictors (X vectors, columns)		= 1
df(total)	= number of cases (N) – 1	= 10 – 1	= 9
df(resid.)	df(total) – df(reg.)	= 9 – 1	= 8

TABLE 3.15 Source table for basic ANOVA through regression

Source	SS	df	MS	F
Regression	40	1	40	26.67
Residual	12	8	1.5	
Total	52	9		

b, is 0 and, simultaneously, that the correlation, r, is 0. The equation for final SS(reg.) is a first cousin to the equation for slope:

$$SS(reg.) = \frac{(SP(YX))^2}{SS(X)} \tag{3.20}$$

Once SS(reg.) is known and SS(Y) is computed, an error term called sum of squares residual, SS(resid.), is computed as the difference between them, as shown in Table 3.13.

SS(reg.) has as many degrees of freedom as there are X columns (columns to code levels of the IV). The df for SS(resid.) is found as the difference between df(total) and df(reg.). Degrees of freedom for the example are in Table 3.14.

Table 3.15 summarizes the information in Tables 3.13 and 3.14 in a source table, rearranged to mimic that of Table 3.8. The values in Table 3.15 are identical to those in Table 3.8; only the names have been changed to protect the innocent. The regression source is the treatment effect, A, and the residual source is the estimate of error, S/A. The F ratio is the same as in Table 3.8:

$$F = \frac{\textbf{MS(reg.)}}{\textbf{MS(resid.)}} = \frac{40}{1.5} = 26.67$$

Table A.1 is used to find the critical value of F for testing null hypotheses, as in the traditional approach to ANOVA. Three null hypotheses are tested at one time through the regression approach: (1) the population correlation between X and Y is equal to 0, (2) the population regression coefficient (slope or b) is equal to 0, and (3) $\mu_1 = \mu_2$. The third hypothesis, of course, is the null hypothesis in the form typically stated for the traditional approach to ANOVA.

3.3.2 Why Bother?

With simple designs, the traditional approach to ANOVA is clearly more efficient. However, as designs become more complicated, the traditional approach often loses its advantage. The traditional computational scheme becomes convoluted for some types of problems and requires something of a leap of faith as to its accuracy. The regression approach, however, with the design of the study clearly laid out in the X columns, remains straightforward. The logic of the process is easy to follow, although the computations require an extra step: preliminary calculations of sums of squares or sums of products for each of the columns, followed by final calculation of SS(reg.) from preliminary calculations. With use of a spreadsheet such as Microsoft Excel or Corel Quattro Pro, the computations are also easy enough to make the approach attractive. Or you could use a multiple regression program available in one of the statistics software packages instead of an ANOVA program. These multiple regression programs have facilities for controlling the entry of variables sequentially, giving you complete control over the analysis.

3.4 ASSUMPTIONS OF ANALYSIS

Certain assumptions are made about the nature of the data when inferential tests are devised. It is simply not possible, most of the time, to develop an inferential test for all possible types of data sets. It may be that the inferential test works well for many types of data sets, including some that violate some of the assumptions. An inferential test is said to be *robust* if it gives the correct answer despite violation of some assumption(s). However, the applicability of an analytic strategy to data sets with various properties has to be investigated. When an inferential test is used for a data set that violates important assumptions, the probabilities that are estimated are not necessarily accurate. The researcher may intend to use an α level of, say, .05, but may actually have an α level of, say, .23. For this reason, it is important to consider the fit between the assumptions of analysis and your data set before using the analysis. Some ANOVA designs have specialized assumptions that are not discussed here but that instead are addressed in the third sections of the chapters devoted to those designs.

3.4.1 Normality of Sampling Distribution of Means

An assumption of ANOVA is that the sampling distribution of means for each level (or combination of levels) of the IV(s) is normal. The assumption is for the sampling distribution, not the raw scores. If the raw scores are normally distributed, the sampling distribution of their means is also normally distributed. However, even if the raw scores are not normally distributed, the Central Limit Theorem assures us that the sampling distribution of means is normally distributed for large enough samples.

Normality of sampling distributions is usually ensured by having sufficiently large and relatively equal sample sizes among levels (or combinations of levels) of the IV. *With relatively equal sample sizes and two-tailed tests (and no outliers), robustness is expected with 20 degrees of freedom for error.* For basic ANOVA, degrees of freedom for error are calculated using Equation 3.7. Single-sided (one-tailed) tests require larger samples.

Rather often, experiments have small sample sizes and fail to provide sufficient df. If error df are low, tests for skewness and kurtosis are applied to the raw scores within each group (Section 2.2.3.3). If these tests also indicate problems, the researcher has to compensate. If sample sizes are unequal or too small and there is excessive skewness or kurtosis or if outliers are present, data transformation may be necessary to achieve normality of distributions of raw scores within each group. (Transformations are discussed in Section 3.5.7.) Recall from Chapter 2 that the transformation is applied to *all* DV scores, not just to the scores in one group. Nonparametric methods are also available for small, non-normally distributed data sets. Siegel and Castellan (1988) and Conover (1980) are rich sources for nonparametric tests. Most statistical computer packages also provide nonparametric analyses.

3.4.2 Independence of Errors

The last term in the GLM is e, the error term. Another assumption of analysis is that the errors are independent of one another—that the size of the error for one case is unrelated to the size of the error for cases near in time, near in space, or whatever. This assumption is easily violated if, for instance, equipment drifts during the course of the study and cases measured near each other in time have more similar errors of measurement than cases measured farther apart in time. In each research setting, care is needed to control the numerous factors that could cause violation of the assumption of independence of errors because violation of the assumption can lead both to larger error terms (due to inclusion of additional factors that are not accounted for in the analysis) and to potentially misleading results if nuisance variables are confounded with levels of treatment.

If cases are entered into the data set in sequential order and the problem is analyzed through regression, there is a formal test of contingencies among errors in time, called the Durbin-Watson statistic. This statistic assesses autocorrelation (self-correlation) among residuals. If the errors (residuals) are independent, autocorrelation is 0. If you suspect violation of this assumption due to contingencies in the sequence of the cases, use of this analysis is appropriate. If violation is found, addition of another IV representing time (say, early, middle, and late in the course of the study) might account for this source of variability in the data set.

The assumption of independence of errors is also violated if an experiment is not properly controlled. For experimental IVs, errors within groups are related (not independent) if all cases within a level are tested together (unless all cases from all levels are tested simultaneously), because cases tested together are subject to the same nuisance variables. For example, a participant may laugh out loud inappropriately in the presence of other members of a group assigned to watch a sad movie. Thus, a mean difference found between groups could be due to the nuisance variables unique to the group rather than to the treatment unique to the group.[14] The seriousness of confounding nuisance variables with levels of treatment by testing groups separately depends on the expected sizes of the effects of the nuisance variables on the DV. If there are potentially important nuisance variables, you should test cases individually or simultaneously for all levels, not in groups defined by levels.

This assumption is rarely applicable for nonexperimental IVs. In the absence of random assignment to levels, there is no justification for causal inference to the treatments, so the assumption of independence loses relevance.

3.4.3 Homogeneity of Variance

The ANOVA model *tests* whether population *means* from different levels of the IV are equal. However, the model *assumes* that population *variances* in different levels of the IV are equal—that is, it is assumed in ANOVA that the variance of DV scores within each level of the design is a separate estimate of the same population variance. Recall from Sections 3.2.2 and 3.2.3.2 that the error term, $MS_{S/A}$, is an average of the variances within each level. If those variances are separate estimates of the same population variance, averaging them is sensible. If the variances are quite different and are not separate estimates of the same population variance, averaging them to produce a single error term is not sensible.

ANOVA is known to be robust to violation of this assumption as long as there are no outliers, sample sizes are large and fairly equal, the sample variances within levels (or combinations of levels) are relatively equal, and a two-tailed hypothesis is tested. The ratio of largest to smallest sample size should be no greater than 4:1. The ratio of largest to smallest variance should be no greater than approximately 10:1 when an F_{max} ratio is calculated in which

$$F_{max} = \frac{s^2_{largest}}{s^2_{smallest}} \tag{3.21}$$

Homogeneity of variance requires that F_{max} be no greater than 10. *If these conditions are met, there is adequate homogeneity of variance.*

If sample sizes are too discrepant, a formal test of homogeneity of variance is useful; some of these tests are described in detail by Winer, Brown, and Michels (1991), Keppel (1991), and others. However, these tests tend to be too sensitive, leading to overly conservative rejection of the use of ANOVA. Many of these tests are

[14] Nested designs, in which groups are necessarily tested together, are discussed in Chapter 11.

also sensitive to non-normality of the DV. One of the tests, Levene's (1960) test of homogeneity of variance, is typically not sensitive to departures from normality, although, like the others, it is usually too sensitive to heterogeneity of variance. Levene's test, which performs ANOVA on the absolute values of the residuals (differences between each score and its group mean) derived from a standard ANOVA, is available in SPSS ONEWAY and GLM. Significance indicates possible violation of homogeneity of variance.

Violations of homogeneity often can be corrected by transformation of the DV scores (see Section 3.5.7); however, interpretation is limited to the transformed scores. Another option is to use a more stringent α level (e.g., for nominal $\alpha = .05$, use .025 with moderate violation and .01 with severe violation of homogeneity) with untransformed DV scores.

Heterogeneity of variance should always be reported and, in any event, is usually of interest in itself. *Why* is spread of scores in groups related to level of treatment? Do some levels of treatment affect all the cases about the same, whereas other levels of treatment affect only some cases or affect some cases much more strongly? This finding may turn out to be one of the most interesting in the study and should be dealt with as an issue in itself, not just as an annoyance in applying ANOVA.

3.4.4 Absence of Outliers

Outliers, which were discussed in some detail in Section 2.2.3.4, are a pervasive problem in data sets and are the subject of a great deal of gossip among statisticians at cocktail parties ("Remember the time that an outlier was found—a 2000-year-old man . . . ?"). The effects of outliers are unpredictable; they can hide real statistical significance or create apparent statistical significance in ANOVA. They severely limit the generalizability of the results. For this reason, it is important to screen data sets for outliers prior to analysis. Remember that they are sought separately in each level (or combination of levels) of a design.

There may be outlying scores on the DV in one or more levels of the IV. An outlier is a score that is unusually far from the mean of its own group and apparently disconnected from the rest of the scores in the group. The z score distribution is used to assess the distance of raw scores from their own group mean. A z score (see Equation 2.7) is calculated as the difference between the score for a deviant case and its group mean divided by the standard deviation for its group. If there are 20 or more df for error, a case with a z (absolute) value of 3.3 or greater is probably an outlier, because the probability of sampling a score of this size if it is truly from the population of interest is .001 or less. If there are fewer than 20 df for error, an absolute value of $z = 2.58$ is appropriate (corresponding to a probability of .01 or less).

Disconnectedness is assessed by visual inspection of the distribution of DV scores in each group. As shown in Figure 3.3, if there is a large distance between the potential outlier and the other scores in the group and the score has a z value of 3.3 or higher, that case is most probably an outlier. If, on the other hand, the score seems to be connected with the rest of the cases because intervening cases fill the gap, then

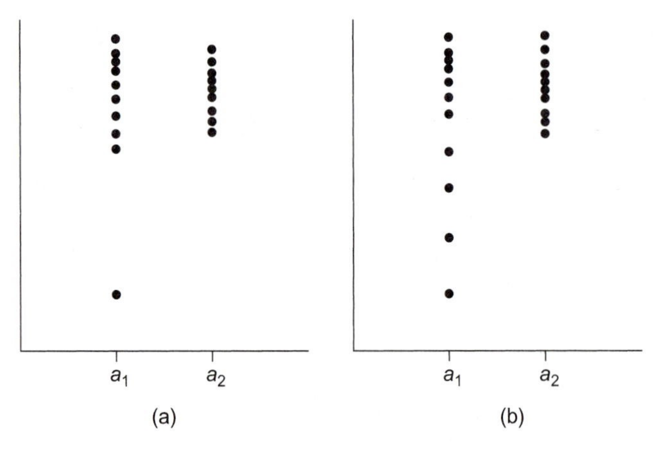

FIGURE 3.3 Assessing outliers through disconnectedness where the low score at a_1 in (a) is more likely to be a true outlier than the same low score in (b).

the case may not be an outlier. Rather, you may be dealing with a non-normal distribution.

If one or more outliers are found, they are dealt with by transforming *all* the DV scores (see Section 3.5.7), by deleting the cases that produce the outlying scores, or by changing only the scores for the outlying cases to make them less deviant (see Section 2.2.3.4). Each of these options has a different consequence. If all scores are transformed, interpretation of the results of ANOVA is limited to the transformed means. If cases are deleted, the sample is changed and so is the scope of generalizability. If scores are changed, you need to reassure your readers that the decision was made prior to and without consideration of its effects on ANOVA. In any case, treatment of outliers is included in the report of the analysis.

3.5 OTHER RECURRING ISSUES

There are several recurring issues in ANOVA, discussed in all the chapters that follow. They are also discussed briefly here to give you a preview of coming attractions.

3.5.1 Effect Size

In some disciplines, there is currently lively debate about the use of the term *significance* to describe mean differences that are likely to replicate. The problem is partially semantic and involves the excess baggage attached to the word *significant.* In ordinary parlance, the term connotes "important," a meaning that is inappropriate when applied in a statistical setting. Mean differences in well-controlled experiments are often statistically significant but quite small and unimportant from any practical standpoint. There is growing awareness, then, of a need to examine effect size, which does indicate importance, as well as reliability. This issue was discussed in Chapter 2 in the context of squared correlation and conceptual diagrams. Discussion of this issue is expanded, and measures of effect size provided, in the chapters that follow.

3.5.2 Power

Power is the probability of correctly rejecting a false null hypothesis. In somewhat plainer English, power is the probability of supporting your research hypothesis if it really is true. If the probability is low, there is very little reason to do the research, because you are unlikely to find evidence in favor of the conclusion you seek, even if it is true. In most settings, a power level of about .80 or so is needed to justify the time and expense of conducting the research.

Power and sample size are interrelated. Larger samples have greater power, all other things being equal. One of the more common things a statistician is asked to do is estimate the sample size needed for the requisite power. The researcher wants to have a big enough sample for adequate power without going overboard and using too many cases. There are numerous programs available for assisting in the assessment of power. These and related concepts are discussed in the chapters that follow.

3.5.3 Comparisons

When there are only two levels of an IV, the results of ANOVA are unambiguous. If the difference is statistically significant, the larger population mean probably (with error set to α) is greater than the smaller one; the two means are estimates of central tendency from two different populations. If the difference is nonsignificant, the two population means probably (with error set to β) are the same; that is, they are separate estimates of central tendency in a single population.

However, when there are more than two levels of an IV, the results of routine (omnibus) ANOVA are ambiguous. The result tells you that there probably is a difference somewhere among the population means but not which means are different. If there are three means, for instance, you could have $\mu_{a_1} \neq \mu_{a_2} \neq \mu_{a_3}$ (all three means different) or $\mu_{a_1} = \mu_{a_2} \neq \mu_{a_3}$ (the first two means equal but unequal to the third) or $\mu_{a_1} = \mu_{a_3} \neq \mu_{a_2}$ (first and third means equivalent but unequal to the second) or . . . well, you get the point.

To resolve the ambiguity, comparisons (also called contrasts) are performed. Comparisons are either done in lieu of ANOVA, in a process called planned comparisons, or after ANOVA, in a process called post hoc comparisons. If you have enough information to plan your comparisons, there is a clear advantage in terms of Type I error rates and power. Post hoc analyses are used if you do not know what to expect, but power is lower.

There are many varieties of comparisons, each with a different tradeoff between power and flexibility. This issue also arises repeatedly in the chapters that follow.

3.5.4 Orthogonality

Orthogonality is perfect *non*relationship between two or more variables. When two variables are orthogonal, their correlation is 0. The issue of orthogonality arises in several places, including in the context of comparisons. The regression approach to ANOVA, for instance, involves setting up a bunch of comparisons, and it "works" easily because orthogonal codes are used in the *X* columns (see Table 3.10); similarly, comparisons in a traditional analysis are enhanced by choosing orthogonal coding coefficients (see Chapter 4).

The issue of orthogonality also arises when designs include more than one IV, as discussed in Chapters 5 (for randomized-groups designs) and 6 (for repeated-measures designs) and in Chapter 7 (for mixed designs). Additional IVs are in factorial arrangement in the designs discussed in these chapters because all combinations of all levels of all IVs are included in the design. As long as the designs are factorial and have equal numbers of scores in every cell, the IVs are orthogonal and each source of variability in the design is assessed independently (except for use of a common error term); that is, the *F* tests for all of the IVs are independent.

However, there are two situations in which the *F* tests become nonorthogonal. The first is when sample sizes are not equal in all cells in the design, and the second is when IVs are added but are not in factorial arrangement with other IVs in the design. Section 5.6.5 has a lengthy discussion of the problem of nonorthogonality associated with unequal *n*. Chapters 9 (Latin-square designs), 10 (screening designs), and 11 (nested designs) discuss analyses for designs when IVs are not in factorial arrangement. The covariates introduced in Chapter 8 are also not orthogonal to the rest of the design.

3.5.5 Coding

Coding is used to perform the specialized work of comparisons, the basis for the regression approach to ANOVA. Codes (also called weighting coefficients) are numerical values used to leave out some levels of an IV, to compare the means from two groups with each other, or to pool groups together for the purposes of testing the pooled mean from two or more groups against the mean for other groups (pooled or not). Consider, for example, the four levels of treatment in Table 3.16, with two control groups (waiting list and placebo) and two treatment groups (Treatment I and Treatment II). The basic idea is to weight scores in groups by numbers to perform some work, as described in the last column of the table.

Often numerical codes simply work to separate levels of an IV from each other, for example, to separate a_1 from a_2 and a_3. There are several schemes for doing this, as illustrated in Table 3.17, namely, dummy variable coding, effects coding, and contrasts coding. There are as many codes needed as there are degrees of freedom for the groups. For instance, there are three groups in the example, a_1, a_2, and a_3, so there are two degrees of freedom and two columns of codes (X_1 and X_2) necessary to separate the groups from each other, no matter which coding scheme is used.

TABLE 3.16 Coding coefficients to delete groups, to compare means of groups, and to pool means of groups

Waiting List	Placebo	Treatment I	Treatment II	Work of Codes (weighting coefficients)
0	0	1	–1	Leave out waiting list and placebo groups and compare mean of Treatment I against mean of Treatment II.
1	1	–1	–1	Pool waiting list and placebo groups and pool Treatment I and Treatment II; compare pooled mean of control groups against pooled mean of treated groups.
0	2	–1	–1	Leave out waiting list and compare the means of the placebo group against the pooled means from the two treatment groups.

TABLE 3.17 Three types of coding—dummy variable, effects, and contrasts—to separate groups from each other

	Dummy Variable Coding		Effects Coding		Contrasts Coding	
Groups	X_1	X_2	X_1	X_2	X_1	X_2
a_1	1	0	1	0	2	0
a_2	0	1	0	1	–1	1
a_3	0	0	–1	–1	–1	–1

In all these schemes, the cases at a_1 have a different set of codes from the cases at a_2 and a_3. Looking at the rows, members of group a_1 have codes of 1,0 for X_1 and X_2 with dummy variable coding and effects coding, but they have a code of 2,0 for contrasts coding. Members of a_2 have 0,1 codes with dummy variable coding and effects coding, but -1,1 for contrasts coding; whereas members of a_3 have 0,0 codes for dummy variable coding, but -1, -1 for effects and contrasts coding. Looking down the columns, and thinking about the information in Table 3.16, you'll see that the X_1 column separates a_1 from a_2 and a_3 in dummy variable coding, compares a_1 against a_3 in effects coding, and compares a_1 against a_2 and a_3 pooled in contrasts coding. What is the X_2 column doing in the three types of coding?

Because contrasts coding is orthogonal, and for other reasons that are too complicated to go into here, we use contrasts coding in the regression examples for most of the book. This is also the type of coding that corresponds to the coding used for comparisons in the traditional approach to ANOVA. This decision is not as limiting as it may seem. There are alternate sets of orthogonal contrasts that may be chosen, especially when the number of levels of the IV is large. Be warned, however, that you will encounter effects coding in Chapter 10 (screening designs) and dummy variable coding from time to time in your study of advanced statistics and in the literature.

3.5.6 Missing Values

In randomized-groups designs, each case provides a single DV score. When that score is missing, the case is deleted from the analysis and the problem of missing values becomes a problem of unequal n. In repeated-measures designs and mixed designs, on the other hand, there are other scores for a case that can be used to estimate the size of the missing value. As the number of DV scores taken for each case increases, it becomes more probable that there will be missing values, but there will also be better estimates of them.

There are several ways to estimate (impute) missing values. Some of them involve honest best guesses about the likely size of a missing value. Others use regression equations to predict the missing score from one or several other scores that are available for a case, as shown in Section 2.3.2.2. Others impute missing values from the average DV score for the case and then average for the level of treatment in which the missing DV occurs. Others match the case that has a missing value to another case that has complete values and then use the value from the complete case as the estimate. Yet others draw a random sample from among the scores of the complete cases. When there are specialized methods of estimating a missing value for a particular analysis, they are discussed in the relevant chapter.

3.5.7 Transformations

Data transformations are often undertaken when a data set violates assumptions, such as normality and homogeneity of variance, or when there are outliers. If you decide to transform, it is important to check that the distribution is improved by transformation. If a variable is only moderately positively skewed, for instance, a square root transformation may make the variable moderately negatively skewed, and thus there is no advantage to transformation. Often you need to try first one transformation and then another until you find the transformation that produces the skewness and kurtosis values nearest 0, homogeneity of variance, and/or the fewest outliers.

Variables differ in the extent to which they diverge from normal. Figure 3.4 presents distributions with several degrees of non-normality, together with the transformations that are likely to render them normal. If the distribution differs moderately from normal, a square root transformation is tried first. If the distribution differs substantially, a logarithm transformation is tried. If the distribution differs severely, the inverse is tried. According to Bradley (1982), the inverse is the best of several alternatives for J-shaped distributions, but even it may not render the distribution acceptably normal. Finally, if the departure from normality is severe and no transformation seems to help, you may want to try dichotomizing the variable.

The direction of the deviation is also considered. When distributions have positive skewness, as discussed previously, the long tail is to the right. When they have negative skewness, the long tail is to the left. If there is negative skewness, the best strategy is to *reflect* the variable and then apply the appropriate transformation for positive skewness. To reflect a variable, find the largest score in the distribution and add 1 to it to form a constant that is larger than any score in the distribution.[15] Then

[15] If the distribution contains negative numbers, add a large enough constant so that the smallest number in the entire distribution is 1 before performing the transformation.

FIGURE 3.4 Distributions with varying degrees of non-normality and their potential transformations

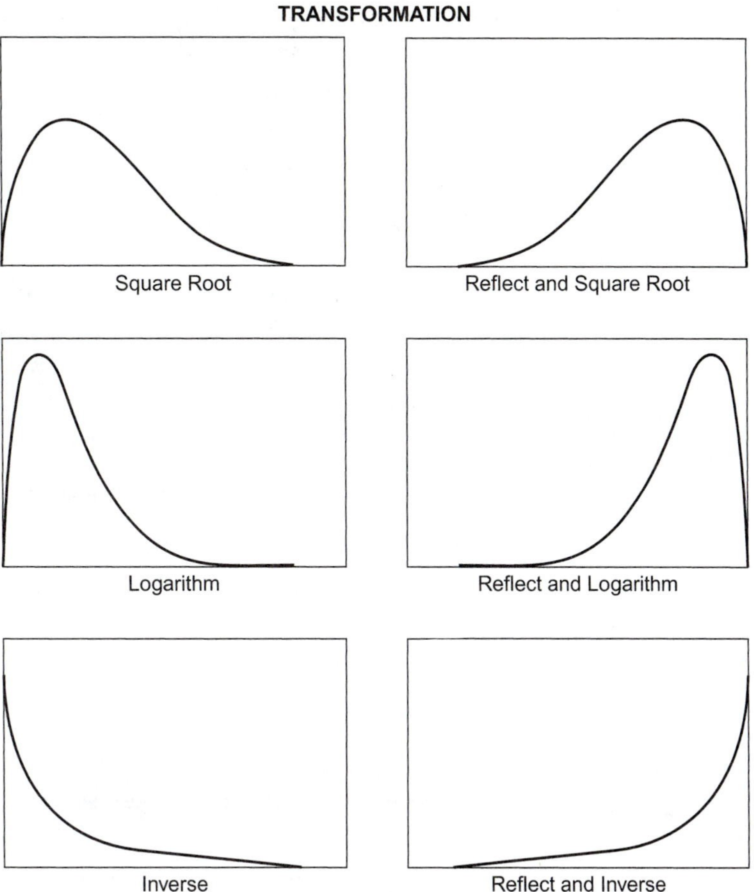

create a new variable by subtracting each score from the constant. In this way, a variable with negative skewness is converted to one with positive skewness prior to transformation. When you interpret a reflected variable, be sure to reverse the direction of the interpretation as well. For instance, if big numbers meant good things prior to reflecting the variable, big numbers mean bad things afterward.

Instructions for transforming variables in four software packages are given in Table 3.18. Notice that a constant is also added if the distribution contains a value less than 1. A constant (to bring the smallest value to at least 1) is added to each score to avoid taking the log, square root, or inverse of 0.

It should be clearly understood that this section merely scratches the surface of the topic of transformations, about which a great deal more is known. The interested reader is referred to Emerson (1991) or the classic Box and Cox (1964) for a more flexible and challenging approach to the problem of transformation.

TABLE 3.18 Original distributions and common transformations to produce normality

	SPSS COMPUTE	SAS DATA
Moderate positive skewness	NEWX=SQRT(*X*)	NEWX=SQRT(*X*);
Substantial positive skewness	NEWX=LG10(*X*)	NEWX=LOG10(*X*);
with zero	NEWX=LG10(*X* + *C*)	NEWX=LOG10(*X* + *C*);
Severe positive skewness, L-shaped	NEWX=1/*X*	NEWX=1/*X*;
with zero	NEWX=1/(*X* + *C*)	NEWX=1/(*X* + *C*);
Moderate negative skewness	NEWX=SQRT(*K* - *X*)	NEWX=SQRT(*K* - *X*);
Substantial negative skewness	NEWX=LG10(*K* - *X*)	NEWX=LOG10(*K* - *X*);
Severe negative skewness, J-shaped	NEWX=1/(*K* - *X*)	NEWX=1=(*K* - *X*);

C = a constant added to each score so that the smallest score is 1.
K = a constant from which each score is subtracted so that the smallest score is 1; usually equal to the largest score +1.

3.6 OVERVIEW OF REMAINING CHAPTERS

Chapters 4–11 describe analyses appropriate to various research designs. Chapter 4 describes basic one-way, randomized-groups, fixed-effects ANOVA. This is the simplest analysis and is the place to begin your study of ANOVA. Chapter 5 provides a generalization of the one-way analysis to a design with two or more IVs. The concept of interaction arises at this point and is the next level of complexity in ANOVA. Chapter 6 describes repeated-measures and Chapter 7 mixed-design ANOVAs, where cases are measured more than one time. Some important assumptions affect these analyses, thus increasing their complexity over the material in Chapters 4 and 5. Because Chapter 7 ends with a summary of the designs of Chapters 5–7, we recommend that you take on these chapters in order (unless you are already familiar with the material in one or more of them). After that, the order in which you approach Chapters 8–11 is best driven by your needs.

In working with the analysis chapters, it is recommended that you select a set of variables appropriate to your own discipline and insert those variable names, as appropriate, into the examples we give. It is also suggested that you apply the various analyses to some interesting data set from your own discipline. Many data banks are readily accessible through computer installations. (One online source of such data sets is http://lib.stat.cmu.edu/DASL/DataArchive.html.)

3.7 PROBLEM SETS

3.1 Pain is measured on a scale of 0 to 30 for two groups of pain sufferers—a control group and a group given drug treatment in the following data set. Check the assumptions for the analysis of this data set.

a_1: Control	a_2: Treatment
14	15
13	12
15	11
14	12
12	12
30	14
16	14
11	10
13	14
14	13
25	10
14	12

3.2 Resolve any problems you found in Problem 3.1 and use the standard computational approach for ANOVA.

3.3 Redo Problem 3.1, using the regression approach and contrasts coding $1, -1$.

Extending the Concepts

3.4 Redo Problem 3.1, using the regression approach with other numbers (say, 3,–3). Remember that every case in the same group has the same code and that the codes for the two groups are different. (*Hint:* Make your life easy and pick small numbers).

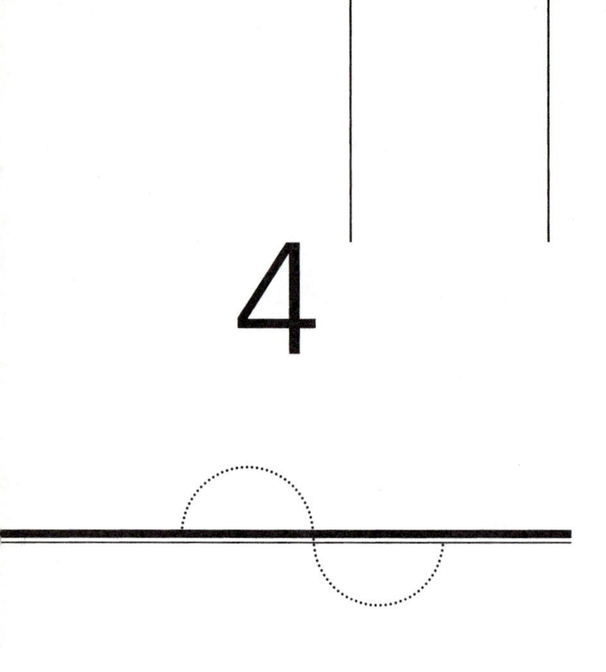

4

One-Way Randomized-Groups Analysis of Variance, Fixed-Effects Designs

4.1 GENERAL PURPOSE AND DESCRIPTION

This simple one-way ANOVA asks whether different groups of cases have statistically significant mean differences on a DV. The different groups form different levels of a single IV; the DV is a continuous variable that potentially has different means in the different groups. For example, do two different types of therapy produce mean differences in self-esteem? The levels of the single IV are the two types of therapy; the DV is self-esteem. The research question is whether the therapies differ in their ability to enhance self-esteem. The null hypothesis is that the two therapies produce about the same average self-esteem. As another example, if the IV is a type of on-the-job personnel training program with three levels (a no training control group, a text-based training program, and a video-based training program) and the DV is the supervisor's overall rating of job performance, ANOVA determines whether there are statistically significant mean differences in supervisor ratings associated with type of training program. The research hypothesis might be that the video-based training program is superior to the text-based program, which in turn is better than no training at all. The null hypothesis is that there are no mean differences in ratings among the three training programs.

ANOVA *compares* means, so the minimum number of levels of the IV is two: One of the levels may be a control and one a treatment, or they may be two different treatments. Although there must be at least two levels of treatment, there may be many levels of treatment. However, designs with very large numbers of levels often become unwieldy.

The levels of treatment differ in either type or amount. Examples of levels that differ in type are the two different types of therapy or the three different types of on-the-job training. Examples of levels that differ in amount are the same therapy for two different lengths of time (say, 1 month or 2 months) or the same type of on-the-job training conducted for one session, two sessions, or three sessions. A special type of technique, called a *trend analysis,* is available when there are three or more levels of treatment that differ in amount.

This chapter discusses completely randomized, fixed-effects, one-way ANOVA, also known as one-way between-subjects ANOVA. This analysis is appropriate when there is a single DV and (1) there is only one IV (one-way), (2) there are different cases in each level of the IV (completely randomized), and (3) all of the levels of the IV are specifically chosen by the researcher (fixed effects) because they are the levels of research interest. Designs with more than one IV are discussed in Chapter 5. Designs in which the same cases are measured repeatedly, or in which cases are matched in some way before random assignment, are discussed in Chapters 6 and 7. Designs in which levels of IVs are not fixed, but are randomly selected by the researcher, are discussed in Chapter 11. Designs with more than one DV (say, self-esteem and level of anxiety) are discussed in Chapters 7 and 8 of Tabachnick and Fidell (2007).

This type of ANOVA is often used to analyze the results of experimental research. If the study is an experiment, there is random assignment of cases to levels of the IV, manipulation of the levels of the IV by the researcher, and control of extraneous variables. The analysis also is frequently and appropriately used for non- and quasi-experimental research in which it is impossible to randomly assign cases to levels of treatment or to manipulate the levels of the IV, but in which interpretation of results is limited in the absence of full experimental control.

4.2 KINDS OF RESEARCH QUESTIONS

The primary question in one-way ANOVA is whether there are statistically significant mean differences on the DV among the different treatment groups. If significant mean differences are found, population means (and often standard deviations or standard errors) are estimated and reported for each group, along with estimates of the strength of the treatment effect. If there are more than two levels of the IV, comparisons are usually desired to pinpoint the location of the mean differences among groups. Comparisons are conducted in addition to or instead of ANOVA.

4.2.1 Effect of the IV

Are there statistically significant mean differences on the DV associated with the different levels of the IV? In other words, holding all else constant,[1] are the mean differences among the different groups of cases larger than would be expected if mean differences are due solely to random fluctuations in scores? For example, is there a mean difference in self-esteem depending on therapy type? Is there a mean difference in job performance rating depending on the training group to which an employee has been assigned?

These questions are answered by applying F tests to the data, as demonstrated in Sections 3.2.3.3 and 4.4. This F test is called an *omnibus test* because it assesses the overall reliability of differences among treatment means but does not reveal the source of the difference(s) if there are more than two treatments.

4.2.2 Specific Comparisons

If there are more than two groups and a mean difference is found, what is the source of the difference? For the job-training example with three levels, is the text-based training program different from no training at all? Is the video-based program different from the text-based program? Is the average performance of trained employees different from that of employees without on-the-job training? Or, if the IV is quantitative, as in length of training, is there a linear relationship between length of training and supervisor ratings? Does supervisor rating increase and then level off with increased training? These questions are answered by applying comparison procedures, as discussed in Section 4.5.5.

Comparisons to pinpoint the source of mean differences can be conducted in lieu of the omnibus F test (planned) or following the omnibus F test (post hoc). The advantages and disadvantages of planned and post hoc tests are also discussed in Section 4.5.5.

4.2.3 Parameter Estimates

If a treatment could be applied to an entire population, what mean would we expect? What mean job performance rating would we expect for a population of employees

[1] Devices discussed in Section 1.1 are used to "hold all else constant" in an experimental design.

who participated in a video-based training program? Population values estimated from samples are called *parameter estimates.* Parameter estimates reported following ANOVA are group means, standard deviations, standard errors, or confidence intervals for the mean (which incorporate both means and standard errors). Means, as well as standard deviations or standard errors, are reported in narrative form, shown in a table, or presented in a graph. Estimation of parameters is illustrated in Section 4.6.2.

4.2.4 Effect Sizes

The size of group mean differences may provide sufficient information to assess the magnitude of treatment effect when the DV is a familiar one, such as size, weight, or a physical measure like number of bacteria. If, however, inspection of means is insufficient, a measure of effect sizes is useful. It is also useful to report effect sizes so that future researchers may include your study in meta-analyses, which combine the results of many published studies. There are two types of effect size measures: One is based on proportion of variance in a DV that can be attributed to differences among treatments, and another expresses differences among group means in standard deviation units. Both types are discussed in Section 4.5.1.

4.2.5 Power

How high is the probability of showing that there are mean differences among groups if, indeed, treatment has an effect? For example, if the video-based training program is indeed superior to the text-based program or to no program at all, what is the probability of demonstrating that superiority? Estimates of power made prior to running a study are used to help determine sample size. Estimates of power as part of the analysis of a completed study are primarily used to determine why the research hypothesis did not find support. Power is discussed in Section 4.5.2.

4.3 ASSUMPTIONS AND LIMITATIONS

4.3.1 Theoretical Issues

The statistical test does not ensure that changes in the DV are caused by the IV. Causal inference depends on the study's design, not on the statistical test used. Considerations discussed in Sections 1.1 and 1.2 determine the legitimacy of causal inference. As a statistical test, ANOVA is also appropriate for nonexperimental IVs, as long as results are interpreted cautiously.

Strict generalizability is limited to populations from which random samples are taken, as discussed in Section 1.7. Results in an experiment generalize to the population of cases available for random assignment to levels of the IV. In nonexperimental studies, it is usually impossible to determine the available population unless there is careful random sampling from the population of interest. Therefore, generalization becomes a logical argument.

4.3.2 Practical Issues

The ANOVA model assumes normality of sampling distributions, homogeneity of variance, independence of errors, and absence of outliers, as described in Section 3.4. Unequal sample sizes, which create nonorthogonality, also pose problems for some assumptions and some comparisons in ANOVA, as discussed in Section 4.5.5.

4.4 FUNDAMENTAL EQUATIONS

One-way randomized-groups designs have one IV with at least two levels. The analysis for two levels is demonstrated in Section 3.2.3. The current section shows allocation of cases to three levels, partition of variance, and calculation of the one-way randomized-groups ANOVA by both the traditional and the regression approach. Analyses with more than three levels are calculated in the same manner as those with three levels.

4.4.1 Allocation of Cases

The cases in an experiment are randomly assigned to one of two or more levels of the IV through any of the methods discussed in Section 1.1.2. Table 4.1 shows how cases are assigned to groups within this design. Notice that every level of A has a different group of cases assigned to it. A is a fixed effect—the three levels are specifically chosen by the researcher because they are the ones of interest, not randomly selected from some larger set of levels. Cases, however, are a random effect because they are randomly assigned to the levels of A from the full set of cases available. This is still considered a fixed-effects design because cases are experimental units, not effects

TABLE 4.1 Assignment of cases in a one-way randomized-groups design

Treatment		
a_1	a_2	a_3
s_1	s_4	s_7
s_2	s_5	s_8
s_3	s_6	s_9

to be tested. For the traditional approach, once DV scores are gathered, they are arranged in such a table, with a columns representing groups (levels of the IV) and with n scores in each group.[2]

Cases in a nonexperimental study are also assigned to the levels of the IV but not on the basis of random assignment. For example, cases could be assigned on the basis of their age group (21–35, 36–50, 51–65) or their religious affiliation.

4.4.2 Partition of Variance

The partition of variance in basic ANOVA is described in Section 3.2.3.1. This partition applies to the one-way randomized-groups design with equal sample sizes in each group, no matter how many levels of A.

4.4.3 Traditional ANOVA Approach (Three Levels)

Section 3.2.3 shows the traditional ANOVA approach with two levels of an IV. The example here is a generalization of that approach to three (or more) levels. Table 4.2 shows a silly, hypothetical, and very small data set in which rating of a ski vacation (the DV) is evaluated as a function of view from the hotel room (the IV). Upon arrival at a mythical ski resort, skiers are randomly assigned to a hotel room that has either a view of the parking lot, a view of the ocean (we said it was mythical), or a view of the ski slopes. We assume proper controls have been instituted to avoid communication among skiers in the experiment. Sample size is inadequate for answering real research questions, but it allows relatively painless hand calculation of the ANOVA.

Equations 3.13–3.16 are used to calculate the sums of squares for this example (see Table 4.3). Table 4.4 shows degrees of freedom for the example. Note that the

TABLE 4.2 Hypothetical data set for a one-way design with three levels

	A: Type of View	
a_1: Parking Lot View	a_2: Ocean View	a_3: Ski Slope View
1	3	6
1	4	8
3	3	6
5	5	7
2	4	5
$A_1 = 12$	$A_2 = 19$	$A_3 = 32$

[2] Recall that n is used for sample size within a single group or level and N is used for total sample size.

TABLE 4.3 Calculation of sums of squares for traditional one-way ANOVA with three levels

$$SS_A = \frac{12^2 + 19^2 + 32^2}{5} - \frac{63^2}{15} = 41.2$$

$$SS_{S/A} = 325 - \frac{12^2 + 19^2 + 32^2}{5} = 19.2$$

$$SS_T = 325 - \frac{63^2}{15} = 1.6$$

TABLE 4.4 Calculation of degrees of freedom for traditional one-way ANOVA with three levels

$$df_A = a - 1 = 3 - 1 = 2$$

$$df_{S/A} = an - a = 15 - 3 = 12$$

$$df_T = an - 1 = 15 - 1 = 14$$

TABLE 4.5 Calculation of mean squares for traditional one-way ANOVA with three levels

$$MS_A = \frac{\mathbf{SS}_A}{\mathbf{df}_A} = \frac{41.2}{\mathbf{2}} = 20.6$$

$$MS_{S/A} = \frac{\mathbf{SS}_{S/A}}{\mathbf{df}_{S/A}} = \frac{19.2}{\mathbf{12}} = 1.6$$

TABLE 4.6 Source table for traditional one-way ANOVA with three levels

Source	SS	df	MS	*F*
A	41.2	2	20.6	12.88
S/A	19.2	12	1.6	
T	60.4	14		

partitions of SS and the partitions of df both sum to their respective totals. Recall that variance (mean square) associated with each source is obtained by dividing SS by df, because SS is the numerator and df is the denominator of variance. Table 4.5 shows calculation of mean squares associated with treatment and error sources of variance. Finally, Table 4.6 is a source table that summarizes the results of Tables 4.3–4.5 and adds the F ratio of Equation 3.12. Note that SS_A and $SS_{S/A}$ add up to SS_T. However, MS_A and $MS_{S/A}$ *do not* add up to MS_T.

Section 3.2.3.3 describes the process for deciding whether the differences among sample means are likely to have occurred by chance. Suppose, for example, our willingness to risk a Type I error is 5%. With numerator df $= 2$ and denominator df $= 12$, the critical value of F in Table A.1 is 3.88. The calculated (obtained) F of 12.88 is larger than 3.88; therefore, the null hypothesis is rejected. We conclude that rating of a ski vacation does, indeed, depend on the view from the hotel room.

4.4.4 Regression Approach (Three Levels)

As discussed in Sections 3.3.1 and 3.5.5, the regression approach to ANOVA uses columns of numbers to code group membership. Because there are only two groups in the example of Section 3.3.1, a single column is sufficient to separate the cases into two groups (1 if a case belonged to one group, −1 if a case belonged to the other group). However, with more than two groups, more than one column is required to separate the groups from each other. For an example, with three groups, the first

TABLE 4.7 Columns for regression approach to analysis of variance with three treatment levels

Level of A	Case	Y	X_1	X_2	YX_1	YX_2	X_1X_2
	s_1	1	1	−1	1	−1	−1
	s_2	1	1	−1	1	−1	−1
a_1	s_3	3	1	−1	3	−3	−1
	s_4	5	1	−1	5	−5	−1
	s_5	2	1	−1	2	−2	−1
	s_6	3	1	1	3	3	1
	s_7	4	1	1	4	4	1
a_2	s_8	3	1	1	3	3	1
	s_9	5	1	1	5	5	1
	s_{10}	4	1	1	4	4	1
	s_{11}	6	−2	0	−12	0	0
	s_{12}	8	−2	0	−16	0	0
a_3	s_{13}	6	−2	0	−12	0	0
	s_{14}	7	−2	0	−14	0	0
	s_{15}	5	−2	0	−10	0	0
Sum		63	0	0	−33	7	0
Squares summed		325	30	10			
N		15					
Means		4.2	0	0			

column could separate group 3 from groups 1 and 2, and the second column could separate groups 1 and 2. So, instead of the single X column of Section 3.3.1, several X columns are necessary, and multiple—rather than bivariate—regression is required.

The number of X columns needed to separate the groups is equal to $\text{df}_A = a - 1$. If, in addition, the numbers in the columns are chosen so that they sum to zero in each column and the columns are orthogonal (cf. Sections 3.5.4 and 4.5.5.2), the sum of the sums of squares for the columns (a of them) equals the sum of squares for the treatment effect, A.

Table 4.7 shows the scores on the DV, Y,[3] with one of the many possible sets of orthogonal contrast codes in the two X columns. The numbers in the X columns are chosen to correspond to research questions relevant to this example. The X_1 column

[3] The column labeling a_i is only for your convenience in identifying group membership; it does not enter into calculations.

TABLE 4.8 Preliminary sums of squares for regression approach to ANOVA with three levels

$$SS(Y) = \sum Y^2 - \frac{(\Sigma Y)^2}{N} = 325 - \frac{63^2}{15} = 60.4$$

$$SS(X_1) = \sum X_1^2 - \frac{(\Sigma X_1)^2}{N} = 30 - \frac{0^2}{15} = 30$$

$$SS(X_2) = \sum X_2^2 - \frac{(\Sigma X_2)^2}{N} = 10 - \frac{0^2}{15} = 10$$

TABLE 4.9 Sums of cross products for regression approach to ANOVA with three levels

$$SP(YX_1) = \sum YX_1 - \frac{(\Sigma Y)(\Sigma X_1)}{N} = -33 - \frac{(63)(0)}{15} = -33$$

$$SP(YX_2) = \sum YX_2 - \frac{(\Sigma Y)(\Sigma X_2)}{N} = 7 - \frac{(63)(0)}{15} = 7$$

$$SP(X_1X_2) = \sum X_1X_2 - \frac{(\Sigma X_1)(\Sigma X_2)}{N} = 0 - \frac{(0)(0)}{15} = 0$$

separates a_3 from a_1 and a_2 (the ski slope view from the parking lot and ocean views), and the X_2 column separates a_1 from a_2 (the parking lot view from the ocean view). Of the many sets of codes that could be devised, these are probably the most interesting for purposes of comparisons, as we will see in Section 4.5.5.1. The remaining columns are found by multiplication.

Table 4.8 shows the sums of squares calculated from Table 4.7. Notice that the second terms in the equations for $SS(X_1)$ and $SS(X_2)$ reduce to zero. This is because the codes we have chosen for columns X_1 and X_2 sum to zero for their columns. This feature is worth the simplification in calculation alone but is also part of the requirement for the columns to be orthogonal.

Table 4.9 shows the sums of cross products that are produced by the columns in Table 4.7. The second terms in the first two sums of cross products of Table 4.9 also reduce to zero, as does the entire sum of cross products for X_1X_2. This holds for columns that are coded orthogonally, as long as sample sizes in each group are equal.

Equation 2.17 for the GLM in regression format needs to be modified to account for additional columns. With two columns, the equation becomes

$$Y' = a + b_1X_1 + b_2X_2 \tag{4.1}$$

where each column, X_j, has its own slope, b_j, and the slope-column products (b_jX_j) are summed over the $a - 1$ columns.

In general, then, the regression equation for any number of levels of A is

$$Y' = a + \sum_{j=1}^{a-1} b_j X_j \tag{4.2}$$

Note again the two uses of a: the intercept and the number of levels of the IV.

TABLE 4.10 Slope and intercept estimates for regression approach to ANOVA with three levels

b_1	$= \dfrac{(\text{SP}(YX_1))(\text{SS}(X_2)) - (\text{SP}(YX_2))(\text{SP}(X_1X_2))}{(\text{SS}(X_1))(\text{SS}(X_2)) - \text{SP}(X_1X_2)^2}$	$= \dfrac{(-33)(10) - (7)(0)}{(30)(10) - 0^2}$	$= -1.1$
b_2	$= \dfrac{(\text{SP}(YX_2))(\text{SS}(X_1)) - (\text{SP}(YX_1))(\text{SP}(X_1X_2))}{(\text{SS}(X_1))(\text{SS}(X_2)) - \text{SP}(X_1X_2)^2}$	$= \dfrac{(7)(30) - (33)(0)}{(30)(10) - 0^2}$	$= 0.7$
a	$= \overline{Y} - \overline{X}_1(b_1) - \overline{X}_2(b_2)$	$= 4.2 - 0(-1.1) - 0(0.7)$	$= 4.2$
Y'	$= a + (b_1)X_1 - (b_2)X_2$		

For group 1, when $X_1 = 1, X_2 = -1$:

$$Y' = 4.2 + (-1.1)(1) + (0.7)(-1) = 2.4$$

For group 2, when $X_1 = 1, X_2 = 1$:

$$Y' = 4.2 + (-1.1)(1) + (0.7)(1) = 3.8$$

For group 3, when $X_1 = -2, X_2 = 0$:

$$Y' = 4.2 + (-1.1)(-2) + (0.7)(0) = 6.4$$

TABLE 4.11 Sums of squares for ANOVA through regression with three treatment levels

SS(reg.)	$= \text{SS}(\text{reg. } X_1) + \text{SS}(\text{reg. } X_2)$	$= \dfrac{\text{SP}(YX_1)^2}{\text{SS}(X_1)} + \dfrac{\text{SP}(YX_2)^2}{\text{SS}(X_2)}$	$= \dfrac{(-33)^2}{30} + \dfrac{7^2}{10}$ $= 36.3 + 4.9$	$= 41.2$
SS(total)		$= \text{SS}(Y)$		$= 60.4$
SS(resid.)		$= \text{SS(total)} - \text{SS(reg.)}$	$= 60.4 - 41.2$	$= 19.2$

Table 4.10 shows equations for estimating slopes and intercept for computation of ANOVA through regression. The equations for slope (b) are quite a bit more complicated than those for the two-level design of Table 3.12 and are facilitated by matrix algebra once the number of columns exceeds two. However, it is not necessary to calculate these slope and intercept values when the purpose of the analysis is to test the reliability of differences among means, which is the goal of ANOVA.

The reduction of some of the terms to zero, with orthogonal contrast coding of columns (and equal sample sizes), continues in the equations for regression coefficients b_1 and b_2. With contrast coding, the intercept, a, is equal to the grand mean. When the codes in the X columns for the group membership for a case are inserted into the prediction equation (as shown in the bottom section of the table), the predicted score for each skier is the mean of that skier's own treatment group.[4]

Sums of squares for the regression analysis are shown in Table 4.11. Degrees of freedom for regression are shown in Table 4.12. Table 4.13 summarizes the

[4] Section 5.6.5 discusses unequal sample sizes, in which columns are no longer orthogonal. Sequential multiple regression is discussed in that section as one way of dealing with nonorthogonality of columns in factorial designs.

TABLE 4.12 Calculation of degrees of freedom for ANOVA through regression with three treatment levels

df(reg.)	= number of predictors (X) vectors		= 2
df(total)	= number of cases (N) − 1	= 15 − 1	= 14
df(resid.)	= df(total) − df(reg.)	= 14 − 2	= 12

TABLE 4.13 Source table for one-way ANOVA through regression with three levels

Source	SS	df	MS	F
Regression	41.2	2	20.6	12.88
Residual	19.2	12	1.6	
Total	60.4	14		

information in Tables 4.11 and 4.12 in a source table, rearranged to mimic that of Table 4.6. The values in Table 4.13 are identical to those in Table 4.6; only the names have been changed. The regression source is the treatment effect, A, and the residual source is the estimate of error, S/A. The F ratio is

$$F = \frac{\textbf{MS(reg.)}}{\textbf{MS(resid.)}} = \frac{20.6}{1.6} = 12.88$$

as in Table 4.6. Table A.1 is used to find the critical value of F for testing the null hypothesis. Three null hypotheses are tested at once through the regression approach: (1) that the population correlation between actual and predicted Y is equal to zero, (2) that each of the population regression coefficients is equal to zero, and (3) that $\mu_1 = \mu_2 = \mu_3$, the null hypothesis in the form typically stated for the traditional approach to ANOVA.

4.4.5 Computer Analyses of Small-Sample One-Way Design

Table 4.14 shows how data are entered for a one-way ANOVA. Data may be entered in a word processing program and saved in text mode, entered in a spreadsheet program and saved in text mode, or entered through facilities provided by the statistical program. Data sets entered through word processors or spreadsheets may include only numbers, without the column headings shown in Table 4.14; data sets entered through statistical programs typically include variable names at the tops of columns.

Each row in this table is a single case. The first column uses 1, 2, or 3 to represent the level of the IV (a_i) for each case. The second column contains the DV score for the case. An additional column, case number, may be added for your convenience in editing the data file.

TABLE 4.14 Hypothetical data set for one-way ANOVA in form suitable for computer analysis

VIEW	RATING
1	1
1	1
1	3
1	5
1	2
2	3
2	4
2	3
2	5
2	4
3	6
3	8
3	6
3	7
3	5

Tables 4.15 and 4.16 show computer analyses of the small-sample example of the one-way randomized-groups small-sample ANOVA through SPSS and SAS, respectively. Table 4.15 shows syntax and output for SPSS ONEWAY.[5] This program, as all SPSS analysis of variance programs, specifies the name of the DV (RATING) followed by the keyword BY, followed by the name of the IV (VIEW). The /MISSING ANALYSIS instruction is the default for handling missing values. Means, standard deviations, standard errors, and 95% confidence interval for the mean, maximum, and minimum values are requested by /STATISTICS DESCRIPTIVES. One-way ANOVA also may be done through SPSS GLM (which is demonstrated in subsequent chapters).

The output begins with a table of descriptive statistics, where 1 is the parking lot view, 2 is the ocean view, and 3 is the ski slope. This table is followed by a source table, titled ANOVA (compare with Table 4.6). The Sig. column provides the probability, to three decimal places, that an observed F ratio this large or larger could have occurred as a result of chance variation. Thus, the probability that this result could have occurred by chance is .001.

Table 4.16 shows SAS ANOVA syntax and output for the small-sample example. SAS PROC GLM may also be used for this analysis and is nearly identical to PROC ANOVA for this application. The IV (`VIEW`) is identified as a categorical (`class`)

[5] If SPSS Windows menus are used, the ONEWAY program is accessed by choosing Analyze > Compare Means > One-Way ANOVA.

TABLE 4.15 One-way randomized-groups ANOVA for Small-Sample Example Through SPSS ONEWAY

```
ONEWAY
  RATING BY VIEW
  /STATISTICS DESCRIPTIVES
  /MISSING ANALYSIS.
```

Oneway

Descriptives

rating

	N	Mean	Std. Deviation	Std. Error	95% Confidence Interval for Mean		Minimum	Maximum
					Lower Bound	Upper Bound		
Parking lot	5	2.4000	1.67332	.74833	.3223	4.4777	1.00	5.00
Ocean	5	3.8000	.83666	.37417	2.7611	4.8389	3.00	5.00
Ski slope	5	6.4000	1.14018	.50990	4.9843	7.8157	5.00	8.00
Total	15	4.2000	2.07709	.53630	3.0497	5.3503	1.00	8.00

ANOVA

rating

	Sum of Squares	df	Mean Square	F	Sig.
Between Groups	41.200	2	20.600	12.875	.001
Within Groups	19.200	12	1.600		
Total	60.400	14			

variable. The `model` instruction follows regression syntax, with the DV (`RATING`) identified on the left side of the equation and the IV on the right. Descriptive statistics are requested by the `means` instruction.[6]

Two source tables are given. The first is for the full model, with all IVs combined; the second identifies IVs (and their interactions) as separate sources of variability. Because there is only one IV in this example, both source tables are the same. S/A is identified as `Error` in the first source table (compare with Table 4.6). The probability that an F value this large or larger could have occurred by chance variation is provided in the `Pr > F` column. This table includes `R-Square`, the proportion of variance accounted for by the whole model (cf. Section 4.5.1). Also shown are the square root of the $MS_{S/A}$ (`Root MSE`, also known as the pooled standard deviation), the grand mean of the DV (`RATING Mean`), and the coefficient of variation (`Coeff Var`., or the square root of the error mean square divided by the grand mean multiplied by 100). Descriptive statistics for each level of the IV are provided after the second source table.

[6] Syntax may be generated by choosing Solutions > Analysis > Analyst. However, that system requires a more complex procedure for obtaining descriptive statistics in ANOVA and GLM.

TABLE 4.16 One-way randomized-groups ANOVA through SAS ANOVA

```
proc anova data=SASUSER.RAND_GRP;
   class VIEW;
   model RATING = VIEW;
   means VIEW;
run;
```

```
                              The ANOVA Procedure

                           Class Level Information

                        Class          Levels    Values

                        VIEW                3    1 2 3

                   Number of Observations Read          15
                   Number of Observations Used          15

Dependent Variable: RATING

                                        Sum of
 Source                      DF        Squares     Mean Square    F Value    Pr > F

 Model                        2    41.20000000     20.60000000      12.88    0.0010
 Error                       12    19.20000000      1.60000000
 Corrected Total             14    60.40000000

                  R-Square     Coeff Var      Root MSE    RATING Mean

                  0.682119      30.11693      1.264911       4.200000

 Source                      DF       ANOVA SS     Mean Square    F Value    Pr > F

 VIEW                         2    41.20000000     20.60000000      12.88    0.0010

                 Level of           ------------RATING-----------

                   VIEW        N            Mean          Std Dev

                   1           5      2.40000000       1.67332005
                   2           5      3.80000000       0.83666003
                   3           5      6.40000000       1.14017543
```

4.5 SOME IMPORTANT ISSUES

Several of the issues discussed in this section apply to other kinds of designs, as well. These issues are discussed here because this is the first opportunity to do so. Subsequent chapters refer back to this section when appropriate.

4.5.1 Effect Size

Finding a statistically significant effect tells you only that your IV probably made a difference—it is unlikely that the observed mean difference among groups occurred by chance. There is nothing in the inferential test that tells how much of a difference your IV made. Although a bigger F ratio often implies a bigger effect, the size of the F ratio depends on error variability and degrees of freedom, as well as on differences among means. For this reason, measures of effect size (also called *strength of association, treatment magnitude,* or *magnitude of effect*) are often used.

Effect size should always be reported if the effect is statistically significant. There is disagreement, however, as to whether effect size should be reported if the mean differences are not significant. Some argue that if an effect is not statistically significant, there is no evidence of an effect, so effect size cannot be reported as different from zero. Thus, these researchers believe that effect size should not be reported when the null hypothesis is retained. Others argue that all treatments produce *some* differences, however small, and that finding a nonsignificant effect is due to lack of power (Section 4.5.2) not lack of a real effect. If that is the case, reporting effect size provides useful information to researchers who want to use the information to design a more powerful study or perform a meta-analysis (in which researchers statistically combine the results of multiple studies of the same variables). We place ourselves firmly on the fence of this controversy.

Occasionally, the DV is so well-understood that a mere statement of parameter estimates (cell means) in appropriate units tells all you need to know about effect size. For example, if a particular prenatal regime produces babies weighing 500 grams (about 1.1 pounds) more at birth, the magnitude of the effect of the treatment is obvious (although some would perhaps prefer a translation into ounces). Estimates of parameters do not always reveal effect size, however. Often the unit of measurement of the DV is arbitrary, derived, transformed, or otherwise not easily understood. Also, different studies often have measured DVs on different scales, making a comparison of their results from parameter estimates difficult. A variety of effect size measures are available to solve these problems.

The most common measures of effect size express treatment magnitude as the proportion of variability in the DV associated with the effect. The easiest way to conceptualize these measures is through a Venn diagram, as shown in Figure 4.1. All of the variability in the DV, from whatever source (measurement error, extraneous variables, etc.), is the circle of interest. The other circle, labeled A, represents the effect of the IV. The extent of overlap is the proportion of variability in the DV that is associated with, related to, predictable from, or accounted for by the effect. Small effects have only a tiny proportion of the variability in the DV associated with them and a correspondingly tiny overlap. Larger effects have larger and larger proportions of the variability in the DV associated with them and larger areas of overlap. If all of the variability in the DV is predictable by the effect, the circles overlap completely.

The simplest measure of effect size is η^2 (also called R^2); η^2 (eta squared) is the sum of squares for the effect divided by the total sum of squares:

$$\eta^2 = R^2 = \frac{\text{SS}_A}{\text{SS}_T} \tag{4.3}$$

For the example of Section 4.4.3 (cf. Table 4.6),

$$\eta^2 = R^2 = \frac{41.2}{60.4} = .68$$

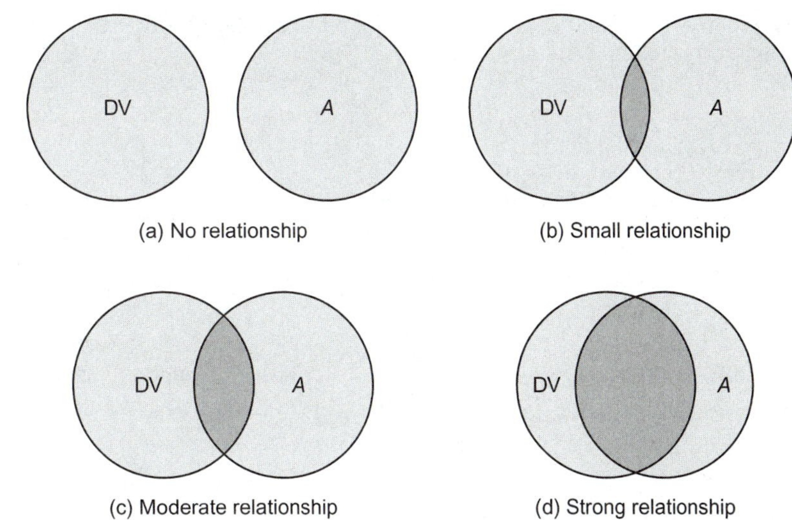

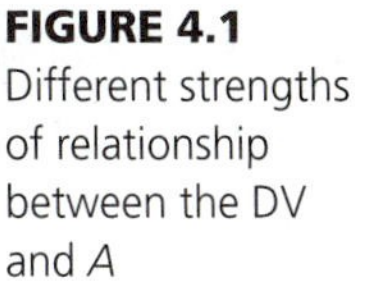
FIGURE 4.1 Different strengths of relationship between the DV and *A*

Thus, about two-thirds of the variability in ratings of a ski vacation is due to the view from the room. You have a very good idea of how highly a vacation is rated by knowing the view from the room.

Confidence intervals around η^2 are available through the Smithson (2003) SPSS or SAS files that are available along with data files for this book. The use of the SPSS file, NONCF3.SPS, is demonstrated in Section 4.6.2. Table 4.17 shows the use of Smithson's SAS file, NONCF2.SAS, to find η^2 and its confidence limits for the small-sample data. Confidence limits for effect sizes are found by adding values to the syntax file NONCF2.SAS. Filled-in values (from Table 4.16) are F (`12.88`), numerator df (`2`), denominator df (`12`), and the proportion for the desired confidence interval (`.95`). Input values are shown in gray in Table 4.17 and replace the default values in the file. The output column labeled `rsq` is η^2; lower and upper confidence limits are labeled `rsqlow` and `rsqupp`, respectively. Thus, $\eta^2 = .68$ (`rsq`), with 95% confidence limits from .22 (`rsqlow`) to .80 (`rsqupp`).

A drawback to η^2 is that the measure is simply a descriptive statistic; it describes the proportion of variability in the DV accounted for by the effect in the particular sample, but, on average, it overestimates the effect size in the population from which the sample was drawn. That is, η^2 is a biased estimate of the parameter. An alternative measure, called $\hat{\omega}^2$ (omega squared), is an unbiased parameter estimate that generalizes to the population. Because it adjusts for the overestimation of effect size by η^2, $\hat{\omega}^2$ is always smaller than η^2. The equation for $\hat{\omega}^2$ starts with the two sums of squares used in η^2, but then something is subtracted from the numerator and added to the denominator.

$$\hat{\omega}^2 = \frac{SS_A - df_A(MS_{S/A})}{SS_T + MS_{S/A}} \tag{4.4}$$

TABLE 4.17 Syntax and output from NONCF2.SAS for effect size (**rsq**) with 95% confidence limits (**rsqlow** and **rsqupp**) for small-sample example

```
.
.
rsq = df1 * F / (df2 + df1 * F);
rsqlow = ncplow / (ncplow + df1 + df2 + 1);
rsqupp = ncpupp / (ncpupp + df1 + df2 + 1);
cards;
12.88    2    12    .95
;
proc print;
run;
```

Obs	F	df1	df2	conf	prlow	prupp	ncplow	ncpupp	rsq	rsqlow	rsqupp
1	12.88	2	12	0.95	0.975	0.025	4.12638	60.2004	0.68220	0.21574	0.80053

For the small-sample example,

$$\hat{\omega}^2 = \frac{41.2 - 2(1.6)}{610.4 + 1.6} = .61$$

which is less than the .68 calculated for η^2.

When you merely want to describe effect size in a sample, η^2 is the measure of choice. When you want to estimate effect size for a population from a sample, $\hat{\omega}^2$ is your choice. However, computation of $\hat{\omega}^2$ becomes very complicated when the design is other than randomized groups with equal sample sizes in all levels of the IV. Also, there are no readily available procedures for calculating confidence intervals around $\hat{\omega}^2$. The descriptive measure η^2, on the other hand, easily generalizes to a variety of ANOVA designs and is often used with complicated designs despite the limits to its generalizability.

Another kind of measure of effect size expresses mean differences in standard deviation units. This is called Cohen's d (1977), or *effect-size index*. In its simplest form, it is the difference between the largest and smallest means divided by the common standard deviation:

$$d = \frac{\overline{Y}_{largest} - \overline{Y}_{smallest}}{\sqrt{\text{MS}_{S/A}}} \tag{4.5}$$

where the subscripts denote largest and smallest means.

This simplest form of the measure requires that there be an even number of means, that they be concentrated at each of the extremes, and that there be equal sample sizes within each level. Thus, the measure only works for some single factor designs. It becomes increasingly complicated when dealing with other arrangements of means within their range or unequal sample sizes. Thus, d is a less convenient measure than $\hat{\omega}^2$ or η^2. Cohen (1977) shows the varieties of d and equations for converting d to η^2.

Effect sizes are also useful in evaluating specific comparisons. These are described in Section 4.5.5.7.

4.5.2 Power and Sample Size

Power, as discussed in Sections 1.6.2 and 3.5.2, is the sensitivity of a research design: the probability that one will be able to reject a false null hypothesis. Power considerations drive many features of research design and experimental control.

4.5.2.1 Designing Powerful Studies

Analyzing the power of a study is most usefully accomplished in the design stage, *before* data are collected. It is too late to change the design after data have been collected and an unfavorable outcome is observed. The most that can be done then is to redesign and rerun the study, usually a frustrating and costly enterprise.

One factor influencing power is the size of the effect in the population. Small effects are harder to detect than larger ones—thus small effects generate less power, all other things being equal. One way to increase power, then, is to be sure to choose strong and very different levels of treatment. This will increase the size of the mean differences that produce MS_A and, therefore, that increase the numerator of *F*.

Error variability also has a profound effect on power; the smaller the error variability, the greater the power. Measures to control extraneous variables and (perhaps) to randomly assign a *homogeneous* sample of cases to different levels of treatment are instituted to minimize $SS_{S/A}$. Controlling extraneous variables is accomplished by holding them constant over levels of treatment, counterbalancing (or partially counterbalancing) their effects, or letting random processes evenly disperse their effects through levels of treatment. In practical terms, this means one must have equipment that is running reliably, experimental technicians who are performing their job consistently, and so on, through the long list of variables that can increase the variability in the DV. Randomly assigning a homogeneous sample of cases to different levels of treatment has the effect of minimizing variability due to differences in cases. This will also decrease the size of $SS_{S/A}$.

Another way to minimize $MS_{S/A}$, the denominator of the *F* ratio, is to increase *its* denominator by increasing $df_{S/A}$. This is reflected in the notion that large sample sizes generate more power. With the same sum of squares for error, mean square for error decreases as degrees of freedom for error increase.

Finally, the level of α chosen for testing hypotheses affects power. The stricter (smaller) the level of α (e.g., using .01 instead of .05), the less the power. Recall from Section 1.6.1 that power is $1 - \beta$, so that α and β errors are negatively related—that is, there is a trade-off between Type I and Type II errors in designing research studies. These relationships have several implications for designing research studies to maximize power:

1. Select levels of the IV that are very different. All other things being equal, it is more useful to choose levels of an IV that are as different from each other as possible if you hope to find a statistically significant outcome—larger effect sizes are more likely to lead to favorable results than are smaller ones. It is harder to find a needle in a haystack than it is to find a nail. If small effects are

to be detected, be prepared to make other compromises, such as testing very large samples.

2. Use the most liberal, legitimate α level for testing hypotheses. Legitimacy depends on the use of the results. Scientific publication has traditionally required use of α levels of .05 or smaller for declaring results to be statistically significant. However, requirements for studies designed for purposes other than scientific publication may justify more liberal α levels. For example, if a choice is to be made between two materials in a manufacturing process, equal in cost and other production factors, with no prior reason for choosing one over the other, the choice could be made on the basis of an observed difference where α is as great as .49. There is, of course, a strong possibility that the "wrong" material is chosen, but that error is still less than would be associated with an arbitrary choice or a flip of a coin. Large Type I errors also may be tolerated if the consequences of Type II errors are grave (e.g., diagnosing a disease that is fatal if not treated) and power cannot be increased in other ways.
3. Control error variability to the extent possible. *Error variability* in randomized-groups designs is the variability of scores within groups given the same treatment. Thus, anything done to control for extraneous influences on the DV tends to increase power. The more homogeneous the sample of cases within treatments, the less that error variability is likely to be. Samples of wheat taken from a single field and then randomly assigned to different treatments are likely to be more similar than samples of wheat taken from a variety of fields. (The problem with this, of course, is that generalizability is limited to the field[s] supplying the samples; cf. Section 1.7.) Individual studies vary in their response to the necessary trade-off between power and generalizability.[7] Assuring that cases are tested under identical conditions (but for levels of the IV) likewise increases homogeneity. Sloppy research hurts power.
4. As part of the research design, compute the sample size necessary to achieve the desired power. The goal is to avoid wasting resources looking for trivial effects or needlessly expending resources in a search for a very large effect that you are unlikely to miss in any event.

4.5.2.2 Estimating Sample Size

A variety of techniques are available for estimating needed sample size, ranging from the quick and dirty to the elaborate. Many view these techniques as tedious when performed by hand. However, several computer programs are also available for estimating sample size, some at minimal cost. SAS GLMPOWER is a new procedure in Version 9 that estimates sample size for a variety of ANOVA designs and that can also be used to solve for power after finding a nonsignificant result. Another program is PC-SIZE (Dallal, 1986), a shareware program that deals with one-way and two-way randomized-groups designs, among others. NCSS PASS 2002 is an example of a comprehensive

[7] Nonhomogeneous samples may be made more homogeneous by moving to a factorial design. For example, create an additional IV with different fields as levels to increase homogeneity within each cell (a combination of both treatment and field), as discussed in Chapter 5.

commercial power analysis and sample size program. Other programs are regularly reviewed in methodological journals, such as *Behavior Research Methods, Instruments, and Computing*. The Internet is a terrific source of power programs—many of them downloadable as free shareware (for example, PC-SIZE) or usable online; try entering "statistical power" into your favorite browser or search engine.

Some decisions must be made before you estimate needed sample size, whether by hand or through computer programs. First, you need to determine your α level and whether you want a one- or two-tailed test of the null hypothesis (some computer programs assume a two-tailed test, but you can trick them into a one-tailed test by doubling the α value that you enter). Second, you need to decide on the power you want to achieve, recognizing that the greater the value you set, the larger sample size you will need. Values in the range of .75 to .90 are typically considered reasonable in scientific research.

You also need to estimate the size of the effect you expect to achieve. These estimates can be either in the form of expected mean differences or in the form of expected effect size. Finally, you need some idea of the within-group (error) variability that is likely to occur given your ability to control extraneous influences on the DV. This last is perhaps the most difficult to estimate. What do you predict to be the distribution of scores on the DV for cases given the same treatment? If you are using a standardized measure as a DV (such as a measure of IQ), the population standard deviation of the scale is probably known (e.g., $\sigma = 15$). Or, previous research may have used the same DV and reported the within-group variability. Or, in the absence of these conveniences, a highly recommended method for estimating variability of a DV is a pilot study. Try out your DV on a small sample of cases, giving them no treatment or only one of the treatments, and then calculate the standard deviation.[8] Or, if all else fails, divide your anticipated range in scores by 5 for a very rough estimate of error variability.

A classic, quick-and-dirty method for estimating sample size is illustrated here primarily because it makes clear the relationships among design decisions:

$$n = \frac{2\sigma^2}{\delta^2}(z_{1-\alpha} + z_{1-\beta})^2 \tag{4.6}$$

where n = sample size for each group

σ^2 = estimated $\text{MS}_{\text{error}} = \text{MS}_{S/A}$ for this design

$z_{1-\alpha}$ = normal deviate corresponding to $1 - \alpha$ (see Table 4.18 for typical values)

$z_{1-\beta}$ = normal deviate corresponding to $1 - \beta$ (see Table 4.18 for typical values)

δ = desired minimal difference between means to be detected

The equation clearly shows the trade-off between α and β error and the increase in sample size necessary for smaller effects and greater error variances. The greater the variability in the DV (σ^2), the larger the retention region for the null hypothesis $(1 - \alpha)$. The more power desired $(1 - \beta)$, the larger the needed sample size; whereas the larger the expected mean difference (δ), the smaller the needed sample size.

Equation 4.6 is meant for simple one-way designs with two levels. It can be used with multilevel designs by defining δ as the smallest mean difference in the set that is

[8] If you decide to use more than one treatment, be sure to calculate variances separately for cases within each treatment, and then find the average standard deviation.

TABLE 4.18 Selected *z* values for sample-size equation

	p	z (2-tail)	z (1-tail)
$1-\alpha$	.99	**2.58**	2.33
	.95	**1.96**	1.65
$1-\beta$	.85	1.44	**1.04**
	.80	1.28	**0.84**
	.75	1.15	**0.67**

to be detected. This method often results in an overestimation of sample size in multilevel designs, because you are solving for the smallest difference, rather than any difference at all. However, you will have at least the power you have selected, and probably much more.

Say, for example, that before running the experiment described in Section 4.4.3 you postulate a mean difference between ski slope view and ocean view of about 2.5 rating points, and $MS_{S/A} = 2$ (from a pilot study). Suppose the desired significance test is two-tailed with α level set at .05, so that $1 - \alpha = .95$ and $z_{1-\alpha} = 1.96$. (Note that this is actually $z_{1-\alpha/2}$ because it is a two-tailed test.) Suppose also that the desired power $(1 - \beta)$ is set at .85, so that $z_{1-\beta} = 1.04$. The one-tailed test is used because the research hypothesis predicted an advantage in favor of the ski slope view.[9] Solving the equation, then,

$$n = \frac{2(2)}{2.5^2}(1.96 + 1.04)^2 = 5.76$$

Fractional numbers are rounded upward, so that there are at least as many cases as required for the stated power. Thus, six skiers need to be randomly assigned to each type of room to achieve a significant result ($p < .05$) with a probability of .85 for the difference between ocean and ski slope views, if such a difference truly exists.

Notice an interesting simplification with these particular choices of α and β. The z values add to 3.00, so that the last part of the expression, squared, is 9.00. As a result, an even quicker and dirtier form of the sample size formula is

$$n = \frac{18\sigma^2}{\delta^2} \tag{4.7}$$

More precise estimates are available with the use of power tables, such as are available in Keppel (1991). Better yet, use of software, such as PC-SIZE, streamlines sample size determination. For the example, assume that the means for the three room views were hypothesized to be 2.5, 4.0, and 6.5. The program requires variability in the form of a standard deviation rather than a variance, so that $(2)^{1/2} = 1.414$ is used. In addition, α is chosen as .05 and power as .85. Table 4.19 shows the setup and results for this problem through PC-SIZE. (Note that selected values appear below the instructions in printed output.)

[9] Null hypotheses are usually bidirectional, indicating a two-tailed test, whereas research hypotheses are usually unidirectional, indicating a one-tailed test.

TABLE 4.19 Sample-Size Determination Through PC-SIZE.

```
                              PC-SIZE
              A Program for Sample Size Determinations
                           Version 2.13
                          (c) 1985, 1986

                    "One of many STATOOLS(tm)..."
                                by

                         Gerard E. Dallal
                         53 Beltran Street
                         Malden, MA  02148

   Please  acknowledge  PC-SIZE  in  any  publication that makes use of its
   calcluations.  A suitable reference is Dallal, G.E. (1986), "PC-SIZE:  A
   Program for Sample Size Determinations," The American Statistician,40,52.

This program operates in seven modes:

    1  --  single factor design
    2  --  two factor design
    3  --  randomized blocks design
    4  --  paired t-test
    5  --  generic F
    6  --  correlation coefficient
    7  --  proportions

Choose a mode [1]:
     1

Enter the level of the test [ .050]:
  .50000E-01
Enter the required power [ .800]:
  .85000
Enter the number of groups [ 2]:
     3

There are three ways to specify the alternative:

   1 -- Specify individual effects.
   2 -- Specify a range throughout which group
           means are uniformly distributed.
   3 -- Specify average squared effect divided
           by the error variance.

Choose a method [2]:
     1
Enter estimate of effect for group  1 [  .00000    ]:
    2.5000
Enter estimate of effect for group 2 [ .00000 ]:
    4.0000
Enter estimate of effect for group 3 [ .00000 ]:
    6.5000
Enter estimate of within cell standard deviation [0.00]:
    1.4140

Level of the test: .0500
Noncentrality parameter: 4.0846 * sample size

    SAMPLE
    SIZE            POWER
    (per cell)

    2               .32311

    3               .66977

    4               .86682
```

Only four skiers in each group are required for power of .87, at the chosen α level and with means and within-cell standard deviation as specified. This provides power to find *any* differences among the three groups; that is, there is enough power to find an omnibus effect. If the desire truly is to find a particular difference between just two of the groups, this software may be used as if only those two groups were run. This use assumes a planned comparison, in which no power is lost by adjusting for post hoc analysis, as per Section 4.5.5.

4.5.2.3 Power of a Nonsignificant Effect

The procedures of Section 4.5.2.2 are usually used to determine needed sample size before gathering data. However, they can be turned around to solve for power after the analysis is run if a nonsignificant result is obtained. This gives you some idea of how far off your study was from producing a significant effect. For example, in Table 4.19, PC-SIZE shows that had we run an experiment with $n = 2$ and achieved the stated means and within-cell standard deviation, the power would have been .32 (indicating a Type II error rate of .68), which means we would need to run more subjects in a second study if significance is to be expected.

SAS GLMPOWER permits power solutions when you provide information about sample size, means, and error standard deviation. Table 4.20 shows syntax and output from SAS GLMPOWER for the hypothetical data analyzed in Section 4.4.3. The `do` loop identifies three levels of the IV (`VIEW`), for which means (shown in gray) are given after the `datalines;` instruction. Note that if sample size is desired instead of power, the instructions change to `ntotal=.;` and `power=.80;` (or whatever your desired power). Type I error rate is set to .05 by default.

Thus, the hypothetical data analyzed in Section 4.4.3 had a probability of .98 of detecting a true mean difference in ratings among the three ocean views. SPSS GLM also gives as part of the analysis the power and effect size (in the form of η^2) for each effect if certain menu selections are made, as demonstrated in Section 5.4.5.

4.5.3 Unequal Sample Sizes

Unequal samples sizes in levels of the IV pose no particular difficulty for computation of one-way omnibus ANOVA. However, they may pose intractable problems if the inequality is caused by differential dropout from some levels of the IV in an experiment designed to have equal sample sizes. For example, if some levels of treatment cause some employees to leave an experiment or rats to die or fabrics to disintegrate, the original random assignment to levels of the IV is no longer random by the time the DV is measured. Therefore, it is no longer an experiment.

Another complication of unequal sample sizes is that orthogonality is lost in both the traditional and regression approaches (and therefore in comparisons), even when the coding itself is orthogonal. For example, note in Table 4.7 that the sums of X_1, X_2, and X_1X_2 are not zero if one or more cases are missing from any group. Suppose the last case in the small sample provided unusable data, resulting in only four scores in a_3. Table 4.21 shows the resulting data set and group means.

TABLE 4.20 Syntax and output of SAS GLMPOWER for results of one-way ANOVA with three levels

```
data SSAMPLE;
do VIEW = 1 to 3;
        input RATING@@;
        output;
end;
datalines;
2.4 3.8 6.4
;
run;
proc glmpower data=SSAMPLE;
class VIEW;
model RATING = VIEW;
power
        stddev = 1.26
        ntotal = 15.
        power = .;
run;
```

```
                    The GLMPOWER Procedure
                   Fixed Scenario Elements

        Dependent Variable                  RATING
        Source                                VIEW
        Error Standard Deviation              1.26
        Total Sample Size                       15
        Alpha                                 0.05
        Test Degrees of Freedom                  2
        Error Degrees of Freedom                12

                       Computed Power

                           Power

                           0.984
```

TABLE 4.21 Hypothetical data set with unequal sample sizes for one-way ANOVA

A: Type of View		
a_1: Parking Lot View	a_2: Ocean View	a_3: Ski Slope View
1	3	6
1	4	8
3	3	6
5	5	7
2	4	
$A_1 = 12$	$A_2 = 19$	$A_3 = 27$
$\overline{Y}_{A_1} = 2.4$	$\overline{Y}_{A_2} = 3.8$	$\overline{Y}_{A_3} = 6.75$
$\Sigma Y_{A_1} = 40$	$\Sigma Y_{A_2} = 75$	$\Sigma Y_{A_3} = 185$

For the example, then, SS_A is found using Equation 3.14, with modification to the denominators for unequal n. With three levels of A, for example, Equation 3.14 becomes

$$SS_A = \left(\frac{A_1^2}{n_1} + \frac{A_2^2}{n_2} + \frac{A_3^2}{n_2}\right) - \frac{T^2}{N} \tag{4.8}$$

For the data of Table 4.21,

$$SS_A = \left(\frac{12^2}{5} + \frac{19^2}{5} + \frac{27^2}{4}\right) - \frac{58^2}{14} = 42.96$$

and $MS_A = SS_A/df_A = 42.96/2 = 21.48$.

For the error term, the sum of squares is found by a modification of Equation 3.15 for unequal n, as follows:

$$SS_{S/A} = \Sigma Y^2 - \left(\frac{A_1^2}{n_1} + \frac{A_2^2}{n_2} + \frac{A_3^2}{n_3}\right) \tag{4.9}$$

For the data of Table 4.20,

$$SS_{S/A} = (40 + 75 + 185) - \left(\frac{12^2}{5} + \frac{19^2}{5} + \frac{27^2}{4}\right) = 16.75$$

Degrees of freedom for error is the number of scores within each group minus 1, or $4 + 4 + 3 = 11$ so that,

$$\mathbf{MS}_{S/A} = \frac{16.75}{4+4+3} = 1.52$$

$$F = \frac{21.48}{1.52} = 14.13$$

Critical F at $\alpha = .05$ from Table A.1, with 2 and 11 df, is 3.98, indicating a statistically significant difference in rating of vacation as a function of view from room.

In this situation, the traditional approach to ANOVA is clearer than the regression approach, because some of the columns do not sum to zero and the computations become much more complex. A feel for the kind of adjustments that are necessary is obtained by looking at the equations for slope in Table 4.10. On the other hand, use of statistical software circumvents the complexity of calculations. As long as software default values remain unchanged, programs protect against the possibility that nonorthogonality caused by unequal sample sizes is great enough to cause the solution to be unstable.

4.5.4 Homogeneity of Variance

Section 3.4.3 discusses the ratio of largest to smallest variance as an indication of compliance with the assumption of homogeneity of variance. To repeat Equation 3.21,

$$F_{\mathbf{max}} = \frac{s^2_{\mathbf{largest}}}{s^2_{\mathbf{smallest}}}$$

Variances (squared standard deviations) in the small-sample example of Section 4.4.3 (with equal n) are $(1.673)^2$, $(0.837)^2$, and $(1.140)^2$ for the three groups, respectively, as shown in Tables 4.15–4.17. Thus,

$$F_{\max} = \frac{1.673^2}{0.837^2} = 4.00$$

Recall that the assumption of homogeneity of variance is robust to violation when (1) $F_{\max} \leq 10$, (2) the ratio of largest to smallest sample size is less than 4:1, (3) two-tailed tests are used, and (4) an omnibus analysis is performed. All of these requirements are met for the small-sample omnibus ANOVA, so the probability associated with the critical value of F is close to the tabled value. Departure from homogeneity of variance is more serious for comparisons than for omnibus analyses, as shown in Section 4.5.5.

Recall from Section 3.4.3 that heterogeneity of variance should always be reported and is often of interest in itself. Differences in dispersion of scores in groups could come from differences in the cases in groups (indicating a failure of random assignment) or from variability in the reactions of various cases to the same treatment. Cases may respond consistently to some levels of treatment and inconsistently to other levels. What is it about the level of treatment or the cases that causes variability in response? Ferreting out the source of this variability often leads the research in unexpected and fruitful directions.

4.5.5 Specific Comparisons

An analysis of variance as described in preceding sections provides only limited information if there are more than two levels of the IV. The overall test of the effect, the omnibus analysis, indicates that there are differences somewhere among the means, but gives no hint as to exactly which means are different from one another. Specific comparisons (also called *contrasts*) are necessary to provide more focused tests of differences among means.

Specific comparisons may either *replace* omnibus ANOVA (planned comparisons) or *follow* omnibus ANOVA (post hoc comparisons). Planned comparisons are far more powerful than post hoc comparisons and are therefore preferred. An analysis "uses up" degrees of freedom, best thought of as a precious, nonrenewable resource. Those degrees of freedom can be used either conducting a one-way omnibus ANOVA or conducting planned comparisons. The one-way omnibus ANOVA uses up $a - 1$ df; those df could, instead, be spent performing planned comparisons that provide specific information regarding the location of mean differences.

If ANOVA is done first, several types of post hoc comparisons are available, all with different goals and different levels of power. The two that are discussed here in greatest detail are the Scheffé procedure for performing numerous and complex comparisons and the Tukey test for comparing all possible pairs of means. The Scheffé procedure is the most flexible, but also the most conservative, of the post hoc procedures. Failure to apply appropriate adjustment for post hoc analyses results

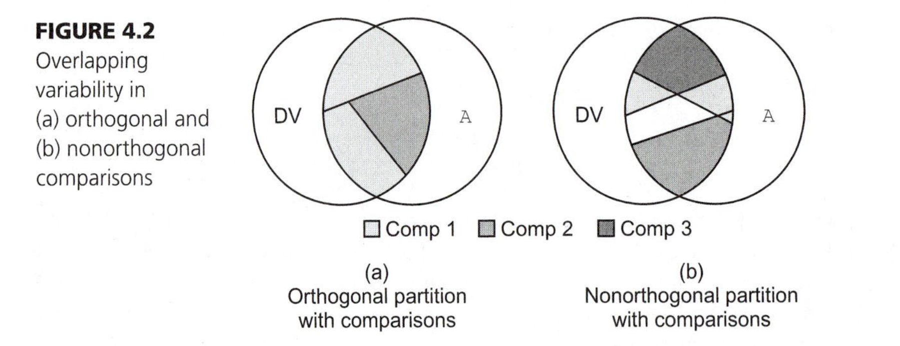

FIGURE 4.2 Overlapping variability in (a) orthogonal and (b) nonorthogonal comparisons

in (grossly) inflated Type I error. Section 4.5.5.5 deals with some adjustment techniques.

The only difference between planned comparisons and several types of post hoc comparisons is which critical F is used to test the F obtained for the comparison; a larger critical value of F is required when the same comparison is done post hoc. That is, F obtained for a comparison is the same whether planned or post hoc; it is the inferential test of the comparison that changes depending on when it is performed.

Gaining the power of planned comparisons imposes restrictions, in addition to the need for preplanning. First, there should only be as many comparisons as there are df_A.[10] For example, in a one-way design with four levels, three comparisons, each with a single df, are available for analysis. Second, it is desirable to use an orthogonal set of weighting coefficients (numerical codes) for the comparisons. Weighting coefficients are chosen to express hypotheses. If possible, these coefficients should be chosen in such a way that the hypotheses tested are independent—the set of coefficients completely analyzes all available variability without reanalyzing any parts of it.

Figure 4.2 uses Venn diagrams to illustrate the difference between orthogonal and nonorthogonal comparisons. In Figure 4.2(a), there are as many comparisons as there are df_A, and the set of comparisons is orthogonal. In this case, the comparisons completely partition the area representing the relationship between the DV and A, with none of the area reanalyzed. Figure 4.2(b) represents a case of nonorthogonal comparisons, where not all of the relationship between the DV and A is analyzed, and some of the area is analyzed by more than one comparison.

4.5.5.1 Weighting Coefficients for Comparisons

Comparison of treatment means begins by assigning a weighting factor (w) to the mean of each level of the IV so that the weights reflect your null hypotheses. Suppose you have a one-way design with a means, and you want to make comparisons. For each comparison: (1) A weight is assigned to each mean. (2) At least two of the means have to have nonzero weights. (3) Weights of zero are assigned to means (groups) that are left out of a comparison. (4) Means that are contrasted with each

[10] If greed sets in and more planned comparisons are desired than the available df, the Bonferroni correction is applied, as discussed in Section 4.5.5.4.3.

other are assigned weights with opposite signs (positive or negative), with the constraint that (5) the weights sum to zero. That is,

$$\sum_{1}^{a} w_j = 0$$

For example, consider an IV with four levels, producing $\overline{Y}_{a_1}$, $\overline{Y}_{a_2}$, $\overline{Y}_{a_3}$, and $\overline{Y}_{a_4}$.

If you want to test the null hypothesis that in the population, $\mu_{a1} - \mu_{a3} = 0$, appropriate weighting coefficients are 1, 0, −1, 0, producing $1\overline{Y}_{a_1} + 0\overline{Y}_{a_2} + (-1)\overline{Y}_{a_3} + \overline{Y}_{a_4} \cdot \overline{Y}_{a_2}$ and $\overline{Y}_{a_4}$ are left out, whereas $\overline{Y}_{a_1}$ is compared with $\overline{Y}_{a_2}$. Or, if you want to test the null hypothesis that $(\mu_{a1} + \mu_{a2})/2 - \mu_{a3} = 0$ (to compare the average of means from the first two groups with the mean of the third group, leaving out the fourth group), weighting coefficients are 1/2, 1/2, −1, 0 (or any multiple of them, such as 1, 1, −2, 0), respectively. Or, if you want to test the null hypothesis that $(\mu_{a1} + \mu_{a2})/2 - (\mu_{a3} + \mu_{a4})/2 = 0$ (to compare the average mean of the first two groups with the average mean of the last two groups), the weighting coefficients are 1/2, 1/2, −1/2, −1/2 (or 1, 1, −1, −1).

Notice that some of these comparisons involve pooling (averaging) means to contrast them with another mean (or another pooled mean). But no matter how complicated the pooling procedure becomes, a comparison always boils down to a test of the difference between two means. For this reason, the *F* test for a comparison has just one degree of freedom in the numerator.

The idea behind the test is that the difference between the weighted means is equal to zero when the null hypothesis is true. The more the sum diverges from zero, the greater the obtained *F*, and the greater the confidence with which the corresponding null hypothesis is rejected.

In the small-sample example, for instance, to test whether the ocean view differs significantly from the parking lot view, the coefficients assigned to the three levels of the IV are −1,1,0 for parking lot, ocean, and ski slope views, respectively. Notice that these are also the X_2 coefficients applied to the three levels in the regression approach to ANOVA (Table 4.7). The other set of coefficients, in X_1, contrasts the combination of parking lot and ocean view with the ski slope view. At this point, you have probably recognized that the weighting coefficients for comparisons are the same as the numerical codes in the columns for the regression approach to ANOVA. Indeed, the regression approach has comparisons embedded as an integral part of the analysis.

4.5.5.2 Orthogonality of Weighting Coefficients for Comparisons

Any pair of comparisons is orthogonal if the sum of the cross products of the weights for the two comparisons is equal to zero. For example, in the three comparisons shown here, the sum of the cross products of weights for comparison 1 and comparison 2 is

$$(1)(-1) + (1)(1) + (-2)(0) = 0$$

Therefore, the first two comparisons are orthogonal. This corresponds to the sum of the X_1X_2 column of Table 4.7, in which codes for each case have been cross-multiplied.

The use of orthogonal coding in the regression approach to ANOVA *builds in* orthogonal comparisons.

	w_1	w_2	w_3
Comparison 1	1	1	−2
Comparison 2	−1	1	0
Comparison 3	1	0	−1

Comparison 3, however, is orthogonal to neither of the first two comparisons. For instance, checking it against comparison 1,

$$(1)(1) + (1)(0) + (-2)(-1) = 3$$

In general, there are as many orthogonal comparisons as there are degrees of freedom for effect. Because $a = 3$ in the example, $df_A = 2$. There are, then, only two orthogonal comparisons when there are three levels of an IV, only three orthogonal comparisons when there are four levels of an IV, and so on. (However, there are always several possible sets of orthogonal comparisons.)

There are advantages to using orthogonal comparisons if a set is available that asks the appropriate research questions. First, there are only as many comparisons as there are degrees of freedom for effect, so the temptation to "overspend" degrees of freedom is avoided. Second, as shown in Figure 4.2, orthogonal comparisons analyze nonoverlapping variance; if one of them is significant, it has no bearing on the significance of any of the others. Last, because they are independent, if all $a - 1$ orthogonal comparisons are performed, the sum of the sum of squares for the comparisons is the same as the sum of squares for the IV in omnibus ANOVA. That is, the sum of squares for the effect has been completely partitioned into the $a - 1$ orthogonal comparisons that comprise it. This is, of course, the tack that is taken in the regression approach to ANOVA.

It should be emphasized that answering the right research question is more important than having an orthogonal set of comparisons. Sometimes an orthogonal set answers the right questions so you can take advantage of its mathematical properties. However, answering the right research questions is always more important than maintaining orthogonality in the set of comparisons. Beware, however, that you cannot simply enter a set of nonorthogonal contrasts into a single computer run, because they all will inappropriately be adjusted for each other. Instead, coefficients for each interesting contrast are entered as part of an orthogonal set of contrasts. You can then ignore the contrasts that are extraneously produced by the program.

4.5.5.3 Obtained *F* for Comparisons

Once the weighting coefficients are chosen, the following equation is used to obtain *F* for the comparison, if sample sizes are equal in each group:

$$F = \frac{n\left(\Sigma w_j \overline{Y}_j\right)^2 / \Sigma w_j^2}{\mathbf{MS}_{S/A}} \tag{4.10}$$

where n = number of scores in each group

$(\Sigma w_j \overline{Y}_j)^2$ = squared sum of the weighted means

Σw_j^2 = sum of the squared coefficients

$MS_{S/A}$ = mean square for error in the ANOVA

(Section 4.5.3 shows an equation to use for unequal n.) The numerator of Equation 4.10 is both sum of squares and mean square for the comparison because a comparison has only one degree of freedom.

In the small-sample example, with $n = 5$ and $MS_{S/A} = 1.6$, if the parking lot and ocean views are pooled and contrasted with the ski slope view, then

$$F = \frac{5((1)(2.4) + (1)(3.8) + (-2)(6.4))^2/(1^2 + 1^2 + (-2)^2)}{1.6} = 22.69$$

This is the same F ratio that is found if the regression approach is used to test X_1 of Table 4.11:

$$F = \frac{\mathbf{SS(reg.}\ X_1\mathbf{)}}{\mathbf{MS(resid.)}} = \frac{36.3}{1.6} = 22.69$$

Once you have obtained F for a comparison, whether by hand calculation or computer, compare the obtained F with a critical F to see if the mean difference is statistically significant. If obtained F exceeds critical F, the null hypothesis represented by the weighting coefficients is rejected. But which critical F is used depends on whether the comparison is planned or post hoc (as well as on other niceties, such as equal n and homogeneity of variance).

4.5.5.4 Planned Comparisons

Choice of tests for planned comparisons depends on the number of comparisons to be done, whether the planned comparisons are orthogonal, and whether the assumption of homogeneity of variance is met. Table 4.22 shows the choices.

4.5.5.4.1 An Orthogonal Set If you are in the enviable position of having (1) planned your comparisons prior to data collection, (2) planned no more comparisons than you have degrees of freedom for the effect, (3) a set of planned comparisons that is orthogonal, and (4) homogeneity of variance, you can read the significance of F for the comparison directly from the output, just as in routine ANOVA. Or, if obtained F is compared with critical F from a table such as Table A.1, each comparison is tested against critical F at chosen alpha, with one degree of freedom in the numerator and df_{error} ($df_{S/A}$) in the denominator. If obtained F is larger than critical F, the null hypothesis represented by the weighting coefficients is rejected.

In the small-sample example, $df_{S/A} = 12$. The critical value from Table A.1 with 1 and 12 df at $\alpha = .05$ is 4.75. With obtained $F = 22.69$, the null hypothesis of no difference between the mean of the ski slope view versus pooled means of the other two views is rejected. Table 4.23 shows how the set of orthogonal comparisons is done

TABLE 4.22 Choosing among tests for planned comparisons

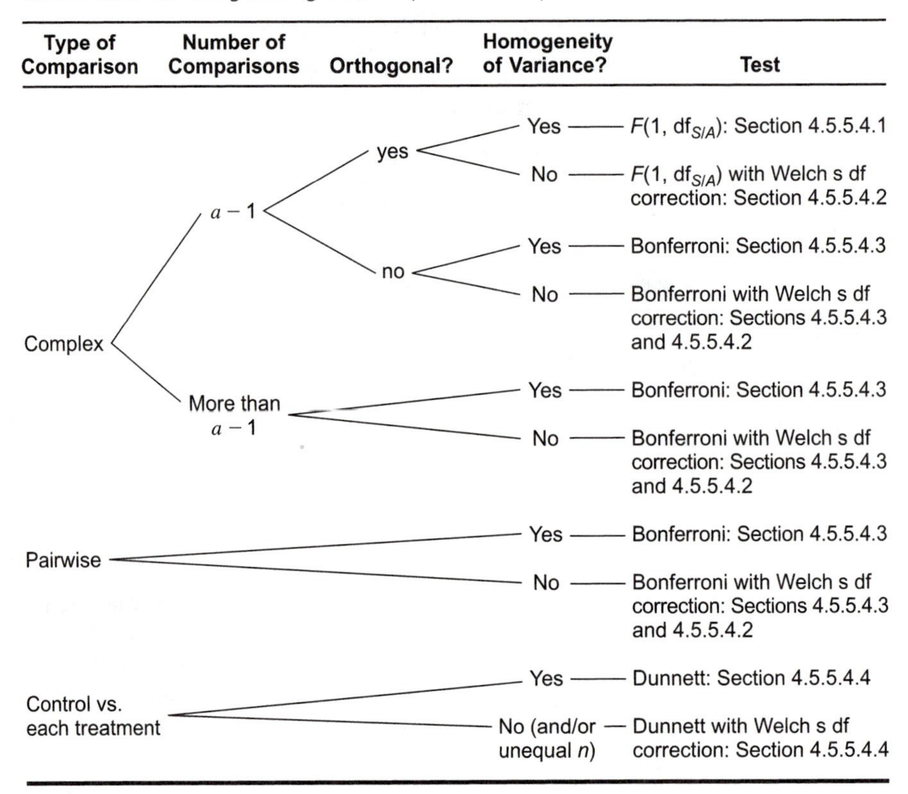

Type of Comparison	Number of Comparisons	Orthogonal?	Homogeneity of Variance?	Test
Complex	$a - 1$	yes	Yes	$F(1, df_{S/A})$: Section 4.5.5.4.1
			No	$F(1, df_{S/A})$ with Welch s df correction: Section 4.5.5.4.2
		no	Yes	Bonferroni: Section 4.5.5.4.3
			No	Bonferroni with Welch s df correction: Sections 4.5.5.4.3 and 4.5.5.4.2
	More than $a - 1$		Yes	Bonferroni: Section 4.5.5.4.3
			No	Bonferroni with Welch s df correction: Sections 4.5.5.4.3 and 4.5.5.4.2
Pairwise			Yes	Bonferroni: Section 4.5.5.4.3
			No	Bonferroni with Welch s df correction: Sections 4.5.5.4.3 and 4.5.5.4.2
Control vs. each treatment			Yes	Dunnett: Section 4.5.5.4.4
			No (and/or unequal n)	Dunnett with Welch s df correction: Section 4.5.5.4.4

through SPSS, SAS, and SYSTAT software by adding contrast coefficients to existing syntax for one-way ANOVA.

With planned comparisons, the results for omnibus ANOVA are not attended; the researcher moves straight to the results for the comparisons. Once the degrees of freedom are spent on the planned comparisons, however, it is perfectly acceptable to snoop the data using the less powerful post hoc procedures (Section 4.5.5.5).

4.5.5.4.2 Welch's Correction for Heterogeneity of Variance All the tests in Table 4.22, including the orthogonal set of comparisons in Section 4.5.5.4.1, assume homogeneity of variance. Therefore, the omnibus ANOVA that provides the error term should also be examined for homogeneity of variance, using the procedures of Section 4.5.4.[11] If the guidelines are violated, Welch's correction is applied to $df_{S/A}$ when looking up critical F to compensate for inflated Type I error rate. The adjusted df_{error} N is based on Welch (1947) and Wang (1971).

[11] Variances can be obtained through any of the software modules that provide descriptive statistics, or they can be calculated by hand.

TABLE 4.23 Syntax for orthogonal comparisons

Program	Syntax	Section of Output	Name of Effect

SPSS ONEWAY	`ONEWAY`
	`RATING BY VIEW`
	`/CONTRAST= 1 1 -2`
	`/CONTRAST= -1 1 0`
	`/MISSING ANALYSIS .`
		Contrast Tests	Assume equal variances

SAS GLM	`Proc glm data=SASUSER.RAND_GRP;`
	`class VIEW;`
	`model RATING=VIEW;`
	`contrast 'X1' VIEW 1 1 -2;`
	`contrast 'X2' VIEW -1 1 0;`
	`run;`
		Contrast	X1, X2

SYSTAT ANOVA	`ANOVA`
	`CATEGORY VIEW`
	`COVAR`
	`DEPEND RATING`
	`ESTIMATE`
	`HYPOTHESIS`
	`EFFECT=VIEW`
	`NOTE 'SLOPE VS OTHERS'`
	`CONTRAST[ 1 1 -2 ]`
	`TEST`
	`HYPOTHESIS`
	`EFFECT=VIEW`
	`NOTE 'PARKING VS OCEAN'`
	`CONTRAST[ -1 1 0 ]`
	`TEST`
		Test of Hypothesis	Hypothesis

[a] t is given rather than F; recall that $t^2 = F$.

$$\text{df}_{error'} = \frac{\left(\frac{w_1^2 s_1^2}{n_1} + \frac{w_2^2 s_2^2}{n_2} + \cdots + \frac{w_a^2 s_a^2}{n_a}\right)^2}{\frac{w_1^4 s_1^4}{n_1^2(n_1 - 1)} + \frac{w_2^4 s_2^4}{n_2^2(n_2 - 1)} + \cdots + \frac{w_a^4 s_a^4}{n_a^2(n_a - 1)}} \tag{4.11}$$

For demonstration purposes, using the variances from the small-sample example, $(1.673)^2$, $(0.837)^2$, and $(1.140)^2$, even though they are not heterogeneous, and assuming the comparison coefficients are 1, 1, -2:

$$\text{df}_{error'} = \frac{\left(\frac{1^2(1.673)^2}{5} + \frac{1^2(0.873)^2}{5} + \frac{(-2)^2(1.140)^2}{5}\right)^2}{\frac{1^4(1.673)^4}{5^2(5-1)} + \frac{1^4(0.873)_2^4}{5^2(5-1)} + \frac{(-2)^4(1.140)^4}{5^2(5-1)}} = \frac{(1.750)^2}{0.292} = 8.65$$

Thus, error degrees of freedom are cut from 12 to 9 to adjust for violation of the assumption of homogeneity of variance. Critical F with df = 1,9 in Table A.1 is 5.32, so the comparison is statistically significant even after this (unnecessary in this example) adjustment to df.

Note Equation 4.11 also incorporates unequal n, if present, which may exacerbate the seriousness of heterogeneity of variance.

4.5.5.4.3 Bonferroni Correction Sometimes the researcher plans more comparisons than degrees of freedom for the effect. For instance, you may plan to compare all pairs of means (pairwise comparisons), but this requires too many degrees of freedom. When there are too many tests, the α level across all tests—the familywise error rate—exceeds the α level for any one test, even if comparisons are planned. The reason for the precipitous rise in error rate is the necessity of making many separate decisions, all of them correctly. When there is only one test, the Type I error rate is .05, so the probability of correctly retaining a true null hypothesis is .95. With two tests, each at $\alpha = .05$, the probability of retaining two true null hypotheses is (.95)(.95) = .90. This comes from the desire to make the correct decision on the first test *and then* the correct decision on the second. The probability of an incorrect decision about one or the other of the hypotheses, or both, is 1 − .90 = .10. Thus, although each test uses $\alpha = .05$, when the familywise error rate is computed, the error rate for both tests combined is .10. With three tests, the joint probability of retaining three true null hypotheses is (.95)(.95)(.95) = .86, and the familywise error rate is .14. When there are five hypothesis tests to be conducted, the familywise Type I error rate is about .23, meaning that there is roughly one chance in four of committing a Type I error in the chain of five decisions. As a rough guide, the familywise error rate is the number of tests to be conducted multiplied by the chosen α level.

When there are too many significance tests, some adjustment of α for the tests are needed. It is common practice to use a Bonferroni-type adjustment where more stringent α levels are used with each test to keep the familywise α across all tests at reasonable levels. For instance, if five comparisons are planned and each is tested at $\alpha = .02$, the alpha across all tests is an acceptable .10 (roughly .02 times 5, the number of tests). Overall familywise error rates of between .05 and .15 are acceptable in many situations. Thus, the simplest form of the Bonferroni correction is applied by dividing the acceptable familywise error rate by the number of comparisons to be made.

However, this familywise error rate need not be apportioned equally among all comparisons. If you want to keep overall α at, say, .10, and you have five tests to perform, you can assign each of them $\alpha = .02$. Or you can assign two of them $\alpha = .04$, with the other three evaluated at $\alpha = .01$, for an overall Type I error rate of roughly .11. Keppel (1991) suggests a modified Bonferroni procedure that takes into account the number of tests legitimately performed as planned comparisons as well as those in excess of the legitimate number. The decision about how to apportion α through the tests is part of planning the analysis and is made prior to data collection.

A problem of inflated Type I error rate also occurs when the researcher plans $a - 1$ comparisons that are nonorthogonal. The problem arises because the researcher is reanalyzing the same sum of squares (see Figure 4.2). The suggested remedy is to apply the Bonferroni correction (with Welch's adjustment if homogeneity of variance is violated), as shown in Table 4.22. Most software programs provide a Bonferroni correction to pairwise comparisons, although not to complex comparisons.[12]

As an aside, it is important to realize that routine ANOVA designs with numerous main effects and interactions suffer from the same problem of inflated Type I error rate across all tests as do planned comparisons in which there are too many tests. Some adjustment of α for separate tests is needed in big ANOVA problems if all effects are evaluated.

4.5.5.4.4 Dunnett's Test for Control Group Versus Each Treatment. Dunnett's test is applicable to a specific design in which there is a single control group and multiple treatment groups and the researcher plans to compare the mean of each treatment, in turn, with the mean of the control. In this situation, there are $a - 1$ comparisons, but they are not orthogonal. Thus, Dunnett's test controls α_{FW} for $a - 1$ nonorthogonal comparisons.

Obtained F for these comparisons uses Equation 4.10, simplified because each comparison involves only two means, coded 1 and -1. If the parking lot view is considered a control condition in the small-sample example, the two F ratios for the Dunnett comparisons are 3.06 for ocean and 25.00 for ski slope.

The critical F ratio for the Dunnett test, F_D, is

$$F_D = q_D^2 \tag{4.12}$$

where q_D is the value from Table A.4, entered with df_{error} ($\text{df}_{S/A}$ for the randomized-groups design), a (the number of means including the control), and your selection of .05 or .01 as α_{FW}.

For the example, with $\text{df}_{S/A} = 12$, $a = 3$, and $\alpha_{FW} = .05$ (two-tailed), the value from Table A.4 is 2.50, which, when squared, produces $F_D = 6.25$. By the criterion of the Dunnett test, the ski slope view differs significantly from the parking lot control, but the ocean view does not.

The procedure demonstrated here is for the two-tailed Dunnett test to be used with a nondirectional alternative hypothesis. This form of the Dunnett test is available in the SAS and SPSS GLM programs. These software implementations also permit a one-sided Dunnett test, with the control group hypothesized to have either a lower (or higher) mean than the treatment groups.

Dunnett's test assumes equal n and a special form of homogeneity of variance. It is assumed that the variance of the control group is equal to the pooled variance of the treatment groups. When these assumptions are not met, you can use Dunnett's test with Welch's correction for error df of the previous section.

[12] An alternative, slightly more powerful method for apportioning familywise Type I error rate is the Sidak correction, available for pairwise comparisons in SAS and SPSS GLM programs.

4.5.5.5 Adjustments for Post Hoc Comparisons

If you are unable to plan your comparisons and choose, instead, to start with routine ANOVA, you almost always follow a significant main effect (with more than two levels) with post hoc comparisons to find the treatment means that are different from one another. Because you have already spent your degrees of freedom in routine ANOVA and are likely to capitalize on chance differences among means that you notice in the data, some form of adjustment of α is necessary.

A staggering array of procedures for dealing post hoc with inflated Type I error rate is available. The tests differ in the number and type of comparisons they permit, the amount of adjustment required of α, and availability in software. The tests that permit more numerous comparisons have correspondingly more stringent adjustments to critical F. Therefore, the fewer comparisons that are possible, the less conservative (i.e., the more powerful) the test. The name of the game is to choose the most liberal (powerful) test that permits you to perform the comparisons of interest.

This section demonstrates the different tests for making comparisons. The most flexible, but also most conservative, adjustment procedure is the Scheffé test, which permits numerous and complex comparisons and controls familywise error rate over all comparisons that can be devised. The Tukey HSD (honestly significant difference) test controls familywise error for all pairwise comparisons among means. Because the Scheffé test permits comparison of all pairs of means, as well as comparisons where means are pooled, the adjustment is more severe. These are the two most popular post hoc procedures, but both require equal n and homogeneity of variance in their simplest forms. Alternative procedures are available when the assumptions are violated, as shown in Table 4.24. (Note that the Dunnett tests in Table 4.25 are for pairwise comparisons and differ from the Dunnett test described in Section 4.5.5.4.4.)

Other comparison procedures available in software include the Duncan test, the Student-Newman-Keuls test, the least-significant difference and modified least-significant difference tests, an alternative Tukey procedure, and so on. In general, these tests are known to be too liberal under certain circumstances and are not recommended.

Different selections of these post hoc adjustment techniques are available in statistical computing packages. Section 4.7 compares one-way ANOVA programs and indicates which techniques are available in each.

4.5.5.5.1 Scheffé Post Hoc Test for Complex Comparisons The test described here (Scheffé, 1953) is the most flexible and most conservative of the methods of post hoc adjustment. The Scheffé adjustment for familywise error rate is applied to critical F. Once this critical F is computed, you use it for all pairwise comparisons and for all combinations of treatment means pooled and contrasted with other treatment means, pooled or not, as desired.[13] Once you pay the "price" in conservatism for this flexibility, you might as well conduct all the comparisons that make sense given your research design.

[13] Some possibilities for pooling are illustrated in "Weighting Coefficients for Comparisons" in Section 4.5.5.1.

TABLE 4.24 Choosing among post hoc analyses

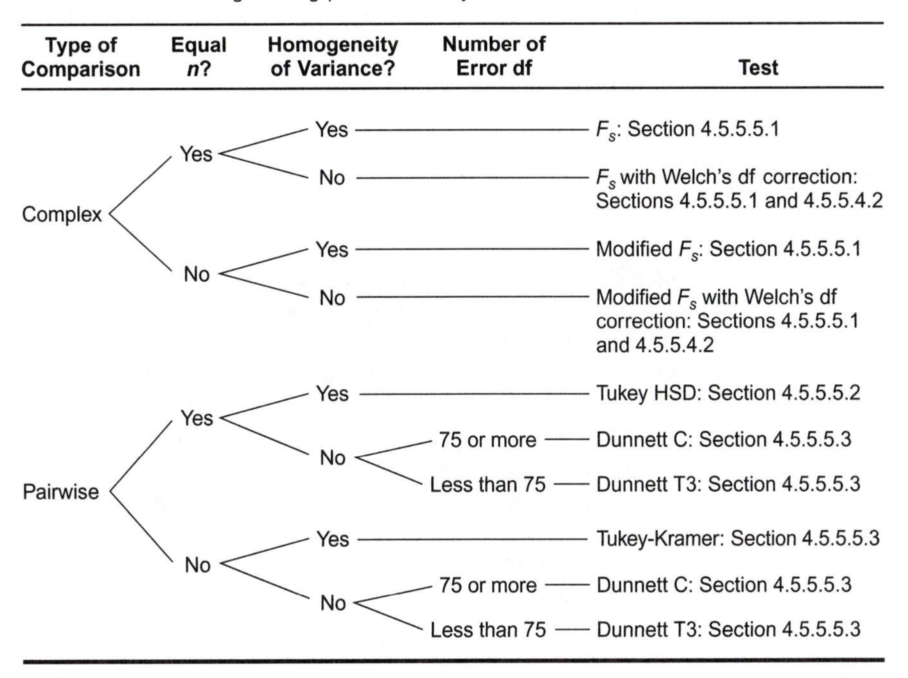

Type of Comparison	Equal *n*?	Homogeneity of Variance?	Number of Error df	Test
Complex	Yes	Yes		F_S: Section 4.5.5.5.1
		No		F_S with Welch's df correction: Sections 4.5.5.5.1 and 4.5.5.4.2
	No	Yes		Modified F_S: Section 4.5.5.5.1
		No		Modified F_S with Welch's df correction: Sections 4.5.5.5.1 and 4.5.5.4.2
Pairwise	Yes	Yes		Tukey HSD: Section 4.5.5.5.2
		No	75 or more	Dunnett C: Section 4.5.5.5.3
			Less than 75	Dunnett T3: Section 4.5.5.5.3
	No	Yes		Tukey-Kramer: Section 4.5.5.5.3
		No	75 or more	Dunnett C: Section 4.5.5.5.3
			Less than 75	Dunnett T3: Section 4.5.5.5.3

The Scheffé adjustment to critical F for a comparison on marginal means is

$$F_S = (a - 1)F_c \tag{4.13}$$

where F_S = Scheffé-adjusted critical F

$(a - 1)$ = degrees of freedom for the effect

F_c (critical F) = tabled F with $a - 1$ degrees of freedom in the numerator and degrees of freedom associated with the error term in the denominator

If obtained F is larger than critical F_S, the null hypothesis represented by the weighting coefficients for the comparison is rejected. For example, had the small-sample comparison of the ski slope view against the pooled means of the ocean and parking lot views been done as part of a post hoc analysis, critical F_S would have been $(3 - 1)(3.88) = 7.76$. The obtained $F = 22.69$ is still statistically significant, meaning that the null hypothesis of equality between the mean of the ski slope view contrasted with the pooled means of the parking lot and ocean views is rejected.

The Scheffé test assumes homogeneity of variance; violation is dealt with by applying Welch's correction to error df (Equation 4.11). The Scheffé test also assumes equal n, violation of which requires modifying Equation 4.10 as follows:

$$F = \frac{(\Sigma w_j \overline{Y}_j)^2 / \Sigma(w_j^2 / n_j)}{\mathbf{MS}_{S/A}} \tag{4.14}$$

Each coefficient is divided by its own group size, instead of multiplying weighted means by a single sample size. For example, using the unequal n data of Section 4.5.3, the comparison of ski slope versus other views with coefficients 1, 1, and −2 produces obtained F:

$$F = \frac{((1)(2.4) + (1)(3.8) - (-2)(6.75))^2/((1^2/5) + (1^2/5) + (-2)^2/4))}{1.52}$$

$$= 25.04$$

4.5.5.5.2 *Tukey HSD Test for Pairwise Comparisons* The Tukey HSD (honestly significant difference) test, or studentized range statistic, is designed to test all means against each other pairwise, meaning that each mean is compared, in turn, with every other mean. This set of comparisons does not include comparisons with pooled or averaged means. Obtained F is calculated using Equation 4.10—this is the same as for any other comparison, except the equation is simplified because the pair of means of interest are given codes 1 and −1 and the remaining means are omitted because their codes are zero. In the small-sample example, the three obtained F ratios are as follows:

Parking lot view versus ocean view:

$$F = \frac{5((1)(2.4) + (-1)(3.8))^2/(1^2 + 1^2)}{1.6} = 3.06$$

Parking lot view versus ski slope view:

$$F = \frac{5((1)(2.4) + (-1)(6.4))^2/(1^2 + 1^2)}{1.6} = 25.00$$

Ocean view versus ski slope view:

$$F = \frac{5((1)(3.8) + (-1)(6.4))^2/(1^2 + 1^2)}{1.6} = 10.56$$

The critical F ratio in the Tukey test is

$$F_T = \frac{q_T^2}{2} \tag{4.15}$$

where q_T is the value from Table A.3. To enter Table A.3, use df_{error} ($df_{S/A}$), a (the number of levels of the IV), and one of two choices for familywise α_{FW} (.05 or .01). The adjustment to critical F for controlling α_{FW} is built into the tabled q_T value.

For the small-sample example, $df_{S/A} = 12$, $a = 3$, and α_{FW} is chosen to be .05. Table A.3 shows that $q_T = 3.77$. Therefore, $F_T = (3.77)^2/2 = 7.11$. Thus, the means for parking lot and ocean views do not differ, but the remaining two pairs of means are significantly different. The critical value for the Tukey test (7.11) is smaller than the corresponding critical value for the Scheffé test (7.76), making the null hypothesis easier to reject because the Tukey procedure permits fewer comparisons.

A useful computational procedure for the Tukey test when there is a large number of means involves solving for the critical mean difference rather than critical F_T.

All pairwise differences between means that are greater than this critical mean difference are statistically significant. The critical mean difference, $\bar{d}_T$, is defined as

$$\bar{d}_T = q_T\sqrt{\frac{\mathbf{MS}_{S/A}}{n}} \quad \textbf{(4.16)}$$

In this example, the critical mean difference is

$$\bar{d}_T = 3.77\sqrt{\frac{1.6}{5}} = 2.13$$

Thus, the difference between means for parking lot view and ocean view $(3.8 - 2.4 = 1.4)$ is not statistically significant, but the other two mean differences exceed 2.13.

Tukey tests are available in SPSS ONEWAY and GLM and in SAS ANOVA and GLM. SPSS ONEWAY and GLM also have an alternative Tukey procedure (TUKEY-b), which has different critical values for means that are different rank distances apart.[14]

SPSS and SAS GLM have numerous additional tests for pairwise comparisons, but some of these, like Scheffé, are far too conservative when applied to pairwise comparisons, and others, such as Student-Newman-Keuls, are often too liberal.

4.5.5.5.3 Pairwise Comparisons When Assumptions Are Violated Tukey tests assume equal n and homogeneity of variance. There are two additional Dunnett tests (not to be confused with the Dunnett planned comparison test for comparing a control with each treatment) that modify the Tukey test to accommodate unequal n and heterogeneity of variance, as shown in Table 4.24. The first, Dunnett C, is more powerful when there are 75 or more error df. The other, Dunnett T3, is more powerful when sample sizes are smaller (Dunnett, 1980). Both are appropriate with equal or unequal sample sizes and are available in SPSS GLM. Both are discussed by Kirk (1995), who has a comprehensive coverage of multiple comparison tests.

When the assumption of homogeneity of variance is met but there is unequal n, the Tukey-Kramer modification to the error MS of Equation 4.16, $\sqrt{\frac{\mathbf{MS}_{S/A}}{n}}$, is applied separately for each comparison:

$$\sqrt{\frac{\mathrm{MS}_{S/A}\left(\frac{1}{n_i}+\frac{1}{n_j}\right)}{2}} \quad \textbf{(4.17)}$$

where n_i and n_j are sample sizes for applicable pair of means.

Both GLM programs, as well as SPSS ONEWAY, apply the Tukey-Kramer modification for unequal sample sizes.

4.5.5.5.4 Summary of Some Critical Values for Comparisons Table 4.25 summarizes the critical values for our example when $\alpha = .05$ (with $a = 3$ and $n = 5$) for

[14] Tukey-b is one of several "multiple range tests" available in software, as shown in Table 4.30.

TABLE 4.25 Critical values for omnibus ANOVA and comparisons

Type of Test	Relevant Statistical Table	Maximum Number of Tests	Critical Value for Rejecting H_0 at $\alpha = .05$
Omnibus ANOVA	A.1	1	3.88
Planned comparisons	A.1	2	4.75
Dunnett test	A.4	2	6.25
Tukey test	A.3	3	7.11
Scheffé test	A.1	6	7.76

the omnibus ANOVA, for orthogonal planned comparisons, for comparisons of a control with each treatment, and for two types of adjustments for post hoc analysis.

Table 4.25 shows the loss of power as more (and less restrictive) tests are permitted. Although the same number of tests are permitted for planned and Dunnett comparisons, the planned comparisons are more restrictive because they are orthogonal. Note that the difference in critical values between Tukey and Scheffé adjustments is small when $a = 3$, but grows as the number of levels of a increases. For example, with $a = 6$ and $n = 5$, the critical value for a Tukey test is $(4.37)^2/2 = 9.55$, but the critical value for a Scheffé test is $(5)(2.62) = 13.10$.

4.5.5.6 Trend Analysis

The functional relationship between the IV and the DV typically is of interest when the levels of the IV differ in amount rather than in type. When differences in levels of the IV are *quantitative* rather than qualitative, one often wants to know how the DV changes as a function of the level of the IV. For instance, a qualitative IV is different types of view from a window, whereas a quantitative IV is the same view at different window sizes. If the IV is quantitative, one might ask, "Does rating of quality of the vacation increase with window size?" Or, "Is there an optimal size above which ratings decline?" Or, "Do ratings increase up to a point and then level off as window sizes become ever larger?"

These questions are answered by a trend analysis (polynomial decomposition) using Table A.5. The weighting coefficients are used in Equation 4.10 in the same way other weighting coefficients are used. These weighting coefficients are orthogonal, and if $a - 1$ of them are used, they completely analyze the variance in the DV associated with treatment. Further, their use may be planned or post hoc (for which a post hoc Scheffé adjustment to critical F is required). In other words, a trend analysis is just like planned orthogonal comparisons or post hoc comparisons with a Scheffé adjustment to critical F, except you look up the coefficients instead of making them up.

A trend analysis checks the *shape* of the relationship between the DV and the levels of the IV. The simplest shape is a linear trend where the means of the DV at

FIGURE 4.3
Shapes represented by linear, quadratic, and cubic trend when $a = 4$

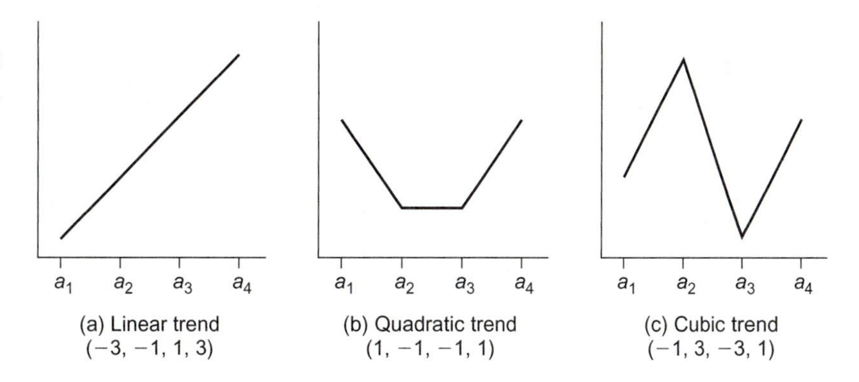

various levels of the IV fall along a straight line. The line may have positive slope (the means increase with increasing levels of the IV) or negative slope (the means decrease with increasing levels of the IV). The closer the means come to falling along a line with nonzero slope, the higher the F value and the greater the chance of rejecting the null hypothesis of equality among the means.[15]

With more than two levels of the IV, higher-order trends become possible. In general, there are $a - 1$ possible trends, although not all of them may be of much interest. If there are four levels of the IV, there are three possible trends: linear, quadratic, and cubic. The shape that is investigated by each trend is shown by plotting the trend coefficients against successive means. Figure 4.3 plots the trend coefficients for the three possible trends when there are four levels of the IV. As can be seen, the shapes become increasingly esoteric as the number of levels of the IV increases. In many applications, only the linear and quadratic trends are interpretable.

Combinations of trends are possible because the coefficients are orthogonal and because each trend is independent of all others. It is rather common to find a combination of linear and quadratic trend, for instance, in the shape of the relationship of means of the DV at various levels of the IV. Figure 4.4 represents several different shapes that would all produce both significant linear and significant quadratic trend (if the error term is small enough).

The interpretation of these various trends is different. If the only significant trend is linear, it means that over the range of sizes of the levels of the IV tested, the DV goes steadily up (or down). Quadratic trend usually implies that there is some optimal level of the IV where the DV scores are at their highest (or lowest). Combinations of linear and quadratic trend usually imply that the process grows (or declines) rapidly and then levels off or that the process is level for a while before it rapidly grows (or declines). Interpretation of yet higher-order trends and combinations of them is left to the vivid imagination of the reader.

[15] If there are only two levels of the IV, the only trend possible is linear ("two points determine a line") and omnibus ANOVA is also the test of linear trend.

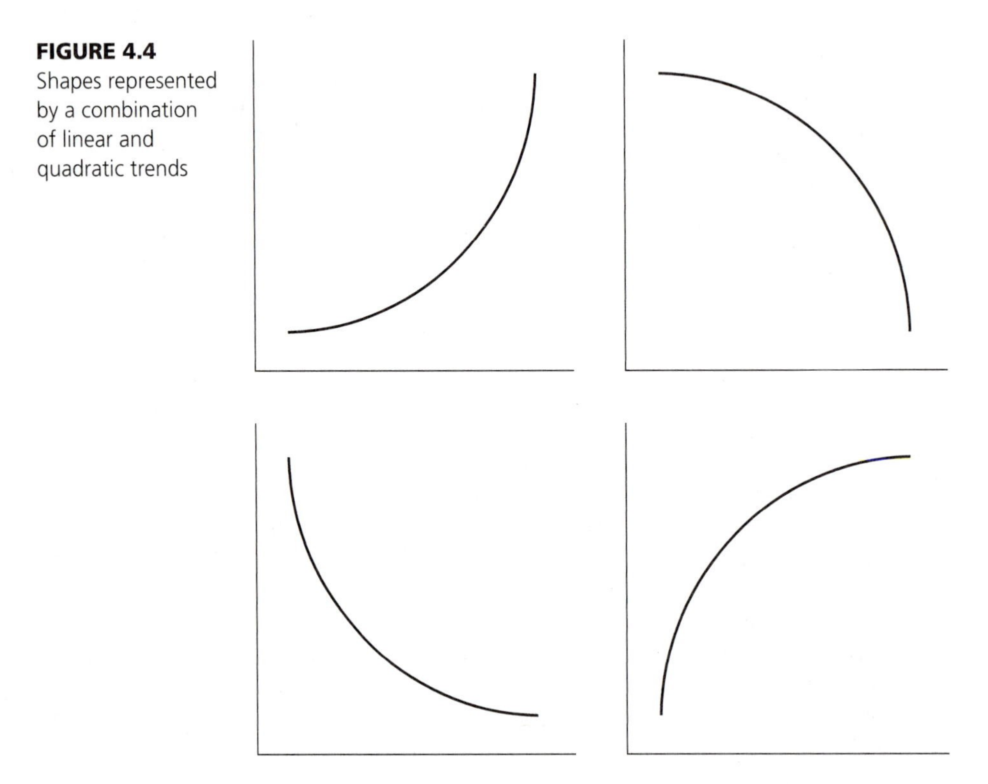

FIGURE 4.4 Shapes represented by a combination of linear and quadratic trends

If the levels of the IV in the ski slope example are *size* of window of ski slope view (say, 250, 500, and 750 cm^2) rather than type of view (parking lot, ocean, or ski slope), a trend analysis is appropriate. A plot of the means for the example appears in Figure 4.5. The means seem to fall in a straight line (represented by the dashed line), but not exactly along it.

The coefficients for linear trend of the three means are $-1, 0, 1$, and $\Sigma w_i^2 = 2$.[16] Applying Equation 4.10 to the small-sample data with $MS_{S/A} = 1.6$ and $n = 5$,

$$F = \frac{5\left((-1)(2.4) + (0)(3.8) + (1)(6.4)\right)^2 /2}{1.6} = 25.00$$

At this point, we know that the rating of the vacation increases linearly with the increasing view of the ski slope.

The quadratic trend tests whether two best-fit lines drawn through the means have significantly different slopes. Figure 4.5 shows that the slope from a_1 to a_2 is not identical to the slope from a_2 to a_3, but we don't know whether the difference is statistically significant. Table A.5 provides the coefficients for quadratic trend, $1, -2, 1$, and $\sigma w_i^2 = 6$. Applying Equation 4.10,

$$F = \frac{5\left((1)(2.4) + (-2)(3.8) + (1)(6.4)\right)^2 /6}{1.6} = 0.75$$

FIGURE 4.5
Mean rating of ski vacation as a function of window size (Graph produced in WordPerfect 8.0.)

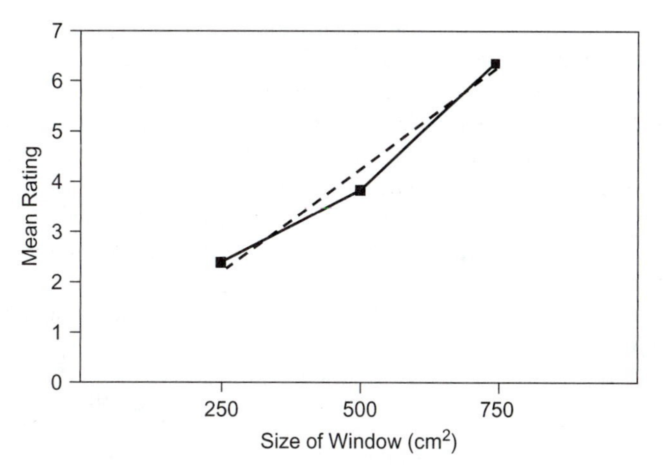

The linear trend is statistically significant by either a planned or post hoc criterion. The quadratic trend is significant by neither criterion. Therefore, we conclude that rating of the vacation increases with increasing size of ski slope view.

SPSS ONEWAY (as demonstrated in Section 4.6.2) has special syntax to produce a trend. SAS GLM requires specification of the coefficients of Table A.5. The coefficients in Table A.5 assume that levels of the IV increase or decrease in equal increments, say from 250 to 500 to 750 cm^2 and that sample sizes are equal in all levels of the IV. Otherwise, coefficients must be devised to reflect the inequality of spacing among levels of the IV and/or inequality in sample sizes. Derivation of coefficients for unequally spaced IVs is difficult by hand,[17] but some statistical programs provide for them by allowing you to enter an integer for each level of the IV (say 200, 250, and 325 cm^2) when levels are unequally spaced and by dealing with unequal sample sizes (see Section 4.5.3). SPSS GLM allows unequal spacing for randomized-groups designs. SAS GLM provides for unequal spacing only for repeated-measures IVs. Section 4.6.2 shows a planned trend analysis (with unequal sample sizes and equal spacing) through SPSS ONEWAY. Trend analysis with unequal spacing is demonstrated in Section 11.6.3.

4.5.5.7 Effect Sizes for Comparisons

Section 4.5.1 describes estimation of effect sizes for the omnibus test of the IV. What is the proportion of variability in the DV associated with differences in levels of A? Effect sizes for comparisons may be regarded in two ways. What proportion of the

[16] Yes, these are the same coefficients used to test the mean of the first versus the mean of the last level of the IV. If means for these levels are significantly different, the best-fit line drawn through the three means is not horizontal, so its slope is not zero. The value of the middle mean does not matter.

[17] See Myers and Well (1991) for an approach to deriving coefficients with unequal spacing between levels of the IV.

variability in the DV is associated with the comparison? Or, what is the proportion of variability in the main effect of A that is associated with the comparison? The numerator in both types of comparisons is the same—the variability associated with the comparison. The denominator reflects the base against which you are evaluating the comparison—either the entire variability of the DV or the variability associated with the omnibus main effect.

Consider the comparison of ski slope view with the average of parking lot and ocean views of Section 4.5.5.3. The SS for the comparison is 36.3. Tables 4.6 and 4.13 show that $SS_T = 60.4$, $SS_A = 41.2$, and $MS_{S/A} = 19.2$.

A variant of Equation 4.3 is used to evaluate the comparison relative to total variability:

$$\eta^2 = \frac{SS_{A\text{comp}}}{SS_T} = \frac{36.3}{60.4} = .60 \quad \textbf{(4.18)}$$

This indicates that 60% of the variance in ratings of a ski vacation is associated with the difference between a ski slope view and the other views.

Alternatively, effect size may be calculated relative to a single analysis. With a series of comparisons, there may be several statistically significant sources of variability, and this measure of effect size can be quite small because the denominator contains other systematic effects than the one of current interest. The denominator (SS_T) is thus larger than it would be in a corresponding one-way design, making the ratio smaller. A solution is to compute partial η^2 in which the denominator consists only of the effect of interest and the error; this produces an estimate of effect size that might be found in a one-way design that includes only the effect of current interest:

$$\text{Partial } \eta^2 = \frac{SS_{A\text{comp}}}{SS_{A\text{comp}} + SS_{S/A}} = \frac{36.3}{36.3 + 19.2} = .65 \quad \textbf{(4.19)}$$

Note that this is a larger value than produced by Equation 4.18 and should be clearly labeled as partial η^2. This is actually the value reported by Smithson's (2003) procedure; in a one-way analysis, η^2 and partial η^2 are the same for the omnibus effect (i.e., $SS_T = SS_A + SS_{S/A}$).

Finally, a different denominator is used to evaluate the variability in the comparison relative to the variability in the main effect of A:

$$\eta^2 = \frac{SS_{A\text{comp}}}{SS_A} = \frac{36.3}{41.2} = .88 \quad \textbf{(4.20)}$$

This indicates that 88% of the difference due to views (levels of A) is in the ski slope view versus the other two views.

4.6 COMPLETE EXAMPLE OF ONE-WAY ANOVA

The experiment analyzed in this section studied the number of facets in the eyes of each fly hatched in one of nine incubators that varied in temperature. This grand old classic study (Hersh, 1924) is described more fully in Appendix B.1. Data files are FLY.*.

The nine incubator temperatures varied from 15 to 31° C, in increments of 2° C. The number of facets was reported in units on a logarithmic scale. The strategy here is a planned trend analysis to evaluate the number of facets as a function of increasing temperature.

4.6.1 Evaluation of Assumptions

4.6.1.1 Accuracy of Input, Independence of Errors, Sample Sizes, and Distributions

Table 4.26 shows setup and output for an SPSS Tables (Basic tables) and Graphs (INTERACTIVE) run on the DV, FACETS. The output is shown separately for each level of the IV. Histograms with a normal curve superimposed are requested in the

TABLE 4.26 Descriptive statistics and histograms through SPSS TABLES and GRAPHS (INTERACTIVE) for fly data

```
* Basic Tables.
TABLES
  /FORMAT BLANK MISSING('.')
  /OBSERVATION FACETS
  /TABLES TEMP > FACETS
  BY (STATISTICS)
  /STATISTICS
  minimum( )
  maximum( )
  mean( )
  stddev( )
  variance( )
  validn( ( NEQUAL5.0 )).
IGRAPH /VIEWNAME='Histogram' /X1 = VAR(FACETS) TYPE = SCALE /Y = $count
/COORDINATE = VERTICAL /PANEL VAR(TEMP)
/X1LENGTH=3.0 /YLENGTH=3.0 /X2LENGTH=3.0 /CHARTLOOK='NONE' /CATORDER VAR(TEMP) (ASCENDING
VALUES OMITEMPTY) /Histogram SHAPE
   = HISTOGRAM CURVE = ON X1INTERVAL AUTO X1START = 0.
EXE.
```

	Minimum	Maximum	Mean	Std Deviation	Variance	Valid N
15.00	1.07	8.07	4.55	1.56	2.43	N=90
17.00	-2.93	8.07	3.51	2.34	5.50	N=54
19.00	-2.93	8.07	2.49	2.30	5.30	N=83
21.00	-4.93	7.07	.32	1.90	3.60	N=100
23.00	-8.93	3.07	-2.43	2.45	5.99	N=86
25.00	-11.93	4.07	-1.63	2.30	5.29	N=122
27.00	-8.93	.07	-4.62	1.72	2.96	N=137
29.00	-11.93	-.93	-6.59	2.53	6.41	N=97
31.00	-15.93	-3.93	-7.82	2.57	6.63	N=54

TABLE 4.26 Continued

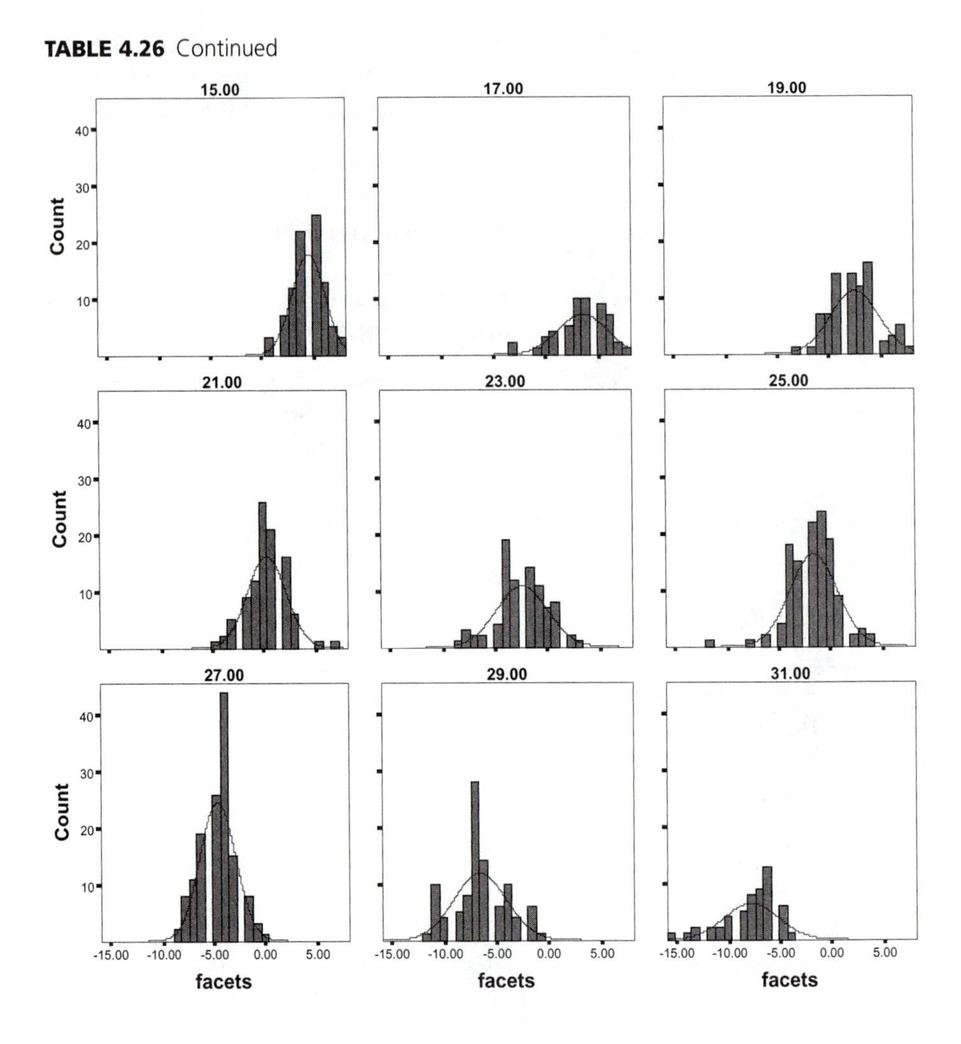

Graphs (INTERACTIVE) run by specifying TEMP as the panel variable. Statistics requested in the Table are minimum, maximum, mean, stddev, variance, and validn.

Sample sizes are unequal, ranging from 54 (for both 17° and 31° C) to 137 (for 27° C). The raw score distributions vary over the nine groups, as well, but normality of sampling distributions is assured by the very large number of df available with 9 groups and 823 observations: $\text{df}_{S/A} = N - a = 823 - 9 = 814$.

Negative values for the DV in some of the groups look suspicious until one realizes that the measurement of number of facets has been subjected to a logarithmic transformation. A constant added to each value before transformation would have avoided this anomaly.

Independence of errors is questionable because flies within a single incubator were not hatched individually. Evidence is needed that there were no systematic differences among incubators other than temperature and that environments within incubators differed only in temperature.

4.6.1.2 Outliers

Table 4.26 provides means, standard deviations, minima, and maxima for each group, which are necessary to identify outliers through standard scores (z scores) within each group. Standard scores for the minimum and maximum values are calculated within each group, and if the absolute value of one of them exceeds $|z| > 3.3$, the case with the maximum (or minimum score) is an outlier. Attention then turns to the case with the next highest (or lowest) score.

The first group has no outliers. Standard scores for the minimum and maximum values are $z = (1.07 - 4.55)/1.56 = -2.23$ and $z = (8.07 - 4.55)/1.56 = 2.26$, respectively. The first possible outlier encountered is in the 21° C group, with $z = (7.07 - 0.32)/1.90 = 3.55$ for the maximum score. Note, however, that the histogram in Table 4.26 shows this score is not notably disconnected from the remaining scores given the scale of the plot (cf. Section 3.4.4). Another possible outlier is the case with the minimum score in the 25° C group, with $z = -4.48$. This score, however, is indeed disconnected from the remaining scores. The next lowest score is not an outlier. Thus, the case with the minimum score in the sixth group is considered an outlier in the data set. Variability noted in temperature measurement over time and within incubators (this was 1924, after all) suggests that this extreme case may have had a different actual hatching temperature, indicating that the case may be deleted. Also, sample sizes in all groups are large enough that there is no problem in deleting this case from subsequent analyses, leaving a total $N = 822$.

4.6.1.3 Homogeneity of Variance

Variances are shown in Table 4.26. The smallest variance is in the group of flies hatched at the lowest temperature, Variance = 2.43, and the largest variance is in the group hatched at the highest temperature, Variance = 6.63. The resulting $F_{max} = 6.63/2.43 = 2.73$ (less than 10) poses no threat to homogeneity of variance, especially with a discrepancy in sample sizes no greater than about 2.5:1.

4.6.2 Planned Trend Analysis for One-Way Randomized-Groups Design

SPSS ONEWAY is selected for the planned analysis, as shown in Table 4.27. The COMPUTE instruction deletes the outlier from group 6. The model is specified in the FACETS BY TEMP, where FACETS (the DV) is analyzed according to the levels of TEMP (the IV). Then the Contrasts submenu is used to specify the trend analysis, POLYNOMIAL 1 (linear) through 5 (quintic). However, only the highest-order trend need be requested; all lower-order trends are included by default.

TABLE 4.27 Planned trend analysis for one-way randomized-groups design (SPSS ONEWAY syntax and selected output)

```
USE ALL.
COMPUTE filter_$=(id NE 535).
VARIABLE LABEL filter_$ 'id NE 535 (FILTER)'.
VALUE LABELS filter_$ 0 'Not Selected' 1 'Selected'.
FORMAT filter_$ (f1.0).
FILTER BY filter_$.
EXECUTE.
ONEWAY
    FACETS BY TEMP
    /POLYNOMIAL=5
    /MISSING ANALYSIS.
```

Oneway

ANOVA

facets

			Sum of Squares	df	Mean Square	F	Sig.
Between Groups	(Combined)		12383.033	8	1547.879	339.084	.000
	Linear Term	Unweighted	11084.807	1	11084.807	2428.276	.000
		Weighted	11967.990	1	11967.990	2621.749	.000
		Deviation	415.043	7	59.292	12.989	.000
	Quadratic Term	Unweighted	19.904	1	19.904	4.360	.037
		Weighted	40.105	1	40.105	8.786	.003
		Deviation	374.938	6	62.490	13.689	.000
	Cubic Term	Unweighted	3.917	1	3.917	.858	.355
		Weighted	.390	1	.390	.085	.770
		Deviation	374.548	5	74.910	16.410	.000
	4th-order Term	Unweighted	5.983	1	5.983	1.311	.253
		Weighted	1.958	1	1.958	.429	.513
		Deviation	372.589	4	93.147	20.405	.000
	5th-order Term	Unweighted	47.139	1	47.139	10.326	.001
		Weighted	80.129	1	80.129	17.553	.000
		Deviation	292.461	3	97.487	21.356	.000
Within Groups			3711.254	813	4.565		
Total			16094.287	821			

The output shows unweighted, weighted, and deviation tests for each trend. The unweighted analysis adjusts all trends for each other and is the analysis of choice. Type I error rate for each of the five trends to be evaluated is set at .01, producing a familywise error rate of about .08 (cf. Section 4.5.5.4.3). Each segment of the source table lists the Sum of Squares, df, Mean Square, F, and Sig. (significance level) for each trend.

TABLE 4.28 Data set output from NONCF3.SPS for effect size (R2) with 95% confidence limits (LR2 and UR2) for linear trend of facets over time, deviations from linear trend, and quintic trend

	fval	df1	df2	conf	lc2	ucdf	uc2	lcdf	power	R2	LR2	UR2
1	2428.28	1	813	.950	2132.156	.9750	2742.153	.0250	1.0000	.75	.72	.77
2	12.9900	7	813	.950	51.1735	.9750	126.7603	.0250	1.0000	.10	.06	.13
3	10.3300	1	813	.950	1.5542	.9750	26.8237	.0250	.8344	.01	.00	.03

The results show a strong linear trend: $F(1, 813) = 2428.276, p < .02$ (α is set at $.10/5 = .02$ to correct for increased familywise error in testing the five trends; cf. Section 4.5.5.4.3). Strength of association between the transformed number of facets and the linear trend of temperature is remarkably large for an experiment, $\eta^2 = 11084.81/16094.28 = .69$ (Equation 4.18). Smithson's (2003) procedure may be used to find partial η^2 and its confidence limits. Table 4.28 shows SPSS output produced by running NONCF3.SPS on NONCF.SAV. A line of values is filled into NONCF.SAV data file for the effect being evaluated. As shown in the first line in Table 4.28, values for the linear trend are F = 2428.28 (fval) and df = 1, 813 (df1, df2). The confidence level is set to .950 (conf). Table 4.28 also shows values for deviations from linear trend (line 2) and quintic trend (line 3). Running NONCF3.SPS fills in the remaining values in the table.

Thus, for the linear trend of facets, partial $\eta^2 = .75$, with 95% confidence limits from .72 to .77. There is also a statistically significant deviation from linear trend: $F(7, 813) = 12.99$. (See Linear Term Deviation in Table 4.27.) Effect size for these deviations is relatively small, however; partial $\eta^2 = .10$, with confidence limits from .06 to .13. For example, the quintic trend (5th order) is reliable but very small: $F(1, 813) = 10.33, p < .02$, partial $\eta^2 < .01$, with confidence limits from .00 to .03. If the remaining possible trends—6th, 7th, and 8th order—are desired, their coefficients may be entered into syntax. Alternatively, the entire 8th-order trend analysis may be done through SPSS UNIVARIATE, although output is less straightforward than that provided by ONEWAY. In any event, trends higher than cubic (which changes direction twice) can be difficult to interpret.

The statistically significant but small deviations from linear trends in this analysis may have resulted from the nature of the manipulation. The recorded temperature for each incubator was the nominal value, with variation of as much as 1° C above or below that nominal value. Variation occurred with time and was not constant throughout a single incubator. Thus, small deviations in trend should not preclude an interpretation of an overall linear trend of decrease in number of facets with increasing temperature.

The pooled standard error of the mean is calculated by

$$SE_M = \sqrt{\frac{\mathbf{MS}_{Within\ Groups}}{\mathbf{df}_{Within\ Groups}}} = 0.075$$

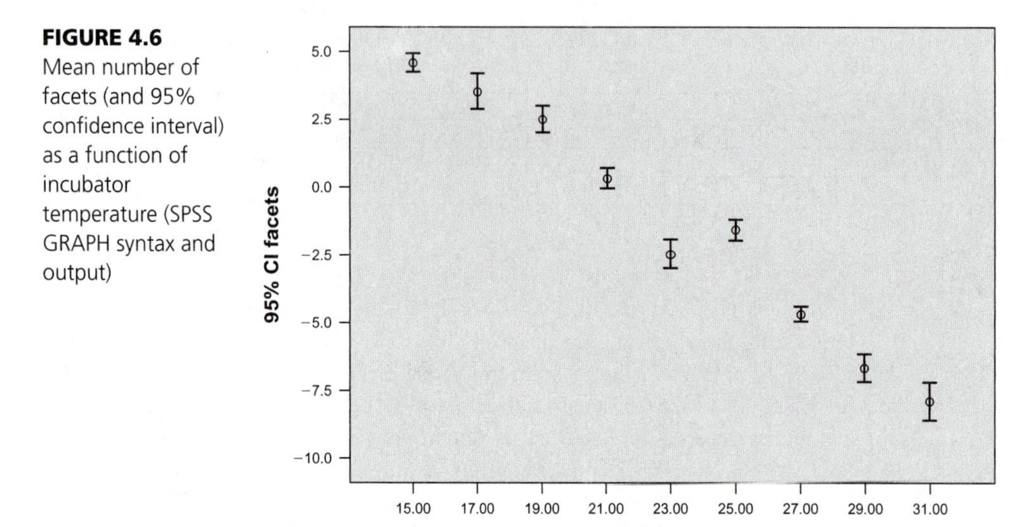

FIGURE 4.6 Mean number of facets (and 95% confidence interval) as a function of incubator temperature (SPSS GRAPH syntax and output)

(See Equation 2.6 for the basic formula for calculating the standard error of the mean.) Recall that a confidence interval may be built as two standard errors above and below each mean.

Figure 4.6 plots the means and 95% confidence intervals for the means (two times the standard error above and below each mean) for the nine groups, illustrating the strong linear trend, as produced by SPSS GRAPH.

Table 4.29 is a checklist for one-way randomized-groups ANOVA. An example of a results section, in journal format, follows for the study just described.

TABLE 4.29 Checklist for one-way randomized-groups ANOVA

1. Issues
   a. Independence of errors
   b. Unequal sample sizes
   c. Normality of sampling distributions
   d. Outliers
   e. Homogeneity of variance
2. Major analyses: Planned comparisons or omnibus *F*, when significant
   a. Effect size(s) and confidence interval(s)
   b. Parameter estimates (means and standard deviations or standard errors or confidence intervals for each group)
3. Additional analyses
   a. Post hoc comparisons
   b. Discussion of departure from homogeneity of variance, if appropriate

Results

A planned trend analysis for a one-way randomized-groups design was performed on a transformed value of the number of facets in eyes of newly hatched *Drosophila melanogaster* Meig heterozygotes. Flies were hatched in nine incubators, varying in temperature from 15° C to 31° C in increments of 2° C. Numbers of facets were transformed into units that make the measurement scale essentially logarithmic.

Total N of 823 was reduced to 822 with the deletion of an outlier. One fly from the 23° C incubator had an unusually small number of facets ($z = -4.48$). The assumption of homogeneity of variance was met, with sample sizes ranging from 54 to 137.

The planned trend analysis at $\alpha = .02$ for each comparison up to quintic showed a strong negative linear trend of incubator temperature: $F(1, 813) = 2428.28$, $p < .05$, partial $\eta^2 = .69$ with 95% confidence limits from .72 to .77, $SE_M = 0.075$. Figure 4.6 shows mean (transformed) number of facets as a function of temperature, with standard errors.

There was a statistically significant deviation from linear trend, as well: $F(7, 813) = 12.99$, $p < .02$, partial $\eta^2 = .10$ with confidence limits from .06 to .13. For example, the quintic trend was statistically significant: $F(1, 813) = 10.33$, $p < .05$, $\eta^2 = .01$ with confidence limits from .00 to .03, $SE_M = 0.075$.

Deviations from the negative linear trend may be explained, at least in part, by the inexact nature of the manipulation. Groups were classified by nominal incubator temperature, which varied over time and within each incubator by as much as 1° C above or below the nominal value. Thus, small deviations in trend should not preclude an interpretation of an overall decrease in the logarithm of the number of facets with increasing temperature.

4.7 COMPARISON OF PROGRAMS

Each package has a number of programs for ANOVA, almost all of which may be used for one-way randomized-groups ANOVA. Table 4.30 reviews one program from each package, typically the simplest one to use for this design. Subsequent chapters review the other programs.

4.7.1 SPSS Package

SPSS has several programs for ANOVA: ONEWAY and a group of GLM (general linear model) programs.[18] Table 4.30 reviews ONEWAY, which is devoted solely to one-way randomized-groups ANOVA and is the easiest of the SPSS ANOVA programs to use. ONEWAY is accessed in Windows through Analyze > Compare Means > One-way ANOVA.

ONEWAY provides a comprehensive set of descriptive statistics, including the 95% confidence interval for the mean for each group. Levene's test for homogeneity of variance is available, along with separate-variance tests for comparisons if homogeneity of variance is violated. A wide variety of post hoc comparison procedures is available. Among programs reviewed in Table 4.30, ONEWAY is the only one with a provision for handling unequal n by specifying equal weighting[19] of sample sizes for post hoc comparisons (the harmonic mean). Trend analysis is easily specified, as illustrated in Section 4.6.2.

4.7.2 SAS System

SAS has two programs for ANOVA: GLM and ANOVA. The ANOVA program runs faster and uses less storage but is limited to equal sample sizes in each group. The one-way omnibus test is accurate in both programs, but ANOVA may produce inaccurate results for comparisons. Therefore, only GLM is reviewed in Table 4.30 and in subsequent chapters.

PROC GLM produces two source tables for omnibus ANOVA—one table (designated the model source of variance) that reports a test of all effects combined and another table that reports a test of each effect separately. These tables provide redundant information in one-way omnibus ANOVA where there is only one effect. Several descriptive statistics are available by request.

This program has a large number of post hoc comparison procedures and provides confidence intervals for differences among means for some of those procedures on request. A measure of effect size, η^2 (labeled R-Square), is provided for the omnibus ANOVA.

[18] SPSS GLM is meant to replace MANOVA but lacks some of its features, as seen in subsequent chapters. The earlier program is available in syntax (batch) format.

[19] These weights are different from the coefficients used to form specific comparisons.

TABLE 4.30 Comparison of SPSS, SAS, and SYSTAT programs for one-way randomized-groups ANOVA

Feature	SPSS ONEWAY	SAS GLM[a]	SYSTAT ANOVA and GLM[a]
Input			
Specify sample size estimate for post hoc comparisons	Harmonic	No	No
Specify α level for post hoc comparisons	Yes	ALPHA	No[b]
Special specification for trend (polynomial) analysis	Polynomial	No	Polynomial
Separate-variance tests for comparisons	Yes	No	Separate
Specify error terms for comparisons	No	Yes	Yes
Output			
ANOVA table	Yes	Yes	Yes
Means for each group	Yes	Yes	LS Mean
Standard deviations for each group	Yes	Yes	No
Standard error for each group	Yes	Yes	No
Pooled within-group standard error	No	Std Err LSMEANS	Yes
Pooled within-group standard deviation	No	No	No
Minimum and maximum values for each group	Yes	No	No
Descriptive statistics for all groups combined	Yes	No	No
Confidence intervals for each mean	Yes	CLM	No
Error coefficient of variation	No	Yes	No
Test(s) for homogeneity of variance	Yes	Yes	No
$R^2(\tilde{\eta}^2)$	No	R-square	Squared multiple R
User-specified comparisons	Yes	Yes	Yes
Tukey HSD test for pairwise differences	TUKEY	TUKEY	TUKEY
Tukey-b multiple range test for pairwise differences	BTUKEY	No	No
Dunnett test for control vs. treatment	DUNNETT	DUNNETT	DUNNETT[c]
Dunnett C and T3 tests for pairwise differences with heterogeneity of variance	C, T3	No	No
Scheffé test for pairwise differences	SCHEFFE	SCHEFFE	SCHEFFE
LSD/Fisher (pairwise *t* tests or confidence intervals)	LSD	LSD	Yes
Bonferroni tests for pairwise differences	BONFERRONI	BON	BONF
Duncan test for pairwise differences	DUNCAN	DUNCAN	No
Student-Newman-Keuls test for pairwise differences	SNK	SNK	No
Gabriel test for pairwise differences	GABRIEL	GABRIEL	No

TABLE 4.30 Continued

Feature	SPSS ONEWAY	SAS GLM[a]	SYSTAT ANOVA and GLM[a]
Ryan-Einot-Gabriel-Welch F step-down procedure for multiple comparisons	FREGW	REGWF	No
Ryan-Einot-Gabriel-Welch Range step-down procedure for multiple comparisons	QREGW	REGWQ	No
Hochberg multiple comparison and range test	GT2	GT2	No
Waller-Duncan multiple comparison test	WALLER	WALLER	No
Tamahane pairwise comparisons for unequal *n*	T2	No	No
Games-Howell pairwise comparisons for unequal *n*	GH	No	No
Sidak tests for pairwise differences	SIDAK	SIDAK	No
Sidak correction for control vs. treatment	No	No	No
Confidence limits for pairwise differences	Yes	CLDIFF	No
Probability values for pairwise differences	Yes	PDIFF	Yes
Confidence limits for predicted values for each case	No	Yes	No
Outliers (studentized residual criterion)	No	P	Yes
Durbin-Watson statistic	No	P	Yes
Plots			
Means for each group	Yes	No	Yes
Standard errors shown on plot of means	No	No	Yes
Confidence interval for each group	No	No	No
Histogram of residuals	No	No	No
Normal probability plot of residuals	No	No	No
Residuals vs. predicted (fitted) values	No	No	No
Residuals vs. sequence of data	No	No	No
Residuals vs. any variable	No	No	No
Saved to file			
Residuals	No	Yes	Yes
Predicted (fitted, adjusted) values (i.e., mean for each case)	No	Yes	Yes
Additional statistics, including those for outliers	No	Yes	Yes
Means, standard deviations, and frequencies	No	Yes	No

[a] Additional features reviewed in subsequent chapters.
[b] Probability values provided for comparisons.
[c] GLM only

The program provides an indication of outliers through the studentized residual criterion and also evaluates autocorrelation of cases through the Durbin-Watson statistic. Autocorrelation causes difficulty only if order of cases is related to treatments, which should not occur when treatments are randomly assigned. Although GLM and ANOVA produce no plots, any of the values saved to file may be plotted through the PROC PLOT.

4.7.3 SYSTAT System

SYSTAT ANOVA is a subset of GLM for analysis of variance with a different setup but largely the same output. The few features that are different are noted in Table 4.30.

SYSTAT ANOVA/GLM provides the mean for each group, accompanied by a pooled standard error of the mean, as long as the option chosen for output length using ANOVA is PRINT=MEDIUM or PRINT=LONG. The standard error of the mean differs among groups only if sample sizes are unequal. Additional descriptive statistics are requested through the STATS program. An effect size measure, η^2 (labeled Multiple R-square), is reported for the omnibus analysis. Outliers (identified by the studentized residual criterion) are highlighted, and the Durbin-Watson autocorrelation statistic is given in the event that the order of cases is not random. Four of the most popular post hoc comparison procedures are available. Separate-variance tests are available for specific comparisons but not for post hoc procedures. SYSTAT is handy for trend analysis when there are many levels of the IV. You can specify any order of a polynomial without needing to enter coefficients.

4.8 PROBLEM SETS

4.1 Do an analysis of variance on the following data set. Assume the IV is minutes of presoaking a dirty garment, and the DV is rating of cleanliness.

Presoak Time

a_1	a_2	a_3	a_4
5 min	10 min	15 min	20 min
2	3	8	6
6	9	11	7
6	9	9	9
6	5	9	6
5	7	6	9
3	5	12	9

a. Test for homogeneity of variance.

b. Provide a source table.

c. Check results by running the data set through at least one computer program.
d. Calculate effect size and find the confidence interval using Smithson's software.
e. Write a results section for the analysis.
f. Do a trend analysis with an appropriate post hoc adjustment.

4.2 Do an analysis of variance on the following pilot data set.

a_1	a_2	a_3	a_4	a_5
6	5	5	6	6
5	6	5	5	7
5	6	7	9	7
	7	5	7	6
	6		7	
			6	

a. Provide a source table.
b. Check results by running the data set through at least one statistical software program.
c. How many cases would you need to run to get a statistically significant main effect at $\alpha = .05$ and $\beta = .20$? Use a statistical program if available, otherwise assume that the minimum difference you are seeking is between a_1 and a_5.

4.3 Do an analysis of variance for the following, with $n = 10$. (Recall that $MS_{S/A}$ is the average of the group variances when sample sizes are equal.)

	a_1	a_2	a_3
Mean	20	22	28
Variance	42	46	43

a. Provide a source table.
b. Calculate effect size.
c. Do a Tukey test at $\alpha = .05$.
d. Do a post hoc test of a_3 versus the other two groups.

4.4 Suppose a researcher is interested in the effects of hours of sleep on performance on a statistics exam. Students are randomly assigned to get 2, 4, 6, or 8 hours of sleep prior to the exam. The DV is exam score. Assume that a_1 is 2 hours of sleep, a_2 is 4 hours of sleep, and so on. Write the coefficients you would need to answer the following questions.

a. Is there a linear trend of exam scores for hours of sleep?
b. Do students who sleep 8 hours do better than those who sleep fewer hours?
c. Is there a difference in average performance between those who sleep 2 hours and those who sleep 4 hours?
d. Do students who sleep 2 or 4 hours do worse than those who sleep 6 or 8 hours?
e. Write an orthogonal set of coefficients that makes the most sense to you given the meaning of the levels of the IV.

4.5 Assume that a researcher runs a drug study for three analgesics against two types of control groups. There are five groups in all (a_1 to a_5, respectively): a waiting list control group, a placebo control group, a group that receives the old analgesic, and two other groups, each assigned to one of the two new analgesics. The DV is pain relief. Write the coefficients you need to answer each of the following questions.

a. Do the two types of control groups produce different levels of pain relief?

b. Are the combined control groups different from the combined analgesic groups?

c. Does the old analgesic differ from the combination of newer analgesics?

d. Do the two newer analgesics differ from each other?

e. Is the response of the first of the new analgesics (a_4) different from the placebo group?

f. Is the response of the second of the new analgesics (a_5) different from the placebo group?

5 Factorial Randomized-Groups, Fixed-Effects Designs

5.1 GENERAL PURPOSE AND DESCRIPTION

The factorial randomized-groups, fixed-effects design is a generalization of the one-way, randomized-groups design to a study in which there is more than one IV. For example, the first IV in a two-way design might be type of on-the-job personnel training with three levels, such as a no-training control group, a text-based training program, and a video-based training program. A second IV might be job classification with two levels, such as clerical personnel and security personnel. The DV might be supervisor's overall job performance rating. This example could also be referred to as a 3×2 factorial design, because there are two IVs—one with three levels, and the other with two levels. A factorial design with three IVs with three, two, and four levels, respectively, is called a $3 \times 2 \times 4$ design; the order in which IVs are listed is arbitrary.

The omnibus ANOVA for this design develops separate statistical tests for each IV and their interaction. For type of on-the-job training, potential mean differences between the video-based training program, the text-based program, and the no-training program are examined. For job classification, the potential mean difference between those trained for clerical jobs and those trained for security jobs is examined. The interaction of type of training and job classification examines potential differences in the *pattern* of response of clerical and security personnel to the different types of on-the-job training. For example, security personnel might respond more favorably to the video-based program, whereas clerical personnel might respond equally to both types of training. Interpretation of interactions is discussed in detail in Section 5.6.1.

The minimum number of IVs for a factorial design is two. A design is factorial when there are two or more IVs and all levels of one IV are used in combination with all levels of the others. The example is factorial, because both security and clerical trainees receive no training, text-based training, or video-based training. If, for example, security personnel never receive text-based training, the design is not fully factorial. It is a randomized-groups design, because different security and clerical trainees are randomly assigned to each of the three levels of training. In general, the distinguishing characteristics of a completely randomized, fixed-effects design are that (1) different cases are randomly assigned to each *combination* of levels of the IVs, (2) each research unit is measured once, and (3) all the levels of all the IVs are fixed (deliberately chosen by the researcher and not randomly chosen from among a range of levels). Such designs are usually also factorial (fully crossed), because cases receive all possible combinations of levels of the IVs. Exceptions to the latter are the dangling group design discussed in Section 5.5.2 and the incomplete designs of Chapters 9 and 10.

There may be any number of IVs with any number of levels each. However, fully crossed, randomized-groups designs become unwieldy when there are many IVs or many levels for each IV. The number of combinations of levels of IVs is a product of the levels of each. For instance, if there are three IVs (a three-way design), the first with three levels, the second with four, and the third with two, then there are $3 \times 4 \times 2 = 24$ combinations to present. Running a study of this size can become a logistical nightmare, and the number of cases required can become excessive. For example, 120 cases are needed if 5 cases are randomly assigned to each combination of levels.

A factorial randomized-groups design is an experiment if at least one of the IVs is manipulated with random assignment to its levels. In the two-way example, type of training is an experimental IV and job classification is nonexperimental. The omnibus ANOVA is also appropriate for nonexperimental and quasi-experimental designs, but interpretation of results is limited to those appropriate for nonexperimental designs.

5.2 KINDS OF RESEARCH QUESTIONS

5.2.1 Main Effects of the IVs

Averaging across all combinations of levels of all other IVs, are there statistically significant mean differences among the levels of this IV? That is, when the effects of all other IVs are held constant by averaging across their effects, do the means at different levels of this IV differ by more than would be expected by chance? For example, is there a mean difference in job performance rating depending on the training group to which an employee is assigned? Is there a mean difference in job performance ratings between clerical and security personnel?

These questions are answered by applying separate F tests to each IV, as demonstrated in Section 5.4. Procedures discussed in Section 1.1.2 are used to hold constant all other effects in an experimental design.

5.2.2 Effects of Interactions among IVs

Holding the effects of other IVs constant, does the pattern of means for one IV change at different levels of another IV? For example, does the pattern of differences in means among training programs depend on whether clerical or security personnel are being trained? This is called the training program-by-job classification interaction.

F tests are conducted separately for each interaction (if there is more than one). In addition, except for the use of a common error term, the F tests are independent of tests of main effects of individual IVs if there are equal numbers of cases at all combinations of levels and if the design is fully crossed. There is an interaction for each combination of IVs; thus, if there are three IVs (A, B, and C), then there is an $A \times B$ interaction, an $A \times C$ interaction, a $B \times C$ interaction, and an $A \times B \times C$ interaction. The greater the number of IVs, the greater the number of interactions generated. If, for example, age of trainee is added to the training program and job classification IVs, there are two additional two-way interactions (training program by age group and job classification by age group), plus the three-way interaction (training program by job classification by age group). The three-way interaction asks whether the difference in effectiveness of training programs for clerical versus security employees depends on their age group. Two-way and higher-order interactions are discussed more fully in Section 5.6.1.

5.2.3 Specific Comparisons

If a difference in means among more than two groups is found, what is the source of the difference? For example, is the text-based training program more effective than no training at all? Is the video-based program better than the text-based program? Is the average performance of trained employees better than that of employees without on-the-job training? Is there a difference between text- and video-based programs for clerical employees? These questions may be answered by applying one of a wide variety of post hoc comparison procedures, some of which are discussed in Section 5.6.4.

Alternatively, some of these specific questions may be answered by conducting a planned set of comparisons instead of testing omnibus main effects and interactions. The advantage of doing so is greater power for these tests when planned than when hypotheses are posed after omnibus differences are observed. Planned comparisons are also discussed in Section 5.6.4.

5.2.4 Parameter Estimates

What population means are expected given a particular treatment or combination of treatments? What job performance rating would we expect for a clerical worker who participated in a video-based training program? Parameter estimates reported in factorial designs (as well as in one-way designs) are means, standard deviations, standard errors, and/or confidence intervals for the mean.

Parameter estimates for main effects are marginal means,[1] or the mean of each level of the IV averaged over all the levels of the other IV(s). When sample sizes are not the same for all cells, there are different ways to average over the other IVs. Section 5.6.5 discusses this and other complications of studies that have unequal sample sizes. Marginal means most often appear in narrative format in a results section, but they may be presented in tabular or graphic format, particularly if the IV has many levels or if those levels represent some underlying quantitative dimension, such as amount of treatment.

Parameter estimates for interactions are cell means, which are reported in narrative format, shown in a table, or plotted in a graph. Graphs are typically used with a statistically significant interaction. Frequently these are line graphs,[2] as shown in the examples in Section 5.6.1.

5.2.5 Effect Sizes

As for one-way ANOVA, effect sizes are typically reported in terms of proportion of variance in the DV associated with a main effect or interaction. Effect sizes are usually reported for all statistically significant effects. Sometimes effect sizes and their confidence intervals are reported for all effects, regardless of statistical significance. Section 5.6.2 demonstrates calculation of effect sizes, as well as software for determining confidence intervals around them.

5.2.6 Power

How high is the probability of showing main effects and interactions if, indeed, IV(s) have effects? For example, what is the probability of finding a difference in performance rating among groups given different training types, if the difference actually exists? What is the probability of finding a difference between clerical and security workers? What is the probability of finding that the pattern of performance ratings among training types is different for clerical and security workers?

[1] Parameter estimates for main effects may also sometimes be standard deviations or standard errors.

[2] Bar graphs are technically more correct when the levels of the IV are qualitatively different, but they are sometimes more difficult to interpret.

Estimates of power made prior to running a study are used to help determine sample size, as they are for a one-way design. Estimates of power made after completing a study help determine why research hypotheses were not supported. Power is discussed in Section 5.6.3.

5.3 ASSUMPTIONS AND LIMITATIONS

5.3.1 Theoretical Issues

Theoretical issues in a factorial design are identical to those in a one-way design. Thus, issues of causal inference and generalizability are of the same concern in factorial designs.

5.3.2 Practical Issues

Whether one-way or factorial, the ANOVA model assumes normality of sampling distributions, homogeneity of variance, independence of errors, and absence of outliers. Unequal sample sizes pose additional problems for factorial ANOVA and are discussed in Section 5.6.5.

5.3.2.1 Normality of Sampling Distributions

Issues of normality of sampling distributions apply to data in the *cells* (combinations of levels of IVs) of a factorial design, where n is the sample size of each cell. However, the criteria are the same as for a one-way design (Section 3.4.1). Similarly, solutions for questionable normality are the same as for a one-way design, except that fewer nonparametric tests are available. If a test of differences between medians is desired, a generalization of the χ^2 test of independence is available in which each IV is a factor and the DV from ANOVA becomes an additional factor when it is transformed into a categorical variable (using a median or some other split). Thus, a two-way ANOVA becomes a three-way χ^2 (with the DV as the additional factor), and so on. Tabachnick and Fidell (2007, Chapter 16) discuss multiway frequency analysis.

5.3.2.2 Homogeneity of Variance

Homogeneity of variance is usually tested in the cells of a factorial design. The guidelines for assessing compliance with the assumption and the solutions for violation are the same as for a one-way design, except that cells formed by the combination of IVs are tested instead of levels of a single IV. In other words, homogeneity of variance is tested in the six cells of a factorial design (formed by combining, say, one IV that has two levels with a second IV that has three levels), the same way as in a one-way design in which the IV has six levels (cf. Sections 3.4.3 and 4.5.5).

Procedures are also available to test homogeneity of variance separately for each main effect and interaction of a factorial design, should these be desired.

5.3.2.3 Independence of Errors

The ANOVA model assumes that errors (deviations of individual scores around their respective means) are independent of one another, both within each cell and between cells of a design. The assumption is identical to that for a one-way design but is applied to cells rather than to levels of a single IV. This assumption is evaluated in the same way as for one-way designs (cf. Section 3.4.2.)

5.3.2.4 Absence of Outliers

Univariate outliers can occur in the DV in one or more cells of the design. Issues surrounding testing and dealing with outliers are the same as for a one-way design (Section 3.4.4), with outliers defined by their cells rather than by levels of a single IV.

5.4 FUNDAMENTAL EQUATIONS

Factorial designs consist of two or more IVs, each of which has at least two levels. The design is factorial if all levels of one IV occur in combination with all levels of the other IVs. Equation 5.1 shows the GLM in traditional deviation format for two-way randomized-groups ANOVA:

$$Y = \mu + \alpha + \beta + \alpha\beta + e \tag{5.1}$$

where α = deviation of a score, Y, from the grand mean, μ, due to the main effect of the first IV, A;
β = deviation of a score due to the main effect of the second IV, B; and
$\alpha\beta$ = is the deviation of a score due to the interaction—the effect of the combination of A and B on Y above and beyond their individual main effects on Y.

Thus, a factorial design with two IVs provides three tests for the price of one: two that you would get from two separate one-way studies, one on each IV, plus a freebie. Say that the first IV is the same as in the one-way design of Section 4.4: The IV is view from the room (parking lot, ocean, and ski slope), and the DV is rating of the vacation. The test of the main effect of type of view is the same test as in the one-way ANOVA. If the second IV is profession of rater (say, administrators, belly dancers, and politicians), you also get a test of possible mean differences in rating of vacation associated with type of profession. The third test (of interaction) is not available if you simply run two studies, one for each IV. The test of interaction asks whether the effect of one IV depends on the level of the other IV. For example, does the effect of view on rating of vacation differ for the different professions?

5.4.1 Allocation of Cases

Cases are randomly assigned to each level of an experimentally manipulated IV. Table 5.1 shows assignment of cases to two treatments, A and B, with two levels each. Twelve different cases are randomly assigned to the four cells representing combinations of levels of A and B, leaving $n = 3$ for each cell. Recall that subscripts for s do not imply sequential assignment but are simply for convenience.

If the design has one manipulated and one nonmanipulated IV, cases are randomly assigned to levels of the manipulated IV within each level of the nonmanipulated IV. This is sometimes called a *treatment by blocks design,* or a blocking design. Table 5.2 shows a design with a manipulated IV, Treatment B, and a nonmanipulated IV, Case Type A. Cases are classified into either a_1 or a_2. Blocking variables test the generalizability of the effects of the manipulated IV through the test of interaction. Does the effect of the manipulated IV depend on the level of the blocking variable? In our current example, type of profession is a blocking variable—the design is said to be blocked on profession; cases are randomly assigned to view but classified as to profession. Common types of blocking variables are environmental, situational, or correlational variables associated with cases.

The six cases classified as a_1 are randomly assigned to the two levels of B, as are the six cases classified as a_2. If there is one nonmanipulated and two manipulated IVs, cases are randomly assigned to every combination of levels of the manipulated IVs within each level of the nonmanipulated IV, and so on.

In a completely nonexperimental design, levels of both A and B are observed; neither is manipulated. Cases are classified into all levels on the basis of their characteristics with respect to A and B. For example, cases are classified into A on the basis of profession and into B on the basis of, say, age. The statistical analysis is identical for the fully randomized design, the treatment by blocks design, and the fully nonexperimental design; only the interpretation with respect to causality differs for the variants of the design.

TABLE 5.1 Assignment of cases in two-way randomized-groups ANOVA with both IVs manipulated

	Treatment B	
Treatment A	b_1	b_2
	s_1	s_7
a_1	s_2	s_8
	s_3	s_9
	s_4	s_{10}
a_2	s_5	s_{11}
	s_6	s_{12}

TABLE 5.2 Assignment of cases in two-way randomized-groups ANOVA with one nonmanipulated IV

	Treatment B	
Case Type A	b_1	b_2
	$s_{a1,b1,1}$	$s_{a1,b2,4}$
a_1	$s_{a1,b1,2}$	$s_{a1,b2,5}$
	$s_{a1,b1,3}$	$s_{a1,b2,6}$
	$s_{a2,b1,1}$	$s_{a2,b2,4}$
a_2	$s_{a2,b1,2}$	$s_{a2,b2,5}$
	$s_{a2,b1,3}$	$s_{a2,b2,6}$

5.4.2 Partition of Sources of Variance

The SS_{bg} of Equation 3.4 (expressed as in Equation 3.3) is further partitioned into SS associated with the first IV, A; SS associated with the second IV, B; and SS associated with the interaction, AB (discussed in Section 5.6.1). That is,

$$SS_{bg} = SS_A + SS_B + SS_{AB} \tag{5.2}$$

Or,

$$n_{jk}\sum_j\sum_k(\overline{Y}_{AB} - GM)^2 = n_j\sum_j(\overline{Y}_A - GM)^2 + n_k\sum_k(\overline{Y}_B - GM)^2$$
$$+\left[n_{jk}\sum_j\sum_k(\overline{Y}_{AB} - GM)^2 - n_j\sum_j(\overline{Y}_A - GM)^2 - n_k\sum_k(\overline{Y}_B - GM)^2\right]$$

The sum of squared differences between cell means ($\overline{Y}_{AB}$) and the grand mean is partitioned into (1) sum of squared differences between means associated with different levels of A ($\overline{Y}_A$) and the grand mean; (2) sum of squared differences between means associated with different levels of B ($\overline{Y}_B$) and the grand mean; and (3) sum of squared differences associated with combinations of A and B ($\overline{Y}_{AB}$) and the grand mean, from which differences associated with A and B are subtracted. Each n is the number of scores composing the relevant marginal or cell mean: n_{jk} is the number of scores in each cell, n_j is the number of scores in each level of A, and n_k is the number of scores in each level of B. Note that the interaction is what is left over after influences associated with each IV are subtracted. That is, any differences among cell means that remain after marginal mean differences are eliminated are associated with the interaction.

The full partition for the factorial randomized-groups design is

$$\sum_i\sum_j\sum_k(Y_{ijk} - GM)^2 = n_j\sum_j(\overline{Y}_A - GM)^2 + n_k\sum_k(\overline{Y}_B - GM)^2$$
$$+\left[n_{jk}\sum_j\sum_k(\overline{Y}_{AB} - GM)^2 - n_j\sum_j(\overline{Y}_A - GM)^2 - n_k\sum_k(\overline{Y}_B - GM)^2\right] \tag{5.3}$$
$$+\sum_i\sum_j\sum_k(Y_{ijk} - \overline{Y}_{AB})^2$$

or

$$SS_T = SS_A + SS_B + SS_{AB} + SS_{S/AB}$$

The notation for the error term is S/AB, which indicates subjects (cases) within the groups formed by combinations of A and B—subjects *nested* within the various AB combinations.

Total degrees of freedom remain the number of scores (expressed as N or abn) minus 1, lost when the grand mean is estimated (cf. Equation 3.6). Degrees of freedom for A are the number of levels of A minus 1, lost because the $\overline{Y}_{a_j}$ average to the grand mean; so $a - 1$ of the means can take on any values whatever, but the last mean is fixed.

$$\text{df}_A = a - 1 \tag{5.4}$$

Similarly, degrees of freedom for B are the number of levels of B minus 1, lost because the $\overline{Y}_{b_k}$ also average to the grand mean; so the first $b - 1$ of them can take on any value, but the last one is fixed.

$$\text{df}_B = b - 1 \tag{5.5}$$

Degrees of freedom for the AB interaction are the degrees of freedom for A times the degrees of freedom for B.

$$\text{df}_{AB} = (a - 1)(b - 1) \tag{5.6}$$

Finally, degrees of freedom for error are the total degrees of freedom minus ab, the number of cell means estimated.

$$\text{df}_{S/AB} = N - ab \tag{5.7}$$

Verifying the equality of degrees of freedom,

$$N - 1 = a - 1 + b - 1 + (a - 1)(b - 1) + N - ab$$

Rearranging terms and multiplying terms within parentheses,

$$N - 1 = N - ab + a - 1 + b - 1 + ab - a - b + 1$$

Figure 5.1 diagrams the partition of sums of squares and degrees of freedom.

Mean squares are formed by dividing sums of squares by degrees of freedom, as for one-way designs, but now there are more of them:

$$\text{MS}_A = \frac{\text{SS}_A}{\text{df}_A} \qquad \text{MS}_B = \frac{\text{SS}_B}{\text{df}_B}$$
$$\text{MS}_{AB} = \frac{\text{SS}_{AB}}{\text{df}_{AB}} \qquad \text{MS}_{S/AB} = \frac{\text{SS}_{S/AB}}{\text{df}_{S/AB}} \tag{5.8}$$

F ratios are formed for each of the three sources of variance to be evaluated: A, B, and AB. Each F ratio is formed by dividing the appropriate MS by the estimate of random-error variation, $\text{MS}_{S/AB}$.

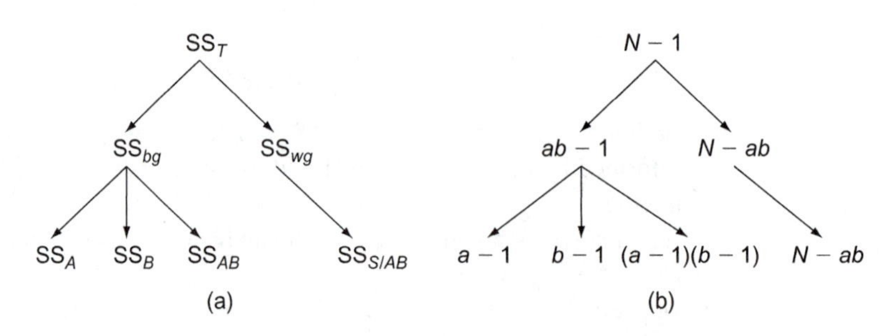

FIGURE 5.1 Partition of (a) sums of squares and (b) degrees of freedom for two-way factorial randomized-groups ANOVA

F_A evaluates differences among marginal means for levels of A, called the main effect of A. The null hypothesis for this test is the same as for a one-way ANOVA, but is now expressed as $\mu_{a_1} = \mu_{a_2} = \cdots = \mu_{a_j}$. The obtained F_A is

$$F_A = \frac{\text{MS}_A}{\text{MS}_{S/AB}} \qquad \text{df} = (a-1),\ (abn - ab) \tag{5.9}$$

Obtained F_A is tested against critical F from Table A.1, with numerator df $= a - 1$ and denominator df $= N - ab$. The null hypothesis is rejected if the obtained F_A is equal to or exceeds critical F.

F_B evaluates differences among marginal means for levels of B, called the main effect of B. The null hypotheses for this test is that $\mu_{b_1} = \mu_b = \cdots = \mu_{b_k}$.

$$F_B = \frac{\text{MS}_B}{\text{MS}_{S/AB}} \qquad \text{df} = (b-1),\ (abn - ab) \tag{5.10}$$

Obtained F_B is tested against critical F from Table A.1, with numerator df $= b - 1$ and denominator df $= N - ab$. The null hypothesis is rejected if the obtained F_B is equal to or exceeds critical F.

F_{AB} evaluates differences that remain among cell means after differences among marginal means are subtracted. This is called the AB interaction. The null hypothesis is that *remaining* cell means are all equal—for example, $\mu_{a_1b_1} = \mu_{a_1b_2} = \mu_{a_2b_1} = \mu_{a_2b_2}$ after effects of A and B are removed—or, alternatively, that the mean differences between the levels of B are the same at the different levels of A—for example, that $(\mu_{a_1b_1} - \mu_{a_1b_2}) = (\mu_{a_2b_1} - \mu_{a_2b_2})$. A deeper understanding of the meaning of interaction is available through the regression approach, as discussed in Section 5.4.4.

$$F_B = \frac{\text{MS}_{AB}}{\text{MS}_{S/AB}} \qquad \text{df} = (a-1)(b-1),\ (abn - ab) \tag{5.11}$$

5.4.3 Traditional ANOVA Approach (3 × 3)

Table 5.3 shows an unrealistically small, silly, hypothetical data set in which the rating of a vacation is evaluated as a function of length of vacation and professional identification of the rater. The first IV, A, is professional identification, with three levels (administrators, belly dancers, and politicians). The second IV, B, is length of vacation, also with three levels (a one-week trip, a two-week trip, or a three-week trip). Each block of nine professionals is randomly assigned to one of the three vacation lengths, resulting in $n = 3$ for each combination of levels of A and B. Thus, vacation length is an experimental IV, and profession is not.

A matrix of sums, such as that in Table 5.4, is very helpful to work with when calculating sums of squares. The table contains sums for the cells (AB), as well as sums for the margins of both A and B. Table 5.5 shows computational equations for sums of squares for A, B, AB, and S/AB.

Note that for each quantity in Table 5.5, the divisor is the number of cases added together to produce the sums that are to be squared in the numerator. In the first

TABLE 5.3 Hypothetical data set for 3 × 3 ANOVA

A: Profession	B: Vacation Length		
	b_1: 1 week	b_2: 2 weeks	b_3: 3 weeks
a_1: Administrators	0	4	5
	1	7	8
	0	6	6
a_2: Belly dancers	5	5	9
	7	6	8
	6	7	8
a_3: Politicians	5	9	3
	6	9	3
	8	9	2

$\sum Y^2 = 0^2 + 1^2 + \cdots + 2^2 = 1046$
$a = 3, b = 3, n = 3$

TABLE 5.4 Cell totals and marginal sums for hypothetical data set for 3 × 3 ANOVA

A: Profession	B: Vacation Length			
	b_1: 1 week	b_2: 2 weeks	b_3: 3 weeks	Marginal sums for A
a_1: Administrators	1	17	19	$A_1 = 37$
a_2: Belly dancers	18	18	25	$A_2 = 61$
a_3: Politicians	19	27	8	$A_3 = 54$
Marginal sums for B	$B_1 = 38$	$B_2 = 62$	$B_3 = 52$	$T = 152$

TABLE 5.5 Computational equations for sums of squares for two-way ANOVA

$$\text{SS}_A = \frac{\sum A^2}{bn} - \frac{T^2}{abn} = \frac{37^2 + 61^2 + 54^2}{9} - \frac{152^2}{27}$$

$$\text{SS}_B = \frac{\sum B^2}{an} - \frac{T^2}{abn} = \frac{38^2 + 62^2 + 52^2}{9} - \frac{152^2}{27}$$

$$\text{SS}_{AB} = \frac{\sum (AB)^2}{n} - \frac{\sum A^2}{bn} - \frac{\sum B^2}{an} + \frac{T^2}{abn} = \frac{1^2 + 17^2 + 19^2 + 18^2 + 18^2 + 25^2 + 19^2 + 27^2 + 8^2}{3}$$
$$- \frac{37^2 + 61^2 + 54^2}{9} - \frac{38^2 + 62^2 + 52^2}{9} + \frac{152^2}{27}$$

$$\text{SS}_{S/AB} = \sum Y^2 - \frac{\sum (AB)^2}{n} = 0^2 + 1^2 + 0^2 + 4^2 + \cdots + 3^2 + 2^2$$
$$- \frac{1^2 + 17^2 + 19^2 + 18^2 + 18^2 + 25^2 + 19^2 + 27^2 + 8^2}{3}$$

$$\text{SS}_T = \sum Y^2 - \frac{T^2}{abn} = 0^2 + 1^2 + 0^2 + 4^2 + \cdots + 3^2 + 2^2 - \frac{152^2}{27}$$

TABLE 5.6 Sums of squares for traditional 3 × 3 ANOVA with three levels of each IV

SS_A	$= 89.55 - 855.7$	$= 33.85$
SS_B	$= 888 - 855.7$	$= 32.30$
SS_{AB}	$= 1026 - 889.55 - 888 + 855.7$	$= 104.15$
$SS_{S/AB}$	$= 1046 - 1026$	$= 20.00$
SS_T	$= 1046 - 855.7$	$= 190.30$

TABLE 5.7 Source table for traditional two-way ANOVA

Source	SS	df	MS	F
A	33.85	2	16.93	15.25
B	32.30	2	16.15	14.55
AB	104.15	4	26.04	23.46
S/AB	20.00	18	1.11	
T	190.30	26		

equation, for example, scores are summed in each level of A $(A_1 + A_2 + A_3)$ before squaring, and there are bn cases in each level of A. Table 5.6 shows the final calculations for sums of squares for the example.

Table 5.7 summarizes the calculation results and shows the mean squares and F values based on Equations 5.3–5.11. Each F ratio is evaluated separately. With $\alpha = .05$, the critical values for the main effects of A and B are the same, because each has 2 df in the numerator and 18 df in the denominator. The critical F from Table A.1 is 3.55. Because both calculated F ratios for main effects are larger than 3.55, the null hypotheses of equality of means on both margins are rejected. We conclude that there is a difference in ratings due to length of ski trip and that ratings differ by profession.

The interaction of length of trip and profession is evaluated with numerator df = 4 and denominator df = 18. With a critical F value of 2.93 (from Table A.1), we also conclude that the ratings of cases with different lengths of vacation depend on profession (or, conversely, that the effect of profession depends on length of vacation). In other words, administrators, belly dancers, and politicians have different patterns in their preferences for length of vacation. A glance at the totals in Table 5.4 suggests that administrators are mild in their preferences for either two- or three-week vacations, belly dancers like vacations of any length but seem to prefer three-week vacations, and politicians seem to prefer two-week vacations. However, these patterns need to be confirmed by further analyses. Put another way, the significant interaction implies that one must know both the length of the vacation and the profession of the case to predict the rating of the vacation. Interpretation of interactions is discussed in Section 5.6.1, where plots of means (parameter estimates) are also constructed.

5.4.4 Regression Approach (3 × 3)

As discussed in Section 4.4.4, the regression approach "works" by using orthogonal weighting coefficients—contrast coefficients—in the columns that code each IV. One column is required for each df associated with main effects and interactions. Two columns are required for the $a - 1$ $(3 - 1 = 2)$ df for the main effect of A and two for the $b - 1$ $(3 - 1 = 2)$ df for the main effect of B; four columns are required for the $(a - 1)(b - 1)$ df for the AB interaction. Altogether, eight columns are required for

TABLE 5.8 Columns for regression approach to ANOVA with three treatment levels in both A and B

		A		***B***		***AB***				***SP***							
						$X_1 * X_3$	$X_1 * X_4$	$X_2 * X_3$	$X_2 * X_4$	$Y * X_1$	$Y * X_2$	$Y * X_3$	$Y * X_4$	$Y * X_5$	$Y * X_6$	$Y * X_7$	$Y * X_8$
	Y	X_1	X_2	X_3	X_4	X_5	X_6	X_7	X_8	X_9	X_{10}	X_{11}	X_{12}	X_{13}	X_{14}	X_{15}	X_{16}
	0	2	0	−1	1	−2	2	0	0	0	0	0	0	0	0	0	0
a_1b_1	1	2	0	−1	1	−2	2	0	0	2	0	−1	1	−2	2	0	0
	0	2	0	−1	1	−2	2	0	0	0	0	0	0	0	0	0	0
	4	2	0	0	−2	0	−4	0	0	8	0	0	−8	0	−16	0	0
a_1b_2	7	2	0	0	−2	0	−4	0	0	14	0	0	−14	0	−28	0	0
	6	2	0	0	−2	0	−4	0	0	12	0	0	−12	0	−24	0	0
	5	2	0	1	1	2	2	0	0	10	0	5	5	10	10	0	0
a_1b_3	8	2	0	1	1	2	2	0	0	16	0	8	8	16	16	0	0
	6	2	0	1	1	2	2	0	0	12	0	6	6	12	12	0	0
	5	−1	1	−1	1	1	−1	−1	1	−5	5	−5	5	5	−5	−5	5
a_2b_1	7	−1	1	−1	1	1	−1	−1	1	−7	7	−7	7	7	−7	−7	7
	6	−1	1	−1	1	1	−1	−1	1	−6	6	−6	6	6	−6	−6	6
	5	−1	1	0	−2	0	2	0	−2	−5	5	0	−10	0	10	0	−10
a_2b_2	6	−1	1	0	−2	0	2	0	−2	−6	6	0	−12	0	12	0	−12
	7	−1	1	0	−2	0	2	0	−2	−7	7	0	−14	0	14	0	−14
	9	−1	1	1	1	−1	−1	1	1	−9	9	9	9	−9	−9	9	9
a_2b_3	8	−1	1	1	1	−1	−1	1	1	−8	8	8	8	−8	−8	8	8
	8	−1	1	1	1	−1	−1	1	1	−8	8	8	8	−8	−8	8	8
	5	−1	−1	−1	1	1	−1	1	−1	−5	−5	−5	5	5	−5	5	−5
a_3b_1	6	−1	−1	−1	1	1	−1	1	−1	−6	−6	−6	6	6	−6	6	−6
	8	−1	−1	−1	1	1	−1	1	−1	−8	−8	−8	8	8	−8	8	−8
	9	−1	−1	0	−2	0	2	0	2	−9	−9	0	−18	0	18	0	18
a_3b_2	9	−1	−1	0	−2	0	2	0	2	−9	−9	0	−18	0	18	0	18
	9	−1	−1	0	−2	0	2	0	2	−9	−9	0	−18	0	18	0	18
	3	−1	−1	1	1	−1	−1	−1	−1	−3	−3	3	3	−3	−3	−3	−3
a_3b_3	3	−1	−1	1	1	−1	−1	−1	−1	−3	−3	3	3	−3	−3	−3	−3
	2	−1	−1	1	1	−1	−1	−1	−1	−2	−2	2	2	−2	−2	−2	−2
Sum	152	0	0	0	0	0	0	0	0	−41	7	14	−34	40	−8	18	34
Sum of sq	1046	54	18	18	54	36	108	12	36								
N	27																

the 3 × 3 factorial design.[3] When necessary, SS(reg.) for separate columns are then combined to form main effects and interactions, just as they are combined to form a main effect in a one-way design that has more than two levels.

Two columns are needed to separate the three groups of professions from each other. The weighting coefficients in one column are arbitrarily chosen to separate administrators from both belly dancers and politicians and the coefficients in the second column to separate belly dancers from politicians. Two columns are also needed to separate the three levels of vacation length from each other, but here the weighting coefficients for trend analysis seem most appropriate. The first column provides a test of the linear trend and the second column of the quadratic trend.

Table 5.8 shows the scores on the DV, Y, in the first column, then the weighting coefficients for the eight columns representing main effects and interaction, and then cross products calculated from the first nine columns. The coding for the X_1 column compares rating of administrators with pooled ratings from belly dancers and politicians. The X_2 column compares ratings of belly dancers with those of politicians. These two columns are orthogonal. Weighting coefficients for trend analysis are found in Table A.5. The X_3 column has the weighting coefficients for the linear trend of vacation length, and the X_4 column has weighting coefficients for the quadratic trend of vacation length. These two columns are orthogonal to each other and to the first two X columns. These two sets of orthogonal weighting coefficients are chosen based on the meaning of the levels of A and B in this example. However, any set of orthogonal weighting coefficients will also be orthogonal to another orthogonal set, as long as sample sizes are equal in the cells.

The X columns for interaction are found by cross-multiplying each X column for A with each for B. First consider a simple case in which there are only two levels of A and two levels of B. As shown in Table 5.9 (where DV scores, Y, are not included), the X_1 column distinguishes a_1 from a_2 and the X_2 column distinguishes b_1 from b_2. Interaction is coded in the X_3 column, which is a cross product of the weighting coefficients in X_1 and the weighting coefficients in X_2. Note that all three columns are orthogonal. Note also that in the interaction column, groups labeled a_1b_1 and a_2b_2 are

TABLE 5.9 Regression approach to a 2 × 2 factorial design

	A	B	AB
Group	X_1	X_2	$X_1*X_2 = X_3$
a_1b_1	1	1	1
a_1b_2	1	−1	−1
a_2b_1	−1	1	−1
a_2b_2	−1	−1	1

[3] This is the same number that would be required for a one-way design with nine levels. Thus, analysis of a factorial design may be seen as simply a set of comparisons developed for a one-way design, with each column representing a single comparison.

coded 1, and those for a_1b_2 and a_2b_1 are coded -1. Thus, the null hypothesis tested by the interaction is that the pooled mean for a_1b_1 and a_2b_2 is the same as the pooled mean for a_1b_2 and a_2b_1. This is the simplest interpretation of interaction as a comparison between pooled means from the two sides of a cross that links the four groups of a 2×2 factorial design. The interaction can also be interpreted as a difference between mean differences—that is, between $(\mu_{a_1b_1} - \mu_{a_1b_2})$ and $(\mu_{a_2b_1} - \mu_{a_2b_2})$.

In the larger 3×3 example, weighting coefficients in the X_5 column are a product of weighting coefficients in X_1 and weighting coefficients in X_3. Weighting coefficients in the X_6 column are a product of weighting coefficients in the X_1 and X_4 columns. Weighting coefficients in the X_7 column are a product of X_2 and X_3, and finally, weighting coefficients in X_8 are a product of weighting coefficients in X_2 and X_4. For the first case in the a_1b_1 group (an administrator's rating of a one-week ski trip), the weighting coefficients in the X_5 to X_8 columns, respectively, are

$$X_5 = X_1X_3 = 2(-1) = -2$$

$$X_6 = X_1X_4 = 2(1) = 2$$

$$X_7 = X_2X_3 = 0(-1) = 0$$

$$X_8 = X_2X_4 = 0(1) = 0$$

Remaining cases are coded in like fashion, multiplying a weighting coefficient from a column representing A by a weighting coefficient from a column representing B.

Thus, the GLM in regression format of Equation 4.2 applies to this factorial design with $j = 8$. However, the equations for the b's (slopes) are most conveniently expressed in matrix format, with matrix algebra used for solution. Mercifully, computation of the solution of the b's (slopes) and a (intercept) is not necessary for testing ANOVA hypotheses.

Table 5.10 shows the general forms of equations for sums of squares and sums of products for the regression approach. Table 5.11 shows the equations for finding the sums of squares for the main effect of type of profession, A. Note that this matches the SS_A of Table 5.6. Table 5.12 shows the equations for finding the sums of

TABLE 5.10 Preliminary sums of squares and cross product terms for regression equations and final SS for regression

$$SS(Y) = \sum Y^2 - \frac{(\Sigma Y)^2}{N}$$

$$SS(X_i) = \sum X_i^2 - \frac{(\Sigma X_i)^2}{N}$$

$$SP(YX_i) = \sum YX_i - \frac{(\Sigma Y)(\Sigma X_i)}{N}$$

$$SS(\text{reg. } X_i) = \frac{SP(YX_i)^2}{SS(X_i)}$$

TABLE 5.11 Equations for the sum of squares for the main effect A

$$SS(\text{reg. } X_1) = \frac{SP(YX_1)^2}{SS(X_1)} = \frac{(-41)^2}{54} = 31.13$$

$$SS(\text{reg. } X_2) = \frac{SP(YX_2)^2}{SS(X_2)} = \frac{7^2}{18} = 2.72$$

$$SS_A = SS(\text{reg. } X_1) + SS(\text{reg. } X_2) = 31.13 + 2.72 = 33.85$$

TABLE 5.12 Equations for the sum of squares for the main effect B

$$SS(\text{reg. } X_3) = \frac{SP(YX_3)^2}{SS(X_3)} = \frac{14^2}{18} = 10.89$$

$$SS(\text{reg. } X_4) = \frac{SP(YX_4)^2}{SS(X_4)} = \frac{(-34)^2}{54} = 21.41$$

$$SS_B = SS(\text{reg. } X_3) + SS(\text{reg. } X_4) = 10.89 + 21.41 = 32.30$$

TABLE 5.13 Equations for the sum of squares for the AB interaction

$$SS(\text{reg. } X_5) = \frac{SP(YX_5)^2}{SS(X_5)} = \frac{40^2}{36} = 44.44$$

$$SS(\text{reg. } X_6) = \frac{SP(YX_6)^2}{SS(X_6)} = \frac{(-8)^2}{108} = 0.59$$

$$SS(\text{reg. } X_7) = \frac{SP(YX_7)^2}{SS(X_7)} = \frac{18^2}{12} = 27.00$$

$$SS(\text{reg. } X_8) = \frac{SP(YX_8)^2}{SS(X_8)} = \frac{34^2}{36} = 32.11$$

$$SS_{AB} = SS(\text{reg. } X_5) + SS(\text{reg. } X_6) + SS(\text{reg. } X_7) + SS(\text{reg. } X_8) = 44.44 + 0.59 + 27.00 + 32.11 = 104.15$$

squares for the main effect of length of vacation, B. Finally, the sum of squares for the interaction of type of profession with vacation length is calculated as per Table 5.13 (with a bit of rounding error in SS_{AB}). The sum of squares for error is found by subtracting all the sums of squares for effects from the total sum of squares:

$$SS_Y = 1046 - \frac{152^2}{27} = 190.30$$

SS_Y is the same as SS_T in Table 5.6, and

$$SS_{S/AB} = SS_Y - SS_A - SS_B - SS_{AB}$$
$$= 190.30 - 33.85 - 32.30 - 104.15 = 20.00$$

5.4.5 Computer Analyses of Small-Sample Factorial Design

Table 5.14 shows how data are entered for this type of two-way ANOVA. Data may be entered in text format (e.g., through a word processor or a spreadsheet and then saved in text format) or through facilities provided by the statistical program. Data entered in text format include numbers only, without the column headings shown in Table 5.14; data entered through statistical programs typically include variable names at the tops of columns. Either way, each row is a single case. One column is needed to code the levels of each IV (in this case, a column for LENGTH and a column for PROFESS), and another column is needed for the DV (RATING). An additional column, case number, may be added for convenience in viewing and editing the data file.

Tables 5.15 and 5.16 show computer analyses of the small-sample example from Section 5.4.3 through SPSS GLM (UNIANOVA)[4] and SAS GLM, respectively. The SPSS GLM (UNIANOVA) syntax lists the DV (RATING), followed by the keyword BY, followed by a list of the IVs (LENGTH and PROFESS). The EMMEANS instructions request marginal and cell means; DESCRIPTIVE in the /PRINT instruction requests descriptive statistics for cells. ETASQ and OPOWER request effect size and observed power statistics. The default is for DESIGN to request the full factorial two-way ANOVA. Remaining syntax (METHOD, INTERCEPT, and CRITERIA) shows the default values produced by SPSS menus.

TABLE 5.14 Hypothetical data set for 3 × 3 ANOVA in a form suitable for computer

PROFESS	LENGTH	RATING	PROFESS	LENGTH	RATING
1	1	0	2	2	7
1	1	1	2	3	9
1	1	0	2	3	8
1	2	4	2	3	8
1	2	7	3	1	5
1	2	6	3	1	6
1	3	5	3	1	8
1	3	8	3	2	9
1	3	6	3	2	9
2	1	5	3	2	9
2	1	7	3	3	3
2	1	6	3	3	3
2	2	5	3	3	2
2	2	6			

[4] This is accessed through SPSS for Windows by choosing Analyze > General Linear Model > Univariate. UNIANOVA is the syntax generated by the menu selection.

TABLE 5.15 Two-way randomized-groups ANOVA for small-sample example through SPSS GLM (UNIANOVA)

```
UNIANOVA
RATING BY PROFESS LENGTH
/METHOD = SSTYPE(3)
/INTERCEPT = INCLUDE
/EMMEANS = TABLES(PROFESS)
/EMMEANS = TABLES(LENGTH)
/EMMEANS = TABLES(PROFESS*LENGTH)
/PRINT = DESCRIPTIVE ETASQ OPOWER
/CRITERIA = ALPHA(.05)
/DESIGN =PROFESS LENGTH PROFESS*LENGTH.
```

Univariate Analysis of Variance

Between-Subjects Factors

		N
profess	1.00	9
	2.00	9
	3.00	9
length	1.00	9
	2.00	9
	3.00	9

Descriptive Statistics

Dependent Variable: rating

profess	length	Mean	Std. Deviation	N
1.00	1.00	.3333	.57735	3
	2.00	5.6667	1.52753	3
	3.00	6.3333	1.52753	3
	Total	4.1111	3.05959	9
2.00	1.00	6.0000	1.00000	3
	2.00	6.0000	1.00000	3
	3.00	8.3333	.57735	3
	Total	6.7778	1.39443	9
3.00	1.00	6.3333	1.52753	3
	2.00	9.0000	.00000	3
	3.00	2.6667	.57735	3
	Total	6.0000	2.87228	9
Total	1.00	4.2222	3.07318	9
	2.00	6.8889	1.83333	9
	3.00	5.7778	2.63523	9
	Total	5.6296	2.70538	27

TABLE 5.15 Continued

Tests of Between-Subjects Effects

Dependent Variable: rating

Source	Type III Sum of Squares	df	Mean Square	F	Sig.	Partial Eta Squared	Noncent. Parameter	Observed Power[a]
Corrected Model	170.296[b]	8	21.287	19.158	.000	.895	153.267	1.000
Intercept	855.704	1	855.704	770.133	.000	.977	770.133	1.000
profess	33.852	2	16.926	15.233	.000	.629	30.467	.997
length	32.296	2	16.148	14.533	.000	.618	29.067	.996
profess * length	104.148	4	26.037	23.433	.000	.839	93.733	1.000
Error	20.000	18	1.111					
Total	1046.000	27						
Corrected Total	190.296	26						

a. Computed using alpha = .05
b. R Squared = .895 (Adjusted R Squared = .848)

Estimated Marginal Means

1. profess

Dependent Variable: rating

profess	Mean	Std. Error	95% Confidence Interval Lower Bound	95% Confidence Interval Upper Bound
1.00	4.111	.351	3.373	4.849
2.00	6.778	.351	6.040	7.516
3.00	6.000	.351	5.262	6.738

2. length

Dependent Variable: rating

length	Mean	Std. Error	95% Confidence Interval Lower Bound	95% Confidence Interval Upper Bound
1.00	4.222	.351	3.484	4.960
2.00	6.889	.351	6.151	7.627
3.00	5.778	.351	5.040	6.516

3. profess * length

Dependent Variable: rating

profess	length	Mean	Std. Error	95% Confidence Interval Lower Bound	95% Confidence Interval Upper Bound
1.00	1.00	.333	.609	−.945	1.612
	2.00	5.667	.609	4.388	6.945
	3.00	6.333	.609	5.055	7.612
2.00	1.00	6.000	.609	4.721	7.279
	2.00	6.000	.609	4.721	7.279
	3.00	8.333	.609	7.055	9.612
3.00	1.00	6.333	.609	5.055	7.612
	2.00	9.000	.609	7.721	10.279
	3.00	2.667	.609	1.388	3.945

TABLE 5.16 Two-way randomized-groups ANOVA for small-sample example through SAS GLM

```
proc glm data=SASUSER.FAC_RAND;
     class LENGTH PROFESS;
     model RATING = LENGTH PROFESS LENGTH*PROFESS / ss3;
     means LENGTH PROFESS LENGTH*PROFESS;
run;
```

The GLM Procedure

Class Level Information

Class	Levels	Values
LENGTH	3	1 2 3
PROFESS	3	1 2 3

Number of Observations Read 27
Number of Observations Used 27

Dependent Variable: RATING

Source	DF	Sum of Squares	Mean Square	F Value	Pr > F
Model	8	170.2962963	21.2870370	19.16	<.0001
Error	18	20.0000000	1.1111111		
Corrected Total	26	190.2962963			

R-Square	Coeff Var	Root MSE	RATING Mean
0.894901	18.72401	1.054093	5.629630

Source	DF	Type III SS	Mean Square	F Value	Pr > F
LENGTH	2	32.2962963	16.1481481	14.53	0.0002
PROFESS	2	33.8518519	16.9259259	15.23	0.0001
LENGTH*PROFESS	4	104.1481481	26.0370370	23.43	<.0001

Level of LENGTH	N	RATING Mean	RATING Std Dev
1	9	4.22222222	3.07318149
2	9	6.88888889	1.83333333
3	9	5.77777778	2.63523138

Level of PROFESS	N	RATING Mean	RATING Std Dev
1	9	4.11111111	3.05959329
2	9	6.77777778	1.39443338
3	9	6.00000000	2.87228132

Level of LENGTH	Level of PROFESS	N	RATING Mean	RATING Std Dev
1	1	3	0.33333333	0.57735027
1	2	3	6.00000000	1.00000000
1	3	3	6.33333333	1.52752523
2	1	3	5.66666667	1.52752523
2	2	3	6.00000000	1.00000000
2	3	3	9.00000000	0.00000000
3	1	3	6.33333333	1.52752523
3	2	3	8.33333333	0.57735027
3	3	3	2.66666667	0.57735027

Cell means and standard deviations are printed as a result of the PRINT= DESCRIPTIVE instruction. Below the segment of output labeled Tests of Between-Subjects Effects is a traditional ANOVA source table with *S/AB* (Table 5.7) labeled Error. The column labeled Sig. shows the probability (to three decimal places) that an *F* ratio this large or larger would be achieved if the null hypothesis were true. The Corrected Model source of variation is a composite of the LENGTH and PROFESS main effects and the LENGTH*PROFESS interaction. The probability that each result occurred by chance if the null hypothesis were true is less than .0005. The Total source of variance includes the sum of squares for intercept (*grand mean* in the coding scheme used by SPSS). The Corrected Total excludes the sum of squares for intercept and is the more useful sum for effect size calculations (cf. Section 5.6.2). The column labeled Partial Eta Squared contains partial η^2 as a measure of effect size (Section 5.6.2). R Squared in footnote b is the proportion of total variance that is predictable from the sum of all effects. Adjusted R Squared adjusts that value for bias due to small sample sizes (cf. Tabachnick and Fidell, 2007). Observed Power is described in Section 5.6.3, and the Noncent. Parameter (noncentrality) is relevant to calculation of power.

Marginal and cell means appear at the end of the output, together with a common standard error for each level of each factor and the 95% confidence interval for the mean. The common standard error is found by dividing the square root of $MS_{S/A}$ (labeled Error in the output) by the square root of the number of cases in each marginal or cell mean.

SAS GLM syntax and output appear in Table 5.16. Discrete IVs are identified in the `class` instruction; the `model` instruction shows (in regression format) that the DV is a function of the IVs and their interaction. The `ss3` instruction requests Type III adjustment for unequal sample sizes (should there be any) and limits the output to just that adjustment (cf. Section 5.6.5). Marginal (`LENGTH` and `PROFESS`) and cell (`LENGTH*PROFESS`) means are requested in the means instruction.

Two source tables are given. The first is for evaluating the entire `Model` (sum of all the effects) with the `Error` (*S/AB*) as well as the `Total` (identified as `Corrected`). This table includes `R-Square`, the proportion of variance predictable from all effects combined. Also shown are the square root of $MS_{S/A}$ (`Root MSE`), the grand mean (`RATING Mean`), and the coefficient of variation (`Coeff Var`; the square root of the error mean square divided by the grand mean multiplied by 100).

The second source table shows the three effects, with the last column (`Pr > F`) indicating the probability of obtained *F* under the null hypothesis. Marginal and cell means are as requested in the `means` instruction.

5.5 OTHER TYPES OF RANDOMIZED-GROUPS DESIGNS

Factorial designs may be expanded to any number of IVs, producing higher-order factorials. A variant on this design is one in which an additional group, typically a control group, is added to an otherwise factorial design.

5.5.1 Higher-Order Factorial Designs

Higher-order randomized-groups factorial designs have more than two IVs, each with two or more levels. In an experiment, at least one of the IVs is experimentally manipulated, with the remaining IVs also experimentally manipulated or not. Quasi-experimental or nonexperimental designs have no experimentally manipulated IVs. Table 5.17 shows allocation of cases in the simplest higher-order factorial design—a $2 \times 2 \times 2$ design. This design can be considered a replication of an $A \times B$ factorial design at both c_1 and c_2. Seven effects are tested—each of the three main effects (A, B, and C), the three two-way interactions (AB, AC, and BC), and the three-way interaction (ABC). Thus, the between-groups sum of squares is partitioned as follows:

$$\text{SS}_{bg} = \text{SS}_A + \text{SS}_B + \text{SS}_C + \text{SS}_{AB} + \text{SS}_{AC} + \text{SS}_{BC} + \text{SS}_{ABC} \qquad \textbf{(5.12)}$$

The error sum of squares remains the within-cell variability, now designated $\text{SS}_{S/ABC}$, so that the partition of the total sum of squares is

$$\text{SS}_T = \text{SS}_A + \text{SS}_B + \text{SS}_C + \text{SS}_{AB} + \text{SS}_{AC} + \text{SS}_{BC} + \text{SS}_{ABC} + \text{SS}_{S/ABC} \qquad \textbf{(5.13)}$$

Table 5.18 shows degrees of freedom and expanded (multiplied out) degrees of freedom for each effect, as well as the corresponding computational equations for SS. Spend some time comparing the expanded df equations and the computational equations: The value that is squared in the numerator of the computational equations is the corresponding value from the expanded df equation. The only exceptions are that the 1's in the df statements are each replaced by T^2 in the computational equations, and the Ns are replaced by ΣY^2. Notice also that each segment of the computational equations contains all the factors in the study (in this case, A, B, C) and n, and the elements that are not in the numerator of a segment are in the denominator. Lastly, notice the pattern in the sequences of addition and subtraction: First, the factor with

TABLE 5.17 Assignment of cases in a three-way randomized-groups design with all IVs experimentally manipulated

Treatment *C*

c_1

	Treatment *B*	
Treatment *A*	b_1	b_2
a_1	s_1 s_2 s_3	s_7 s_8 s_9
a_2	s_4 s_5 s_6	s_{10} s_{11} s_{12}

c_2

	Treatment *B*	
Treatment *A*	b_1	b_2
a_1	s_{13} s_{14} s_{15}	s_{19} s_{20} s_{21}
a_2	s_{16} s_{17} s_{18}	s_{22} s_{23} s_{24}

TABLE 5.18 Degrees of Freedom and SS Equations for Three-Way ANOVA

Source	Degrees of Freedom and Expanded df	Sum of Squares Equations
A	$a - 1$	$\frac{\sum A^2}{bcn} - \frac{T^2}{abcn}$
B	$b - 1$	$\frac{\sum B^2}{acn} - \frac{T^2}{abcn}$
C	$c - 1$	$\frac{\sum C^2}{abn} - \frac{T^2}{abcn}$
AB	$(a-1)(b-1)$ $= ab - a - b + 1$	$\frac{\sum AB^2}{cn} - \frac{\sum A^2}{bcn} - \frac{\sum B^2}{acn} + \frac{T^2}{abcn}$
AC	$(a-1)(c-1)$ $= ac - a - c + 1$	$\frac{\sum AC^2}{bn} - \frac{\sum A^2}{bcn} - \frac{\sum C^2}{abn} + \frac{T^2}{abcn}$
BC	$(b-1)(c-1)$ $= bc - b - c + 1$	$\frac{\sum BC^2}{an} - \frac{\sum B^2}{acn} - \frac{\sum C^2}{abn} + \frac{T^2}{abcn}$
ABC	$(a-1)(b-1)(c-1)$ $= abc - ab - ac - bc$ $+ a + b + c - 1$	$\frac{\sum ABC^2}{n} - \frac{\sum AB^2}{cn} - \frac{\sum AC^2}{bn} - \frac{\sum BC^2}{an}$ $+ \frac{\sum A^2}{bcn} + \frac{\sum B^2}{acn} + \frac{\sum C^2}{abn} - \frac{T^2}{abcn}$
S/ABC	$N - abc$	$\sum Y^2 - \frac{\sum ABC^2}{n}$
T	$N - 1$	$\sum Y^2 - \frac{T^2}{abcn}$

the most elements (in this case, ABC) is added. Next, all two-way combinations of those elements are subtracted. Then, all one-way elements are added. And finally, the T element is subtracted.[5] The segments of the computational equation, in short, follow the same pattern of addition and subtraction as the expanded df statement follows. Once this pattern is apparent to you, you can generate the computational equations for any factorial design from the expanded degrees of freedom.

Figure 5.2 diagrams the partition of sums of squares and degrees of freedom for the three-way factorial randomized-groups ANOVA.

Suppose we are investigating the eagerness (the DV) of a bull in the presence of a cow. Factor A is deprivation time for the bull (short vs. long), B is familiarity of the cow (familiar vs. unfamiliar), and C is decoration of the pen (unpainted wood vs. floral wallpaper). Table 5.19 shows a hypothetical data set for this experiment.

[5] In a four-way study, after the four-way interaction ($ABCD$) is included, the three-way interactions are subtracted, then the two-way combinations are added, then the one-way elements are subtracted, and finally, the T element is added.

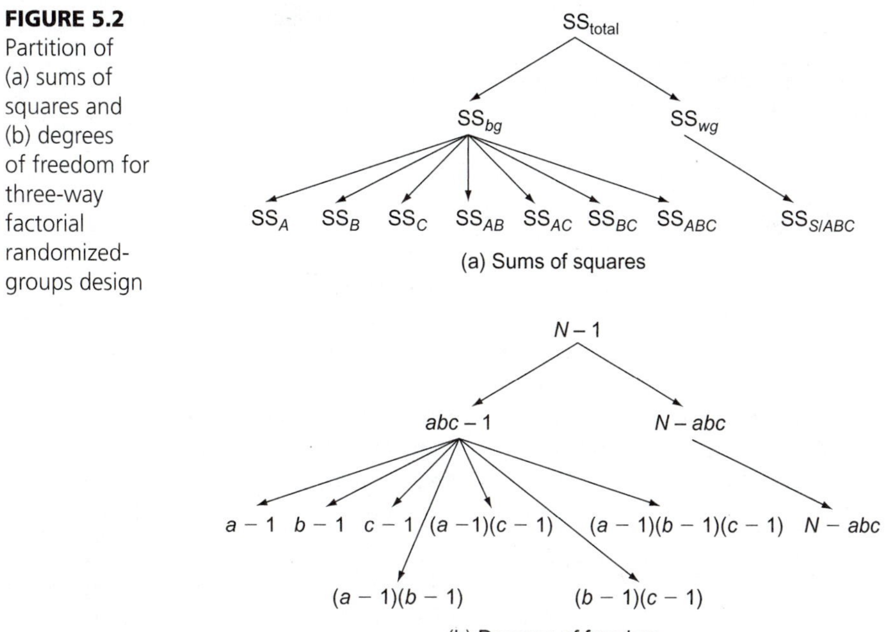

FIGURE 5.2 Partition of (a) sums of squares and (b) degrees of freedom for three-way factorial randomized-groups design

TABLE 5.19 Hypothetical data set for three-way randomized-groups ANOVA

Deprivation	Familiarity	Decor	Eagerness	Deprivation	Familiarity	Decor	Eagerness
1	1	1	4	2	1	1	0
1	1	1	3	2	1	1	5
1	1	1	2	2	1	1	4
1	1	1	3	2	1	1	1
1	1	1	0	2	1	1	3
1	1	2	3	2	1	2	6
1	1	2	0	2	1	2	7
1	1	2	2	2	1	2	6
1	1	2	1	2	1	2	6
1	1	2	4	2	1	2	6
1	2	1	0	2	2	1	9
1	2	1	2	2	2	1	6
1	2	1	2	2	2	1	6
1	2	1	0	2	2	1	6
1	2	1	2	2	2	1	7
1	2	2	6	2	2	2	6
1	2	2	5	2	2	2	10
1	2	2	5	2	2	2	9
1	2	2	5	2	2	2	9
1	2	2	3	2	2	2	8

TABLE 5.20 Three-way randomized-groups ANOVA through SAS GLM syntax and selected output

```
proc   glm  data = SASUSER.THREEWY;
       class DEPRIVAT FAMILIAR DECOR EAGERNES;
       model EAGERNES = DEPRIVAT|FAMILIAR|DECOR;
       means DEPRIVAT|FAMILIAR|DECOR;

run;                                    The GLM Procedure
```

Dependent Variable: eagerness eagernes

Source	DF	Sum of Squares	Mean Square	F Value	Pr > F
Model	7	241.6000000	34.5142857	17.59	<.0001
Error	32	62.8000000	1.9625000		
Corrected Total	39	304.4000000			

R-Square	Coeff Var	Root MSE	eagernes Mean
0.793693	32.57890	1.400893	4.300000

Source	DF	Type III SS	Mean Square	F Value	Pr > F
deprivat	1	115.6000000	115.6000000	58.90	<.0001
familiar	1	40.0000000	40.0000000	20.38	<.0001
deprivat*familiar	1	14.4000000	14.4000000	7.34	0.0108
décor	1	44.1000000	44.1000000	22.47	<.0001
deprivat*décor	1	2.5000000	2.5000000	1.27	0.2674
familiar*décor	1	2.5000000	2.5000000	1.27	0.2674
depriv*familia*décor	1	22.5000000	22.5000000	11.46	0.0019

Level of deprivat	N	----------eagernes-------- Mean	Std Dev
1	20	2.60000000	1.84676103
2	20	6.00000000	2.55466549

Level of familiar	N	----------eagernes-------- Mean	Std Dev
1	20	3.30000000	2.22663305
2	20	5.30000000	2.99297423

Level of deprivat	Level of familiar	N	----------eagernes---------- Mean	Std Dev
1	1	10	2.20000000	1.47572957
1	2	10	3.00000000	2.16024690
2	1	10	4.40000000	2.36643191
2	2	10	7.60000000	1.57762128

Level of decor	N	----------eagernes---------- Mean	Std Dev
1	20	3.25000000	2.59300679
2	20	5.35000000	2.64127162

Level of deprivat	Level of decor	N	----------eagernes----------- Mean	Std Dev
1	1	10	1.80000000	1.39841180
1	2	10	3.40000000	1.95505044
2	1	10	4.70000000	2.75075747
2	2	10	7.30000000	1.56702124

TABLE 5.20 Continued

Level of familiar	Level of decor	N	eagernes Mean	eagernes Std Dev
1	1	10	2.50000000	1.71593836
1	2	10	4.10000000	2.46981781
2	1	10	4.00000000	3.16227766
2	2	10	6.60000000	2.27058485

Level of deprivat	Level of familiar	Level of decor	N	eagernes Mean	eagernes Std Dev
1	1	1	5	2.40000000	1.51657509
1	1	2	5	2.00000000	1.58113883
1	2	1	5	1.20000000	1.09544512
1	2	2	5	4.80000000	1.09544512
2	1	1	5	2.60000000	2.07364414
2	1	2	5	6.20000000	0.44721360
2	2	1	5	6.80000000	1.30384048
2	2	2	5	8.40000000	1.51657509

Table 5.20 shows the syntax and output for a SAS GLM ANOVA run on these data. The syntax `DEPRIVAT|FAMILIAR|DECOR` requests the three-way interaction and all lower-order effects.

`Error Mean Square=1.9625` from the first source table serves as the error term for all effects. The source table with `Type III SS` shows that all three main effects are statistically significant. Tables of means follow the source table in the same order as the effects in the source table. For example, the mean eagerness of bulls after short deprivation (level 1) is 2.6, and the mean eagerness after long deprivation is 6.0. Among the interactions, only the two-way interaction of deprivation by familiarity ($p = 0.0108$) and the three-way interaction of deprivation by familiarity by decoration ($p = 0.0019$) are statistically significant. Interpretation of interactions for three-way designs is discussed in Section 5.6.1.2.

Generating computational equations for factorial designs with more than three IVs follows a simple pattern:

1. List all IVs. There is a main effect for each IV, with degrees of freedom equal to the number of levels of that IV minus 1.
2. Form all combinations of two-way interactions among IVs. Degrees of freedom for each interaction are the product of the degrees of freedom for each of the IVs forming the interaction.
3. Form all combinations of three-way interactions among IVs. Again, degrees of freedom are the products of the degrees of freedom for component IVs.
4. Continue forming combinations of interactions until the highest-order interaction is listed (e.g., *ABCDE*).
5. The error term is the within-cell variability: *S* before the slash, the highest-order interaction after the slash, (e.g., $S/ABCDE$).

6. Expand degrees of freedom equations to develop the segments needed and their patterns of addition and subtraction (cf. Table 5.18).
7. Form the computational equations with all elements in either the numerator or denominator of each segment, and follow the rules of substitution for 1s and *N*s.

Computer analysis of higher-order factorials requires a column in the data file for each IV and a column for the DV. Then the IVs are listed in the appropriate syntax instruction for factors used by the statistical program. Interactions are formed and tested automatically by SPSS UNIANOVA. SAS requires that they be specified in the model instruction (cf. Tables 5.16 and 5.20).

5.5.2 Factorial Design with a Single Control

Sometimes factorial designs have two or more IVs, representing levels of experimental treatments but only a single control group. Consider a study with two types of drugs and two levels of dosage, say, 5 mg and 10 mg. If the researcher also wants a control group with no drug treatment (no *A* or *B*), there is only one such group, because 0 mg of the first drug is the same as 0 mg of the second drug. Or consider an experiment in which the researcher is interested in the effects on examination scores of type (classical and heavy metal) and loudness (40 dB and 80 dB) of background music during study. If the researcher also wants a 0 dB control group, there is only one because 0 dB of classical music sounds just like 0 dB of heavy metal.

This is sometimes called a dangling group design. Table 5.21 shows a layout for the simplest form of such a design. The factorial portion of this design can have any number of IVs, with two or more levels of each IV; the distinguishing feature is the single control group.

The first step in the analysis is to turn the design into a one-way design with, in this example, five levels. The error term is $SS_{S/A}$, found as in Sections 3.2.3.2 and 4.4.3. Degrees of freedom for the error term are found by multiplying the total number of groups (in this example, $a = 5$) by $n - 1$. This error term is easily found by computer by using a column of numbers that identifies group membership; this is the appropriate error term for all the tests.

TABLE 5.21 Factorial design with two levels each of two drug treatments and a single control

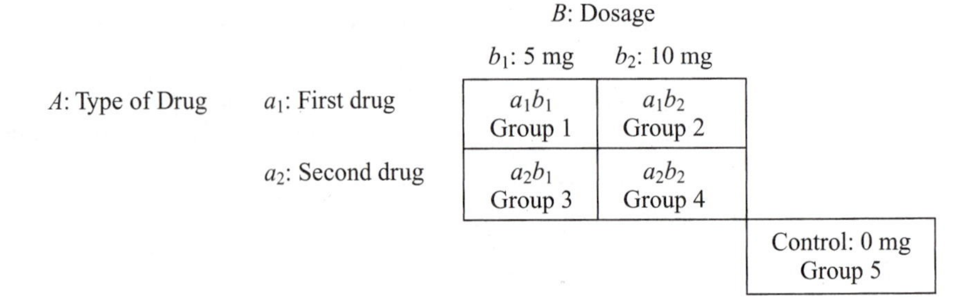

		B: Dosage		
		b_1: 5 mg	b_2: 10 mg	
A: Type of Drug	a_1: First drug	a_1b_1 Group 1	a_1b_2 Group 2	
	a_2: Second drug	a_2b_1 Group 3	a_2b_2 Group 4	
				Control: 0 mg Group 5

The second step is to use comparisons, preferably planned beforehand (see Section 5.6.4), to test relevant hypotheses. One likely test is of the control condition mean against the pooled means for four drug conditions (with weighting coefficients of 1, 1, 1, 1, −4). An example of this step of the analysis is in Section 5.6.4.1.2. The other three comparisons might well be an omnibus analysis of the 2 × 2 factorial part of the design into *A*, *B*, and *AB*. The weighting coefficients for these comparisons are in Table 5.9, to be used with the procedures on comparisons on marginal means of Section 5.6.4.1.2. These four comparisons are a complete orthogonal partition of the sum of squares for the design. Analysis of these designs is also discussed by Keppel (1991), Himmelfarb (1975), and Brown, Michels, and Winer (1991).

5.6 SOME IMPORTANT ISSUES

Several of the issues discussed in this section also apply to designs other than randomized groups. For example, the interpretation of interactions is the same for any factorial design. They are discussed here only because this is the first opportunity to do so. Subsequent chapters refer back to this section when appropriate.

5.6.1 Interpreting Interactions

Presence of an interaction indicates that the effect of one IV on the DV depends on the level of another IV. Absence of an interaction indicates that the effect of one IV on the DV is similar for all levels of another IV. The null hypothesis is that there is no interaction. Rejection of the null hypothesis indicates that there probably is, in reality, an interaction.

5.6.1.1 Two-Way Designs

Interactions in a two-way design may be expressed in either direction or from a prediction perspective. For example, the interaction between length of stay and profession in the 3 × 3 example of Section 5.4.3 may be stated either as, "The effect of different lengths of stay on ratings of a vacation depends on type of profession," or as, "The effect of different professions on ratings of a vacation depends on length of stay." Or, from a prediction perspective, one could say, "Prediction of the rating of the vacation requires knowledge of both the type of profession and the length of stay."

Tests of interactions are also tests of differences among mean *differences*. For example, the interaction in the two-way design may also be stated as, "The differences between mean ratings of vacations of various lengths are different for the different professions," or "The differences between mean ratings for different professions are different for the three lengths of stay." The means for the small-sample example of Section 5.4.3 are in Table 5.22.

TABLE 5.22 Cell and marginal means for hypothetical data set for 3 × 3 ANOVA of Table 5.3

	B: Vacation Length			
A: Profession	b_1: 1 week	b_2: 2 weeks	b_3: 3 weeks	Marginal means for A
a_1: Administrators	0.33	5.67	6.33	4.11
a_2: Belly dancers	6.00	6.00	8.33	6.78
a_3: Politicians	6.33	9.00	2.67	6.00
Marginal means for B	4.22	6.89	5.78	5.63

Interactions are best illustrated in graphs of means in which levels of one IV define the abscissa, the DV is on the ordinate, and the second IV is represented by separate lines. As an example, a plot of the cell means of Table 5.22 is in Figure 5.3. This plot shows differences among professions in vacation ratings as a function of length of stay. Politicians prefer two-week vacations and do not much like three-week vacations. Belly dancers like vacations of any length but prefer three-week vacations. Administrators really dislike one-week vacations and may prefer three-week vacations to two-week vacations. This is just an eyeball description of the patterns. Specific comparisons (Section 5.6.4) are necessary to make definite statements about preferences.

The interaction could also be plotted in the other direction, with profession defining the abscissa and length of stay represented by separate lines, as shown in

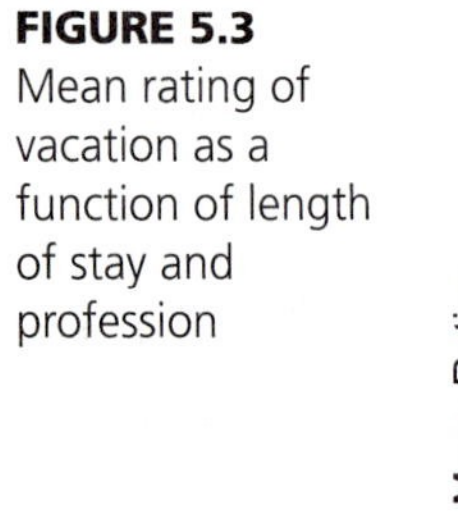

FIGURE 5.3
Mean rating of vacation as a function of length of stay and profession

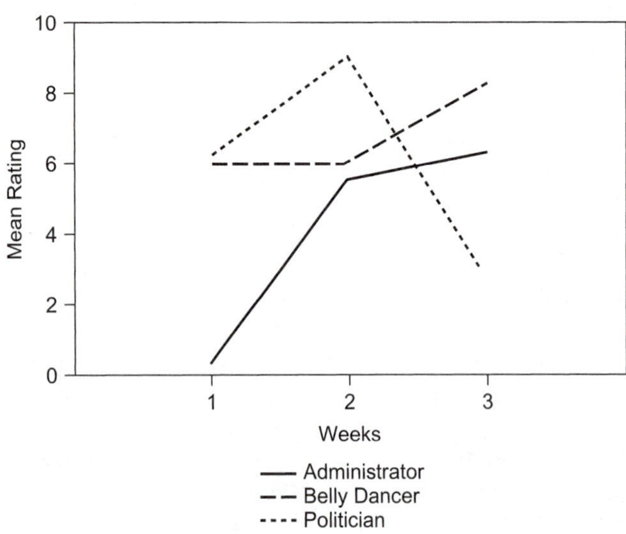

FIGURE 5.4
Mean rating of vacation as a function of length of stay and profession

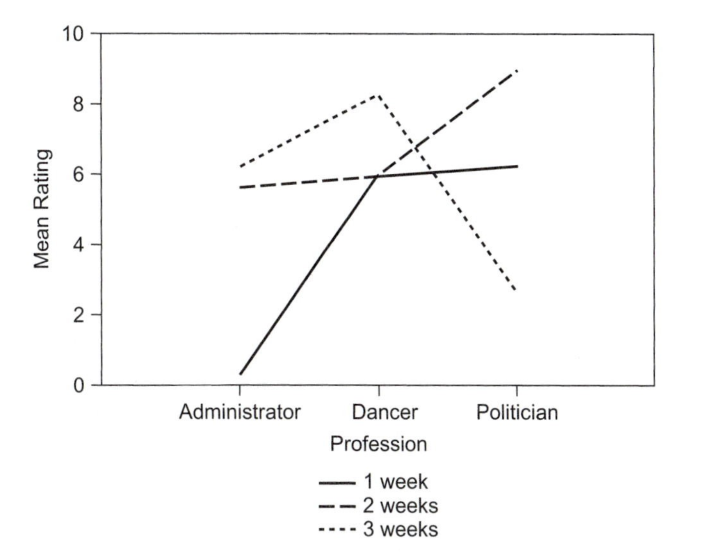

Figure 5.4.[6] This figure is a bit more difficult to interpret verbally, because it suggests an interpretation in terms of weeks rather than of profession. Three-week vacations are disliked by politicians but most preferred by belly dancers; two-week vacations are preferred by politicians but produce moderate ratings by the other two groups; one-week vacations receive moderate ratings by politicians and belly dancers but are heartily disliked by administrators.

Both orientations are equally legitimate. The choice is yours as a researcher, but ease of interpretation is the final criterion. The following guidelines may be useful:

1. If one IV is quantitative (e.g., weeks) and the other is qualitative (e.g., professions), interpretation is usually easier if the quantitative IV defines the abscissa.
2. If one IV is a blocking variable (e.g., profession), interpretation is often easier if the blocking IV is plotted as separate lines.
3. If one IV has more levels than the other, interpretation is often easier if the IV with more levels defines the abscissa.

In practice, the researcher frequently produces both plots and chooses the one that is more easily "verbalized."

When there is interaction, the lines are not parallel. The null hypothesis for an interaction can also be expressed as parallelism of the lines, with rejection of the null hypothesis when there is significant departure from parallelism. Suppose that belly

[6] By convention, interaction is plotted as a line graph to emphasize parallelism of lines. The use of a line graph with profession defining the abscissa does not imply that profession is a quantitative variable. Strictly speaking, bar graphs are more appropriate when IVs are qualitative, but line graphs are more easily interpreted.

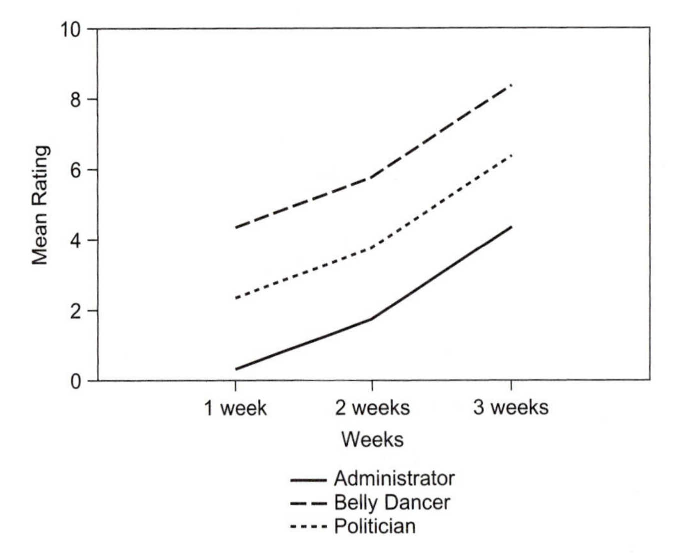

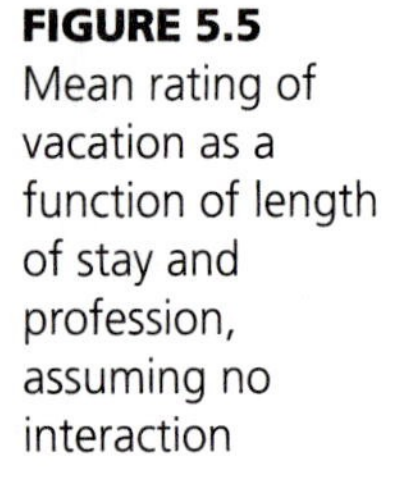

FIGURE 5.5 Mean rating of vacation as a function of length of stay and profession, assuming no interaction

dancers consistently rate their vacations higher than politicians, and politicians rate their vacations higher than administrators. Suppose also that everybody prefers a two-week vacation to a one-week vacation and a three-week vacation to a two-week vacation—all by about the same amounts. In this case, there is no interaction and the lines are parallel, as shown in Figure 5.5,[7] which shows that differences in ratings among professions are the same for all levels of length of vacation or that differences in ratings among different lengths of vacation are the same for all professions. There are no differences among mean differences.

Notice that Figure 5.5 makes it easy to interpret main effects, assuming that they are statistically significant. The line for the belly dancers is consistently higher than the line for politicians, which, in turn, is consistently higher than the line for administrators. On average, three-week vacations get higher ratings than two-week vacations, and so on. Differences in means for main effects are often visible, even when lines are not parallel. For example, Figure 5.3 also shows higher ratings by belly dancers than by administrators and, on average, higher ratings for two-week vacations than for one-week vacations.

Caution is needed when interpreting significant main effects in the presence of a significant interaction. One way of looking at the presence of interaction is that the effects of one IV on the DV do not generalize over the levels of the other IV. Once you know that the effects of one IV are tempered by or contingent on the levels of another IV, interpretation of either IV alone is problematic. Some (e.g., Keppel, 1991) recommend only interpreting significant main effects either when there is no interaction

[7] Lines do not have to be precisely parallel for you to conclude that there is no significant interaction. Failing to reject the null hypothesis for interaction means that lines are parallel within some tolerable range.

TABLE 5.23 Problematic nature of main effects when interaction is significant

		Type of Drug		Main Effect of Kid
		Stimulant	Tranquilizer	
Type of Kid	ADHD	Low	High	Medium
	Normal	High	Low	Medium
Main Effect of Drug		Medium	Medium	

or when the interaction is significant but effect size is small relative to main effects and there is an ordinal pattern to the means. If the means are ordinal (as in Figure 5.5), they line up in the same order at all levels of the other IV, so the lines do not cross each other even if they are not parallel. The means in Figures 5.3 and 5.4 are disordinal because the lines cross. In any event, main effects should be interpreted cautiously or not at all when there is strong, significant interaction.

Consider, for example, the tests of main effects in an experiment involving ADHD (hyperactive) kids and "normal" kids who are given either stimulants or tranquilizers. ADHD kids have a paradoxical response to these two drugs and become *more* active when given tranquilizers and *less* active when given stimulants. The other kids respond the other way to the two drugs. Table 5.23 shows the typical pattern of behavior.

Results of ANOVA on data like these are probably a significant interaction but a nonsignificant main effect for type of drug and for type of kid. This is because the means on the margins are very similar when averaged over both high and low scores. Taken at face value, the tests of main effects imply that ADHD kids and normal kids have the same behavior, as do kids given stimulants and tranquilizers; these are clearly misleading conclusions but are indicative of the potential problem of interpreting main effects in the presence of interaction.

5.6.1.2 Higher-Order Factorials

Numerous interactions are tested in higher-order factorials. For example, a three-way design with *A*, *B*, and *C* provides tests of all two-way interactions (*AB*, *AC*, and *BC*), plus the three-way interaction (*ABC*). Any of these or all of these may be statistically significant. If the three-way interaction is not significant, but one or more of the two-way interactions are significant, they are plotted and interpreted as if a two-way design had been analyzed. For example, in the three-way design of Section 5.5.1, the two-way interaction between deprivation and familiarity is statistically significant (cf. Table 5.20). Figure 5.6 plots the means involved in that interaction. The plot shows that the effect of deprivation time on eagerness is greater in the presence of unfamiliar than familiar cows.

However, when the three-way interaction is significant as well, interpretation of lower-order interactions and main effects is also problematic. From a prediction

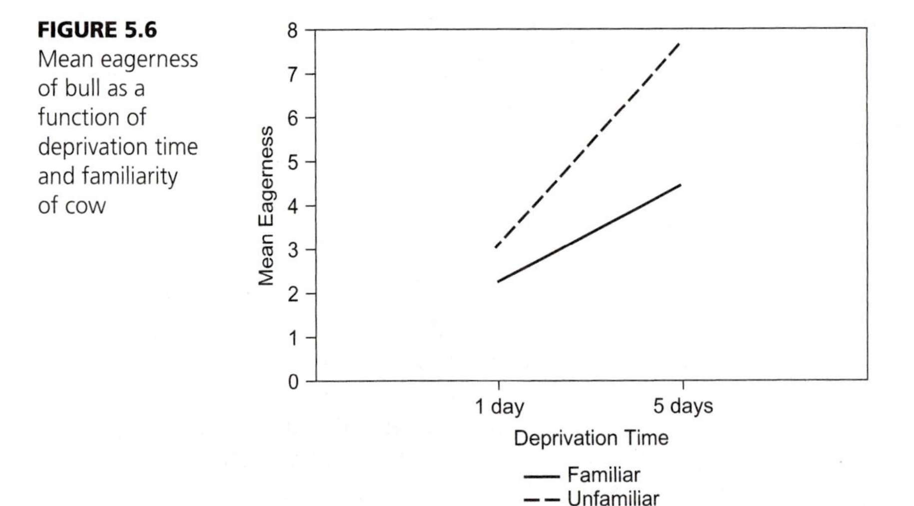

FIGURE 5.6 Mean eagerness of bull as a function of deprivation time and familiarity of cow

perspective, a three-way interaction means that the researcher must know the status on all three IVs in order to accurately predict a score for a case on the DV. That is, the generalizability of two-way interactions and main effects is contingent on the level of the third variable, so that

1. the *AB* interaction varies with the level of *C*, or
2. the *AC* interaction varies with the level of *B*, or
3. the *BC* interaction varies with the level of *A*.

Figure 5.7 plots all the cell means for the three-way interaction found in Section 5.5.1.

The interaction between deprivation and familiarity shown in Figure 5.7 is different for wood- and wallpaper-decorated pens. There is little interaction in the presence of floral wallpaper. At c_2 (floral wallpaper), deprivation and familiarity are said to be additive (i.e., there is no interaction), because eagerness is a simple sum of the effects of longer deprivation and unfamiliar cow (note the nearly horizontal line for familiar cows). There is an interaction in the presence of wood, however, in which the effect of deprivation on eagerness occurs only in the presence of unfamiliar cows. Notice that this figure also suggests main effects of all three IVs. Plotted points are higher on average for wallpaper than for wood decoration, are higher on average for five days than for one day of deprivation, and are higher on average for unfamiliar than for familiar cows.

If there were no three-way interaction, the relationship between deprivation and familiarity would be the same in both types of pens. For example, both plots of Figure 5.7 might have the same interaction as found in the presence of wood. Or both plots might have the same lack of interaction found in the presence of wallpaper. Full interpretation of a three-way interaction requires further analysis through specific comparisons, just as does full interpretation of a two-way interaction or main effects with more than two levels.

FIGURE 5.7
Mean eagerness of bull as a function of familiarity of cow, days of deprivation, and decoration of pen

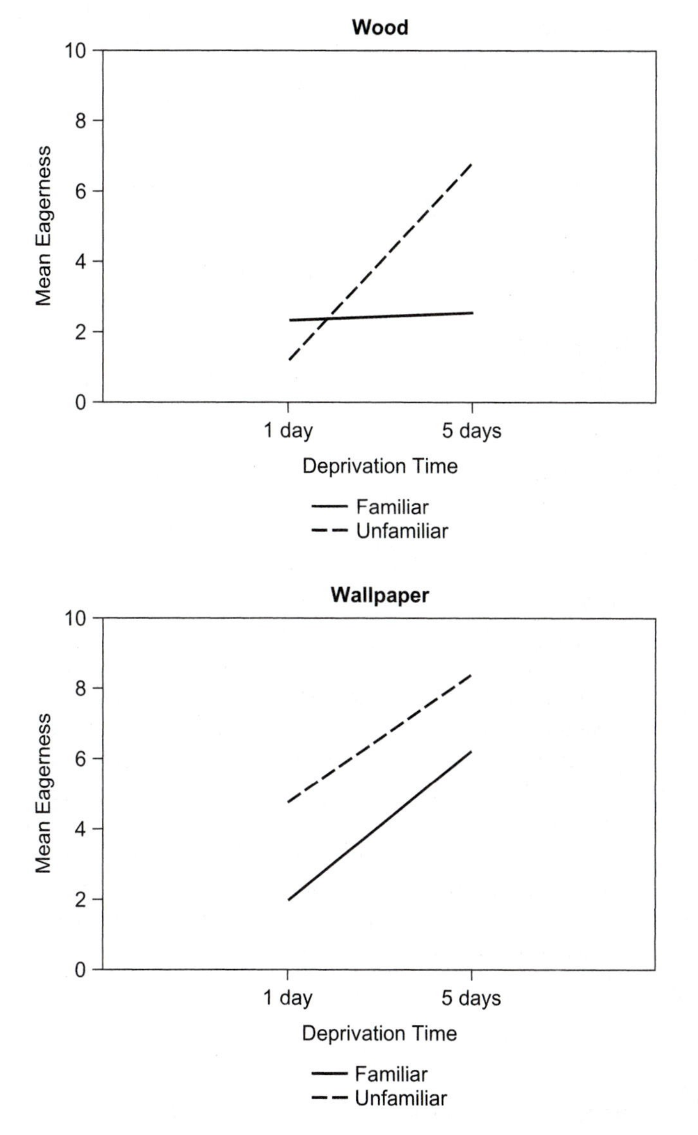

5.6.2 Effect Size

Recall that effect size is the proportion of variability in the DV that is associated with or accounted for by an IV (cf. Section 4.5.1). In a Venn diagram (e.g., Figure 4.2) in which the circle for the DV represents all of the variability in the DV, effect size is the overlapping area between the DV and an IV (a main effect or an interaction). Effect sizes in a factorial design are reported for all statistically significant effects. Sometimes researchers also choose to report effect size for nonsignificant effects.

The generalization of η^2 for factorial designs is

$$\eta^2_{\text{effect}} = R^2_{\text{effect}} = \frac{\text{SS}_{\text{effect}}}{\text{SS}_T} \tag{5.14}$$

where $\text{SS}_{\text{effect}}$ is the sum of squares for A or B, or AB, or whatever.

The three significant effects for the two-way small-sample example of Table 5.7 thus produce for profession:

$$\eta^2_A = R^2_A = \frac{33.85}{190.30} = .18$$

and for length of stay:

$$\eta^2_B = R^2_B = \frac{32.30}{190.30} = .17$$

and for interaction of length of stay with profession:

$$\eta^2_{AB} = R^2_{AB} = \frac{104.15}{190.30} = .55$$

The two main effects—length of stay and profession of rater—account for approximately the same proportion of variance in the rating of a vacation. The interaction, however, accounts for a much greater proportion of variance. In a factorial design with equal n, these effects are orthogonal and additive, so that .90 (.17 + .18 + .55) of the variability in rating of vacation is predictable from the combination of IVs.

With factorial designs and several statistically significant sources of variability, this measure of effect size can be quite small, because the denominator contains systematic effects other than the one of current interest. Thus, the denominator (SS_T) is larger than it would be if a corresponding one-way ANOVA had been run on one of the factors (e.g., A), making the ratio smaller. A solution is to compute partial η^2, in which the denominator consists only of the effect of interest and the error; this produces a better guess of the effect size than might be found in a one-way design with just factor A:

$$\text{Partial } \eta^2 = \frac{\text{SS}_{\text{effect}}}{\text{SS}_{\text{effect}} + \text{SS}_{\text{error}}} \tag{5.15}$$

where SS_{error} is the error sum of squares ($\text{SS}_{S/AB}$, $\text{SS}_{S/ABC}$, or whatever). For instance, for the main effect of A (profession) in the two-way example,

$$\text{Partial } \eta^2_A = \eta^2_A = \frac{33.85}{33.85 + 20.00} = .63$$

which is a much larger estimate than that produced by standard η^2. This is the Partial Eta Squared produced by SPSS GLM (cf. Table 5.15). However, when partial η^2 is used, effect sizes from various sources are *not* additive.

Smithson's (2003) software may be used to find confidence limits around partial η^2. Use of the SPSS files is demonstrated in Section 5.7.2.1. For SAS, confidence limits are found by adding values to the syntax file NONCF2.SAS. Table 5.24 shows F,

TABLE 5.24 Partial syntax and output from NONCF2.SAS for effect size (rsq) with 95% confidence limits (rsqlow and rsqupp) for length, profess, and their interaction, respectively

```
    rsq = df1 * F / (df2 + df1 * F);
    rsqlow = ncplow / (ncplow + df1 + df2 + 1);
    rsqupp = ncpupp / (ncpupp + df1 + df2 + 1);
    cards;
15.23  2  18  .95
14.53  2  18  .95
23.43  4  18  .95
;
proc print;
run;
```

Obs	F	df1	df2	conf	prlow	prupp	ncplow	ncpupp	rsq	rsqlow	rsqupp
1	15.23	2	18	0.95	0.975	0.025	7.1494	65.0041	0.62856	0.25398	0.75583
2	14.53	2	18	0.95	0.975	0.025	6.5990	62.4871	0.61751	0.23910	0.74846
3	23.43	4	18	0.95	0.975	0.025	32.5523	174.840	0.83888	0.58598	0.88374

numerator df, denominator df, and desired proportion for confidence limits for PROFESS, LENGTH, and the interaction, respectively. Values are from Table 5.16. Table 5.24 shows that for PROFESS, partial $\eta^2 = .63$, with confidence limits from .25 to .76. For LENGTH, partial η^2 (rsq) $= .62$, with confidence limits from .24 (rsqlow) to .75 (rsqupp). For the interaction, partial $\eta^2 = .84$, with confidence limits from .59 to .88. Effect sizes for comparisons may be calculated with respect to the total SS or with respect to the effect they are decomposing. These are demonstrated in context of the various types of comparisons in Section 5.6.4.

As in the one-way analysis, η^2 (or partial η^2) is a descriptive statistic, not a parameter estimate. The corresponding parameter estimate, ω^2, may also be applied to factorial designs. As in one-way designs, ω^2 adjusts for overestimation by η^2 and is always smaller than η^2. For factorial designs in which only one effect is significant,

$$\hat{\omega}^2 = \frac{\text{SS}_{\text{effect}} - \text{df}_{\text{effect}}(\text{MS}_{\text{error}})}{\text{SS}_T + \text{MS}_{\text{error}}} \tag{5.16}$$

where MS_{error} is $\text{MS}_{S/AB}$, $\text{MS}_{S/ABC}$, or whatever is appropriate for the design.

Although there is more than one significant effect in the design so application of Equation 5.16 is problematic, ω^2 for the main effect of A in the two-way example is

$$\hat{\omega}^2 = \frac{33.85 - 2(1.11)}{190.30 + 1.11} = .17$$

which is smaller than the .18 produced by η^2.

The partial form of ω^2 is available when there is more than one significant effect (Keren and Lewis, 1979):

$$\text{Partial } \hat{\omega}^2 = \frac{\text{df}_{\text{effect}}\left(\text{MS}_{\text{effect}} - \text{MS}_{\text{error}}\right)/N}{\left(\text{df}_{\text{effect}}\left(\text{MS}_{\text{effect}} - \text{MS}_{\text{error}}\right)\right)/N + \text{MS}_{\text{error}}} \tag{5.17}$$

Applied to the main effect of A,

$$\text{Partial } \hat{\omega}^2 = \frac{2(16.93 - 1.11)/27}{2(16.93 - 1.11)/27 + 1.11} = .53$$

This estimate is smaller than partial η^2 but much larger than unadjusted ω^2.

A drawback of ω^2 (whether adjusted or not) is that it becomes much more complicated when there are several significant effects in the design and when the layout is other than a randomized-groups design with equal sample sizes in all cells. The descriptive measures η^2 or partial η^2, on the other hand, easily generalize to a variety of ANOVA designs. In addition, confidence limits are readily available only for partial η^2.

Cohen's d (1977), described in Section 4.5.1, works well for main effects if means are evenly spaced throughout their range. However, it also becomes complicated for other arrangements of means within their range and when applied to interactions in factorial designs (see Cohen, 1977, Chapter 8). Thus, d is a less convenient measure than ω^2 or η^2, especially for factorial designs.

5.6.3 Power and Sample Size

Considerations for designing a powerful factorial research study are the same as those for designing a powerful one-way study, as discussed in Section 4.5.2.

5.6.3.1 Estimating Sample Size

Some techniques discussed in Section 4.5.2 are available for estimating sample size in factorial designs. However, the quick-and-dirty methods of Equations 4.6 and 4.7 are problematic with factorial designs because of the many possible mean differences to be evaluated.

One approach is to solve Equation 4.6 or 4.7 separately for each main effect and each interaction in a factorial design and then use the largest resulting sample size. For the interaction, the difference between means, δ, is defined as the smallest difference between mean differences to be detected.[8] Say, for example, that a pilot study for a 2×2 experiment produces $MS_{S/AB} = 1$ and the cell and marginal means of Table 5.25 (see Table 5.26).

TABLE 5.25 Hypothetical means from a pilot study for a 2×2 design

	b_1	b_2	Mean A
a_1	5.6	5.8	5.7
a_2	6.6	7.5	7.05
Mean B	6.1	6.65	

[8]Interpretation of interactions as differences among mean differences is discussed in Section 5.4.

Estimation of sample size for either of the main effects follows procedures of Section 4.5.2.2, using $MS_{S/AB}$ as the estimate of error variance and realizing that the resulting sample size is divided among levels of the other IV. For example, the main effect of B has $\delta = 6.65 - 6.1 = 0.55$ in the pilot study, which may be used as an estimate of the smallest difference to be detected. Entering into Equation 4.7 (which assumes $\alpha = .05$ and $\beta = .15$), we have

$$n = \frac{18\sigma^2}{\delta^2} = \frac{18(1)}{0.55^2} = 59.5$$

This value is divided by the number of levels of A to produce estimates of cell size, so that $n = 59.5/2 = 29.75$, or 30 per cell.

For the interaction, the difference among mean differences is $\delta = (7.5 - 6.6) - (5.8 - 5.6) = 0.7$. When entered in Equation 4.7,

$$n = \frac{18(1)}{0.57} = 36.73$$

This value must be multiplied by 2, because each difference is based on two cells, so that $n = 2(36.73) = 73.46$. Rounding up, $n = 74$ per cell.

PC-SIZE (Dallal, 1986) and NCSS PASS work well for estimating sample size for main effects in a factorial design, but interaction estimates are complicated. SAS GLMPOWER, however, handles all effects in factorial designs quite well. Table 5.26 shows the syntax and output for the 2 × 2 example, assuming $\alpha = .05$ and $\beta = .15$. The `do` loop shows two levels each for A and B. Hypothesized means, shown in gray, follow the `datalines;` instruction. The `model DV=A|B` instruction indicates that the $A \times B$ interaction is to be considered, as well as the lower-order main effects. The lack of a number following `ntotal` indicates that total sample size is desired as output.

The output provides separate sample size estimates for each effect. Choose the largest sample size if you want all effects to be statistically significant. Thus, a total sample of 296 cases (the same as estimated from the quick-and-dirty method) is needed for the proposed study.

5.6.3.2 Power of a Nonsignificant Effect

SAS GLMPOWER also facilitates power estimates for a completed factorial design when information about the design, sample size, means, and pooled error standard deviation are entered. Power estimates are often used at this point to investigate the reason for a nonsignificant finding. Table 5.27 shows output from SAS GLMPOWER for the data in the 3 × 3 ANOVA of Section 5.4.3. The square root of the error MS ($1.11^{1/2} = 1.05$) provides the `stddev` value. Note that it is the `power` instruction that now has no number. All of the effects in the small-sample example had power in excess of 99%. Note that these are the same values available from SPSS GLM in Table 5.15.

TABLE 5.26 Sample size determination through SAS GLMPOWER (syntax and selected output)

```
data TWOWAY;
do A = 1 to 2;
    do B = 1 to 2;
        input DV @@;
        output;
  end;
end;
datalines;
 5.6  5.8
 6.6  7.5
;
 run;
proc glmpower data=TWOWAY;
class A B;
model DV = A|B;
power
  stddev = 1
  ntotal = .
  power = .85;
run;
```

Computed N Total

Index	Source	Test DF	Error DF	Actual Power	N Total
1	A	1	20	0.882	24
2	B	1	120	0.859	124
3	A*B	1	292	0.851	296

5.6.4 Specific Comparisons

A significant main effect in an omnibus analysis is ambiguous when the main effect has more than two levels. Just as in a one-way design with three or more levels, ANOVA tells you that the means are not equal, but it gives no clue as to which means are different from each other. And a significant interaction—even in the 2×2 design, in which both IVs have two levels—almost always warrants closer inspection. Various comparison procedures provide more focused tests of differences among means.

As for a one-way design, comparisons may be either planned or post hoc (Section 4.5.5). Comparisons done post hoc are more conservative (less powerful), so it is better to plan them if you can. Or, as advocated by Keppel (1991), you can mix the two by following a sequential plan in which the choice of post hoc comparisons depends on the outcome of omnibus tests.

5.6.4.1 Types of Comparisons

Comparisons test differences among means or combinations of means by applying weighting coefficients to them and forming F ratios that are tested using critical

TABLE 5.27 Output of SAS GLMPOWER for results of 3 × 3 factorial ANOVA (syntax and selected output)

```
data TWOWAY;
do A = 1 to 3;
    do B = 1 to 3;
        input DV @@;
        output;
    end;
end;
datalines;
  0.33  5.67  6.33
  6.00  6.00  8.33
  6.33  9.00  2.67
;
 run;
proc glmpower data=TWOWAY;
class A B;
model DV = A|B;
power
  stddev = 1.05
  ntotal = 27.
  power = .;
run;
```

Computed Power

Index	Source	Test DF	Power
1	A	2	0.997
2	B	2	0.996
3	A*B	4	>.999

values of F based on either planned or post hoc rules of assessment. Three kinds of comparisons are available for factorial designs: comparisons of marginal means for main effects, comparisons applied to cell means for investigating special designs (such as the dangling control group), and comparisons intended to dissect an interaction (as discussed in Section 5.6.4.4).

5.6.4.1.1 Comparisons on Marginal Means When a main effect with more than two levels is statistically significant, or when there is a plan to investigate a main effect, comparisons on marginal means are often desirable. Remember, however, that comparisons on marginal effects are problematic when there is significant interaction, as discussed in Section 5.6.1. Significant interaction implies that the effect of one IV differs depending on the level of another IV. If you already know that a main effect has limited generalizability, its interpretation on the margin is often misleading.

To conduct a comparison on the margin (e.g., $SS_{A\text{comp}}$ or $SS_{B\text{comp}}$), you would generate coefficients and apply them to marginal means, as is done in a one-way design. The means for each level, averaged across all levels of other IVs, are weighted by coefficients, with the constraint that the coefficients sum to 0 for each comparison. Two comparisons are orthogonal if, in addition, the sum of the cross products of weighting coefficients is 0.

As discussed in Section 5.4, there are two approaches to ANOVA—the traditional approach and the regression approach. The regression approach, as illustrated in Section 5.4.4, builds marginal comparisons into the X columns themselves. The first column of Table 5.8, X_1, applies the coefficients $2\ -1\ -1$ to the three levels of A. The sum of squares for regression for this column *is* the sum of squares for that comparison on the marginal means for A ($SS_{A\text{comp1}}$). The second column, X_2, applies the coefficients $0\ 1\ -1$ to the levels of A. The sum of squares for regression for this column is the sum of squares for that comparison on the marginal means ($SS_{A\text{comp2}}$).

The traditional ANOVA approach to the same comparisons is available through the following equation. Because comparisons each have 1 df, the sum of squares is the same as the mean square, so the F ratio for marginal comparisons is obtained in one step:

$$F = \frac{n_{\overline{Y}}\left(\Sigma w_j \overline{Y}_j\right)^2 / \Sigma w_j^2}{MS_{\text{error}}} \tag{5.18}$$

where $n_{\overline{Y}}$ = number of scores in each of the means to be compared,
$(\Sigma w_j \overline{Y}_j)^2$ = squared sum of the weighted means,
Σw_j^2 = sum of the squared coefficients, and
MS_{error} = mean square for error in the ANOVA.

By using Equation 5.18 for the traditional ANOVA approach to data of the small-sample example, the coefficients for the comparison of administrators against belly dancers and politicians combined ($2\ -1\ -1$, just as for the regression approach) are applied to the marginal means of A (4.11, 6.78, 6.00), type of profession:

$$F = \frac{9((2)(4.11) + (-1)(6.78) + (-1)(6.00))^2/(2^2 + (-1)^2 + (-1)^2)}{1.11}$$
$$= \frac{31.193}{1.11} = 28.10$$

The $n_{\overline{Y}}$ for this analysis is the number of scores in each of the marginal means for A, or 9. Note that the numerator (which is both $SS_{A\text{comp1}}$ and $MS_{A\text{comp1}}$) is equivalent (when computations are carried to four decimal places) to SS(reg. X_1) of Table 5.11. The same procedure is available for the second comparison on A; when the coefficients defining that comparison are used in Equation 5.18, it yields SS(reg. X_2) in the numerator (which is also both $SS_{A\text{comp2}}$ and $MS_{A\text{comp2}}$). If the trend coefficients for X_3 and X_4 in Table 5.8 are applied to marginal means of B (4.22, 6.89, 5.78—for length of stay), the numerators of Equation 5.18 become SS(reg. X_3)—or $SS_{B\text{comp1}}$—and SS(reg. X_4)—or $SS_{B\text{comp2}}$—respectively. Comparisons of marginal means are interpreted as in a one-way design.

Effect size for this comparison may be calculated in two ways. First, η^2 may be calculated for the comparison as a proportion of the total sum of squares for the factorial design:

$$\eta^2 = \frac{SS_{A\text{comp}}}{SS_T} = \frac{31.19}{190.30} = .16 \tag{5.19}$$

Alternatively, η^2 may be calculated for the comparison as a proportion of the main effect of which it is a component:

$$\eta^2 = \frac{SS_{A\text{comp}}}{SS_A} = \frac{31.19}{33.85} = .92 \quad \textbf{(5.20)}$$

The interpretation of these two versions of η^2 is quite different, so specify which you used when you report your results.

5.6.4.1.2 ***Comparisons on Cell Means*** Usually, when weighting coefficients are applied to cell means, the researcher is attempting to understand an interaction. (Procedures appropriate to dissecting an interaction in a factorial design are discussed in Section 5.6.4.4.) Sometimes, however, there is an unusual design, and coefficients are applied to cell means to analyze features of the design.

One example of an unusual design in which comparisons might be applied to cell means is the dangling group design of Section 5.5.2. In this design, the researcher very likely wants to know whether the mean for the control group differs from the combined mean of the four treated groups. To accomplish this comparison, each of the four treated groups is assigned a coefficient of 1 and the control group a coefficient of −4. Table 5.28 shows the arrangement with the assigned coefficients.

Suppose the means for the five groups are as shown in Table 5.29. The $n_{\bar{Y}}$ of Equation 5.18 is the cell sample size. In this example, if we assume that $n = 5$ and that $MS_{S/A} = 2.3$, then

$$F = \frac{5((1)(5.6) + (1)(5.8) + (1)(6.6) + (1)(7.5) + (-4)(4))^2/(1^2 + 1^2 + 1^2 + 1^2 + (-4)^2)}{2.3}$$

$$= 9.81$$

TABLE 5.28 Assignment of weighting coefficients in a factorial design with two levels each of two drug treatments and a single control

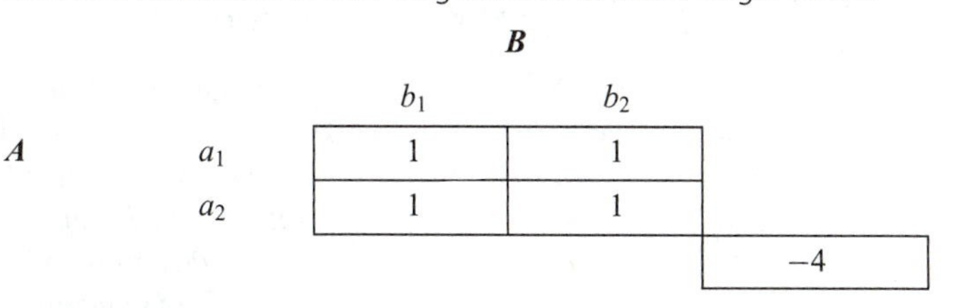

		B		
		b_1	b_2	
A	a_1	1	1	
	a_2	1	1	
				−4

TABLE 5.29 Hypothetical means for a factorial design with two levels each of two drug treatments and a single control

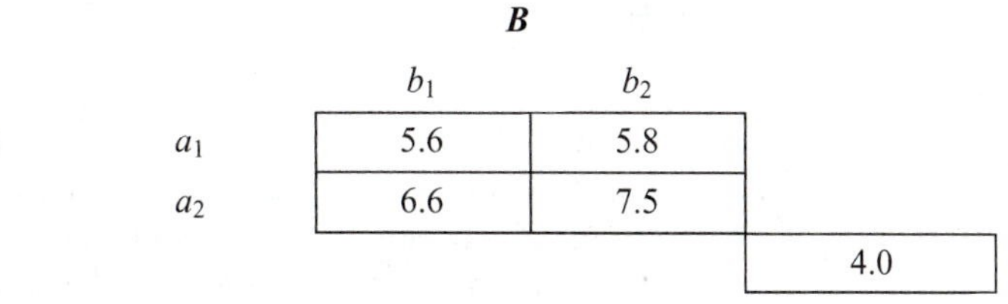

		B		
		b_1	b_2	
A	a_1	5.6	5.8	
	a_2	6.6	7.5	
				4.0

Obtained F is evaluated as either planned or post hoc, depending on when the decision to do the comparison was made. Effect size is calculated as the proportion of total variance associated with the effect:

$$\eta^2 = \frac{SS_{comp}}{SS_T} \tag{5.21}$$

A solution requires calculation of SS_T, which, in turn, requires the raw scores not given in this example.

5.6.4.2 Critical F for Planned Comparisons

If you have planned your comparisons prior to data collection and if you have planned no more comparisons than you have degrees of freedom for the effect to be tested, critical F is obtained from the tables, just as in routine ANOVA. Each comparison is tested against critical F at routine α with 1 df in the numerator and degrees of freedom associated with MS_{error} in the denominator. If obtained F is larger than critical F, the null hypothesis represented by the weighting coefficients is rejected.

In the 3×3 small-sample example, error df $= 18$ (see Table 5.7), and the critical value in Table A.1 is 4.41 at $\alpha = .05$. The obtained $F = 28.10$ for the first marginal comparison, A_{comp1}, clearly exceeds the critical value, leading to the conclusion that there is a significant mean difference in ratings of vacations by administrators versus belly dancers and politicians.

In the dangling control group example, the denominator df $= a(n - 1) = 5(5 - 1) = 20$. The critical value from Table A.1 with 1 and 20 df and $\alpha = .05$ is 4.35. With obtained $F = 9.81$, the null hypothesis of no mean difference between the control group and the pooled treated groups is rejected.

As for the one-way design, the Bonferroni correction is applied if more comparisons are planned than are justified by the available df (Section 4.5.5.4.3).

5.6.4.3 Adjustments for Post Hoc Comparisons

Criteria for post hoc tests in the factorial design are the same as for the one-way design. The researcher chooses the most powerful post hoc method that allows the comparisons of interest. Both Scheffé and Tukey adjustments are available for marginal tests in factorial designs or for unusual designs. The Dunnett test is typically not applied in factorial designs. An exception is the design with a single control, which is treated as a one-way design for purposes of the Dunnett test (Section 4.5.5.4.3). Additional post hoc tests available in software are summarized in Table 4.24.

5.6.4.3.1 Scheffé The Scheffé method for computing critical F for a comparison on marginal means is

$$F_S = (k - 1)F_c \tag{5.22}$$

where F_S = adjusted critical F,
$(k - 1)$ = degrees of freedom for the effect with k levels (a or b), and
F_c = tabled F with effect degrees of freedom in the numerator and degrees of freedom associated with the error term in the denominator.

If obtained F is larger than F_S, the null hypothesis represented by the weighting coefficients for the comparison is rejected. For example, for comparisons on the marginal means of A, critical $F_S = 2(3.55) = 7.10$. The obtained $F_{A\text{compl}} = 28.10$ for comparing the means for administrators against the pooled mean for belly dancers and politicians exceeds even this strong adjustment. The same critical value is used for post hoc tests on marginal means of B, because in this example the number of levels of B is also three.

The adjustment for tests of cell means instead of marginal means for unusual designs is more drastic, because there are usually numerous levels and, therefore, many possible comparisons. To find the total number of levels, you transform the factorial segment of the design into a one-way design with $a \times b$ levels and then add the remaining cells. Thus, the 2×2 part of the dangling control groups example yields four levels, plus the level for the control group, for a total of five. If the dangling control groups comparison is done as part of a post hoc analysis, the critical value F_S is based on F_c with 4 and 20 df, so that $F_S = (5 - 1)(2.87) = 11.48$. The obtained $F = 9.81$, in this case, is not statistically significant.

5.6.4.3.2 ***Tukey*** There are two ways to use the Tukey test on main effects in a factorial design. The first, and probably most sensible, option is to do a separate Tukey test for each statistically significant main effect. If done by hand, obtained F is calculated by assigning coefficients to marginal means (as in Section 5.6.4.1.1), whereas critical F_T is based on the number of marginal means. In Section 5.6.4.1.1, obtained F for the difference in mean ratings between administrators and the other two groups is 28.10. Critical $F_T = (3.61)^2/2 = 6.52$ (Equation 4.15) for the three marginal means of A. Therefore, the mean difference for the comparison is significant by the Tukey criterion. This approach to the Tukey test is available in the GLM programs of both SPSS[9] and SAS.

The second approach is simply to ignore the factorial arrangement and treat the means on the margins as if from a one-way design. That is, record the cells into a one-way design, as illustrated later in Table 5.29 and elsewhere. The Tukey test is then identical to that described in Section 4.5.3.5.2. This approach may also be taken with software for other pairwise comparisons that are available for one-way ANOVA, such as Bonferroni or Sidak.

If this approach is taken, the 3×3 example of Section 5.4 is treated as a one-way design with 9 groups and 36 possible pairwise comparisons. Obtained Fs are calculated by assigning coefficients of 1 and -1 separately to each pair of means and critical F_T is based on nine means, so that, in the example, $F_T = (4.96)^2/2 = 12.3$ at $\alpha = .05$. With so many possible comparisons, it is usually best to put the mean differences in order and test the smallest mean difference first, followed by the next smallest, and so on until the first statistically significant difference is found. The larger differences are also significant. (Alternatively, you can test the largest difference first and work down until you find a nonsignificant difference.) Or, you may find the minimum significant mean difference, $\bar{d}_T$, as per Equation 4.16.

[9] SPSS GLM also has a Dunnett C test, which does not require homogeneity of variance but which uses the same test distribution as the Tukey test.

5.6.4.4 Analyzing Interactions

In omnibus ANOVA, there are two approaches to pinpointing the location of the mean differences when the interaction is significant: a simple-effects analysis or an interaction-contrasts analysis. There is a helpful notational scheme that clearly labels these various types of comparison.

Simple-effects analysis includes both simple main effects and simple comparisons. Both break a factorial design down into a series of one-way designs by examining one IV separately at each level of another IV. A *simple main effect* holds one IV constant at a chosen level and examines mean differences among all levels of the other IV. For instance, you could compute $SS_{A \text{ at } b1}$ (to test the cell means for A at the first level of B) or $SS_{B \text{ at } a1}$ (to test the cell means for B at the first level of A). Then, if either A or B has more than two levels, *simple comparisons* are possible. For instance, you could compute $SS_{A\text{comp1 at } b1}$ (to compare the two means formed by applying weighting coefficients to the cell means of A at the first level of B) or $SS_{B\text{comp1 at } a1}$ (to compare the two means formed by applying weighting coefficients to the cell means of B at the first level of A). For the example, $SS_{A \text{ at } b1}$ represents the sum of squares for the mean difference in ratings among administrators, belly dancers, and politicians for a one-week vacation. $SS_{A\text{comp1 at } b1}$ represents the sum of squares for the difference between administrators and the other two groups for a one-week vacation.

An interaction-contrasts analysis examines interaction by forming smaller interactions with weighting coefficients applied to either A or B or both. When weighting coefficients are applied to A, wherein all levels of B are used, one assesses $F_{A\text{comp} \times B}$. When weighting coefficients are applied to B, wherein all levels of A are used, one assesses $F_{A \times B\text{comp}}$. These are sometimes called *partial factorials.* If weighting coefficients are applied to both A and B, one assesses $F_{A\text{comp} \times B\text{comp}}$. For our example, $F_{A\text{comp1} \times B}$ is a test of the interaction of administrators versus the other two groups with vacations of all three lengths: Do administrators show the same pattern of responses to vacations of the same length as the other two groups? $F_{A \times B\text{comp1}}$ is a test for the interaction of the linear trend of vacation length for all three professions: Do all three professions have the same linear trend? Finally, $F_{A\text{comp1} \times B\text{comp1}}$ is a test of the similarity of the linear trend of vacation length for administrators versus the other two groups: Do administrators have the same linear trend as the other two groups?

Either simple effects or interaction contrasts can be part of a comprehensive sequential plan, such as that recommended by Keppel (1991, Chapters 11, 12, 20), where the decision about which post hoc analyses to use depends on the outcome of the omnibus ANOVA. The general strategy for a sequential analysis considers the statistical significance of all interaction(s) and main effects. If the interaction is significant, and at least one main effect is *not* significant, use simple-effects analysis. If the interaction is significant, but so are all of the main effects, use interaction contrasts. If the interaction is significant, and none of the main effects is significant, use either simple effects or interaction contrasts, whichever is the easier to explain.

The sequential plan usually follows the results of omnibus ANOVA, so post hoc adjustments are required. However, either simple effects or interaction contrasts also could be planned. If you plan no more than three or four comparisons and eschew omnibus ANOVA, adjustment is unnecessary. If, however, you plan more comparisons than that, a Bonferroni correction (cf. Section 4.5.5.4.2) is prudent.

Simple-effects and interaction-contrasts analyses are demonstrated through software in the following sections, with analysis by hand also discussed to some extent. Briefly, simple-effects analyses are done by hand by performing a one-way ANOVA, with follow-up comparisons, as described in Chapter 4, on the appropriate row or column of data. (Equations and hand-worked examples for the traditional ANOVA approach are also demonstrated in Keppel [1991].) Interaction-contrasts analyses are most easily done by hand, by using the regression strategy in conjunction with a spreadsheet program, or by creating the needed coefficients, as illustrated in Table 5.35 and using them in syntax for one of the software packages.

5.6.4.4.1 Simple Effects In simple-effects analysis, one IV is held constant at a selected level, and mean differences are examined on the levels of the other IV, as shown in Figure 5.8. For instance, the level of A is held constant at administrator, while mean differences are examined in ratings of one-, two-, and three-week vacations (Figure 5.8(a)). The researcher asks whether administrators have mean differences in their ratings of one-, two-, and three-week vacations by assessing $F_{B \text{ at } a1}$. Or, length of vacation is held constant at one week, while mean differences are explored

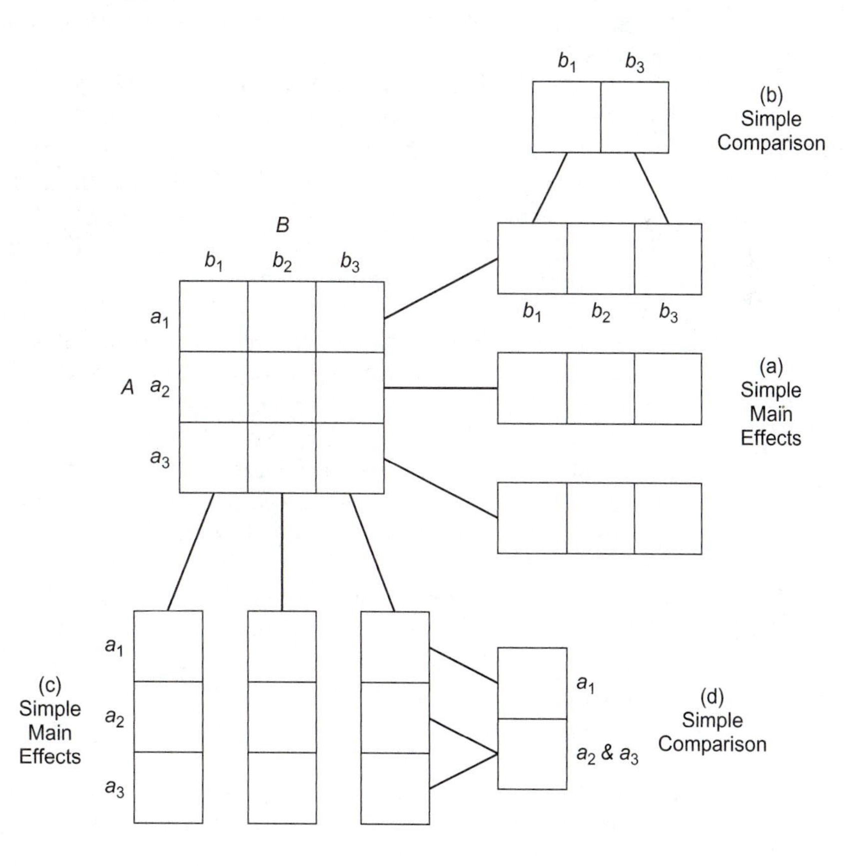

FIGURE 5.8 Simple-effects analysis exploring (a) differences among B for each level of A, followed by (b) simple comparison between levels of B for one level of A; (c) differences among A for each level of B, followed by (d) simple comparison between levels of A for one level of B

among professions (Figure 5.8(c)). The researcher asks whether the three types of professions have different mean ratings of one-week vacations by assessing $F_{A \text{ at } b1}$. Each test of these simple main effects has 2 df in the numerator, because, in this example, A and B both have three levels. A simple main effects analysis could then be followed by simple comparisons (Figures 5.8(b) and 5.8(d)), where the numerator again has 1 df. These simple comparisons may be among single cells, as in Figure 5.8(b), or among combined cells, as in Figure 5.8(d).

In the small-sample 3×3 example of Section 5.4.3, both main effects and the interaction are significant. Because all effects are significant, the guidelines suggest that interaction contrasts are appropriate rather than simple effects. For didactic purposes, however, the simple effects of A (profession) will be analyzed at each level of B (vacation length).

Simple-effects analysis partitions one of the main effects from the omnibus ANOVA in addition to the interaction. When A is held constant at a single level to analyze B, both the sum of squares for interaction and the sum of squares for B are reapportioned into SS for simple effects of B. When levels of B are held constant and A is analyzed, both the sum of squares for interaction and the sum of squares for A are reapportioned into SS for simple effects of A. In simple-effects analyses, because the interaction sum of squares is confounded (intermingled) with one or the other of the main effects, it seems best to confound them with a nonsignificant main effect where possible.

Hand calculations for simple effects are straightforward. You do a one-way ANOVA on each "slice" of the design. For example, to do simple effects on A at each level of B, you do three one-way ANOVAs—A at b_1, A at b_2, and A at b_3—by applying the equations described in Section 4.4.3. This provides you with $MS_{A \text{ at } bj}$ and $MS_{S/A \text{ at } bj}$.[10]

To form the F ratio, divide $MS_{A \text{ at } bj}$ by either $MS_{S/A \text{ at } bj}$ or $MS_{S/AB}$, the error term for the factorial design. Usually, $MS_{S/AB}$ is used because it is considered more stable, as it is based on more numerous estimates of the same within-cell error variability (all the cells instead of just some of them). However, if there are rather larger differences in the variability in the cells (failure or near failure of homogeneity of variance at some levels of B), it makes more sense to use the error term developed specifically for the simple effect. Whichever error term you choose, for small data sets, calculating the ANOVAs by hand may prove more efficient than using software.

Table 5.30 shows syntax and location of output for simple-effects analysis through SPSS GLM, SAS GLM, and SYSTAT GLM with $MS_{S/AB}$ as the error term. Also shown is SPSS MANOVA, which has a special, simpler syntax for simple effects.

Simple main effects in SPSS GLM are done through LMATRIX syntax, wherein the researcher provides weighting coefficients for the effect to be analyzed (profess) and for the interaction. For example, the first LMATRIX is identified (within quotes) as testing the simple main effect of profess at one week. The LMATRIX instruction has two parts, one for each degree of freedom in the profess main effect, separated

[10] These one-way ANOVAs can also be performed through software by selecting data for just one level of B at a time and doing the analysis, with comparisons if desired.

TABLE 5.30 Syntax for simple main effects analysis

Program	Syntax	Section of Output	Name of Effect
SPSS GLM	UNIANOVA RATING BY PROFESS LENGTH /METHOD = SSTYPE(3) /INTERCEPT = INCLUDE /CRITERIA = ALPHA(.05) /LMATRIX = "profess at one week" profess 2 -1 -1 profess*length 2 0 0 -1 0 0 -1 0 0 ; profess 0 1 -1 profess*length 0 0 0 1 0 0 -1 0 0 /LMATRIX = "profess at two weeks" profess 2 -1 -1 profess*length 0 2 0 0 -1 0 0 -1 0; profess 0 1 -1 profess*length 0 0 0 0 1 0 0 -1 0 /LMATRIX = "profess at three weeks" profess 2 -1 -1 profess*length 0 0 2 0 0 -1 0 0 -1; profess 0 1 -1 profess*length 0 0 0 0 0 1 0 0 -1 /DESIGN profess, length, profess*length.	**Custom Hypothesis Test #1**, etc.	Test Results
SPSS MANOVA	MANOVA rating BY profess length(1,3) /PRINT = SIGNIF(BRIEF) /DESIGN length profess WITHIN length(1), profess WITHIN Length(2), profess WITHIN length(3).	Tests for rating using UNIQUE sums of squares	Source of Variation
SAS GLM	data SASUSER.FAC_RANG; set SASUSER.FAC_RAND; IF PROFESS=1 AND LENGTH=1 THEN GROUP=1; IF PROFESS=1 AND LENGTH=2 THEN GROUP=2; IF PROFESS=1 AND LENGTH=3 THEN GROUP=3; IF PROFESS=2 AND LENGTH=1 THEN GROUP=4; IF PROFESS=2 AND LENGTH=2 THEN GROUP=5; IF PROFESS=2 AND LENGTH=3 THEN GROUP=6; IF PROFESS=3 AND LENGTH=1 THEN GROUP=7; IF PROFESS=3 AND LENGTH=2 THEN GROUP=8; IF PROFESS=3 AND LENGTH=3 THEN GROUP=9; PROC GLM; CLASS GROUP; MODEL RATING = GROUP; CONTRAST 'PROFESS AT 1 WK' GROUP 2 0 0 -1 0 0 -1 0 0, GROUP 0 0 0 1 0 0 -1 0 0; CONTRAST 'PROFESS AT 2 WK' GROUP 0 2 0 0 -1 0 0 -1 0, GROUP 0 0 0 0 1 0 0 -1 0 ; CONTRAST 'PROFESS AT 3 WK' GROUP 0 0 2 0 0 -1 0 0 -1, GROUP 0 0 0 0 0 1 0 0 -1; RUN;	Contrast	PROFESS AT 1 WK., etc.

SYSTAT GLM	IF (PROFESS=1) AND (LENGTH=1) THEN LET GROUP=1
	IF (PROFESS=1) AND (LENGTH=2) THEN LET GROUP=2
	IF (PROFESS=1) AND (LENGTH=3) THEN LET GROUP=3
	IF (PROFESS=2) AND (LENGTH=1) THEN LET GROUP=4
	IF (PROFESS=2) AND (LENGTH=2) THEN LET GROUP=5
	IF (PROFESS=2) AND (LENGTH=3) THEN LET GROUP=6
	IF (PROFESS=3) AND (LENGTH=1) THEN LET GROUP=7
		Test of Hypothesis	Hypothesis

TABLE 5.30 Continued

```
IF (PROFESS=3) AND (LENGTH=2)  THEN LET GROUP=8
IF (PROFESS=3) AND (LENGTH=3)  THEN LET GROUP=9
GLM
CATEGORY GROUP
COVAR
DEPEND RATING
ESTIMATE
HYPOTHESIS
EFFECT=group
NOTE 'PROFESS AT WEEK1'
CONTRAST[ 2 0 0 -1 0 0 -1 0 0 0 0 0 1 0 0 -1 0 0  ]
TEST
HYPOTHESIS
EFFECT=group
NOTE 'PROFESS AT WEEK2'
CONTRAST [ 0 2 0 0 -1 0 0 -1 0 0 0 0 0 1 0 0 -1 0 ]
TEST
HYPOTHESIS
EFFECT=group
NOTE 'PROFESS AT WEEK3'
CONTRAST [ 0 0 2 0 0 -1 0 0 -1 0 0 0 0 0 1 0 0 -1 ]
TEST
```

by a semicolon. Within each part, coefficients are provided for the first degree of freedom for the main effect of profess (2 −1 −1) and for the profess*length interaction at one week (2 0 0 −1 0 0 −1 0 0), and then for the second degree of freedom for the main effect of profess (0 1 −1) and its profess*length interaction at one week (0 0 0 1 0 0 −1 0 0). LMATRIX instructions then follow for the 2 df for a two-week vacation and for the 2 df for a three-week vacation. The results of both simple comparisons are combined in the output because both are within a single LMATRIX instruction. The output for the first LMATRIX instruction is labeled Custom Hypothesis Test #1, and so on.

Table 5.31 helps clarify the specification for the interaction, where A is type of profession and B is length of vacation. Because profess is named before length, the nine cells of the design are laid out in the order specified in the table. The coefficients are assigned to select the appropriate level of B for each comparison. Note, for instance, that when the comparison is *at* b_1, coefficients are only present in the b_1 columns.

TABLE 5.31 Order of cells for specification of weight coefficients for simple-effects analyses

	a_1b_1	a_1b_2	a_1b_3	a_2b_1	a_2b_2	a_2b_3	a_3b_1	a_3b_2	a_3b_3
$A_{\text{compl at } b1}$	2	0	0	−1	0	0	−1	0	0
$A_{\text{comp2 at } b1}$	0	0	0	1	0	0	−1	0	0
$A_{\text{compl at } b2}$	0	2	0	0	1	0	0	1	0
$A_{\text{comp2 at } b2}$	0	0	0	0	1	0	0	−1	0
$A_{\text{compl at } b3}$	0	0	2	0	0	−1	0	0	−1
$A_{\text{comp2 at } b3}$	0	0	0	0	0	1	0	0	−1

Table 5.30 also shows syntax for the simple-effects analysis through SAS GLM. The data step recodes the design into a nine-level one-way layout, with the IV labeled GROUP. SAS GLM requires separate CONTRAST instructions for simple effects of profession at each level of length. A set of coefficients is derived for each df of the simple-effects analysis. Two sets of coefficients, separated by a comma, are necessary, because there are three levels of profession. Here the set includes (2 −1 −1) and (0 1 −1), applied to the appropriate cells of the design representing the level of length selected, with zeros applied to all other cells. These are the same coefficients used for SPSS GLM.

SPSS MANOVA, available only in syntax, uses the WITHIN keyword in the DESIGN paragraph to indicate the levels within which simple effects are sought, in this case length. Thus, simple effects of profess are requested WITHIN each level of length: 1, 2, and 3. Notice that main effect of length is also specified in the DESIGN paragraph.

All programs test the three simple main effects as if they were planned. If these analyses were done post hoc, however, the printed significance levels would be inappropriate and a Scheffé adjustment would be needed to control for inflated familywise Type I error. For these tests, the Scheffé adjustment is

$$F_S = (a - 1)\, F_{(a-1),(\text{df}_{\text{error}})} \tag{5.23}$$

where a is the number of levels in the main effect being analyzed and $\text{df}_{\text{error}} = N - ab$.

For the example, using $\alpha = .05$,

$$F_S = (3 - 1)\, F_{2,18} = 7.10$$

All three simple main effects survive the Scheffé adjustment, with F values of 30.70, 9.10, and 22.30, respectively, indicating that there are differences among professions for all three levels of length of vacation. However, these simple effects each have 2 df (because there are three levels of A). Therefore, a further decomposition of the interaction (and main effect of A) into separate degrees of freedom comparisons is in order.

As shown in Figure 5.8(d), simple comparisons are possible on A at each level of B because A has three levels. A set of orthogonal comparisons tests (1) the mean difference between the administrators and the other two professions and then (2) the mean difference between belly dancers and politicians. The weighting coefficients for the first comparison are 2 −1 −1, and the coefficients for the second comparison are 0 1 −1. These comparisons are done by hand by applying Equation 5.18 to cell means, where $n_{\bar{Y}}$ is the number of scores per cell and where weights are applied to cell rather than to marginal means. F ratios are formed with $\text{MS}_{S/A \text{ at } bj}$ or $\text{MS}_{S/AB}$ as the error term in the denominator. Here, again, you may find it more efficient to do hand calculations for small data sets than to set up the rather complex syntax required by software programs.

Table 5.32 shows syntax and location of output for simple comparisons through SPSS GLM, SAS GLM, and SYSTAT GLM, with $\text{MS}_{S/AB}$ as the error term. Note that syntax for the SPSS and SAS is very similar to that used for simple main effects, except that in SPSS and SAS there is a separate set of contrast coefficients for each

TABLE 5.32 Syntax for simple comparisons in a two-way factorial design

Program	Syntax	Section of Output	Name of Effect
SPSS GLM	UNIANOVA RATING BY PROFESS LENGTH /METHOD = SSTYPE(3) /INTERCEPT = INCLUDE /CRITERIA = ALPHA(.05) /LMATRIX = "admin vs others at one week" profess 2 -1 -1 profess*length 2 0 0 -1 0 0 -1 0 0 /LMATRIX = "belly vs politic at one week" profess 0 1 -1 profess*length 0 0 0 1 0 0 -1 0 0 /LMATRIX = "admin vs others at two weeks" profess 2 -1 -1 profess*length 0 2 0 0 -1 0 0 -1 0 /LMATRIX = "belly vs politic at two weeks" profess 0 1 -1 profess*length 0 0 0 0 1 0 0 -1 0 /LMATRIX = "admin vs others at three weeks" profess 2 -1 -1 profess*length 0 0 2 0 0 -1 0 0 -1 /LMATRIX = "belly vs politic at three weeks" profess 0 1 -1 profess*length 0 0 0 0 0 1 0 0 -1 /DESIGN profess, length, profess*length.	**Custom Hypothesis Tests #1**, etc.	Contrast Results
SAS GLM	proc glm[a] class GROUP; model RATING = GROUP; contrast 'admin vs others at one week' GROUP 2 0 0 -1 0 0 -1 0 0; contrast 'belly vs politic at one week' GROUP 0 0 0 1 0 0 -1 0 0; contrast 'admin vs others at two weeks' GROUP 0 2 0 0 -1 0 0 -1 0; contrast 'belly vs politic at two weeks' GROUP 0 0 0 0 1 0 0 -1 0; contrast 'admin vs others at three weeks' GROUP 0 0 2 0 0 -1 0 0 -1; contrast 'belly vs politic at three weeks' GROUP 0 0 0 0 0 1 0 0 -1; run;	Contrast	admin vs others at one week, etc.

SYSTAT ANOVA	IF (PROFESS=1) AND (LENGTH=1) THEN LET GROUP=1
	IF (PROFESS=1) AND (LENGTH=2) THEN LET GROUP=2
	IF (PROFESS=1) AND (LENGTH=3) THEN LET GROUP=3
	IF (PROFESS=2) AND (LENGTH=1) THEN LET GROUP=4
	IF (PROFESS=2) AND (LENGTH=2) THEN LET GROUP=5
	IF (PROFESS=2) AND (LENGTH=3) THEN LET GROUP=6
	IF (PROFESS=3) AND (LENGTH=1) THEN LET GROUP=7
	IF (PROFESS=3) AND (LENGTH=2) THEN LET GROUP=8
		Test of Hypothesis	Hypothesis

```
IF (PROFESS=3) AND (LENGTH=3)  THEN LET GROUP=9
ANOVA
CATEGORY GROUP
COVAR
DEPEND RATING
ESTIMATE
HYPOTHESIS
NOTE 'admin vs others at one week'
EFFECT=GROUP
CONTRAST [ 2 0 0 -1 0 0 -1 0 0 ]
TEST
HYPOTHESIS
NOTE 'belly vs politic at one week'
EFFECT=GROUP
CONTRAST [ 0 0 0 1 0 0 -1 0 0 ]
TEST
HYPOTHESIS
NOTE 'admin vs others at two weeks'
EFFECT=GROUP
CONTRAST [ 0  2  0  0 -1 0 0 -1 0 ]
TEST
HYPOTHESIS
NOTE 'belly vs politic at two weeks'
EFFECT=GROUP
CONTRAST [ 0  0  0  0 1 0 0 -1 0 ]
TEST
HYPOTHESIS
NOTE 'admin vs others at three weeks'
EFFECT=GROUP
CONTRAST [ 0  0 2 0 0 -1 0 0 -1 ]
TEST
HYPOTHESIS
NOTE 'belly vs politic at three weeks'
EFFECT=GROUP
CONTRAST [ 0  0  0  0  0 1 0 0 -1 ]
TEST
```

[a] Preceded by coding into one-way design, as per Table 5.30.

TABLE 5.33 Columns for regression approach to simple-effects analysis

		***A* at b_1**	
		Acomp1 at b_1	Acomp2 at b_1
	Y	X_1	X_2
	0	2	0
a_1b_1	1	2	0
	0	2	0
	4	0	0
a_1b_2	7	0	0
	6	0	0
	5	0	0
a_1b_3	8	0	0
	6	0	0
	5	−1	1
a_2b_1	7	−1	1
	6	−1	1
	5	0	0
a_2b_2	6	0	0
	7	0	0

		***A* at b_1**	
		Acomp1 at b_1	Acomp2 at b_1
	Y	X_1	X_2
	9	0	0
a_2b_3	8	0	0
	8	0	0
	9	−1	−1
a_3b_1	9	−1	−1
	9	−1	−1
	9	0	0
a_2b_3	8	0	0
	8	0	0
	9	0	0
a_3b_1	9	0	0
	9	0	0

comparison. SAS requires recoding of the two IVs into a single IV in a one-way layout.

It may have occurred to you that the nine coefficients in the syntax look suspiciously like coefficients in the columns of the regression approach, as indeed they are. Simple main effects and simple comparisons are done through regression by the simple expedient of applying zeros to the levels of the variable currently omitted from analysis. The columns necessary to compute $SS_{A\text{comp1 at } b1}$ and $SS_{A\text{comp2 at } b1}$ are shown in Table 5.33. The sum of SS(reg.) for these two columns is $SS_{A \text{ at } b1}$.

Table 5.34 shows how simple-effects analysis partitions a main effect and the interaction and is then further decomposed into simple comparisons. Degrees of freedom and sums of squares in parentheses are those that are replaced by finer-grained analyses. A_{comp1} and A_{comp2} refer to the first and second comparisons on A, respectively. In the table, the 6 df from combining the main effect of A and the AB interaction are partitioned, first into simple main effects with 2 df each and then into the 6 single df comparisons. Table 5.34 also shows the results of the tests of simple comparisons produced by the syntax in Table 5.32.

If planned with a Bonferroni adjustment of α to .02 for all comparisons, ratings of administrators differ from the others for one-week and two-week, but not for three-week, vacations, whereas belly dancers differ from politicians for two-week and

TABLE 5.34 Source table for simple main effects analysis, followed by simple comparisons

Source	SS	df	MS	F
B: Length of Vacation	32.30	2	16.15	14.55
A: Profession	[(33.85)]	[(2)]		
AB	[(104.15)]	[(4)]		
A at b_1	(68.22)	(2)		
A_{comp1} at b_1	68.06	1	68.06	61.25
A_{comp2} at b_1	0.17	1	0.17	0.16
A at b_2	(20.22)	(2)		
A_{comp1} at b_2	6.72	1	6.72	6.05
A_{comp2} at b_2	13.50	1	13.50	12.15
A at b_3	(49.56)	(2)		
A_{comp1} at b_3	1.39	1	1.39	1.25
A_{comp2} at b_3	48.17	1	48.17	43.35
S/AB	20.00	18	1.11	
T	190.30	26		

three-week vacations. If tested post hoc, $F_S = 2(F_{2,18}) = 7.70$, and ratings of administrators differ from the others for one-week, but not for two- or three-week, vacations, whereas belly dancers still differ from politicians for both two- and three-week vacations.

Effect sizes may be calculated with respect to total SS or to the combination of the interaction and the effect being decomposed, in this case the combination of AB and A. For the simple effect of A at b_1 (assuming that a further decomposition into simple comparisons is not done),

$$\eta^2 = \frac{SS_{A \text{ at } b_1}}{SS_T} = \frac{68.22}{190.30} = .36 \tag{5.24}$$

or

$$\eta^2 = \frac{SS_{A \text{ at } b_1}}{SS_A + SS_{AB}} = \frac{68.22}{33.85 + 104.15} = .49 \tag{5.25}$$

For the simple comparison A_{comp1} at b_1,

$$\eta^2 = \frac{SS_{A_{\text{comp}} \text{ at } b_1}}{SS_T} = \frac{68.06}{190.30} = .36 \tag{5.26}$$

or

$$\eta^2 = \frac{SS_{A_{\text{comp}} \text{ at } b_1}}{SS_A + SS_{AB}} = \frac{68.06}{33.85 + 104.15} = .49 \tag{5.27}$$

As usual, it is important to report which form of η^2 you are using.

Suppose you are planning an analysis in which you expect a main effect of *B* and an *AB* interaction but no main effect of *A*. This might be an expectation when *A* is a control variable that you hypothesize to interact with *B* (your primary IV) but to have no effect on its own. Further, suppose your interest in *B* is on mean differences between the first level and the other two (which is especially appropriate if the first level of *B* is a control condition) and then on mean differences between the last two levels. The simple comparisons of this section are appropriate as an analytic strategy: Do one simple comparison in which b_1 is compared with the combination of b_2 and b_3 at each level of *A*, and then b_2 is compared with b_3 at each level of *A*. This analysis "spends" 6 df and produces a familywise error rate of approximately .12 if each simple comparison is evaluated at the .02 alpha level.

It should be noted that simple interactions are also possible when there are more than two IVs. If there are three IVs, for instance, you could ask if *AB* interactions are present at each level of *C* (or if *AC* interactions are present at each level of *B* or if *BC* interactions are present at each level of *A*). When significant interactions are found, simple main effects might follow to investigate, for example, the simple main effect of *A* at each *BC* combination and then, perhaps, the simple comparison of *A* at each *BC* combination. Partial η^2s and their confidence intervals are available through Smithson's (2003) files for SPSS and SAS. Recall that all that is required for these procedures are *F* ratios, degrees of freedom for effect and error, and the desired confidence level.

5.6.4.4.2 Interaction Contrasts Interaction contrasts are more appropriate than simple-effects analysis when main effects are statistically significant, or when you expect them to be. Interaction contrasts are not confounded with main effects because their sum of squares comes from the omnibus interaction. Thus, interaction contrasts are always appropriate. However, most researchers (and their readers) find simple effects easier to understand, so we recommend using interaction contrasts only when the main effects are also significant.

In interaction-contrasts analysis, the interaction between two IVs is examined through one or more smaller interactions, as depicted in Figure 5.9. Smaller interactions are obtained by deleting levels of one or more of the IVs or by combining levels with use of appropriate weighting coefficients or both. For instance, the significant interaction in a 3 × 3 design might be reduced to examination of a possible interaction between just two levels of *A* and two levels of *B*, as in Figure 5.9(a). Or, you could combine two of the levels of *A* and two of the levels of *B* to test the interaction of the four cells that are produced, as in Figure 5.9(b). Or, you could select just two levels of one IV while using all levels of the other, as in Figure 5.9(c).

Interaction contrasts are formed by applying weighting coefficients to either *A* or *B* or both to evaluate $F_{A\text{comp} \times B}$ (where coefficients are applied only to *A*) or $F_{A \times B\text{comp}}$, as in Figure 5.9(c) (where coefficients are applied only to *B*), or $F_{A\text{comp} \times B\text{comp}}$, as in Figures 5.9(a) and 5.9(b) (where coefficients are applied to both *A* and *B*). When the weighting coefficients are applied to just one IV, the analysis is sometimes called a partial factorial. Interactions can be formed from single cells, as in Figures 5.9(a) and 5.9(c), or by combining cells of one or both IVs, as in Figure 5.9(b).

FIGURE 5.9 Interaction-contrasts analysis exploring small interaction formed by partitioning a large (3 × 3) interaction: (a) 2 × 2 with selected single levels of both *A* and *B*, (b) 2 × 2 with combined levels of *A* (lighter gray levels are combined) and *B* (patterned levels are combined), and (c) 2 × 3 with all levels of *A* and selected levels of *B* (partial factorial)

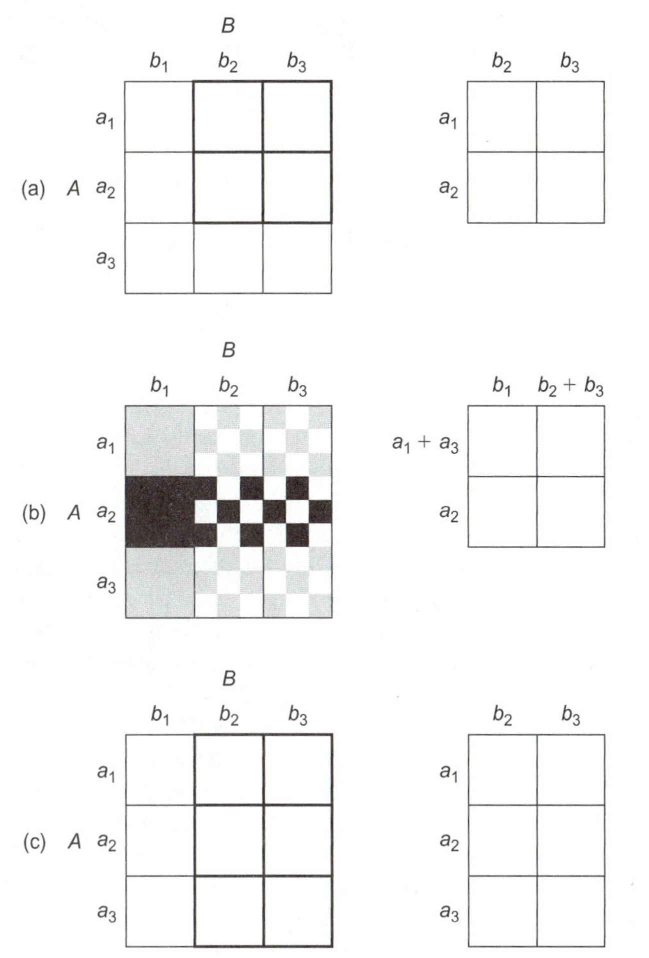

Weighting coefficients for interaction contrasts are found by cross multiplying the weighting coefficients for each of the two IVs. First, appropriate weighting coefficients are chosen for each comparison (say, 2 −1 −1 for *A* and 1 0 −1, for *B*); then they are cross multiplied to find the coefficients that will be applied to the means of the cells and that are given in parentheses below the coefficients in Table 5.35. For the current example, this analysis asks whether administrators have the same linear trend of vacation length as belly dancers and politicians have.

It is at this point that the traditional approach to ANOVA becomes tricky, whereas the regression approach remains straightforward. Simple interaction contrasts that involve weighting coefficients of 0, 1, and −1 are fairly straightforward when done by hand, but contrasts that involve fractional weighting coefficients or weighting coefficients larger than 1 become complex. Keppel (1991, pp. 258–266) did as clear a presentation as is possible, but the logic of the procedure is still difficult to understand. The regression approach, on the other hand, remains clear.

TABLE 5.35 Weighting coefficients and their cross products for the first interaction contrast on the small-sample example (cell means in parentheses)

	$b_1 = -1$	$b_2 = 0$	$b_3 = 1$
$a_1 = 2$	−2 (0.33)	0 (5.67)	2 (6.33)
$a_2 = -1$	1 (6.00)	0 (6.00)	−1 (8.33)
$a_3 = -1$	1 (6.33)	0 (9.00)	−1 (2.67)

Consider the numbers in the X_5 column of Table 5.8, for instance. They are identical to the weighting coefficients derived for the interaction contrast in Table 5.35. This is because the numbers in the X_5 column are a product of the coefficients in the X_1 and X_3 columns, which are the same as the coefficients for A and B, respectively. The regression approach, then, has interaction contrasts built in.

Table 5.36 shows syntax and location of output for interaction contrasts through SPSS GLM, SAS GLM, and SYSTAT ANOVA. The weighting coefficients for the first interaction contrast are derived in Table 5.35; the weighting coefficients for the other three interaction contrasts are found through the same procedures, as in columns X_6, X_7, and X_8 in Table 5.8. Because the sets of coefficients for A and B on the margins are orthogonal, the set of interaction contrasts is also orthogonal. However, a nonorthogonal set could be chosen if it were more meaningful. Note that SYSTAT ANOVA requires first recoding the two IVs into a one-way layout.

Results of the interaction-contrasts analyses produced by syntax of Table 5.36 are in Table 5.37, which shows how interaction-contrasts analysis partitions the interaction when orthogonal weighting coefficients are used on the two main effects and when there is equal n in each cell. Degrees of freedom and sums of squares for the large interaction (in parentheses) are replaced by finer-grained analysis.

Note that the sums of squares for the contrasts match the sums of squares for the interaction columns, X_5 to X_8, of Table 5.13, where factorial ANOVA is done through multiple regression. If each sum of squares for the interaction columns is tested individually, rather than combined for the omnibus test of interaction, results are identical to these interaction contrasts. Thus, this multiple-regression approach to factorial ANOVA builds interaction contrasts into the fabric of the analysis.

Effect sizes for interaction contrasts are calculated with respect to either total SS or SS_{AB} (the effect being decomposed). For A_{compl} by B_{linear},

$$\eta^2 = \frac{SS_{A_{\text{compl}} \times B_{\text{linear}}}}{SS_T} = \frac{44.44}{190.30} = .23 \quad \textbf{(5.28)}$$

or

$$\eta^2 = \frac{SS_{A_{\text{compl}} \times B_{\text{linear}}}}{SS_{AB}} = \frac{44.44}{104.15} = .43 \quad \textbf{(5.29)}$$

TABLE 5.36 Syntax for interaction contrasts in a two-way factorial design

Program	Syntax	Section of Output	Name of Effect
SPSS GLM	UNIANOVA RATING BY PROFESS LENGTH /METHOD = SSTYPE(3) /INTERCEPT = INCLUDE /CRITERIA = ALPHA(.05) /LMATRIX = "admin vs others by linear" profess*length 2 0 -2 -1 0 1 -1 0 1 /LMATRIX = "admin vs. others by quadratic" profess*length 2 -4 2 -1 2 -1 -1 2 -1 /LMATRIX = "belly vs. polit by linear" profess*length 0 0 0 -1 0 1 1 0 -1 /LMATRIX = "belly vs. polit by quadratic" profess*length 0 0 0 1 -2 1 -1 2 -1 /DESIGN profess, length, profess*length .	**Custom Hypothesis Tests #1, etc.**	Test Result
SAS GLM	proc glm data=SASUSER.FAC_RAND; class PROFESS LENGTH; model RATING = PROFESS LENGTH PROFESS*LENGTH; contrast 'admin vs others by linear' PROFESS*LENGTH 2 0 -2 -1 0 1 -1 0 1; contrast 'admin vs. others by quadratic' PROFESS*LENGTH 2 -4 2 -1 2 -1 -1 2 -1; contrast 'belly vs. polit by linear' PROFESS*LENGTH 0 0 0 -1 0 1 1 0 -1; contrast 'belly vs. polit by quadratic' PROFESS*LENGTH 0 0 0 1 -2 1 -1 2 -1; run;	Contrast	admin vs others by linear, etc.

(Continued)

TABLE 5.36 Continued

Program	Syntax	Section of Output	Name of Effect

SYSTAT ANOVA	IF (PROFESS=1) AND (LENGTH=1) THEN LET GROUP=1
	IF (PROFESS=1) AND (LENGTH=2) THEN LET GROUP=2
	IF (PROFESS=1) AND (LENGTH=3) THEN LET GROUP=3
	IF (PROFESS=2) AND (LENGTH=1) THEN LET GROUP=4
	IF (PROFESS=2) AND (LENGTH=2) THEN LET GROUP=5
	IF (PROFESS=2) AND (LENGTH=3) THEN LET GROUP=6
	IF (PROFESS=3) AND (LENGTH=1) THEN LET GROUP=7
	IF (PROFESS=3) AND (LENGTH=2) THEN LET GROUP=8
	IF (PROFESS=3) AND (LENGTH=3) THEN LET GROUP=9
	ANOVA
	CATEGORY GROUP
	COVAR
	DEPEND RATING
	ESTIMATE
	HYPOTHESIS
	NOTE 'admin vs others by linear'
	EFFECT=GROUP
	CONTRAST [2 0 -2 -1 0 1 -1 0 1]
	TEST
	HYPOTHESIS
	NOTE 'admin vs. others by quadratic'
	EFFECT=GROUP
	CONTRAST [2 -4 2 -1 2 -1 -1 2 -1]
	TEST
	HYPOTHESIS
	NOTE 'belly vs. polit by linear'
	EFFECT=GROUP
	CONTRAST [0 0 0 -1 0 1 1 0 -1]
	TEST
	HYPOTHESIS
	NOTE ’belly vs. polit by quadratic’
	EFFECT=GROUP
	CONTRAST [0 0 0 1 -2 1 -1 2 -1]
	TEST
		Test of Hypothesis	Hypothesis

TABLE 5.37 Source table for interaction-contrasts analysis

Source	SS	df	MS	F
A	32.30	2	16.15	14.55
B	33.85	2	16.93	15.25
AB	(104.15)	(4)		
$A_{\text{comp1}} \times B_{\text{linear}}$	44.44	1	44.44	40.00
$A_{\text{comp1}} \times B_{\text{quad}}$	0.59	1	0.59	0.53
$A_{\text{comp2}} \times B_{\text{linear}}$	27.00	1	27.00	24.30
$A_{\text{comp2}} \times B_{\text{quad}}$	32.11	1	32.11	28.90
S/A	20.00	18	1.11	
	190.30	26		

TABLE 5.38 Means compared in interaction contrast of linear trend of administrators versus others

	One-Week Vacation	Three-Week Vacation
Administrators	0.33	6.33
Others	$(6.00 + 6.33)/2 = 6.16$	$(8.33 + 2.67)/2 = 5.50$

Interaction contrasts may be done as part of a planned analysis, replacing the omnibus test of the interaction, or as a post hoc analysis to dissect a significant interaction. The Scheffé adjustment for post hoc analysis for inflated Type I error rate is

$$F_S = (a - 1)(b - 1)\, F_{(a-1)(b-1),\ (\text{df}_{\text{error}})} \tag{5.30}$$

In this case, with $\alpha = .05$:

$$F_S = (3 - 1)(3 - 1)(2.93) = 11.72$$

Whether these interaction contrasts are done as planned or post hoc analyses, there is a difference in linear trend between administrators and others ($F = 40.00$), a difference in linear trend between belly dancers and politicians ($F = 24.30$), and a difference in quadratic trend between belly dancers and politicians ($F = 28.90$).

Interpretation of a significant interaction contrast involves plotting the means compared in the interaction. The means for the significant linear trend of administrators versus others are shown in Table 5.38. The first interaction contrast is plotted in Figure 5.10.

The plot of means suggests that administrators are far less satisfied with one-week vacations than are the others as a group. If further clarification is desired, a significant interaction contrast is followed up with one or more simple comparisons. For example, a simple comparison is run between administrators and others at the one-week level of vacation. Or, a simple comparison of one and three weeks of vacation is run for administrators. Procedures for simple comparisons like these are in Table 5.36, previously illustrated as a follow-up to simple main effects analysis. These simple comparisons need a Scheffé adjustment for post hoc testing.

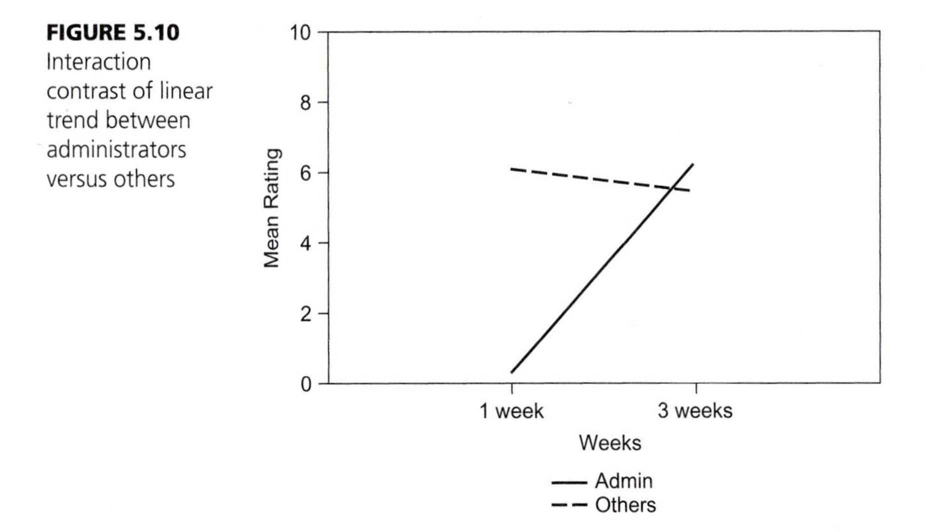

FIGURE 5.10 Interaction contrast of linear trend between administrators versus others

5.6.4.5 Extending Comparisons through the Regression Approach

Table 5.39 presents a verbal layout of the eight columns of the regression approach to our current problem. At this point, the marginal comparisons for both A and B are available, as are the four orthogonal interaction contrasts.

However, additional information can be gleaned from the table by adding together various columns, as shown in Table 5.40. $A_{\text{comp1}} \times B$ asks whether there is an interaction between administrators and others at the three levels of B. $A_{\text{comp2}} \times B$ asks whether there is an interaction between belly dancers and politicians at the three levels of B. $A \times B_{\text{linear}}$ asks whether the linear trends are the same for the three professions, while $A \times B_{\text{quad}}$ asks whether the quadratic trends are the same for the three professions.

TABLE 5.39 Columns for randomized-groups ANOVA through regression

X_1	X_2	X_3	X_4	X_5	X_6	X_7	X_8
				$X_1 \times X_3$	$X_1 \times X_4$	$X_2 \times X_3$	$X_2 \times X_4$
A_{comp1}	A_{comp2}	B_{linear}	B_{quad}	$A_{\text{comp1}} \times B_{\text{linear}}$	$A_{\text{comp1}} \times B_{\text{quad}}$	$A_{\text{comp2}} \times B_{\text{linear}}$	$A_{\text{comp2}} \times B_{\text{quad}}$
A		B		$A \times B$			

TABLE 5.40 Partial factorials through regression

$A_{\text{comp1}} \times B$	$X_5 + X_6$	$A_{\text{comp1}} \times B_{\text{linear}} + A_{\text{comp1}} \times B_{\text{quad}}$
$A_{\text{comp2}} \times B$	$X_7 + X_8$	$A_{\text{comp2}} \times B_{\text{linear}} + A_{\text{comp2}} \times B_{\text{quad}}$
$A \times B_{\text{linear}}$	$X_5 + X_7$	$A_{\text{comp1}} \times B_{\text{linear}} + A_{\text{comp2}} \times B_{\text{linear}}$
$A \times B_{\text{quad}}$	$X_6 + X_8$	$A_{\text{comp1}} \times B_{quad} + A_{\text{comp2}} \times B_{\text{quad}}$

It should be noted that if you have the error term, you do not need to work out the complete regression problem to compute the sum of squares for a particular element of the design. For example, if you want the sum of squares for the interaction contrast of $A_{\text{comp1}} \times B_{\text{linear}}$, you need only compute the preliminary sum of squares for X_5 and the sum of products for $Y \times X_5$. The sum of squares regression for the interaction contrast is obtained by applying Equation 3.20.

As with all comparisons, effect size is calculated with respect to total sum of squares or the sum of squares for the effect being decomposed; in this case, SS_A is the denominator of η^2 for A_{comp1} and A_{comp2}, and SS_{AB} is the denominator of η^2 for, say, $A \times B_{\text{linear}}$ or $A_{\text{comp1}} \times B_{\text{linear}}$.

5.6.4.6 Comparisons in Higher-Order Designs

The highest-order interaction sets the stage for comparisons in higher-order designs. If the highest-order interaction is not statistically significant, the analysis is treated as a next lower level design. For example, if the highest-order interaction is nonsignificant in a three-way design, the three two-way interactions are examined and treated as if from three separate two-way designs. Thus, any two-way interaction that is statistically significant is dealt with as in Sections 5.6.4.4 and 5.6.4.5. So, for example, a significant $A \times C$ interaction (with values in the *ac* cells summed over levels of B) is followed by either simple main effects or interaction contrasts that, in turn, are followed by simple comparisons. Any significant main effects that are not involved in a significant interaction (as B might be in this example) are followed by simple comparisons on marginal means.

If the higher-order interaction is statistically significant, either the simple-effects or interaction comparisons strategy is again adopted. Now, however, the first stage in a simple-effects analysis is simple interactions. For example, a significant three-way interaction is followed by tests of the two-way interactions at each level of the third IV—for example, $A \times B$ at c_1, $A \times B$ at c_2, and so on. The strategy that involves least confounding (intermingling of SS) is to test a nonsignificant (or the least substantial) two-way interaction or, if comparisons are planned beforehand, a test that is expected to be nonsignificant. For example, the choice of $A \times B$ at c_i assumes that the $A \times B$ interaction is not significant, or has the smallest effect size of the three. (An interaction-contrasts approach is preferable if all three interactions are, or are expected to be, significant and substantial.)

Any significant two-way interaction (at one level of the third IV) is then treated as in a two-way design. That is, two-way interactions are followed up with simple main effects or interaction comparisons, which in turn are followed by simple comparisons. In the example of Table 5.20, the appropriate simple-effects strategy is based on one of the two nonsignificant two-way interactions: DEPRIVATION by DECOR at each level of FAMILIARITY, or FAMILIARITY by DECOR at each level of DEPRIVATION. A simple-effects interaction analysis is demonstrated in Section 7.7.2.2 for a three-way mixed randomized-groups repeated-measures design.

If all lower-order interactions, as well as the highest-order interaction, are statistically significant, interaction contrasts are appropriate, as per Section 5.6.4.4.2.

These are formed by applying contrasts coefficients to each of the IVs that have more than two levels. For example, you might base interaction contrasts on a mix of comparisons and trends in a 3 × 3 × 3 design. You would have eight tests of the form $A_{comp} \times B_{comp} \times C_{comp}$. For example, two of the comparisons might be $A_{comp1} \times B_{linear} \times C_{comp1}$ and $A_{comp2} \times B_{quad} \times C_{comp1}$, and so on.

5.6.5 Unequal Sample Sizes

All of the problems of unequal n discussed in Section 4.5.3 for one-way designs carry over to factorial designs, and that is just the beginning. The first, and largest, problem is still the logical one. If unequal sample sizes are caused by one of the treatment levels in a design meant to have equal sample sizes, interpretation is problematic. Results, if interpreted, generalize only to cases that provide data. Such cases may be substantially different from cases that are unable (or unwilling) to provide data.

The next two problems are statistical. First, orthogonality of effects is lost for all the main effects and all the lower-order interactions in higher-order designs. The only effect that is unrelated to all others is the highest-order interaction, because the sum of squares for that interaction is found by subtracting the sums of squares for the main effects (cf. Equation 5.3 and Table 5.5) in the traditional ANOVA equations.[11] Nonorthogonality is most easily seen for the simple two-way factorial design of Table 5.9, where both A and B have only two levels. In Table 5.41, the same design is represented with equal $n = 3$ (Table 5.41(a)) and one cell (a_1b_2) with only two cases (Table 5.41(b)). Notice that in Table 5.41(a), the correlation between column X_1 and column X_2 is 0 because 1 and -1 are paired equally often with each other. However, in Table 5.9(b), where there is unequal n, the correlation between column X_1 and column X_2 is no longer 0. In this case, the sum of squares for the X_1 column (SS_A) and the sum of squares for the X_2 column (SS_B) are nonorthogonal. The analysis of a design with unequal n has to take this nonorthogonality (overlapping sum of squares) into account.

Notice that with unequal n, the X_3 (interaction) column also is nonorthogonal with the other two, but again, its overlap with the main effects is eliminated in the traditional analysis by subtracting sums of squares associated with both A and B. Figure 5.11 demonstrates the overlap in Venn diagrams in the traditional analysis.

The second statistical problem with unequal sample sizes has to do with ambiguity regarding the marginal means. The data in Table 5.41 for a_1, at both b_1 and b_2, are duplicated in Table 5.42. There are two ways to compute $\overline{Y}_{a1}$. The first is to compute the mean of the scores in the column. The second is to compute the mean of the two cell means. When there is equal n, these two methods of computing the marginal mean yield the same value (5.67), as shown in Table 5.42(a). When there is unequal

[11]All effects, including the highest-order interaction, are nonorthogonal in the regression approach and require matrix calculations. Also, most adjustments require a sequential series of regression equations rather than the single one demonstrated in Section 3.2.

TABLE 5.41 Effects of equal and unequal n on the orthogonality of columns in ANOVA

(a) Equal n in all four cells

	Case	Y	X_1 (A)	X_2 (B)	$X_3 = X_1X_2$ (AB)
	1	5	1	1	1
a_1b_1	2	3	1	1	1
	3	4	1	1	1
	4	7	1	−1	−1
a_1b_2	5	7	1	−1	−1
	6	8	1	−1	−1
	7	3	−1	1	−1
a_2b_1	8	5	−1	1	−1
	9	4	−1	1	−1
	10	4	−1	−1	1
a_2b_2	11	1	−1	−1	1
	12	3	−1	−1	1

(b) Unequal n when case 6 is lost in the a_1b_2 cell:

	Case	Y	X_1 (A)	X_2 (B)	$X_3 = X_1X_2$ (AB)
	1	5	1	1	1
a_1b_1	2	3	1	1	1
	3	4	1	1	1
	4	7	1	−1	−1
a_1b_2	5	7	1	−1	−1
	7	3	−1	1	−1
a_2b_1	8	5	−1	1	−1
	9	4	−1	1	−1
	10	4	−1	−1	1
a_2b_2	11	1	−1	−1	1
	12	3	−1	−1	1

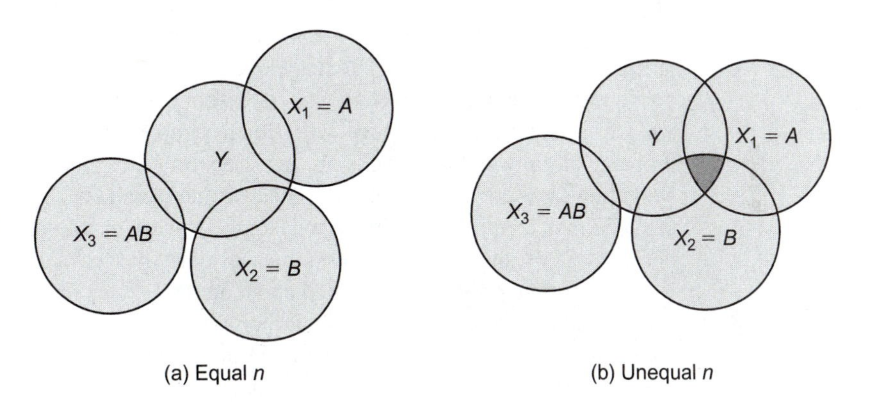

FIGURE 5.11 Relationships among effects in orthogonal and nonorthogonal two-way designs: (a) no overlapping variance among main effects with equal n, and (b) overlap among main effects with unequal n

n, these two methods of computing the marginal mean yield different values (5.2 and 5.5), as shown in Table 5.42(b). ANOVA, which "lives" to test for mean differences, is in a quandary with unequal n. The researcher has to specify which marginal mean is to be tested. This ambiguity exists for all means in a larger factorial design except for those associated with the cells of the highest-order interaction.

Again, there are two approaches to calculating marginal means in a factorial design. One approach is to find the cell means for each level of one IV at a level of the other IV, add them up, and divide by the number of cell means. This is the procedure

TABLE 5.42 Ambiguity of marginal means in unequal *n* designs

	(a) Equal n in every cell			(b) Unequal n in one cell		
	Case	***Y***	**Cell Means**	**Case**	***Y***	**Cell Means**
	1	5		1	5	
a_1b_1	2	3	$\overline{Y}_{a_1b_1} = 4$	2	3	$\overline{Y}_{a_1b_1} = 4$
	3	4		3	4	
	4	7		4	7	
a_1b_2	5	7	$\overline{Y}_{a_1b_2} = 7.33$	5	7	$\overline{Y}_{a_1b_2} = 7$
	6	8				
Marginal Means	$\overline{Y}_{a_1} =$	$(5 + \cdots + 8)/6$	$(4 + 7.33)/2$	$\overline{Y}_{a_2} =$	$(5 + \cdots + 7)/5$	$(4 + 7)/2$
		$= 5.67$	$= 5.67$		$= 5.2$	$= 5.5$

used in Table 5.42 when, for instance, the 4 and 7 are averaged to produce a marginal mean of 5.5. This procedure produces unweighted means, because cells are weighted equally, regardless of the number of cases each cell has. The other approach is to add up all the scores in that level of the IV, ignoring the other IV(s), and divide by the number of scores. In Table 5.42, this procedure produces the marginal mean of 5.2. This procedure produces weighted[12] means, because cell means are weighted by their cell sample sizes. In this case, the marginal mean is closer to the cell mean with the larger sample size. With two potential candidates for the mean, which marginal mean do we want to analyze? Which mean reflects population values?

In an experiment designed with equal sample cell sizes, with some random dropout, the unweighted means are more appropriate: There is no justification for weighting smaller cells less heavily, because their small size occurred by chance. On the other hand, in nonexperimental designs, or when there are one or more nonexperimental IVs, the inequality in sample sizes may be caused by population size differences. If type of job training and gender are IVs, for example, there really may be more women than men in the particular job for which training is being offered. A design that weights cell means by their cell sample sizes makes more sense, because in the population, the marginal mean would, indeed, be closer to the mean for women.

With unequal sample sizes, however, there is the further complication of overlapping variance. Some way to subtract out the overlap is needed. In the unweighted-means approach, the overlap is subtracted from the effects; none of the effects takes credit for it. The sums of squares for effects plus error do not add up to the total SS because total SS includes the area of overlap. In the weighted-means approach, the researcher *decides* which effect deserves the credit (gets the overlapping variance) and gives it higher priority. The sums of squares for effects plus error then add up to total SS.

[12] Note that these weights are sample sizes and are different from the coefficients used to weight means for specific comparisons.

There are some guidelines for assigning priority to IVs. The decision is based on the roles the various IVs are playing in the design. If there are two or more IVs, give the one of greater interest the higher priority, because in that position, it will also most probably have higher power. On the other hand, if one or more of the IVs is a blocking variable, you may want to adjust the treatment IV (e.g., vacation length) for differences due to the blocking variable (e.g., profession) before testing mean differences on the treatment IV. The blocking IV (profession) becomes a covariate, as discussed in Chapter 8. For example, in the type of job training by gender example, you might want to adjust for gender differences before looking for mean differences due to type of job training. Or, if your IVs occur in a time sequence, assign the ones that happen earlier the higher priority so that later events are adjusted for earlier ones. All of these considerations also apply to lower-order interactions in higher-order factorials. Intermediate interactions overlap among themselves and with main effects. In any design, however, the highest-order interaction is independent of all other effects, whether sample sizes are equal or unequal, because the sum of squares for the highest-order interaction is formed by subtracting sums of squares for all lower-order effects in traditional ANOVA calculations.

Researchers usually use the unweighted-means approach in experiments, where overlapping sums of squares are simply unanalyzed unless there is compelling reason to assign priority. If one of the experimental IVs is of overriding interest, the only strategy that is likely to make a meaningful difference in its power is assigning it highest priority. All other strategies are likely to yield the same, possibly sad, results.

Some software programs are limited to the unweighted-means approach or to algorithms that are consistent with that strategy. SAS ANOVA, for example, cannot handle unequal n. However, a weighted-means analysis can be produced by both software packages if the regression programs are used in conjunction with the setups for the regression strategy. Much more detail about this approach is available in Tabachnick and Fidell (2007, Chapter 5).

Table 5.43 shows SPSS REGRESSION syntax and selected output for a sequential regression analysis of the data in Table 5.42(b).[13] The METHOD instruction specifies that X_1 (A) is entered first (given highest priority and all overlapping sums of squares); then X_2 (B) is entered to test the nonoverlapping sums of squares for B. Finally, X_3 ($A \times B$) is entered in the final step of the sequence. If, on the other hand, B is to be given priority over A, it is entered first into the equation. The CHANGE statistic is selected to produce the sequential tests of effects; remaining options are produced by default through the SPSS Windows menus.

The tests of effects are in the F Change column. Thus $F_A(1, 9) = 3.86$ with $p = .081$ indicates no main effect of A. Similarly, there is no main effect of B, with $F_B(1, 8) = 0.36, p = .566$. However, there is a significant interaction, with $F_{AB}(1, 7) = 10.11, p = .015$.

Both SPSS and SAS GLM offer several options for dealing with unequal n and use similar notation and procedures. Table 5.44, adapted from Tabachnick and Fidell (2007), summarizes the various options in the programs.

[13] Note that the X_3 column may be entered in the data set or created through transformation.

TABLE 5.43 Weighted-means ANOVA through sequential regression (SPSS REGRESSION syntax and selected output)

```
REGRESSION
 /MISSING LISTWISE
 /STATISTICS COEFF OUTS R ANOVA CHANGE
 /CRITERIA=PIN(.05) POUT(.10)
 /NOORIGIN
 /DEPENDENT Y
 /METHOD=ENTER X1 /METHOD=ENTER X2 /METHOD=ENTER X3 .
```

Model Summary

Model	R	R Square	Adjusted R Square	Std. Error of the Estimate	Change Statistics: R Square Change	F Change	df1	df2	Sig. F Change
1	.548[a]	.300	.223	1.568	.300	3.864	1	9	.081
2	.575[b]	.330	.163	1.627	.030	.358	1	8	.566
3	.852[c]	.726	.609	1.113	.396	10.111	1	7	.015

a. Predictors: (Constant), ×1
b. Predictors: (Constant), ×1, ×2
c. Predictors: (Constant), ×1, ×2, ×3

TABLE 5.44 Terminology for strategies for adjustment of unequal sample sizes

Research Type	Overall and Spiegel (1969)	SPSS GLM	SAS GLM
1. Experiments designed to be equal n, with random dropout; all cells equally important	Method 1	METHOD=SSTYPE(3): (default) METHOD=SSTYPE(4)[a]	TYPE III and TYPE IV[a]
2. Nonexperimental research in which sample sizes reflect importance of cells; Main effects have equal priority[b]	Method 2	METHOD=SSTYPE(2)	TYPE II
3. Like number 2, except all effects have unequal priority	Method 3	METHOD=SSTYPE(1)	TYPE I
4. Research in which cells are given unequal weight on basis of prior knowledge	N/A	N/A	N/A

[a] Type III and IV differ only if there are missing cells.

[b] The programs take different approaches to adjustment for interaction effects.

Overall and Spiegel's (1969) Method 1 is the unweighted-means analysis. The remaining methods are various ways of applying weighted-means analyses, as discussed ad nauseam in Tabachnick and Fidell (2007, Section 6.5.4.2). Method 1, the unweighted-means analysis in which all effects are adjusted for each other, is usually appropriate for experimental work. Method 3 is most commonly used for nonexperimental work in which differences in cell sizes represent population differences and all effects are adjusted for higher-priority effects. This is the weighted-means method that requires a priority order of effects.

5.7 COMPLETE EXAMPLE OF TWO-WAY RANDOMIZED-GROUPS ANOVA

This complete example of factorial randomized-groups ANOVA is from a study of the bonding strength of periodontal dressings, tested in vitro at one of two times. The five dressings were all commercially available and were tested at 3 or 24 hours after application. Researchers were interested in differences among the dressings in bonding strength, in differences between the 3- and 24-hour bonding strength, and in any interaction between time and dressing. The dependent variable was tensile bond strength, measured in units of kilonewtons per millimeter (kN/mm). There were ten observations in each of the ten combinations of dressing and setting time, resulting in $N = 100$. The study is described in more detail in Appendix B.2. Data files are DENTAL.*.

5.7.1 Evaluation of Assumptions

5.7.1.1 Sample Sizes, Normality, and Independence of Errors

There are 10 cases in all cells of the design, so that $df_{error} = N - ab = 90$, which is well over the 20 required to support normality of sampling distributions. Independence of errors is assumed because trials were run individually. Table 5.45 shows an SPSS FREQUENCIES run of the data for diagnostic purposes. The file is first sorted and split so that descriptive statistics are produced separately for each group. Histograms are requested, with a normal curve superimposed. All statistics are requested. SESKEW and SEKURT are standard errors for skewness and kurtosis, respectively. The /FORMAT=NOTABLE instruction suppresses the table of frequencies and percentages for each DV value.

5.7.1.2 Homogeneity of Variance

Output of Table 5.45 contains indications that variances are heterogeneous rather than homogeneous, as is assumed by the analysis (cf. Section 3.4.3). The standard deviations range from 37.642 to 227.17, and the ratio of largest to smallest variance far exceeds 10: $F_{max} = 51606.40/1416.89 = 36.41$. This suggests that a square root or log transform of the DV should be considered.

Table 5.46 shows syntax and results of these transformations. The COMPUTE instructions provide square root and log transforms of the DV (STRENGTH) to produce SSTREN and LSTREN, respectively. Only variances, minima, and maxima are requested in this FREQUENCIES run.

With the square root transformation of the DV, $F_{max} = 16.1436/1.6609 = 9.72$, which is a bit less than 10. With the log-transformed DV, $F_{max} = .045/.002 = 22.5$.

TABLE 5.45 Diagnostic run for two-way ANOVA through SPSS FREQUENCIES

```
SORT CASES BY DRESSING HOURS .
SPLIT FILE
  SEPARATE BY DRESSING HOURS .
FREQUENCIES
  VARIABLES=STRENGTH
  /FORMAT=NOTABLE
  /STATISTICS=STDDEV VARIANCE MINIMUM MAXIMUM MEAN SKEWNESS SESKEW KURTOSIS
  SEKURT
  /HISTOGRAM NORMAL
  /ORDER  ANALYSIS .
```

Frequencies

Type of dressing = care, Time after application = 3

Statistics[a]

Bond strength KN/sq m

N	Valid	10
	Missing	0
Mean		354.90
Std. Deviation		147.934
Variance		21884.322
Skewness		.368
Std. Error of Skewness		.687
Kurtosis		−.276
Std. Error of Kurtosis		1.334
Minimum		140
Maximum		596

a. Type of dressing = care, Time after application = 3

Histogram

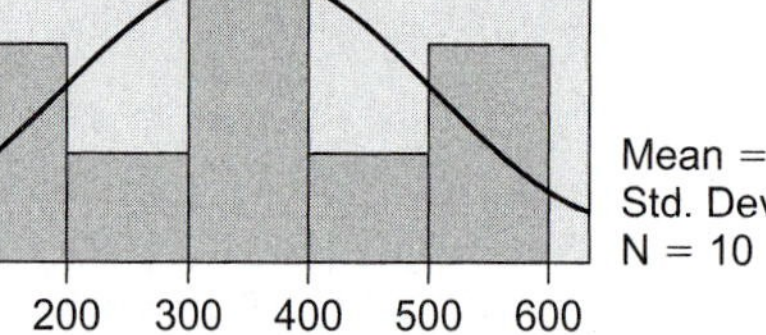

Type of dressing = care, Time after application = 24

Statistics[a]

Bond strength KN/sq m

N	Valid	10
	Missing	0
Mean		683.30
Std. Deviation		116.259
Variance		13516.233
Skewness		.115
Std. Error of Skewness		.687
Kurtosis		.439
Std. Error of Kurtosis		1.334
Minimum		476
Maximum		882

a. Type of dressing = care, Time after application = 24

Histogram

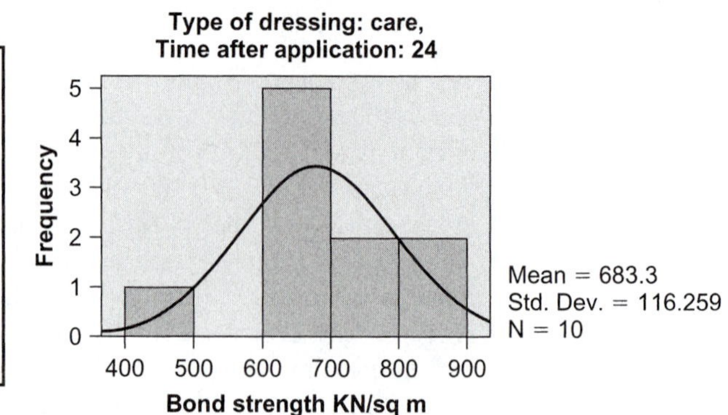

TABLE 5.45 Continued

Type of dressing = coe, Time after application = 3

Statistics[a]

Bond strength KN/sq m

N	Valid	10
	Missing	0
Mean		573.60
Std. Deviation		61.312
Variance		3759.156
Skewness		−.179
Std. Error of Skewness		.687
Kurtosis		−.786
Std. Error of Kurtosis		1.334
Minimum		469
Maximum		665

a. Type of dressing = coe, Time after application = 3

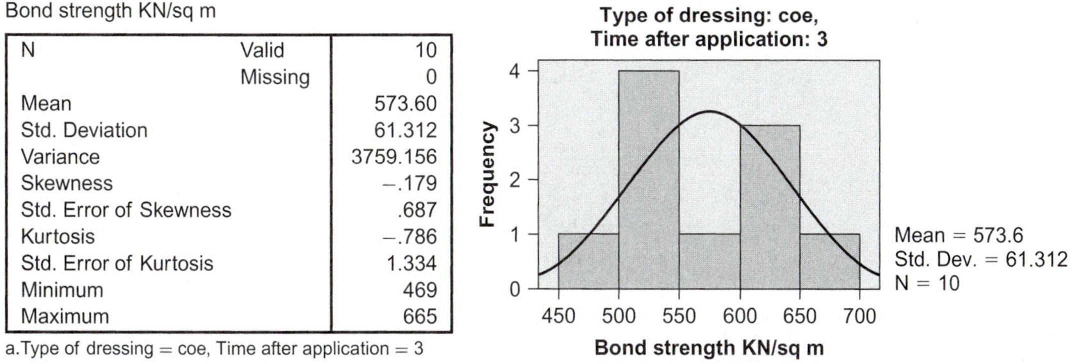

Type of dressing = coe, Time after application = 24

Statistics[a]

Bond strength KN/sq m

N	Valid	10
	Missing	0
Mean		781.60
Std. Deviation		104.812
Variance		10985.600
Skewness		−.252
Std. Error of Skewness		.687
Kurtosis		−.957
Std. Error of Kurtosis		1.334
Minimum		602
Maximum		910

a. Type of dressing = coe, Time after application = 24

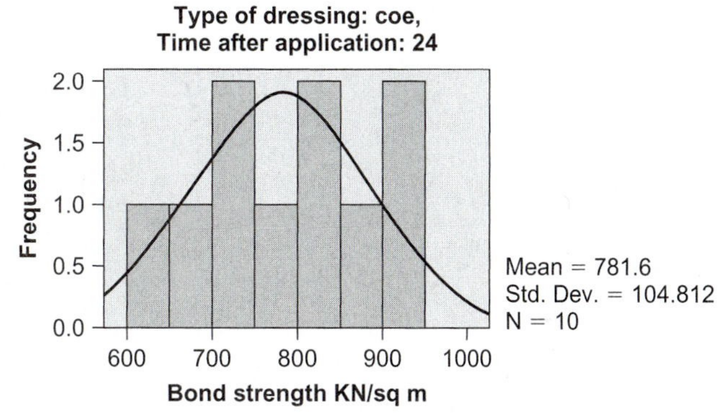

Type of dressing = peripac, Time after application = 3

Statistics[a]

Bond strength KN/sq m

N	Valid	10
	Missing	0
Mean		112.00
Std. Deviation		43.369
Variance		1880.889
Skewness		−.369
Std. Error of Skewness		.687
Kurtosis		−.521
Std. Error of Kurtosis		1.334
Minimum		34
Maximum		168

a. Type of dressing = peripac, Time after application = 3

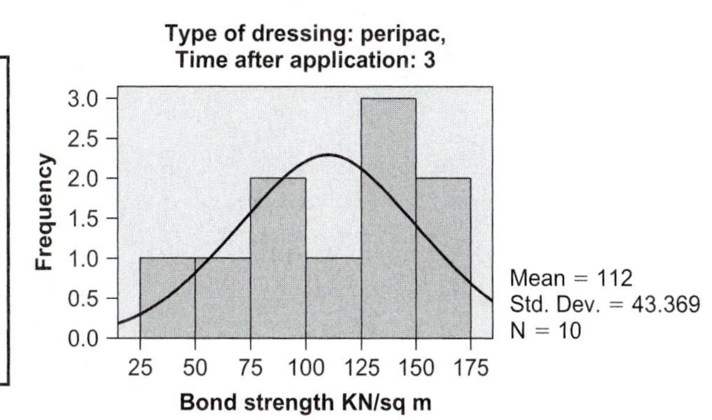

TABLE 5.45 Continued

Type of dressing = peripac, Time after application = 24

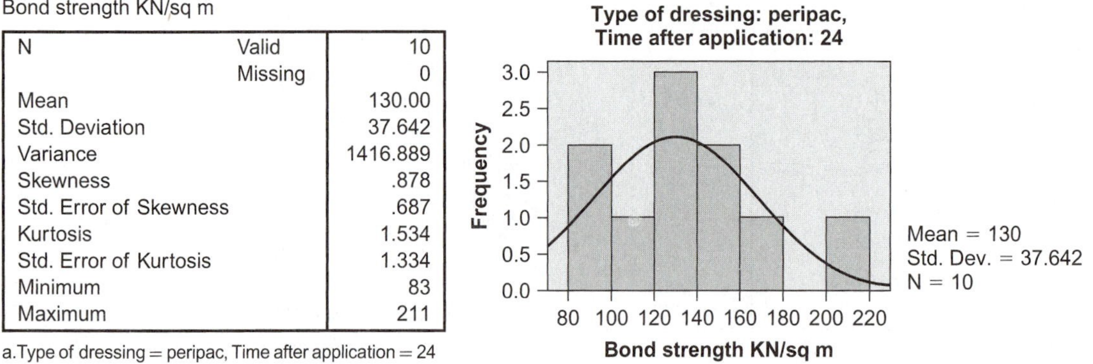

Statistics[a]

Bond strength KN/sq m

N	Valid	10
	Missing	0
Mean		130.00
Std. Deviation		37.642
Variance		1416.889
Skewness		.878
Std. Error of Skewness		.687
Kurtosis		1.534
Std. Error of Kurtosis		1.334
Minimum		83
Maximum		211

a.Type of dressing = peripac, Time after application = 24

Histogram

Type of dressing = putty, Time after application = 3

Statistics[a]

Bond strength KN/sq m

N	Valid	10
	Missing	0
Mean		837.90
Std. Deviation		189.874
Variance		36052.322
Skewness		.595
Std. Error of Skewness		.687
Kurtosis		−.398
Std. Error of Kurtosis		1.334
Minimum		631
Maximum		1190

a.Type of dressing = putty, Time after application = 3

Histogram

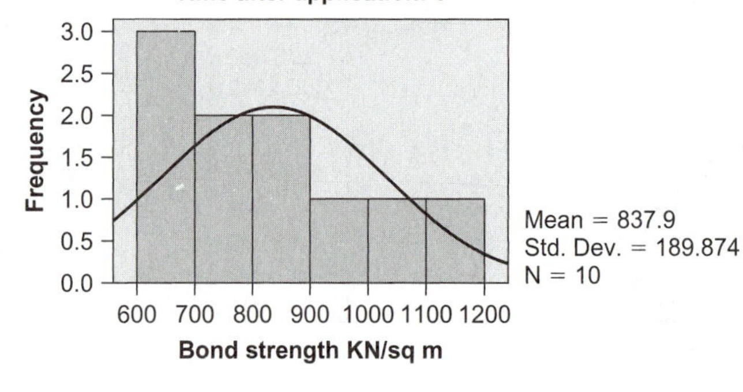

Type of dressing = putty, Time after application = 24

Statistics[a]

Bond strength KN/sq m

N	Valid	10
	Missing	0
Mean		852.20
Std. Deviation		227.170
Variance		51606.400
Skewness		.395
Std. Error of Skewness		.687
Kurtosis		−1.029
Std. Error of Kurtosis		1.334
Minimum		553
Maximum		1225

a.Type of dressing = putty, Time after application = 24

Histogram

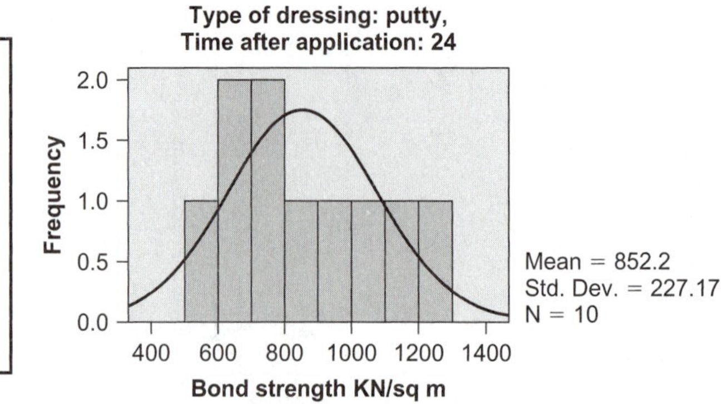

TABLE 5.45 Continued

Type of dressing = zone, Time after application = 3

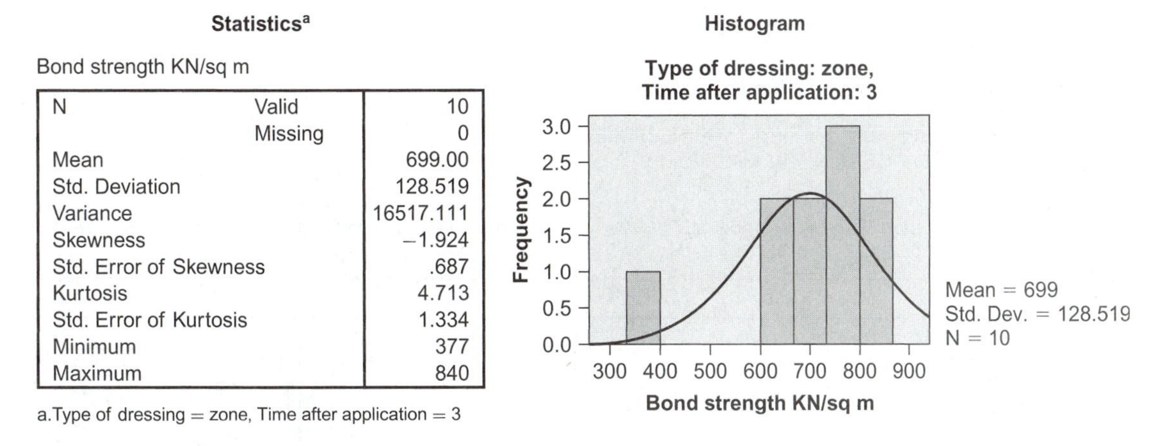

Statistics[a]

Bond strength KN/sq m

N	Valid	10
	Missing	0
Mean		699.00
Std. Deviation		128.519
Variance		16517.111
Skewness		−1.924
Std. Error of Skewness		.687
Kurtosis		4.713
Std. Error of Kurtosis		1.334
Minimum		377
Maximum		840

a. Type of dressing = zone, Time after application = 3

Type of dressing = zone, Time after application = 24

Statistics[a]

Bond strength KN/sq m

N	Valid	10
	Missing	0
Mean		1193.90
Std. Deviation		198.367
Variance		39349.433
Skewness		−.810
Std. Error of Skewness		.687
Kurtosis		−.513
Std. Error of Kurtosis		1.334
Minimum		827
Maximum		1399

a. Type of dressing = zone, Time after application = 24

Histogram

Type of dressing: zone, Time after application: 24

Mean = 1193.9
Std. Dev. = 198.367
N = 10

Bond strength KN/sq m

5.7.1.3 Outliers

Runs producing Tables 5.43 and 5.44 provide maximum and minimum values for each treatment to check for outliers. All scores are well within 3.3 standard deviations of their means and are apparently connected to other scores in their groups, so that outliers pose no problem in this data set.

5.7.2 Randomized-Groups Analysis of Variance

There is a choice for the omnibus ANOVA at this point: either the ANOVA on the square root transform of strength or the ANOVA on the original, untransformed data

TABLE 5.46 SPSS FREQUENCIES to request variances with square root and log transforms

```
COMPUTE SSTREN=SQRT(STRENGTH).
COMPUTE LSTREN=LG10(STRENGTH).
SORT CASES BY DRESSING HOURS .
SPLIT FILE
 SEPARATE BY DRESSING HOURS .
FREQUENCIES
 VARIABLES=SSTREN LSTREN /FORMAT=NOTABLE
 /STATISTICS=VARIANCE MINIMUM MAXIMUM
 /ORDER ANALYSIS .
```

Type of dressing = care, Time after application = 3

Statistics[a]

		SSTREN	LSTREN
N	Valid	10	10
	Missing	0	0
Variance		5.012	.006
Minimum		21.82	2.68
Maximum		29.70	2.95

a.Type of dressing = care, Time after application = 24

Type of dressing = care, Time after application = 24

Statistics[a]

		SSTREN	LSTREN
N	Valid	10	10
	Missing	0	0
Variance		5.012	.006
Minimum		21.82	2.68
Maximum		29.70	2.95

a.Type of dressing = care, Time after application = 24

Type of dressing = coe, Time after application = 3

Statistics[a]

		SSTREN	LSTREN
N	Valid	10	10
	Missing	0	0
Variance		1.661	.002
Minimum		21.66	2.67
Maximum		25.79	2.82

a.Type of dressing = coe, Time after application = 3

Type of dressing = coe, Time after application = 24

Statistics[a]

		SSTREN	LSTREN
N	Valid	10	10
	Missing	0	0
Variance		3.594	.004
Minimum		24.54	2.78
Maximum		30.17	2.96

a.Type of dressing = coe, Time after application = 24

Type of dressing = peripac, Time after application = 3

Statistics[a]

		SSTREN	LSTREN
N	Valid	10	10
	Missing	0	0
Variance		4.993	.045
Minimum		5.83	1.53
Maximum		12.96	2.23

a.Type of dressing = peripac, Time after application = 3

Type of dressing = peripac, Time after application = 24

Statistics[a]

		SSTREN	LSTREN
N	Valid	10	10
	Missing	0	0
Variance		2.612	.015
Minimum		9.11	1.92
Maximum		14.53	2.32

a.Type of dressing = peripac, Time after application = 24

Type of dressing = putty, Time after application = 3

Statistics[a]

		SSTREN	LSTREN
N	Valid	10	10
	Missing	0	0
Variance		10.454	.009
Minimum		25.12	2.80
Maximum		34.50	3.08

a.Type of dressing = putty, Time after application = 3

Type of dressing = putty, Time after application = 24

Statistics[a]

		SSTREN	LSTREN
N	Valid	10	10
	Missing	0	0
Variance		14.982	.013
Minimum		23.52	2.74
Maximum		35.00	3.09

a.Type of dressing = putty, Time after application = 24

TABLE 5.46 Continued

Type of dressing = zone, Time after application = 3

Statistics[a]

		SSTREN	LSTREN
N	Valid	10	10
	Missing	0	0
Variance		7.149	.010
Minimum		19.42	2.58
Maximum		28.98	2.92

a.Type of dressing = zone, Time after application = 3

Type of dressing = zone, Time after application = 24

Statistics[a]

		SSTREN	LSTREN
N	Valid	10	10
	Missing	0	0
Variance		8.835	.006
Minimum		28.76	2.92
Maximum		37.40	3.15

a.Type of dressing = zone, Time after application = 24

with α set at a somewhat more stringent level (e.g., .01 rather than .05) given the strong heterogeneity. Because all analyses produce essentially the same results at $\alpha =$ .01, the preference is for the simplest (untransformed) analysis. However, the report of results will include a statement of the finding of heterogeneity of variance, as well as an attempt to interpret it, if warranted.

5.7.2.1 Omnibus Tests

Table 5.47 shows the SPSS GLM ANOVA for the original untransformed scores. Both of the main effects and the interaction are highly significant, $p < .001$. Tables of Estimated Marginal Means show standard errors of means for each effect, to be reported in the results section. However, treatment magnitude varies widely among the three effects. Applying Equation 5.17 for dressing, hours, and dressing*hours:

$$\text{Partial } \hat{\omega}_{D^2} = \frac{4(2098498.5 - 19696.8)/100}{4(2098498.5 - 19696.8)/100 + 19696.8} = .81$$

$$\text{Partial } \hat{\omega}_{H^2} = \frac{4(1131244.96 - 19696.8)/1001}{4(1131244.96 - 19696.8)/100 + 19696.8} = .36$$

$$\text{Partial } \hat{\omega}_{DH^2} = \frac{4(851580.34 - 19696.8)/100}{4(851580.34 - 19696.8)/100 + 19696.8} = .10$$

A great deal of systematic variance is accounted for in these partial analyses.

Alternatively, Smithson's (2003) files may be used to find, for each of the effects, η^2s and their confidence intervals. Table 5.48 shows SPSS output produced by running NONCF3.SPS on NONCF.SAV. One line of values is filled into NONCF.SAV data file for each effect: dressing, hours, and the dressing by hours interaction. The Table 5.47 values for the main effect of dressing, for example, are F = 106.540 (fval) and df = 4, 90 (df1, df2). The confidence level is set to .950 (conf).

Thus, for the main effect of dressing, partial $\eta^2 = .83$, with 95% confidence limits from .75 to .86. For the main effect of hours, partial $\eta^2 = .39$, with 95% confidence limits from .23 to .51. For the interaction, partial $\eta^2 = .32$, with 95% confidence limits from .15 to .43. Effect sizes are not much different from those produced by $\hat{\omega}^2$ and these values have the advantage of indicating confidence intervals as well as point estimations.

TABLE 5.47 Two-way analysis of variance of strength through SPSS GLM (UNIANOVA)

```
UNIANOVA
  STRENGTH BY DRESSING HOURS
  /METHOD = SSTYPE(3)
  /INTERCEPT = INCLUDE
  /PLOT = PROFILE( HOURS*DRESSING )
  /EMMEANS = TABLES(DRESSING)
  /EMMEANS = TABLES(HOURS)
  /EMMEANS = TABLES(DRESSING*HOURS)
  /PRINT = DESCRIPTIVE ETASQ
  /CRITERIA = ALPHA(.05)
  /DESIGN = DRESSING HOURS DRESSING*HOURS .
```

Univariate Analysis of Variance

Between-Subjects Factors

		N
Type of dressing	care	20
	coe	20
	peripac	20
	putty	20
	zone	20
Time after application	3	50
	24	50

Tests of Between-Subjects Effects

Dependent Variable: Bond strength KN/sq m

Source	Type III Sum of Squares	df	Mean Square	F	Sig.	Partial Eta Squared
Corrected Model	10376820.2[a]	9	1152980.027	58.536	.000	.854
Intercept	38668498.6	1	38668498.56	1963.183	.000	.956
dressing	8393994.940	4	2098498.735	106.540	.000	.826
hours	1131244.960	1	1131244.960	57.433	.000	.390
dressing * hours	851580.340	4	212895.085	10.809	.000	.324
Error	1772715.200	90	19696.836			
Total	50818034.0	100				
Corrected Total	12149535.4	99				

a. R Squared = .854 (Adjusted R Squared = .840)

TABLE 5.47 Continued

Estimated Marginal Means

1. Type of dressing

Dependent Variable: Bond strength KN/sq m

Type of dressing	Mean	Std. Error	95% Confidence Interval	
			Lower Bound	Upper Bound
care	519.100	31.382	456.754	581.446
coe	677.600	31.382	615.254	739.946
peripac	121.000	31.382	58.654	183.346
putty	845.050	31.382	782.704	907.396
zone	946.450	31.382	884.104	1008.796

2. Time after application

Dependent Variable: Bond strength KN/sq m

Time after application	Mean	Std. Error	95% Confidence Interval	
			Lower Bound	Upper Bound
3	515.480	19.848	476.049	554.911
24	728.200	19.848	688.769	767.631

3. Type of dressing * Time after application

Dependent Variable: Bond strength KN/sq m

Type of dressing	Time after application	Mean	Std. Error	95% Confidence Interval	
				Lower Bound	Upper Bound
care	3	354.900	44.381	266.729	443.071
	24	683.300	44.381	595.129	771.471
coe	3	573.600	44.381	485.429	661.771
	24	781.600	44.381	693.429	869.771
peripac	3	112.000	44.381	23.829	200.171
	24	130.000	44.381	41.829	218.171
putty	3	837.900	44.381	749.729	926.071
	24	852.200	44.381	764.029	940.371
zone	3	699.000	44.381	610.829	787.171
	24	1193.900	44.381	1105.729	1282.071

TABLE 5.48 Data set output from NONCF3.SPS for effect size (**R2**), with 95 confidence limits (**LR2** and **UR2**) for dressing, hours, and the interaction, respectively

fval	df1	df2	conf	lc2	ucdf	uc2	lcdf	power	R2	LR2	UR2
106.540	4	90	.950	287.1586	.9750	583.4209	.0250	1.0000	.83	.75	.86
57.4330	1	90	.950	28.2117	.9750	96.2872	.0250	1.0000	.39	.23	.51
10.8090	4	90	.950	16.6357	.9750	73.0663	.0250	.9998	.32	.15	.43

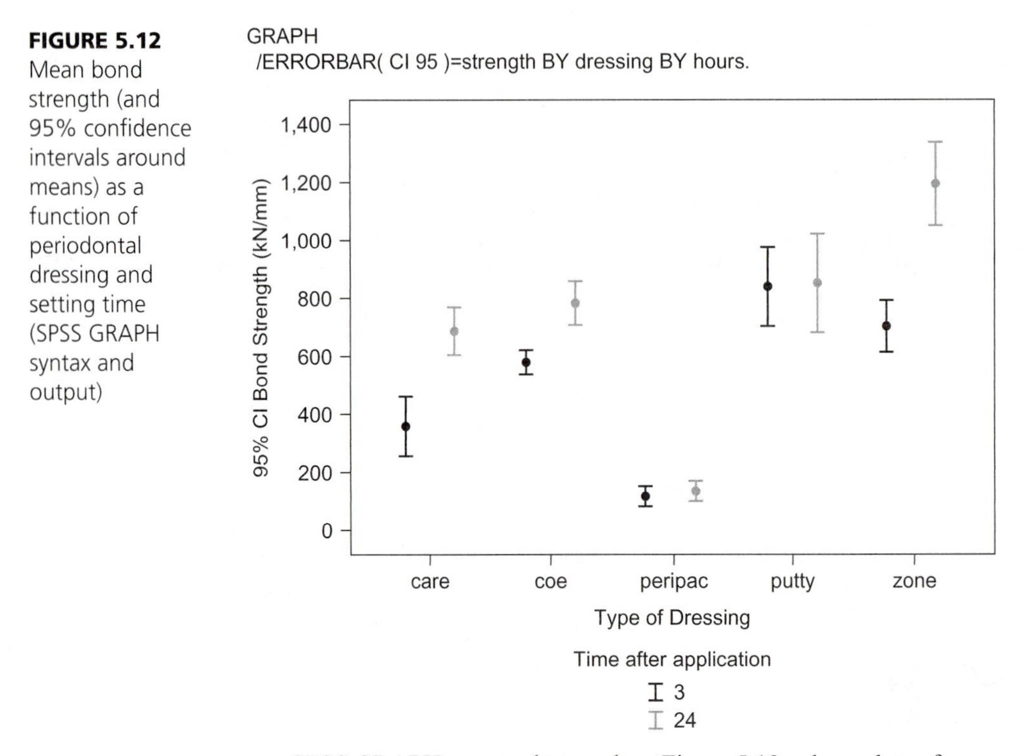

FIGURE 5.12 Mean bond strength (and 95% confidence intervals around means) as a function of periodontal dressing and setting time (SPSS GRAPH syntax and output)

SPSS GRAPH was used to produce Figure 5.12, where plots of means and 95% confidence intervals aid in interpreting the interaction. All of the dressings show greater strength at 24 hours than at 3 hours, but the difference is small for Putty and Peripac. The Zone dressing tested at 24 hours appears superior to all others. The most straightforward way to interpret these differences (and lack of them) is pairwise comparison of all 10 treatments.

5.7.2.2 Comparisons among Cell Means

Omnibus ANOVA is followed by post hoc pairwise comparisons of means and by post hoc pairwise comparisons of variance. SPSS GLM offers pairwise tests that do not assume homogeneity of variance. The Dunnett C test—requested as POSTHOC = GROUP(C)—is convenient because it is based on the studentized range, as per the Tukey test. Table 5.49 displays SPSS GLM (UNIANOVA) syntax and selected output, showing the pairwise Dunnett C comparisons. Critical $\alpha = .01$ is used instead of .05 because of the large number of comparisons. A new variable, group, is manually added to the data set, formed out of the 10 combinations of Dressing and Hours.

Significant differences at $\alpha = .01$ are marked with an asterisk (*) in this output. The dressing with the highest mean, Zone after 24 hours, is stronger than any other dressing, except Putty after either 3 or 24 hours. Therefore, a decision in favor of any of the three combinations is statistically supported—Putty, if quicker setting time is important. There are no mean differences between 3 and 24 hours for Putty or for Peripac, but time differences are statistically significant for the three other dressings.

TABLE 5.49 SPSS GLM (UNIANOVA) syntax and partial output for pairwise comparisons among means with Dunnett C adjustment

```
UNIANOVA
  STRENGTH BY GROUP
 /METHOD = SSTYPE(3)
 /INTERCEPT = INCLUDE
 /POSTHOC = GROUP ( C )
 /CRITERIA = ALPHA(.01)
 /DESIGN = GROUP .
```

Multiple Comparisons

Dependent Variable: Bond strength KN/sq m

Dunnett C

(I) GROUP	(J) GROUP	Mean Difference (I-J)	Std. Error	95% Confidence Interval	
				Lower Bound	Upper Bound
care24	care3	328.40*	59.498	86.98	569.82
	coe24	−98.30	49.499	−299.15	102.55
	coe3	109.70	41.564	−58.95	278.35
	peri24	553.30*	38.643	396.50	710.10
	peri3	571.30*	39.239	412.08	730.52
	putty24	−168.90	80.699	−496.35	158.55
	putty3	−154.60	70.405	−440.28	131.08
	zone24	−510.60*	72.709	−805.63	−215.57
	zone3	−15.70	54.803	−238.07	206.67
care3	care24	−328.40*	59.498	−569.82	−86.98
	coe24	−426.70*	57.332	−659.33	−194.07
	coe3	−218.70*	50.639	−424.18	−13.22
	peri24	224.90*	48.271	29.03	420.77
	peri3	242.90*	48.750	45.09	440.71
	putty24	−497.30*	85.727	−845.15	−149.45
	putty3	−483.00*	76.116	−791.85	−174.15
	zone24	−839.00*	78.252	−1156.52	−521.48
	zone3	−344.10*	61.969	−595.55	−92.65
coe24	care24	98.30	49.499	−102.55	299.15
	care3	426.70*	57.332	194.07	659.33
	coe3	208.00*	38.399	52.19	363.81
	peri24	651.60*	35.217	508.70	794.50
	peri3	669.60*	35.870	524.05	815.15
	putty24	−70.60	79.115	−391.62	250.42

TABLE 5.49 Continued

(I) GROUP	(J) GROUP	Mean Difference (I-J)	Std. Error	95% Confidence Interval Lower Bound	Upper Bound
coe24 (cont.)	putty3	−56.30	68.584	−334.59	221.99
	zone24	−412.30*	70.947	−700.18	−124.42
	zone3	82.60	52.443	−130.20	295.40
coe3	care24	−109.70	41.564	−278.35	58.95
	care3	218.70*	50.639	13.22	424.18
	coe24	−208.00*	38.399	−363.81	−52.19
	peri24	443.60*	22.751	351.28	535.92
	peri3	461.60*	23.749	365.24	557.96
	putty24	−278.60	74.408	−580.52	23.32
	putty3	−264.30*	63.096	−520.32	−8.28
	zone24	−620.30*	65.657	−886.71	−353.89
	zone3	−125.40	45.029	−308.11	57.31
peri24	care24	−553.30*	38.643	−710.10	−396.50
	care3	−224.90*	48.271	−420.77	−29.03
	coe24	−651.60*	35.217	−794.50	−508.70
	coe3	−443.60*	22.751	−535.92	−351.28
	peri3	18.00	18.160	−55.69	91.69
	putty24	−722.20*	72.817	−1017.67	−426.73
	putty3	−707.90*	61.212	−956.28	−459.52
	zone24	−1063.90*	63.849	−1322.98	−804.82
	zone3	−569.00*	42.349	−740.84	−397.16
peri3	care24	−571.30*	39.239	−730.52	−412.08
	care3	−242.90*	48.750	−440.71	−45.09
	coe24	−669.60*	35.870	−815.15	−524.05
	coe3	−461.60*	23.749	−557.96	−365.24
	peri24	−18.00	18.160	−91.69	55.69
	putty24	−740.20*	73.135	−1036.96	−443.44
	putty3	−725.90*	61.590	−975.81	−475.99
	zone24	−1081.90*	64.211	−1342.45	−821.35
	zone3	−587.00*	42.893	−761.04	−412.96
putty24	care24	168.90	80.699	−158.55	496.35
	care3	497.30*	85.727	149.45	845.15
	coe24	70.60	79.115	−250.42	391.62
	coe3	278.60	74.408	−23.32	580.52

TABLE 5.49 Continued

(I) GROUP	(J) GROUP	Mean Difference (I-J)	Std. Error	95% Confidence Interval	
				Lower Bound	Upper Bound
putty24 (cont.)	peri24	722.20*	72.817	426.73	1017.67
	peri3	740.20*	73.135	443.44	1036.96
	putty3	14.30	93.626	−365.60	394.20
	zone24	−341.70	95.371	−728.68	45.28
	zone3	153.20	82.537	−181.71	488.11
putty3	care24	154.60	70.405	−131.08	440.28
	care3	483.00*	76.116	174.15	791.85
	coe24	56.30	68.584	−221.99	334.59
	coe3	264.30*	63.096	8.28	520.32
	peri24	707.90*	61.212	459.52	956.28
	peri3	725.90*	61.590	475.99	975.81
	putty24	−14.30	93.626	−394.20	365.60
	zone24	−356.00*	86.834	−708.34	−3.66
	zone3	138.90	72.505	−155.30	433.10
zone24	care24	510.60*	72.709	215.57	805.63
	care3	839.00*	78.252	521.48	1156.52
	coe24	412.30*	70.947	124.42	700.18
	coe3	620.30*	65.657	353.89	886.71
	peri24	1063.90*	63.849	804.82	1322.98
	peri3	1081.90*	64.211	821.35	1342.45
	putty24	341.70	95.371	−45.28	728.68
	putty3	356.00*	86.834	3.66	708.34
	zone3	494.90*	74.744	191.62	798.18
zone3	care24	15.70	54.803	−206.67	238.07
	care3	344.10*	61.969	92.65	595.55
	coe24	−82.60	52.443	−295.40	130.20
	coe3	125.40	45.029	−57.31	308.11
	peri24	569.00*	42.349	397.16	740.84
	peri3	587.00*	42.893	412.96	761.04
	putty24	−153.20	82.537	−488.11	181.71
	putty3	−138.90	72.505	−433.10	155.30
	zone24	−494.90*	74.744	−798.18	−191.62

Based on observed means.

*The mean difference is significant at the .05 level.

5.7.2.3 Interpreting Heterogeneity of Variance

Some variances in strength are far larger than others, although this finding did not compromise the major analyses. Pairwise F tests of variances are conducted post hoc with α set to .001, because there are 45 pairs where

$$F = \frac{s^2_{\text{larger}}}{s^2_{\text{smaller}}}$$

and df = 9 and $9(n - 1)$ for each F test. This test is similar to the F_{max} test, except that a tabled critical value is used and all pairs of variances are tested, not just the largest and smallest pair. The critical value from Appendix A.6 for these variance tests is 10.37.

Table 5.50 shows (with an asterisk) which of the variances differed.

In general, this analysis shows that Peripac, the dressing with least strength, is also less variable than the stronger dressings (Zone after 24 hours and Putty after 3 and 24 hours). Peripac is also less variable in strength than some of the intermediate dressings. The weakest dressing produced the most consistent results.

Table 5.51 is a checklist for factorial randomized-groups ANOVA. An example of a results section, in journal format, follows for the study just described.

TABLE 5.50 Significant (*) differences among variances in strength of dressings after 3 and 24 hours

	Zone 3 hr	Zone 24 hr	Putty 3 hr	Putty 24 hr	Coe 3 hr	Care 3 hr
Peripac 3 hr		*	*	*	*	*
Peripac 24 hr	*	*	*	*		*

TABLE 5.51 Checklist for factorial randomized-groups ANOVA

1. Issues
   a. Independence of errors
   b. Unequal sample sizes
   c. Normality of sampling distributions
   d. Outliers
   e. Homogeneity of variance
2. Major analyses
   a. Main effects or planned comparisons. If statistically significant:
      (1) Effect size(s) and their confidence limits
      (2) Parameter estimates (marginal means and standard deviations or standard errors or confidence intervals)
   b. Interactions or planned comparisons. If significant:
      (1) Effect size(s) and their confidence limits.
      (2) Parameter estimates (cell means and standard deviations or standard errors or confidence intervals in table or plot).
3. Additional analyses
   a. Post hoc comparisons.
   b. Interpretation of departure from homogeneity of variance.

Results

A 5 × 2 randomized-groups ANOVA was performed on bond strength of periodontal dressings tested in vitro. Dressings, commercially available, were Zone, Perio Putty, Coe Pak, PerioCare+, and Peripac. Half of the 20 samples of each dressing were tested for tensile bond strength after 3 hours and half tested were tested after 24 hours. Severe heterogeneity of variance was noted; however, no substantive distortion of the results was evident. Obtained probability levels were far enough beyond the preset α = .05 to compensate for any inflation in Type I error due to heterogeneity. Therefore, no additional adjustment for heterogeneity was made. Figure 5.12 plots means and standard errors for all combinations of dressing and time.

Strong, statistically significant differences among dressing were evident, averaged over the two time periods, $F(4, 90)$ = 106.54, p < .001, SE_M = 31.28, partial η^2 = .83, with a confidence interval ranging from .75 to .86. Tensile strength varied from a mean high of 946.56 kN/mm for Zone dressing to a mean low of 121.00 kN/mm for Peripac.

A smaller, but still statistically significant, effect of time was also evident, averaged over dressings, $F(1, 90)$ = 57.43, p < .001, SE_M = 19.85, partial η^2 = .39, with a confidence interval ranging from .23 to .61. Average strength was less after 3 hours (M = 515.48 kN/mm) than after 24 hours (M = 728.2 kN/mm). These main effects were modified by a small but statistically significant interaction between dressing and time, $F(1, 90)$ 10.81, p < .001, SE_M = 44.38, partial η^2 = .32, with a confidence interval ranging from .15 to .43.

All pairwise comparisons among means were tested using a Dunnett C correction at α = .01. These comparisons showed that the dressing with the strongest mean (Zone after 24 hours) was superior to all other dressings except Perio Putty after either 3 or 24 hours. These findings suggest a choice in favor of Perio Putty if setting time is important.

However, pairwise comparisons among variances in strength among the 10 sets of samples showed greatest variance for the strongest dressings; strength measurements for Zone after 24 hours and Putty after 3 and 24 hours were less consistent than were strength measurements for the weakest dressings (Peripac after 3 and after 24 hours). Peripac after 3 hours was also significantly less variable in strength than PerioCare+ after 3 hours and Coe Pak after 3 hours. Peripac after 24 hours was significantly less variable in strength than Zone after 3 hours and PerioCare+ after 3 hours. All of the samples using Zone after 24 hours and using Putty, however, were stronger than any of the samples using Peripac.

5.8 COMPARISON OF PROGRAMS

Table 5.52 reviews programs from SPSS, SAS, and SYSTAT that are appropriate for factorial randomized-groups ANOVA. Only the features applicable to these designs are reviewed here; other features are reviewed elsewhere.

5.8.1 SPSS Package

GLM (UNIANOVA) is the SPSS program available for factorial analysis of variance.[14] It offers the standard features, such as source tables, and the major descriptive

[14] In SPSS for Windows, on the General Linear Models submenu, UNIANOVA program is identified as General Linear Model, Univariate.

TABLE 5.52 Comparison of SPSS, SAS, and SYSTAT programs for factorial randomized-groups ANOVA

Feature	SPSS GLM	SAS GLM[a]	SYSTAT ANOVA and GLM[a]
Input			
Maximum number of IVs	No limit	No limit	No limit
Options for unequal n	Yes	Yes	Yes[b]
Specify α level for post hoc comparisons	Yes	ALPHA	No[d]
Special specification for trend (polynomial) analysis	Yes	No	Polynomial
Unequal spacing of levels for trend analysis	Yes	No	Yes
Streamlined specification for simple effects	No	SLICE	No
Specify error terms for effects and/or contrasts	Yes	Yes	Yes
Specify number of Bonferroni comparisons	No	No	No
Custom hypothesis matrices	Yes	Yes	Yes
Case weighting	Yes	Yes	No
Specify tolerance	EPS	SINGULAR	Yes
Output			
ANOVA table	Yes	Yes	Yes
Cell means	Yes	Yes	No
Confidence interval for cell means	No	No	No
Unadjusted marginal means	No[c]	Yes	No
Cell standard deviations	Yes	Yes	No
Confidence interval for adjusted cell means	EMMEANS	Yes	No
Standard errors for adjusted cell means	EMMEANS	Data file	Yes
Adjusted marginal means	EMMEANS	LS MEANS	Yes
Standard errors for adjusted marginal means	EMMEANS	STDERR	Pooled only
Descriptive statistics for all groups combined	Yes	No	No
Power analysis and effect sizes	Yes	No	No
Tests for pairwise contrasts that do not require homogeneity of variance	Yes	No	No
t test for all pairs of cell means	No	Yes	Yes
Tukey comparisons for main effects	Yes	Yes	No
Bonferroni pairwise comparisons for main effects	Yes	Yes	No
Scheffé pairwise comparisons for main effects	Yes	Yes	No
LSD pairwise comparisons for main effects	Yes	Yes	No

TABLE 5.52 Continued

Feature	SPSS GLM	SAS GLM[a]	SYSTAT ANOVA and GLM[a]
Output (cont.)			
Sidak pairwise comparisons for main effects	Yes	Yes	No
Additional post hoc tests with adjustments[f]	Yes	Yes	Yes
User-specific contrasts	Yes	Yes	Yes
Multiple R and/or R^2	Yes	Yes	Yes
Levene's test for homogeneity of variance	Yes	No	No
Predicted values and residuals	Yes	Yes	Data file
Confidence limits for predicted values	No	Yes	No
Outliers (studentized residual criterion)	Yes	P	Yes
Durbin-Watson statistic	No	P	Yes
Plots			
Means	Yes	No	Yes
Box-Cox plots for homogeneity of variance (group means vs. group standard deviations)	Yes	No	No
Plots of residuals	Yes	No	No
Histogram of residuals	No	No	No
Normal probability plot of residuals	No	No	No
Residuals vs. predicted (fitted) values	No	No	No
Residuals vs. order of data	No	No	No
Residuals vs. any variable	No	No	No
Saved to file			
Residuals	Yes	Yes	Yes
Predicted (fitted, adjusted) values (i.e., mean for each case)	Yes	Yes	Yes
Additional statistics, including those for outliers	Yes	Yes	Yes
Means, standard deviations, and frequencies	No	Yes	Yes

[a]Additional features reviewed in other chapters.

[b]Requires multiple runs for any but Method 1.

[c]Unadjusted marginal means only for first-listed factor.

[d]Probability values provided for comparisons.

[f]See Table 4.30 for additional details about post hoc tests.

statistics and has options for unequal sample sizes. SPSS GLM is the only program reviewed that provides achieved power and effect sizes (partial η^2). GLM has a vast array of post hoc tests (20 in all), including several choices of pairwise comparisons that do not require homogeneity of variance. GLM also provides tests for homogeneity of variance, predicted values and residuals, plots of residuals, plots of means versus variances (Box-Cox plots, which are handy for diagnosis of homogeneity of variance), and plots of means to aid interpretation. SPSS MANOVA, an earlier program available only in syntax, has special simplified language for simple-effects analysis.

5.8.2 SAS System

SAS GLM has all the general features of ANOVA programs, along with several unique features. However, GLM's capacity to do a wide variety of multivariate general linear modeling makes it more difficult to use than some of the more specialized ANOVA programs. For instance, modeling syntax is necessary to specify IVs and the DV. Although SAS ANOVA is less computer-intensive, it is no easier to use than GLM and is limited to data sets with equal sample sizes in all cells.

GLM has an especially rich array of options for planned and post hoc comparisons, and all sorts of matrices may be hypothesized and printed out for the more sophisticated analyst. The program produces predicted values and residuals for each case, as well as confidence limits for those values, on request. Information about outliers is available, along with the Durbin-Watson statistic if case order is meaningful.

5.8.3 SYSTAT System

SYSTAT ANOVA is a subset of the GLM program; both produce the same output and have the same features, but ANOVA uses a simpler syntax for identifying IVs and the DV. GLM uses modeling syntax, similar to SAS.

The program has several of the more popular post hoc comparison procedures, along with user-specified weighting coefficients. However, it is somewhat short on descriptive statistics, among them standard deviations within cells. (Instead, a standard error is printed that is the square root of the pooled within-cell variance divided by the square root of the cell size.) This means that other modules within the system are necessary to evaluate homogeneity of variance.

Studentized residual outliers are automatically printed, as is the Durbin-Watson statistic. Plots of means are provided but not in the usual "factorial" format. SYSTAT ANOVA and GLM produce an unweighted-means analysis by default if sample sizes are unequal. Any other option for dealing with unequal n requires multiple runs.

5.9 PROBLEM SETS

5.1 Do an analysis of variance on the following data set. Assume that A is type of fabric, B is minutes of presoaking a garment, and the DV is rating of cleanliness.

	Presoak Time			
	b_1	b_2	b_3	b_4
	5 min	10 min	15 min	20 min
	2	3	8	6
a_1: Cotton	6	9	11	7
	6	9	9	9
	6	5	9	6
a_2: Polyester	5	7	6	9
	3	5	12	9

a. Test for homogeneity of variance.

b. Provide a source table.

c. Check results by running the data set through at least one statistical computer program.

d. Calculate effect size(s) and their confidence limits.

e. Write a results section for the analysis.

f. Use one of the statistical software programs to do a trend analysis and apply an appropriate post hoc adjustment.

5.2 Table 5.37 shows several statistically significant interaction contrasts for the small-sample data of Table 5.3. One of these is plotted in Figure 5.10. Plot one of the remaining significant contrasts.

5.3 Do an analysis of variance (through a statistical software program) on the following pilot data set, assuming these are pilot data for an experiment in which cell sample sizes will be equal.

	b_1	b_2	b_3	b_4	b_5
	6	5	5	6	6
a_1	5	6	5	5	7
			6		
	5	6	7	9	7
a_2	8	7	5	7	6
		6		7	
				6	

a. Provide a source table.

b. Show appropriate marginal means.

c. How many cases would you need to run to get a statistically significant main effect of B at $\alpha = .05$ and $\beta = .20$?

5.4 Suppose a researcher is interested in the effects of hours of sleep and type of beverage consumed on test performance in a statistics course. Students in the course are randomly assigned to get either 4, 6, or 8 hours of sleep and to get either coffee, fruit juice, or beer in the 24 hours prior to the exam. Consider type of beverage factor A and hours of sleep factor B. For each question below, indicate the type of analysis required (marginal comparison, simple main effect, simple comparison, or interaction contrast) and use A and B symbols to specify the analysis (e.g., $A_{\text{comp at } b3}$).

a. Do students who get 8 hours of sleep perform better than those who get 4 or 6 hours?

b. Do those who drink coffee perform better than those who drink fruit juice?

c. For those sleeping 6 hours, do the different beverages produce different exam scores?

d. For those drinking fruit juice, is there a linear trend of hours of sleep?

e. For coffee drinkers, is there a difference in exam scores between those who sleep 4 hours and those who sleep 8 hours?

f. For those drinking beer, is there a difference in scores depending on hours of sleep?

g. For those sleeping 4 hours, is there a difference between fruit juice drinkers and coffee drinkers?

h. Do coffee drinkers and fruit juice drinkers have a different linear trend of hours of sleep?

Extending the Concepts

5.5 Do a three-way ANOVA using the data set of Table 5.19.

a. Test for homogeneity of variance.

b. Provide a source table.

c. Check results by running the data set through at least one statistical software program.

d. Calculate effect size(s).

e. Write a results section for the analysis.

f. Use one of the statistical software programs to test the simple effect of deprivation at each level of familiarity. Assume this was part of the sequential analysis plan. (*Hint:* If using SPSS, recode into eight groups or use coefficients in ONEWAY contrasts as per Chapter 4.)

5.6 Develop the IV coding columns for a $2 \times 3 \times 3$ ANOVA using the regression approach.

6

Repeated-Measures Designs

6.1 GENERAL PURPOSE AND DESCRIPTION

In repeated-measures designs, each case is measured more than once on the same DV. In randomized-groups designs, each case is measured a single time.[1] Repeated-measures designs are often used to track changes in a DV over time. For instance, a marketing firm might be interested in changes in sales of a new product in the six months after introduction into a city. Sales figures are collected in each of several sales districts (the cases) for each month and are analyzed using repeated-measures ANOVA. Are sales higher at the end? Is there a difference between the first and second month of sales? When did sales really pick up? Are sales still going up, or have they leveled off?

Although the IV in repeated-measures designs is often a variable, such as time or distance, whose levels are in a fixed order, IVs whose levels are measured in any order are also used. For instance, a hospital might be interested in the efficacy of several

[1] There are designs in which cases are measured once, but on several DVs, and analyzed using MANOVA (multivariate analysis of variance) or in which cases are measured several times, each time on several DVs, and analyzed using repeated-measures MANOVA (see Tabachnick and Fidell, 2007).

different disinfectants for cleaning pieces of equipment. The levels of the IV are the different disinfectants. The cases are the same pieces of equipment cleaned repeatedly, each time using a different disinfectant. The order in which the disinfectants are used is completely arbitrary. In this application of repeated measures, special controls for the potential effects of order are necessary, as described in Section 6.5.1 and Chapter 7.

Two or more repeated-measures IVs are often used in factorial combination in the same study. For instance, different disinfectants are used in combination with different temperatures of washing solution (high and low) to find, if possible, the particular combination of disinfectant and temperature that is the most efficacious. A repeated-measures IV can also be used in combination with one or more randomized-groups IVs in a mixed design (see Chapter 7).

Repeated-measures designs are favored in many settings because they are often more efficient and more powerful than randomized-groups designs. They are more efficient for at least two reasons: (1) They use far fewer cases than a corresponding randomized-groups design. For instance, a randomized-groups design with three levels probably requires about 10 cases per level, for a total of 30 cases. In a repeated-measures design, the same 10 cases are measured three times. This feature of repeated measures is especially valuable when cases are expensive. (2) It often takes considerable time and effort to access and prepare a case. Once a case is ready, it is often more efficient to measure it several times unless there is some reason not to.

Repeated-measures analysis is often more powerful than randomized-groups analysis because differences associated with the cases themselves are assessed and subtracted from the error term. In ANOVA, anything measured more than one time is eligible for analysis as a source of variability. In repeated-measures designs, cases are measured more than once. Therefore, differences between cases are eligible for analysis. For instance, in the marketing example, differences in sales districts are assessed and subtracted from the error term; in the hospital example, differences in the pieces of equipment are assessed and removed from the error term. The error term in a repeated-measures design, then, frequently is smaller than the error term in a corresponding randomized-groups design, making the overall ANOVA more powerful. Repeated-measures designs are often chosen to capitalize on this feature of the analysis.

However, there are no free lunches, so increases in efficiency and power with this design are sometimes offset by a restrictive assumption that makes analysis by traditional ANOVA problematic. The assumption—sphericity—is discussed in Sections 6.3.2.3 and 6.6.2. When the assumption is violated, routine analysis is untenable. There are several alternatives, with some involving more complicated statistical analyses and some involving reconceptualization of the design.

6.2 KINDS OF RESEARCH QUESTIONS

The research questions asked in repeated-measures designs are basically the same as those asked in randomized-groups designs. Are there significant mean differences associated with the levels of an IV? Do combinations of levels of IVs interact to produce changes of their own in the means? If so, parameters and effect size are estimated. If an IV has more than two levels, comparisons are appropriate in lieu of, or after, ANOVA.

6.2.1 Effect of the IVs

The main effect of each IV in the design is available for analysis. Are there significant mean differences associated with different levels of each IV? Are there mean differences in the sales figures for the six months? Do different disinfectants produce different levels of cleanliness? There is, however, another main effect available for analysis in a one-way design, should it prove interesting, and that is for cases. Do cases have different average scores on the DV? Do the sales districts differ? Are pieces of equipment different in cleanliness? In some settings, differences due to cases are extremely interesting, while in others, they are merely a nuisance. These questions are answered as in randomized-groups ANOVA, by applying *F* tests to the data, as demonstrated in Section 6.4.1.

6.2.2 Effect of Interactions among IVs

Does the effect of one IV depend on the level of another IV? Are there particular combinations of levels of IVs that are better (or worse) than other combinations? Is the effectiveness of type of disinfectant different at various levels of temperature of the washing solution? Section 6.4.2 demonstrates application of the *F* test to answer questions about interaction.

6.2.3 Parameter Estimates

What mean DV score is predicted for each level of the IV in the population? For each combination of levels of IVs? What average cleanliness score is predicted for each type of disinfectant? Parameter estimates reported for statistically significant IVs are sample means and either standard deviations, standard errors, or confidence intervals around the means. Statistically significant mean differences are often shown graphically, as in Figure 2.10.

6.2.4 Effect Sizes

How much of the variability in the DV is associated with differences in levels of the IV? With differences in combinations of levels of IVs? How much of the total variability in sales is associated with month? How much of the total variability in cleanliness is associated with the type of disinfectant? Computation of effect size for repeated-measures IVs is discussed in Section 6.6.4.

6.2.5 Power

What is the probability of finding a mean difference on the IV if the DV really does change with different levels of the IV? More formally, if the alternative hypothesis is true (the null hypothesis is false), what are the chances of concluding in its favor? If there really are differences in disinfectants, what is the probability of finding a mean difference in cleanliness? Are enough cases (pieces of equipment) being tested so that the outcome is likely to be statistically significant? Computation of power in repeated-measures designs is discussed in Section 6.6.3.

6.2.6 Specific Comparisons

If there are more than two levels of an IV, which means are different from which other means? Do some levels of the IV produce similar means whereas other levels produce different mean DV scores? Are all the disinfectants different from each other? Are some of them the same while others are different? Is the best one significantly better than the next best one? As in randomized-groups designs, comparisons are either planned and performed in lieu of ANOVA or, with lower power, performed post hoc. Comparisons are a nasty topic in repeated-measures designs, as discussed in Section 6.6.6.

6.3 ASSUMPTIONS AND LIMITATIONS

6.3.1 Theoretical Issues

If the repeated-measures IV is passage of time (whose levels are chosen, but not manipulated), the study is not an experiment, and causal inference is inappropriate. When the levels are manipulated (e.g., type of music), issues of causal inference are resolved by attention to the details of the study's design, as in randomized-groups designs.

Control of extraneous variables may be especially difficult because of the passage of time during repeated measurement. Have extraneous variables been adequately controlled? What conditions, if any, change during the course of the study? Is equipment stable? Do testing conditions remain the same? Do the researchers behave consistently over time? If cases are people, are they just as perky and involved at the last measurement as at the first? Is their attention stable, or does it waver over time? Do people become more skilled at performing the DV over time, regardless of the IV? If the levels of the IV can be presented in any order and if there is reason to expect change over repeated measurement, you are wise to implement a Latin-square design as an additional control over these changes, as discussed in Chapter 9.

There is also the issue of generalizability of findings: How are cases obtained, and what population do they represent? Because repeated-measures designs use fewer cases than randomized-groups designs use, generalization is more problematic, making the representativeness of the sample more critical than for randomized-groups design. Also, the pattern of effects of the IV can only be generalized to cases that have been exposed to all levels of the IV. Thus, generalization is usually more restricted in a repeated-measures design than in a randomized-groups design. These and other questions are resolved by attention to the details of the research design, and not to the statistics.

6.3.2 Practical Issues

Practical issues of normality of sampling distributions, homogeneity of variance, and absence outliers are similar in both repeated-measures and randomized-groups designs. However, the assumption of independence of errors has greater relevance in this design; additivity (absence of an interaction between cases and levels of the IV) is an additional assumption. Although logically different, this assumption plays out as the assumption of sphericity, special to repeated measurement. This assumption is critical to the analysis and is often violated with both real and made-up data sets.

All the analyses in this chapter also assume equal spacing between adjacent levels of IVs (e.g., 10, 20, 30 seconds), as opposed to unequal spacing (e.g., 5, 20, 30 seconds). Analysis of unequally spaced levels of IVs is discussed in Section 11.6.3.

6.3.2.1 Normality of Sampling Distributions

In randomized-group designs, normality of sampling distributions of means is expected if there are relatively equal numbers of cases in every cell and 20 df for error. In repeated measures, there is only one value in every combination of case and IV level (i.e., one case in each cell, as shown in Table 6.1(a)), and the number of cases at each level of the IV is equal because the same cases participate at all levels. If there is only one IV, *normality of the sampling distribution of means is anticipated if there are 20 df for the error term.*

However, a complexity arises with small N studies and factorial repeated measures. In factorial repeated measures, several error terms are developed—a different one for each IV and for each combination of IVs in the design. Some error terms may have sufficient df to assure normality, but others may not. If df are sparse for the error

TABLE 6.1 Conversion of raw scores to scores tested by each IV in a factorial repeated-measures design

(a) Raw scores

	a_1		a_2	
	b_1	b_2	b_1	b_2
s_1	3	7	4	6
s_2	1	3	2	4
s_3	5	9	6	8

(b) Scores tested in main effect of A

	a_1	a_2
	$b_1 + b_2$	$b_1 + b_2$
s_1	$3 + 7 = 10$	$4 + 6 = 10$
s_2	$1 + 3 = 4$	$2 + 4 = 6$
s_3	$5 + 9 = 14$	$6 + 8 = 14$

(c) Scores tested in main effect of B

	b_1	b_2
	$a_1 + a_2$	$a_1 + a_2$
s_1	$3 + 4 = 7$	$7 + 6 = 13$
s_2	$1 + 2 = 3$	$3 + 4 = 7$
s_3	$5 + 6 = 11$	$9 + 8 = 17$

terms associated with some effects, look at the skewness and kurtosis values associated with the DV for those effects, as described in Section 2.2.3.3. If the values are satisfactory, normality of sampling distributions is anticipated. To look at skewness and kurtosis, you have to generate the DV scores for each IV or combination of IVs, because the scores that are actually tested are not the raw scores, but rather the sums of raw scores, as shown in Tables 6.1(b) and 6.1(c). It is the distributions of summed scores (10, 4, 14) for a_1 and summed scores (10, 6, 14) for a_2 that need to be fairly normally distributed to test the main effect of *A*; similarly, to test the main effect of *B*—b_1 (7, 3, 11) versus b_2 (13, 7, 17).

The principle involved with generating these scores is the same no matter how complex the design becomes: You sum the numbers over the IV that is currently "left out" for each level of the IV in which you are interested. For instance, if you are creating a data set for the main effect of *A*, you sum the scores for *B* at a_1 and repeat the process at a_2, and so on, as illustrated in Table 6.1(b). If the design has three IVs and you are interested in the main effect of *A*, you sum over both *B* and *C*. When you are interested in the $A \times B$ interaction, you sum over scores for levels of *C*. Both programs have convenient facilities for combining scores in this manner.

If non-normality threatens the analysis, a data transformation of the DV scores may be desirable; if so, *all* the scores at all levels of the repeated-measures IV are transformed. Various transformations are discussed in Section 2.2.3.3. Nonparametric methods are available for small, non-normally distributed data sets if there is only one IV. The Wilcoxon signed ranks test is available when $a = 2$; the Friedman test is available for $a > 2$. Most statistical computer packages provide these nonparametric analyses.

6.3.2.2 Homogeneity of Variance

There is only one DV score in each cell of a design in which all cases participate in all levels of all IVs, so the assumption of homogeneity of variance within each cell is not relevant. However, homogeneity of variance for all the scores contributing to a level of an IV or a combination of levels of IVs is required. Variances are computed for scores at each level of an IV or each combination of levels of IVs and examined for homogeneity of variance. Because DV scores for each level or each combination of levels of IVs are in a separate column of the data set in a repeated-measures design (as shown later in Table 6.13), the variance for each column is computed separately and then the variances are compared. The guidelines and procedures in Section 4.5.5 or 5.3.2.2 are also appropriate for a repeated-measures design.

6.3.2.3 Independence of Errors, Additivity, Homogeneity of Covariance, and Sphericity

Although assumptions of independence of errors, homogeneity of covariance, and additivity are conceptually different, they are practically combined into an assumption of sphericity in the repeated-measures design. The reason is that the error term for the design confounds random error (the independence of error assumption) and the $A \times S$ (IV by cases) interaction (related to the additivity assumption). One's fondest hope in repeated measures is that there is no genuine (population) $A \times S$ interaction, so the error term is only random error. This is a forlorn hope with many real data sets, because of contingencies in the responses of the same cases as they are measured repeatedly over time.

Violation of independence of errors occurs because of (1) reassessment of the same cases, (2) effects over time, and (3) differences in the similarities of levels of treatment. The error components for each case are almost certainly associated with each other over levels of A because they are the same cases. This violates the assumption of independence of errors. Also, cases are likely to have idiosyncratic responses to different levels of treatment. Some may respond strongly to some levels, whereas others respond strongly or not at all to other levels. This produces $A \times S$ interaction, which violates additivity.[2]

A second source of violation is potential nonindependence of errors due to passage of time. Not only cases but also equipment and the testing situation are likely to drift over time. Measurements made close to each other in time are probably more similar than measurements made farther apart in time.

The third source of violation is patterns of similarity in the different levels of A. If some levels of A are similar to each other but different from other levels, there are likely to be higher correlations in the error components among those that are more similar. These sources contribute to violation of homogeneity of covariance (a part of compound symmetry and related to sphericity).

Section 6.6.2 discusses more technically these assumptions and the tests of them. It also discusses the alternative analytic strategies available when a data set violates the assumptions.

6.3.2.4 Absence of Outliers

Outliers have the same devastating effect on repeated-measures designs that they have on randomized-groups designs. Outliers are sought among the same sets of DV scores as those used to assess homogeneity of variance: separately in each level or combination of levels of repeated-measures IVs. Therefore, potential outliers are sought for each DV column in the data set (as shown later in Table 6.13). The guidelines of Section 3.4.5 or 4.3.2.4 are appropriate for this design, as well. If an outlier is found, deletion of the case, transformation of all the DV scores, or adjustment of the single score is appropriate.

6.3.2.5 Missing Data

If data are missing in randomized-groups designs, there is very little you can do, because there is no other information from which to estimate the size of the missing

[2] If you can identify some characteristic of the cases (such as age, gender, or socioeconomic status) that is associated with differences in response to different levels of A, you can block cases on it. This creates a mixed design (Chapter 7) and makes the formerly pesky interaction directly available for analysis and interpretation.

value. In repeated measures, however, there are other DV scores for the case from which to estimate (impute) the missing value. If only one or two DV scores are missing, it is often better to estimate them rather than lose all the other information for those cases. Various estimation procedures are described in Section 6.6.5.

6.4 FUNDAMENTAL EQUATIONS

There are several varieties of repeated-measures designs. The simplest is a design with a single IV. The next most complicated design has two IVs in factorial combination. These are covered in the sections that follow. Designs, of course, sometimes have three or more IVs, but the analytical principles remain the same and are easily generalized from the two-IV situation. Designs that have at least one randomized-group IV in addition to at least one repeated-measures IV are discussed in Chapter 7.

6.4.1 One-Way Repeated-Measures ANOVA

6.4.1.1 Allocation of Cases

All cases participate at all levels of *A* in a one-way repeated-measures design. Table 6.2 shows how cases or samples are used in a one-way design with three levels. Notice that every level of *A* has the same cases (samples) and each case participates at every level of *A*. Once DV scores are gathered, they are arranged in a table with *a* columns representing levels of the IV, one row for each case, and *s* scores within each level.

6.4.1.2 Partition of Sources of Variance

Anything that is measured more than one time is potentially a source of variance in ANOVA. In repeated-measures designs, cases (samples) are measured more than one time and are therefore analyzed as a potential source of variability in the DV. The

TABLE 6.2 Assignment of cases in a one-way repeated-measures design

Treatment		
a_1	a_2	a_3
s_1	s_1	s_1
s_2	s_2	s_2
s_3	s_3	s_3
s_4	s_4	s_4

other identifiable source of variability is factor A. Once these two sources of variability are computed, the remaining variability serves as the error term. Total variability (SS_T), then, is partitioned into variability due to treatment, cases, and error.

$$SS_T = SS_A + SS_S + SS_{AS} \quad \textbf{(6.1)}$$

Notice that this partition of total variability is identical to that of the main effects and interaction in a two-way randomized-groups design, A and B and $A \times B$. That is, there is a main effect of A, a main effect of S, and the interaction of A with S. Levels of S (cases) cross levels of A; all combinations of A and S appear in the design. In a two-way randomized-groups design, however, there is an additional source of variability, S/AB (cases nested within AB combinations), that serves as the error term. In the repeated-measures design, as shown in Table 6.2, there is only one score at each AS combination.[3] Therefore, there is no variability in scores within each cell. In the absence of the traditional error term, $A \times S$ is used as the error term. This term is both the interaction of cases with treatment (the idiosyncratic responses of cases to levels of treatment) and random variability (error). More about that later in this section.

Degrees of freedom in ANOVA partition the same way as sums of squares:

$$df_T = df_A + df_S + df_{AS} \quad \textbf{(6.2)}$$

Total degrees of freedom is the total number of scores, as, minus 1, lost when the grand mean is estimated. Therefore,

$$df_T = as - 1 \quad \textbf{(6.3)}$$

Note that s is the notation used for number of cases, so that $as = N$, the total number of scores.

Degrees of freedom for treatment are identical to those in a corresponding randomized-groups design:

$$df_A = a - 1 \quad \textbf{(6.4)}$$

Degrees of freedom for samples or cases are

$$df_S = s - 1 \quad \textbf{(6.5)}$$

And degrees of freedom for the error term are

$$df_{AS} = (a - 1)(s - 1) \quad \textbf{(6.6)}$$

Following is a complete partition of the degrees of freedom (and, by extension, sum of squares):

$$as - 1 = a - 1 + s - 1 + as - a - s + 1$$

Figure 6.1 is a graphic representation of the partition, showing the partition of SS_T in a randomized-groups design and then in a repeated-measures design. Notice that SS_A is the same in the two designs. Notice also that SS_S and SS_{AS} come out of the error term ($SS_{S/A}$) in the randomized-groups design. This explains why the error term for repeated measures is often smaller than the error term for randomized

[3] This is the reason that repeated-measures designs are sometimes labeled *one-score-per-cell.*

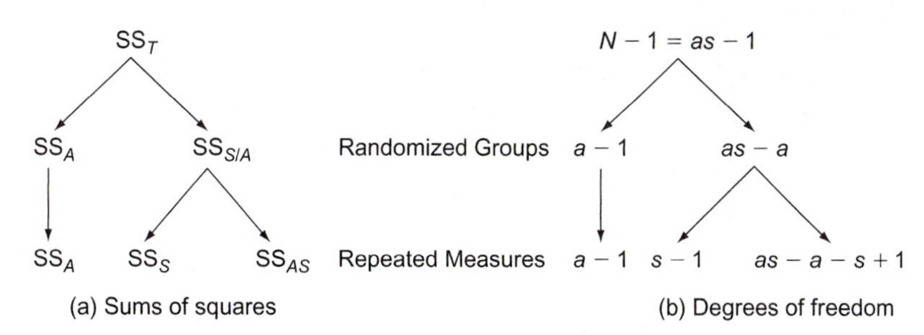

FIGURE 6.1 Partition of (a) sums of squares and (b) degrees of freedom in a randomized-groups design and then in a repeated-measures design.

groups: Differences due to cases are assessed and removed as a source of variability. Notice, however, that the degrees of freedom for the repeated-measures error term are also reduced.

Variances (mean squares) are produced by dividing SS by df. Thus, equations for mean squares are

$$\mathrm{MS}_A = \frac{\mathrm{SS}_A}{\mathrm{df}_A} \tag{6.7}$$

and

$$\mathrm{MS}_S = \frac{\mathrm{SS}_S}{\mathrm{df}_S} \tag{6.8}$$

and

$$\mathrm{MS}_{AS} = \frac{\mathrm{SS}_{AS}}{\mathrm{df}_{AS}} \tag{6.9}$$

In many research applications, MS_S, the variability in cases, is not a useful quantity.

The F distribution is a ratio of two variances. In repeated-measures ANOVA, MS_A and MS_{AS} provide the variances for the F ratio to test the null hypothesis that $\mu_{a_1} = \mu_{a_2} = \cdots = \mu_{aj}$.

$$F_A = \frac{\mathrm{MS}_A}{\mathrm{MS}_{AS}} \qquad \mathrm{df} = (a-1), (a-1)(s-1) \tag{6.10}$$

If the null hypothesis is true and there are no treatment effects and no $A \times S$ interaction, the F ratio boils down to a ratio of two estimates of the same error and is likely to be around 1.

Once F is obtained, it is tested against critical F (Table A.1) with numerator df $= a - 1$ and denominator df $= (a - 1)(s - 1)$ at desired alpha level. If obtained F equals or exceeds critical F, the null hypothesis is rejected in favor of the hypothesis that there is a difference among the means in the a levels.

6.4.1.3 Traditional One-Way Repeated-Measures ANOVA

Table 6.3 shows a data set in which number of science fiction books read per month serves as the DV. There are five readers in this small, inadequate, and hypothetical

TABLE 6.3 Hypothetical data set for a one-way repeated-measures design with three levels

	A: Month			
	a_1: Month 1	a_2: Month 2	a_3: Month 3	Case Totals
s_1	1	3	6	$S_1 = 10$
s_2	1	4	8	$S_2 = 13$
s_3	3	3	6	$S_3 = 12$
s_4	5	5	7	$S_4 = 17$
s_5	2	4	5	$S_5 = 11$
Treatment Totals	$A_1 = 12$	$A_2 = 19$	$A_3 = 32$	$T = 63$

data set. (Note that this is not an experiment, because time cannot be manipulated.) The last column contains the sum of the DV scores for each case over the three months. These values are used to assess differences among the cases because they reflect overall differences in preferences for reading, or skill levels, or whatever.

By using the notational scheme developed for randomized groups, the equation for SS_A, sum of squares attributable to treatment, is

$$SS_A = \frac{\Sigma A^2}{s} - \frac{T^2}{as} \tag{6.11}$$

The equation for SS_S, sum of squares attributable to cases, is

$$SS_S = \frac{\Sigma S^2}{a} - \frac{T^2}{as} \tag{6.12}$$

The equation for SS_{AS}, sum of squares attributable to error, calculated as the $A \times S$ interaction is

$$SS_{A \times S} = \Sigma\,(Y)^2 - \frac{\Sigma A^2}{s} - \frac{\Sigma S^2}{a} + \frac{T^2}{as} \tag{6.13}$$

And the equation for SS_T is

$$SS_T = \Sigma Y^2 - \frac{T^2}{as} \tag{6.14}$$

Notice that the elements in these equations are *Y*, *A*, *S*, and *T*, each divided by the number of scores added together to produce the total that is squared. Notice also that the symbols in the denominator are the symbols *not* in the numerator. Verify that these elements are a complete partition of SS_T.

Table 6.4 shows sums of squares for the example. Take a moment to compare these results with those from the randomized-groups example of Chapter 4. Notice first that the data in Table 4.2 are the same as the data in Table 6.3. The same data set was used partly out of sloth and partly to illustrate a very important point: Computation of the

TABLE 6.4 Calculation of sums of squares for traditional one-way repeated-measures ANOVA with three levels

$$SS_A = \frac{12^2 + 19^2 + 32^2}{5} - \frac{63^2}{3(5)} = 41.2$$

$$SS_S = \frac{10^2 + 13^2 + 12^2 + 17^2 + 11^2}{3} - \frac{63^2}{3(5)} = 9.7$$

$$SS_{AS} = 325 - \frac{12^2 + 19^2 + 32^2}{5} - \frac{10^2 + 13^2 + 12^2 + 17^2 + 11^2}{3} + \frac{63^2}{3(5)} = 9.5$$

$$SS_T = 325 - \frac{63^2}{3(5)} = 60.4$$

main effect of A and computation of SS_T are identical for the two designs. When computing main effects (and interactions), the way cases are allocated is irrelevant. Then notice that the sum of SS_S (9.7) and SS_{AS} (9.5) is equal to $SS_{S/A}$ (19.2) from the randomized-groups design (Table 4.3). Variability due to cases (along with their df) is subtracted from the error term in a corresponding randomized-group design, making, it is hoped, the error terms in the repeated-measures design, MS_{AS}, smaller. SS are converted to MS by dividing by df. Table 6.5 shows degrees of freedom for the example. It is reassuring that the partitions of SS and the partitions of df both sum to their respective totals. If they do not sum to SS_T and df_T, there has been a computational error.

Table 6.6 shows calculation of mean squares for treatment and error sources of variance. Compare these values with those in Table 4.5. MS_A has not changed, because neither the SS nor the df have changed from the randomized-groups design. The error term for the repeated-measures design, MS_{AS}, on the other hand, is smaller than the error term for the randomized-groups design, $MS_{S/A}$ ($1.2 < 1.6$). This is because there are some differences due to cases when the data are analyzed as repeated measures. If there are no differences due to cases, the repeated-measures design is usually equal to or lower in power than the randomized-groups design (because of the loss of df for error), although it may still be more efficient to run the cases through several levels of A once they are set up to be measured (e.g., volunteers are already in the research cubicle).

TABLE 6.5 Calculation of degrees of freedom for traditional one-way repeated-measures ANOVA with three levels

$$df_A = a - 1 = 3 - 1 = 2$$

$$df_S = c - 1 = 5 - 1 = 4$$

$$df_{AS} = (a - 1)(s - 1) = (3 - 1)(5 - 1) = 8$$

$$df_T = as - 1 = 15 - 1 = 14$$

TABLE 6.6 Calculation of mean squares for traditional one-way repeated-measures ANOVA with three levels

$$MS_A = \frac{SS_A}{df_A} = \frac{41.2}{2} = 20.6$$

$$MS_S = \frac{SS_S}{df_S} = \frac{9.7}{4} = 2.4$$

$$MS_{AS} = \frac{SS_{AS}}{df_{AS}} = \frac{9.5}{8} = 1.2$$

TABLE 6.7 Source table for one way repeated-measures ANOVA

Source	SS	df	MS	F
A	41.2	2	20.6	17.17
S	9.7	4		
$A \times S$	9.5	8	1.2	
T	60.4	14		

Finally, Table 6.7 is a source table that summarizes the results of Tables 6.4–6.6 and that includes the F ratio. The F ratio is tested with 2 and 8 df. If our willingness to risk a Type I error is 5%, the critical value of F in Table A.1 is 4.46. Because the F ratio of 17.17 is larger than 4.46, the null hypothesis is rejected in favor of the hypothesis that there is a difference in mean numbers of science fiction books read over the three months of the study. Note that the effect of S (cases) is not tested as part of the analysis. If they were to be tested, they would be a random effect (cf. Chapter 11)—that is, cases in this study are assumed to be randomly selected (at least in an experiment) from some larger population of cases.

6.4.1.4 The Regression Approach to One-Way Repeated Measures

The regression approach uses columns to code IV levels, as in the randomized-groups analysis. In general, df_A such columns are required to compute SS_A. In this example, $df_A = 2$, so two columns are required to code levels of A. Because levels of A represent passage of time, it makes sense to use the codes for trend analysis for this example, which are -1 0 1 for linear trend and 1 -2 1 for quadratic trend. Table 6.8 shows the scores on the DV, Y, with trend coefficients in the two X columns. Column X_1 contains the linear trend coefficients, and column X_2 the quadratic. Sums for cases are in column X_3, repeated three times, once at each level of A. This shortcut using case totals only works in the one-way repeated-measures design. Remaining columns are found by multiplication. Table 6.9 shows the preliminary sums of squares calculated from Table 6.8. Table 6.10 shows the sums of cross products produced by the columns in Table 6.8. Finally, sums of squares for the regression analysis are shown in Table 6.11. Degrees of freedom for regression are in Table 6.12. If this information is rearranged into the format of Table 6.7, the results are identical.

6.4.1.5 Computer Analyses of Small-Sample One-Way Repeated-Measures Design

Data are entered in the format of Table 6.3 for the SPSS and SAS GLM programs. Each line of data represents a case and contains sequential DV values for each level of the repeated-measures IV. Each line may also include a case identification number, as shown in Table 6.13.

Tables 6.14 and 6.15 show computer analyses of the small-sample example of the one-way repeated-measures ANOVA through SPSS GLM, and SAS GLM, respectively.

TABLE 6.8 Regression approach to one-way repeated-measures ANOVA with three treatment levels

		A		*S*[a]			
	Y	X_1	X_2	X_3	YX_1	YX_2	YX_3
	1	−1	1	10	−1	1	10
	1	−1	1	13	−1	1	13
a_1	3	−1	1	12	−3	3	36
	5	−1	1	17	−5	5	85
	2	−1	1	11	−2	2	22
	3	0	−2	10	0	−6	30
	4	0	−2	13	0	−8	52
a_2	3	0	−2	12	0	−6	36
	5	0	−2	17	0	−10	85
	4	0	−2	11	0	−8	44
	6	1	1	10	6	6	60
	8	1	1	13	8	8	104
a_3	6	1	1	12	6	6	72
	7	1	1	17	7	7	119
	5	1	1	11	5	5	55
Sum	63	0	0	189	20	6	823
Sum sq	325	10	30	2469			
N	15						

[a]Alternatively, cases can be divided into $s - 1 = 4$ orthogonal columns, as in the two-way design of Table 6.26.

TABLE 6.9 Preliminary sums of squares for regression approach to repeated-measures ANOVA

$$SS(Y) = \sum Y^2 - \frac{(\Sigma Y)^2}{N} = 325 - \frac{63^2}{15} = 60.4$$

$$SS(X_1) = \sum X_1^2 - \frac{(\Sigma X_1)^2}{N} = 10 - \frac{0^2}{15} = 10.0$$

$$SS(X_2) = \sum X_2^2 - \frac{(\Sigma X_2)^2}{N} = 30 - \frac{0^2}{15} = 30.0$$

$$SS(X_3) = \sum X_3^2 - \frac{(\Sigma X_3)^2}{N} = 2469 - \frac{189^2}{15} = 87.6$$

TABLE 6.10 Sums of cross products for regression approach to repeated-measures ANOVA

$$\text{SP}(YX_1) = \sum YX_1 - \frac{(\Sigma Y)(\Sigma X_1)}{N} = 20 - \frac{(63)(0)}{15} = 20.0$$

$$\text{SP}(YX_2) = \sum YX_2 - \frac{(\Sigma Y)(\Sigma X_2)}{N} = 6 - \frac{(63)(0)}{15} = 6.0$$

$$\text{SP}(YX_3) = \sum YX_3 - \frac{(\Sigma Y)(\Sigma X_3)}{N} = 823 - \frac{(63)(189)}{15} = 29.2$$

TABLE 6.11 Sums of squares for repeated-measures ANOVA through regression

$$\text{SS}_A = \text{SS(reg. } X_1) + \text{SS(reg. } X_2) = \frac{[\text{SP}(YX_1)]^2}{\text{SS}(X_1)} + \frac{[\text{SP}(YX_2)]^2}{\text{SS}(X_2)} = \frac{(20)^2}{10} + \frac{(6)^2}{30} = 41.2$$

$$\text{SS}_S = \text{SS(reg. } X_3) = \frac{[\text{SP}(YX_3)]^2}{\text{SS}(X_3)} = \frac{(29.2)^2}{87.6} = 9.7$$

$$\text{SS}_T = \text{SS}_Y = 60.4$$

$$\text{SS(resid.)} = \text{SS}_T - \text{SS}_A - \text{SS}_S = 60.4 - 41.2 - 9.7 = 9.5$$

TABLE 6.12 Calculation of degrees of freedom for ANOVA through regression with three treatment levels

df_A = number of predictors (X vectors): $(a - 1) = 3 - 1 = 2$

df_S = number of cases -1: $(s - 1) = 5 - 1 = 4$

df_T = number of scores -1 $(as - 1) = 15 - 1 = 14$

$\text{df(resid.)} = \text{df}_T - \text{df}_A - \text{df}_S = 14 - 2 - 4 = 8$

TABLE 6.13 Hypothetical data for one-way ANOVA in a form suitable for SPSS and SAS analysis

CASE	MONTH1	MONTH2	MONTH3
1	1	3	6
2	1	4	8
3	3	3	6
4	5	5	7
5	2	4	5

Table 6.14 shows syntax and output for SPSS GLM > Repeated Measures. The repeated measure is defined as month with three levels. Once the three levels are defined as MONTH_1, MONTH_2, and MONTH_3, the /MEASURE instruction identify books as the DV. Options selected for printing are means (EMMEANS), descriptive statistics, estimates of effect size (ETASQ), and obtained power (OPOWER). The rest of the syntax in Table 6.14 is produced by default by the SPSS Windows menu program.

TABLE 6.14 One-way repeated-measures ANOVA for small-sample example through SPSS GLM (syntax and output)

```
GLM
  MONTH_1 MONTH_2 MONTH_3
  /WSFACTOR = month 3 Polynomial
  /MEASURE = books
  /METHOD = SSTYPE(3)
  /EMMEANS = TABLES(month)
  /PRINT = DESCRIPTIVE ETASQ OPOWER
  /CRITERIA = ALPHA(.05)
  /WSDESIGN = month .
```

General Linear Model

Within-Subjects Factors

Measure: books

month	Dependent Variable
1	MONTH_1
2	MONTH_2
3	MONTH_3

Descriptive Statistics

	Mean	Std. Deviation	N
MONTH_1	2.4000	1.67332	5
MONTH_2	3.8000	.83666	5
MONTH_3	6.4000	1.14018	5

Multivariate Tests[c]

Effect		Value	F	Hypothesis df	Error df	Sig.	Partial Eta Squared	Noncent. Parameter	Observed Power[a]
month	Pillai's Trace	.870	10.035[b]	2.000	3.000	.047	.870	20.070	.631
	Wilks' Lambda	.130	10.035[b]	2.000	3.000	.047	.870	20.070	.631
	Hotelling's Trace	6.690	10.035[b]	2.000	3.000	.047	.870	20.070	.631
	Roy's Largest Root	6.690	10.035[b]	2.000	3.000	.047	.870	20.070	.631

a. Computed using alpha = .05
b. Exact statistic
c. Design: Intercept
Within Subjects Design: month

Mauchly's Test of Sphericity[b]

Measure: books

					Epsilon[a]		
Within Subjects Effect	Mauchly's W	Approx. Chi-Square	df	Sig.	Greenhouse-Geisser	Huynh-Feldt	Lower-bound
month	.509	2.027	2	.363	.671	.885	.500

Tests the null hypothesis that the error covariance matrix of the orthonormalized transformed dependent variables is proportional to an identity matrix.

a. May be used to adjust the degrees of freedom for the averaged tests of significance. Corrected tests are displayed in the Tests of Within-Subjects Effects table.

b. Design: Intercept
Within Subjects Design: month

TABLE 6.14 Continued

Tests of Within-Subjects Effects

Measure: books

Source		Type III Sum of Squares	df	Mean Square	F	Sig.	Partial Eta Squared	Noncent. Parameter	Observed Power[a]
month	Sphericity Assumed	41.200	2	20.600	17.408	.001	.813	34.817	.993
	Greenhouse-Geisser	41.200	1.341	30.718	17.408	.006	.813	23.349	.949
	Huynh-Feldt	41.200	1.770	23.276	17.408	.002	.813	30.814	.986
	Lower-bound	41.200	1.000	41.200	17.408	.014	.813	17.408	.870
Error(month)	Sphericity Assumed	9.467	8	1.183					
	Greenhouse-Geisser	9.467	5.365	1.765					
	Huynh-Feldt	9.467	7.080	1.337					
	Lower-bound	9.467	4.000	2.367					

a. Computed using alpha = .05

Tests of Within-Subjects Contrasts

Measure: books

Source	month	Type III Sum of Squares	df	Mean Square	F	Sig.	Partial Eta Squared	Noncent. Parameter	Observed Power[a]
month	Linear	40.000	1	40.000	20.000	.011	.833	20.000	.909
	Quadratic	1.200	1	1.200	3.273	.145	.450	3.273	.286
Error(month)	Linear	8.000	4	2.000					
	Quadratic	1.467	4	.367					

a. Computed using alpha = .05

Tests of Between-Subjects Effects

Measure: books

Transformed Variable: Average

Source	Type III Sum of Squares	df	Mean Square	F	Sig.	Partial Eta Squared	Noncent. Parameter	Observed Power[a]
Intercept	264.600	1	264.600	108.740	.000	.965	108.740	1.000
Error	9.733	4	2.433					

a. Computed using alpha = .05

TABLE 6.14 Continued

Estimated Marginal Means

month

Measure: books

month	Mean	Std. Error	95% Confidence Interval	
			Lower Bound	Upper Bound
1	2.400	.748	.322	4.478
2	3.800	.374	2.761	4.839
3	6.400	.510	4.984	7.816

Confirmation of the DV used at each level of month is shown first, followed by means, standard deviations, and number of scores for each level. Multivariate tests of the mean difference for month is next; this is one of the options when sphericity is violated (Section 6.6.2) and is default output. By using Wilks' Lambda (a multivariate test criterion described by Tabachnick and Fidell, 2007), the difference in means is statistically significant at $\alpha = .047$. Observed Power for the multivariate test follows (.631). Next is Mauchly's Test of Sphericity. Because the test is not significant (Sig. = .363), sphericity (see Section 6.6.2) is acceptable. The next table contains results of the Huynh-Feldt adjustment, as described in Section 6.6.2, and the test of month: Sphericity Assumed. This latter test is the univariate ANOVA, with Type III Sum of Squares = 41.200, df = 2, Mean Square = 20.600, F = 17.408, and Sig. = .001. If Mauchly's Test of Sphericity is significant, a second option is to use the results of the Huynh-Feldt test (sig. = .002) instead. The table also includes effect size (month: Sphericity Assumed, Partial Eta Squared = .813) and Observed Power (.993), which is greater than the power of the Multivariate Test (.631). That is followed by a table of tests of linear and quadratic trend, the default output (useful as the third option when sphericity fails), in a table labeled Tests of Within-Subjects Contrasts. In this example, the Linear trend is statistically significant (.011), but the Quadratic trend is not (.145). The next table, Tests of Between-Subjects Effects, contains Sums of Squares, df, and Mean Square for cases, labeled Error. Finally, means and the 95% confidence bounds are given for parameter estimation. This voluminous output, most of it produced by default, contains the result of routine repeated-measures ANOVA, as well as tests of the three options available when sphericity is violated. In this example, all four options indicate a difference in means over the three months.

Table 6.15 shows SAS GLM syntax and output for the one-way repeated-measures example.[4] The `model` statement reflects the absence of a randomized-group IV to the right of the equal sign. The repeated-measures IV (`MONTH`) is defined in the `repeated` statement. Means are also requested.

This default output for SAS GLM begins with separate ANOVAs on each DV (not shown) and follows with the multivariate test, which matches that of SPSS

[4] SAS PROC ANOVA uses the same syntax for this analysis and produces the same output.

TABLE 6.15 One-way repeated-measures ANOVA through SAS GLM (syntax and selected output)

```
proc glm data=SASUSER.SS_RPTD;
     model  MONTH1 MONTH2 MONTH3 =;
     repeated MONTH / mean;
run;
```

```
                              The GLM Procedure
                      Repeated Measures Analysis of Variance

                       Repeated Measures Level Information

                   Dependent Variable       MONTH1    MONTH2    MONTH3

                      Level of MONTH             1         2         3

  MANOVA Test Criteria and Exact F Statistics for the Hypothesis of no MONTH Effect
                        H = Type III SSCP Matrix for MONTH
                             E = Error SSCP Matrix

                           S=1      M=0      N=0.5

    Statistic                      Value     F Value    Num DF     Den DF     Pr > F

    Wilks' Lambda             0.13003802       10.04         2          3     0.0469
    Pillai's Trace            0.86996198       10.04         2          3     0.0469
    Hotelling-Lawley Trace    6.69005848       10.04         2          3     0.0469
    Roy's Greatest Root       6.69005848       10.04         2          3     0.0469

              Univariate Tests of Hypotheses for Within Subject Effects

                                                                    Adj Pr > F
Source        DF   Type III SS   Mean Square  F Value   Pr > F      G-G        H-F
MONTH          2   41.20000000   20.60000000    17.41   0.0012   0.0060     0.0021
Error(MONTH)   8    9.46666667    1.18333333

                   Greenhouse-Geisser Epsilon     0.6706
                   Huynh-Feldt Epsilon            0.8850

                     Means of Within Subjects Effects

                    Level of
                    MONTH          N            Mean           Std Dev

                    1              5      2.40000000        1.67332005
                    2              5      3.80000000        0.83666003
                    3              5      6.40000000        1.14017543
```

GLM. The `Univariate Test of Hypotheses for Within Subject Effects` shows the univariate test of `MONTH` and its error term, together with probability following the `G-G` (Greenhouse-Geisser) and `H-F` (Huynh-Feldt) adjustments to *F*. Sum of squares for cases is not given, nor is total sum of squares. The program also produces Mauchly's test of sphericity and polynomial trend analysis, if requested. Means and standard deviations produced by the `mean` instruction follow.

6.4.2 Factorial Repeated-Measures Designs

Factorial repeated-measures designs have two or more repeated-measures IVs in factorial combination. Cases provide a DV at each combination of levels of the IVs. Beyond exhaustion of cases (and researchers), there is no limit to the number of IVs and their levels—and, therefore, combinations of levels—that can be used.

The complexity that arises with this analysis is the development of a separate error term for each effect in the design. Each error term is an interaction of cases with the effect being tested. When testing the main effect of B, for instance, the error term is the $B \times S$ interaction. When testing the $A \times B$ interaction, the error term is the $A \times B \times S$ interaction, and so on.

6.4.2.1 Allocation of Cases

Allocation of cases for the simplest 2×2 design is found in Table 6.16.

6.4.2.2 Partition of Sources of Variance

The partition of SS for a two-way repeated-measures factorial design is

$$SS_T = SS_A + SS_B + SS_{AB} + SS_S + SS_{AS} + SS_{BS} + SS_{ABS} \qquad \textbf{(6.15)}$$

In this analysis, SS_A, SS_B, and SS_{AB} represent the main effect of A, the main effect of B, and their interaction, respectively. These effects are both logically and computationally equivalent to the same effects in a two-way factorial randomized-groups design. In the last four terms of the equation, SS_S represents differences among cases, and the remaining interactions are the various error terms.

As usual, each SS has a corresponding df. The df for each effect is formed in the usual way. For a main effect, 1 is subtracted from the number of levels—for example, $df_S = s - 1$. For an interaction, it is the df for one effect times the df for the other(s)—for example, $df_{AS} = (a - 1)(s - 1)$ and $df_{ABS} = (a - 1)(b - 1)(s - 1)$.

A graphic representation of the partition appears in Figure 6.2. As usual, each MS is formed by dividing an SS by its corresponding df (e.g., $MS_{ABS} = SS_{ABS}/df_{ABS}$).

TABLE 6.16 Allocation of cases in a two-way factorial repeated-measures design

	b_1		b_2	
	a_1	a_2	a_1	a_2
s_1	s_1	s_1	s_1	s_1
s_2	s_2	s_2	s_2	s_2
s_3	s_3	s_3	s_3	s_3

FIGURE 6.2 Partition of (a) sums of squares and (b) degrees of freedom in a two-way factorial repeated-measures design

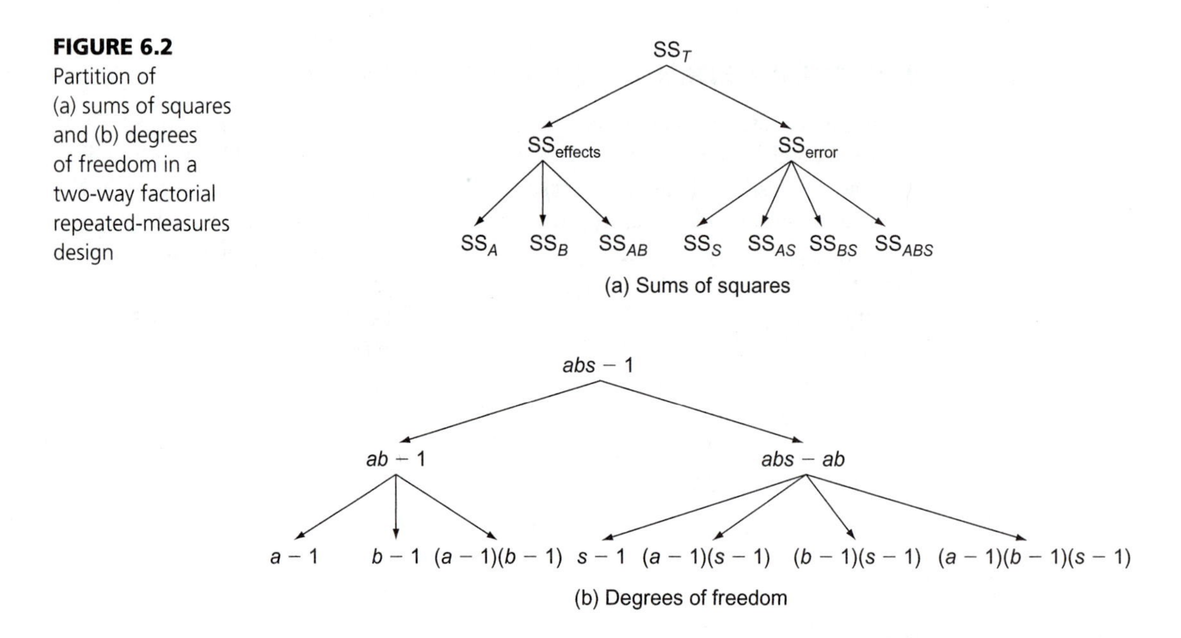

The error term for each effect is the interaction of that effect with cases. The sources A, B, and $A \times B$, then, have $A \times S$, $B \times S$, and $A \times B \times S$ as error terms, respectively.[5]

$$F_A = \frac{MS_A}{MS_{AS}} \tag{6.16}$$

$$F_B = \frac{MS_B}{MS_{BS}} \tag{6.17}$$

and

$$F_{AB} = \frac{MS_{AB}}{MS_{ABS}} \tag{6.18}$$

The use of separate error terms makes sense, because each error term is, computationally and logically, an interaction of cases with an effect. These interactions represent the *special* way cases reacted to the effect, above and beyond the direct effects of treatment. In other words, each error term is an average of the idiosyncratic reactions

[5] There is another difference between this analysis and the one-way repeated-measures analysis—the absence of a clear choice for the error term for S should a test of S be desired.

of the various cases to the effect in question. Why should idiosyncratic reactions to levels of B or $A \times B$ combinations be involved in a test of the significance of A? Or why should idiosyncratic reactions to A be involved in a test of B, and so on?

6.4.2.3 Traditional ANOVA for Two-Way Repeated Measures

A small data set for a two-way repeated-measures design is in Table 6.17. The same five cases (and their data) as in the one-way design are used; however, in this example, they are assigned to read science fiction novels *and* mystery novels for three months for each type.[6] Month is A, with three levels, and type of novel is B, with two levels. There are, then, six combinations of month and type of novel. Number of novels read each month again serves as the DV.

The equations for the main effects of interest are the same as those in randomized-groups analysis (Chapter 5) and appear later in Table 6.19. The computational equations for interactions are the same, too, but to use them, you have to develop the necessary values. You must sum together the values over cases for the $A \times B$ interaction. You sum together the values over levels of A for the $B \times S$ interaction, and you sum together the values over levels of B for the $A \times S$ interaction. This is shown in Table 6.18 for the data set of Table 6.17.

Computational equations for sums of squares for this ANOVA appear in Table 6.19. And the sums of cross-products are shown in Table 6.20, with a bit of adjustment for rounding error. Table 6.21 has the equations and computations for degrees of freedom for this example. Finally, Table 6.22 summarizes the results of the

TABLE 6.17 Hypothetical data set for a two-way factorial repeated-measures design

	b_1: Science Fiction			b_2: Mystery			
	Month 1 a_1	Month 2 a_2	Month 3 a_3	Month 1 a_1	Month 2 a_2	Month 3 a_3	Case Totals
s_1	1	3	6	5	4	1	$S_1 = 20$
s_2	1	4	8	8	8	4	$S_2 = 33$
s_3	3	3	6	4	5	3	$S_3 = 24$
s_4	5	5	7	3	2	0	$S_4 = 22$
s_5	2	4	5	5	6	3	$S_5 = 25$
Treatment Totals	$A_1B_1 = 12$	$A_2B_1 = 19$	$A_3B_1 = 32$	$A_1B_2 = 25$	$A_2B_2 = 25$	$A_3B_2 = 11$	$T = 124$

[6] In this example, where type of novel is an IV whose levels can be reordered (either science fiction first or mystery first), the researcher would be well-advised to use counterbalancing and Latin-square analysis, as described in Chapter 9.

calculations and shows the mean squares and F values for each effect. Take a moment to closely examine the F ratios. Notice the pattern of effects and error terms in the table.

Each effect in the table is tested against critical F with its own df. Critical F for the test of A (months) at $\alpha \leq .05$ with 2 and 8 df is 4.46; because 2.77 is less than 4.46, the null hypothesis of no mean difference in number of books read per month is

TABLE 6.18 Computations necessary for computing interaction terms in a two-way factorial repeated-measures design

(a) Sums required to compute the $A \times B$ interaction

	A: Month			
B: Type of Novel	a_1: Month 1	a_2: Month 2	a_3: Month 3	Novel Totals
b_1: Science Fiction	12	19	32	$B_1 = 63$
b_2: Mystery	25	25	11	$B_2 = 61$
Month Totals	$A_1 = 37$	$A_2 = 44$	$A_3 = 43$	$T = 124$

(b) Sums required to compute the $A \times S$ interaction

	A: Month			
	Month 1 a_1	Month 2 a_2	Month 3 a_3	Case Totals
s_1	$1 + 5 = 6$	$3 + 4 = 7$	$6 + 1 = 7$	$S_1 = 20$
s_2	$1 + 8 = 9$	$4 + 8 = 12$	$8 + 4 = 12$	$S_2 = 33$
s_3	$3 + 4 = 7$	$3 + 5 = 8$	$6 + 3 = 9$	$S_3 = 24$
s_4	$5 + 3 = 8$	$5 + 2 = 7$	$7 + 0 = 7$	$S_4 = 22$
s_5	$2 + 5 = 7$	$4 + 6 = 10$	$5 + 3 = 8$	$S_5 = 25$
Month Totals	$A_1 = 37$	$A_2 = 44$	$A_3 = 43$	$T = 124$

(c) Sums required to compute the $B \times S$ interaction

	B: Type of Novel		
	b_1: Science Fiction	b_2: Mystery	Case Totals
s_1	$1 + 3 + 6 = 10$	$5 + 4 + 1 = 10$	$S_1 = 20$
s_2	$1 + 4 + 8 = 13$	$8 + 8 + 4 = 20$	$S_2 = 33$
s_3	$3 + 3 + 6 = 12$	$4 + 5 + 3 = 12$	$S_3 = 24$
s_4	$5 + 5 + 7 = 17$	$3 + 2 + 0 = 5$	$S_4 = 22$
s_5	$2 + 4 + 5 = 11$	$5 + 6 + 3 = 14$	$S_5 = 25$
Novel Totals	$B_1 = 63$	$B_2 = 61$	$T = 124$

TABLE 6.19 Computational equations for sums of squares for two-way repeated-measures ANOVA

$$SS_A = \frac{\sum A^2}{bs} - \frac{T^2}{abs} = \frac{37^2 + 44^2 + 43^2}{2(5)} - \frac{124^2}{3(2)(5)}$$

$$SS_B = \frac{\sum B^2}{as} - \frac{T^2}{abs} = \frac{63^2 + 61^2}{3(5)} - \frac{124^2}{3(2)(5)}$$

$$SS_{AB} = \frac{\sum (AB)^2}{s} - \frac{\sum A^2}{bs} - \frac{\sum B^2}{as} + \frac{T^2}{abs} = \frac{12^2 + 25^2 + 19^2 + 25^2 + 32^2 + 11^2}{5} - \frac{37^2 + 44^2 + 43^2}{2(5)} - \frac{63^2 + 61^2}{3(5)} + \frac{124^2}{3(2)(5)}$$

$$SS_S = \frac{\sum S^2}{ab} - \frac{T^2}{abs} = \frac{20^2 + 33^2 + 24^2 + 22^2 + 25^2}{3(2)} - \frac{124^2}{3(2)(5)}$$

$$SS_{AS} = \frac{\sum (AS)^2}{b} - \frac{\sum A^2}{bs} - \frac{\sum S^2}{ab} + \frac{T^2}{abs} = \frac{6^2 + 9^2 + 7^2 + 8^2 + \cdots + 8^2}{2} - \frac{37^2 + 44^2 + 43^2}{2(5)} - \frac{20^2 + 33^2 + \cdots + 25^2}{3(2)} + \frac{124^2}{3(2)(5)}$$

$$SS_{BS} = \frac{\sum (BS)^2}{a} - \frac{\sum B^2}{as} - \frac{\sum S^2}{ab} + \frac{T^2}{abs} = \frac{10^2 + 13^2 + 12^2 + 17^2 + \cdots + 14^2}{3} - \frac{63^2 + 61^2}{3(5)} - \frac{20^2 + 33^2 + \cdots + 25^2}{3(2)} + \frac{124^2}{3(2)(5)}$$

$$SS_{ABS} = \sum Y^2 - \frac{\sum (AB)^2}{s} - \frac{\sum (AS)^2}{b} - \frac{\sum (BS)^2}{a} + \frac{\sum A^2}{bs} + \frac{\sum B^2}{as} + \frac{\sum S^2}{ab} - \frac{T^2}{abs}$$

$$= 1^2 + 1^2 + 3^2 + \cdots + 0^2 + 3^2 - \frac{12^2 + \cdots + 11^2}{5} - \frac{6^2 + \cdots + 8^2}{2} - \frac{10^2 + \cdots + 14^2}{3} + \frac{37^2 + 44^2 + 43^2}{2(5)} + \frac{63^2 + 61^2}{3(5)} + \frac{20^2 + \cdots + 25^2}{3(2)} - \frac{124^2}{3(2)(5)}$$

$$SS_T = \sum Y^2 - \frac{T^2}{abs} = 1^2 + 1^2 + 3^2 + 5^2 + \cdots + 0^2 + 3^2 - \frac{124^2}{3(2)(5)}$$

retained. Critical F for the test of B (type of novel) at $\alpha \leq .05$ with 1 and 4 df is 7.71; because 0.02 is less than 7.71, the null hypothesis of no mean difference in number of science fiction and mystery novels read is retained. Finally, critical F for the test of interaction (months by type of novel) at $\alpha \leq .05$ with 2 and 8 df is 4.46; because 26.20 is greater than 4.46, the null hypothesis of no mean difference in number of books read per month by type of book is rejected. A glance at the numbers in Table 6.17 suggests that for science fiction books, the number of books read increases over the three months, whereas for mystery novels, the number decreases over the three months. This informal observation needs to be verified by the procedures of Section 6.6.6.

TABLE 6.20 Sums of squares for traditional factorial repeated-measures ANOVA

$$SS_A = 515.40 - 512.53 = 2.87$$

$$SS_B = 512.67 - 512.53 = 0.14$$

$$SS_{AB} = 580.00 - 515.40 - 512.67 + 512.53 = 64.46$$

$$SS_S = 529.00 - 512.53 = 16.47$$

$$SS_{AS} = 536.00 - 515.40 - 529.00 + 512.53 = 4.13$$

$$SS_{BS} = 562.67 - 512.67 - 529.00 + 512.53 = 33.53$$

$$SS_{ABS} = 644.00 - 580.00 - 536.00 - 562.67 + 515.40 + 512.67 + 529.00 - 512.53 = 9.87$$

$$SS_T = 644.00 - 512.53 = 131.47$$

TABLE 6.21 Degrees of freedom for traditional factorial repeated-measures ANOVA

df_A	$= a - 1$	$3 - 1$	$= 2$
df_B	$= b - 1$	$2 - 1$	$= 1$
df_{AB}	$= (a - 1)(b - 1)$	$(3 - 1)(2 - 1)$	$= 2$
df_S	$= s - 1$	$5 - 1$	$= 4$
df_{AS}	$= (a - 1)(s - 1)$	$(3 - 1)(5 - 1)$	$= 8$
df_{BS}	$= (b - 1)(s - 1)$	$(2 - 1)(5 - 1)$	$= 4$
df_{ABS}	$= (a - 1)(b - 1)(s - 1)$	$(3 - 1)(2 - 1)(5 - 1)$	$= 8$
df_T	$= abs - 1$	$3(2)(5) - 1$	$= 29$

TABLE 6.22 Source table for two-way repeated-measures ANOVA

Source	SS	df	MS	F
A	2.87	2	1.44	$\frac{1.44}{0.52} = 2.77$
B	0.13	1	0.13	$\frac{0.14}{8.38} = 0.02$
AB	64.47	2	32.23	$\frac{32.23}{1.23} = 26.20$
S	16.47	4	4.12	
AS	4.13	8	0.52	
BS	33.53	4	8.38	
ABS	9.87	8	1.23	
T	131.47	29		

6.4.2.4 Regression Approach to Factorial Repeated Measures

In Table 6.23, the data set of Table 6.17 is rearranged into the format appropriate for regression. The levels of A appear as the first column and the levels of B the second. Case designations are in the third column. These three columns are for your convenience in understanding the rearrangement of the data set and the use of coding in columns that follow. The fourth column contains DV (Y) values. The total df for this problem is 29, so it takes 29 columns to code the design (excluding the 4 columns already mentioned) and another 29 columns for the cross multiplication of the DV scores with each design column. The X_1 column contains the linear trend coefficients for A and the X_2 column the quadratic trend coefficients. Column X_3 contains contrast coefficients for B, and columns X_4 and X_5 are $A_{\text{linear}} \times B$ and $A_{\text{quad}} \times B$, respectively. The next four columns, X_6–X_9, contain contrast coding for cases; the choice of coefficients for cases is arbitrary, provided that the sum of each column is 0 and the products of all pairs of columns sum to 0. In this case, the coefficients used are (4 −1 −1 −1 −1) for Sc1 (subject contrast 1), (0 1 1 −1 −1) for Sc2, (0 1 −1 0 0) for Sc3, and (0 0 0 1 −1) for Sc4.

The remaining columns are found by multiplication. Columns X_{10} to X_{29} contain all possible cross products of columns X_1 to X_9. The X_{10} column contains codes for $A_{\text{linear}} \times$ Sc1 (cross of the linear trend coefficients for A by the code for the first df for cases). Column X_{29} contains codes for $A_{\text{quad}} \times B \times$ Sc4 (cross of the quadratic coefficients for A by B by the code for the last contrast for cases). Columns X_{30} to X_{58} are Y crossed with each of the preceding columns; this is necessary for computation of sums of products.

To develop the several error terms for the factorial design, it is necessary to separate the cases from each other, which differs from the one-way repeated-measures design, in which cases can be combined into the $S(X_3)$ column of Table 6.8. This is accomplished in columns X_6 to X_9 of Table 6.23, separately for each of the four df for cases and labeled Sc1, Sc2, Sc3, and Sc4. If a different set of orthogonal coefficients is used, SS and SP for each column are different, but the total for the effect is the same, just as in any other orthogonal partition of a main effect. The df for cases in the one-way design can also be separated into different columns for the omnibus design[7] and must be separated into different columns for comparisons (cf. Section 6.6.6).

Preliminary sums of squares for the regression approach are in Table 6.24. And the sums of cross products appear in Table 6.25. Finally, sums of squares for the regression analysis are shown in Table 6.26.

Degrees of freedom for each effect in the regression approach is the number of columns (number of SS(reg. X_i)s) for the effect in the design. When SS and df are combined into a source table, the results are identical (except for minor rounding error) to the values shown for the traditional ANOVA approach, as shown in Table 6.27.

[7] If you're masochistic.

TABLE 6.23 Regression format for factorial repeated-measures design

				Al	Aq	B	$Al{*}B$	$Aq{*}B$	Sc1	Sc2	Sc3	Sc4	$Al{*}$Sc1	$Al{*}$Sc2	$Al{*}$Sc3	$Al{*}$Sc4
			Y	X_1	X_2	X_3	X_4	X_5	X_6	X_7	X_8	X_9	X_{10}	X_{11}	X_{12}	X_{13}
a_1	b_1	s_1	1	−1	1	1	−1	1	4	0	0	0	−4	0	0	0
		s_2	1	−1	1	1	−1	1	−1	1	1	0	1	−1	−1	0
		s_3	3	−1	1	1	−1	1	−1	1	−1	0	1	−1	1	0
		s_4	5	−1	1	1	−1	1	−1	−1	0	1	1	1	0	−1
		s_5	2	−1	1	1	−1	1	−1	−1	0	−1	1	1	0	1
	b_2	s_1	5	−1	1	−1	1	−1	4	0	0	0	−4	0	0	0
		s_2	8	−1	1	−1	1	−1	−1	1	1	0	1	−1	−1	0
		s_3	4	−1	1	−1	1	−1	−1	1	−1	0	1	−1	1	0
		s_4	3	−1	1	−1	1	−1	−1	−1	0	1	1	1	0	−1
		s_5	5	−1	1	−1	1	−1	−1	−1	0	−1	1	1	0	1
a_2	b_1	s_1	3	0	−2	1	0	−2	4	0	0	0	0	0	0	0
		s_2	4	0	−2	1	0	−2	−1	1	1	0	0	0	0	0
		s_3	3	0	−2	1	0	−2	−1	1	−1	0	0	0	0	0
		s_4	5	0	−2	1	0	−2	−1	−1	0	1	0	0	0	0
		s_5	4	0	−2	1	0	−2	−1	−1	0	−1	0	0	0	0
	b_2	s_1	4	0	−2	−1	0	2	4	0	0	0	0	0	0	0
		s_2	8	0	−2	−1	0	2	−1	1	1	0	0	0	0	0
		s_3	5	0	−2	−1	0	2	−1	1	−1	0	0	0	0	0
		s_4	2	0	−2	−1	0	2	−1	−1	0	1	0	0	0	0
		s_5	6	0	−2	−1	0	2	−1	−1	0	−1	0	0	0	0
a_3	b_1	s_1	6	1	1	1	1	1	4	0	0	0	4	0	0	0
		s_2	8	1	1	1	1	1	−1	1	1	0	−1	1	1	0
		s_3	6	1	1	1	1	1	−1	1	−1	0	−1	1	−1	0
		s_4	7	1	1	1	1	1	−1	−1	0	1	−1	−1	0	1
		s_5	5	1	1	1	1	1	−1	−1	0	−1	−1	−1	0	−1
	b_2	s_1	1	1	1	−1	−1	−1	4	0	0	0	4	0	0	0
		s_2	4	1	1	−1	−1	−1	−1	1	1	0	−1	1	1	0
		s_3	3	1	1	−1	−1	−1	−1	1	−1	0	−1	1	−1	0
		s_4	0	1	1	−1	−1	−1	−1	−1	0	1	−1	−1	0	1
		s_5	3	1	1	−1	−1	−1	−1	−1	0	−1	−1	−1	0	−1
		sum	124	0	0	0	0	0	0	0	0	0	0	0	0	0
		sqd	644	20	60	30	20	60	120	24	12	12	80	16	8	8
		ss	131.47	20	60	30	20	60	120	24	12	12	80	16	8	8
		sp		6	−8	2	34	20	−24	10	9	−3	−1	5	1	−2
		reg		1.8	1.07	0.13	57.8	6.67	4.8	4.17	6.75	0.75	0.01	1.56	0.12	0.5

			Aq*Sc1 X_{14}	Aq*Sc2 X_{15}	Aq*Sc3 X_{16}	Aq*Sc4 X_{17}	B*Sc1 X_{18}	B*Sc2 X_{19}	B*Sc3 X_{20}	B*Sc4 X_{21}	Al*B*Sc1 X_{22}	Al*B*Sc2 X_{23}	Al*B*Sc3 X_{24}	Al*B*Sc4 X_{25}	Aq*B*Sc1 X_{26}
a_1	b_1	s_1	4	0	0	0	4	0	0	0	−4	0	0	0	4
		s_2	−1	1	1	0	−1	1	1	0	1	−1	−1	0	−1
		s_3	−1	1	−1	0	−1	1	−1	0	1	−1	1	0	−1
		s_4	−1	−1	0	1	−1	−1	0	1	1	1	0	−1	−1
		s_5	−1	−1	0	−1	−1	−1	0	−1	1	1	0	1	−1
	b_2	s_1	4	0	0	0	−4	0	0	0	4	0	0	0	−4
		s_2	−1	1	1	0	1	−1	−1	0	−1	1	1	0	1
		s_3	−1	1	−1	0	1	−1	1	0	−1	1	−1	0	1
		s_4	−1	−1	0	1	1	1	0	−1	−1	−1	0	1	1
		s_5	−1	−1	0	−1	1	1	0	1	−1	−1	0	−1	1
a_2	b_1	s_1	−8	0	0	0	4	0	0	0	0	0	0	0	−8
		s_2	2	−2	−2	0	−1	1	1	0	0	0	0	0	2
		s_3	2	−2	2	0	−1	1	−1	0	0	0	0	0	2
		s_4	2	2	0	−2	−1	−1	0	1	0	0	0	0	2
		s_5	2	2	0	2	−1	−1	0	−1	0	0	0	0	2
	b_2	s_1	−8	0	0	0	−4	0	0	0	0	0	0	0	8
		s_2	2	−2	−2	0	1	−1	−1	0	0	0	0	0	−2
		s_3	2	−2	2	0	1	−1	1	0	0	0	0	0	−2
		s_4	2	2	0	−2	1	1	0	−1	0	0	0	0	−2
		s_5	2	2	0	2	1	1	0	1	0	0	0	0	−2
a_3	b_1	s_1	4	0	0	0	4	0	0	0	4	0	0	0	4
		s_2	−1	1	1	0	−1	1	1	0	−1	1	1	0	−1
		s_3	−1	1	−1	0	−1	1	−1	0	−1	1	−1	0	−1
		s_4	−1	−1	0	1	−1	−1	0	1	−1	−1	0	1	−1
		s_5	−1	−1	0	−1	−1	−1	0	−1	−1	−1	0	−1	−1
	b_2	s_1	4	0	0	0	−4	0	0	0	−4	0	0	0	−4
		s_2	−1	1	1	0	1	−1	−1	0	1	−1	−1	0	1
		s_3	−1	1	−1	0	1	−1	1	0	1	−1	1	0	1
		s_4	−1	−1	0	1	1	1	0	−1	1	1	0	−1	1
		s_5	−1	−1	0	−1	1	1	0	1	1	1	0	1	1
		sum	0	0	0	0	0	0	0	0	0	0	0	0	0
		sqd	240	48	24	24	120	24	12	12	80	16	8	8	240
		ss	240	48	24	24	120	24	12	12	80	16	8	8	240
		sp	3	1	−3	6	−2	−16	−7	15	11	5	7	0	−5
		reg	0.04	0.02	0.38	1.5	0.03	10.67	4.08	18.75	1.51	1.56	6.12	0	0.10

(Continued)

TABLE 6.23 Continued

			Aq*B*Sc2 X_{27}	Aq*B*Sc3 X_{28}	Aq*B*Sc4 X_{29}	Y*Al X_{30}	Y*Aq X_{31}	Y*B X_{32}	Y*Al*B X_{33}	Y*Aq*B X_{34}	Y*Sc1 X_{35}	Y*Sc2 X_{36}	Y*Sc3 X_{37}	Y*Sc4 X_{38}	Y*Al*Sc1 X_{39}
a_1	b_1	s_1	0	0	0	−1	1	1	−1	1	4	0	0	0	−4
		s_2	1	1	0	−1	1	1	−1	1	−1	1	1	0	1
		s_3	1	−1	0	−3	3	3	−3	3	−3	3	−3	0	3
		s_4	−1	0	1	−5	5	5	−5	5	−5	−5	0	5	5
		s_5	−1	0	−1	−2	2	2	−2	2	−2	−2	0	−2	2
	b_2	s_1	0	0	0	−5	5	−5	5	−5	20	0	0	0	−20
		s_2	−1	−1	0	−8	8	−8	8	−8	−8	8	8	0	8
		s_3	−1	1	0	−4	4	−4	4	−4	−4	4	−4	0	4
		s_4	1	0	−1	−3	3	−3	3	−3	−3	−3	0	3	3
		s_5	1	0	1	−5	5	−5	5	−5	−5	−5	0	−5	5
a_2	b_1	s_1	0	0	0	0	−6	3	0	−6	12	0	0	0	0
		s_2	−2	−2	0	0	−8	4	0	−8	−4	4	4	0	0
		s_3	−2	2	0	0	−6	3	0	−6	−3	3	−3	0	0
		s_4	2	0	−2	0	−10	5	0	−10	−5	−5	0	5	0
		s_5	2	0	2	0	−8	4	0	−8	−4	−4	0	−4	0
	b_2	s_1	0	0	0	0	−8	−4	0	8	16	0	0	0	0
		s_2	2	2	0	0	−16	−8	0	16	−8	8	8	0	0
		s_3	2	−2	0	0	−10	−5	0	10	−5	5	−5	0	0
		s_4	−2	0	2	0	−4	−2	0	4	−2	−2	0	2	0
		s_5	−2	0	−2	0	−12	−6	0	12	−6	−6	0	−6	0
a_3	b_1	s_1	0	0	0	6	6	6	6	6	24	0	0	0	24
		s_2	1	1	0	8	8	8	8	8	−8	8	8	0	−8
		s_3	1	−1	0	6	6	6	6	6	−6	6	−6	0	−6
		s_4	−1	0	1	7	7	7	7	7	−7	−7	0	7	−7
		s_5	−1	0	−1	5	5	5	5	5	−5	−5	0	−5	−5
	b_2	s_1	0	0	0	1	1	−1	−1	−1	4	0	0	0	4
		s_2	−1	−1	0	4	4	−4	−4	−4	−4	4	4	0	−4
		s_3	−1	1	0	3	3	−3	−3	−3	−3	3	−3	0	−3
		s_4	1	0	−1	0	0	0	0	0	0	0	0	0	0
		s_5	1	0	1	3	3	−3	−3	−3	−3	−3	0	−3	−3
		sum	0	0	0	6	−8	2	34	20	−24	10	9	−3	−1
		sqd	48	24	24										
		ss	48	24	24										
		sp	5	−1	0										
		reg	0.52	0.04	0										

			Y*Al*Sc2 X_{40}	Y*Al*Sc3 X_{41}	Y*Al*Sc4 X_{42}	Y*Aq*Sc1 X_{43}	Y*Aq*Sc2 X_{44}	Y*Aq*Sc3 X_{45}	Y*Aq*Sc4 X_{46}	Y*B*Sc1 X_{47}	Y*B*Sc2 X_{48}	Y*B*Sc3 X_{49}	Y*B*Sc4 X_{50}
a_1	b_1	s_1	0	0	0	4	0	0	0	4	0	0	0
		s_2	−1	−1	0	−1	1	1	0	−1	1	1	0
		s_3	−3	3	0	−3	3	−3	0	−3	3	−3	0
		s_4	5	0	−5	−5	−5	0	5	−5	−5	0	5
		s_5	2	0	2	−2	−2	0	−2	−2	−2	0	−2
	b_2	s_1	0	0	0	20	0	0	0	−20	0	0	0
		s_2	−8	−8	0	−8	8	8	0	8	−8	−8	0
		s_3	−4	4	0	−4	4	−4	0	4	−4	4	0
		s_4	3	0	−3	−3	−3	0	3	3	3	0	−3
		s_5	5	0	5	−5	−5	0	−5	5	5	0	5
a_2	b_1	s_1	0	0	0	−24	0	0	0	12	0	0	0
		s_2	0	0	0	8	−8	−8	0	−4	4	4	0
		s_3	0	0	0	6	−6	6	0	−3	3	−3	0
		s_4	0	0	0	10	10	0	−10	−5	−5	0	5
		s_5	0	0	0	8	8	0	8	−4	−4	0	−4
	b_2	s_1	0	0	0	−32	0	0	0	−16	0	0	0
		s_2	0	0	0	16	−16	−16	0	8	−8	−8	0
		s_3	0	0	0	10	−10	10	0	5	−5	5	0
		s_4	0	0	0	4	4	0	−4	2	2	0	−2
		s_5	0	0	0	12	12	0	12	6	6	0	6
a_3	b_1	s_1	0	0	0	24	0	0	0	24	0	0	0
		s_2	8	8	0	−8	8	8	0	−8	8	8	0
		s_3	6	−6	0	−6	6	−6	0	−6	6	−6	0
		s_4	−7	0	7	−7	−7	0	7	−7	−7	0	7
		s_5	−5	0	−5	−5	−5	0	−5	−5	−5	0	−5
	b_2	s_1	0	0	0	4	0	0	0	−4	0	0	0
		s_2	4	4	0	−4	4	4	0	4	−4	−4	0
		s_3	3	−3	0	−3	3	−3	0	3	−3	3	0
		s_4	0	0	0	0	0	0	0	0	0	0	0
		s_5	−3	0	−3	−3	−3	0	−3	3	0	3	3
		sum	5	1	−2	3	1	−3	6	−2	−16	−7	15

(Continued)

TABLE 6.23 Continued

			$Y^*A1^*B^*$Sc1 X_{51}	$Y^*A1^*B^*$Sc2 X_{52}	$Y^*A1^*B^*$Sc3 X_{53}	$Y^*A1^*B^*$Sc4 X_{54}	$Y^*Aq^*B^*$Sc1 X_{55}	$Y^*Aq^*B^*$Sc2 X_{56}	$Y^*Aq^*B^*$Sc3 X_{57}	$Y^*Aq^*B^*$Sc4 X_{58}
a_1	b_1	s_1	−4	0	0	0	4	0	0	0
		s_2	1	−1	−1	0	−1	1	1	0
		s_3	3	−3	3	0	−3	3	−3	0
		s_4	5	5	0	−5	−5	−5	0	5
		s_5	2	2	0	2	−2	−2	0	−2
	b_2	s_1	20	0	0	0	−20	0	0	0
		s_2	−8	8	8	0	8	−8	−8	0
		s_3	−4	4	−4	0	4	−4	4	0
		s_4	−3	−3	0	3	3	3	0	−3
		s_5	−5	−5	0	−5	5	5	0	5
a_2	b_1	s_1	0	0	0	0	−24	0	0	0
		s_2	0	0	0	0	8	−8	−8	0
		s_3	0	0	0	0	6	−6	6	0
		s_4	0	0	0	0	10	10	0	−10
		s_5	0	0	0	0	8	8	0	8
	b_2	s_1	0	0	0	0	32	0	0	0
		s_2	0	0	0	0	−16	16	16	0
		s_3	0	0	0	0	−10	10	−10	0
		s_4	0	0	0	0	−4	−4	0	4
		s_5	0	0	0	0	−12	−12	0	−12
a_3	b_1	s_1	24	0	0	0	24	0	0	0
		s_2	−8	8	8	0	−8	8	8	0
		s_3	−6	6	−6	0	−6	6	−6	0
		s_4	−7	−7	0	7	−7	−7	0	7
		s_5	−5	−5	0	−5	−5	−5	0	−5
	b_2	s_1	−4	0	0	0	−4	0	0	0
		s_2	4	−4	−4	0	4	−4	−4	0
		s_3	3	−3	3	0	3	−3	3	0
		s_4	0	0	0	0	0	0	0	0
		s_5	3	0	3	3	3	0	3	3
		sum	11	5	7	0	−5	5	−1	0

TABLE 6.24 Preliminary sums of squares for regression approach to factorial repeated-measures ANOVA

$$SS_Y = \sum Y^2 - \frac{(\sum Y)^2}{N} = 644 - \frac{124^2}{30} = 131.47$$

$$SS(X_1) = \sum X_1^2 - \frac{(\sum X_1)^2}{N} = 20 - \frac{0^2}{30} = 20$$

$$SS(X_2) = \sum X_2^2 - \frac{(\sum X_2)^2}{N} = 60 - \frac{0^2}{30} = 60$$

$$SS(X_3) = \sum X_3^2 - \frac{(\sum X_3)^2}{N} = 30 - \frac{0^2}{30} = 30$$

$$SS(X_4) = \sum X_4^2 - \frac{(\sum X_4)^2}{N} = 20 - \frac{0^2}{30} = 20$$

$$SS(X_5) = \sum X_5^2 - \frac{(\sum X_5)^2}{N} = 60 - \frac{0^2}{30} = 60$$

$$SS(X_6) = \sum X_6^2 - \frac{(\sum X_6)^2}{N} = 120 - \frac{0^2}{30} = 120$$

$$SS(X_7) = \sum X_7^2 - \frac{(\sum X_7)^2}{N} = 24 - \frac{0^2}{30} = 24$$

$$SS(X_8) = \sum X_8^2 - \frac{(\sum X_8)^2}{N} = 12 - \frac{0^2}{30} = 12$$

$$SS(X_9) = \sum X_9^2 - \frac{(\sum X_9)^2}{N} = 12 - \frac{0^2}{30} = 12$$

$$SS(X_{10}) = \sum X_{10}^2 - \frac{(\sum X_{10})^2}{N} = 80 - \frac{0^2}{30} = 80$$

$$\vdots \qquad \vdots \qquad \vdots \qquad \vdots$$

$$SS(X_{29}) = \sum X_{29}^2 - \frac{(\sum X_{29})^2}{N} = 24 - \frac{0^2}{30} = 24$$

TABLE 6.25 Sums of cross products for regression approach to factorial repeated-measures ANOVA

$$SP(YX_1) = \sum YX_1 - \frac{(\sum Y)(\sum X_1)}{N} = 6 - \frac{(124)(0)}{30} = 6$$

$$SP(YX_2) = \sum YX_2 - \frac{(\sum Y)(\sum X_2)}{N} = -8 - \frac{(124)(0)}{30} = -8$$

$$SP(YX_3) = \sum YX_3 - \frac{(\sum Y)(\sum X_3)}{N} = 2 - \frac{(124)(0)}{30} = 2$$

TABLE 6.25 Continued

$$SP(YX_4) = \sum YX_4 - \frac{(\sum Y)(\sum X_4)}{N} = 34 - \frac{(124)(0)}{30} = 34$$

$$SP(YX_5) = \sum YX_5 - \frac{(\sum Y)(\sum X_5)}{N} = 20 - \frac{(124)(0)}{30} = 20$$

$$SP(YX_6) = \sum YX_6 - \frac{(\sum Y)(\sum X_6)}{N} = -24 - \frac{(124)(0)}{30} = -24$$

$$SP(YX_7) = \sum YX_7 - \frac{(\sum Y)(\sum X_7)}{N} = 10 - \frac{(124)(0)}{30} = 10$$

$$SP(YX_8) = \sum YX_8 - \frac{(\sum Y)(\sum X_8)}{N} = 9 - \frac{(124)(0)}{30} = 9$$

$$SP(YX_9) = \sum YX_9 - \frac{(\sum Y)(\sum X_9)}{N} = -3 - \frac{(124)(0)}{30} = -3$$

$$SP(YX_{10}) = \sum YX_{10} - \frac{(\sum Y)(\sum X_{10})}{N} = -1 - \frac{(124)(0)}{30} = -1$$

$$\vdots$$

$$SP(YX_{29}) = \sum YX_{29} - \frac{(\sum Y)(\sum X_{29})}{N} = 0 - \frac{(124)(0)}{30} = 0$$

TABLE 6.26 Sums of squares for factorial repeated-measures ANOVA through regression

SS_A	$SS(\text{reg. } X_1) + SS(\text{reg. } X_2)$	$\frac{6^2}{20} + \frac{(-8)^2}{60}$	$1.80 + 1.07$	2.87
SS_B	$SS(\text{reg. } X_3)$	$\frac{2^2}{30}$	0.13	0.13
SS_{AB}	$SS(\text{reg. } X_4) + SS(\text{reg. } X_5)$	$\frac{34^2}{20} + \frac{20^2}{60}$	$57.80 + 6.67$	64.47
SS_S	$SS(\text{reg. } X_6) + SS(\text{reg. } X_7)$ $+ SS(\text{reg. } X_8) + SS(\text{reg. } X_9)$	$\frac{(-24)^2}{120} + \frac{10^2}{24} + \frac{9^2}{12} + \frac{(-3)^2}{12}$	$4.80 + 4.17$ $+ 6.75 + 0.75$	16.47
SS_{AS}	$SS(\text{reg. } X_{10}) + SS(\text{reg. } X_{11})$ $+ SS(\text{reg. } X_{12}) + SS(\text{reg. } X_{13})$ $+ SS(\text{reg. } X_{14}) + SS(\text{reg. } X_{15})$ $+ SS(\text{reg. } X_{16}) + SS(\text{reg. } X_{17})$	$\frac{(-1)^2}{80} + \frac{5^2}{16} + \frac{1^2}{8} + \frac{(-2)^2}{8}$ $+ \frac{3^2}{240} + \frac{1^2}{48} + \frac{(-3)^2}{24} + \frac{6^2}{24}$	$0.01 + 1.56$ $+ 0.12 + 0.50$ $+ 0.04 + 0.02$ $+ 0.38 + 1.50$	4.13
SS_{BS}	$SS(\text{reg. } X_{18}) + SS(\text{reg. } X_{19})$ $+ SS(\text{reg. } X_{20}) + SS(\text{reg. } X_{21})$	$\frac{(-2)^2}{120} + \frac{(-16)^2}{24} + \frac{(-7)^2}{12} + \frac{15^2}{12}$	$0.03 + 10.67$ $+ 4.08 + 18.75$	33.53
SS_{ABS}	$SS(\text{reg. } X_{22}) + SS(\text{reg. } X_{23})$ $+ SS(\text{reg. } X_{24}) + SS(\text{reg. } X_{25})$ $+ SS(\text{reg. } X_{26}) + SS(\text{reg. } X_{27})$ $+ SS(\text{reg. } X_{28}) + SS(\text{reg. } X_{29})$	$\frac{11^2}{80} + \frac{5^2}{16} + \frac{7^2}{8} + \frac{0^2}{0}$ $+ \frac{(-5)^2}{240} + \frac{5^2}{48} + \frac{(-1)^2}{24} + \frac{0^2}{24}$	$1.51 + 1.56$ $+ 6.12 + 0$ $+ 0.10 + 0.52$ $+ 0.04 + 0$	9.87
SS_T	SS_Y	$644 - \frac{124^2}{30}$	$644 - 512.53$	131.47

TABLE 6.27 Source table for the regression approach to factorial-repeated-measures ANOVA

Source	SS	df	MS	F
A	2.87	2	1.44	$\frac{1.44}{0.52} = 2.77$
B	0.13	1	0.13	$\frac{0.14}{8.38} = 0.02$
AB	64.47	2	32.23	$\frac{32.23}{1.23} = 26.20$
S	16.47	4	4.12	
AS	4.13	8	0.52	
BS	33.53	4	8.38	
ABS	9.87	8	1.23	
T	131.47	29		

6.4.2.5 Computer Analyses of Small-Sample Factorial Repeated-Measures Design

Table 6.28 shows the data format for the two-way repeated-measures ANOVA through SPSS and SAS. Repeated-measures IVs are created out of the numerous DV values for each case by naming them and their levels. The choice of names for the DVs and the order in which IVs and their levels are named are critical, because they have to correspond. Consider the two-way factorial data set of Table 6.17. There are six DV values from three months and two types of novels. If the data set is organized as in Table 6.17, the three months are collected below each type of novel. The index for months is said to change more rapidly than the index for type of novel because while novel is held constant (first at science fiction and then at mystery), DV values for all three months are presented. The data could also be rearranged so that the index for type of novel changes more rapidly than the index for month: month1 (science fiction, mystery), month2 (science fiction, mystery), and month3 (science fiction,

TABLE 6.28 Hypothetical data set for two-way repeated-measures design for analysis through SPSS and SAS

CASE	SCIFI_M1	SCIFI_M2	SCIFI_M3	MSTRY_M1	MSTRY_M2	MSTRY_M3
1	1	3	6	5	4	1
2	1	4	8	8	8	4
3	3	3	6	4	5	3
4	5	5	7	3	2	0
5	2	4	5	5	6	3

mystery). Both of these arrangements represent the same six DVs, but in different orders.

Tables 6.29 and 6.30 present syntax and highly edited output from SPSS GLM and SAS GLM, respectively. Except for being lengthier due to increases in number of effects and errors, the output is the same as for the one-way design. It is the syntax that is tricky for factorial repeated-measures analysis.

The syntax for SPSS GLM (Table 6.29) lists all six conditions as DVs. Then the /WSFACTOR instruction defines those six DVs as two repeated-measures IVs: the first with two levels and the second with three levels. Then /MEASURE names books as the DV. The correspondence between levels of the IVs and DVs becomes clear when you use the SPSS Windows menu system to click over the DVs that correspond to each combination of levels. The remaining syntax in Table 6.29 is produced by the SPSS Windows menu system.

The first table in the output shows the correspondence between the levels of novel and month and the six DVs. The next table shown is for Mauchly's Test of Sphericity; both month and the novel*month interaction have sphericity.[8] Finally, the three effects are tested against their corresponding error terms. There is no main effect of novel (Sig. = .906) or of month (Sig. = .122), but there is a novel*month interaction (Sig. = .000). Multivariate output and trend analysis are omitted here.

TABLE 6.29 Factorial repeated measures through SPSS GLM (syntax and selected output)

```
GLM
  SCIFI_M1 SCIFI_M2 SCIFI_M3 MSTRY_M1 MSTRY_M2 MSTRY_M3
  /WSFACTOR = novel 2 Polynomial month 3 Polynomial
  /MEASURE = books
  /METHOD = SSTYPE(3)
  /CRITERIA = ALPHA(.05)
  /WSDESIGN = novel month novel*month .
```

General Linear Model

Within-Subjects Factors

Measure: books

novel	month	Dependent Variable
1	1	scifi_m1
	2	scifi_m2
	3	scifi_m3
2	1	mstry_m1
	2	mstry_m2
	3	mstry_m3

[8] The effect of novel has no test for sphericity and no adjustments, because it only has two levels.

TABLE 6.29 Continued

Mauchly's Test of Sphericity[b]

Measure: books

Within Subjects Effect	Mauchly's W	Approx. Chi-Square	df	Sig.	Epsilon[a] Greenhouse-Geisser	Huynh-Feldt	Lower-bound
novel	1.000	.000	0	.	1.000	1.000	1.000
month	.796	.684	2	.710	.831	1.000	.500
novel*month	.252	4.135	2	.127	.572	.651	.500

Tests the null hypothesis that the error covariance matrix of the orthonormalized transformed dependent variables is proportional to an identity matrix.

a. May be used to adjust the degrees of freedom for the averaged tests of significance. Corrected tests are displayed in the Tests of Within-Subjects Effects table.

b. Design: Intercept
Within Subjects Design: novel+month+novel*month

Tests of Within-Subjects Effects

Measure: books

Source		Type III Sum of Squares	df	Mean Square	F	Sig.
novel	Sphericity Assumed	.133	1	.133	.016	.906
	Greenhouse-Geisser	.133	1.000	.133	.016	.906
	Huynh-Feldt	.133	1.000	.133	.016	.906
	Lower-bound	.133	1.000	.133	.016	.906
Error(novel)	Sphericity Assumed	33.533	4	8.383		
	Greenhouse-Geisser	33.533	4.000	8.383		
	Huynh-Feldt	33.533	4.000	8.383		
	Lower-bound	33.533	4.000	8.383		
month	Sphericity Assumed	2.867	2	1.433	2.774	.122
	Greenhouse-Geisser	2.867	1.661	1.726	2.774	.136
	Huynh-Feldt	2.867	2.000	1.433	2.774	.122
	Lower-bound	2.867	1.000	2.867	2.774	.171
Error(month)	Sphericity Assumed	4.133	8	.517		
	Greenhouse-Geisser	4.133	6.645	.622		
	Huynh-Feldt	4.133	8.000	.517		
	Lower-bound	4.133	4.000	1.033		
novel*month	Sphericity Assumed	64.467	2	32.233	26.135	.000
	Greenhouse-Geisser	64.467	1.144	56.344	26.135	.004
	Huynh-Feldt	64.467	1.303	49.479	26.135	.003
	Lower-bound	64.467	1.000	64.467	26.135	.007
Error(novel*month)	Sphericity Assumed	9.867	8	1.233		
	Greenhouse-Geisser	9.867	4.577	2.156		
	Huynh-Feldt	9.867	5.212	1.893		
	Lower-bound	9.867	4.000	2.467		

TABLE 6.30 Factorial repeated-measures ANOVA through SAS GLM (syntax and selected output)

```
proc glm data=SASUSER.SS_RPTD2;
     model SCIFIM1 SCIFIM2 SCIFIM3 MSTRYM1 MSTRYM2 MSTRYM3 = ;
     repeated NOVEL 2, MONTH 3 / mean;
run;
```

The GLM Procedure
Repeated Measures Analysis of Variance
Univariate Tests of Hypotheses for Within Subject Effects

Source	DF	Type III SS	Mean Square	F Value	Pr > F
NOVEL	1	0.13333333	0.13333333	0.02	0.9057
Error(NOVEL)	4	33.53333333	8.38333333		

						Adj Pr > F	
Source	DF	Type III SS	Mean Square	F Value	Pr > F	G-G	H-F
MONTH	2	2.86666667	1.43333333	2.77	0.1216	0.1365	0.1216
Error(MONTH)	8	4.13333333	0.51666667				

Greenhouse-Geisser Epsilon	0.8306
Huynh-Feldt Epsilon	1.3481

						Adj Pr > F	
Source	DF	Type III SS	Mean Square	F Value	Pr > F	G-G	H-F
NOVEL*MONTH	2	64.46666667	32.23333333	26.14	0.0003	0.0044	0.0027
Error(NOVEL*MONTH)	8	9.86666667	1.23333333				

(Continued)

```
Greenhouse-Geisser Epsilon    0.5721
Huynh-Feldt Epsilon           0.6514
```

Means of Within Subjects Effects

Level of NOVEL	N	Mean	Std Dev
1	15	4.20000000	2.07708587
2	15	4.06666667	2.25092574

Level of MONTH	N	Mean	Std Dev
1	10	3.70000000	2.16281709
2	10	4.40000000	1.71269768
3	10	4.30000000	2.58413966

Level of NOVEL	Level of MONTH	N	Mean	Std Dev
1	1	5	2.40000000	1.67332005
1	2	5	3.80000000	0.83666003
1	3	5	6.40000000	1.14017543
2	1	5	5.00000000	1.87082869
2	2	5	5.00000000	2.23606798
2	3	5	2.20000000	1.64316767

Syntax and highly selected output from SAS GLM are shown in Table 6.30. SAS GLM sets up in a similar fashion to the one-way repeated SAS GLM example of Table 6.15. Both the repeated factors, as well as their number of levels, are listed in the `repeated` instruction; order is important, as is the comma between the two factors. Notice that in the syntax, `NOVEL` (with two levels) is named before `MONTH` (with three levels), which corresponds to the order of DVs in the data set. Means are requested.

Separate source tables, with appropriate error terms, are shown for each effect tested, as well as epsilon values for evaluating homogeneity of variance or sphericity (cf. Section 6.6.2). Multivariate output is not shown. Means and standard deviations are at the end of Table 6.30.

6.5 TYPES OF REPEATED-MEASURES DESIGNS

6.5.1 Time as a Variable

In many repeated-measures designs, the repeating variable is time or trials. Cases are measured at two or more sequential times during the course of the study. The design is useful for assessing the time course of change, or the growth or decline of a process over a time period. When tracking change is the goal, it is often useful to use planned comparisons with trend coefficients to model the change. Is the change linear? If so, what is the slope? Is there a quadratic component to the change? Does the process first increase and then decline (or the reverse)? How long does the change last? Is there any permanent change? Is the DV the same at the end of the time period as it was at the beginning?

Passage of time is a variable with inherent order. The first trial is always before the second and so on. Because trials cannot be given in different orders, control features, such as randomization, counterbalancing, and Latin square (Chapter 9), are not available to control extraneous variables. Extraordinary care is needed, therefore, to eliminate or control extraneous variables over time to guarantee that there are no influences on the process other than the IVs.

In addition, the design virtually guarantees violation of homogeneity of covariance/sphericity (Section 6.6.2). Events that are closer in time are likely to be more highly correlated than events farther apart in time. This gives even greater advantage to a planned trend analysis over an omnibus test of the effect of the IV, because homogeneity of covariance/sphericity is not an issue with trend analysis.

Mere passage of time is not a manipulated variable. Although the researcher can choose when and how often to measure, the levels of the repeated measure are not manipulated; thus, strictly speaking, the study is not an experiment. However, if a second, manipulated IV is added to the design, it becomes a true experiment. Often the second IV is a randomized-groups IV in which one group gets a treatment of some sort and the other serves as control. Or different amounts or types of treatment (e.g., science fiction or mystery novels) are given to cases in different groups, and the consequences for the DV are studied over time. This is the popular mixed design in which changes associated with two or more treatments are studied over a period of time. Does one treatment produce a larger or more permanent change than another? The mixed design is the topic of Chapter 7.

6.5.2 Simultaneous Repeated Measures

Sometimes the levels of the repeated-measures IV are administered simultaneously or intermixed within an experimental or other research session. For example, the levels of the IV might be verbs versus nouns that appear haphazardly within a passage to be memorized. This is one of the few situations in which a repeated-measures IV may be used in an experiment without concern for counterbalancing or violation of sphericity.

6.5.3 Matched Randomized Blocks

Another variant on the repeated-measures design actually uses different cases in the different cells of the design but analyzes the results as if produced by the same cases. That is, the design is really randomized groups (with one case per cell), but the analysis is repeated measures. This is the matched randomized-blocks design.[9] When applied in experiments, this design allows random assignment of cases to levels, yet it has the advantage of greater power than the randomized-groups design offers.

The first step is to measure all the cases on some variable. In some studies, the variable on which cases are equated is the DV administered before treatment (a pretest). Sometimes several variables are measured during the first step, and cases are equated on several variables simultaneously. The second step is to create equivalent blocks (sets) of cases based on their scores on the measured variable(s), with as many cases in each block as there are levels of the repeated-measures IV. If there are three levels of the repeated-measures IV, three "equivalent" cases are put into each block of cases, and so on. The third step is to randomly assign one case in each block to each level of the repeated-measures IV, as shown in Table 6.31. During the analytic phase, the cases in each block are treated as one in a routine repeated-measures analysis. The researcher argues that for these research purposes, these cases are equivalent.

[9] Some sources label the design *matched randomized,* others label it *randomized blocks.*

TABLE 6.31 Matched randomized-blocks design with $a = 4$

	A			
Block	a_1	a_2	a_3	a_4
block 1 ($s_1 = s_9 = s_{11} = s_{12}$)	s_9	s_1	s_{12}	s_{11}
block 2 ($s_2 = s_3 = s_4 = s_8$)	s_4	s_8	s_3	s_2
block 3 ($s_5 = s_6 = s_7 = s_{10}$)	s_5	s_7	s_6	s_{10}

This design is useful when cases need to be naïve during each measurement or when cases or samples are used up by a single measurement. Because cases are actually different, they do not have previous experience with another level of the IV before their own trial. However, the researcher is putting a lot of faith into the matching process, and if cases are matched on several variables, it is common to have them come "unglued" on some variables while trying to match them on others. It is also common to have such a reduced set of cases by the time they are matched that the generalizability of the findings is in doubt. An obvious practical requirement for the design is that all cases are measured ahead of time so the researcher can determine which are equivalent before randomly assigning them to levels of the repeated-measures IV.

This design has the advantages of the greater power usually associated with the repeated-measures design (assuming the matching is effective) without the expectation that sphericity will be violated.

6.6 SOME IMPORTANT ISSUES

6.6.1 Carryover Effects: Control of Extraneous Variables

In planning a repeated-measures study, it is extremely important to consider the impact of repeated experience with levels of the IV(s) and the DV on the performance of the cases or samples. If cases are people, is it a problem if they gain some knowledge of the purpose and nature of the study by experiencing several levels of the IV? Are cases changed by the measurement process? For instance, does measurement "use up" a case or sample or strengthen or weaken it somehow? If cases have memory, do they become more or less skilled in performing the DV over time, independent of the level of the IV? Does their attention or motivation change over time? Do some levels of the IV make permanent or semipermanent changes in the case, whereas others do not?

The researcher contemplating a repeated-measures study considers all these questions. Changes in cases over time, regardless of the level of the IV (becoming more or less skilled in the performance of the DV, or shifting attention or motivation, or becoming stronger or weaker), are called *carryover effects*. When the IV is passage of time (trials), carryover effects *are* the focus of study: The researcher specifically wants to investigate changes over time. The area of learning in psychology, for instance, *is* the study of carryover effects. In many situations, however, the researcher has another research question, and carryover effects are potentially a dangerous nuisance. Consider, for instance, the factorial repeated-measures design of Section 6.4.2, in which people are assigned to read both science fiction and mystery novels, each for three months. It is entirely possible that people's reading habits change over a three-month interval, regardless of type of novel. In Table 6.17, the sums of novels read increase over three months and then decrease a pattern that might be expected of those participating in a research study as the novelty wears off.[10] If control is not exercised over the order in which types of novels are read, these carryover effects could be misinterpreted as an effect of type of novel.

There are several ways to exercise control over carryover effects. Two of the most popular are complete counterbalancing and partial counterbalancing (using Latin-square designs). Complete counterbalancing involves presenting the levels of the IV in all possible orders to different groups of cases. In the example in which there are two levels of B, some cases read b_1 (science fiction novels) first, whereas others read b_2 (mystery novels) first. This "decouples" carryover effects from the effects of type of novel. Partial counterbalancing is used when there are several levels of the IV; complete counterbalancing is impractical. In partial counterbalancing, some, but not all, of the potential orders are used with different cases. Partial counterbalancing through a Latin-square design is the worthy topic of Chapter 9. Counterbalancing has the added advantage of identifying order as a systematic source of variance and removing it from the error term.

A third way to decouple carryover effects from effects of the IV is to present the levels of the IV in a different random order for each case. This strategy is often used when there are many levels of the IV. With numerous levels of the IV presented over many trials, random processes should more or less equalize the different levels at each trial, thereby evenly distributing the carryover effects. Any order effects, however, remain in the error term(s).

A fourth way to minimize carryover effects is to wait until performance of the DV stabilizes before running the study. As cases learn to perform a DV, there is often a rapid change in performance over the first few trials, followed by a period of time when scores are more or less stable. Cases are given several practice trials before the actual study begins. Sometimes researchers give a constant number of practice trials

[10] Sorry.

to all cases, whereas other times they establish a criterion for stability of performance to be met before the trials of the study begin.

When the IV is passage of time, the presence of carryover effects is assessed as the test of trials. When the repeated measure is something other than passage of time and controls such as complete or partial counterbalancing are used so that treatment effects and effects of trials/position/order are unconfounded, the test of trials/position/order is the test of the presence of carryover effects. This analysis is described in Chapter 9.

The most dangerous type of carryover is differential carryover, which results from the (relatively) long-lasting effects of some levels of the IV. For instance, in drug studies, a substantial washout period is usually required between delivery of different types (or dosages) of drugs to ensure that the first drug is thoroughly metabolized before the second is introduced. However, if one drug produces a permanent physiological change, no amount of delay will help, and the apparent effects of all the levels of the IV that follow are contaminated. When some levels of an IV produce long-lasting changes and others do not, there is differential carryover. Designs to assess differential carryover are described in Section 9.6.3.4.

6.6.2 Assumptions of Analysis: Independence of Errors, Sphericity, Additivity, and Compound Symmetry

Independence of errors is an untenable assumption in repeated-measures ANOVA. Errors are necessarily correlated, because measures are made on the same cases; any consistency in individual differences among cases produces correlated errors. In a logical sense, these correlations among errors are removed by calculating variance due to individual differences, SS_S, separately from what would have been the error term in a randomized-groups design (see Figure 6.1).

In practice, the assumption of independence of errors is replaced by the assumption of sphericity (defined in Table 6.32) when there are more than two levels of the repeated-measures IV. The F test for the effect of treatment is too liberal if the assumption of sphericity is violated; thus, the probability of Type I error is greater than the nominal value. In general, sphericity is violated when the correlations among pairs of levels of A are unequal. This is most likely to occur when the IV is time; correlations tend to decrease over time, and variances of difference scores between pairs of levels increase over time.

There is a somewhat complicated relationship among sphericity, additivity, and compound symmetry (the combination of homogeneity of variance and homogeneity of covariance). Each of these is described in Table 6.32, in which it is assumed that $a = 3$.

Additivity is the absence of an $A \times S$ interaction. Recall that in a repeated-measures ANOVA, it is this interaction that serves as the error term, as discussed in Section 6.4.1. If there truly is an interaction between treatments and cases, this is a

TABLE 6.32 Definitions of sphericity, compound symmetry, and additivity

Assumption	Verbal Definition	Algebraic Definition
Sphericity	Variances of difference scores between pairs of levels of *A* are equal.	$\sigma^2_{Y_{a1}-Y_{a2}} \cong \sigma^2_{Y_{a1}-Y_{a3}} \cong \sigma^2_{Y_{a2}-Y_{a3}}$
Compound symmetry		
Homogeneity of variance	Variances in different levels of *A* are equal.	$\sigma^2_{a1} \cong \sigma^2_{a2} \cong \sigma^2_{a3}$
Homogeneity of covariance	Covariances (and correlations) between pairs of levels of *A* are equal; variances of difference scores are equal.	$\sigma_{a1,a2} \cong \sigma_{a1,a3} \cong \sigma_{a2,a3}$
Additivity	There is no $A \times S$ interaction; difference scores are equivalent for all cases. Variances of difference scores are 0.	$(Y_{1,a1} - Y_{1,a2}) \cong (Y_{2,a1} - Y_{2,a2}) \cong \cdots \cong (Y_{s,a1} - Y_{s,a2})$ $(Y_{1,a1} - Y_{1,a3}) \cong (Y_{2,a1} - Y_{2,a3}) \cong \cdots \cong (Y_{s,a1} - Y_{s,a3})$ $(Y_{1,a2} - Y_{1,a3}) \cong (Y_{2,a2} - Y_{2,a3}) \cong \cdots \cong (Y_{s,a2} - Y_{s,a3})$

distorted error term, because it includes a systematic source of variance (the interaction), as well as random error.[11]

Because the interaction means that different cases have different patterns of response to treatment, a better, more powerful, and generalizable design would take the interaction into account by blocking on cases that have similar patterns of response to the levels of IV. For example, if men show one consistent pattern of response over the levels of the repeated-measures IV and women show a different consistent pattern of response, gender should be made an additional randomized-groups IV. This provides an explicit test of the former nonadditivity (treatment by gender interaction) and removes it from the error term.

The assumption of sphericity is that the variances of *difference* scores between *pairs* of levels of *A* are equal. This explains why the assumption is not relevant when there are only two levels of *A*—there is only one variance of difference scores and nothing for it to equal. Note that with complete additivity, there is zero variance in difference scores (Table 6.32). Because all zeros are equal, there is also sphericity. Thus, additivity is the most restrictive assumption.

After sphericity, the next most restrictive assumption is compound symmetry, which means that both the variances in levels of *A* and the correlations between pairs of levels of *A* are equal. In this situation, the variances in differences scores are not 0 (as they are with additivity), but they are equal. With either additivity or compound symmetry, then, the assumption of sphericity is met. However, it is possible to have sphericity without having either additivity or compound symmetry, as demonstrated by Myers and Well (1991).

[11] Use of this design when there is $A \times S$ interaction is also suspect, because the interaction shows up in the numerator of F_A as well (cf. Chapter 11).

If your data meet requirements for either additivity or compound symmetry, you can be confident about sphericity. However, if requirements for additivity or compound symmetry are not met, you may still have sphericity and meet statistical requirements for an honest (noninflated) *F* test of treatment. In practice, researchers rely on the results of a combination of tests for homogeneity of variance and the Mauchly (1940) test for sphericity.

SPSS GLM offers the Mauchly test of sphericity by default; SAS GLM and ANOVA produce it by request. In addition, the programs that do repeated-measures ANOVA display epsilon factors that are used to adjust degrees of freedom should the assumption of sphericity be violated.

If the Mauchly test is nonsignificant, or if the adjustment based on epsilon (described in what follows) does not alter the nominal probability of rejecting the null hypothesis (e.g., .05 or .01) and conditions for homogeneity of variance are met, the *F* test for routine repeated-measures ANOVA is appropriate.

The Mauchly test, however, is sensitive to non-normality of the DV, as well as to heterogeneity of covariance. Therefore, the test is sometimes significant when there is non-normality rather than failure of sphericity. If the Mauchly test is significant, closer examination of the distribution of the DV is in order. If it is markedly skewed, repeat the sphericity test after a normalizing transformation of the DV (cf. Chapter 2). If, as a result, the test is now nonsignificant, the problem with the data set is probably non-normality rather than failure of sphericity. If the test is still significant, however, there is probably nonsphericity. The Mauchly test also has low power for small samples and is overly sensitive with very large samples. Thus, with large samples, it is sometimes statistically significant when departure from sphericity is slight. Thus, it is always worthwhile to consider the magnitude of the epsilon factor, even when the test of sphericity is explicitly provided.

There are four options when the assumption of sphericity is not tenable: (1) Use specific comparisons on the IV in question (usually trend analysis), instead of the omnibus test; (2) use an adjusted *F* test; (3) use a multivariate test of the repeated-measures effect(s); or (4) use a maximum likelihood procedure that lets you specify that the structure of the variance-covariance matrix is something other than compound symmetry.

The variance-covariance matrix shows the variances for each level of a repeated-measures IV in the diagonal positions and the covariances (cf. Section 2.3.2.1) for each pair of levels in the off-diagonal positions.

	Y_1	Y_2	Y_3
Y_1	σ^2_{a1}	$\sigma_{a1,a2}$	$\sigma_{a1,a3}$
Y_2	$\sigma_{a2,a1}$	σ^2_{a2}	$\sigma_{a2,a3}$
Y_3	$\sigma_{a3,a1}$	$\sigma_{a3,a2}$	σ^2_{a3}

As shown in Table 6.32, compound symmetry assumes that all the variances are the same and all the covariances are the same.

The first option—comparisons—takes advantage of the fact that sphericity is not required when $a = 2$. This option is often good because questions about trends in the

DV over time are usually the ones researchers want answered anyway and because the assumption of homogeneity of covariance is most likely to be violated when the IV is related to time. Does the DV increase (or decrease) steadily over time? Does the DV first increase and then decrease (or the reverse)? Are both patterns present, superimposed on each other? Are there other, more complicated patterns in the data?

The second option involves using a more stringent F test of the effects of the IV in question. Two standard adjustments are routinely offered: Greenhouse-Geisser (1959) and Huynh-Feldt (1976). Both compute an adjustment factor, epsilon (ϵ), based on the degree of failure of sphericity in the data.[12] Once ϵ is estimated, it is used to reduce df associated with both numerator and denominator of the F test. That is, the adjustment factor is less than or equal to 1 and is used to multiply both the df for treatment and the df for error to produce new (fractional) df for treatment and error. Reducing df makes the F test more conservative. Both Greenhouse-Geisser and Huynh-Feldt compute an ϵ value, but Greenhouse-Geisser usually produces a stronger adjustment (larger ϵ value) than Huynh-Feldt. The more liberal Huynh-Feldt adjustment is preferred because it seems to produce results closer to nominal α levels.

The third option is to use a multivariate test that does not require sphericity, instead of using routine ANOVA. This option is more often associated with analysis of nonexperimental studies. Description of multivariate tests is beyond the scope of this book, but is described in Tabachnick and Fidell (2007, Chapters 7–8) and elsewhere.

A fourth option is to use a maximum likelihood strategy instead of ANOVA, in which the variance-covariance matrix is user-specified or is left unspecified. SAS MIXED and SPSS MIXED MODELS produce this type of analysis. The appropriate variance-covariance matrix structure for a time-related repeated-measures IV, for example, is first-order autoregressive, in which correlations among pairs of levels decrease the farther apart they are in time. This is requested as AR(1) in SAS MIXED or SPSS MIXED MODELS. Another maximum-likelihood strategy is multilevel modeling (Tabachnick and Fidell, 2007, Chapter 15), in which a hierarchy is set up with occasions as the first level of analysis and cases the second. Analysis of these models is also available in SAS MIXED and SPSS MIXED MODELS.

Before sophisticated software was available, trend analysis (or other comparisons) was preferred on strictly computational grounds. A trend analysis is still preferred if the researcher has questions about the shape of the patterns in the DV over time. However, a disadvantage of trend analysis (or any set of comparisons) in a repeated-measures design is that each comparison has its own error term (cf. Section 6.6.6). This reduces the number of error degrees of freedom—and consequently the power—available for the test of the comparison.

ANOVA programs provide trend analyses, multivariate tests, and the Huynh-Feldt adjustment, so the researcher can easily choose any of these three options. Indeed, many combinations of these options are produced by default alongside the univariate

[12] The adjustment is based on compound symmetry and is computed from the variance-covariance matrix of the data. The adjustment increases as the individual variances and covariances deviate from their own mean. Computation of ϵ is illustrated in Myers and Well (1991), among others.

F test, as shown in Tables 6.14 and 6.15. The fourth option, maximum-likelihood analysis, is not included in standard ANOVA programs (see Section 6.7).

6.6.3 Power, Sample Size, and Relative Efficiency

A basic strategy for increasing power is to decrease the size of the error term in the F ratio. In repeated-measures analysis, sums of squares for differences among cases are assessed and subtracted from the error term. A repeated-measures design, then, is often more powerful than a corresponding randomized-groups design. More precisely, if the ratio of sum of squares to df for S is *larger* than the ratio of sum of squares to df for the error term in a randomized-groups design, repeated-measures is more powerful. If the ratio of sum of squares to df for S is *the same* as the ratio of sum of squares to df for the error term in a randomized-groups design, the two designs have equal power. However, if the ratio of sum of squares to df for S is *smaller* than the ratio of sum of squares to df for the error term in a randomized-groups design, the repeated-measures design is less powerful, because df are lost without a corresponding proportion of sum of squares being removed. Thus, if power is the main issue, the repeated-measures design offers an advantage only when systematic differences among cases are anticipated.

Relative efficiency is a useful measure of the gain in power by using a one-way repeated-measures design over a one-way randomized-groups design. Note from Figure 6.1 that the error term in a repeated-measures ANOVA, SS_{AS}, is the error term from the randomized-groups ANOVA, $SS_{S/A}$, with individual differences, SS_S, subtracted out:

$$SS_{AS} = SS_{S/A} - SS_S \tag{6.19}$$

By taking into account the differences in df for the error terms in the two designs, relative efficiency (Fisher, 1935) is

$$\text{Relative efficiency} = \frac{MS_{S/A}}{MS_{AS}}\left(\frac{df_{AS}+1}{df_{AS}+3}\right)\left(\frac{df_{S/A}+3}{df_{S/A}+1}\right) \tag{6.20}$$

$MS_{S/A}$ is found either by running the analysis as a one-way randomized-groups design or by applying the following equation to the results of the repeated-measures ANOVA:

$$MS_{S/A} = \frac{(s-1)\,MS_S + s\,(a-1)\,MS_{AS}}{as-1} \tag{6.21}$$

For the small-sample example, using $MS_{S/A}$ from Table 4.6,

$$\text{Relative efficiency} = \frac{1.6}{1.2}\left(\frac{8+1}{8+3}\right)\left(\frac{12+3}{12+1}\right) = 1.26$$

Thus, the error term (corrected for differences in df) for the randomized-groups design is about 1.26 times as large as that for the error term of the repeated-measures design—a 26% gain in efficiency for the repeated-measures analysis.

Relative efficiency can also be used to show how many cases would be necessary in each level of a randomized-blocks design to match the efficiency of a repeated-measures design:

$$n = s(\text{Relative efficiency}) = 5(1.26) = 6.3 \qquad \textbf{(6.22)}$$

Rounding up, this shows that 7 cases per level of A are necessary in a randomized-groups design to get the same power as 5 cases in the one-way repeated-measures design (or in the matched randomized blocks design).

The basic procedures for calculating power and determining the number of cases (see Sections 4.5.2 and 5.5.3) apply. The one-way repeated-measures design is a two-way (treatment and case) layout, but sample-size estimations are typically of interest only for treatment effects. In factorial repeated measures, sample-size requirements are calculated for all effects (e.g., A, B, and $A \times B$), and the result that calls for the largest sample size is used.

Section 4.5.2 describes the ins and outs of estimating power for a study, including use of Dallal's (1986) PC-SIZE program. Notice in Table 4.19 that one of the options for estimating power is the paired t test, your choice for a repeated-measures design. SPSS GLM provides observed power (OPOWER) for all significance tests on request, as demonstrated in Table 6.14.

6.6.4 Effect Size

Two problems arise with estimating effect size in repeated-measures designs. The first involves inclusion of sum of squares for cases in sum of squares total. Some argue that variability due to individual differences among cases should not be included in estimates of total variance. The second problem involves the error term in repeated-measures designs that includes both random variability and unwanted $A \times S$ interaction. An estimate of random variability is needed for effect size computations, but it is impossible to separate random variability from the $A \times S$ interaction term in the repeated-measures design. If the test of sphericity indicates that there is no violation of the assumption, MS_{AS} is taken as an assessment of random variability. On the other hand, if there is evidence of $A \times S$ interaction or violation of sphericity, some accommodation is needed when estimating effect size.

Because of doubt regarding the wisdom of including sum of squares for cases in sum of squares total, and when there is little evidence of $A \times S$ interaction, we recommend using the form of partial $\hat{\omega}^2$ offered by Keppel (1991):

$$\text{Partial } \hat{\omega}_A^2 = \frac{\text{df}_A(\text{MS}_A - \text{MS}_{AS})}{\text{df}_A(\text{MS}_A - \text{MS}_{AS}) + as\text{MS}_{AS}} \qquad \textbf{(6.23)}$$

This equation is for the effect of A. When estimating effect size for other main effects and interactions in a factorial arrangement, substitute into the equation the error terms, df, and levels for those effects.

A more conservative approach, and the one recommended by Myers and Well (1991) when $A \times S$ interaction is present, is to compute a lower and upper bound for the value of $\hat{\omega}^2$. The lower bound is

$$\hat{\omega}^2_L = \frac{\text{df}_A(\text{MS}_A - \text{MS}_{AS})}{\text{df}_A(\text{MS}_A - \text{MS}_{AS}) + as\text{MS}_{AS} + s\text{MS}_S} \tag{6.24}$$

and the upper bound is the same as partial $\hat{\omega}^2$ (Equation 6.23). Equations 6.23 and 6.24 differ only in inclusion of the effect of cases in the denominator.

Partial $\hat{\omega}^2$ and its lower bound are computed for A (month) for the one-way repeated-measures example of Table 6.7.

$$\text{Partial } \hat{\omega}^2_A = \frac{2(20.600 - 1.183)}{2(20.600 - 1.183) + 3\,(5)(1.183)} = .69$$

$$\hat{\omega}^2_L = \frac{2(20.600 - 1.183)}{2(20.600 - 1.183) + 3\,(5)\,1.183 + 5\,(2.433)} = .56$$

In the example, there is apparent sphericity, so the partial $\hat{\omega}^2$ value of .69 is appropriate. If Mauchly's test is significant and there is other evidence of failure of sphericity, partial $\hat{\omega}^2$ is estimated instead to fall within the range of .56 and .69; that is, between 56% and 69% of the variability in number of novels read is associated with month.

The same considerations apply to the simpler estimate of effect size, η^2—that is, partial η^2 is appropriate if there is sphericity, whereas ${\eta^2}_L$ is used as a lower bound if there is doubt about sphericity. As usual,

$$\text{Partial } \eta^2 = \frac{\text{SS}_A}{\text{SS}_A + \text{SS}_{\text{error}}} = \frac{\text{SS}_A}{\text{SS}_A + \text{SS}_{AS}} \tag{6.25}$$

which, for the data in the one-way repeated small-sample example, is

$$\text{Partial } \eta^2 = \frac{41.2}{41.2 + 9.5} = .81$$

This is the value reported by SPSS GLM when ETASQ is requested.

For a lower-bound value,

$$\eta^2_L = \frac{\text{SS}_A}{\text{SS}_T} = \frac{\text{SS}_A}{\text{SS}_A + \text{SS}_S + \text{SS}_{AS}} \tag{6.26}$$

For the small-sample one-way data,

$$\eta^2_L = \frac{41.2}{41.2 + 9.7 + 9.5} = .68$$

The lower-bound value of η^2, then, is the same as the η^2 for a corresponding one-way randomized-groups design (Section 4.5.1). As usual, the values for η^2 and partial η^2 are larger than the corresponding $\hat{\omega}^2$ and partial $\hat{\omega}^2$, because η^2 is a descriptive statistic not a parameter estimate (cf. Section 4.5.1).

Effect sizes and their confidence intervals are also available through the Smithson (2003) software for SPSS and SAS. Procedures for using the software are identical to those demonstrated for randomized-groups designs in Chapters 4 and 5. This procedure is also demonstrated in Section 6.7.

Recall from previous chapters that estimates of effect size are also appropriate for comparisons and are done with respect to the effect(s) of which they are a part. Lower-bound values are not needed when calculating effect sizes for comparisons, because sphericity is not an issue with single degree of freedom tests. Partial $\hat{\omega}^2$ for components of a trend analysis is demonstrated in Section 6.7.2. An example of η^2 for an interaction contrast, as a component of the full interaction, is in Section 6.6.6.3.

6.6.5 Missing Data

Each case in repeated-measures designs participates at all levels (or all combinations of levels) of the effect(s). If a case is missing, then, it is a problem of overall reduced n and lower power, rather than a problem of unequal sample size. However, a common problem in repeated-measures designs, particularly in factorial designs with numerous combinations of levels of treatment, is to have a missing DV score. If a case is missing a DV measure, one option (the default for the software) is to delete the case. Although expedient, this option is unpopular with researchers who have measured numerous other DV scores for the case. A second option is to use available information (the other DV scores) to estimate (impute) the missing value.

Several estimation procedures are available. One is to replace the missing value with a value estimated from the mean for that level of A and for that case. The following equation takes into account both the mean for the case and the mean for the level of A, as well as the grand mean:

$$Y_{ij}^* = \frac{sS_i' + aA_j' - T'}{(a-1)(s-1)} \tag{6.27}$$

where Y_{ij}^* = predicted score to replace missing score,

s = number of cases,

S_i' = sum of the known values for that case,

a = number of levels of A,

A_j' = sum of the known values of A, and

T' = sum of all known values.

Say that the final score, s_5 in a_3, is missing from Table 6.3. The remaining scores for that case (in a_1 and a_2) sum 6 (4 + 2). The remaining scores for a_3 sum to 27 (6 + 8 + 6 + 7). The remaining scores for the entire table sum to 58 (63 − 5). Plugging these values into Equation 6.27,

$$Y_{5,3}^* = \frac{5(6) + 3(27) - 58}{(2)(4)} = 6.625$$

The error term has df reduced by the number of imputed values—here, $\text{df}_{AS} = (a-1)(s-1) - 1 = 7$.

Most of the other popular methods for imputing missing values, such as regression or expectation maximization (EM; see Tabachnick and Fidell, 2007) available in SPSS MVA, do not take into account the commensurate nature of the measurement; that the DV is measured on the same scale for all its occurrences. They also do not take into account the consistency afforded by the repeated measurement; that is, individual differences. Multiple imputation as implemented in SOLAS MDA (Statistical Solutions, Inc., 1997) and SAS MI/MIANALYZE is a procedure that is indeed applicable to longitudinal data (or any other repeated measures) but is difficult to implement. The procedure requires you to create several data sets with different imputations of missing values and derive summary statistics from analysis of all of them. Or, if you happen to have BMDP5V (Dixon, 1992), the program imputes and prints out missing values for repeated-measures analysis.

Whatever estimation procedure is used, it is reported along with the other findings in your results section.

6.6.6 Specific Comparisons

Repeated-measures designs present the same issues of planned versus post hoc comparisons, degree of adjustment, and type of comparison as randomized-groups designs. Comparisons in repeated-measures designs can also be either planned ahead of time and performed in lieu of omnibus ANOVA or done after omnibus ANOVA, with the power advantage to planning. Also, as in randomized-groups designs, the degree of adjustment to α depends both on when the comparison is performed and on the number of comparisons that can be performed, with the Scheffé procedure the most conservative. Finally, either user-generated coefficients or trend coefficients are available for comparisons, although when the repeated measure IV is time, trend coefficients are likely to be more appropriate.

Repeated-measures comparisons differ from randomized-groups comparisons in development of a separate error term for each comparison. The need for this arises from the nature of the error terms as, in part, interaction. The error terms are the idiosyncratic reactions of cases to different levels of treatment, plus random error. If a comparison omits one of the levels of treatment, why should the error term for that comparison contain idiosyncratic reactions to the omitted level of treatment? When there is no $A \times S$ interaction, the error terms are all estimates of the same random variability and about the same size as MS_{AS}, so which is used has little practical consequence. When there is interaction, however, the sizes of the various error terms differ, with direct consequences for the power of various comparisons. In practice, there is no harm in developing a special error term for each comparison. If MS_{AS} includes interaction, you are safe; if not, you may have gone to some extra effort, but the analysis is still justified.

6.6.6.1 Tests of Repeated-Measures Main Effects

This section discusses procedures for conducting a comparison on an IV with more than two levels. If there is only one IV, the procedure of comparing designs with one repeated-measures IV, discussed in Section 6.6.6.1.1, is appropriate. If there are two or more IVs, the procedure of performing marginal comparisons, also discussed in Section 6.6.6.1.2, is appropriate. Remember that interpretation of a main effect is problematic in the presence of statistically significant interaction, as discussed in Section 5.6.1.1.

6.6.6.1.1 Comparisons for Designs with One Repeated-Measures IV When there are more than two levels of A, comparisons to pinpoint the source of differences in means are appropriate. The logic is the same as for a one-way randomized-groups design (Section 4.5.3). However, procedures differ because of the development of a separate error term for each comparison.

The most straightforward way to perform these comparisons by hand is to build the comparisons into a new data set and do standard analysis on the new set. (This method also works by computer if transformations are used to convert the data or if programs operate on a special data set developed for the comparison.) There are two reasons that this approach works for repeated-measures IVs but not for randomized-groups IVs: (1) Because each line of data comes from a single case, there is a rationale for determining which DV scores to combine over levels of the IV. (2) The analysis develops the mean square for the new error term right along with the comparison mean square for the IV; that is, you get $MS_{A\text{comp}}$ and $MS_{A\text{comp} \times S}$ simultaneously.

Consider the data of Table 6.3. Suppose you want two comparisons, the first month versus the second and then the first two months versus the third. The contrast coefficients are 1 −1 0 and 1 1 −2. Table 6.33 shows the new data set needed for each comparison. These data sets are then analyzed by standard procedures (Section 6.4.1.3) to test $MS_{A\text{comp}}$ against $MS_{A\text{comp} \times S}$.

The data set for the 1 −1 0 comparison simply deletes data for the third level of A. The data set for the 1 1 −2 comparison pools the DV scores for the first two levels of A and doubles the DV scores at a_3. The DV scores with positive coefficients are combined in one column and the DV scores with negative coefficients are combined in the second column. Because each comparison has 1 df, new data sets for comparisons

TABLE 6.33 Development of data sets for comparisons in one-way repeated-measures designs

(a) Data set for the 1 −1 0 comparison

Case	a_1: Month 1	a_2: Month 2	Case Totals
s_1	1	3	4
s_2	1	4	5
s_3	3	3	6
s_4	5	5	10
s_5	2	4	6

(b) Data set for the 1 1 −2 comparison

Case	$a_1 + a_2$ Month 1 + Month 2	$2(a_3)$ 2(Month 3)	Case Totals
s_1	$1 + 3 = 4$	$2(6) = 12$	16
s_2	$1 + 4 = 5$	$2(8) = 16$	21
s_3	$3 + 3 = 6$	$2(6) = 12$	18
s_4	$5 + 5 = 10$	$2(7) = 14$	24
s_5	$2 + 4 = 6$	$2(5) = 10$	16

always have only two columns.[13] This procedure readily generalizes to more complicated comparisons. For instance, the linear trend coefficients for an IV with four levels are -3 -1 1 3. To accomplish this comparison, create two columns of data. The first column contains $3(\text{DV}_{a1}) + \text{DV}_{a2}$ for each case, and the second column contains $\text{DV}_{a3} + 3(\text{DV}_{a4})$ for each case. These two columns are then analyzed in routine repeated-measures ANOVA (Section 6.4.1.3).

SAS and SPSS compute the separate error terms needed for each comparison. Table 6.34 summarizes the syntax and error term for trend analysis and user-specified comparisons. Note, however, that the SAS GLM notation in Table 6.34 can only be used for pairwise comparisons.[14]

The SAS GLM `contrast(1)` indicates that the first level of month is to be contrasted with each other level. So, for example, if you want to compare level 2 with level 3, you would need to change the number in parentheses to either 2 or 3. Complex user-specified contrasts—for example, (1 1 -2)—cannot be readily specified in SAS GLM. The user-specified contrast matrix in SPSS GLM is entered in syntax; it cannot be generated through the menu system.

Once you have obtained F for a comparison, whether by hand or computer, compare the obtained F with a critical F to see if there is a mean difference. If obtained F exceeds critical F, the null hypothesis of equality of means for the comparison is rejected. But which critical F is used depends on whether the comparison is planned or performed post hoc. If planned, the probability levels reported in the output are correct, and the comparison is tested against tabled F with 1 and $\text{df}_{A\text{comp} \times S}$. If post hoc, the probability levels in the output are incorrect; instead, Equation 4.13 is used to compute F_S: $(a - 1)F_{\text{df}A,\text{df}AS}$.[15]

Note that the regression approach of Table 6.8 has contrast coefficients built into each column. SS(reg.) for a column *is* the sum of squares (and mean square) for a comparison, whatever coefficients are used. In the example of Table 6.8, linear trend coefficients are used in the X_1 column, so SS(reg. X_1) is $\text{SS}_{\text{linear}}$ for A. However, this approach, in which totals are used in the X_3 column, does not automatically develop the special error term necessary to test the comparison. To do the comparison, you need to either expand the df for cases and develop the special error term by cross multiplying the columns for cases with the column for linear trend or develop the new DV values and submit them to routine regression analysis.

6.6.6.1.2 Marginal Comparisons in Factorial Repeated-Measures Designs Comparisons for main effects in factorial designs investigate differences in means at the margins of the data set. To perform this analysis by hand, you again have to develop the data set for the comparison, this time both by collapsing over the levels of the other IV(s) (NOVEL, in this example) and by using the contrast coefficients to

[13] This is why comparisons are considered to have $a = 2$, and sphericity is not an issue for them.

[14] Both programs can be run on a data set specifically developed for a comparison, as in Table 6.33.

[15] SPSS has a set of macros (available online at www.spss.com) that do LSD, Bonferroni, and Dunn-Sidak post hoc tests for a one-way repeated-measures design.

TABLE 6.34 Syntax and error terms for trend analyses and user-specified comparisons for one-way repeated-measures designs

Type of Comparison	Program	Syntax for One-Way Example	Name of Effect/ Name of Error in Output
Trend analysis	SPSS GLM (default)	GLM month_1 month_2 month_3 /WSFACTOR month 3 POLYNOMIAL /METHOD = SSTYPE(3) /CRITERIA = ALPHA(.05) /WSDESIGN = month.	MONTH Linear/ Error(MONTH) Linear
	SAS GLM	proc glm data=SASUSER.SS_RPTD; model MONTH1 MONTH2 MONTH3 = ; repeated MONTH 3 polynomial/ summary; run;	MONTH_1 MEAN/ Error
	SYSTAT GLM (default with PRINT = MEDIUM	GLM MODEL MONTH_1 MONTH_2 MONTH_3 = CONSTANT/ REPEAT =3,NAMES='MONTH' ESTIMATE	Polynomial Test of Order 1 (Linear)/ Error
User-specified coefficients	SPSS GLM	GLM month_1 month_2 month_3 /WSFACTOR month 3 SPECIAL (1 1 1 1 –1 0 1 1 –2) /METHOD = SSTYPE(3) /CRITERIA = ALPHA(.05) /WSDESIGN = month.	MONTH L1/ Error(MONTH) L1
	SAS GLM	proc glm data=SASUSER.SS_RPTD; model MONTH1 MONTH2 MONTH3 = ; repeated MONTH 3 contrast(1)/ summary; run;	MONTH_2 MEAN/ Error
	SYSTAT GLM	GLM MODEL MONTH_1 MONTH_2 MONTH_3 = CONSTANT/ REPEAT = 3, NAMES ='MONTH' ESTIMATE HYPOTHESIS WITHIN="MONTH" STANDARDIZE=WITHIN CONTRAST [1 -1 0] TEST	Test of Hypothesis/ Error

transform the DV scores. The new data set for the 1 −1 0 comparison on MONTH is in Table 6.35. Once the data set is developed, procedures of Section 6.4.1.3 are used to compute $MS_{A\text{comp}}$ and $MS_{A\text{comp} \times S}$ for the ANOVA .

To do either trend analysis or user-specified comparisons by computer, use the syntax in Table 6.36. SAS and SPSS develop the special error terms necessary for these comparisons. Trend analyses in these programs also include tests for trends of the interaction. Recall that SAS does not perform user-specified comparisons for repeated-measures effects, so that special data sets are necessary for each comparison.

TABLE 6.35 Development of data set for marginal comparison (1 −1 0) in two-way repeated-measures designs

	a_1: **Month 1**	a_2: **Month 2**	
Case	SciFi + Mystery	SciFi + Mystery	Case Totals
s_1	$1 + 5 = 6$	$3 + 4 = 7$	13
s_2	$1 + 8 = 9$	$4 + 8 = 12$	21
s_3	$3 + 4 = 7$	$3 + 5 = 8$	15
s_4	$5 + 3 = 8$	$5 + 2 = 7$	15
s_5	$2 + 5 = 7$	$4 + 6 = 10$	17

TABLE 6.36 Syntax and error terms for trend analyses and user-specified comparisons for factorial repeated-measures designs

Type of Comparison	**Program**	**Syntax for Two-Way Example**	**Name of Effect/Name of Error in Output**
Trend analysis	SPSS GLM (default)	GLM scifi_m1 scifi_m2 scifi_m3 mstry_m1 mstry_m2 mstry_m3 /WSFACTOR = novel 2 polynomial month 3 polynomial /MEASURE = books /METHOD = SSTYPE(3) /CRITERIA = ALPHA(.05) /WSDESIGN = novel month novel*month .	MONTH Linear/ Error(MONTH) Linear
	SAS GLM	proc glm data=SASUSER.SS_RPTD2; model SCIFIM1 SCIFIM2 SCIFIM3 MSTRYM1 MSTRYM2 MSTRYM3 = ; repeated NOVEL 2, MONTH 3 polynomial / summary; run;	MONTH_1 MEAN/ Error
	SYSTAT GLM (default with PRINT= MEDIUM)	GLM MODEL SCIFI_M1 SCIFI_M2 SCIFI_M3, MSTRY_M1 MSTRY_M2 MSTRY_M3 = , CONSTANT/ REPEAT =2,3, NAMES='NOVEL','MONTH' ESTIMATE	Polynomial Test of Order 1 (Linear) MONTH/ Error
User-specified coefficients	SPSS GLM	GLM scifi_m1 scifi_m2 scifi_m3 mstry_m1 mstry_m2 mstry_m3 /WSFACTOR = novel 2 month 3 special (1 1 1 1 −1 0 1 1 −2) /METHOD = SSTYPE(3) /CRITERIA = ALPHA(.05) /WSDESIGN = novel month novel*month .	MONTH L1/ Error(MONTH) L1

TABLE 6.36 Continued

Type of Comparison	Program	Syntax for Two-Way Example	Name of Effect/Name of Error in Output
	SYSTAT GLM	GLM MODEL SCIFI_M1 SCIFI_M2 SCIFI_M3, MSTRY_M1 MSTRY_M2 MSTRY_M3 = CONSTANT/ REPEAT =2,3, NAMES='NOVEL','MONTH' ESTIMATE HYPOTHESIS WITHIN="MONTH" STANDARDIZE=WITHIN CONTRAST [1 -1 0] TEST	Test of Hypothesis/ Error

6.6.6.2 Tests of Repeated-Measures Interactions: Simple-Effects Analysis

In simple-effects analysis, one IV is held constant at one level, while mean differences are examined between the levels of the other IV (Figure 5.8). For instance, only data from a_1, the first level of month, are used, and mean differences are examined in number of science fiction versus mystery novels read; the researcher asks whether there are differences in number of science fiction and mystery novels read in the first month by assessing $F_{B \text{ at } a1}$. The same analyses can be performed for months 2 ($F_{B \text{ at } a2}$) and 3 ($F_{B \text{ at } a3}$). Or, type of novel is held constant at science fiction, and mean differences are explored among months using $F_{A \text{ at } b1}$; the researcher asks whether there are mean differences in the three months when science fiction novels are read. The same analysis can be performed for mystery novels, using $F_{A \text{ at } b2}$. The test of the simple main effect of B (novels) has 1 df in the numerator of the F test because there are only two types of novels. The test of the simple main effect of A (month) has 2 df in the numerator of the F test because there are three months. Because month has 2 df, the simple main effects analysis could be followed by simple comparisons (see Figures 5.7 and 5.8). However, the guidelines of Section 5.6.4.4.1 are also appropriate for factorial repeated measures, because simple-effects analyses confound main effects with interaction.

Hand calculations for simple main effects and simple comparisons are straightforward—you do a one-way repeated-measures ANOVA on each "slice" of the design. For the example, Table 6.37 contains the data sets necessary (a) to test for mean differences between novels for the first month, (b) to test for mean differences between months for science fiction novels, and (c) to compare the first two months against the third for science fiction novels.

The easiest way to do simple main effects analyses by computer is to select just the DVs you need in order to create one-way repeated-measures syntax and then, for simple comparisons, add appropriate syntax. Table 6.38 shows syntax for the simple main effect of month for science fiction novels and the simple comparison through SPSS of

TABLE 6.37 Simple effects and simple comparisons by hand for factorial repeated-measures designs

(a) Data set for the simple main effect of novels at the first month

	Month 1		
Cases	Science Fiction	Mystery	Case Totals
s_1	1	5	6
s_2	1	8	9
s_3	3	4	7
s_4	5	3	8
s_5	2	5	7
Novel Totals	12	25	37

(b) Data set for the simple main effect of months for science fiction novels

		Science Fiction		
Case	Month 1	Month 2	Month 3	Case Totals
s_1	1	3	6	10
s_2	1	4	8	13
s_3	3	3	6	12
s_4	5	5	7	17
s_5	2	4	5	11
Month Totals	12	19	32	63

(c) Data set for the simple comparison of the first two months versus the third for science fiction novels

	Science Fiction		
Case	Month 1 + Month 2	2(Month 3)	Case Totals
s_1	$1 + 3 = 4$	$2(6) = 12$	16
s_2	$1 + 4 = 5$	$2(8) = 16$	21
s_3	$3 + 3 = 6$	$2(6) = 12$	18
s_4	$5 + 5 = 10$	$2(7) = 14$	24
s_5	$2 + 4 = 6$	$2(5) = 10$	16
Month Totals	$12 + 19 = 31$	$2(32) = 64$	95

the first two months versus the third for science fiction novels. The syntax and output are identical to that for a one-way repeated-measures design. The only difference is in selection of only some of the available DVs. For some user-specified comparisons, SAS GLM, as before, requires a special comparison data set for the simple comparison.

TABLE 6.38 Syntax for simple main effects and simple comparisons for factorial repeated-measures designs

Type of Comparison	Program	Syntax	Labels for Effect/Error
Simple main effect	SPSS GLM	GLM SCIFI_M1 SCIFI_M2 SCIFI_M3 /WSFACTOR = month 3 Polynomial /METHOD = SSTYPE(3) /CRITERIA = ALPHA(.05) /WSDESIGN = month .	month Linear/ Error(month) Linear
	SAS GLM	proc glm data=SASUSER.SS_RPTD2; model SCIFIM1 SCIFIM2 SCIFIM3 = ; repeated MONTH 3; run;	MONTH/ Error (MONTH)
	SYSTAT GLM	GLM MODEL SCIFI_M1 SCIFI_M2 SCIFI_M3 = CONSTANT/, REPEAT =3,NAMES='MONTH' ESTIMATE	MONTH/ Error
Simple comparison (For 1 1 –2)	SPSS GLM	GLM SCIFI_M1 SCIFI_M2 SCIFI_M3 /WSFACTOR = month 3 special (1 1 1 1 1 –2 1 –1 0) /METHOD = SSTYPE(3) /CRITERIA = ALPHA(.05) /WSDESIGN = month .	month L1/ Error(month) L1
	SYSTAT GLM	GLM MODEL SCIFI_M1 SCIFI_M2 SCIFI_M3 = CONSTANT/REPEAT=3, NAMES='MONTH' ESTIMATE HYPOTHESIS WITHIN="MONTH" STANDARDIZE=WITHIN CONTRAST [1 1 -2] TEST	Test of Hypothesis/ Error

Both SPSS and SAS test the effects as if they were planned. If these analyses are done post hoc, the printed significance levels are inappropriate, and Scheffé adjustment is needed to control for inflated familywise Type I error.

When there are three or more IVs, simple interactions, simple main effects, and simple comparisons are all possible. Researchers back a significant three-way interaction ($A \times B \times C$) into a series of simple two-way interactions ($MS_{B \times S \text{ at } a1}$, $MS_{B \times S \text{ at } a2}$, etc.), followed by simple main effects (followed, perhaps, by simple comparisons) analyses of those that have statistically significant mean differences.

6.6.6.3 Tests of Repeated-Measures Interactions: Interaction Contrasts

The second option for pinpointing the source of an interaction is interaction-contrasts analysis. As you recall, interaction contrasts are often more appropriate than simple main effects analysis in the presence of large or disordinal (crossing) effects, because only the sum of squares for omnibus interaction is partitioned. In interaction-contrasts analysis, the interaction between two IVs is examined through smaller interactions (cf. Figure 5.9), which are obtained by deleting levels of one or more of the IVs or by combining levels using appropriate weighting coefficients.

If contrast coefficients are applied to all the IVs (because all have more than two levels), the analysis is interaction contrasts. If contrasts are applied to some, but not all, of the IVs, the analysis is a partial factorial. Because novel has only two levels in this two-way example, the analysis is partial factorial. Linear trend analysis is chosen for the three months.

Interaction-contrasts analysis is performed by hand by creating the necessary data set and redoing the ANOVA. The appropriate data set for the partial factorial of novel with the linear trend of month is in Table 6.39. For the linear trend, the second level of month is simply deleted for both science fiction novels and mystery novels. Once the data are derived, a two-way, repeated-measures ANOVA is performed in which the interaction term is the desired partial factorial. Alternatively, the regression approach is available, in which SS(reg. X_4) provides the desired result, as shown in Table 6.23.

The syntax of Table 6.36 automatically produces the partial factorial of NOVEL $\times$ MONTH$_{\text{linear}}$ for SAS GLM and SPSS GLM. In SAS, the effect and its error are labeled `NOVEL_1*MONTH_1` (`MEAN` and `Error`).[16] In SPSS GLM, the effect is labeled novel*month Linear, with Error (novel*month Linear).

Effect size for the linear interaction contrast, relative to the full interaction, may be done as per Equation 5.29,

$$\eta^2 = \frac{SS_{A\text{comp1} \times B\text{comp2}}}{SS_{AB}} = \frac{57.8}{64.467} = .90$$

TABLE 6.39 Hypothetical data set for the partial factorial of novel with month$_{\text{linear}}$.

	b_1: Science Fiction		b_2: Mystery		
Case	a_1: Month 1	a_3: Month 3	a_1: Month 1	a_3: Month 3	Case Totals
s_1	1	6	5	1	$S_1 = 13$
s_2	1	8	8	4	$S_2 = 21$
s_3	3	6	4	3	$S_3 = 16$
s_4	5	7	3	0	$S_4 = 15$
s_5	2	5	5	3	$S_5 = 15$
Treatment Totals	$A_1B_1 = 12$	$A_3B_1 = 32$	$A_1B_2 = 25$	$A_3B_2 = 11$	$T = 80$

[16] Recall that contrasts in SAS GLM for repeated-measures designs are restricted to built-in comparisons, which include trend analysis. There is no provision for user-specified comparisons unless separate data sets developed for the comparison are used.

Thus, 90% of the AB interaction is attributable to its linear trend. This effect size and its confidence interval can be found by using Smithson's (2003) software for SPSS and SAS.

Interaction contrasts may be done as part of a planned analysis, replacing the omnibus test of the interaction, or as a post hoc analysis to dissect a significant interaction that is observed. The Scheffé adjustment for post hoc analysis for inflated Type I error rate is

$$F_S = (a - 1)(b - 1)F_{(a-1)(b-1),(b-1)(s-1)} \tag{6.28}$$

In this case,

$$F_S = (3 - 1)(2 - 1)(4.46) = 8.92$$

The obtained F of 25.13 exceeds critical F_c, whether planned ($F_{1,8} = 5.32$) or post hoc ($F_S = 8.92$), so there is an interaction of linear trend with novel—that is, the linear trend over months is different for science fiction and mystery novels. Interpretation of the partial interaction is most easily achieved through a plot of the means, which is somewhat difficult to achieve for repeated-measures designs. The relevant plot is produced by SPSS GLM (edited), as shown in Figure 6.3. The number of science fiction novels read increases steadily over the three months (positive slope), while the number of mystery novels read decreases over the three months (negative slope). The plot also suggests a quadratic trend of the interaction—that is, the curves for the two types of novels have different shapes.

FIGURE 6.3 SPSS GLM syntax and (edited) profile plot of means over three months for science fiction and mystery novels

```
GLM
 SCIFI_M1 SCIFI_M2 SCIFI_M3 MSTRY_M1 MSTRY_M2 MSTRY_M3
 /WSFACTOR = novel 2 Polynomial month 3 Polynomial
 /MEASURE = books
 /METHOD = SSTYPE(3)
 /PLOT = PROFILE( month*novel )
 /CRITERIA = ALPHA(.05)
 /WSDESIGN = novel month novel*month .
```

Profile Plots

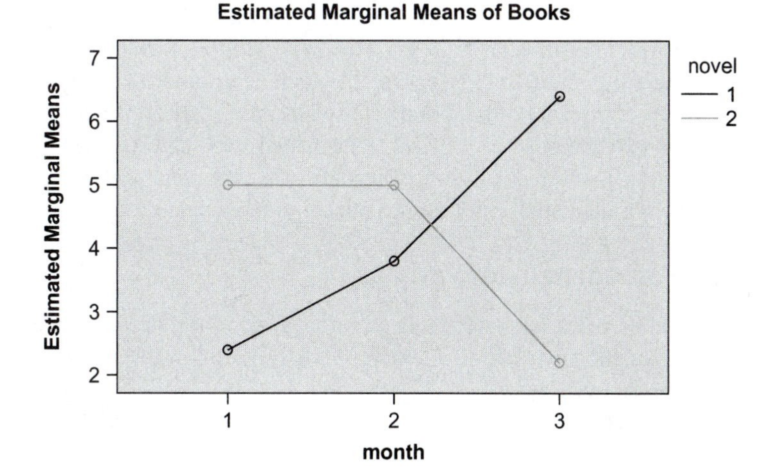

6.7 COMPLETE EXAMPLE OF TWO-WAY REPEATED-MEASURES ANOVA

Data for this example were extracted from a study by Damos (1989) in which 20 participants were required to identify figures under various angles of rotation (0°, 60°, 120°, and 180°), each in standard and mirror-image orientation. The DV selected for use here is reaction time, and data selected are from the last set of 20 sessions run over a period of two weeks. Thus, there are two repeated-measures IVs: orientation with two levels (standard and mirror) and rotation with four levels. Previous research had shown a linear trend of increasing reaction time the greater the angle of rotation. Questions to be addressed in the current research are as follows: (1) Are there trends other than linear? (2) Are there differences associated with mirror-image as opposed to standard orientation? (3) Are there differences in trend over angle of rotation for standard and mirror-image orientation—that is, what is the trend of the rotation by orientation interaction? Data files are REACTIME.*.

6.7.1 Evaluation of Assumptions

There are no missing data in this set. Therefore, the issues to be addressed are normality of sampling distributions, homogeneity of variance, sphericity, and absence of outliers.

6.7.1.1 Normality of Sampling Distributions

The error term for the main effect of rotation (the rotation by participation interaction) has df $= (b - 1)(s - 1) = (4 - 1)(20 - 1) = 57$, which is sufficient to assume normality for purposes of ANOVA. Likewise, the interaction error term (orientation by rotation by participant interaction) has df $= (a - 1)(b - 1)(s - 1) = (2 - 1)(4 - 1)(20 - 1) = 57$. The error term for the main effect of orientation (standard vs. mirror) has df $= (a - 1)(s - 1) = (2 - 1)(20 - 1) = 19$, which is close enough to the criterion of 20 to be acceptable, as long as other assumptions are met.

The planned trend analysis reduces error df for each trend comparison to 19 as well, because each trend is evaluated with df $= (a - 1)(1)(s - 1) = (2 - 1)(1)(20 - 1)$ when df for B is reduced to 1 for each comparison. Again, this is not of concern unless other assumptions appear to be violated.

6.7.1.2 Homogeneity of Variance

Ratios of variances for this factorial design are checked for cells representing the orientation by rotation interaction. SAS MEANS is used to print out a full set of cell descriptive statistics.

Table 6.40 shows syntax and output from the SAS MEANS run. The `vardef=df` statement instructs SAS MEANS to calculate variances and standard deviations using

TABLE 6.40 Descriptive statistics for two-way repeated ANOVA (SAS DATA and MEANS syntax and output)

```
proc means data=SASUSER.REACTIME vardef=df
         N MIN MAX MEAN STD VAR SKEWNESS KURTOSIS;
     var SO S60 S120 S180 MO M60 M120 M180;
run;
```

Variable	N	Minimum	Maximum	Mean	Std Dev	Variance	Skewness
SO	20	419.0400000	550.9100000	482.9825000	41.5966443	1730.28	0.2594316
S60	20	418.8000000	539.3600000	484.0725000	36.2295931	1312.58	0.0122305
S120	20	432.3500000	693.6700000	506.1235000	58.8820937	3467.10	1.6999893
S180	20	422.8400000	770.8000000	546.4735000	80.8858206	6542.52	1.2617860
MO	20	440.5100000	592.3900000	508.9835000	53.1946684	2829.67	0.1993758
M60	20	435.4200000	581.9700000	511.3695000	44.6091112	1989.97	-0.0246631
M120	20	430.7400000	748.6800000	523.8220000	73.6641852	5426.41	1.4852998
M180	20	423.8000000	890.0600000	546.4510000	99.4082426	9882.00	2.2065031

Variable	Kurtosis
SO	-1.3603214
S60	-1.1656553
S120	4.4887456
S180	2.5805090
MO	-1.6264543
M60	-1.0082450
M120	3.4826521
M180	7.1502194

df rather than n in the denominator (cf. Equations 2.1 and 2.2). Statistics requested are sample size (`N`), minimum, maximum, standard deviation (`Std Dev`), variance, skewness, and kurtosis. Several of these are also useful for evaluating outliers.

A glance at the `Variance` column assures us that there is no cause for concern about heterogeneity of variance. The ratio of largest to smallest variance, $F_{max} = 9882.00/1312.58 = 7.53$, is less than 10.

6.7.1.3 Sphericity

We are interested only in trend analysis in this study, both of the main effect of rotation and of the rotation by orientation interaction. Therefore, the outcome of a test of sphericity for the omnibus effects is irrelevant. This is fortunate, because the nature of the rotation IV is such that we might expect violation of sphericity (the greater the difference in angles of rotation, the lower the expected correlation).

6.7.1.4 Outliers

Table 6.40 provides information to evaluate the presence of outlying scores on the variables used for the tests. Scores are within 3.3 standard deviations of their means, with one exception—the maximum score for mirror-image orientation with 180° rotation has $z = (890.060 - 546.451)/99.408 = 3.46$. A look at all the variables, however, indicates fairly general positive skewness, as well as kurtosis, and the nature of the DV, reaction time, is such that a transformation is usually applied. There is a physiological lower limit to reaction time, but very long ones can be produced by a variety of influences, many of which are unrelated to the cognitive processing strategies of interest in this research. Therefore, the decision is made to apply a square root (minimal) transform to all scores.

Table 6.41 shows SAS DATA syntax to perform the transformation and create a file, `REACSQRT`, that contains the transformed, as well as the original, data. Table 6.41 also shows `means` syntax and output for the transformed data. This transformation successfully pulls in the outlier and reduces skewness and kurtosis in the entire data set.

6.7.2 Planned Trend Analysis of Two-Way Repeated-Measures Design

Table 6.42 shows SAS GLM syntax and highly selected output for the planned trend analysis. Trend analysis is requested in the `polynomial` instruction, along with `summary` to display the results of the analysis. The `nom` instruction suppresses multivariate output.

The output shown in this table (much has been omitted) begins with a translation of the variables that are used to form levels of the two IVs, reassuring us that our syntax correctly identified them. All output relating to individual tests of DVs, as well as the omnibus ANOVA, are omitted here. Marginal and cell means and standard deviations are shown at the end of the output. Table 6.42 first shows the test of type of orientation (`type_1`), which is solely a linear test, because there are only two levels:

TABLE 6.41 Descriptive statistics for square root transformation of repeated-measures data (SAS DATA and MEANS syntax and output)

```
data SASUSER.REACSQRT;
   set SASUSER.REACTIME;
   S_S0 = SQRT(S0);
   S_S60 = SQRT(S60);
   S_S120 = SQRT(S120);
   S_S180 = SQRT(S180);
   S_M0 = SQRT(M0);
   S_M60 = SQRT(M60);
   S_M120 = SQRT(M120);
   S_M180 = SQRT(M180);
run;
proc means data=SASUSER.REACSQRT vardef=df
         N MIN MAX MEAN STD VAR SKEWNESS KURTOSIS;
     var S_S0 S_S60 S_S120 S_S180 S_M0 S_M60 S_M120 S_M180;
run;
```

Variable	N	Minimum	Maximum	Mean	Std Dev	Variance	Skewness
S_S0	20	20.4704665	23.4714720	21.9576297	0.9431169	0.8894694	0.2170095
S_S60	20	20.4646036	23.2241254	21.9869736	0.8242959	0.6794637	-0.0377583
S_S120	20	20.7930277	26.3376157	22.4635189	1.2623367	1.5934938	1.4753388
S_S180	20	20.5630737	27.7632851	23.3195153	1.6776252	2.8144263	0.9994462
S_M0	20	20.9883301	24.3390633	22.5315384	1.1757543	1.3823981	0.1642077
S_M60	20	20.8667199	24.1240544	22.5929337	0.9888039	0.9777332	-0.0918186
S_M120	20	20.7542767	27.3620175	22.8370978	1.5522365	2.4094382	1.2347343
S_M180	20	20.5864033	29.8338734	23.2957358	1.9893652	3.9575740	1.8231569

Variable	Kurtosis
S_S0	-1.3601786
S_S60	-1.1264405
S_S120	3.5694429
S_S180	1.9965889
S_M0	-1.6503563
S_M60	-0.9522283
S_M120	2.5758677
S_M180	5.4112779

TABLE 6.42 Trend analysis of two-way repeated design (SAS GLM syntax and selected output)

```
proc glm data=SASUSER.REACSQRT;
  model S_s0 S_s60 S_s120 S_s180 S_m0 S_m60 S_m120 S_m180 =;
  repeated type 2, angle 4 polynomial / mean summary nom;
run;
```

The GLM Procedure
Repeated Measures Analysis of Variance

Repeated Measures Level Information

Dependent Variable	S_S0	S_S60	S_S120	S_S180	S_M0	S_M60	S_M120	S_M180
Level of type	1	1	1	1	2	2	2	2
Level of angle	1	2	3	4	1	2	3	4

The GLM Procedure
Repeated Measures Analysis of Variance
Analysis of Variance of Contrast Variables

type_N represents the contrast between the nth level of type and the last

Contrast Variable: type_1

Source	DF	Type III SS	Mean Square	F Value	Pr > F
Mean	1	46.79768435	46.79768435	9.91	0.0053
Error	19	89.68845068	4.72044477		

angle_N represents the nth degree polynomial contrast for angle

Contrast Variable: angle_1

Source	DF	Type III SS	Mean Square	F Value	Pr > F
Mean	1	50.39520738	50.39520738	11.47	0.0031
Error	19	83.50809801	4.39516305		

Contrast Variable: angle_2

Source	DF	Type III SS	Mean Square	F Value	Pr > F
Mean	1	7.48959774	7.48959774	11.10	0.0035
Error	19	12.81538178	0.67449378		

Contrast Variable: angle_3

Source	DF	Type III SS	Mean Square	F Value	Pr > F
Mean	1	0.00129924	0.00129924	0.00	0.9585
Error	19	8.89360669	0.46808456		

Contrast Variable: type_1*angle_1

Source	DF	Type III SS	Mean Square	F Value	Pr > F
Mean	1	4.10243072	4.10243072	10.49	0.0043
Error	19	7.43242632	0.39118033		

Contrast Variable: type_1*angle_2

Source	DF	Type III SS	Mean Square	F Value	Pr > F
Mean	1	0.92196416	0.92196416	4.81	0.0410
Error	19	3.64387122	0.19178270		

Contrast Variable: type_1*angle_3

Source	DF	Type III SS	Mean Square	F Value	Pr > F
Mean	1	0.00989138	0.00989138	0.10	0.7544
Error	19	1.86472067	0.09814319		

Means of Within Subjects Effects

Level of type	N	Mean	Std Dev
1	80	22.43190940	1.32069823
2	80	22.81432640	1.47999689

Level of angle	N	Mean	Std Dev
1	40	22.24458406	1.09144922
2	40	22.28995367	0.94947512
3	40	22.65030832	1.40923074
4	40	23.30762554	1.81640431

Level of type	Level of angle	N	Mean	Std Dev
1	1	20	21.95762975	0.94311688
1	2	20	21.98697363	0.82429586
1	3	20	22.46351889	1.26233666
1	4	20	23.31951533	1.67762519
2	1	20	22.53153838	1.17575427
2	2	20	22.59293371	0.98880394
2	3	20	22.83709775	1.55223650
2	4	20	23.29573576	1.98936522

standard and mirror. This is followed by the test of the three polynomials for angle: linear = `angle_1`, quadratic = `angle_2`, and cubic = `angle_3`. Finally, there are the tests of the three trends of the interaction: `type_1*angle_1` = type by linear trend of angle of rotation, and so on.

Table 6.42 shows that two of the three trends of the interaction are statistically significant at $\alpha = .05$. Smithson's (2003) procedure is used to find effect sizes and confidence limits for all three trends. Table 6.43 shows edited SAS syntax (NONCF2.SAS) with values from Table 6.42 in gray and output for this run.

Effect size for the linear trend is strong—partial $\eta^2 = .36$, with confidence limits from .04 to .58. However, the quadratic trend of the interaction is problematic; although partial $\eta^2 = .20$, the confidence limit descends to 0, with an upper limit of .46. Little, if any, systematic variance is associated with the cubic trend, with partial $\eta^2 = .01$ and limits from 0 to .18.

Parameter estimates for the interaction are cell means and standard errors, produced in Table 6.44 from data in Table 6.41. In each case, the standard error of the mean is the standard deviation divided by the square root of n (i.e., $\sqrt{20} = 4.472$).

The interaction is interpreted by plotting the cell means of Table 6.42 in Figure 6.4. This is most easily done in a spreadsheet or word processing program, because the absence of randomized-groups variables makes plotting difficult except for SPSS GLM.

The plot shows that the linear trend of the interaction has a shallower slope for mirror orientation trials; that is, the angle of rotation has less influence on (square root of) reaction time with mirror-image than with standard trials. Thus, the difference between standard and mirror-image trials diminishes with greater angles of rotation. The quadratic trend of the interaction shows that the greater acceleration for standard trials is only for angles of rotation greater than 60°. In other words, curves are parallel up to 60°; increase in reaction time between 0° and 60° is the same for standard and mirror-image trials.

There is also a statistically significant difference in square root of reaction time due to standard versus mirror-image orientation—$F(1, 19) = 9.91, p = .0053$. However, interpretation of the main effect is problematic due to the presence of strong

TABLE 6.44 Mean square root of reaction time as a function of type of presentation and angle of rotation

	Angle of Rotation			
Type	0°	60°	120°	180°
Standard				
M	21.96	21.99	22.46	23.32
SE_M	0.21	0.18	0.28	0.38
Mirror				
M	22.53	22.59	22.84	23.30
SE_M	0.26	0.22	0.35	0.44

TABLE 6.43 Syntax and output from NONCF2.SAS for effect size (`rsq`) with 95% confidence limits (`rsqlow` and `rsqupp`) for trend analysis

```
.
.
.
    rsq = df1 * F / (df2 + df1 * F);
    rsqlow = ncplow / (ncplow + df1 + df2 + 1);
    rsqupp = ncpupp / (ncpupp + df1 + df2 + 1);
    cards;
10.49  1 19 .95
4.81   1 19 .95
0.10   1 19 .95

;
proc print;
run;
```

Obs	F	df1	df2	conf	prlow	prupp	ncplow	ncpupp	rsq	rsqlow	rsqupp
1	10.49	1	19	0.95	0.975	0.025	0.98594	29.3346	0.35571	0.044844	0.58279
2	4.81	1	19	0.95	0.975	0.025	0.00000	18.0316	0.20202	0.000000	0.46197
3	0.10	1	19	0.95	0.975	0.025	0.00000	4.7232	0.00524	0.000000	0.18362

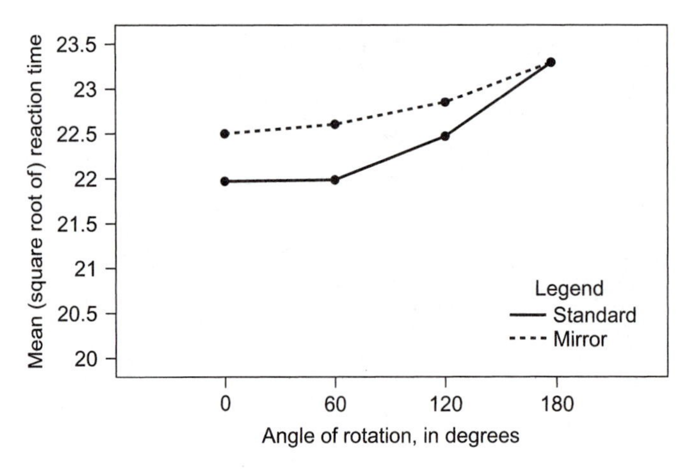

FIGURE 6.4 Mean square root of reaction time as a function of angle of rotation and standard versus mirror orientation (Graph produced in Quattro Pro.)

interaction. If main effects are to be interpreted, marginal means are found in Table 6.42: 22.432 for standard orientation and 22.814 for mirror orientation.

Angle of rotation shows the expected significant linear trend (`angle_1`), with $F(1, 19) = 11.10, p = .0031$. There is also a significant quadratic trend (`angle_2`), with $F(1, 19) = 11.47, p = .0035$. No significant cubic trend is evident, $p = .9585$. Again, interpretation of trends in angle of rotation is problematic in the face of significant trends of the interaction.

Table 6.45 is a checklist for factorial repeated-measures ANOVA, followed by an example of a results section, in journal format, for the study just described.

TABLE 6.45 Checklist for factorial repeated-measures ANOVA.

1. Issues
   a. Missing data
   b. Normality of sampling distributions
   c. Homogeneity of variance
   d. Sphericity
   e. Outliers
2. Major analyses
   a. Main effects or planned comparisons. If statistically significant:
      (1) Effect size(s) and their confidence limits
      (2) Parameter estimates (marginal means and standard errors or standard deviations or confidence intervals)
   b. Interactions or planned comparisons. If significant:
      (1) Effect size(s) and their confidence limits
      (2) Parameter estimates (cell means in table and/or plot)
3. Additional analyses
   a. Post hoc comparisons
   b. Interpretation of departure from homogeneity of variance

Results

A 2 × 4 repeated-measures design was analyzed using planned trend analysis on reaction time for 20 participants. Factors were two types of orientation (standard and mirror image) and four angles of rotation (0°, 60°, 120°, and 180°). Reaction times were positively skewed, and an outlying score was noted among the mirror-image trials at 180°. A square root transformation of reaction times normalized the distributions adequately, and the extreme score was no longer an outlier.

Cell means are shown in Table 6.44 (with standard errors) and are plotted in Figure 6.4. Statistically significant effects included the main effect of type of orientation, linear and quadratic trends of angle of rotation, and the linear and quadratic trends of the interaction. Interpretation, therefore, is limited to interaction.

Increase in square root of reaction time as a function of increasing angle of rotation was greater for mirror-image than for standard trials, producing a statistically significant linear trend of the interaction, $F(1, 19) = 10.49$, $p = .0043$, partial $\eta^2 = .36$, with confidence limits from .04 to .58. Thus, the greater the angle of rotation, the less the advantage of standard orientation. The quadratic trend of the interaction was also statistically significant, $F(1, 19) = 4.81$, $p = .041$, partial $\eta^2 = .20$, with confidence limits from 0 to .46, suggesting that the linear trend of the interaction is limited to trials with greater angles of rotation. The cubic trend of the interaction was not statistically significant, $F < 1$, partial $\eta^2 = .01$, with confidence limits from 0 to .18. The difference between 0° and 60° angles is the same for standard and mirror-image trials.

6.8 COMPARISON OF PROGRAMS

The programs reviewed in Table 5.52 handle repeated-measures ANOVA. Table 6.46 summarizes features of the programs that are unique to repeated-measures designs. Other features are reviewed elsewhere, particularly in Sections 4.8 and 5.8.

6.8.1 SPSS Package

GLM is a program in the SPSS package for repeated-measures ANOVA. The Windows menu system is especially easy to use for these analyses. The program has

TABLE 6.46 Comparison of SPSS, SYSTAT, and SAS programs for repeated-measures ANOVA

Feature	SPSS GLM	SAS GLM	SAS MIXED[a]	SPSS MIXED MODELS	SYSTAT ANOVA and GLM
Input					
Special specification for trend (polynomial) analysis	Polynomial (default)	Yes	No	No	PRINT= MEDIUM
Simple syntax to specify unequal spacing of levels for trend analysis	Yes	Yes	No	No	Yes
Specify error terms for effects and/or contrasts	Yes	Yes	N/A	N/A	Yes
Uses maximum likelihood methods rather than ANOVA	No	No	Yes	Yes	No
Specify structure of variance-covariance matrix	No	No	Yes	Yes	No
Handles missing data without deleting cases	No	No	Yes	Yes	No
Specify exact multivariate tests for repeated-measures effects	No	MSTAT= EXACT	N/A	N/A	No
Requires data set format with one DV value per line.	No	Yes	No	Yes	No
Output					
Standard ANOVA source table	Yes	Yes	No	No	Yes
Mauchly test of sphericity	Yes	PRINTE	No	No	No
Greenhouse-Geisser and Huynh-Feldt epsilon	Yes	Yes	N/A	N/A	Yes
Lower-bound epsilon	Yes	No	N/A	N/A	No
Epsilon-adjusted df	Yes	No	N/A	N/A	No
Greenhouse-Geisser and Huynh-Feldt probabilities	Yes	Yes	N/A	N/A	Yes
Multivariate tests of repeated-measures effects	Yes	Yes	No	No	Yes
Printed residuals and predicted values	Yes	Yes	PREDICTED	No	No
Residuals and predicted values saved to file	Yes	Yes	Yes	Yes	Yes
Plots					
Plots of residuals against predicted values[b]	Yes	No	No	No	Yes
Profile plots of means	Yes	No	No	No	No

[a]Additional features described in Section 11.8.

[b]Saved values may be plotted through other programs in package.

all the popular features for repeated-measures designs, with Mauchly's sphericity test, trend analysis, and multivariate tests printed out by default. Also printed are details for four univariate tests for each effect: sphericity assumed, Greenhouse-Geisser adjusted, Huynh-Feldt adjusted, and with a lower-bound adjustment (usually considered far too conservative), although a simpler output format is available if requested.

Residuals may be printed or saved to file, and a residuals plot is available by request. Section 5.7 shows additional features common to SPSS GLM analyses, such as effect size, observed power, and user-specified contrasts.

SPSS MIXED MODELS is available as an alternative to repeated-measures ANOVA in the event that sphericity is violated and the structure of the variance-covariance matrix either is known or assumed to be related to distance between levels of the IV (cf. Section 6.6.2). This strategy requires data to be entered in a format different from that used by other programs: Each line has a column to identify the case, a column for each repeated-measures IV to identify its level, and a single column to identify the DV value for that combination of IV levels.

6.8.2 SAS System

SAS does repeated-measures ANOVA through PROC GLM or ANOVA. The two procedures work identically for repeated-measures designs, as described in this chapter. Multivariate output is produced by default, but trend analysis requires additional syntax. Exact tests of multivariate effects may be specified instead of the usual F approximations. Sufficient information is available to evaluate sphericity and adjust for its failure, particularly if Mauchly's test is requested. SAS does only limited user-specified comparisons for repeated-measures designs; others require a special data set developed for each comparison.

SAS MIXED is also available as an alternative to repeated-measures ANOVA in the event that sphericity is violated and the structure of the variance-covariance matrix either is known or can be assumed to be related to distance between levels of the IV (cf. Section 6.6.2). Use of this program requires data to be in the "one-line-per-response" format, as per SPSS MIXED MODELS.

6.8.3 SYSTAT System

SYSTAT GLM is a good, easy-to-use repeated-measures ANOVA program, particularly if trend analysis is to be used with unequal spacing of levels. Mauchly's sphericity test is not available, but Greenhouse-Geisser and Huynh-Feldt epsilon values and adjusted probabilities are shown. If these probabilities differ from the unadjusted probabilities, steps can be taken to deal with violation of sphericity (Section 6.6.2). A request for medium-level printing of output produces trend analysis, in addition to the usual univariate repeated-measures ANOVA and multivariate output. Plots of residuals versus predicted values are shown by default.

6.9 PROBLEM SETS

6.1 Put together the data set of Table 6.17 for testing the normality assumption as per Table 6.1.

6.2 Use computer software to do an analysis of variance, including trend analysis, of the following data set. Assume the IV is trials, and the DV is time to ski a run.

Trial			
1	2	3	4
11	7	5	4
12	8	6	6
9	8	5	5
8	5	3	2
13	8	9	6
9	6	6	4
10	5	5	3
7	3	1	0
10	5	5	0
4	3	2	1

a. Test the assumptions of normality, homogeneity of variance, and sphericity.

b. Check for outliers.

c. Calculate effect size using a form of η^2 appropriate given the results of the sphericity test.

d. Write a results section assuming the trend analysis was planned.

6.3 Reanalyze the data (either by computer or by hand) in Table 6.17 (two-way repeated measures) as a two-way randomized-groups design, and then compare values.

6.4 Do a two-way repeated-measures ANOVA for the following matched randomized-blocks design. *A* comprises two kinds of altimeters (digital and analog). *B* is position (high, low, right, left). The DV is accuracy of altimeter reading.

	***A*: Altimeter Type**							
	a_1: Digital				a_2: Analog			
B: Position	High	Low	Left	Right	High	Low	Left	Right
	84	45	47	90	80	49	45	73
	69	28	12	68	33	43	18	46
	99	40	35	71	36	22	56	38
	70	35	29	59	51	20	42	50
	91	18	19	85	37	44	38	39

a. Test appropriate assumptions for factorial repeated-measures ANOVA.
b. Calculate effect size(s).
c. Do any post hoc analyses you feel are appropriate.
d. Write a results section.

6.5 Complete Problem 6.4 using the regression strategy of Table 6.23.

6.6 What is the relative efficiency of the analysis of Problem 6.2 of the repeated-measures design compared with a randomized-groups design?

Extending the Concepts

6.7 Develop the variance sources, df, and error terms for a three-way repeated-measures design.

7 Mixed Randomized-Repeated Designs

7.1 GENERAL PURPOSE AND DESCRIPTION

In this popular design, there are different cases at different levels of one (or more) randomized-groups IV(s), and each case is measured repeatedly on one (or more) repeated-measures IV(s). The simplest mixed design is a two-way factorial with one IV of each type. For example, a marketing firm introduces a new product with several different promotional schemes in different cities and tracks sales over the first six months. The different promotional schemes, randomly assigned to cities, are the levels of the randomized-groups IV, sales districts within cities are the cases, and months are the levels of the repeated-measures IV. The DV is sales figures. As another example, a pharmaceutical company tests the effectiveness over several hours of a new over-the-counter analgesic for patients with three different types of pain (arthritic, headache, and lower back). Type of pain is the randomized-group IV, trials (time) is the repeated measure, and patients are the cases. The DV is pain relief. This design is sometimes called a *split-plot design*[1] due to its origins in an agricultural setting.

[1] The design is also sometimes referred to simply as *repeated measures,* even though there are one or more randomized-groups IVs.

A piece of land was split into several different plots (strips) for treatment with different fertilizers (or whatever), and repeated measures were made along the length of each strip.

This design provides an analysis of the effects of each IV separately and of their interaction(s). There is no limit (beyond the endurance of the researcher) to the number of IVs of each type that are included in the design.[2] However, if the study is an experiment, at least one of the IVs is manipulated. In the marketing example, promotional schemes are manipulated and randomly assigned. In the pharmaceutical example, the randomized-groups IV (type of pain) is a selected characteristic of the cases, and the repeated measure is trials; neither is manipulated.

As in purely repeated-measures designs, the repeated-measures IVs can be of a type where the levels (or combination of levels) are in a predetermined order (trials/time, distance) or where the order in which levels are presented is arbitrary. Control of extraneous variables over the time period is essential for either type of IV. However, additional controls are possible when the order of presentation of the repeated measure is flexible. These controls are the subject of Chapter 9.

There are two major advantages to this type of design. The first is the test of the generalizability of the repeated-measures IV(s) over the levels of the randomized-groups IV(s) when the randomized-groups IV is a selected characteristic of cases. If a repeated-measures IV produces a different pattern of results at different levels of the randomized-groups IV, there is significant interaction and knowledge that the effects of the repeated-measures IV apply differently to different groups. For example, does the new analgesic show the same pattern of pain relief over time for those with arthritic pain, headache pain, and lower-back pain (no interaction), or is there a different pattern of relief over time for different types of pain (interaction)?

The second advantage is the (usually) increased power due to the (usually) smaller error terms associated with the repeated-measures segment of the design. As in the purely repeated-measures design, differences among cases are assessed and removed from the repeated-measures error terms.[3] Because these error terms are used with both the repeated-measures IV(s) and the interaction(s) with repeated-measures, power for these effects is increased.

A disadvantage of the design is addition of the complexities of repeated measures to those of randomized groups. For instance, the design has all the assumptions of randomized-groups analyses, plus the assumptions of repeated-measures analyses (e.g., sphericity). It also has all the complexity of comparisons in randomized groups, plus the complexity of comparisons in repeated measures (where separate error terms are used). For example, some simple main effects are randomized groups, whereas other simple main effects are repeated measures.

[2] One of our Ph.D. dissertations had two randomized-groups and three repeated-measures IVs, and this was before computerized statistical software!

[3] The sum of squares for differences among cases is then partitioned into the randomized-groups effects and their error term.

7.2 KINDS OF RESEARCH QUESTIONS

7.2.1 Effects of the IVs

Are the means different at different levels of the randomized-groups IV(s)? At different levels of the repeated-measures IV(s)? Do the different promotional schemes produce different average sales? Are there mean differences over the six months of the promotion? Are there mean differences in pain relief associated with arthritic pain, headache pain, and lower-back pain? Are pain assessments different over the several hours of measurement? These questions are answered in the usual ANOVA fashion by applying F tests to the data, as demonstrated in Section 7.4.

7.2.2 Effects of Interactions among IVs

Does the effect of one IV depend on the level of the other IV? For example, does one promotional scheme work well in the early months, whereas another works later or not at all? Is the pattern of pain relief over hours the same for the different types of pain? For example, does the new analgesic work more rapidly for headache pain than for arthritic pain or lower-back pain? F tests to answer these questions are in Section 7.4.

7.2.3 Parameter Estimates

What means are anticipated in the population for each level of the statistically significant effects in the design? What sales are predicted for each promotional scheme over the months? What average pain relief is predicted for each type of pain over the hours of the study? These are calculated as in any factorial ANOVA, as described in Section 5.2.4.

7.2.4 Effect Sizes

How much of the variability in the DV is associated with each statistically significant effect in the design? How much of the variability in sales is associated with type of sales promotion? With months? With the interaction between type of sales promotions and months? How much of the variability in pain relief is associated with type of pain? With hours? With their interaction? Section 5.6.2 describes ω^2 (Equation 5.14) and partial ω^2 (Equation 5.15) for factorial designs. These equations are appropriate for mixed randomized-repeated designs with the use of the proper error term, as described in Section 7.4.

7.2.5 Power

Are there enough cases to reveal differences when the differences are real? Are enough sales districts included in each city to reveal real differences in promotional schemes? Are enough cases included for each type of pain to show that the analgesic really works better for one type of pain than for another? Are consistent individual differences among cases strong enough to counteract the decrease in error df? The basic procedures described in Section 5.6.3 for calculating power and determining sample size apply.

7.2.6 Specific Comparisons

If an IV of either type has more than two levels, which levels are different from which other levels? Which sales promotion scheme is most effective? Is it more effective than the next most effective one? When are sales at their peak? Does the analgesic work better for one type of pain than for another? Is there a linear or quadratic trend to pain relief over hours? Sections 7.6.1–7.6.5 describe procedures for specific comparisons in mixed randomized-repeated designs.

7.3 ASSUMPTIONS AND LIMITATIONS

7.3.1 Theoretical Issues

The same issues of causal inference and generalizability accrue to mixed designs as to randomized-groups or repeated-measures designs. If causal inference is intended, the levels of an IV are manipulated, with random assignment of cases and control of extraneous variables. Similarly, results generalize to the kinds of cases included in the design, although, as mentioned before, the generalizability of various effects is often tested directly in this design with the use of selected characteristics of cases to compose (one or more of) the randomized-groups IV(s). For instance, suppose levels of A represent items or people of different age (young and old), and levels of B represent tests at two different times (early and late). A statistically significant interaction implies that young and old cases show a different pattern of responses over the two time periods. Or, if levels of B represent two different doses of drugs, the interaction reveals that young and old cases have different reactions to the two doses. If cases are people, this result implies that prescribing should be contingent on age.

7.3.2 Practical Issues

7.3.2.1 Normality of Sampling Distributions

In the mixed design, normality of sampling distributions for the randomized-groups IV(s) applies to case averages (where scores have been combined over the levels of the repeated measure). For the randomized-groups IV(s), this requirement is satisfied if (1) there are roughly equal sample sizes in each level (or combination of levels) of the randomized-groups IV(s), (2) there are approximately 20 df for the randomized-groups error term, and (3) two-tailed tests are used. If these criteria are not met, follow the procedures for randomized-groups designs as described in Chapters 4 and 5.

Criteria for the repeated-measures segments of the design are as indicated in Chapter 6. Recall that you have to generate the necessary DV scores if a test of their skewness is required. With marked skewness and inadequate df for error, data transformation is often desirable. There are no nonparametric alternatives for analysis of a mixed design.

7.3.2.2 Homogeneity of Variance

Homogeneity of variance in the mixed design is required at each level (or combination of levels) of the randomized-groups IV(s). Follow the procedures discussed in Chapters 4 (Sections 4.3.2.2 and 4.5.5) and 5 (Sections 5.3.2.2 and 5.8.1.2) for assessing homogeneity of variance and for dealing with its violation for the randomized-groups portion of the design. Homogeneity of variance for the repeated-measures portion of the design is handled as described in Section 6.3.2.2.

7.3.2.3 Independence of Errors, Additivity, and Sphericity

Independence of errors is required for the randomized-groups error term, as discussed in Section 4.3.2.3. However, each repeated-measures error term confounds random variability with the cases by effects interaction, so assumptions of independence of error and additivity are intertwined. The requirement is for sphericity (as discussed in Sections 6.3.2.3 and 6.6.2), and data are tested through Mauchly's test, as in purely repeated-measures designs.[4]

By request, SPSS GLM and SAS GLM can provide Mauchly's test for each repeated-measures effect in the design. If the test fails (i.e., there is heterogeneity) for one or more effect, the usual three alternatives are available for analysis of that effect: single df comparisons, Huynh-Feldt corrections, or multivariate tests.

Note that Mauchly's test is for the average variance-covariance matrix over the levels of the randomized-groups IV. It is possible to have homogeneity of variance-covariance matrices for some groups but not for others. Because groups are sometimes deleted during development of special error terms for comparisons, the assumption may be met for some comparisons but not for others. If you use SPSS GLM, each run for a comparison provides Mauchly's test for that comparison, and if there is a violation, you have the same options of single df comparisons, Huynh-Feldt corrections, or multivariate tests.

[4] In the mixed design, there is a variance-covariance matrix (cf. Section 6.6.2) of the repeated-measures DVs at each level (or combination of levels) of the randomized-groups IV(s). Error terms are developed from the average variance-covariance matrix over the several groups, and the reasonableness of those terms for omnibus ANOVA depends on the reasonableness of averaging. Tests of homogeneity of variance-covariance matrices are available—for example, Box's *M* test in SPSS GLM. However, these tests are overly sensitive to small departures from the assumption that may have little or no effect on nominal probabilities of Type I error rate. Although violation of homogeneity of variance-covariance matrices affects Type I error rate in the repeated-measures portion of the design, those biases are expected to be detected by Mauchly's test of sphericity.

7.3.2.4 Absence of Outliers

Outliers have the same unpredictable effect on analysis of the mixed design as in any other ANOVA and are identified using the same criteria of large z score and disconnectedness as in any other design. *Search for outliers separately for each cell of the design.* Thus, each repeated-measures DV is evaluated separately (as described in Section 6.3.2.4) in each level (or combination of levels) of the randomized-groups IV(s). This is demonstrated in Section 7.7. If an outlier is found, you have the same options as in any repeated-measures design.

7.3.2.5 Missing Data and Unequal Sample Sizes

Because portions of the mixed design are repeated measures, you may be able to estimate one or two data points using procedures discussed in Section 6.3.2.5. Unequal sample sizes in levels of a single randomized-groups IV pose no problem. However, unequal sample sizes create the usual problems of nonorthogonality if the randomized-groups portion of the design is itself factorial—that is, if there is more than one randomized-groups IV, as discussed in Section 5.6.5.

7.4 FUNDAMENTAL EQUATIONS

In many ways, the mixed design combines material from Chapters 4, 5, and 6. Different cases appear in the different levels (or combination of levels) of the randomized-groups IV(s), but they provide a DV value at each level (or combination of levels) of the repeated-measures IV(s). Therefore, the organization of data for both traditional ANOVA and the regression approach is literally a combination of the organizations for randomized groups and repeated measures. The error terms are also a literal combination of randomized-groups and repeated-measures error terms.

7.4.1 Allocation of Cases

The simplest mixed design has two IVs, one of each type, as shown in Table 7.1. The first three cases are in level a_1, and the last three cases are in level a_2. Therefore, A is a randomized-groups IV. All cases provide a DV value at both b_1 and b_2, so B is a repeated-measures IV.

Remember crossing and nesting from Section 1.2.5? In this design, cases nest in levels of A: s_1, s_2, and s_3 are nested in a_1, and s_4, s_5, and s_6 are nested in a_2. Therefore, one effect in the analysis is S/A. Meanwhile, cases cross levels of B, so another effect in the analysis is $B \times S$. Notice, however, that cases cross levels of B within each level of A. There is one $B \times S$ cross at a_1 and another $B \times S$ cross at a_2. So each $B \times S$ cross is really nested within levels of A: $B \times S/A$. Understanding this is important for

TABLE 7.1 Allocation of cases in a two-way mixed design

		Repeated Measures	
		b_1	b_2
Randomized Groups	a_1	s_1	s_1
		s_2	s_2
		s_3	s_3
	a_2	s_4	s_4
		s_5	s_5
		s_6	s_6

understanding the error terms that are developed during the analysis. However, the main effects of A and B and the $A \times B$ interaction are logically and computationally identical to any other factorial design.

7.4.2 Partition of Sources of Variance

The partition of the sum of squares is best conceptualized as a split between the randomized-groups and the repeated-measures segments of the design:

$$\mathrm{SS}_T = \mathrm{SS}_{\text{randomized}} + \mathrm{SS}_{\text{repeated}} \tag{7.1}$$

This fundamental split is differences between cases (randomized groups) and differences within cases (repeated measures). The randomized-groups segment sums to SS_S from the purely repeated-measures design. This effect is missing from the mixed design because it is further partitioned into A and S/A, the main effect of A and its error term. That is, differences in cases are due both to effects produced by levels of A and to otherwise unexplained variability in cases, just as in a randomized-groups design.

$$\mathrm{SS}_{\text{randomized}} = \mathrm{SS}_A + \mathrm{SS}_{S/A} \tag{7.2}$$

The repeated-measures segment of the design is partitioned into B, $A \times B$, and $B \times S/A$ (the main effect of B, the $A \times B$ interaction, and the error term for these two sources of variability).

$$\mathrm{SS}_{\text{repeated}} = \mathrm{SS}_B + \mathrm{SS}_{AB} + \mathrm{SS}_{B\times S/A} \tag{7.3}$$

As expected, the error term for the randomized-groups segment is a nested term, and the error term for the repeated-measures segment is a crossed term.

Degrees of freedom follow the partition of the sums of squares:

$$\mathrm{df}_T = \mathrm{df}_{\text{randomized}} + \mathrm{df}_{\text{repeated}} = \mathrm{df}_A + \mathrm{df}_{S/A} + \mathrm{df}_B + \mathrm{df}_{AB} + \mathrm{df}_{B\times S/A} \tag{7.4}$$

Total degrees of freedom is the total number of scores, *abs*, minus 1, lost when the grand mean is estimated; therefore,

$$\mathrm{df}_T = abs - 1 \tag{7.5}$$

Degrees of freedom for the randomized-groups part of the design are $as - 1$, the same as the total in a one-way randomized-groups ANOVA, but substituting s for n. Degrees of freedom for the randomized-groups part are further partitioned into df for the main effect of A, $a - 1$ as usual, and $a(s - 1)$, corresponding to the error term in a one-way randomized-groups design.

$$\text{df}_{\text{randomized}} = as - 1 = a - 1 + a(s - 1) \tag{7.6}$$

Degrees of freedom for the repeated portion of the design are $abs - as$ and are further partitioned into the main effect of B, $b - 1$; the $A \times B$ interaction, $(a - 1)(b - 1)$; and the interaction of B with S, which, in turn, is nested within A, $a(b - 1)(s - 1)$:

$$\text{df}_{\text{repeated}} = abs - as = b - 1 + (a - 1)(b - 1) + a(b - 1)(s - 1) \tag{7.7}$$

Verifying the partition for the repeated portion of the design:

$$abs - as = b - 1 + ab - a - b + 1 + abs - ab - as + a \tag{7.8}$$

Figure 7.1 diagrams the partition of sums of squares and degrees of freedom for the two-way randomized-repeated design.

There are a couple things to notice about the analysis for a mixed design. The first is that repeated measures are kind of "sticky." The $A \times B$ interaction, for instance, contains both a randomized-groups effect and a repeated-measures effect, but it is analyzed in the repeated-measures part of the design. That is generally true: Any interaction term that contains one (or more) repeated-measures main effect(s) is analyzed as part of the repeated-measures segment of the design. The second is that the error terms

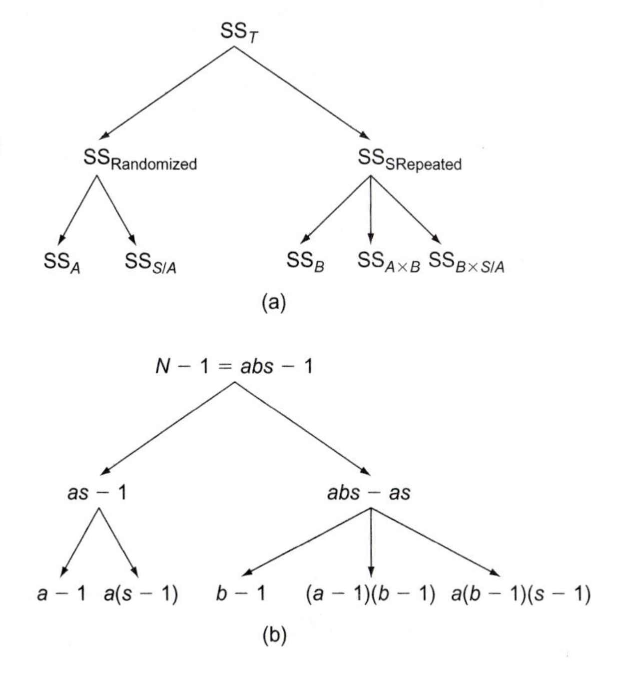

FIGURE 7.1 Partition of (a) sums of squares and (b) degrees of freedom in a mixed randomized-repeated design

for the repeated-measures effects in the design are always cases crossed with (in interaction with) effects, which, in turn, are nested in the randomized-groups effects in the designs. This generalizes to extensions of the design, as shown in Section 7.5.2.

7.4.3 Traditional ANOVA for the Mixed Design

An unrealistically small, hypothetical 3 × 3 data set for a two-way mixed design is in Table 7.2. There are three levels of each IV to illustrate comparisons. Type of novel (A) is the randomized-groups IV, with three levels (science fiction, mystery, and romance). A total of 15 readers participate in the study—5 readers each randomly assigned to read only novels of one type for 3 months. Month is the repeated-measures IV, B. Number of novels read each month still serves as the DV. Both the IVs are fixed—that is, specific levels are selected by the researcher. Readers (cases) are a random factor but are not tested explicitly.

TABLE 7.2 Hypothetical data set for a mixed design

		***B*: Month**			
***A*: Novel**		b_1: Month 1	b_2: Month 2	b_3: Month 3	Case Totals
	s_1	1	3	6	$A_1S_1 = 10$
	s_2	1	4	8	$A_1S_2 = 13$
a_1: Science Fiction	s_3	3	3	6	$A_1S_3 = 12$
	s_4	5	5	7	$A_1S_4 = 17$
	s_5	2	4	5	$A_1S_5 = 11$
		$A_1B_1 = 12$	$A_1B_2 = 19$	$A_1B_3 = 32$	$A_1 = 63$
	s_6	3	1	0	$A_2S_6 = 4$
	s_7	4	4	2	$A_2S_7 = 10$
a_2: Mystery	s_8	5	3	2	$A_2S_8 = 10$
	s_9	4	2	0	$A_2S_9 = 6$
	s_{10}	4	5	3	$A_2S_{10} = 12$
		$A_2B_1 = 20$	$A_2B_2 = 15$	$A_2B_3 = 7$	$A_2 = 42$
	s_{11}	4	2	0	$A_3S_{11} = 6$
	s_{12}	2	6	1	$A_3S_{12} = 9$
a_3: Romance	s_{13}	3	3	3	$A_3S_{13} = 9$
	s_{14}	6	2	1	$A_3S_{14} = 9$
	s_{15}	3	3	2	$A_3S_{15} = 8$
		$A_3B_1 = 18$	$A_3B_2 = 16$	$A_3B_3 = 7$	$A_3 = 41$
		$B_1 = 50$	$B_2 = 50$	$B_3 = 46$	$T = 146$

The equations for the main effects and interaction of interest—A, B, and $A \times B$—are identical to those in both factorial randomized-groups analysis (Chapter 5) and factorial repeated-measures analysis (Chapter 6). In general, computations for main effects and interactions are "blind" to the way cases are used in a design. The analysis is similar to repeated measures, in that case totals (AS) are needed to assess overall differences among cases—a source of variability that is then partitioned into A and S/A. The last source of variability is $B \times S/A$, the error term for the repeated-measures portion of the analysis. Computational equations for sums of squares for mixed ANOVA appear in Table 7.3. As always, A and B are fixed IVs; cases (S) are random, presumably randomly selected and assigned from some larger population of cases. Sums of squares are in Table 7.4. The equations and computations for degrees of freedom for this example are in Table 7.5. Finally, Table 7.6 summarizes the results of the calculations and shows the mean squares and F values for each

TABLE 7.3 Computational equations for sums of squares for mixed-design ANOVA

$$\mathrm{SS}_A = \frac{\sum A^2}{bs} - \frac{T^2}{abs} = \frac{63^2 + 42^2 + 41^2}{3(5)} - \frac{146^2}{3(3)(5)}$$

$$\mathrm{SS}_{S/A} = \frac{\sum (AS)^2}{b} - \frac{\sum A^2}{bs} = \frac{10^2 + 13^2 + 12^2 + \cdots + 8^2}{3} - \frac{63^2 + 42^2 + 41^2}{3(5)}$$

$$\mathrm{SS}_B = \frac{\sum B^2}{as} - \frac{T^2}{abs} = \frac{50^2 + 50^2 + 46^2}{3(5)} - \frac{146^2}{3(3)(5)}$$

$$\mathrm{SS}_{AB} = \frac{\sum (AB)^2}{s} - \frac{\sum A^2}{bs} - \frac{\sum B^2}{as} + \frac{T^2}{abs} = \frac{12^2 + 19^2 + 32^2 + 20^2 + 15^2 + 7^2 + 18^2 + 16^2 + 7^2}{5} - \frac{63^2 + 42^2 + 41^2}{3(5)} - \frac{50^2 + 50^2 + 46^2}{3(5)} + \frac{146^2}{3(3)(5)}$$

$$\mathrm{SS}_{B \times S/A} = \sum Y^2 - \frac{\sum (AB)^2}{s} - \frac{\sum (AS)^2}{s} + \frac{\sum A^2}{bs} = 1^2 + 1^2 + 3^2 + 5^2 + 2^2 + 3^2 + 4^2 + 5^2 + \cdots + 1^2 + 2^2 - \frac{12^2 + 19^2 + 32^2 + 20^2 + 15^2 + 7^2 + 18^2 + 16^2 + 7^2}{5} - \frac{10^2 + 13^2 + \cdots + 8^2}{3} + \frac{63^2 + 42^2 + 41^2}{3(5)}$$

$$\mathrm{SS}_T = \sum Y^2 - \frac{T^2}{abs} = 1^2 + 1^2 + 3^2 + 5^2 + 2^2 + 3^2 + \cdots + 1^2 + 2^2 - \frac{146^2}{3(3)(5)}$$

TABLE 7.4 Sums of squares for mixed-design ANOVA

$$\mathrm{SS}_A = 494.27 - 473.69 = 20.58$$

$$\mathrm{SS}_{S/A} = 520.37 - 494.27 = 26.40$$

$$\mathrm{SS}_B = 474.40 - 473.69 = 0.71$$

$$\mathrm{SS}_{AB} = 566.40 - 494.27 - 474.40 + 473.69 = 71.42$$

$$\mathrm{SS}_{B \times S/A} = 630 - 566.40 - 520.67 + 494.27 = 37.20$$

$$\mathrm{SS}_T = 630 - 473.69 = 156.31$$

TABLE 7.5 Degrees of freedom for mixed-design ANOVA

df_A	$= a - 1$	$= 3 - 1$	$= 2$
$df_{S/A}$	$= a(s - 1)$	$= 3(5 - 1)$	$= 12$
df_B	$= b - 1$	$= 3 - 1$	$= 2$
df_{AB}	$= (a - 1)(b - 1)$	$= (3 - 1)(3 - 1)$	$= 4$
$df_{B \times S/A}$	$= a(b - 1)(s - 1)$	$= 3(3 - 1)(5 - 1)$	$= 24$
df_T	$= abs - 1 = N - 1 = 3(3)(5) - 1$		$= 44$

TABLE 7.6 Source table for mixed-design ANOVA

Source	SS	df	MS	F
		Randomized Groups		
A	20.58	2	10.29	$\frac{10.29}{2.20} = 4.68$
S/A	26.40	12	2.20	
		Repeated Measures		
B	0.71	2	0.36	$\frac{0.36}{1.55} = 0.23$
AB	71.42	4	17.86	$\frac{17.86}{1.55} = 11.52$
$B \times S/A$	37.20	24	4.55	
T	156.31	44		

effect. As you examine the F ratios, notice the pattern of effects and error terms. The error term for A is S/A, and the error term for B and $A \times B$ is $B \times S/A$.

Each effect in the table is tested against critical F with its own df. Critical F for the test of A (type of novel) at $\alpha \leq .05$ with 2 and 12 df is 3.88; the null hypothesis of no mean differences among science fiction, mystery, and romance novels read is rejected because 4.68 is greater than 3.88. A glance at the numbers in Table 7.2 suggests that, overall, more science fiction novels are read, a conclusion tested in Section 7.6.1.1 (main-effects analysis for randomized groups). Critical F for the test of B (month) at $\alpha \leq .05$ with 2 and 24 df is 3.40; the null hypothesis of no mean difference in number of novels read per month is retained because 0.23 is less than 3.40. Finally, critical F for the type of novel by month interaction at $\alpha \leq .05$ with 4 and 24 df is 2.78; the null hypothesis of no interaction of number of novels read per month by type of novel is rejected because 11.52 is greater than 2.78. The numbers in Table 7.2 suggest that the number of science fiction novels read increases over the three months, whereas the numbers of mystery novels and romance novels read stay steady or decrease. This informal observation is tested by the procedures of Section 7.6.3 or 7.6.4 (simple effects trend analysis or interaction contrasts).

As an aside, it should be noted that the analysis of the randomized-groups part of this design, A and S/A, is identical to a one-way randomized-groups ANOVA of the case totals.

7.4.4 Regression Approach to the Mixed Design

The mixed nature of this design carries over into the regression approach to the analysis. Columns are needed for Y and for each df in the design—44 in this example. The 44 columns that partition the variability represent each df for main effects, for interaction, and for cases, which are still treated like another main effect. An additional 44 columns (not shown) are needed for the cross of each of the first 44 columns with Y. The last 44 columns produce the sums of products that are summarized under the first 44 columns as SP. The data set of Table 7.2 is rearranged for regression analysis in Table 7.7.

There are many things to notice about Table 7.7. Levels of A (3), B (3), and case designations are provided and repeated for clarification. Cases 1 to 5 repeat three times, once for each level of B at a_1; cases 6 to 10 also repeat three times, once for each level of B at a_2; and cases 11 to 15 repeat three times at a_3. The DV scores, Y, are in a single column, with scores for the same cases organized by level of B. Columns X_1 and X_2 contain contrast coefficients for the two df for A; the first coefficients compare a_1 with pooled results from a_2 and a_3; the second, a_2 versus a_3. These two columns are an orthogonal partition of the sum of squares for A. X_3 and X_4 contain codes for B_{linear} and $B_{\text{quadratic}}$, respectively. (Note that signs for the quadratic trend coefficients are $-1\ 2\ -1$ instead of $1\ -2\ 1$. This has no effect on the results.) These two columns are an orthogonal partition of the sum of squares for B. Columns X_5 to X_8 are columns for $A_{\text{comp1}} \times B_{\text{linear}}$, $A_{\text{comp1}} \times B_{\text{quadratic}}$, $A_{\text{comp2}} \times B_{\text{linear}}$, and $A_{\text{comp2}} \times B_{\text{quadratic}}$, respectively. These four columns are an orthogonal partition of the sum of squares for the $A \times B$ interaction.

The two error terms, one nested and one crossed, are set up in columns X_9 to X_{44}. Columns X_9–X_{20} literally represent the nesting of the five cases at each level of A. The same arbitrary but orthogonal coefficients for cases are used and repeated at each level of A. For example, X_9 contrasts case 1 with the other cases in a_1, and X_{13} contrasts case 6 with the other cases in a_2. X_{12} contrasts cases 4 and 5 in a_1, and so on. Notice that the nesting arrangement for cases is preserved in the coding where 0's "blank out" the cases that are not in that level of A. These 12 columns (X_9–X_{20}) are an orthogonal partition of sum of squares S/A (which has 12 df). The last 24 columns (X_{21}–X_{44}) are the cross of B_{linear} with cases (X_{21}–X_{32}) and then $B_{\text{quadratic}}$ with cases (X_{33}–X_{44}). These 24 columns are an orthogonal partition of the sum of squares for $B \times S/A$ (which has 24 df).

The remaining 44 columns are Y crossed with each of the 44 previous columns, necessary for computation of sums of products but not shown. This problem is set up using a spreadsheet, with sums, sum of squared values (sumsq), sum of squares (SS), sum of products (SP), and sum of squares for regression (SS(reg.)) computed using the spreadsheet software. Equations for computing preliminary sums of squares are in Table 6.19 and equations for computing sums of products are in Table 6.20. Table 7.8 summarizes the computations needed to produce the sum of squares for the various effects in the design.

These results are the same as those produced by the traditional ANOVA approach, within rounding error, as shown by comparing SS and df in Tables 7.6 and 7.8. The advantage is in the clarity of the layout of the design and in the separation of each df into a separate source of variability for follow-up analyses (cf. Section 7.6.5).

TABLE 7.7 Organization of data for the regression approach to mixed-design ANOVA

				$A1$	$A2$	Bl	Bq	$A1*B$l	$A1*B$q	$A2*B$l	$A2*B$q	$A1$,Sc1	$A1$,Sc2	$A1$,Sc3
			Y	X_1	X_2	X_3	X_4	X_5	X_6	X_7	X_8	X_9	X_{10}	X_{11}
a_1	b_1	s_1	1	2	0	−1	−1	−2	−2	0	0	4	0	0
		s_2	1	2	0	−1	−1	−2	−2	0	0	−1	1	1
		s_3	3	2	0	−1	−1	−2	−2	0	0	−1	1	−1
		s_4	5	2	0	−1	−1	−2	−2	0	0	−1	−1	0
		s_5	2	2	0	−1	−1	−2	−2	0	0	−1	−1	0
	b_2	s_1	3	2	0	0	2	0	4	0	0	4	0	0
		s_2	4	2	0	0	2	0	4	0	0	−1	1	1
		s_3	3	2	0	0	2	0	4	0	0	−1	1	−1
		s_4	5	2	0	0	2	0	4	0	0	−1	−1	0
		s_5	4	2	0	0	2	0	4	0	0	−1	−1	0
	b_3	s_1	6	2	0	1	−1	2	−2	0	0	4	0	0
		s_2	8	2	0	1	−1	2	−2	0	0	−1	1	1
		s_3	6	2	0	1	−1	2	−2	0	0	−1	1	−1
		s_4	7	2	0	1	−1	2	−2	0	0	−1	−1	0
		s_5	5	2	0	1	−1	2	−2	0	0	−1	−1	0
a_2	b_1	s_6	3	−1	1	−1	−1	1	1	−1	−1	0	0	0
		s_7	4	−1	1	−1	−1	1	1	−1	−1	0	0	0
		s_8	5	−1	1	−1	−1	1	1	−1	−1	0	0	0
		s_9	4	−1	1	−1	−1	1	1	−1	−1	0	0	0
		s_{10}	4	−1	1	−1	−1	1	1	−1	−1	0	0	0
	b_2	s_6	1	−1	1	0	2	0	−2	0	2	0	0	0
		s_7	4	−1	1	0	2	0	−2	0	2	0	0	0
		s_8	3	−1	1	0	2	0	−2	0	2	0	0	0
		s_9	2	−1	1	0	2	0	−2	0	2	0	0	0
		s_{10}	5	−1	1	0	2	0	−2	0	2	0	0	0
	b_3	s_6	0	−1	1	1	−1	−1	1	1	−1	0	0	0
		s_7	2	−1	1	1	−1	−1	1	1	−1	0	0	0
		s_8	2	−1	1	1	−1	−1	1	1	−1	0	0	0
		s_9	0	−1	1	1	−1	−1	1	1	−1	0	0	0
		s_{10}	3	−1	1	1	−1	−1	1	1	−1	0	0	0
a_3	b_1	s_{11}	4	−1	−1	−1	−1	1	1	1	1	0	0	0
		s_{12}	2	−1	−1	−1	−1	1	1	1	1	0	0	0
		s_{13}	3	−1	−1	−1	−1	1	1	1	1	0	0	0
		s_{14}	6	−1	−1	−1	−1	1	1	1	1	0	0	0
		s_{15}	3	−1	−1	−1	−1	1	1	1	1	0	0	0
	b_2	s_{11}	2	−1	−1	0	2	0	−2	0	−2	0	0	0
		s_{12}	6	−1	−1	0	2	0	−2	0	−2	0	0	0
		s_{13}	3	−1	−1	0	2	0	−2	0	−2	0	0	0
		s_{14}	2	−1	−1	0	2	0	−2	0	−2	0	0	0
		s_{15}	3	−1	−1	0	2	0	−2	0	−2	0	0	0
	b_3	s_{11}	0	−1	−1	1	−1	−1	1	−1	1	0	0	0
		s_{12}	1	−1	−1	1	−1	−1	1	−1	1	0	0	0
		s_{13}	3	−1	−1	1	−1	−1	1	−1	1	0	0	0
		s_{14}	1	−1	−1	1	−1	−1	1	−1	1	0	0	0
		s_{15}	2	−1	−1	1	−1	−1	1	−1	1	0	0	0
sum			146	0	0	0	0	0	0	0	0	0	0	0
sumsq			630	90	30	30	90	60	180	20	60	60	12	6
ss			156.31	90	30	30	90	60	180	20	60	60	12	6
sp				43	1	−4	4	64	−22	−2	−4	−13	−3	1
ss(reg)				20.54	0.03	0.53	0.18	68.27	2.69	0.2	0.27	2.82	0.75	0.17

TABLE 7.7 Continued

			$A1$,Sc4 X_{12}	$A2$,Sc1 X_{13}	$A2$,Sc2 X_{14}	$A2$,Sc3 X_{15}	$A2$,Sc4 X_{16}	$A3$,Sc1 X_{17}	$A3$,Sc2 X_{18}	$A3$,Sc3 X_{19}	$A3$,Sc4 X_{20}	$B1$*Sc1 X_{21}	$B1$*Sc2 X_{22}
		s_1	0	0	0	0	0	0	0	0	0	−4	0
		s_2	0	0	0	0	0	0	0	0	0	1	−1
	b_1	s_3	0	0	0	0	0	0	0	0	0	1	−1
		s_4	1	0	0	0	0	0	0	0	0	1	1
		s_5	−1	0	0	0	0	0	0	0	0	1	1
		s_1	0	0	0	0	0	0	0	0	0	0	0
		s_2	0	0	0	0	0	0	0	0	0	0	0
a_1	b_2	s_3	0	0	0	0	0	0	0	0	0	0	0
		s_4	1	0	0	0	0	0	0	0	0	0	0
		s_5	−1	0	0	0	0	0	0	0	0	0	0
		s_1	0	0	0	0	0	0	0	0	0	4	0
		s_2	0	0	0	0	0	0	0	0	0	−1	1
	b_3	s_3	0	0	0	0	0	0	0	0	0	−1	1
		s_4	1	0	0	0	0	0	0	0	0	−1	−1
		s_5	−1	0	0	0	0	0	0	0	0	−1	−1
		s_6	0	4	0	0	0	0	0	0	0	0	0
		s_7	0	−1	1	1	0	0	0	0	0	0	0
	b_1	s_8	0	−1	1	−1	0	0	0	0	0	0	0
		s_9	0	−1	−1	0	1	0	0	0	0	0	0
		s_{10}	0	−1	−1	0	−1	0	0	0	0	0	0
		s_6	0	4	0	0	0	0	0	0	0	0	0
		s_7	0	−1	1	1	0	0	0	0	0	0	0
a_2	b_2	s_8	0	−1	1	−1	0	0	0	0	0	0	0
		s_9	0	−1	−1	0	1	0	0	0	0	0	0
		s_{10}	0	−1	−1	0	−1	0	0	0	0	0	0
		s_6	0	4	0	0	0	0	0	0	0	0	0
		s_7	0	−1	1	1	0	0	0	0	0	0	0
	b_3	s_8	0	−1	1	−1	0	0	0	0	0	0	0
		s_9	0	−1	−1	0	1	0	0	0	0	0	0
		s_{10}	0	−1	−1	0	−1	0	0	0	0	0	0
		s_{11}	0	0	0	0	0	4	0	0	0	0	0
		s_{12}	0	0	0	0	0	−1	1	1	0	0	0
	b_1	s_{13}	0	0	0	0	0	−1	1	−1	0	0	0
		s_{14}	0	0	0	0	0	−1	−1	0	1	0	0
		s_{15}	0	0	0	0	0	−1	−1	0	−1	0	0
		s_{11}	0	0	0	0	0	4	0	0	0	0	0
		s_{12}	0	0	0	0	0	−1	1	1	0	0	0
a_3	b_2	s_{13}	0	0	0	0	0	−1	1	−1	0	0	0
		s_{14}	0	0	0	0	0	−1	−1	0	1	0	0
		s_{15}	0	0	0	0	0	−1	−1	0	−1	0	0
		s_{11}	0	0	0	0	0	4	0	0	0	0	0
		s_{12}	0	0	0	0	0	−1	1	1	0	0	0
	b_3	s_{13}	0	0	0	0	0	−1	1	−1	0	0	0
		s_{14}	0	0	0	0	0	−1	−1	0	1	0	0
		s_{15}	0	0	0	0	0	−1	−1	0	−1	0	0
sum			0	0	0	0	0	0	0	0	0	0	0
sumsq			6	60	12	6	6	60	12	6	6	40	8
ss			6	60	12	6	6	60	12	6	6	40	8
sp			6	−22	2	0	−6	−11	1	0	1	5	5
ss(reg)			6	8.07	0.33	0	6	2.02	0.08	0	0.17	0.625	3.125

TABLE 7.7 Continued

			Bl*Sc3 X_{23}	Bl*Sc4 X_{24}	Bl*Sc1 X_{25}	Bl*Sc2 X_{26}	Bl*Sc3 X_{27}	Bl*Sc4 X_{28}	Bl*Sc1 X_{29}	Bl*Sc2 X_{30}	Bl*Sc3 X_{31}	Bl*Sc4 X_{32}	Bq*Sc1 X_{33}
a_1	b_1	s_1	0	0	0	0	0	0	0	0	0	0	−4
		s_2	−1	0	0	0	0	0	0	0	0	0	1
		s_3	1	0	0	0	0	0	0	0	0	0	1
		s_4	0	−1	0	0	0	0	0	0	0	0	1
		s_5	0	1	0	0	0	0	0	0	0	0	1
	b_2	s_1	0	0	0	0	0	0	0	0	0	0	8
		s_2	0	0	0	0	0	0	0	0	0	0	−2
		s_3	0	0	0	0	0	0	0	0	0	0	−2
		s_4	0	0	0	0	0	0	0	0	0	0	−2
		s_5	0	0	0	0	0	0	0	0	0	0	−2
	b_3	s_1	0	0	0	0	0	0	0	0	0	0	−4
		s_2	1	0	0	0	0	0	0	0	0	0	1
		s_3	−1	0	0	0	0	0	0	0	0	0	1
		s_4	0	1	0	0	0	0	0	0	0	0	1
		s_5	0	−1	0	0	0	0	0	0	0	0	1
a_2	b_1	s_6	0	0	−4	0	0	0	0	0	0	0	0
		s_7	0	0	1	−1	−1	0	0	0	0	0	0
		s_8	0	0	1	−1	1	0	0	0	0	0	0
		s_9	0	0	1	1	0	−1	0	0	0	0	0
		s_{10}	0	0	1	1	0	1	0	0	0	0	0
	b_2	s_6	0	0	0	0	0	0	0	0	0	0	0
		s_7	0	0	0	0	0	0	0	0	0	0	0
		s_8	0	0	0	0	0	0	0	0	0	0	0
		s_9	0	0	0	0	0	0	0	0	0	0	0
		s_{10}	0	0	0	0	0	0	0	0	0	0	0
	b_3	s_6	0	0	4	0	0	0	0	0	0	0	0
		s_7	0	0	−1	1	1	0	0	0	0	0	0
		s_8	0	0	−1	1	−1	0	0	0	0	0	0
		s_9	0	0	−1	−1	0	1	0	0	0	0	0
		s_{10}	0	0	−1	−1	0	−1	0	0	0	0	0
a_3	b_1	s_{11}	0	0	0	0	0	0	−4	0	0	0	0
		s_{12}	0	0	0	0	0	0	1	−1	−1	0	0
		s_{13}	0	0	0	0	0	0	1	−1	1	0	0
		s_{14}	0	0	0	0	0	0	1	1	0	−1	0
		s_{15}	0	0	0	0	0	0	1	1	0	1	0
	b_2	s_{11}	0	0	0	0	0	0	0	0	0	0	0
		s_{12}	0	0	0	0	0	0	0	0	0	0	0
		s_{13}	0	0	0	0	0	0	0	0	0	0	0
		s_{14}	0	0	0	0	0	0	0	0	0	0	0
		s_{15}	0	0	0	0	0	0	0	0	0	0	0
	b_3	s_{11}	0	0	0	0	0	0	4	0	0	0	0
		s_{12}	0	0	0	0	0	0	−1	1	1	0	0
		s_{13}	0	0	0	0	0	0	−1	1	−1	0	0
		s_{14}	0	0	0	0	0	0	−1	−1	0	1	0
		s_{15}	0	0	0	0	0	0	−1	−1	0	−1	0
sum			0	0	0	0	0	0	0	0	0	0	0
sumsq			4	4	40	8	4	4	40	8	4	4	120
ss			4	4	40	8	4	4	40	8	4	4	120
sp			4	−1	−2	0	1	−3	−9	5	−1	−4	1
ss(reg)			4	0.25	0.1	0	0.25	2.25	2.025	3.125	0.25	4	0.01

TABLE 7.7 Continued

			Bq*Sc2 X_{34}	Bq*Sc3 X_{35}	Bq*Sc4 X_{36}	Bq*Sc1 X_{37}	Bq*Sc2 X_{38}	Bq*Sc3 X_{39}	Bq*Sc4 X_{40}	Bq*Sc1 X_{41}	Bq*Sc2 X_{42}	Bq*Sc3 X_{43}	Bq*Sc4 X_{44}
		s_1	0	0	0	0	0	0	0	0	0	0	0
		s_2	−1	−1	0	0	0	0	0	0	0	0	0
	b_1	s_3	−1	1	0	0	0	0	0	0	0	0	0
		s_4	1	0	−1	0	0	0	0	0	0	0	0
		s_5	1	0	1	0	0	0	0	0	0	0	0
		s_1	0	0	0	0	0	0	0	0	0	0	0
		s_2	2	2	0	0	0	0	0	0	0	0	0
a_1	b_2	s_3	2	−2	0	0	0	0	0	0	0	0	0
		s_4	−2	0	2	0	0	0	0	0	0	0	0
		s_5	−2	0	−2	0	0	0	0	0	0	0	0
		s_1	0	0	0	0	0	0	0	0	0	0	0
		s_2	−1	−1	0	0	0	0	0	0	0	0	0
	b_3	s_3	−1	1	0	0	0	0	0	0	0	0	0
		s_4	1	0	−1	0	0	0	0	0	0	0	0
		s_5	1	0	1	0	0	0	0	0	0	0	0
		s_6	0	0	0	−4	0	0	0	0	0	0	0
		s_7	0	0	0	1	−1	−1	0	0	0	0	0
	b_1	s_8	0	0	0	1	−1	1	0	0	0	0	0
		s_9	0	0	0	1	1	0	−1	0	0	0	0
		s_{10}	0	0	0	1	1	0	1	0	0	0	0
		s_6	0	0	0	8	0	0	0	0	0	0	0
		s_7	0	0	0	−2	2	2	0	0	0	0	0
a_2	b_2	s_8	0	0	0	−2	2	−2	0	0	0	0	0
		s_9	0	0	0	−2	−2	0	2	0	0	0	0
		s_{10}	0	0	0	−2	−2	0	−2	0	0	0	0
		s_6	0	0	0	−4	0	0	0	0	0	0	0
		s_7	0	0	0	1	−1	−1	0	0	0	0	0
	b_3	s_8	0	0	0	1	−1	1	0	0	0	0	0
		s_9	0	0	0	1	1	0	−1	0	0	0	0
		s_{10}	0	0	0	1	1	0	1	0	0	0	0
		s_{11}	0	0	0	0	0	0	0	−4	0	0	0
		s_{12}	0	0	0	0	0	0	0	1	−1	−1	0
	b_1	s_{13}	0	0	0	0	0	0	0	1	−1	1	0
		s_{14}	0	0	0	0	0	0	0	1	1	0	−1
		s_{15}	0	0	0	0	0	0	0	1	1	0	1
		s_{11}	0	0	0	0	0	0	0	8	0	0	0
		s_{12}	0	0	0	0	0	0	0	−2	2	2	0
a_3	b_2	s_{13}	0	0	0	0	0	0	0	−2	2	−2	0
		s_{14}	0	0	0	0	0	0	0	−2	−2	0	2
		s_{15}	0	0	0	0	0	0	0	−2	−2	0	−2
		s_{11}	0	0	0	0	0	0	0	−4	0	0	0
		s_{12}	0	0	0	0	0	0	0	1	−1	−1	0
	b_3	s_{13}	0	0	0	0	0	0	0	1	−1	1	0
		s_{14}	0	0	0	0	0	0	0	1	1	0	−1
		s_{15}	0	0	0	0	0	0	0	1	1	0	1
sum			0	0	0	0	0	0	0	0	0	0	0
sumsq			24	12	12	120	24	12	12	120	24	12	12
ss			24	12	12	120	24	12	12	120	24	12	12
sp			−3	2	−3	−8	−2	3	−3	−7	11	9	−4
ss(reg)			0.38	0.33	0.75	0.53	0.17	0.75	0.75	0.41	5.04	6.75	1.33

TABLE 7.8 Sums of squares for the regression approach to mixed-design ANOVA

Source	Columns		SS	df (number of columns)
A	SS(reg. $X1$) + SS(reg. $X2$)	20.54 + 0.03	20.57	2
S/A	SS(reg. $X9$) + SS(reg. $X10$) + ⋯ + SS(reg. $X20$)	2.82 + 0.75 + ⋯ + 0.17	26.40	12
B	SS(reg. $X3$) + SS(reg. $X4$)	0.53 + 0.18	0.71	2
$A \times B$	SS(reg. $X5$) + SS(reg. $X6$) + SS(reg. $X7$) + SS(reg. $X8$)	68.27 + 2.69 + 0.20 + 0.27	71.46	4
$B \times S/A$	SS(reg. $X21$) + SS(reg. $X22$) + ⋯ + SS(reg. $X44$)	0.62 + 3.12 + ⋯ + 1.33	37.20	24
Total	SS(Y)		156.31	44

7.4.5 Computer Analyses of Small-Sample Mixed Design

The data set for mixed-design ANOVA is laid out as in Table 7.2. One column is needed to code the levels of each randomized-groups IV, and the DV scores are given at each level of the repeated-measures IV. As in the purely repeated-measures design, the name of the repeated-measures IV and its levels are created "on the fly" through the statistical software in SPSS and SAS. If there is more than one repeated-measures IV, the DV scores are set up as in Table 6.17, with the same cautions about the necessity of consistency in the order in which DVs are presented in the data set and identified to the software.

Syntax and output from SPSS GLM for the example are in Table 7.9. Syntax is default except for the requests for /EMMEANS (marginal and cell means, standard errors, and confidence intervals).

After identifying the elements in the design, multivariate tests are given of the two effects with repeated measures, month and month*novel, should you wish to take this approach to testing their significance. Mauchly's Test of Sphericity follows, nonsignificant in this example, indicating no violation of sphericity or homogeneity of variance-covariance matrix assumptions. Univariate tests of month and month*novel follow, together with adjustments, including the Huynh-Feldt adjustment, which you would use if Mauchly's test is significant and you wish to take this approach to evaluating significance. This is followed by trend analysis of month and month*novel, your third alternative to evaluating significance. Then, there is output regarding the effect of the randomized-groups IV, novel. Finally, marginal and cell statistics are shown in the section labeled Estimated Marginal Means. See Section 6.5 for more detailed interpretation of the repeated-measures portion of the output.

Table 7.10 contains syntax and selected output from SAS GLM for the mixed-design example. Note that means are requested separately for the repeated and randomized portions of the design.

TABLE 7.9 SPSS GLM syntax and output for mixed-design ANOVA

```
GLM
  MONTH1 MONTH2 MONTH3 BY NOVEL
  /WSFACTOR = month 3 Polynomial
  /MEASURE = books
  /METHOD = SSTYPE(3)
  /EMMEANS = TABLES(NOVEL)
  /EMMEANS = TABLES(month)
  /EMMEANS = TABLES(NOVEL*month)
  /CRITERIA = ALPHA(.05)
  /WSDESIGN = month
  /DESIGN = NOVEL .
```

General Linear Model

Within-Subjects Factors

Measure: books

month	Dependent Variable
1	month1
2	month2
3	month3

Between-Subjects Factors

		Value Label	N
novel	1.00	science fiction	5
	2.00	mystery	5
	3.00	romance	5

Multivariate Tests[c]

Effect		Value	F	Hypothesis df	Error df	Sig.
month	Pillai's Trace	.062	.361[a]	2.000	11.000	.705
	Wilks' Lambda	.938	.361[a]	2.000	11.000	.705
	Hotelling's Trace	.066	.361[a]	2.000	11.000	.705
	Roy's Largest Root	.066	.361[a]	2.000	11.000	.705
month * novel	Pillai's Trace	.859	4.518	4.000	24.000	.007
	Wilks' Lambda	.155	8.463[a]	4.000	22.000	.000
	Hotelling's Trace	5.354	13.384	4.000	20.000	.000
	Roy's Largest Root	5.336	32.019[b]	2.000	12.000	.000

a. Exact statistic

b. The statistic is an upper bound on F that yields a lower bound on the significance level.

c. Design: Intercept+novel
Within Subjects Design: month

TABLE 7.9 Continued

Mauchly's Test of Sphericity[b]

Measure: books

Within Subjects Effect	Mauchly's W	Approx. Chi-Square	df	Sig.	Epsilon[a]		
					Greenhouse-Geisser	Huynh-Feldt	Lower-bound
month	.789	2.608	2	.272	.826	1.000	.500

Tests the null hypothesis that the error covariance matrix of the orthonormalized transformed dependent variables is proportional to an identity matrix.

a. May be used to adjust the degrees of freedom for the averaged tests of significance. Corrected tests are displayed in the Tests of Within-Subjects Effects table.

b. Design: Intercept+novel
Within Subjects Design: month

Tests of Within-Subjects Effects

Measure: books

Source		Type III Sum of Squares	df	Mean Square	F	Sig.
month	Sphericity Assumed	.711	2	.356	.229	.797
	Greenhouse-Geisser	.711	1.651	.431	.229	.755
	Huynh-Feldt	.711	2.000	.356	.229	.797
	Lower-bound	.711	1.000	.711	.229	.641
month * novel	Sphericity Assumed	71.422	4	17.856	11.520	.000
	Greenhouse-Geisser	71.422	3.303	21.624	11.520	.000
	Huynh-Feldt	71.422	4.000	17.856	11.520	.000
	Lower-bound	71.422	2.000	35.711	11.520	.002
Error(month)	Sphericity Assumed	37.200	24	1.550		
	Greenhouse-Geisser	37.200	19.818	1.877		
	Huynh-Feldt	37.200	24.000	1.550		
	Lower-bound	37.200	12.000	3.100		

Tests of Within-Subjects Contrasts

Measure: books

Source	month	Type III Sum of Squares	df	Mean Square	F	Sig.
month	Linear	.533	1	.533	.320	.582
	Quadratic	.178	1	.178	.124	.731
month * novel	Linear	68.467	2	34.233	20.540	.000
	Quadratic	2.956	2	1.478	1.031	.386
Error(month)	Linear	20.000	12	1.667		
	Quadratic	17.200	12	1.433		

TABLE 7.9 Continued

Tests of Between-Subjects Effects

Measure: books

Transformed Variable: Average

Source	Type III Sum of Squares	df	Mean Square	F	Sig.
Intercept	473.689	1	473.689	215.313	.000
novel	20.578	2	10.289	4.677	.031
Error	26.400	12	2.200		

Estimated Marginal Means

1. novel

Measure: books

			95% Confidence Interval	
novel	Mean	Std. Error	Lower Bound	Upper Bound
science fiction	4.200	.383	3.366	5.034
mystery	2.800	.383	1.966	3.634
romance	2.733	.383	1.899	3.568

2. month

Measure: books

			95% Confidence Interval	
month	Mean	Std. Error	Lower Bound	Upper Bound
1	3.333	.353	2.565	4.102
2	3.333	.362	2.544	4.122
3	3.067	.313	2.385	3.748

3. novel * month

Measure: books

				95% Confidence Interval	
novel	month	Mean	Std. Error	Lower Bound	Upper Bound
science fiction	1	2.400	.611	1.069	3.731
	2	3.800	.627	2.434	5.166
	3	6.400	.542	5.220	7.580
mystery	1	4.000	.611	2.669	5.331
	2	3.000	.627	1.634	4.366
	3	1.400	.542	.220	2.580
romance	1	3.600	.611	2.269	4.931
	2	3.200	.627	1.834	4.566
	3	1.400	.542	.220	2.580

TABLE 7.10 Syntax and selected output from SAS GLM for mixed ANOVA

```
proc glm data=SASUSER.SS_MIXED;
   class NOVEL;
   model MONTH1 MONTH2 MONTH3 = NOVEL;
   repeated month 3 / mean;
   lsmeans NOVEL;
run;
```

The GLM Procedure
Repeated Measures Analysis of Variance

Repeated Measures Level Information

Dependent Variable	MONTH1	MONTH2	MONTH3
Level of month	1	2	3

MANOVA Test Criteria and Exact F Statistics for the Hypothesis of no month Effect
H = Type III SSCP Matrix for month
E = Error SSCP Matrix

S=1 M=0 N=4.5

Statistic	Value	F Value	Num DF	Den DF	Pr > F
Wilks' Lambda	0.93839102	0.36	2	11	0.7049
Pillai's Trace	0.06160898	0.36	2	11	0.7049
Hotelling-Lawley Trace	0.06565385	0.36	2	11	0.7049
Roy's Greatest Root	0.06565385	0.36	2	11	0.7049

MANOVA Test Criteria and F Approximations for the Hypothesis of no month*NOVEL Effect
H = Type III SSCP Matrix for month*NOVEL
E = Error SSCP Matrix

S=2 M=-0.5 N=4.5

Statistic	Value	F Value	Num DF	Den DF	Pr > F
Wilks' Lambda	0.15515386	8.46	4	22	0.0003
Pillai's Trace	0.85905715	4.52	4	24	0.0073
Hotelling-Lawley Trace	5.35362220	14.40	4	12.235	0.0001
Roy's Greatest Root	5.33645857	32.02	2	12	<.0001

NOTE: F Statistic for Roy's Greatest Root is an upper bound.
NOTE: F Statistic for Wilks' Lambda is exact.

The GLM Procedure

Repeated Measures Analysis of Variance
Tests of Hypotheses for Between Subjects Effects

Source	DF	Type III SS	Mean Square	F Value	Pr > F
NOVEL	2	20.57777778	10.28888889	4.68	0.0315
Error	12	26.40000000	2.20000000		

Univariate Tests of Hypotheses for Within Subject Effects

Source	DF	Type III SS	Mean Square	F Value	Pr > F	Adj Pr > F G-G	Adj Pr > F H-F
month	2	0.71111111	0.35555556	0.23	0.7967	0.7554	0.7967
month*NOVEL	4	71.42222222	17.85555556	11.52	<.0001	0.0001	<.0001
Error(month)	24	37.20000000	1.55000000				

Greenhouse-Geisser Epsilon	0.8257
Huynh-Feldt Epsilon	1.1003

Means of Within Subjects Effects

Level of month	N	Mean	Std Dev
1	15	3.33333333	1.44749373
2	15	3.33333333	1.34518542
3	15	3.06666667	2.68505564

Least Squares Means

NOVEL	MONTH1 LSMEAN	MONTH2 LSMEAN	MONTH3 LSMEAN
1	2.40000000	3.80000000	6.40000000
2	4.00000000	3.00000000	1.40000000
3	3.60000000	3.20000000	1.40000000

Like SPSS, SAS gives a routine univariate source table, plus two of the alternatives: multivariate F tests (`Wilks' Lambda`) and probability of F after Huynh-Feldt correction (`Adj Pr > F, H-F`). Although no formal test of sphericity is given unless requested, the epsilon values near 1 for Greenhouse-Geisser and Huynh-Feldt adjustments suggest no or very little violation of the assumption. Marginal means are shown for the repeated-measures IV but not for the randomized-groups IV. Marginal means for `NOVEL`, then, are found by averaging over `MONTH` for each of `NOVEL`.

7.5 TYPES OF MIXED DESIGNS

7.5.1 The Pretest-Posttest Design

One variant on the mixed design requires that the first measurement occur before any treatment is given, as shown in Table 7.11. The first trial is a pretest, and all subsequent trials are part of the posttest.

The mean DV scores on the first trial (pretest) are compared to see whether the randomized groups are equivalent *before* treatment is given.[5] Once reassured that the groups are equivalent, interaction is examined to determine whether the groups have a different pattern of responses over time. A significant interaction in this design implies that the group with treatment has a different pattern of mean DV scores over trials than the control group has—that is, treatment produces some change.[6] If the control group also changes between the pretest and the posttest, one has reason to suspect the influence of uncontrolled extraneous variables.

7.5.2 Expanding Mixed Designs

As noted in Section 7.4.2, any interaction term that contains one (or more) repeated-measures main effect(s) is analyzed as part of the repeated-measures segment of the design. Also noted in that section, error terms for the purely randomized-groups

TABLE 7.11 Pretest-posttest mixed design

	b_1: Pretest	Treatment	b_2: Posttest
a_1: Treatment Group	Yes	Yes	Yes
a_2: Control Group	Yes	No	Yes

[5] If the groups are not equivalent, the pretest can be used as a covariate to adjust each case's score on the posttest by differences on the pretest, as described in Chapter 8.

[6] Analysis of difference scores between posttest and pretest produces the same result.

TABLE 7.12 Partition of randomized groups and repeated measures terms in mixed designs with their error terms

Design	Randomized-Groups IVs	Repeated-Measures IVs	Sources of Variability		
			Randomized Groups	Repeated Measures	Error Terms
Two-way mixed	A	B	A		S/A
				$B, A \times B$	$B \times S/A$
Three-way mixed	A, B	C	$A, B, A \times B$		S/AB
				$C, A \times C, B \times C,$ $A \times B \times C$	$C \times S/AB$
	A	B, C	A		S/A
				$B, A \times B$	$B \times S/A$
				$C, A \times C$	$C \times S/A$
				$B \times C, A \times B \times C$	$B \times C \times S/A$
Four-way mixed	A, B	C, D	$A, B, A \times B$		S/AB
				$C, A \times C, B \times C,$ $A \times B \times C$	$C \times S/AB$
				$D, A \times D, B \times D,$ $A \times B \times D$	$D \times S/AB$
				$C \times D, A \times C \times D,$ $B \times C \times D, A \times B \times C \times D$	$C \times D \times S/AB$

effects in the design are always cases nested in levels (or combinations of levels) of the randomized-groups IV(s). The error terms for the repeated-measures effects in the design are always cases crossed with (in interaction with) those effects, nested in the randomized-groups effects in the designs.

These generalizations are summarized in Table 7.12, where C is the third IV. The table shows the breakdowns for a two-way mixed design, a three-way design with two randomized-groups IVs (A and B) and one repeated measure (C), a three-way design with one randomized-groups IV (A) and two repeated measures (B and C), and finally a four-way design with two randomized-groups IVs (A and B) and two repeated measures (C and D).

Once you understand the pattern, you should be able to generalize to a mixed design of any complexity. If a main effect or interaction contains no repeated factors, the error term is cases within levels of randomized-groups factors. If a main effect or interaction contains one or more repeated factors, the error term is an interaction between the repeated factor(s) and the cases nested within randomized-groups factors.

7.6 SOME IMPORTANT ISSUES

Issues of unequal n, power, effect size, and the like are the same for the mixed design as they are for the randomized-groups design (Chapters 4 and 5) and the repeated-measures design (Chapter 6). Consult these earlier chapters for a discussion of these important issues. The difficulty with the mixed design involves comparisons. The same logic drives comparisons on the margins and tests of interaction (with either simple-effects analysis or interaction contrasts) as in the other analyses, but in the mixed design, the process is computationally more complicated.

The complexity is that comparisons involving the randomized-groups IV(s) use the randomized-groups computations but different error terms for some comparisons, whereas comparisons on the repeated-measures IV(s) use repeated-measures computations and a different error term for each comparison. However, the full array of comparisons is available: comparisons on the margins, simple effects (both simple main effects and simple comparisons), and interaction contrasts.

7.6.1 Comparisons on the Margins

Comparisons on the margins are appropriate when the interaction is not statistically significant or when it is significant but small and ordinal (not crossing). The marginal comparisons are conducted differently, however, depending on whether they are for a randomized-groups or a repeated-measures effect, as shown in Table 7.13. In this table (like in the example of Section 7.4), A (type of novel) is a randomized-groups IV and B (month) is a repeated measure.

Contrast coefficients are chosen for A to compare the average number of science fiction books read against the pooled average of mystery and romance novels (2 −1 −1). The desired error term is S/A, because the contrast is concerned only with A, the randomized-groups IV. Linear contrast coefficients (−1 0 1) are chosen for B

TABLE 7.13 Comparisons on the margins

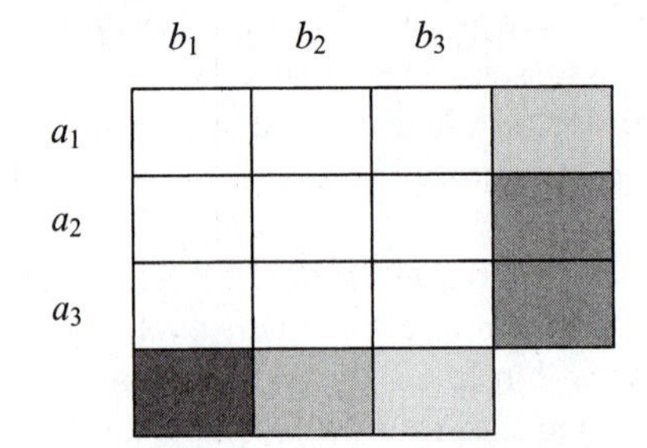

Notes: A_{compl} (a_1 vs. the average of a_2 and a_3) is tested against S/A and is a randomized-groups effect. B_{linear} is tested against $B_{\text{linear}} \times S/A$ and is a repeated-measures effect.

to determine whether the average number of novels read increases (or decreases) steadily over the months. The desired error term is $B_{\text{linear}} \times S/A$, a term developed specifically for this analysis.

7.6.1.1 The Randomized-Groups Margin

The effect of A and the $A \times B$ interaction are statistically significant in the small-sample example. The $A \times B$ interaction is large, by Equation 6.16,

$$\text{Partial } \eta^2 = \frac{\text{SS}_{AB}}{\text{SS}_{AB} + \text{SS}_{B\times S/A}} = \frac{71.24}{71.24 + 37.20} = .66$$

and disordinal, as shown in Figure 7.2. Thus, 66% of the variance in number of books read is associated with the interaction between month and novel type.

Although under these circumstances, comparisons on the margins of A are inappropriate and potentially misleading, they are conducted here for didactic purposes. If the comparison is calculated by hand, the sum of squares, mean square, and F for the comparison are computed using an appropriately modified version of Equation 5.18.

For the comparison between science fiction versus other novels:

$$F = \frac{n_{\overline{Y}}\left(\Sigma w_j \overline{Y}_j\right)^2 / \Sigma w_j^2}{\text{MS}_{S/A}} \tag{7.8}$$

$$F = \frac{(3(5)(2(4.20) - 2.80 - 2.73)^2)/(2^2 + (-1)^2 + (-1)^2)}{2.20} = 9.36$$

If this comparison is planned, it is tested with 1 and 12 df, and critical F at $\alpha = .05$ is 4.75. If it is tested post hoc, F_S is computed using an appropriately modified version of Equation 5.20. For this example, F_S is 2(3.88) = 7.76. Thus, the comparison

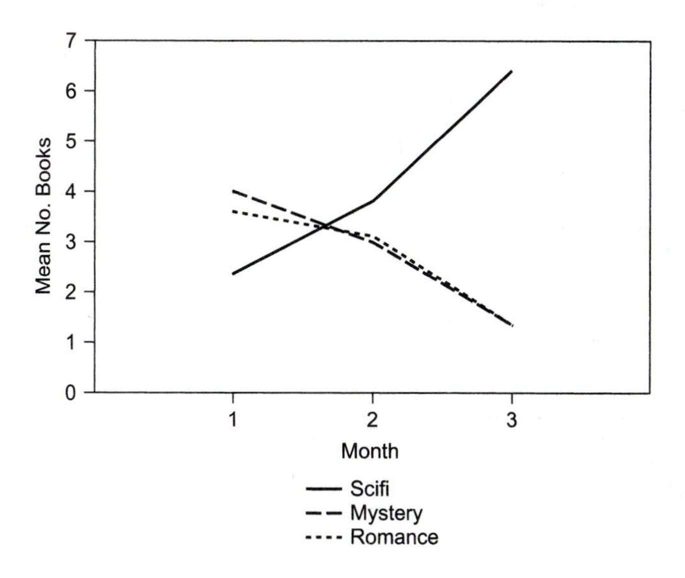

FIGURE 7.2 Mean number of books read as a function of type of novel and month

TABLE 7.14 Syntax for comparisons on the margin of the randomized-groups IV in a mixed design

Program	Syntax	Section of Output	Name of Effect
SPSS GLM	GLM MONTH1 MONTH2 MONTH3 BY NOVEL /WSFACTOR = month 3 Polynomial /METHOD = SSTYPE(3) /CRITERIA = ALPHA(.05) /WSDESIGN = month /LMATRIX "comp1" NOVEL 2 -1 -1 /DESIGN = NOVEL .	**Contrast Results (K Matrix)** **Test Results**	Contrast

SAS GLM	proc glm data=SASUSER.SS_MIXED;
	class NOVEL;
	model MONTH1 MONTH2 MONTH3 = NOVEL;
	contrast 'marg_com' novel 2 -1 -1;
	repeated month 3;
	run;
		Tests of Hypotheses for Between Subjects Effects: Contrast	marg_com

SYSTAT GLM[a]	LET BOOKS = MONTH1+MONTH2+MONTH3 ANOVA CATEGORY NOVEL COVAR DEPEND BOOKS ESTIMATE HYPOTHESIS EFFECT=NOVEL CONTRAST[2 -1 -1] TEST	Test of Hypothesis: Source	Hypothesis

[a] Note that a new DV, books, is created as the sum of the original DVs.

is significant if planned or performed post hoc; science fiction novels are preferred to mystery and romance novels.

Syntax for performing this marginal comparison on the randomized-groups IV through SPSS GLM, SAS GLM, and SYSTAT GLM appears in Table 7.14.

7.6.1.2 The Repeated-Measures Margin

Comparisons are formed on the margin of the repeated-measures IV by hand using the procedures of Section 6.6.6.1.2. Essentially, the data set is revised to build in the desired comparison, and the results are then reanalyzed. Because the error term is developed right along with the effect, the appropriate error term is available for use.

Table 7.15 shows the data set for the $(-1\ 0\ 1)$ comparison, which simply deletes b_2 from the intact data set of Table 7.2. Routine ANOVA (Section 7.4) produces Table 7.16.[7]

In this analysis, the "main effect" of B is the linear trend comparison on months, B_{comp} (month 1 versus month 3), using the appropriate error term, $B_{comp} \times S/A$. Critical value for df $= 1, 12$ at $\alpha = .05$ for a planned comparison is again 4.75, which,

[7] You will have your chance at this analysis in the problem sets.

TABLE 7.15 Hypothetical data set for a mixed design with month 2 eliminated for analysis with coefficients (−1 0 1)

		B: Month		
A: Novel		b_1: Month 1	b_3: Month 3	Case Totals
a_1: Science Fiction	s_1	1	6	$A_1S_1 = 7$
	s_2	1	8	$A_1S_2 = 9$
	s_3	3	6	$A_1S_3 = 9$
	s_4	5	7	$A_1S_4 = 12$
	s_5	2	5	$A_1S_5 = 7$
		$A_1B_1 = 12$	$A_1B_3 = 32$	$A_1 = 44$
a_2: Mystery	s_6	3	0	$A_2S_6 = 3$
	s_7	4	2	$A_2S_7 = 6$
	s_8	5	2	$A_2S_8 = 7$
	s_9	4	0	$A_2S_9 = 4$
	s_{10}	4	3	$A_2S_{10} = 7$
		$A_2B_1 = 20$	$A_2B_3 = 7$	$A_2 = 27$
a_3: Romance	s_{11}	4	0	$A_3S_{11} = 4$
	s_{12}	2	1	$A_3S_{12} = 3$
	s_{13}	3	3	$A_3S_{13} = 6$
	s_{14}	6	1	$A_3S_{14} = 7$
	s_{15}	3	2	$A_3S_{15} = 5$
		$A_3B_1 = 18$	$A_3B_3 = 7$	$A_3 = 25$
		$B_1 = 50$	$B_3 = 46$	$T = 96$

TABLE 7.16 Source table for mixed-design ANOVA with month 2 eliminated

Source	SS	df	MS	F
Randomized Groups				
A	21.80	2	10.90	6.54
S/A	20.00	12	1.67	
Repeated Measures				
B_{comp}	0.53	1	0.53	0.32
$A \times B_{comp}$	68.47	2	34.23	20.54
$B_{comp} \times S/A$	20.00	12	1.67	
T	130.80	29		

of course, F_{Bcomp} does not exceed. The Scheffé-adjusted critical value, should this comparison be done post hoc, is

$$F_S = (b - 1)F_{(b-1),\, a(b-1)(s-1)} \tag{7.10}$$

which is $(2)(3.40) = 6.80$.

SAS GLM, SYSTAT GLM, and SPSS GLM (default) all have simplified syntax for trend (orthogonal polynomial) analysis on a repeated-measures factor, as shown in the first portion of Table 7.17. The second portion of the table shows syntax that generalizes to any marginal comparison on a repeated-measures factor, using the

TABLE 7.17 Syntax for comparisons on the margin of the repeated-measures IV in a mixed design

Type of Comparison	Program	Syntax	Section of Output	Name of Effect
Polynomial (linear trend)	SPSS GLM	GLM MONTH1 MONTH2 MONTH3 BY NOVEL /WSFACTOR = month 3 Polynomial /MEASURE = books /METHOD = SSTYPE(3) /CRITERIA = ALPHA(.05) /WSDESIGN = month /DESIGN = NOVEL .	**Tests of Within-Subjects Contrasts**	MONTH Linear (See Table 7.9)
	SAS GLM	proc glm data=SASUSER.SS_MIXED; class NOVEL; model MONTH1 MONTH2 MONTH3 = NOVEL ; repeated MONTH 3 polynomial/summary; run;	MONTH.N: MONTH_1	MEAN
	SYSTAT GLM (default with PRINT= MEDIUM)	ANOVA CATEGORY NOVEL COVAR DEPEND MONTH1 MONTH2 MONTH3/, REPEAT =3,NAMES='MONTH' ESTIMATE	Single Degree of Freedom Polynomial Contrasts: Polynomial Test of Order 1 (Linear)	MONTH
User-specified (-1 0 1)	SPSS GLM	GLM MONTH1 MONTH2 MONTH3 BY NOVEL /WSFACTOR = month 3 special (1 1 1 -1 0 1 1 -2 1) /MEASURE = books /METHOD = SSTYPE(3) /CRITERIA = ALPHA(.05) /WSDESIGN = month /DESIGN = NOVEL .	**Tests of Within-Subjects Contrasts**	month L1

TABLE 7.17 Continued

Type of Comparison	Program	Syntax	Section of Output	Name of Effect

	SAS GLM	`proc glm`
		`data=SASUSER.SS_MIXED;`
		`class NOVEL;`
		`model MONTH1 MONTH2 MONTH3`
		`= NOVEL;`
		`manova m = -1*MONTH1 +`
		`0*MONTH2 +`
		`1*MONTH3 H=INTERCEPT;`
		`run;`
			MANOVA Test...no Overall Intercept Effect	Wilks' Lambda

	SYSTAT GLM	`GLM`
		`CATEGORY NOVEL`
		`MODEL MONTH1 MONTH2 MONTH3=,`
		`CONSTANT + NOVEL/ REPEAT=3,`
		`NAMES='month'`
		`ESTIMATE`
		`HYPOTHESIS`
		`ALL`
		`WITHIN='month'`
		`STANDARDIZE=WITHIN`
		`CONTRAST [-1 0 1]`
		`TEST`
			Test for effect called CONSTANT: Test of Hypothesis	Source: Hypothesis

coefficients (−1 0 1), which happen to be the ones for linear trend. Because SPSS requires a full matrix of orthogonal effects, the coefficients for the quadratic effect are included in the syntax. SAS and SYSTAT do not require the full matrix; adjustments are handled internally.

7.6.2 Simple Main Effects Analysis

In simple main effects analysis, one IV, either randomized or repeated, is held constant at some value while mean differences are examined on the levels of the other IV, as shown in Figure 5.8. The difference between the mixed design and the completely randomized design of Chapter 5 is that simple main effects may be done on a repeated or randomized factor, and error terms differ depending on which IV is examined.

Table 7.18 shows that each simple main effect of the randomized-groups IV, A, is tested against a separate randomized-groups error term.[8] For example, in Table 7.18(a), the test of A at the first level of B uses S/A at b_1 as an error term. Similarly, each simple main effect of the repeated-measures IV, B, is tested against a

[8] A single error term, $SS_{\text{within cell}} = SS_{S/A} + SS_{B \times S/A}$, may be used if the df for that error term exceed 30. Tests using that error term are biased with fewer df, as noted by Brown, Michels, and Winer (1991).

TABLE 7.18 Simple main effects for randomized groups and repeated measures

(a) Simple main effect of A at b_1 is tested against S/A at b_1 and is a randomized-groups effect

(b) Simple main effect of B at a_1 is tested against $B \times S$ at a_1 and is a repeated-measures effect

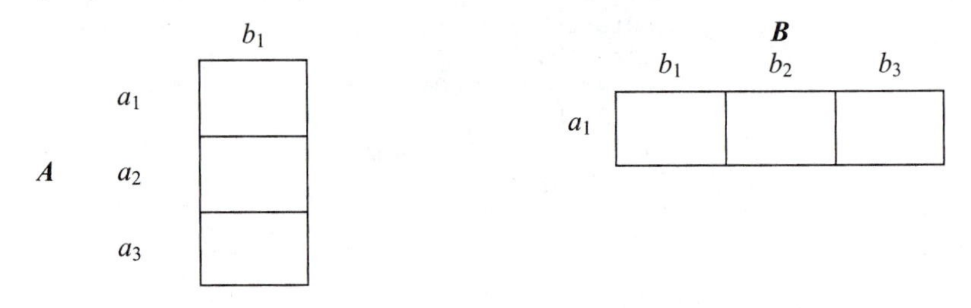

different repeated-measures error term. For example, in Table 7.18(b), $B \times S$ at a_1 provides the error term for the test of B at the first level of A.

As in any simple main effects analysis, significant effects with more than two levels may be followed up by simple comparisons among levels (see Figure 5.8). Again, each comparison among levels has its own error term.

If comparisons are done post hoc to evaluate a significant interaction, then F_S for interaction is appropriate for all simple main effects, simple comparisons, and interaction contrasts. F_S for interaction for this design is an appropriately revised version of Equation 5.19:

$$F_S = (a-1)(b-1)F_{(a-1)(b-1),a(b-1)(s-1)} \tag{7.11}$$

In the small-sample example of Section 7.4,

$$F_S = 2(2)F_{4,24} = 4(2.78) = 11.12$$

7.6.2.1 Simple Main Effects for the Randomized-Groups IV

Hand calculations for simple main effects of the randomized-groups IV are done as one-way ANOVAs, separately on each "slice" (level of B). Equations of Sections 3.2.3.2 and 4.4.3 are used to calculate SS_A and MS_A separately at b_1, b_2, and b_3—that is, use Equation 3.10 three times, one for each level of B. Then each numerator MS is divided by $MS_{S/A \text{ at } bj}$ calculated at that level of B. Finally, you calculate and evaluate each F, based on df_A in the numerator and $df_{S/A}$ at the level B. The result for the first level of B in the small-sample example of Section 7.4 is $F(2, 12) = 1.86, p > .05$.

By computer, you also do a one-way randomized-groups ANOVA with just one of the DV measures, such as the first month (Table 7.19). SPSS GLM (UNIANOVA) may be used as an alternative to ONEWAY.

7.6.2.2 Simple Main Effects for the Repeated-Measures IV

A one-way repeated-measures ANOVA is performed on the set of DVs (levels of B) separately for each level of A. Equations 6.11–6.14 are used to calculate SS by hand

TABLE 7.19 Syntax for tests of simple main effects of the randomized-groups IV in a mixed design

Program	Syntax	Section of Output	Name of Effect
SPSS Compare Means	ONEWAY MONTH1 BY NOVEL /MISSING ANALYSIS .	ANOVA	Between Groups
SAS GLM	proc glm data=SASUSER.SS_MIXED; class NOVEL; model MONTH1 = NOVEL ; run;	Dependent Variable: MONTH1 Source	NOVEL
SYSTAT ANOVA	ANOVA CATEGORY NOVEL COVAR DEPEND MONTH1 ESTIMATE	Analysis of Variance: Source	NOVEL

TABLE 7.20 Syntax for simple main effects on the repeated-measures IV in a mixed design

Program	Syntax	Section of Output	Name of Effect
SPSS GLM	[DATA SELECT CASES IF (NOVEL = 1)] GLM MONTH1 MONTH2 MONTH3 /WSFACTOR = month 3 Polynomial /METHOD = SSTYPE(3) /CRITERIA = ALPHA(.05) /WSDESIGN = month .	**Tests of Within-Subjects Effects**	MONTH
SAS GLM	proc glm data=SASUSER.SS_MIXED; by NOVEL; class NOVEL; model MONTH1 MONTH2 MONTH3 = ; repeated MONTH 3; run;	NOVEL=1 Source:	MONTH
SYSTAT ANOVA	SELECT (NOVEL= 1) ANOVA CATEGORY COVAR DEPEND MONTH1 MONTH2 MONTH3/ , REPEAT=3,NAMES='MONTH' ESTIMATE	Within Subjects: Source	MONTH

at each level of A, substituting B for A. The result for the first level of novel (science fiction) for the small-sample example is $F(2, 8) = 17.35, p < .01$.

Case selection is used to limit the analysis to the desired level of A for each, as shown in Table 7.20. SAS GLM has special simple-effects language that may be used with the repeated-measures simple effect only.

7.6.3 Simple Comparisons

When there are more than two levels, simple comparisons follow significant simple main effects to examine the specific sources of the simple main effect. As usual, they differ from simple comparisons in randomized-groups or fully repeated-measures factorial designs only in the choice of error terms.

Table 7.21 shows the simple comparisons to be demonstrated for the small-sample design of Section 7.4. Comparison on the randomized-groups factor, A, is the first group (science fiction) versus the other two groups (mystery and romance) for the first month, as shown in Table 7.21(a). Linear trend, in Table 7.21(b), is analyzed over the repeated-measures factor, B, for science fiction novels.

7.6.3.1 Simple Comparisons for the Randomized-Groups IV

Simple comparisons for the randomized-groups IV follow naturally from the analysis of simple main effects and use the same error term. Therefore, Equation 7.8 is applied, with $\text{MS}_{S/A \text{ at } b1}$ as the error term for any comparison following the simple main effect of A at b_1.

For the comparison of group 1 (science fiction) versus groups 2 and 3 (mystery and romance),

$$F = \frac{((5)(2(2.40) - 4.00 - 3.60)^2)/(2^2 + (-1)^2 + (-1)^2)}{1.87} = 3.49$$

Critical F for a planned comparison with 1 and 12 df is 4.75, indicating that there is no difference between number of science fiction books read and the number of other books read during the first month.

The analysis by computer is done by adding contrast coefficients to the one-way analysis, using only the data for the first level of month as the DV (Table 7.22). Note that the special simple-effects language of SAS GLM cannot be used for this simple comparison.

7.6.3.2 Simple Comparisons for the Repeated-Measures IV

Simple comparisons on the repeated-measures factor all have different error terms, as do all repeated-measures comparisons. Therefore, a data set is developed for each

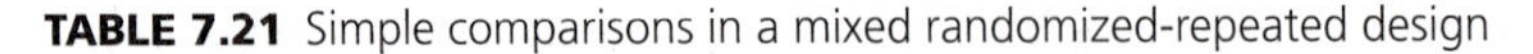

TABLE 7.21 Simple comparisons in a mixed randomized-repeated design

(a) A_{comp} at b_1 is tested against S/A at b_1 and is a randomized-groups effect

b_1

a_1

a_2

a_3

(b) B_{linear} at a_1 is tested against $B_{linear} \times S/A$ at a_1 and is a repeated-measures effect

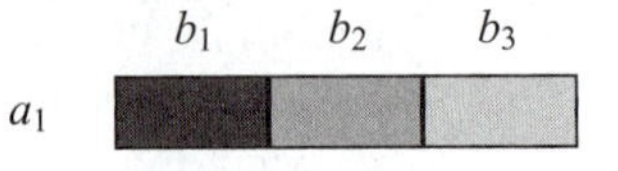

TABLE 7.22 Syntax for simple comparisons on the randomized-groups IV in a mixed design

Program	Syntax	Section of Output	Name of Effect
SPSS GLM	ONEWAY MONTH1 BY NOVEL /CONTRAST= 2 -1 -1 /MISSING ANALYSIS .	**Contrast Tests**	Assume equal variances[a]
SAS GLM	proc glm data=SASUSER.SS_MIXED; class NOVEL; model MONTH1 = NOVEL ; contrast 'SCIFI_VS' NOVEL 2 -1 -1; run;	Dependent Variable: MONTH1 Contrast	SCIFI_VS
SYSTAT ANOVA	ANOVA CATEGORY NOVEL COVAR DEPEND MONTH1 ESTIMATE HYPOTHESIS EFFECT=NOVEL CONTRAST[2 -1 -1] TEST	Test of Hypothesis: Source	Hypothesis

[a] t is given rather than F; recall that $t^2 = F$ with 1 df in the numerator and df_{error} in the denominator.

comparison, as in Table 6.33, based on the coefficients chosen. Indeed, the analysis is exactly the same as for the factorial repeated-measures design, because only one level of novels (A) is used. That is, the design is treated as one-way repeated, with a data set generated that reflects the coefficients. For example, the data set of Table 6.33(b) is used if months 1 and 2 are to be contrasted with month 3. A data set with columns b_1 and b_3 (omitting b_2) is developed at the first level of A to test the linear trend of months for science fiction novels. The result for the small-sample example with b_2 eliminated is $F(1, 4) = 20.00, p < .05$.

Computer analysis simply requires selection of the level of A to be used. Contrasts, either user-specified or polynomial, are then specified (Table 7.23).

7.6.4 Interaction Contrasts

As discussed in Section 5.6.4.4.2, interaction contrasts are appropriate when both main effects, as well as the interaction, are statistically significant. The current data set has one significant main effect, along with a significant interaction. Therefore, the tests for simple main effects of A at each level of B (followed by simple comparisons) are most appropriate. However, interaction contrasts are acceptable, although they are not as easily interpreted.

Table 7.24 describes the interaction between the linear trend of months (B_{linear}) and science fiction versus other novels (A_{comp}). Figure 7.3 plots the combined means relevant to this test.

TABLE 7.23 Syntax for simple comparisons on the repeated-measures IV in a mixed design

Type of Comparison	Program	Syntax	Section of Output	Name of Effect
Polynomial (linear)	SPSS GLM	GLM MONTH1 MONTH2 MONTH3 /WSFACTOR = month 3 Polynomial /METHOD = SSTYPE(3) /CRITERIA = ALPHA(.05) /WSDESIGN = month .	**Tests of Within-Subjects Contrasts**	month Linear

	SAS GLM	proc glm data=SASUSER.SS_MIXED;
		where NOVEL=1;
		model MONTH1 MONTH2 MONTH3 =;
		repeated month 3 polynomial /
		summary;
		run;
			Contrast Variable: MONTH_1	MEAN

	SYSTAT ANOVA	SELECT (NOVEL= 1) ANOVA CATEGORY COVAR DEPEND MONTH1 MONTH2 MONTH3/, REPEAT=3,NAMES='MONTH' ESTIMATE HYPOTHESIS WITHIN="MONTH" CONTRAST /POLYNOMIAL TEST	Univariate F Tests: Source	1
User-specified (for 1, 1, -2)	SPSS GLM	[DATA SELECT CASES IF (NOVEL = 1)] GLM MONTH1 MONTH2 MONTH3 /WSFACTOR = month 3 special (1 1 1 1 1 -2 1 -1 0) /MEASURE = books /METHOD = SSTYPE(3) /CRITERIA = ALPHA(.05) /WSDESIGN = month .	**Tests of Within-Subjects Contrasts**	month L1

	SAS GLM	proc glm data=SASUSER.SS_MIXED;
		where NOVEL=1;
		model MONTH1 MONTH2 MONTH3 =;
		manova m = 1*MONTH1 + 1*MONTH2
		– 2*MONTH3 H=INTERCEPT;
		run;
			MANOVA Test...no Overall Intercept Effect	Wilks' Lambda

	SYSTAT ANOVA	SELECT (NOVEL=1) ANOVA CATEGORY COVAR DEPEND MONTH1 MONTH2 MONTH3/ REPEAT=3, NAMES="MONTH" ESTIMATE HYPOTHESIS WITHIN="MONTH" CONTRAST[1 1 -2] TEST	Test of Hypothesis: : Source	Hypoth-esis

TABLE 7.24 Interaction contrasts: $A_{comp} \times B_{linear}$ is tested against $B_{linear} \times S/A$ and is a repeated-measures (interaction) effect

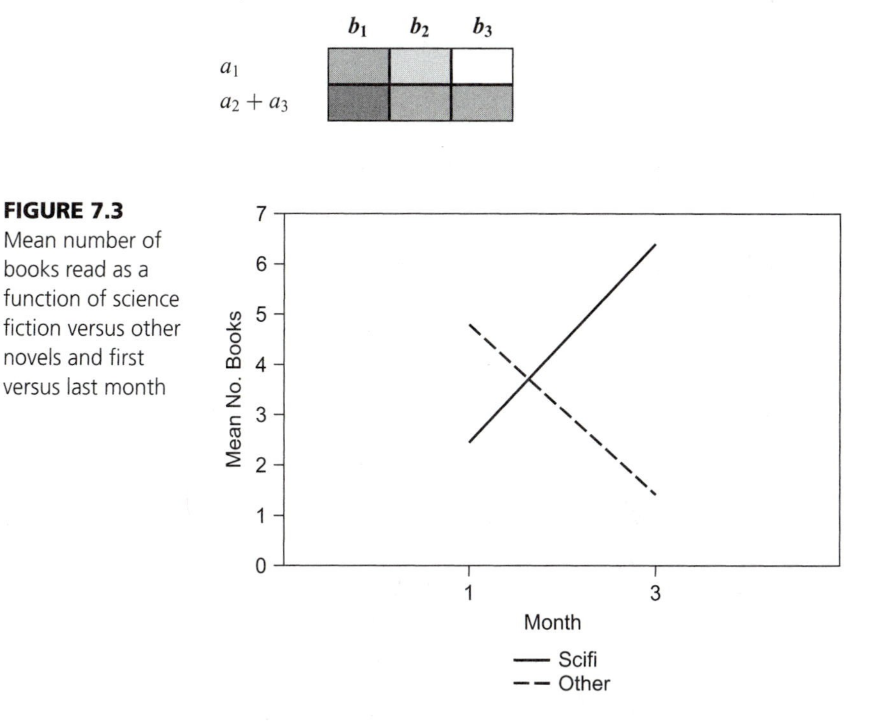

FIGURE 7.3 Mean number of books read as a function of science fiction versus other novels and first versus last month

Month 2 is deleted from the plot because it has a coefficient of 0 when coding for the linear trend. Interaction contrasts done by hand require development of a data set representing the cells of interest: a_1b_1, a_1b_3, $a_{2+3}b_1$, and $a_{2+3}b_3$. Again, month 2 (b_2) is omitted because it has a 0 coefficient in coding for the linear trend with three levels. A standard mixed ANOVA is performed, and the calculated interaction tests $A_{comp} \times B_{linear}$. This yields $F(1, 12) = 40.96, p < .001$. Alternatively, the regression approach is used for interaction comparisons, as demonstrated in the following section.

Table 7.25 shows syntax for an interaction contrast through software.

7.6.5 Comparisons through the Regression Approach

The regression approach is preferable in many ways to the traditional ANOVA approach for comparisons in the mixed design. Each degree of freedom represents a different partition of the variability in the design and a different comparison or interaction contrast.[9] In Table 7.26, the information of Table 7.8 is further partitioned into its elements to illustrate the detailed analysis that is readily available.

[9] Simple-effects analyses (main effects and comparisons) can also be done through regression, but you have to reorganize the data set in order to represent the comparisons you are seeking and then recompute the regression.

TABLE 7.25 Syntax for interaction contrasts in a mixed design

Type of Comparison	Program	Syntax	Location of Output	Name of Effect
User-specified on both randomized IV (2 –1 –1) and repeated IV (–1 0 1)	SPSS GLM	GLM MONTH1 MONTH2 MONTH3 BY NOVEL /MMATRIX = "LINEAR" MONTH1 -1 MONTH2 0 MONTH3 1 /MEASURE = books /WSFACTOR = month 3 /WSDESIGN = month /LMATRIX = "comp1" NOVEL 2 –1 –1 /DESIGN = NOVEL.	**Test Results**	Contrast

	SAS GLM	proc glm data=SASUSER.SS_MIXED;
		class NOVEL;
		model MONTH1 MONTH2 MONTH3 = NOVEL;
		contrast 'Acomp1' NOVEL 2 –1 –1;
		manova m = –1*MONTH1 + 0*MONTH2 +
		1*MONTH3;
		run;
			MANOVA Test ....no Overall Acomp1 Effect	Wilks' Lambda[a]

	SYSTAT ANOVA	ANOVA CATEGORY NOVEL COVAR DEPEND MONTH1 MONTH2 MONTH3/ REPEAT =3, NAMES='MONTH' ESTIMATE HYPOTHESIS CMATRIX [–1 0 1] EFFECT=NOVEL CONTRAST [2 –1 –1] TEST	Test of hypothesis: Source	Hypothesis

<table>
<tr><td>User-specified on randomized IV (2 −1 −1); polynomial (linear) on repeated IV</td><td>SPSS GLM</td><td>Same as user-specified on both randomized and repeated, above</td><td></td><td></td></tr>
<tr><td></td><td>SAS GLM</td><td><pre>proc glm data=SASUSER.SS_MIXED;
   class NOVEL;
   model MONTH1 MONTH2 MONTH3 = NOVEL;
   contrast 'comp' NOVEL 2 -1 -1;
   repeated month 3 polynomial /
   summary;
run;</pre></td><td>Contrast Variable: month_1</td><td>comp</td></tr>
<tr><td></td><td>SYSTAT ANOVA</td><td>Same as user-specified on both randomized and repeated, above</td><td></td><td></td></tr>
</table>

[a] This is a multivariate test, but produces the same F ratio and df as other programs.

TABLE 7.26 Sums of squares for the regression approach to mixed-design ANOVA

Source	Columns	SS	Name of Effect
A	SS(reg. X_1) + SS(reg. X_2)	20.54 + 0.03	
	SS(reg. X_1) = $SS_{A\text{comp1}}$	20.54	A_{comp1} on the A margin
	SS(reg. X_2) = $SS_{A\text{comp2}}$	0.03	A_{comp2} on the A margin
S/A	SS(reg. X_9) + SS(reg. X_{10}) + ⋯ + SS(reg. X_{20})	2.82 + 0.75 + ⋯ + 0.17	
B	SS(reg. X_3) + SS(reg. X_4)	0.53 + 0.18	
	SS(reg. X_3) = $SS_{B\text{linear}}$	0.53	B_{linear} on the B margin
	SS(reg. X_4) = $SS_{B\text{quad}}$	0.18	B_{quad} on the B margin
$A \times B$	SS(reg. X_5) + SS(reg. X_6) + SS(reg. X_7) + SS(reg. X_8)	68.27 + 2.69 + 0.20 + 0.27	
	SS(reg. X_5) = $SS_{A\text{comp1} \times B\text{linear}}$	68.27	Interaction contrast
	SS(reg. X_6) = $SS_{A\text{comp1} \times B\text{quad}}$	2.69	Interaction contrast
	SS(reg. X_7) = $SS_{A\text{comp2} \times B\text{linear}}$	0.20	Interaction contrast
	SS(reg. X_8) = $SS_{A\text{comp2} \times B\text{quad}}$	0.27	Interaction contrast
	SS(reg. X_5) + SS(reg. X_6) = $SS_{A\text{comp1} \times B\text{linear}} + SS_{A\text{comp1} \times B\text{quad}} = SS_{A\text{comp1} \times B}$	68.27 + 2.69	Partial factorial (interaction of A_{comp1} with B)
	SS(reg. X_7) + SS(reg. X_8) = $SS_{A\text{comp2} \times B\text{linear}} + SS_{A\text{comp2} \times B\text{quad}} = SS_{A\text{comp2} \times B}$	0.20 + 0.27	Partial factorial (interaction of A_{comp2} with B)
	SS(reg. X_5) + SS(reg. X_7) = $SS_{A\text{comp1} \times B\text{linear}} + SS_{A\text{comp2} \times B\text{linear}} = SS_{A \times B\text{linear}}$	68.27 + 0.20	Partial factorial (interaction of A with B_{linear})
	SS(reg. X_6) + SS(reg. X_8) = $SS_{A\text{comp1} \times B\text{quad}} + SS_{A\text{comp2} \times B\text{quad}} = SS_{A \times B\text{quadr}}$	2.69 + 0.27	Partial factorial (interaction of A with B_{quad})
$B \times S/A$	SS(reg. X_{21}) + SS(reg. X_{22}) + ⋯ + SS(reg. X_{44})	0.62 + 3.12 + ⋯ + 1.33	
	SS(reg. X_{21}) + ⋯ + SS(reg. X_{32})	0.62 + ⋯ + 4.00	Error term, $B_{\text{linear}} \times S/A$
	SS(reg. X_{33}) + ⋯ + SS(reg. X_{44})	0.01 + ⋯ + 1.33	Error term, $B_{\text{quadr}} \times S/A$
Total	SS(Y)		

Examination of Table 7.26 reveals that comparisons on the margins and interaction contrasts, with their corresponding error terms, are readily available from the regression breakdown. Further, partial factorial interactions are easily obtained by adding together the SS(reg.) for the columns that combine to produce an effect. For instance, if you combine SS(reg. X_5) and SS(reg. X_6), you are "going over" B_{linear} and B_{quad} for A_{comp1}. The result is $A_{\text{comp1}} \times B$. The evaluation of this sum of squares asks whether the comparison on A has the same pattern over the three levels of B. Or, if you combine SS(reg. X_5) and SS(reg. X_7), you are "going over" A_{comp1} and A_{comp2} for B_{linear}. The result is $A \times B_{\text{linear}}$. The evaluation of this sum of squares asks whether there is a different linear trend on B among the three different levels of A. The relevant sums of squares for the specialized error terms are also obtained by adding together sums of squares for the relevant columns.

The advantage to this approach is the absolute clarity of the various elements of the design. You select the orthogonal partition for the randomized-groups IV(s), the orthogonal partition for the repeated-measures IV(s) (often a trend analysis), and the orthogonal partition for cases, and then you obtain the desired results through cross multiplying appropriate columns and evaluating their sums of squares. Addition of sums of squares regression for appropriate columns gives you the numbers needed to test various elements of the design. With modern spreadsheets, this is not even all that painful.[10]

7.7 COMPLETE EXAMPLE OF MIXED RANDOMIZED-REPEATED ANOVA

The data set for this example was available on the Internet (with free use authorized) and is from a statement by Texaco, Inc., to the Air and Water Pollution Subcommittee of the Senate Public Works Committee on June 26, 1973. The data were used to compare the noise produced by an Octel filter with that produced by a standard filter. Eighteen cars were tested in all, nine with each type of filter. Each type of filter was tested with cars of three sizes: small, medium, and large. Noise for each car was measured on the right and left sides. Thus, this example is a 3 × 2 × 2 with two randomized-groups IVs (car size and filter type) and one repeated-measures IV (side). Data files are FILTER.*

The goal of the original research was to show that the Octel filter was no noisier than the standard filter. General questions about the effects of size, the side of car, and the interactions among size, side, and filter type are also addressed in the current analysis.

[10] Well. . . .

7.7.1 Evaluation of Assumptions

7.7.1.1 Normality of Sampling Distributions, Missing Data, and Unequal Sample Sizes

There are no missing data. Sample sizes are equal in each cell but small, with cells including three cases in each combination of car size, filter type, and side. Note that a corresponding completely randomized-groups design (different cases for each side) would require 36 cases rather than the 18 required for this mixed design.

Sample sizes within cells for the randomized portion of the design are too small to ensure normality of sampling distributions; with only three cases per cell, $df_{S/AB} = ab(s-1) = (3)(2)(3-1) = 12$ and $df_{C \times S/AB} = ab(c-1)(s-1) = 12$. Similarly, the within-cell sample size is too small to evaluate normality either statistically or graphically. Instead, the noise scores at each level of the repeated-measures IV are examined. Table 7.27 shows overall descriptive statistics for noise at each side (left and right), as produced by SPSS DESCRIPTIVES. Labels are provided to ease interpretation of output.

Skewness and kurtosis are well within guidelines of Section 2.2.3.3, at least for the test of side. The critical z value for tests of skewness and kurtosis using $\alpha = .01$ is 2.58. Each skewness value is divided by its standard error, according to Equations 2.10 and 2.11. For the right side of the car,

$$z_s = \frac{-0.461}{0.536} = -0.860$$

$$z_k = \frac{-0.946}{1.038} = -0.911$$

and for the left side of the car,

$$z_s = \frac{-0.528}{0.536} = -0.985$$

$$z_k = \frac{1.410}{1.038} = -1.358$$

TABLE 7.27 Descriptive statistics of noise at each side (SPSS DESCRIPTIVES syntax and output)

```
DESCRIPTIVES
  VARIABLES=right left
  /STATISTICS=MEAN STDDEV VARIANCE MIN MAX KURTOSIS SKEWNESS .
```

Descriptive Statistics

	N	Minimum	Maximum	Mean	Std.	Variance	Skewness		Kurtosis	
	Statistic	Statistic	Statistic	Statistic	Statistic	Statistic	Statistic	Std. Error	Statistic	Std. Error
RIGHT	18	76.50	84.50	80.9722	2.43427	5.926	-.461	.536	-.946	1.038
LEFT	18	76.00	85.50	81.0556	3.41230	11.644	-.528	.536	-1.410	1.038
Valid N (listwise)	18									

Therefore, there is no concern with deviation from normality, as long as no outliers are found.

7.7.1.2 Homogeneity of Variance

Cells in this design are formed by the three cases in each combination of size, filter, and side. The data are split into six groups, using size and filter, and then the variances within the groups for both left and right noise measurements are examined. Table 7.28 shows SPSS DESCRIPTIVES for each cell of the design.

TABLE 7.28 Descriptive statistics for each cell of the design

```
SORT CASES BY SIZE FILTER .
SPLIT FILE
  LAYERED BY SIZE FILTER .
DESCRIPTIVES
  VARIABLES=right left
  /STATISTICS=MEAN STDDEV VARIANCE MIN MAX .
```

Descriptive Statistics

SIZE	FILTER		N	Minimum	Maximum	Mean	Std. Deviation	Variance
small	standard	RIGHT	3	81.00	82.00	81.6667	.57735	.333
		LEFT	3	83.50	83.50	83.5000	.00000	.000
		Valid N (listwise)	3					
	octel	RIGHT	3	82.00	82.00	82.0000	.00000	.000
		LEFT	3	82.50	82.50	82.5000	.00000	.000
		Valid N (listwise)	3					
medium	standard	RIGHT	3	84.00	84.50	84.1667	.28868	.083
		LEFT	3	84.50	85.50	85.0000	.50000	.250
		Valid N (listwise)	3					
	octel	RIGHT	3	82.00	82.50	82.1667	.28868	.083
		LEFT	3	81.50	82.50	82.1667	.57735	.333
		Valid N (listwise)	3					
large	standard	RIGHT	3	78.50	79.00	78.6667	.28868	.083
		LEFT	3	76.00	77.00	76.3333	.57735	.333
		Valid N (listwise)	3					
	octel	RIGHT	3	76.50	77.50	77.1667	.57735	.333
		LEFT	3	76.00	77.50	76.8333	.76376	.583
		Valid N (listwise)	3					

The largest variance in the data set is .583, for the large size car, Octel filter, with noise measured on the left side. However, three of the cells have no variance at all, and it is impossible to evaluate homogeneity of variance with 0 in the denominator of F_{max} (Equation 3.21). The decision is made to test all effects at $\alpha = .025$ to counteract any positive bias (increase in Type I error) that may be associated with heterogeneity of variance.

7.7.1.3 Outliers

Table 7.28 provides information for evaluating outliers. The first cell (small size, standard filter, right side), for example, has a Minimum score of 81.00, a Mean of 81.667, and a Std. Deviation (standard deviation) of .577. Therefore, according to Equation 2.7,

$$z = \frac{81.0 - 81.667}{0.577} = -1.155$$

At $\alpha = .01$ for the small sample, the critical value of $z = 2.58$; thus, the score is not an outlier.

Cells with no variance have no outliers, because all scores are at the mean. None of the other minimum or maximum scores are outliers within their cells.

7.7.1.4 Independence of Errors and Sphericity

Independence of errors for the randomized-groups portion of the design is ensured by the (presumed) random assignment of cars to filter types. There is no opportunity to violate sphericity, because there are only two levels of the repeated-measures IV, side.

7.7.2 Three-Way Mixed Randomized-Repeated ANOVA

7.7.2.1 Major Analysis

Table 7.29 shows syntax and output for SPSS GLM (Repeated Measures).

Marginal means for side are in the Descriptive Statistics table and differ little (80.972 for the right sides of the cars and 81.056 for the left sides of the cars).

The Tests of Within-Subjects Effects and Tests of Between-Subjects Effects show that most effects are statistically significant at $p \leq .025$, with the exception of the main effect of side ($F = 0.243$, $p = .631$) and the side by filter type interaction ($F = 0.027$, $p = .875$).

The highest order (side*SIZE*FILTER) interaction is statistically significant—$F(2, 12) = 9.43$, $p = .003$. Effect size is calculated as per Equation 5.13:

$$\text{Partial } \eta^2 = \frac{4.847}{4.847 + 3.083} = .61$$

with Smithson (2003) confidence limits ranging from .12 to .76 (see Table 5.48 for demonstration).

This suggests the need for a simple-effects analysis in this three-way design in which the simple interaction between two of the IVs is plotted at each level of the third IV. As for any simple-effects analysis, the analyzed effect should be the one that is, in itself, nonsignificant, because the interaction test is confounded. The only nonsignificant two-way interaction in Table 7.29 is side by filter type. This suggests that interactions between side and filter type be analyzed separately for each level of car size.

Table 7.30 shows means and standard errors for the 12 cells participating in the three-way interaction. Means and standard deviations are from Table 7.29. Standard deviations are converted to standard errors by dividing standard deviations by the square root of 3, the cell size: $\sqrt{3} = 1.732$.

The three-way interaction is plotted in Figure 7.4 through SPSS GLM. Note that three graphs are generated, one for each car size. Figure 7.4 suggests different patterns of the interaction for each car size. The main effect of filter type is also of particular interest in this design, despite the strong, slightly disordinal interaction,

TABLE 7.29 Mixed randomized-repeated ANOVA (SPSS GLM syntax and selected output)

```
GLM
  RIGHT LEFT BY SIZE FILTER
  /WSFACTOR = side 2 Polynomial
  /MEASURE = noise
  /METHOD = SSTYPE(3)
  /PLOT = PROFILE( side*FILTER*SIZE )
  /PRINT = DESCRIPTIVE
  /CRITERIA = ALPHA(.05)
  /WSDESIGN = side
  /DESIGN = SIZE FILTER SIZE*FILTER .
```

Within-Subjects Factors

Measure: MEASURE_1

side	Dependent Variable
1	RIGHT
2	LEFT

Between-Subjects Factors

SIZE	FILTER			Value Label	N
small	standard	SIZE	1	small	3
		FILTER	1	standard	3
	octel	SIZE	1	small	3
		FILTER	2	octel	3
medium	standard	SIZE	2	medium	3
		FILTER	1	standard	3
	octel	SIZE	2	medium	3
		FILTER	2	octel	3
large	standard	SIZE	3	large	3
		FILTER	1	standard	3
	octel	SIZE	3	large	3
		FILTER	2	octel	3

TABLE 7.29 Continued

Descriptive Statistics

	SIZE	FILTER	Mean	Std. Deviation	N
RIGHT	small	standard	81.6667	.57735	3
		octel	82.0000	.00000	3
		Total	81.8333	.40825	6
	medium	standard	84.1667	.28868	3
		octel	82.1667	.28868	3
		Total	83.1667	1.12546	6
	large	standard	78.6667	.28868	3
		octel	77.1667	.57735	3
		Total	77.9167	.91742	6
	Total	standard	81.5000	2.41091	9
		octel	80.4444	2.48048	9
		Total	80.9722	2.43427	18
LEFT	small	standard	83.5000	.00000	3
		octel	82.5000	.00000	3
		Total	83.0000	.54772	6
	medium	standard	85.0000	.50000	3
		octel	82.1667	.57735	3
		Total	83.5833	1.62532	6
	large	standard	76.3333	.57735	3
		octel	76.8333	.76376	3
		Total	76.5833	.66458	6
	Total	standard	81.6111	4.02941	9
		octel	80.5000	2.79508	9
		Total	81.0556	3.41230	18

Tests of Within-Subjects Effects

Measure: noise

Source		Type III Sum of Squares	df	Mean Square	F	Sig.
side	Sphericity Assumed	.062	1	.062	.243	.631
	Greenhouse-Geisser	.062	1.000	.062	.243	.631
	Huynh-Feldt	.062	1.000	.062	.243	.631
	Lower-bound	.062	1.000	.062	.243	.631
side * SIZE	Sphericity Assumed	9.875	2	4.938	19.216	.000
	Greenhouse-Geisser	9.875	2.000	4.938	19.216	.000
	Huynh-Feldt	9.875	2.000	4.938	19.216	.000
	Lower-bound	9.875	2.000	4.938	19.216	.000

(continued)

TABLE 7.29 Continued

Source		Type III Sum of Squares	df	Mean Square	F	Sig.
side * FILTER	Sphericity Assumed	.007	1	.007	.027	.872
	Greenhouse-Geisser	.007	1.000	.007	.027	.872
	Huynh-Feldt	.007	1.000	.007	.027	.872
	Lower-bound	.007	1.000	.007	.027	.872
side * SIZE * FILTER	Sphericity Assumed	4.847	2	2.424	9.432	.003
	Greenhouse-Geisser	4.847	2.000	2.424	9.432	.003
	Huynh-Feldt	4.847	2.000	2.424	9.432	.003
	Lower-bound	4.847	2.000	2.424	9.432	.003
Error (side)	Sphericity Assumed	3.083	12	.257		
	Greenhouse-Geisser	3.083	12.000	.257		
	Huynh-Feldt	3.083	12.000	.257		
	Lower-bound	3.083	12.000	.257		

Tests of Between-Subjects Effects

Measure: noise
Transformed Variable: Average

Source	Type III Sum of Squares	df	Mean Square	F	Sig.
Intercept	236277.007	1	236277.007	1620185	.000
SIZE	260.514	2	130.257	893.190	.000
FILTER	10.563	1	10.563	72.429	.000
SIZE * FILTER	8.042	2	4.021	27.571	.000
Error	1.750	12	.146		

TABLE 7.30 Mean noise as a function of car size, filter type, and side of car

	Filter Type			
	Octel		Standard	
Car size	Right	Left	Right	Left
Small				
M	82.00	82.50	81.67	83.50
SE_M	0	0	0.33	0
Medium				
M	82.17	82.17	84.17	85.00
SE_M	0.17	0.33	0.17	0.29
Large				
M	77.17	76.83	78.67	76.33
SE_M	0.33	0.44	0.17	0.33

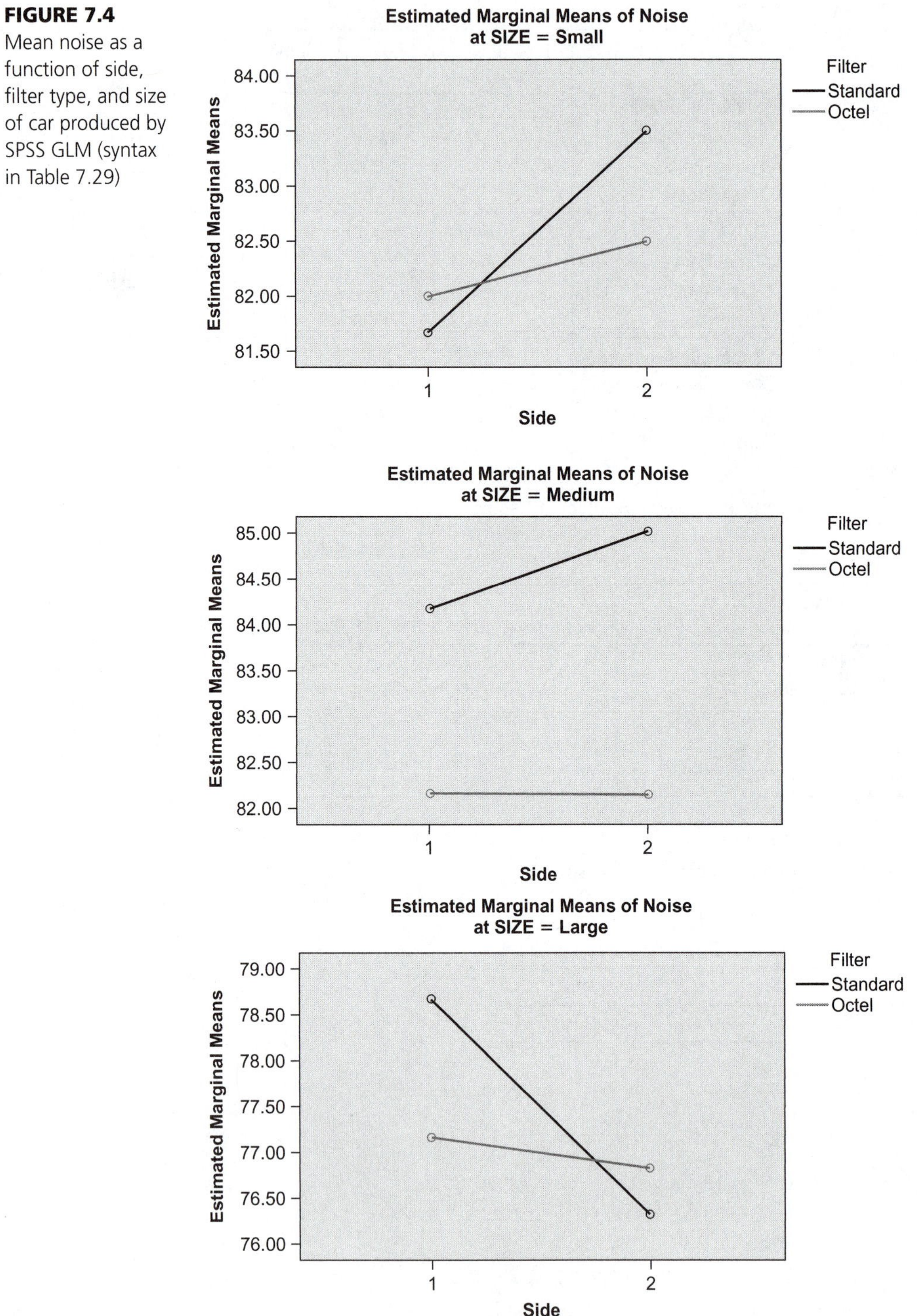

FIGURE 7.4 Mean noise as a function of side, filter type, and size of car produced by SPSS GLM (syntax in Table 7.29)

because the original goal was to show that the Octel filter is no noisier than the standard filter. This is supported by the statistically significant main effect of filter type—$F(1, 12) = 72.43, p < .001$, partial $\omega^2 = (10.563)/(10.563 + 1.75) = .86$. Smithson's (2003) software provides confidence limits around ω^2 of .58 to .92. A look at the means of Table 7.29 shows *lower* means for the Octel filter, on average, and Figure 7.4 suggests no condition in which the Octel filter is appreciably noisier. This observation is tested as a set of simple comparisons.

7.7.2.2 Simple Interaction Comparisons

Simple interactions between side and filter type are analyzed separately for small, medium, and large cars. Because there are only two levels for filter type (standard and Octel) and side (right and left), there is no need for the special language of Table 7.25. This analysis is achieved by sorting and splitting the data for car size and then running the mixed-design ANOVA.

Scheffé-adjusted critical F for these simple interactions requires adding the C effect (side) to Equation 7.11:

$$F_S = (a - 1)(b - 1)(c - 1)F_{(a-1)(b-1)(c-1),\, ab(c-1)(s-1)} \qquad \textbf{(7.12)}$$

For the current hypothesis at $\alpha = .025$,

$$F_S = (2)(1)(1)F_{(2)(1)(1),(3)(2)(1)(2)} = (2)F_{(2,12)} = (2)(5.10) = 10.20$$

By this criterion, one of the three simple interactions, side by filter for small cars, is statistically significant—$F(1,\ 4) = 16.00$. Noise is greater on the left side than on right side of the car using the standard filter, but no such difference is noted for the Octel filter. The basis for calculating effect size for the simple interaction is not clear for this three-way analysis. One reasonable strategy is to look at the proportion of variance in small-car noise that is associated with the filter by side interaction. Thus,

$$\eta^2 = \frac{\text{SS}_{A\times B}}{\text{SS}_{T\text{at small}}} = \frac{1.333}{0.333 + 0.333 + 4.083 + 1.333 + 0.333} = .21$$

where SS_T includes all sources of variance for the small-car analysis from Table 7.31. Thus, 21% of the variance in noise among small cars is attributable to the side by filter interaction. Software for calculating confidence limits around ω^2 is inappropriate for this analysis.

7.7.2.3 Simple Comparisons

Two simple comparisons are of interest in light of the simple main effect of filter (favoring the Octel filter): the higher levels of noise for the Octel filter (shown in Figure 7.4) for small cars (on the right) and large cars (on the left). Table 7.32 shows

TABLE 7.31 Simple interactions of filter by side at each car size through SPSS GLM (syntax and selected output)

```
SORT CASES BY SIZE .
SPLIT FILE
 LAYERED BY SIZE .
GLM
 RIGHT LEFT BY FILTER
 /WSFACTOR = side 2 Polynomial
 /METHOD = SSTYPE(3)
 /PRINT = DESCRIPTIVE
 /CRITERIA = ALPHA(.05)
 /WSDESIGN = side
 /DESIGN = FILTER .
```

Tests of Within-Subjects Effects

Measure: MEASURE_1

SIZE	Source		Type III Sum of Squares	df	Mean Square	F	Sig.
small	side	Sphericity Assumed	4.083	1	4.083	49.000	.002
		Greenhouse-Geisser	4.083	1.000	4.083	49.000	.002
		Huynh-Feldt	4.083	1.000	4.083	49.000	.002
		Lower-bound	4.083	1.000	4.083	49.000	.002
	side * FILTER	Sphericity Assumed	1.333	1	1.333	16.000	.016
		Greenhouse-Geisser	1.333	1.000	1.333	16.000	.016
		Huynh-Feldt	1.333	1.000	1.333	16.000	.016
		Lower-bound	1.333	1.000	1.333	16.000	.016
	Error(side)	Sphericity Assumed	.333	4	.083		
		Greenhouse-Geisser	.333	4.000	.083		
		Huynh-Feldt	.333	4.000	.083		
		Lower-bound	.333	4.000	.083		
medium	side	Sphericity Assumed	.521	1	.521	3.571	.132
		Greenhouse-Geisser	.521	1.000	.521	3.571	.132
		Huynh-Feldt	.521	1.000	.521	3.571	.132
		Lower-bound	.521	1.000	.521	3.571	.132
	side * FILTER	Sphericity Assumed	.521	1	.521	3.571	.132
		Greenhouse-Geisser	.521	1.000	.521	3.571	.132
		Huynh-Feldt	.521	1.000	.521	3.571	.132
		Lower-bound	.521	1.000	.521	3.571	.132
	Error(side)	Sphericity Assumed	.583	4	.146		
		Greenhouse-Geisser	.583	4.000	.146		
		Huynh-Feldt	.583	4.000	.146		
		Lower-bound	.583	4.000	.146		

(*continued*)

TABLE 7.31 Continued

SIZE	Source		Type III Sum of Squares	df	Mean Square	F	Sig.
large	side	Sphericity Assumed	5.333	1	5.333	9.846	.035
		Greenhouse-Geisser	5.333	1.000	5.333	9.846	.035
		Huynh-Feldt	5.333	1.000	5.333	9.846	.035
		Lower-bound	5.333	1.000	5.333	9.846	.035
	side * FILTER	Sphericity Assumed	3.000	1	3.000	5.538	.078
		Greenhouse-Geisser	3.000	1.000	3.000	5.538	.078
		Huynh-Feldt	3.000	1.000	3.000	5.538	.078
		Lower-bound	3.000	1.000	3.000	5.538	.078
	Error(side)	Sphericity Assumed	2.167	4	.542		
		Greenhouse-Geisser	2.167	4.000	.542		
		Huynh-Feldt	2.167	4.000	.542		
		Lower-bound	2.167	4.000	.542		

Tests of Between-Subjects Effects

Measure: MEASURE_1

Transformed Variable: Average

SIZE	Source	Type III Sum of Squares	df	Mean Square	F	Sig.
small	Intercept	81510.083	1	81510.083	978121.0	.000
	FILTER	.333	1	.333	4.000	.116
	Error	.333	4	.083		
medium	Intercept	83416.688	1	83416.688	364000.1	.000
	FILTER	17.521	1	17.521	76.455	.001
	Error	.917	4	.229		
large	Intercept	71610.750	1	71610.750	572886.0	.000
	FILTER	.750	1	.750	6.000	.070
	Error	.500	4	.125		

SPSS GLM (Univariate) syntax and selected output for (a) the comparison between the filters for small cars, right side; and (b) the comparison between filters for large cars, left side. These are the only cases in which the sample Octel filters produced more noise than the sample of standard filters. SELECT CASES is used to select the large-size cars for the first and the small-size cars for the second comparison. As expected, neither of the simple comparisons approaches statistical significance.

Table 7.33 is a checklist for mixed randomized-repeated ANOVA. An example of a results section, in journal format, follows for the study just described.

TABLE 7.32 Simple comparisons between filters conducted through SPSS UNIANOVA

(a) Right side, small cars

```
USE ALL.
COMPUTE filter_$=(SIZE=3 ).
VARIABLE LABEL filter_$ 'SIZE=3  (FILTER)'.
VALUE LABELS filter_$  0 'Not Selected' 1 'Selected'.
FORMAT filter_$ (f1.0).
FILTER BY filter_$.
EXECUTE .
UNIANOVA
  LEFT  BY FILTER
  /METHOD = SSTYPE(3)
  /INTERCEPT = INCLUDE
  /CRITERIA = ALPHA(.05)
  /DESIGN = FILTER .
```

Tests of Between-Subjects Effects

Dependent Variable: LEFT

SIZE	Source	Type III Sum of Squares	df	Mean Square	F	Sig.
large	Corrected Model	.375[a]	1	.375	.818	.417
	Intercept	35190.042	1	35190.042	76778.273	.000
	FILTER	.375	1	.375	.818	.417
	Error	1.833	4	.458		
	Total	35192.250	6			
	Corrected Total	2.208	5			

a. R Squared = .170 (Adjusted R Squared = -.038)

(b) Left side, large cars

```
USE ALL.
COMPUTE filter_$=(SIZE=1 ).
VARIABLE LABEL filter_$ 'SIZE=1  (FILTER)'.
VALUE LABELS filter_$  0 'Not Selected' 1 'Selected'.
FORMAT filter_$ (f1.0).
FILTER BY filter_$.
EXECUTE .
UNIANOVA
  RIGHT  BY FILTER
  /METHOD = SSTYPE(3)
  /INTERCEPT = INCLUDE
  /CRITERIA = ALPHA(.05)  /DESIGN = FILTER .
```

TABLE 7.32 Continued

Tests of Between-Subjects Effects

Dependent Variable: RIGHT

SIZE	Source	Type III Sum of Squares	df	Mean Square	F	Sig.
small	Corrected Model	.167[a]	1	.167	1.000	.374
	Intercept	40180.167	1	40180.167	241081.0	.000
	FILTER	.167	1	.167	1.000	.374
	Error	.667	4	.167		
	Total	40181.000	6			
	Corrected Total	.833	5			

a. R Squared = .200 (Adjusted R Squared = .000)

TABLE 7.33 Checklist for mixed randomized-repeated ANOVA

1. Issues
   a. Independence of errors
   b. Unequal sample sizes
   c. Normality of sampling distributions
   d. Outliers
   e. Homogeneity of variance
   f. Sphericity
2. Major analyses
   a. Main effects or planned comparisons. If statistically significant:
      (1) Effect size(s) and their confidence limits
      (2) Parameter estimates (marginal means and standard deviations or standard errors or confidence intervals)
   b. Interactions or planned comparisons. If significant:
      (1) Effect size(s) and their confidence limits
      (2) Parameter estimates (cell means and standard deviations or standard errors or confidence intervals in table and/or plot)
3. Additional analyses
   a. Post hoc comparisons
   b. Interpretation of departure from homogeneity of variance

Results

A $2 \times 3 \times 2$ mixed randomized-repeated ANOVA was performed on automobile noise. Half of the 18 cars were tested with standard filters and half with Octel filters. Each type of filter was tested with cars of three sizes: small, medium, and large. Noise was measured on the left and right sides of each car. There were three cars in each combination of size and filter type. Normality of sampling distributions and homogeneity of variance could not be evaluated in the presence of small samples and lack of variability in some cells. Therefore, α for testing all effects was set to .025, to compensate for any inflation of Type I error rate due to non-normality or heterogeneity of variance.

There was a statistically significant three-way interaction among car size, filter type, and side of car, $F(2, 12) = 9.43$, $p = .003$, partial $\eta^2 = .61$ with 95% confidence limits from .12 to .76. Means and standard errors for the 12 cells are in Table 7.30. Means are plotted in Figure 7.4.

The interaction between side and filter type differs for the three car sizes. Therefore, interactions were compared separately for small, medium, and large cars, with a Scheffé adjustment for post hoc testing. Only for small cars was the interaction statistically significant, $F(1, 4) = 16.00$, $p < .025$. Relative to noise in small cars, $\eta^2 = .21$ for the interaction. Figure 7.4 shows a difference between left and right sides for the standard filter but not for the Octel filter.

Two additional comparisons are of interest in this research, because a major goal was to demonstrate that the Octel filter is no noisier than a standard filter. The main effect of filter type shows that, on average, the standard filter is noisier than the Octel filter, $F(1, 12) = 72.43$, $p < .001$, partial $\eta^2 = .86$ with confidence limits from .59 to .92. However, Figure 7.4 suggests that the Octel filter is noisier for large cars measured on the left side and small cars measured on the right side. Therefore, Scheffé-adjusted simple comparisons between the filters were made for these two conditions. Neither of the simple comparisons approached statistical significance; $F(1, 4) = 1.00$ and 0.82 for the right side (small cars) and left side (large cars), respectively.

The main effect of size and the interaction between size and side were also statistically significant, but were not of interest in light of the significant three-way interaction and the emphasis on filters.

7.8 COMPARISON OF PROGRAMS

Features of the SPSS, SAS, and SYSTAT programs for factorial randomized-groups designs and factorial repeated-measures designs apply to mixed randomized-repeated designs and are detailed in Sections 5.7 and 6.6. The only features that are unique to mixed designs have to do with specific comparisons; these are summarized in Table 7.34.

7.8.1 SPSS Package

SPSS GLM handles mixed designs with no difficulty, lacking only a special syntax for simple main effects analysis. These are easily performed, however, by limiting the analysis to one level of the IV to be held constant. Simple-effects analysis on the randomized-groups IV also may be done through SPSS ONEWAY, as demonstrated in Table 7.17. SPSS MANOVA is also useful for interaction contrasts, because a whole set of orthogonal contrasts may be done in a single run.

TABLE 7.34 Comparison of SPSS, SAS, and SYSTAT programs for mixed randomized-repeated measures ANOVA-specific comparisons

Feature	SPSS GLM	SAS GLM	SYSTAT ANOVA/GLM
Input			
Specify comparisons on the randomized-groups margin	LMATRIX	CONTRAST	CONTRAST
Specify polynomial comparisons on the repeated-measures margin	Yes (default)	Yes	PRINT=MEDIUM
User-specified comparisons on the repeated-measures margin	SPECIAL	No	No
Special syntax for simple main effects on repeated-measures IV	No	BY	No
Specify polynomial comparisons for repeated-measures simple comparisons	Yes (default)	Yes	Yes
User-specified coefficients for repeated-measures simple comparisons	SPECIAL	No	CONTRAST
User-specified interaction contrasts for both (1) randomized and (2) repeated-measures IVs	(1) LMATRIX (2) MMATRIX,	(1) CONTRAST, (2) MANOVA	(1) CONTRAST (2) CMATRIX
Output			
Univariate results for fully user-specified interaction contrasts	Yes	No	Yes
Multivariate results for fully user-specified interaction contrasts	No	Yes	No

7.8.2 SAS System

SAS GLM is a less satisfying program for mixed analysis, because it does not permit user-specified contrasts on the repeated-measures factor, either for marginal tests or for simple comparisons. User-specified interaction contrasts are available; however, the output is in the unfamiliar form of a multivariate test. Nevertheless, the results match those of the univariate tests of the other programs. One handy feature of SAS GLM is the special syntax that permits a full set of simple main effects on the repeated-measures factor in a single run.

7.8.3 SYSTAT System

SYSTAT ANOVA (and GLM) also has some idiosyncrasies in handling mixed designs. For example, user-specified coefficients may be used for repeated-measures simple comparisons but not for comparisons on the repeated-measures margin. Marginal means are not available for the randomized-groups factor but may be calculated easily from the cell means printed out with a request for PRINT=MEDIUM.

7.9 PUTTING IT ALL TOGETHER FOR FACTORIAL DESIGNS

This section revisits material from Chapters 5 (randomized groups), 6 (repeated measures), and 7 (mixed randomized, repeated measures) factorial designs. The goal is to make some overall patterns apparent by comparing the three analyses. A two-way factorial design is used for the examples, each IV with three levels. Consider A qualitative (e.g., type of novel) and B quantitative (e.g., months). As more IVs are added, the analysis becomes more complex, but the basic rules generalize.

7.9.1 Allocation of Cases

Table 7.35 compares the allocation of cases (three in this example) for (a) the randomized-groups design, (b) the repeated-measures design, and (c) the mixed design (where A is the randomized-groups IV and B is the repeated-measures IV).

There are 27 cases in the randomized-groups design, each providing a single DV score. There are three cases in the repeated-measures design, each providing nine DV scores, one at each combination of levels of A and B. The mixed randomized, repeated-measures design is genuinely a blend of both designs. In this example, the design requires nine cases, each providing three scores, one at each level of B.

If cases are rare or expensive to run, the repeated-measures design has clear advantages. However, if cases need to be naïve or if cases are changed by the measurement process, the randomized-groups design is preferable. If cases need to be naïve with respect to one IV (A) but can provide more than one response to the other (B), the mixed design is preferable. Finally, in many settings, it is desirable to block on a variable that represents characteristics of cases (e.g., demography) to produce more homogeneous groups for development of the error term; either the randomized-groups design or the mixed design is available for this goal.

7.9.2 Assumptions of Analysis

Table 7.36 summarizes the assumptions of analysis for the three designs. Consult each chapter for alternatives if an assumption is violated.

7.9.3 Error Terms

The computations of main effects (A and B, in this example) and the interactions of main effects ($A \times B$, in this example) are the same for all three designs. It is the error terms that change depending on how cases are used in the design. Table 7.37 summarizes the error terms used for each effect and for various types of comparisons for randomized-groups, repeated-measures, and mixed designs. For clarity, and because

TABLE 7.35 Allocation of cases in 3 × 3 randomized-groups, repeated-measures, and mixed designs

(a) Randomized-groups design

	b_1	b_2	b_3
a_1	s_1 s_2 s_3	s_{10} s_{11} s_{12}	s_{19} s_{20} s_{21}
a_2	s_4 s_5 s_6	s_{13} s_{14} s_{15}	s_{22} s_{23} s_{24}
a_3	s_7 s_8 s_9	s_{16} s_{17} s_{18}	s_{25} s_{26} s_{27}

(b) Repeated-measures design

	a_1			a_2			a_3		
	b_1	b_2	b_3	b_1	b_2	b_3	b_1	b_2	b_3
s_1 s_2 s_3									

(c) Mixed randomized, repeated-measures design

		b_1	b_2	b_3
a_1	s_1 s_2 s_3			
a_2	s_4 s_5 s_6			
a_3	s_7 s_8 s_9			

it is often the desirable analysis when a repeated measure is quantitative, a comparison on A is referred to as A_{comp}, and a comparison on B is referred to as B_{linear}.

There are several things to notice in this table. First, in the randomized-groups design, there is a single error term throughout. Second, in the repeated-measures design, each effect is tested by that effect in interaction with cases. Third, in the

TABLE 7.36 Assumptions of analysis for randomized-groups, repeated-measures, and mixed designs

	Randomized Groups	Repeated Measures	Mixed
Normality of sampling distributions of means	Met with 20 df for error, roughly equal n in each cell, and 2-tailed tests. In 3×3 design, requires 4 cases per cell. If df low, check normality in the cells.	Met with 20 df for each error term. In 3×3 design, requires 11 cases. If df low, check normality on the margin for each effect.	*Randomized groups:* Met with 20 df for error, roughly equal n in each group, and 2-tailed tests. In 3×3 design, requires 8 cases per level of A. *Repeated measures:* Met with 20 df for each error term.
Homogeneity of variance	Met if F_{max} across cells 10 or less, with sample-size ratio $\leq$ 4:1; or Levene's test nonsignificant with sample size ratio > 4:1.	Met if F_{max} on the margin for each effect 10 or less.	*Randomized groups:* Met if F_{max} for case totals 10 or less; with sample-size ratio $\leq$ 4 :1; or Levene's test nonsignificant with sample size ratio > 4:1. *Repeated measures:* Met if F_{max} on the margin for each effect 10 or less.
Independence of errors	Depends on control of extraneous variables during testing.	Met if Mauchly's test of sphericity not significant for each effect with more than 1 df.	*Randomized groups:* Depends on control of extraneous variables. *Repeated measures:* Met if Mauchly's test of sphericity not significant for each effect with more than 1 df.
Additivity	N/A		
Absence of outliers	Search in each cell of the design.	Search in each separate column of the DV.	Search in each separate column of the DV at each level of the randomized-groups IV.

mixed design, things are more complicated, but there are still some rules. Purely randomized-groups effects (in this example, only A) are tested against cases nested within those effects. Any effect with a repeated-measures component (B and $A \times B$, in this example) is tested against the repeated-measures effect(s) in interaction with cases, nested within levels of the randomized-groups design. Fourth, for both repeated measures and the repeated portion of a mixed design, when a comparison is applied to the repeated measure, it is also applied to the error term. For example, the error term for B_{linear} at a_1 is $B_{linear} \times S$ at a_1.

7.9.4 Setup for Regression

Table 7.38 contains the setup for doing a randomized-groups analysis through multiple regression. In this example, A is qualitative and has 2 df. Therefore, two columns are required to code the three levels of A. Any pair of two sets of orthogonal contrast

TABLE 7.37 Error terms used for various effects in randomized-groups, repeated-measures, and mixed designs

Source	Randomized Groups (Chapter 5)	Repeated Measures (Chapter 6)	Mixed (Chapter 7)
A	S/AB	$A \times S$	S/A
B	S/AB	$B \times S$	$B \times S/A$
$A \times B$	S/AB	$A \times B \times S$	$B \times S/A$
	Marginal Comparisons		
A_{comp}	S/AB	$A_{comp} \times S$	S/A
B_{linear}	S/AB	$B_{linear} \times S$	$B_{linear} \times S/A$
	Simple Effects		
Simple Main Effects			
A at b_j	S/AB[a]	$A \times S$ at b_j	S/A at b_j
B at a_i	S/AB[b]	$B \times S$ at a_i	$B \times S$ at a_i
Simple Comparisons			
A_{comp} at b_j	S/AB[a]	$A_{comp} \times S$ at b_j	S/A at b_j
B_{linear} at a_i	S/AB[b]	$B_{linear} \times S$ at a_i	$B_{linear} \times S$ at a_i
	Interaction Contrasts		
$A_{comp} \times B$	S/AB	$A_{comp} \times B \times S$	$B \times S/A$
$A \times B_{linear}$	S/AB	$A \times B_{linear} \times S$	$B_{linear} \times S/A$
$A_{comp} \times B_{linear}$	S/AB	$A_{comp} \times B_{linear} \times S$	$B_{linear} \times S/A$

[a] Use S/A at b_j if failure of homogeneity of variance at b_j.
[b] Use S/B at a_i if failure of homogeneity of variance at a_i.

TABLE 7.38 Columns for randomized-groups ANOVA through regression (also first columns for repeated-measures and mixed designs)

A		***B***		***A × B***			
A_{comp1}	A_{comp2}	B_{linear}	B_{quad}	$A_{comp1} \times B_{linear}$	$A_{comp1} \times B_{quad}$	$A_{comp2} \times B_{linear}$	$A_{comp2} \times B_{quad}$
X_1	X_2	X_3	X_4	X_5	X_6	X_7	X_8
				$X_1 \times X_3$	$X_1 \times X_4$	$X_2 \times X_3$	$X_2 \times X_4$

coefficients can be used; call them A_{comp1} and A_{comp2}. B is quantitative, also with 2 df. Two columns are required to code these three levels, as well—say B_{linear} and B_{quad}. There are 4 df for the interaction, so four columns are required to code for interaction. It is important to realize that these eight columns are also required to code the main effects and interaction for repeated-measures and mixed designs. The additional columns in these designs are required to code various elements of the error terms.

TABLE 7.39 Additional columns required to produce the error terms for repeated-measures designs

S		*A* × *S*				*B* × *S*			
Sc_1	Sc_2	$A_{comp1} \times Sc_1$	$A_{comp1} \times Sc_2$	$A_{comp2} \times Sc_1$	$A_{comp2} \times Sc_2$	$B_{linear} \times Sc_1$	$B_{linear} \times Sc_2$	$B_{quad} \times Sc_1$	$B_{quad} \times Sc_2$
X_9	X_{10}	X_{11}	X_{12}	X_{13}	X_{14}	X_{15}	X_{16}	X_{17}	X_{18}
		$X_1 \times X_9$	$X_1 \times X_{10}$	$X_2 \times X_9$	$X_2 \times X_{10}$	$X_3 \times X_9$	$X_3 \times X_{10}$	$X_4 \times X_9$	$X_4 \times X_{10}$

A × *B* × *S*							
$A_{comp1} \times B_{linear} \times Sc_1$	$A_{comp1} \times B_{linear} \times Sc_2$	$A_{comp1} \times B_{quad} \times Sc_1$	$A_{comp1} \times B_{quad} \times Sc_2$	$A_{comp2} \times B_{linear} \times Sc_1$	$A_{comp2} \times B_{linear} \times Sc_2$	$A_{comp2} \times B_{quad} \times Sc_1$	$A_{comp2} \times B_{quad} \times Sc_2$
X_{19}	X_{20}	X_{21}	X_{22}	X_{23}	X_{24}	X_{25}	X_{26}
$X_5 \times X_9$	$X_5 \times X_{10}$	$X_6 \times X_9$	$X_6 \times X_{10}$	$X_7 \times X_9$	$X_7 \times X_{10}$	$X_8 \times X_9$	$X_8 \times X_{10}$

Cases are distinguished from each other to complete the analysis for a repeated-measures design. In this example, there are a very inadequate three cases, requiring two columns, say Sc_1 and Sc_2. Table 7.39 contains those columns, plus the interaction of cases with each of the preceding eight columns. The interaction of cases with the main effects and interactions produces the required error terms.

After the SS(reg.) for each column is determined, various columns can be combined to produce desired error terms for comparisons. For instance, if X_{11} and X_{12} are added, the result is $A_{comp1} \times S$. If X_{17} and X_{18} are added, the result is $B_{quad} \times S$. To produce $A_{comp2} \times B \times S$, add together X_{23}–X_{26}. To produce $A_{comp1} \times B_{linear} \times S$, add together X_{19} and X_{20}.

In the repeated-measures design, cases are crossed with all effects. In the mixed design, cases cross the repeated-measures effects but are nested in the randomized-groups effects. This pattern in the way cases participate in the design is used to develop the error terms for the mixed design, as shown for the example in Table 7.40. Assume that Sc_1 and Sc_2 represent the first three of the nine cases nested in a_1, Sc_3

TABLE 7.40 Additional columns required to produce the error terms for mixed designs

S/*A*					
Sc_1 at a_1	Sc_2 at a_1	Sc_3 at a_2	Sc_4 at a_2	Sc_5 at a_3	Sc_6 at a_3
X_9	X_{10}	X_{11}	X_{12}	X_{13}	X_{14}

B × *S*/*A*											
$B_{linear} \times Sc_1$	$B_{linear} \times Sc_2$	$B_{linear} \times Sc_3$	$B_{linear} \times Sc_4$	$B_{linear} \times Sc_5$	$B_{linear} \times Sc_6$	$B_{quad} \times Sc_1$	$B_{quad} \times Sc_2$	$B_{quad} \times Sc_3$	$B_{quad} \times Sc_4$	$B_{quad} \times Sc_5$	$B_{quad} \times Sc_6$
X_{15}	X_{16}	X_{17}	X_{18}	X_{19}	X_{20}	X_{21}	X_{22}	X_{23}	X_{24}	X_{25}	X_{26}
$X_3 \times X_9$	$X_3 \times X_{10}$	$X_3 \times X_{11}$	$X_3 \times X_{12}$	$X_3 \times X_{13}$	$X_3 \times X_{14}$	$X_4 \times X_9$	$X_4 \times X_{10}$	$X_4 \times X_{11}$	$X_4 \times X_{12}$	$X_4 \times X_{13}$	$X_4 \times X_{14}$

and Sc_4 represent the next three cases nested in a_2, and Sc_5 and Sc_6 represent the last three cases nested in a_3. The S/A error term has $a(s - 1) = 6$ df and requires six columns. Cases then cross levels of B, while they nest in levels of A. The $B \times S/A$ error term has $a(b - 1)(s - 1) = 12$ df and requires 12 columns.

In Table 7.40, if X_{15} through X_{20} are added, the result is $B_{\text{linear}} \times S/A$; whereas if X_{21} through X_{26} are added, the result is $B_{\text{quad}} \times S/A$. Columns associated with cases are not meaningful unless there is some rationale to the comparisons applied to the cases.

It should be noted that once the columns for the regression approach are created, the analysis can be performed through one of the multiple-regression software programs, as shown in Table 5.43. Most of the programs have the capacity for performing a sequential analysis (see Tabachnick and Fidell, 2007), where the researcher determines the order of entry of the variables. If there is equal n in each cell, the analysis produces the same results, no matter the order. However, with unequal n, the researcher can determine which effect gets overlapping variability (Figure 5.11) by entering them into the equation first. Columns for continuous variables (serving either as covariates or simply as continuous IVs) can also be added. In an ordinary ANCOVA (Chapter 8), covariates are entered first and "take" overlapping variability. Through use of sequential regression, the priority of the covariates (or a continuous IV) can be adjusted by entering them into the equation later. Using this method, the researcher can take absolute control of the progress of the solution.

7.9.5 Developing Computational Equations from Degrees of Freedom

Once you understand the pattern of crossing and nesting in your design, you can write the degrees of freedom for each effect. The degrees of freedom can be readily expanded and "translated" into the computational equations for an effect. Table 7.41 shows the expansion for the two-way mixed design.

TABLE 7.41 Development of computational equations from degrees of freedom for the mixed design

Source	Degrees of Freedom	Expansion of df	Computational Equation
A	$(a - 1)$	$a - 1$	$\frac{\sum A^2}{bn} - \frac{T^2}{abn}$
S/A	$a(s - 1)$	$as - a$	$\frac{\sum (AS)^2}{b} - \frac{\sum A^2}{bn}$
B	$(b - 1)$	$b - 1$	$\frac{\sum B^2}{an} - \frac{T^2}{abn}$
$A \times B$	$(a - 1)(b - 1)$	$ab - a - b + 1$	$\frac{\sum (AB)^2}{n} - \frac{\sum A^2}{bn} - \frac{\sum B^2}{an} + \frac{T^2}{abn}$
$B \times S/A$	$a(b - 1)(s - 1)$	$a(bs - b - s + 1)$ $= abs - ab - as + a$	$\sum Y^2 - \frac{\sum (AB)^2}{n} - \frac{\sum (AS)^2}{b} + \frac{\sum A^2}{bn}$

The elements in the design are A, B, and S. Notice that each element in the computational equations contains all three, either in the numerator or in the denominator. The letters and their combinations that are developed by expansion of the degrees of freedom refer to the numerators in the computational equations. When *abs* is developed in the $B \times S/A$ error term, it refers to a particular S at a particular AB combination. That is the same as Y, which is substituted into the computational equation. When a 1 is developed, T is substituted into the computational equation. The values in the numerator are squared, and the letters that do not appear in the numerator are in the denominator. Once this pattern is apparent to you, you can develop the computational equations for any design from the degrees of freedom.

7.10 PROBLEM SETS

7.1 Do a one-way randomized-groups ANOVA for the case totals in Table 7.2 and compare with the results for mixed design. That is, use A as the single IV, and use the case totals (summed over B) as DV scores.

7.2 In the following two-way mixed design, the randomized-groups IV is type of music, and the repeated-measures IV is time of day. The DV is rating of the music.

			Time of Day			
			9 a.m. b_1	11 a.m. b_2	1 p.m. b_3	4 p.m. b_4
		s_1	8	7	5	3
Rock	a_1	s_2	9	7	5	1
		s_3	8	5	4	1
		s_4	6	6	4	2
		s_5	2	4	8	10
Classical	a_2	s_6	6	5	9	10
		s_7	2	4	6	10
		s_8	3	5	6	9

a. Use hand calculations to check the assumptions of normality, homogeneity of variance, and no outliers.
b. Do the omnibus ANOVA for this design using hand calculation.
c. Calculate effect sizes and find their 95% confidence limits.
d. Use a computer program to check the preceding results and test sphericity.
e. Write a results section.

7.3 Do the contrast in Table 7.15 using the regression approach (linear trend of month).

7.4 Do a test by hand of the simple main effect of novel at each month in the small-sample example.

Extending the Concepts

7.5 Generate error terms for the following designs:

a. Three randomized-groups IVs and one repeated-measures IV
b. One randomized-groups IV and three repeated-measures IVs

7.6 Consider the following three-way design with one randomized group and two repeated factors. The randomized-groups IV is pregnancy state. Four glucose blood tests were made at hourly intervals, either after fasting or after a glucose load (100 g glucose). The DV is glucose reading.

	Fasting				**Post-Glucose**			
Hours	1	2	3	4	1	2	3	4
	60	57	50	45	97	110	98	66
Pregnant	80	79	76	69	103	105	107	108
	74	64	70	66	109	101	103	100
	60	62	61	65	130	134	121	125
	45	83	66	71	59	85	60	73
Nonpregnant	89	90	89	77	90	144	100	121
	103	104	93	90	111	129	96	124
	77	82	82	78	86	115	93	89

a. Use computer software to do the omnibus ANOVA and test all assumptions of the analysis.
b. Calculate effect size(s) and use computer software to find their confidence limits.
c. Do appropriate post hoc analyses suggested by the results of the omnibus ANOVA.
d. Write a results section.

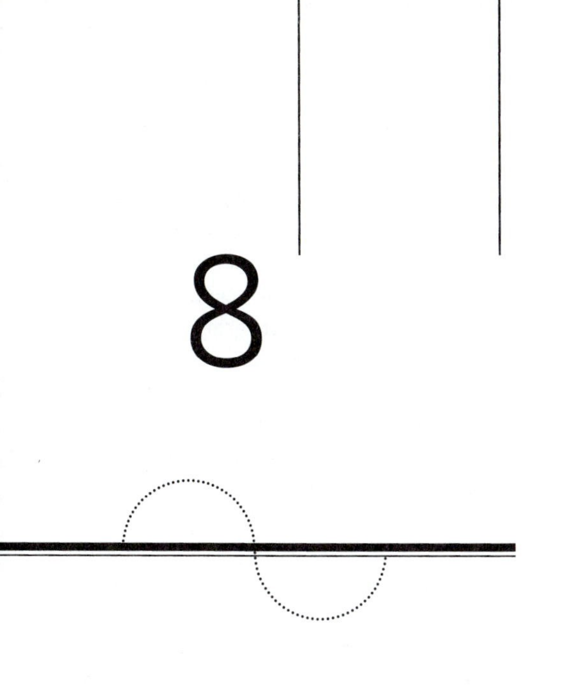

Analysis of Covariance

8.1 GENERAL PURPOSE AND DESCRIPTION

Analysis of covariance (ANCOVA) is an extension of analysis of variance in which main effects and interactions of IVs are assessed after DV scores are adjusted for differences associated with one or more covariates (continuous variables that are measured before the DV and correlated with it).[1] The major questions for ANCOVA are the same as for ANOVA: Are mean differences among levels of the IV on the adjusted DV likely to have occurred by chance? Is there a mean difference among levels of an IV on a posttest (the DV) after posttest scores are adjusted for differences in pretest scores (the covariate)? For example, skill in assembling computers (the covariate) is assessed *before* random assignment to one of three training groups: a no training control group, a videotaped demonstration of the assembly procedures, and hands-on training. The same test of skill then serves as the DV when administered after training. During analysis, differences in skill among cases on the pretest are removed from the error term, making the adjusted ANOVA a potentially more powerful test of the effects of treatment. Statistical analysis tests the null hypothesis that the three groups have the same mean DV score after adjustment for preexisting differences in skill.

[1] Much of the material for this chapter is adapted from Tabachnick and Fidell (2007).

The analysis develops a regression between the DV and the covariate(s). This regression is used to assess the "noise"—the undesirable variance in the DV (individual differences among cases) estimated from scores on covariates (CVs). Although in most experiments, cases are randomly assigned to groups, random assignment in no way ensures equality among groups—it only guarantees, within probability limits, that there are no preexisting *systematic* differences among groups. Random differences among cases tend to spread out scores among cases within groups (thereby increasing the error term) and may also create mean differences that are not associated with treatment. Either of these effects can obscure differences among groups due to levels of experimental treatment. ANCOVA removes from the error term some differences among cases (making the error term smaller) and adjusts the means to the levels predicted if all cases had the same score on the covariate at the beginning.

Both ANCOVA and repeated-measures ANOVA (cf. Chapter 6) are used to increase the power of a study by assessing and then removing from the error term variability associated with differences among cases. In repeated-measures ANOVA, variance due to individual differences among cases is estimated from consistencies within cases over treatments. In ANCOVA, variance due to individual differences is estimated from the linear regression between the DV and the CV(s); the DV scores are statistically adjusted to what they would be if all cases scored identically on the CV(s), and then ANOVA is performed on the adjusted DV scores. Tests of main effects and interactions are more powerful when there are preexisting differences among cases on the DV as assessed by the CV. Otherwise, power is the same as, or lower than, in randomized-groups designs without CVs.

In the classical experimental use of ANCOVA, cases are measured on one or more CVs before the cases are randomly assigned to levels of one or more IVs. A CV is usually a continuous variable and is often measured on the same scale as the DV. However, a CV might also be some characteristic of the case (e.g., IQ, educational level, anxiety level, thickness of the fiber in an example testing fiber strength) that is measured on a different scale from the DV. In experiments, CVs are not systematically related to levels of IV(s)[2] because they are assessed prior to treatment.

ANCOVA is also used in nonexperimental studies when cases cannot be randomly assigned to treatment or when levels of IVs are not manipulated. Group means are adjusted to what they would be if all cases had identical scores on the CV(s) and differences among cases on CVs are removed, so that, presumably, the only mean differences that remain are related to the grouping IV(s). One problem is that differences could also be due to attributes that have not been used as CVs. A further problem is that these CVs are frequently related to the IV(s) as well as the DV. In nonexperimental studies, then, interpretation of ANCOVA is fraught with difficulty. This application of ANCOVA is primarily for descriptive model building—the CV enhances prediction of the DV—but there is no implication of causality. If the research question involves causality, ANCOVA is no substitute for running an experiment.[3]

[2] However, IV-CV relationships may occur by chance.

[3] A third application of ANCOVA is to use stepdown analysis to interpret IV differences when several DVs are used in MANOVA, as described in Tabachnick and Fidell (2007, Chapter 7) and elsewhere.

Although interpretations differ, the statistical operations are identical in any application of ANCOVA. In randomized-groups ANCOVA, like randomized-groups ANOVA, variance in scores is partitioned into variance due to differences between levels of the IV and variance due to differences within levels. Squared differences between scores and various means are summed, and these sums of squares (when divided by appropriate degrees of freedom) provide estimates of variance attributable to main effects, interactions, and error. In ANCOVA, however, the regression of one or more CVs on the DV is estimated first, DV scores are adjusted, and then ANOVA is performed on adjusted values.

Lee (1975) presented an intuitively appealing illustration of the manner in which ANCOVA reduces error variance in a one-way randomized-groups design with three levels of the IV (Figure 8.1). The vertical axis on the right side of the figure illustrates scores and group means in traditional ANOVA. The error term, computed from the sum of squared deviations of DV scores around their associated group means, is substantial because there is considerable spread in scores within each group.

When the same scores are analyzed in ANCOVA, a regression line is first found that relates the DV to the CV. The error term is based on the (sum of squared) deviations of the DV scores from the regression line running through each group mean instead of from the means themselves. Consider the score in the lower left corner of Figure 8.1. The score is near the regression line (a small deviation for error in ANCOVA) but far from the mean for its own group (a large deviation for error in ANOVA). As long as the slope of the regression lines is not 0, ANCOVA produces a smaller sum of squares for error than ANOVA produces. If the slope is 0, sum of squares for error is the same as in ANOVA, but mean square for error is slightly larger because each CV uses up 1 df from the error term.

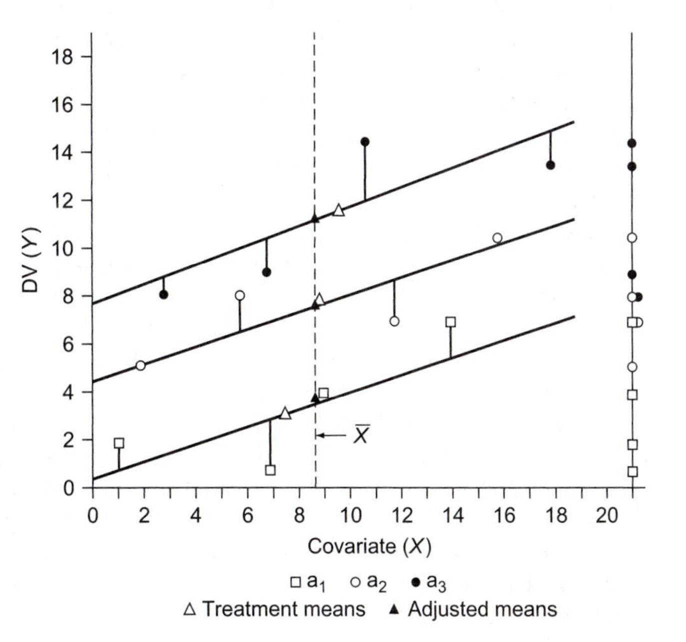

FIGURE 8.1 Plot of hypothetical data: The straight lines with common slope are those that best fit the data for the three treatments. The data points are also plotted along the single vertical line on the right. (From *Experimental Design and Analysis,* by Wayne Lee. Copyright © 1975 by W. H. Freeman and Company. Used with permission.)

ANCOVA is extremely flexible. CVs can be measured on the same scale as the DV or on a different scale. CVs can be either continuous or dichotomous. CVs can be measured once or repeatedly. There may be one or more CVs, measured on the same or different scales, each measured once or repeatedly. CVs can also be used in all ANOVA designs—one-way and factorial randomized-groups designs, repeated-measures designs, mixed randomized-repeated designs, partial factorial designs, random-effects designs, and so on. Specific comparisons and trend analysis are available in ANCOVA, along with estimates of effect size. A price for this flexibility includes some additional assumptions and the necessity of testing those assumptions, along with a somewhat less straightforward interpretation of results.

8.2 KINDS OF RESEARCH QUESTIONS

As with ANOVA, the question in ANCOVA is whether mean differences on the adjusted DV between levels (or combinations of levels) of the IV(s) are larger than expected by chance. In ANCOVA, however, there are additional questions associated with the CV(s).

8.2.1 Effect of the IV(s)

Holding all else constant, are differences in adjusted means that are associated with different levels of an IV larger than expected by chance? For example, is skill in computer assembly affected by different types of training after removing prior individual differences in assembly skill? If there is a second IV (say, size of computer to be assembled) with two levels (desktop and laptop), mean differences in adjusted skill in assembly are tested for this IV as well. The procedures described in Section 8.4 answer this question by testing the null hypothesis that this IV also has no systematic effect on the adjusted DV.

In experimental design, "holding all else constant" is accomplished through several procedures. One of them, the topic for this chapter, is to measure the influence of extraneous variables (CVs) and hold their influence constant by statistically adjusting for differences on them. A second procedure is to institute strict experimental controls over extraneous variables to hold them constant while levels of IV(s) are manipulated. A third procedure is to turn a potentially effective extraneous variable into a blocking IV and study its effects by crossing the levels of the manipulated IV with it in factorial design. Another procedure is to randomize the levels of the extraneous variable (such as time of day) throughout the levels of the primary IV in the hopes of distributing the effects of the extraneous variable equally among treatments. A major consideration in experimental design is which of these procedures, or which combination of these procedures, is more effective or more feasible for which extraneous variables in a given research program.

8.2.2 Effect of Interactions among IVs

Holding all else constant, does change in response over levels of one IV depend on levels of another IV? That is, do IVs interact in their effect on response?[4] For example, when size of computer is added to type of training as a second IV, are the differences in skill produced by the three levels of training the same for both desktop and laptop computers after adjusting for preexisting differences in assembly skill? With more than two IVs, numerous interactions are generated.

Although interpreted differently from main effects, tests of interactions are statistically similar, as demonstrated in Section 8.5. Except for common error terms in some designs, each interaction is tested separately from main effects and other interactions. These tests are independent when sample sizes in all groups are equal and the design is balanced.

8.2.3 Effects of Covariates

Analysis of covariance is based on a linear regression between CV(s) and the DV, but there is no guarantee that the regression actually provides adjustment. The regression is evaluated statistically by testing the CV(s) as a source of variance in DV scores. In the computer assembly example, is pretest skill (the CV) related to posttest skill (the DV), ignoring or adjusting for effects of differential treatment? Evaluation of CV(s) is illustrated in Section 8.4.4.

8.2.4 Parameter Estimates

If a main effect or interaction is statistically significant, what adjusted population parameter (adjusted mean) is estimated for each level of the IV? What is the group mean for each level of the IV after adjustment for CVs, and what are the standard errors of those means? For example, if there is a main effect of type of training, what is the mean adjusted posttest assembly skill score in each of the three groups? If there is significant interaction, what then are the mean adjusted posttest assembly skill scores in each of the six cells that are formed by combining type of training with size of computer? Section 8.6.5 shows how to calculate adjusted means, and Section 8.7 demonstrates their use, along with standard errors, in an example.

8.2.5 Effect Sizes

If a main effect or interaction is associated with changes in the DV, the next logical question is, by how much? How much of the variance in adjusted DV scores is associated with each main effect or interaction? For example, if the mean difference for type of training is statistically significant, one may ask, What proportion of variance

[4] Chapter 5 contains an extensive discussion of interaction.

in the adjusted assembly skill scores is attributed to the IV? Estimation of adjusted effect size and its confidence interval is described in Section 8.6.3 and demonstrated in Section 8.7.

8.2.6 Power

How high is the probability of showing that there are mean differences among groups if, indeed, treatment has an effect? For example, if the hands-on program is superior to the video-based training program or to no program at all, what is the probability of demonstrating that superiority? Estimates of power made prior to running a study are used to help determine sample size. Estimates of power made as part of the analysis of a completed study are primarily used to determine why the research hypothesis did not find support and how many cases need to be added to demonstrate statistical significance in a replication of the study. Power is discussed in Section 8.6.4.

8.2.7 Specific Comparisons

When statistically significant effects are found for an IV with more than two levels, what is the source of the differences? Which adjusted group means differ significantly from each other? Or, if the levels of the IV differ quantitatively, is there a simple trend over levels of the IV? For example, after adjusting for individual differences in skill, (1) are the two treatment groups more effective in increasing assembly skill than the no training control, and (2) is the hands-on training more effective than video-based training?

Tests for these two questions could either be planned and performed in lieu of ANCOVA or, with some loss in sensitivity, be done post hoc after finding a main effect of the IV in ANCOVA. Planned and post hoc comparisons in ANCOVA are discussed in Section 8.6.6.

8.3 ASSUMPTIONS AND LIMITATIONS

8.3.1 Theoretical Issues

As with ANOVA, the statistical test in no way ensures that changes in the DV are caused by an IV. The inference of causality is a logical, rather than a statistical, problem that depends on (1) the manner in which cases are assigned to levels of the IV(s), (2) manipulation of levels of the IV(s) by the researcher, and (3) controls used in the research. Statistics are available to test hypotheses from both nonexperimental and experimental research, but only in the latter case is attribution of causality justified. Nonexperimental use of CVs is both controversial and fraught with difficulty of interpretation. For example, if age is a CV, the interpretation of the results of treatment includes the caveat, "If everyone were the same age." But in a nonexperimental study, the reality may be that different treatments inherently attract cases of different ages.

Choice of CVs is a logical exercise as well. As a general rule, one wants a very small number of CVs, all correlated with the DV and none correlated with each other. The goal is maximum adjustment of the DV with minimum loss of degrees of freedom for error. Calculating the regression of the DV on the CV(s) costs 1 df for error for each CV. Thus, *the gain in power from a smaller sum of squares for error may be offset by the loss in degrees of freedom.* When there is substantial correlation between a DV and a CV, the reduction in sum of squares for error more than offsets the loss of 1 df. With multiple CVs, however, a point of diminishing returns is quickly reached, especially if the CVs are correlated (see Section 8.6.1).

A frequent caution in experimental work is that the CVs must be independent of treatment. Scores on CVs are gathered before treatment is administered. Violation of this precept results in removal of that portion of the IV-DV relationship that is associated with the CV. In this situation, adjusted group means may be closer together than unadjusted means, and these adjusted means are difficult to interpret.

In nonexperimental work, sources of bias are many and subtle and can produce either underadjustment or overadjustment of the DV. At best, the nonexperimental use of ANCOVA allows you to look at IV-DV relationships (noncausal) adjusted for the effects of CVs, as measured. Do not expect ANCOVA to permit causal inference of treatment effects with non–randomly assigned groups. If random assignment is absolutely impossible, or if it breaks down because of nonrandom loss of cases, be sure to thoroughly ground yourself in the literature regarding use of ANCOVA, starting with Cook and Campbell (1979).

Limitations to generalizability apply to ANCOVA just as they do to ANOVA or any other statistical test. One can generalize only to those populations from which a random sample is taken. ANCOVA may, in some very limited sense, adjust for a failure to randomly assign the sample to groups, but it does not affect the relationship between the sample and the population to which one can generalize.

8.3.2 Practical Issues

The ANCOVA model assumes reliability of CVs, linearity between pairs of CVs and between CVs and the DV, lack of multicollinearity among CVs, and homogeneity of regression, in addition to the usual ANOVA requirements. Experiments are typically designed with only a single CV, simplifying evaluation of linearity and outliers and eliminating the need to consider multicollinearity. However, procedures when there are multiple CVs are described in this chapter, as well.

8.3.2.1 Absence of Outliers

There are two kinds of outliers—univariate and multivariate. Univariate outliers are cases with extreme scores on a single variable (whether DV or CV), apparently disconnected from other cases in their group, as discussed in Sections 2.2.3.4 and 3.4.4.

In ANCOVA, *univariate outliers are sought separately on the DV and each of the CV(s)*. They are also sought separately at each level, or at each combination of levels, of the IV(s) by looking for disconnected cases with large *z* scores.[5] Once potential univariate outliers are located, if there is more than one CV, the search for multivariate outliers begins.

Multivariate outliers, which are sought among the CVs,[6] are harder to spot because they are cases with unusual *combinations* of scores. Cases with such unusual combinations are sometimes called *discrepant.* The *z* scores on each CV may be reasonable separately, but their combination is not. For instance, a 5-foot-tall person is not unusual, nor is a person who weighs 200 pounds; however, a 5-foot-tall person who weighs 200 pounds is unusual. If height and weight are being used as CVs in a study, such a person is likely to be a multivariate outlier. The problem with outliers, in general, is that they have undue impact on the analysis. In ANCOVA, that translates as unreasonable adjustment of the DV.

Multivariate outliers may be detected by statistical or graphical methods. Of these two, statistical methods are preferred.[7] A commonly used statistical procedure is computation of *Mahalanobis distance* for each case. Mahalanobis distance is the distance of a case from the centroid of the remaining cases in its cell, where the *centroid* is a point created by the means of all the CVs. For instance, if there are two CVs, height and weight, the centroid is the point that represents the intersection of the mean on height and the mean on weight. The Mahalanobis distance for a case is the distance of the point for the case's own two scores from the point representing both means. The farther the scores from the centroid, the greater the Mahalanobis distance.

The Mahalanobis distance of each case, in turn, from all other cases in its cell, is available in SPSS DISCRIMINANT (by requesting a display of casewise results, as demonstrated in Section 8.7.2.1.3). A very conservative probability estimate for a case being an outlier, say $p < .001$, is appropriate with Mahalanobis distance.

Mahalanobis distances are also found through SPSS and SAS by running REGRESSION separately for each group, with the CVs as predictors (IVs) and some other variable, arbitrarily chosen, as the predicted variable (the DV). Use of regression may identify a slightly different set of cases as outliers, however, because different error terms are developed in each of the separate runs, unlike SPSS DISCRIMINANT.

With SAS there is the further complication that a value of leverage, h_{ii}, is reported instead of Mahalanobis distance. However, this is easily converted to Mahalanobis distance:

$$\text{Mahalanobis distance} = (n - 1)\left(h_{ii} - \frac{1}{n}\right) \tag{8.1}$$

[5] A dichotomous CV (such as a grouping variable) is considered a univariate outlier when fewer than 10% of the cases are in one of the two groups.

[6] Multivariate outliers can also occur between the DV and the CV(s). In this situation, however, they are likely to produce heterogeneity of regression (Section 8.3.2), leading to rejection of ANCOVA. If CVs are serving as a convenience in most analyses, rejection of ANCOVA because there are multivariate outliers is hardly convenient.

[7] The discussion of statistical methods for detecting outliers is adapted from Tabachnick and Fidell (2007).

Or, as is sometimes more useful,

$$h_{ii} = \frac{\text{Mahalanobis distance}}{n - 1} + \frac{1}{n}$$

The latter form is handy if you want to find a critical value for leverage at $\alpha = .001$ by translating the critical value for Mahalanobis distance.

Some outliers may hide behind others. When one or two cases identified as outliers are deleted, the data become more consistent, but other cases become extreme. Sometimes it is a good idea to screen for outliers several times, each time dealing with cases identified as outliers on the last run, until finally no new outliers are identified. It is also a good idea to see if univariate outliers are also multivariate outliers before deciding what to do with them. Often the same cases show up in both analyses; sometimes they do not. It is usually better to decide what to do with outliers when the total extent of the problem is known.

Recent research has pointed out unreliability of methods of outlier detection based on Mahalanobis distance or other leverage measures (e.g., Egan and Morgan, 1998; Hadi and Simonoff, 1993; Rousseeuw and van Zomeren, 1990). Unfortunately, alternative methods are computationally challenging and currently unavailable in statistical packages. Therefore, results of any method of outlier detection should be viewed cautiously.

8.3.2.2 Absence of Multicollinearity

If there are two or more CVs, they should not be highly correlated with each other. Extremely highly correlated variables are multicollinear. Perfectly correlated variables cause singularity. *One or more CVs in a set of highly correlated CVs should be eliminated* because (1) they add very little adjustment to the DV, (2) they each remove a degree of freedom from the error term, and (3) they may cause potential computational difficulties if they are singular or multicollinear. Software programs typically guard against multicollinearity among CVs that are high enough to produce computational difficulties. Long before that point is reached, however, there is a net loss to the analysis by using highly correlated CVs.

The size of correlation is evaluated through bivariate or multiple-regression procedures, where each CV, in turn, is used as the DV, with the remaining CVs treated as IVs. The squared multiple correlation is the variance in the CV currently used as DV that is associated with the other CVs. For purposes of ANCOVA, *any CV with a squared multiple correlation (SMC) in excess of .50 may be considered redundant and deleted from further analysis.* Section 8.6.1.1 shows a check for multicollinearity among CVs.

8.3.2.3 Normality of Sampling Distributions

As in all ANOVA, it is assumed that the sampling distributions of means are normal within each group. As discussed before, it is the sampling distributions of means, and not the raw scores within each cell, that need to be normally distributed. Section 3.4.1

provides guidelines for dealing with normality of sampling distributions; these guidelines apply to CV(s) as well as to the DV. However, the alternative of using a nonparametric method when the assumption is violated is not available for ANCOVA.

8.3.2.4 Independence of Errors

The assumption that deviations of individual DV scores around their group means are independent applies equally to randomized-groups ANCOVA and ANOVA, as do the issues surrounding violation of the assumption. Independence of errors in repeated-measures IVs is dealt with through tests of sphericity, as discussed in Sections 6.3.2.3 and 6.6.2.

8.3.2.5 Homogeneity of Variance

Just as in ANOVA, it is assumed in ANCOVA that the variance of DV scores within each cell of the design is a separate estimate of the same population variance. In ANCOVA, the CV(s) are also evaluated for homogeneity of variance. If a CV fails the test, either a more stringent test of A is required (e.g., $\alpha = .025$ instead of .05) or the CV is dropped from the analysis. Sections 3.4.3 and 4.5.4 provide guidelines and formal tests for evaluating homogeneity of variance, as well as remedies for violation of the assumption.

8.3.2.6 Linearity

The ANCOVA model assesses the linear relationship between each CV and the DV and among all pairs of CVs if there is more than one. Nonlinear relationships, if present, are not captured by the analysis, so error terms are not reduced as fully as they might be, optimum matching of groups is not achieved, and group means are incompletely adjusted. In short, the power of the statistical test is reduced, producing conservative errors in statistical decision making.

If nonlinearity is suspected, *examine scatterplots of the DV with each CV and all CVs with one another* in each cell of the design. All programs, including spreadsheets, have modules for producing scatter plots.

If there seems to be a curvilinear relationship between the DV and a CV,[8] it may be corrected by using a transformation of the CV, by raising the CV to a higher-order power, or by simply eliminating a CV that produces nonlinearity. Nonlinear relationships between pairs of CVs can be handled by the same procedures, although it may not be worthwhile to preserve a CV that complicates interpretation.

[8] Tests for deviation from linearity are available (cf. Myers and Well, 1991), but there seems to be little agreement as to the appropriate test or the seriousness of significant deviation in the case of ANCOVA. Therefore, formal tests are not recommended.

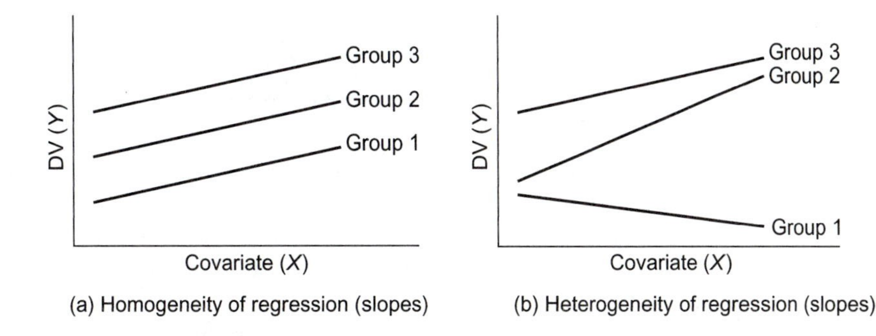

FIGURE 8.2 DV-CV regression lines for three groups plotted on the same coordinates for conditions of (a) homogeneity and (b) heterogeneity of regression (slopes)

8.3.2.7 Homogeneity of Regression

Adjustment of scores in ANCOVA is made on the basis of an average within-cell regression coefficient. The assumption is that the slope of the regression between the DV and the CV(s) within each cell is an estimate of the same population regression coefficient—that is, the slopes are equal for all cells.

Heterogeneity of regression implies that there is a different DV-CV(s) slope in some cells of the design or, stated another way, that there is an interaction between IV(s) and CV(s). If the IV(s) and CV(s) interact, the relationship between the CV(s) and the DV is different at different levels of the IV(s), and the CV adjustment that is needed for various cells is different. Figure 8.2 illustrates, for three groups, perfect homogeneity of regression (equality of slopes) and extreme heterogeneity of regression (inequality of slopes).

Test the assumption of homogeneity of regression using procedures described in Section 8.6.2. ANCOVA is inappropriate if the assumption is violated; alternatives to ANCOVA are discussed in Section 8.6.7.

8.3.2.8 Reliability of Covariates

It is assumed in ANCOVA that CVs are measured without error (i.e., they are perfectly reliable). The issue is whether there is consistency in assessment of the CV score over the course of the study so that the assessment is the same at two points in time or with two types of measurement. In the case of some variables (such as diameter or category of religious affiliation, for instance), the assumption is usually justified. With other variables (such as those measured psychometrically or with unstable instrumentation), the assumption is not so easily made. In addition, some CVs (such as attitude) may be reliable at the point of measurement but fluctuate over short periods.

In experimental research, unreliable CVs lead to loss of power and a conservative statistical test through underadjustment of the error term. In nonexperimental applications, unreliable CVs can lead to either underadjustment or overadjustment of the means, as well. Thus, the issue is more serious in nonexperimental research.

8.4 FUNDAMENTAL EQUATIONS

The simplest ANCOVA has one randomized-groups IV with at least two levels and one CV. This section shows allocation of cases to levels, partition of sources of variance, and computation of a one-way randomized-groups ANCOVA by the traditional method. Analyses with more than three levels are calculated in the same manner as those with three levels.

Analysis using the regression method is not as straightforward as for ANOVA because the CV is not orthogonal with the coding of the IV(s). Therefore, only the traditional approach to the analysis is demonstrated in this chapter.

8.4.1 Allocation of Cases to Conditions

Cases are assigned to groups in the same manner as they are in ANOVA, with the exception that CVs are measured before treatments are applied. Measurement of the CV(s) may be made either before or after random assignment, as long as the measurement process is precisely the same for all groups and knowledge of the CV score does not affect the randomization process. Assignment is made as in Table 4.1.

8.4.2 Partition of Sources of Variance

Equations for ANCOVA are an extension of those for ANOVA. Averaged squared deviations from means (SS) are partitioned into SS associated with different levels of the IV (between-groups SS) and SS associated with differences in scores within groups (unaccounted for or error SS). There are three partitions, however: one for the DV, one for the CV, and one for the cross products of CV with DV. Recall that in ANOVA, SS for the DV is partitioned by summing and squaring differences between scores and various means.

$$\sum_i \sum_j (Y - \mathbf{GM})^2 = n \sum_j (\overline{Y}_j - \mathbf{GM})^2 + \sum_i \sum_j (Y_{ij} - \overline{Y}_j)^2 \qquad \textbf{(8.2)}$$

or

$$\text{SS}_T = \text{SS}_A + \text{SS}_{S/A}$$

The total sum of squared differences between scores on Y (the DV) and the grand mean (GM) is partitioned into two components: (1) sum of squared differences between group means ($_j$) and the grand mean (i.e., systematic or between-groups variability) and (2) sum of squared differences between individual scores (Y_{ij}) and their respective group means (i.e., error).

In ANCOVA, the two additional partitions involve the CV and the CV-DV relationship. First, the differences between CV scores and their GM are partitioned into between-groups and within-groups sums of squares:

$$SS_{T(x)} = SS_{A(x)} + SS_{S/A(x)} \tag{8.3}$$

The total sum of squared differences on the CV (X) is also partitioned into differences between groups and differences within groups.

Finally, the covariance (the linear relationship between the DV and the CV) is partitioned into sums of products associated with covariance between groups and sums of products associated with covariance within groups.

$$SP_T = SP_A + SP_{S/A} \tag{8.4}$$

A *sum of squares* involves taking deviations of scores from means (e.g., $X_{ij} - \overline{X}_j$ or $Y_{ij} - \overline{Y}_j$), squaring them, and then summing the squares over all cases; a *sum of products* involves taking two deviations from the same case (e.g., both $X_{ij} - \overline{X}_j$ and $Y_{ij} - \overline{Y}_j$), multiplying them together (instead of squaring), and then summing the products over all cases. As discussed in Section 3.2.3.1, the means that are used to produce the deviations are different for the different sources of variance in the research design.

The partitions for the CV (Equation 8.2) and the partitions for the association between the CV and DV (Equation 8.4) are used to adjust the sums of squares for the DV. The adjusted sum of squares for treatment is

$$SS'_A = SS_A - \left(\frac{(SP_T)^2}{SS_{T(x)}} - \frac{(SP_{S/A})^2}{SS_{S/A(x)}} \right) \tag{8.5}$$

The adjusted between-groups sum of squares (SS'_A) is found by subtracting from the unadjusted between-groups sum of squares a term based on sums of squares associated with the CV, X, and sums of products for the linear relationship between the DV and the CV.

The adjusted sum of squares for error is

$$SS'_{S/A} = SS_{S/A} - \frac{(SP_{S/A})^2}{SS_{S/A(x)}} \tag{8.6}$$

The adjusted within-groups sum of squares ($SS'_{S/A}$) is found by subtracting from the unadjusted within-groups sum of squares a term based on within-groups sums of squares and products associated with the CV and with the linear relationship between the DV and the CV.

The explanation of Equation 8.6 can be expressed in an alternate, somewhat more intelligible form. The adjustment for each score is the deviation of that score from the grand mean of the DV minus the deviation of the corresponding CV from the grand mean on the CV, weighted by the regression coefficient for predicting the DV from the CV. Symbolically, for an individual score,

$$(Y - Y') = (Y - GM_y) - B_{y.x}(X - GM_x) \tag{8.7}$$

The adjustment for any subject's score ($Y - Y'$) is obtained by subtracting from the unadjusted deviation score ($Y - GM_y$) the individual's deviation on the CV ($X - GM_x$) weighted by the regression coefficient, $B_{y.x}$.

FIGURE 8.3 Partitions of (a) total adjusted sum of squares into sum of squares due to adjusted differences between the groups and to adjusted differences among cases within the groups and (b) total adjusted degrees of freedom into degrees of freedom associated with groups and with cases within the groups

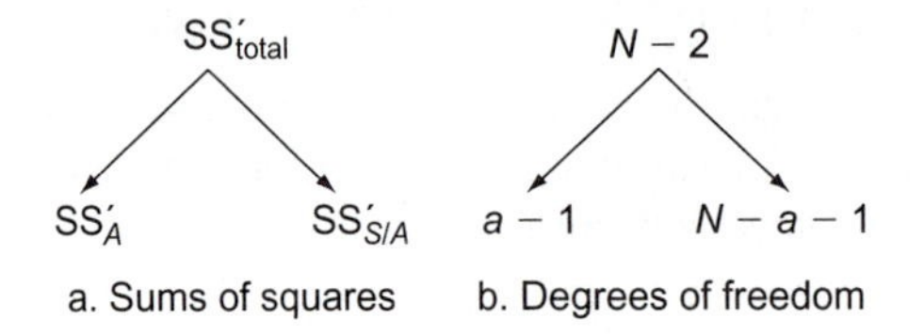

a. Sums of squares b. Degrees of freedom

The df_A are the same as for ANOVA:

$$\text{df}'_A = a - 1 \tag{8.8}$$

However, calculation of the regression coefficient for the DV-CV relationship (B) requires 1 df, so that with a single CV,

$$\text{df}'_T = N - 2 \tag{8.9}$$

and

$$\text{df}'_{S/A} = N - a - 1 = a(n - 1) - 1 = an - a - 1 \tag{8.10}$$

Figure 8.3 diagrams the partition of adjusted sums of squares and degrees of freedom for ANCOVA with a single CV.

Once adjusted sums of squares are found, mean squares are found by dividing by appropriately adjusted degrees of freedom. Thus,

$$\text{MS}'_A = \frac{\text{SS}'_A}{\text{df}'_A} \tag{8.11}$$

and

$$\text{MS}'_{S/A} = \frac{\text{SS}'_{S/A}}{\text{df}'_{S/A}} \tag{8.12}$$

MS'_T is not a useful quantity for ANCOVA.

Thus, the F ratio is

$$F = \frac{\text{MS}'_A}{\text{MS}'_{A/S}} \qquad \text{df} = (a - 1), (a(n - 1) - 1) \tag{8.13}$$

8.4.3 Traditional Approach with Three Levels and One Covariate

The unrealistically small, hypothetical data set of Table 8.1 is the same as that in Section 4.4, with the addition of a CV. Rating of a ski vacation (the DV) is evaluated as a function of view from the hotel room (the IV) adjusted for rating of past ski vacations (the CV). Skiers first provide a rating of their past ski vacations and then are randomly assigned to a hotel room with a view of the parking lot, a view of the ocean,

TABLE 8.1 Hypothetical data set for a one-way randomized-groups ANCOVA with three levels and one CV

A: Type of View					
a_1: Parking Lot View		a_2: Ocean View		a_3: Ski Slope View	
X	Y	X	Y	X	Y
2	1	2	3	3	6
1	1	4	4	4	8
4	3	4	3	2	6
4	5	4	5	5	7
1	2	3	4	3	5
$A_{x1} = 12$	$A_1 = 12$	$A_{x2} = 17$	$A_2 = 19$	$A_{x3} = 17$	$A_3 = 32$

$T_x = 46$ $\Sigma X^2 = 162$ $\Sigma XY = 215$
$T_y = 63$ $\Sigma Y^2 = 325$

or a view of the ski slope (the IV). This particular vacation is assessed at the end of the trip.

Raw score equations are more useful than deviation equations for computation of values in Equations 8.5 and 8.6. Unadjusted SS_A and $SS_{S/A}$ are as defined in Equations 3.13 and 3.14.

Treatment sum of squares for the CV is

$$SS_{A(x)} = \frac{\Sigma A_x^2}{n} - \frac{T_x^2}{an} \quad \textbf{(8.14)}$$

Error sum of squares for the CV is

$$SS_{S/A(x)} = \sum X^2 - \frac{\Sigma A_x^2}{n} \quad \textbf{(8.15)}$$

and total sum of squares for the CV is

$$SS_{T(x)} = \sum X^2 - \frac{T_x^2}{an} \quad \textbf{(8.16)}$$

Sums of products between the DV and CV are also required. Error sum of products is

$$SP_{S/A} = \sum XY - \frac{\Sigma A_x A}{n} \quad \textbf{(8.17)}$$

and total sum of products is

$$SP_{\text{total}} = \sum XY - \frac{T_x T}{an} \quad \textbf{(8.18)}$$

When applied to the data in Table 8.1, the seven sums of squares and sums of products are as in Table 8.2. These values permit calculation of the adjusted sums of squares, shown in Table 8.3. Table 8.4 shows degrees of freedom for the example. Note that 1 df is lost from the adjusted error term because of the single CV. Variance (mean square) for each source is obtained by dividing each adjusted SS by adjusted df. Table 8.5 shows calculation of adjusted mean squares for treatment and

TABLE 8.2 Sums of squares and sums of products for traditional one-way ANCOVA with three levels and one CV

$$SS_A = \frac{12^2 + 19^2 + 32^2}{5} - \frac{63^2}{15} = 41.2$$

$$SS_{A(x)} = \frac{12^2 + 17^2 + 17^2}{5} - \frac{46^2}{15} = 3.3$$

$$SS_{S/A} = 325 - \frac{12^2 + 19^2 + 32^2}{5} = 19.2$$

$$SS_{S/A(x)} = 162 - \frac{12^2 + 17^2 + 17^2}{5} = 17.6$$

$$SS_{T(x)} = 162 - \frac{46^2}{15} = 20.9$$

$$SP_{S/A} = 215 - \frac{(12)(12) + (17)(19) + (17)(32)}{5} = 12.8$$

$$SP_T = 215 - \frac{(46)(63)}{15} = 21.8$$

TABLE 8.3 Adjusted sums of squares for traditional one-way ANCOVA with three levels and one CV

$$SS'_A = 41.2 - \left[\frac{(21.8)^2}{20.9} - \frac{(12.8)^2}{17.6}\right] = 27.8$$

$$SS'_{S/A} = 19.2 - \left[\frac{(12.8)^2}{17.6}\right] = 9.9$$

TABLE 8.4 Degrees of freedom for traditional one-way ANCOVA with three levels and one CV

$$df'_A = a - 1 = 3 - 1 = 2$$

$$df'_{S/A} = an - a - 1 = 15 - 3 - 1 = 11$$

TABLE 8.5 Mean squares for traditional one-way ANCOVA with three levels

$$MS'_A = \frac{SS'_A}{df'_A} = \frac{27.8}{2} = 13.9$$

$$MS'_{S/A} = \frac{SS'_{S/A}}{df'_{S/A}} = \frac{9.9}{11} = 0.9$$

TABLE 8.6 Source table for traditional one-way ANCOVA with three levels

Source	Adjusted SS	Adjusted df	Adjusted MS	F
A	27.8	2	13.9	15.44
S/A	9.9	11	0.9	

error sources of variance. Finally, Table 8.6 is a source table that summarizes the results of Tables 8.3–8.5 and adds the F ratio of Equation 8.13.

The critical value for the main effect of A at $\alpha = .05$ with df = 2 and 11 is 3.98. Because 15.44 is larger than 3.98, there is a mean difference in ratings of ski vacations for the three levels of treatment after adjustment for individual differences in general rating of ski vacations. The test of the relationship between the DV and the CV is reported in Section 8.4.4; the test of the IV (treatment)–CV relationship is described in Section 8.6.2. The total SS is omitted from this source table because of ambiguity, as is discussed in subsequent sections. Note that the adjusted $SS'_T = SS'_A + SS'_{S/A} = 37.7$.

8.4.4 Computer Analyses of Small-Sample ANCOVA

Data for a one-way ANCOVA are entered as for a one-way ANOVA, with the addition of a column for each CV. With one CV, there is a column for the DV, a column coding the IV, and a column for the CV scores. Order of columns (or rows) does not matter. Tables 8.7 and 8.8 show computer analyses of the small-sample ANCOVA example of Section 8.4.3 by SPSS GLM and SAS GLM.

Table 8.7 shows syntax and output for the small-sample example through SAS GLM. Both the CV and IV are on the right side of the `model` equation, but only the IV is identified as a `class` (categorical) variable. This signals GLM that PRE_RATE is a CV. The `solution` instruction requests parameter estimates (*B* weights). Both `means` (unadjusted means) and `lsmeans` (means adjusted for the CV) are requested.

TABLE 8.7 One-way analysis of covariance for small-sample example through SAS GLM

```
proc glm data=SASUSER.SSANCOVA;
     class VIEW;
     model RATING = PRE_RATE VIEW / solution;
     lsmeans VIEW/ ;
     means VIEW ;
run;
```

The GLM Procedure

Class Level Information

Class	Levels	Values
VIEW	3	1 2 3

Number of Observations Read	15
Number of Observations Used	15

The GLM Procedure

Dependent Variable: RATING

Source	DF	Sum of Squares	Mean Square	F Value	Pr > F
Model	3	50.50909091	16.83636364	18.72	0.0001
Error	11	9.89090909	0.89917355		
Corrected Total	14	60.40000000			

R-Square	Coeff Var	Root MSE	RATING Mean
0.836243	22.57732	0.948248	4.200000

Source	DF	Type I SS	Mean Square	F Value	Pr > F
PRE_RATE	1	22.70254777	22.70254777	25.25	0.0004
VIEW	2	27.80654314	13.90327157	15.46	0.0006

Source	DF	Type III SS	Mean Square	F Value	Pr > F
PRE_RATE	1	9.30909091	9.30909091	10.35	0.0082
VIEW	2	27.80654314	13.90327157	15.46	0.0006

Parameter		Estimate	Standard Error	t Value	Pr > |t|
Intercept		3.927272727 B	0.87774042	4.47	0.0009
PRE_RATE		0.727272727	0.22602966	3.22	0.0082
VIEW	1	−3.272727273 B	0.64090469	−5.11	0.0003
VIEW	2	−2.600000000 B	0.59972445	−4.34	0.0012
VIEW	3	0.000000000 B	.	.	.

NOTE: The X'X matrix has been found to be singular, and a generalized inverse was used to solve the normal equations. Terms whose estimates are followed by the letter 'B' are not uniquely estimable.

TABLE 8.7 Continued

Least Squares Means

VIEW	RATING LSMEAN
1	2.88484848
2	3.55757576
3	6.15757576

The GLM Procedure

Level of VIEW	N	RATING Mean	RATING Std Dev	PRE_RATE Mean	PRE_RATE Std Dev
1	5	2.40000000	1.67332005	2.40000000	1.51657509
2	5	3.80000000	0.83666003	3.40000000	0.89442719
3	5	6.40000000	1.14017543	3.40000000	1.14017543

TABLE 8.8 One-way analysis of covariance for small-sample example through SPSS GLM

```
UNIANOVA
 RATING BY VIEW  WITH PRE_RATE
 /METHOD = SSTYPE(3)
 /INTERCEPT = INCLUDE
 /EMMEANS = TABLES(VIEW) WITH(PRE_RATE=MEAN)
 /PRINT = DESCRIPTIVE ETASQ OPOWER PARAMETER
 /CRITERIA = ALPHA(.05)
 /DESIGN = PRE_RATE VIEW .
```

Univariate Analysis of Variance

Between-Subjects Factors

		Value Label	N
View from room	1.00	parking lot	5
	2.00	ocean	5
	3.00	ski slope	5

Descriptive Statistics

Dependent Variable: rating

View from room	Mean	Std. Deviation	N
parking lot	2.4000	1.67332	5
ocean	3.8000	.83666	5
ski slope	6.4000	1.14018	5
Total	4.2000	2.07709	15

TABLE 8.8 Continued

Tests of Between-Subjects Effects

Dependent Variable: RATING

Source	Type III Sum of Squares	df	Mean Square	F	Sig.	Partial Eta Squared	Noncent. Parameter	Observed Power[a]
Corrected Model	50.509[b]	3	16.836	18.724	.000	.836	56.173	1.000
Intercept	6.455	1	6.455	7.179	.021	.395	7.179	.685
PRE_RATE	9.309	1	9.309	10.353	.008	.485	10.353	.833
VIEW	27.807	2	13.903	15.462	.001	.738	30.925	.993
Error	9.891	11	.899					
Total	325.000	15						
Corrected Total	60.400	14						

a. Computed using alpha = .05
b. R Squared = .836 (Adjusted R Squared = .792)

Parameter Estimates

Dependent Variable: rating

Parameter	B	Std. Error	t	Sig.	95% Confidence Interval		Partial Eta Squared	Noncent. Parameter	Observed Power[a]
					Lower Bound	Upper Bound			
Intercept	3.927	.878	4.474	.001	1.995	5.859	.645	4.474	.982
pre_rate	.727	.226	3.218	.008	.230	1.225	.485	3.218	.833
[view=1.00]	-3.273	.641	-5.106	.000	-4.683	-1.862	.703	5.106	.996
[view=2.00]	-2.600	.600	-4.335	.001	-3.920	-1.280	.631	4.335	.976
[view=3.00]	0[b]	.	.	.	.	.	.	.	.

a. Computed using alpha = .05
b. This parameter is set to zero because it is redundant.

Estimated Marginal Means

View from room

Dependent Variable: rating

View from room	Mean	Std. Error	95% Confidence Interval	
			Lower Bound	Upper Bound
parking lot	2.885[a]	.450	1.894	3.875
ocean	3.558[a]	.431	2.610	4.506
ski slope	6.158[a]	.431	5.210	7.106

a. Covariates appearing in the model are evaluated at the following values: pre_rate = 3.0667.

The usual two sets of SAS source tables are printed out, one showing the full model and error term and another showing the effects of interest—the IV (VIEW) and the CV (PRE_RATE). The latter tests the reliability of the relationship between the DV and the CV. Results for Type I and Type III sums of squares differ substantially because of nonorthogonality between the CV and the IV. The Type I results adjust effects in the order in which they are listed in the model instruction; the effect listed first is unadjusted, the second is adjusted for the first, and so on. In this example, VIEW is adjusted for PRE_RATE, but PRE_RATE is not adjusted for VIEW. The Type I result for the CV (PRE_RATE), then, is based on the correlation between PRE_RATE and RATING, without taking into account differences among groups. The Type III results, in which all effects are adjusted for each other, are typically of greater interest in ANCOVA (cf. Section 8.6.1).

The test of the main effect of VIEW matches that of Table 8.6, but for rounding error, with an F Value of 15.46 and Pr > F less than .05. Of lesser interest is the test of the CV (if we did not expect a relationship, we would not use ANCOVA), which is statistically significant, with an F Value of 10.35 and Pr > F less than .05. The next section shows the B weight for the CV, PRE_RATE, as in Equation 8.6; its use is discussed in Section 8.6.5.

Mean RATING for the three VIEW groups, adjusted for PRE_RATE, are in the table labeled Least Squares Means. Unadjusted means and standard deviations for both RATING and PRE_RATE at each level of VIEW are in the last table of the output.

SPSS GLM UNIANOVA (General Linear Model > Univariate . . . in the menu system) identifies the CV after WITH in the GLM syntax, after both DV and IV are identified, as shown in Table 8.8. Adjusted means are requested by the /EMMEANS instructions and unadjusted means by the DESCRIPTIVE instruction. ETASQ and OPOWER request effect size and power statistics. PARAMETER requests B weights. The default method for computing sums of squares—SSTYPE(3)—is equivalent to Type III SS in SAS GLM.

The output first shows descriptive statistics, means, standard deviations, and sample sizes. Next is a source table, containing, among other things, the test of VIEW against Error (adjusted S/A) that has the same F ratio, within rounding error, as in Table 8.6. Of secondary interest is the test of PRE_RATE against adjusted S/A as error that shows a statistically significant relationship between the DV and CV, after adjusting for the main effect of A. The Corrected Model and Corrected Total sums of squares correspond to those of SAS GLM and are based on a model in which the CV is not adjusted for the IV. Partial Eta Squared and Observed Power are described in Sections 8.6.4 and 8.6.5, respectively.

The Parameter Estimates table shows the B weight for the CV, PRE_RATE, as in Equation 8.7; its use is discussed in Section 8.6.5. Note that t^2 for that parameter is equal to F for PRE_RATE in the source table. Estimated Marginal Means are the three group means on the DV after adjustment for the CV (cf. Section 8.6.5). The section also shows standard errors and confidence intervals for those means.

TABLE 8.9 Data set for regression approach to ANCOVA

Y Rating	X_1 Pre_rate	X_2 View_1	X_3 View_2	Y Rating	X_1 Pre_rate	X_2 View_1	X_3 View_2
1	2	1	1	5	4	1	−1
1	1	1	1	4	3	1	−1
3	4	1	1	6	3	−2	0
5	4	1	1	8	4	−2	0
2	1	1	1	6	2	−2	0
3	2	1	−1	7	5	−2	0
4	4	1	−1	5	3	−2	0
3	4	1	−1				

8.4.5 Computer Analysis Using Regression Approach to ANCOVA

Although hand computation of ANCOVA through regression requires matrix algebra because the column for the CV is not orthogonal to the column(s) for the IVs, the analysis is easily performed through a standard multiple-regression program. Table 8.9 contains the small-sample data set organized appropriately for a regression analysis. Notice that, as usual, there are as many columns for the IV (VIEW) as degrees of freedom. In this case, VIEW has three levels and 2 df, so the View_1 column is the comparison of the third level of view (ski slope) against the two other views pooled, and the View_2 column is the comparison of the parking lot view with the ocean view. The CV is in the Pre-rate column, and the DV is in the Rating column.

Analysis of these data through SPSS REGRESSION is illustrated in Table 8.10. The key to producing the relevant output is use of NEXT on the Linear Regression menu to specify that PRE_RATE (the CV) enters before VIEW_1 and VIEW_2 (which enter as a block—although, if they enter separately, tests of the individual comparisons are provided) and a request for R ANOVA CHANGE through the STATISTICS menu. The rest of the syntax is produced by default.

The test of the IV is found in the Model Summary table, where the F Change value of 15.462 is the omnibus test of VIEW. The test of the CV that agrees with the results of the GLM programs is in the table labeled Coefficients in the t column, because $F = t^2 = (3.218)^2 = 10.36$. This is a test of the CV in the equation that also includes VIEW_1 and VIEW_2. The test of the relationship between the CV alone and the DV is the test of Model 1 in the Model Summary table (where the F Change value is 7.829) or in the test of Model 1 in the ANOVA table. In the ANOVA table, the test of Model 2 ($F = 18.724$) is a test of the relationship between the DV and all three variables together (pre_rate, view_1, and view_2), a test that is also provided by SAS GLM and SPSS GLM.

TABLE 8.10 ANCOVA for the small-sample example through SPSS REGRESSION (syntax and output)

```
REGRESSION
 /MISSING LISTWISE
 /STATISTICS COEFF OUTS R ANOVA CHANGE
 /CRITERIA=PIN(.05) POUT(.10)
 /NOORIGIN
 /DEPENDENT RATING
 /METHOD=ENTER PRE_RATE/METHOD=ENTER VIEW_1 VIEW_2 .
```

Regression

Variables Entered/Removed[b]

Model	Variables Entered	Variables Removed	Method
1	pre_rate[a]	.	Enter
2	view_1, view_2[a]	.	Enter

a. All requested variables entered.
b. Dependent Variable: rating

Model Summary

					Change Statistics				
Model	R	R Square	Adjusted R Square	Std. Error of the Estimate	R Square Change	F Change	df1	df2	Sig. F Change
1	.613[a]	.376	.328	1.70288	.376	7.829	1	13	.015
2	.914[b]	.836	.792	.94825	.460	15.462	2	11	.001

a. Predictors: (Constant), pre_rate
b. Predictors: (Constant), pre_rate, view_1, view_2

ANOVA[c]

Model		Sum of Squares	df	Mean Square	F	Sig.
1	Regression	22.703	1	22.703	7.829	.015[a]
	Residual	37.697	13	2.900		
	Total	60.400	14			
2	Regression	50.509	3	16.836	18.724	.000[b]
	Residual	9.891	11	.899		
	Total	60.400	14			

a. Predictors: (Constant), pre_rate
b. Predictors: (Constant), pre_rate, view_1, view_2
c. Dependent Variable: rating

TABLE 8.10 Continued

Coefficients[a]

Model		Unstandardized Coefficients		Standardized Coefficients	t	Sig.
		B	Std. Error	Beta		
1	(Constant)	1.006	1.223		.823	.425
	pre_rate	1.041	.372	.613	2.798	.015
2	(Constant)	1.970	.735		2.679	.021
	pre_rate	.727	.226	.428	3.218	.008
	view_1	-.979	.177	-.690	-5.524	.000
	view_2	-.336	.320	-.137	-1.050	.316

a. Dependent Variable: rating

At this point, you have probably realized that the regression approach offers near-complete flexibility in analysis of ANOVA-like models. For instance, the CV is just a continuous variable that could as easily be considered a continuous IV. Further, the order in which IVs and CVs are entered (and, thus, adjusted for each other) is at the discretion of the researcher. The trick is to set up the columns of data appropriate for regression and then use the facilities of one of the multiple-regression programs to enter the variables, alone or in blocks, in any order desired for the analysis.

8.5 TYPES OF DESIGNS USING COVARIATES

Section 8.4 demonstrates one-way randomized-groups ANCOVA with a single CV. ANCOVA may also be used in factorial randomized-groups designs and, under some circumstances, in repeated-measures designs.

8.5.1 Randomized-Groups Factorial

When factorial randomized-groups designs (Chapter 5) incorporate one or more CVs, the CV(s) adjusts the DV (cf. Equation 8.5) for all main effects and interactions. For example, if length of vacation (one or two weeks) is added to view from window in factorial arrangement, the research questions are: (1) Does length of vacation affect rating of vacation after adjusting for individual differences in vacation ratings? (2) Does view affect rating of vacation after adjusting for individual differences in vacation ratings? and (3) Does the effect of view on rating depend on length of vacation after adjusting for individual differences in vacation ratings?

Computationally, adjustment of sums of squares for effects proceeds as in Section 8.4. For a two-way design, for instance, to compute the main effect of *B*, substitute *B* for *A* in Equations 8.3–8.6, and for the interaction, substitute *AB* in those equations while substituting *S*/*AB* for *S*/*A*. Adjusted mean squares are formed and tested in the usual fashion by dividing adjusted sums of squares by their adjusted degrees of freedom. Remember that 1 df is subtracted from the error term for each CV to produce adjusted df. Hypotheses are tested and interpreted in the usual fashion, as described in Chapter 5.

Similarly, CVs may be applied in any of the randomized-groups Latin-square designs of Chapter 9 and the screening designs of Chapter 10. After computed sums of squares, degrees of freedom, and mean squares are adjusted for the CV(s), hypothesis tests proceed as in those chapters.

8.5.2 Repeated Measures

Both ANCOVA and repeated measures have the goal of reducing the error term by systematic variance associated with individual differences among cases. Although the two designs assess individual differences in separate ways, the effect on the analysis is the same. Thus, a purely repeated-measures design, in which all cases participate in all combinations of levels of all IVs, usually does not benefit from use of CVs measured only once; rather CVs lower power by reducing the degrees of freedom for error by 1 df each. However, in other designs with repeated measures, CVs may be helpful.

8.5.2.1 Same Covariate(s) for All Cells

There are two approaches to analysis of designs with at least one repeated-measures IV—the traditional approach and the general linear model (GLM) approach. In the traditional approach, the repeated-measures effects in the designs are not adjusted for CVs. In the GLM approach, the repeated measures are adjusted by the interaction of the CV(s) with the repeated-measures effects. SPSS MANOVA takes the traditional approach to ANCOVA. Programs labeled GLM in SAS and SPSS use the general linear model approach.

Table 8.11 contains a hypothetical data set for a mixed ANCOVA with a single CV measured once for each case. The randomized-groups IV is type of music: a_1 = rock and a_2 = classical. The repeated-measures IV is time of day—b_1 = 9 p.m., b_2 = 11 p.m., b_3 = 1 a.m., and b_4 = 3 a.m. (labeled T1–T4 in the data set)—at which selections are rated (the DV). The CV (EASY) is a rating of an easy-listening music selection, collected before random assignment of listeners to type of music.

8.5.2.1.1 Traditional Analysis Mixed designs (with at least one randomized-groups IV) benefit by adjustment to the randomized-groups effect(s) and error term in the analysis. All randomized-groups main effects are adjusted for CV(s), as are interactions involving only randomized-groups IVs. There is no adjustment to repeated-measures main effects, nor to any interaction involving a repeated measure, nor to error terms for these effects.

TABLE 8.11 Data set for analysis of mixed randomized-repeated ANCOVA with a nonvarying CV

MUSIC	EASY	T1	T2	T3	T4
1	6	8	7	5	3
1	6	9	7	5	1
1	4	8	5	4	1
1	3	6	6	4	2
2	6	2	4	8	10
2	7	6	5	9	10
2	3	2	4	6	10
2	5	3	5	6	9

Table 8.12 shows the traditional analysis of the data set of Table 8.11 through SPSS MANOVA. Syntax is similar to SPSS GLM, except that the levels of the IV (MUSIC) must be indicated in parentheses after the label. The instruction to /PRINT=SIGNIF(AVERF, EFSIZE) CELLINFO(MEANS) requests univariate results for the repeated-measures portion of the design, printing of the effect size (partial η^2), and unadjusted cell means. Adjusted marginal means for MUSIC are requested by the /PMEANS instruction. Adjusted cell means are not available with a repeated-measures IV. Adjusted marginal means for the repeated-measures IV, TRIALS, do not differ from unadjusted means in this analysis because repeated-measures effects are not adjusted for the covariate.

The output begins with a separate printout of unadjusted cell means and marginal means for TRIALS (labeled For entire sample), for each level of the repeated-measures factor (T1–T4). The Tests of Between-Subjects Effects source table shows the test of the CV, labeled REGRESSION, and the test of the IV, MUSIC. The randomized-groups error term (S/A) is labeled WITHIN+RESIDUAL. Effect-size measures for the CV and randomized-groups IV follow. Then information about the CV is shown in the section labeled Regression analysis for WITHIN+RESIDUAL error term. The regression analysis provides redundant information about the significance of the adjustment by the CV; recall that $F = t^2$. Both the CV and the main effect of MUSIC are statistically significant. Note that in the ANOVA tables, the DV is misleadingly labeled T1 (instead of rating), and the CV is labeled TEASY (instead of easy).

The adjusted marginal means (doubled) for the randomized-groups IV (MUSIC) then follow, along with additional redundant output that is omitted here. Note that these adjusted means require a request for AVERF in the PRINT paragraph.

The repeated-measures portion of the output begins with information about the sphericity test, along with adjustment factors should the assumption be violated (see Section 6.6.2). The Tests involving 'TRIALS' Within-Subject Effect source table includes tests for TRIALS and the MUSIC BY TRIALS, with the error term ($B \times S/A$) also labeled WITHIN+RESIDUAL. The interaction is

TABLE 8.12 ANCOVA for a mixed design with a CV measured once (SPSS MANOVA syntax and selected output)

```
MANOVA
 T1 T2 T3 T4 BY MUSIC(1,2) WITH (EASY)
 /WSFACTOR = TRIALS(4)
 /MEASURE = RATING
 /PRINT = SIGNIF(AVERF, EFSIZE) CELLINFO(MEANS)
 /PMEANS = TABLES(MUSIC)
 /WSDESIGN = TRIALS
 /DESIGN .
```

```
- - - - - - - - - - - - - - - - - - - - - - - - - - - - - - - - - - - - -

Cell Means and Standard Deviations
Variable .. T1

 FACTOR           CODE                           Mean  Std. Dev.          N

 MUSIC            rock                          7.750      1.258          4
 MUSIC            classic                       3.250      1.893          4
For entire sample                               5.500      2.828          8

- - - - - - - - - - - - - - - - - - - - - - - - - - - - - - - - - - - - -

Variable .. T2

 FACTOR           CODE                           Mean  Std. Dev.          N

 MUSIC            rock                          6.250       .957          4
 MUSIC            classic                       4.500       .577          4
For entire sample                               5.375      1.188          8

- - - - - - - - - - - - - - - - - - - - - - - - - - - - - - - - - - - - -

Variable .. T3

 FACTOR           CODE                           Mean  Std. Dev.          N

 MUSIC            rock                          4.500       .577          4
 MUSIC            classic                       7.250      1.500          4
For entire sample                               5.875      1.808          8

- - - - - - - - - - - - - - - - - - - - - - - - - - - - - - - - - - - - -

Variable .. T4

 FACTOR           CODE                           Mean  Std. Dev.          N

 MUSIC            rock                          1.750       .957          4
 MUSIC            classic                       9.750       .500          4
For entire sample                               5.750      4.334          8

- - - - - - - - - - - - - - - - - - - - - - - - - - - - - - - - - - - - -

Variable .. EASY

 FACTOR           CODE                           Mean  Std. Dev.          N

 MUSIC            rock                          4.750      1.500          4
 MUSIC            classic                       5.250      1.708          4
For entire sample                               5.000      1.512          8
```

TABLE 8.12 Continued

```
- - - - - - - - - - - - - - - - - - - - - - - - - - - - - - - - - - - - - - -
Variable .. EASY

 FACTOR          CODE                              Mean  Std. Dev.         N

 MUSIC           rock                             4.750      1.500         4
 MUSIC           classic                          5.250      1.708         4
For entire sample                                 5.000      1.512         8
- - - - - - - - - - - - - - - - - - - - - - - - - - - - - - - - - - - - - - -
Variable .. EASY

 FACTOR          CODE                              Mean  Std. Dev.         N

 MUSIC           rock                             4.750      1.500         4
 MUSIC           classic                          5.250      1.708         4
For entire sample                                 5.000      1.512         8
- - - - - - - - - - - - - - - - - - - - - - - - - - - - - - - - - - - - - - -
Variable .. EASY

 FACTOR          CODE                              Mean  Std. Dev.         N

 MUSIC           rock                             4.750      1.500         4
 MUSIC           classic                          5.250      1.708         4
For entire sample                                 5.000      1.512         8
- - - - - - - - - - - - - - - - - - - - - - - - - - - - - - - - - - - - - - -

* * * * * * A n a l y s i s   o f   V a r i a n c e -- design  1 * * * * * *

Tests of Between-Subjects Effects.

 Tests of Significance for T1 using UNIQUE sums of squares
 Source of Variation            SS      DF       MS          F  Sig of F

 WITHIN+RESIDUAL              3.55       5      .71
 REGRESSION                  11.33       1    11.33      15.96      .010
 MUSIC                        6.44       1     6.44       9.07      .030
- - - - - - - - - - - - - - - - - - - - - - - - - - - - - - - - - - - - - - -
 Effect Size Measures
                          Partial
 Source of Variation      ETA Sqd

 Regression                  .761
 MUSIC                       .645
- - - - - - - - - - - - - - - - - - - - - - - - - - - - - - - - - - - - - - -
 Regression analysis for WITHIN+RESIDUAL error term
 --- Individual Univariate .9500 confidence intervals
 Dependent variable .. T1

 COVARIATE            B         Beta   Std. Err.     T-Value   Sig. of t

 TEASY           .42742       .68387        .107       3.995        .010

 COVARIATE    Lower -95%   CL- Upper     ETA Sq.

 TEASY             .152         .702        .761
```

TABLE 8.12 Continued

```
- - - - - - - - - - - - - - - - - - - - - - - - - - - - - - - - - - - - - - -
 Adjusted and Estimated Means
 Variable .. T1
  CELL          Obs. Mean    Adj. Mean    Est. Mean   Raw Resid.   Std. Resid.

     1            10.125       10.339       10.125         .000          .000
     2            12.375       12.161       12.375         .000          .000

- - - - - - - - - - - - - - - - - - - - - - - - - - - - - - - - - - - - - - -
- - - - - - - - - - - - - - - - - - - - - - - - - - - - - - - - - - - - - - -


* * * * * * A n a l y s i s   o f   V a r i a n c e -- design   1 * * * * * *


Tests involving 'TRIALS' Within-Subject Effect.

 Mauchly sphericity test, W =        .32890
 Chi-square approx. =               5.25111 with 5 D. F.
 Significance =                      .386
 Greenhouse-Geisser Epsilon =        .61441
 Huynh-Feldt Epsilon =              1.00000
 Lower-bound Epsilon =               .33333

AVERAGED Tests of Significance that follow multivariate tests are equivalent
to univariate or split-plot or mixed-model approach to repeated measures.
Epsilons may be used to adjust d.f. for the AVERAGED results.


- - - - - - - - - - - - - - - - - - - - - - - - - - - - - - - - - - - - - - -
- - - - - - - - - - - - - - - - - - - - - - - - - - - - - - - - - - - - - - -


* * * * * * A n a l y s i s   o f   V a r i a n c e -- design   1 * * * * * *


Tests involving 'TRIALS' Within-Subject Effect.

 AVERAGED Tests of Significance for RATING using UNIQUE sums of squares
 Source of Variation          SS      DF        MS         F  Sig of F

 WITHIN+RESIDUAL           15.63      18       .87
 TRIALS                     1.25       3       .42       .48      .700
 MUSIC BY TRIALS          179.62       3     59.87     68.98      .000
- - - - - - - - - - - - - - - - - - - - - - - - - - - - - - - - - - - - - - -

 Effect Size Measures

                         Partial
 Source of Variation     ETA Sqd

 TRIALS                     .074
 MUSIC BY TRIALS            .920
```

statistically significant, but the main effect of TRIALS is not. Finally, effect size measures are provided for the repeated-measures analysis: TRIALS and MUSIC BY TRIALS. Multivariate output is deleted here.

SPSS MANOVA does not provide adjusted marginal or cell means for repeated measures in ANCOVA. The means for the main effect of the repeated-measures factor, TRIALS, are the same as the unadjusted means. Adjusted marginal means for the main effect of MUSIC and adjusted cell means are found as per Section 8.6.5. Note that the adjusted means in Section 8.6.5 are *not* the same adjusted means that result from the GLM approach.

8.5.2.1.2 General Linear Model Analysis In the GLM approach, all the repeated-measures sums of squares (for main effects, interactions, and error terms) are adjusted by the interaction of the repeated-measure IV(s) with the CV. Table 8.13 shows an analysis of the data set of Table 8.11 through SAS GLM.

The portions of this GLM output that refer to repeated-measures IV(s) differ from those in the traditional approach because the GLM approach adjusts repeated measures. For instance, there are adjusted means for the DV over the four trials in the section labeled Least Squares Means that differ from estimates in the traditional approach. On the other hand, the adjusted marginal means for the two levels of music (means of 5.17 and 6.08 for rock and classical music, respectively) have the same adjustment as in the traditional approach. In the source table, the test of the randomized-groups IV (music) and the CV (easy) are also the same as in the traditional approach. However, the tests of the repeated-measures effects, trials and the trials*music interaction, are different because of their adjustment by the trials*easy interaction.

The trials*easy interaction is reported as an additional source of variance. This is a test of differences in the relationship between the DV and CV over the four trials: Does the relationship between rating of easy music and the rating of either rock or classic music depend on trials? Thus, this is a test of homogeneity of regression for the repeated-measures IV, as discussed in Section 8.6.2. There is no trials*easy interaction in this data set—that is, there is no violation of the homogeneity of regression assumption for the repeated-measures IV. Nevertheless, adjustment for the effect in this example reduces sum of squares for error enough to strengthen tests of trials and the trials*music interaction, even after the degrees of freedom for this interaction are deducted from the repeated-measures error term.[9]

It is not clear that the GLM strategy makes much sense or enhances power in ANCOVA. There is usually no a priori reason to expect a different relationship between the DV and CV for different levels of the repeated-measures IV(s). If no such relationship is present, the loss of degrees of freedom for estimating this effect could more than offset the reduction in sum of squares for error, resulting in a less powerful test of the repeated-measures effects. When using a GLM program, you

[9] The multivariate output, which also includes the trials*easy interaction, is omitted.

TABLE 8.13 ANCOVA with the GLM approach for a mixed design with a constant CV (SAS GLM syntax and selected output)

```
proc glm data=SASUSER.MUSIC;
  class music;
  model t1 t2 t3 t4 = music easy;
  repeated trials 4 / mean;
  lsmeans music;
run;
```

Repeated Measures Analysis of Variance

Repeated Measures Level Information

Dependent Variable	t1	t2	t3	t4
Level of trials	1	2	3	4

Tests of Hypotheses for Between Subjects Effects

Source	DF	Type III SS	Mean Square	F Value	Pr > F
music	1	6.43598790	6.43598790	9.07	0.0297
easy	1	11.32661290	11.32661290	15.96	0.0104
Error	5	3.54838710	0.70967742		

Univariate Tests of Hypotheses for Within Subject Effects

Source	DF	Type III SS	Mean Square	F Value	Pr > F	Adj Pr > F G-G	Adj Pr > F H-F
trials	3	4.1000861	1.3666954	1.80	0.1901	0.2149	0.1901
trials*music	3	181.1990927	60.3996976	79.62	<.0001	<.0001	<.0001
trials*easy	3	4.2459677	1.4153226	1.87	0.1788	0.2051	0.1788
Error(trials)	15	11.3790323	0.7586022				

Greenhouse-Geisser Epsilon	0.6634
Huynh-Feldt Epsilon	1.5417

Means of Within Subjects Effects

Level of trials	N	Mean	Std Dev
1	8	5.50000000	2.82842712
2	8	5.37500000	1.18773494
3	8	5.87500000	1.80772153
4	8	5.75000000	4.33424899

Least Squares Means

music	t1 LSMEAN	t2 LSMEAN	t3 LSMEAN	t4 LSMEAN
1	7.93548387	6.32661290	4.64919355	1.76612903
2	3.06451613	4.42338710	7.10080645	9.73387097

could rerun the analysis without the CV to obtain the traditional (unadjusted) repeated-measures portion of the source table, as demonstrated in Section 8.8.2.

8.5.2.2 Varying Covariates(s) over Cells

There are two common designs in which CVs differ for cells: matched randomized-blocks designs, wherein cases in the cells of a repeated-measures IV actually are different cases (cf. Section 6.5.3), and designs where the CV is reassessed prior to administration of each level of the repeated-measures IV(s). When the CVs differ for levels of the repeated-measures IV(s), they are potentially useful in enhancing power for all effects.

A hypothetical data set for a one-way repeated-measures ANCOVA with a varying CV appears in Table 8.14. There are two levels of the repeated-measures IV, *T*, and two levels of the CV, *X*, measured before each level of *T*.

8.5.2.2.1 Traditional Analysis Table 8.15 shows SPSS MANOVA syntax and selected output, using the traditional approach to this repeated-measures ANCOVA. The critical syntax is the listing of the two CVs, B1_X and B2_X without parenthesis after the WITH instruction.

Unadjusted cell means and standard deviations (not shown) begin the output and are followed by tests of the intercept (also not shown); both are not of interest in this analysis. The meaningful output is in the tables labeled `Tests involving 'TRIALS' Within-Subject Effect` and `Regression analysis for WITHIN+RESIDUAL error term`. The DV is misleadingly labeled `T2` in both tables. The results show a statistically significant effect of `TRIALS` but no adjustment by the CV (labeled `T4`) with `t-Value` $= -.434$, $p = .677$.

TABLE 8.14 Hypothetical data for one-way repeated-measures ANCOVA with a varying CV

Case	*T1_X*	*T1_Y*	*T2_X*	*T2_Y*
1	4	9	3	15
2	8	10	6	16
3	13	14	10	20
4	1	6	3	9
5	8	11	9	15
6	10	10	9	9
7	5	7	8	12
8	9	12	9	20
9	11	14	10	20

TABLE 8.15 Repeated-measures ANCOVA with a varying CV (SPSS MANOVA syntax and selected output)

```
MANOVA
 B1_Y B2_Y WITH B1_X B2_X
 /WSFACTOR = TRIALS(2)
 /PRINT = SIGNIF(EFSIZE), CELLIFO(MEANS)
 /WSDESIGN TRIALS
 /DESIGN .
```

```
* * * * * * A n a l y s i s   o f   V a r i a n c e -- design   1 * * * * * *

Tests involving 'TRIALS' Within-Subject Effect.
 Tests of Significance for T2 using UNIQUE sums of squares
 Source of Variation            SS       DF        MS          F  Sig of F

 WITHIN+RESIDUAL             26.08        7      3.73
 REGRESSION                    .70        1       .70        .19      .677
 TRIALS                      99.16        1     99.16      26.62      .001
 - - - - - - - - - - - - - - - - - - - - - - - - - - - - - - - - - - - - - -

 Effect Size Measures
                             Partial
 Source of Variation         ETA Sqd

 Regression                     .026
 TRIALS                         .792
 - - - - - - - - - - - - - - - - - - - - - - - - - - - - - - - - - - - - - -

 Regression analysis for WITHIN+RESIDUAL error term
 --- Individual Univariate .9500 confidence intervals
 Dependent variable .. T2

 COVARIATE             B         Beta   Std. Err.     t-Value   Sig. of t

 T4              -.21805     -.16198        .502       -.434        .677

 COVARIATE   Lower -95%   CL- Upper     ETA Sq.

 T4              -1.405        .969        .026
 - - - - - - - - - - - - - - - - - - - - - - - - - - - - - - - - - - - - - -
```

8.5.2.2.2 General Linear Model Analysis SPSS and SAS GLM programs have no special syntax for specifying a CV that changes with each level of a repeated-measures IV. Instead, the problem is set up as a randomized-groups design with one IV representing cases where measurements for each trial are on a separate line. For this example, each case has two lines, one for each trial. Trial number, CV score, and DV score are on each line. The repeated-measures design is simulated by considering both cases and trials randomized-groups IVs. Table 8.16 shows the previous data set rearranged for this analysis.

Table 8.17 shows SAS GLM setup and output for this analysis. Because no interactions are requested, the CASE by T interaction is the error term. Type III sums of squares (`ss3`) are requested to limit output.

TABLE 8.16 Hypothetical data for repeated-measures ANCOVA with varying CV (one-line-per-trial setup)

CASE	T	X	Y	CASE	T	X	Y
1	1	4	9	5	2	9	15
1	2	3	15	6	1	10	10
2	1	8	10	6	2	9	9
2	2	6	16	7	1	5	7
3	1	13	14	7	2	8	12
3	2	10	20	8	1	9	12
4	1	1	6	8	2	9	20
4	2	3	9	9	1	11	14
5	1	8	11	9	2	10	20

TABLE 8.17 Repeated-measures ANCOVA: Varying CV through SAS GLM

```
proc glm data=SASUSER.SPLTPLOT;
   class CASE T;
   model Y = CASE T X / ss3 ;
   lsmeans T/ ;
   means T ;
run;
```

Class Level Information

Class	Levels	Values
CASE	9	1 2 3 4 5 6 7 8 9
T	2	1 2

Number of Observations Read 18
Number of Observations Used 18

Dependent Variable: Y

Source	DF	Sum of Squares	Mean Square	F Value	Pr > F
Model	10	295.5359231	29.5535923	7.93	0.0059
Error	7	26.0751880	3.7250269		
Corrected Total	17	321.6111111			

R-Square	Coeff Var	Root MSE	Y Mean
0.918923	15.17056	1.930033	12.72222

Source	DF	Type III SS	Mean Square	F Value	Pr > F
CASE	8	105.1901182	13.1487648	3.53	0.0569
T	1	99.1581454	99.1581454	26.62	0.0013
X	1	0.7025898	0.7025898	0.19	0.6771

Least Squares Means

T	Y LSMEAN
1	10.3575606
2	15.0868839

Level of T	N	Y Mean	Y Std Dev	X Mean	X Std Dev
1	9	10.3333333	2.78388218	7.66666667	3.74165739
2	9	15.1111111	4.42844342	7.44444444	2.78886676

Used this way, the SAS GLM source table provides a test of CASE as an IV. Results are consistent with those of SPSS MANOVA. However, there is no test for homogeneity of covariance using this setup, which would be relevant if there were more than two levels of the repeated-measures IV.

8.6 SOME IMPORTANT ISSUES

Analysis of covariance is complex and burdened with assumptions beyond those of ANOVA. Thus, there are a number of issues to consider, including alternative designs to enhance power.

8.6.1 Multiple Covariates

The use of multiple CVs complicates the analysis in terms of calculation, testing of assumptions (cf. Section 8.6.2), and interpreting coefficients. CVs in ANCOVA can themselves be interpreted as predictors of the DV. From a sequential regression perspective (Section 5.6.5), each CV is simply a high-priority, continuous IV, with discrete IVs (main effects and interactions) evaluated after the relationship between the CV(s) and the DV is removed.

Significance tests for CVs assess their relationship with the DV. If a CV is statistically significant, it is related to the DV. For the small-sample example of Table 8.1, the CV, PRE_RATE, is related to the DV, RATING, with $F(1, 11) = 10.35, p < .05$, and $B = 0.727$. This relationship is interpreted in the same way as any IV in multiple regression (Section 4.4.4).

With multiple CVs, all CVs enter the multiple regression equation at once, and, as a set, they are treated as a standard multiple regression with a regression coefficient for each CV. Within the set of CVs, the significance of each CV is assessed as if it entered the equation last; only the unique relationship between the CV and the DV is tested for significance after overlapping variability with other CVs, in their relationship with the DV, is removed. Therefore, although a CV may be significantly correlated with the DV when considered individually, it may add no adjustment to the DV when considered last. When interpreting the utility of a CV, it is necessary to consider correlations among CVs, correlations between each CV and the DV, and significance levels for each CV as reported in ANCOVA source tables.

Unstandardized regression coefficients, provided by most statistical software programs, have the same meaning as regression coefficients, described in Section 4.4.4. In ANCOVA, however, interpretation of the coefficients depends on type of sum of squares used. When Type III sums of squares (standard multiple regression, which is appropriate for experiments) are used, the significance of the regression coefficients for CVs is assessed as if the CV entered the regression equation after all other CVs and after all main effects and interactions. With other types of SS, however, CVs enter the

FIGURE 8.4 Partition of total adjusted degrees of freedom into degrees of freedom associated with groups and associated with cases within the groups, where c = number of CVs

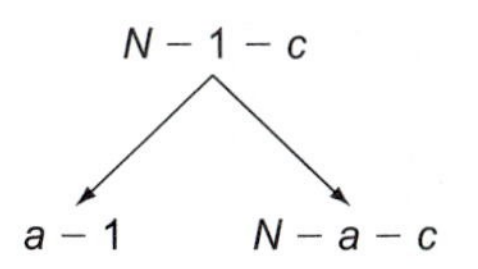

equation differently. Type I SS in GLM programs, for example, evaluates all effects, including CVs, as entering the equation in the order they are listed in the model instructions. In any event, the coefficient is evaluated at whatever point the CV enters the equation.

Each CV "costs" 1 df, so that degrees of freedom are reduced by the number of CVs (c), and the partition of df is as shown in Figure 8.4, which is a generalized version of Figure 8.3(b).

Matrix algebra is required to find the c regression coefficients (Bs) and to adjust the sums of squares for the analysis. Brown, Michels, and Winer (1991) show equations for ANCOVA with multiple CVs and a numerical example with two CVs.

Software programs differ considerably in how they present ANCOVA with multiple CVs. The GLM programs provide tests of each CV individually but no test of the combined effect of the CVs on the DV (although SAS provides information to compute the combined effect, as shown in what follows). Of the software demonstrated here, only SPSS MANOVA provides both individual and combined tests of multiple CVs. Regression coefficients are available in SPSS GLM with a request for parameter estimates, but they are unavailable in SAS GLM.

Table 8.18 shows age group and skiing experience added to prerating as CVs for the small-sample example. SPSS MANOVA analysis of that data set is in Table 8.19.

Note that the additional two CVs have weakened the test of the IV, `VIEW` (cf. Table 8.8). The `REGRESSION` source of variation shows that the set of CVs adjusts the DV, $F(3, 9) = 4.72, p = .03$. However, none of the CVs individually adjusts the DV once the other CVs are taken into account. Even `PRE_RATE`, which alone was statistically significant in its adjustment of `RATING`, no longer is. This suggests that the CVs are too highly correlated (cf. "Relationships among Covariates," on the following page) to be fruitfully combined in the analysis; we would have been better off with only one of them. Later in this section, choosing among potential CVs is discussed. SPSS MANOVA also provides bivariate correlations and r^2 for each CV with the DV.

Table 8.20 shows syntax for the same analysis in SAS GLM. Note that the IV, `VIEW`, is entered before the CVs. This permits calculation of a test for the combined CVs, as discussed on the following page.

Tests for each of the CVs are provided, along with the test of the IV, `VIEW`, in the portion of the output associated with `Type III SS`, in which all effects are adjusted for one another. The results for `VIEW` and each of the CVs are the same as in Table 8.19.

TABLE 8.18 Small-sample data with two additional covariates

View	Prerating	Age Group	Experience	Rating
1	2	3	4	1
1	1	1	2	1
1	4	6	8	3
1	4	7	9	5
1	1	2	3	2
2	2	2	3	3
2	4	5	6	4
2	4	2	7	3
2	4	5	8	5
2	3	4	5	4
3	3	5	6	6
3	4	4	7	8
3	2	3	2	6
3	5	6	9	7
3	3	4	5	5

The test of combined CVs is easily calculated from this output as long as `TYPE I SS` results are available and the IV is entered before the CVs in the model instruction. Sum of squares for combined CVs is

$$SS_{covariates} = SS_{\texttt{model}} - SS_{\texttt{VIEW(Type I SS)}}$$

$$= 52.9386 - 41.2 = 11.7386$$

Degrees of freedom $= c = 3$, so that:

$$MS_{covariates} = \frac{SS_{covariates}}{df_{covariates}} = \frac{11.73864363}{3} = 3.9129$$

and

$$F_{covariates} = \frac{MS_{covariates}}{MS_{error}} = \frac{3.9129}{0.8290} = 4.72$$

as in Table 8.19.

8.6.1.1 Relationships among Covariates

Interpreting significance tests of CVs is aided by knowledge of relationships among them. Sometimes tests of CVs show that they are statistically significant when considered together but not when considered individually because of strong correlations among them, as in Table 8.19. Occasionally correlations among tests of CVs are high enough to cause multicollinearity (Section 8.3.2.2). In any event, correlations among multiple CVs are reported as part of the results of an ANCOVA.

TABLE 8.19 ANCOVA with multiple CVS through SPSS MANOVA (syntax and selected output)

```
MANOVA
 RATING BY VIEW(1,3) WITH AGEGRP EXPER PRE_RATE
 /DESIGN .
```

```
* * * * * * A n a l y s i s   o f   V a r i a n c e -- design  1 * * * * * *

 Tests of Significance for RATING using UNIQUE sums of squares
 Source of Variation          SS       DF         MS         F  Sig of F

 WITHIN CELLS               7.46        9        .83
 REGRESSION                11.74        3       3.91      4.72      .030
 VIEW                      23.27        2      11.64     14.03      .002

 (Model)                   52.94        5      10.59     12.77      .001
 (Total)                   60.40       14       4.31

 R-Squared =            .876
 Adjusted R-Squared =   .808
 - - - - - - - - - - - - - - - - - - - - - - - - - - - - - - - - - - - - - -

 Correlations between Covariates and Predicted Dependent Variable

           COVARIATE

 VARIABLE      AGEGRP      EXPER       PRE_RATE

 RATING          .592       .520           .655
 - - - - - - - - - - - - - - - - - - - - - - - - - - - - - - - - - - - - - -

 Squared Correlations between Covariates and Predicted Dependent Variable

 VARIABLE   AVER. R-SQ

 AGEGRP          .350
 EXPER           .271
 PRE_RATE        .429
 - - - - - - - - - - - - - - - - - - - - - - - - - - - - - - - - - - - - - -

 Regression analysis for WITHIN CELLS error term
 --- Individual Univariate .9500 confidence intervals
 Dependent variable .. RATING

 COVARIATE             B        Beta   Std. Err.     t-Value   Sig. of t

 AGEGRP           .33797      .28494        .263       1.286        .231
 EXPER            .33105      .38478        .414        .800        .444
 PRE_RATE        -.38753     -.22814        .871       -.445        .667

 COVARIATE   Lower -95%  CL- Upper

 AGEGRP           -.256        .932
 EXPER            -.605       1.267
 PRE_RATE        -2.357       1.582
```

TABLE 8.20 ANCOVA with multiple CVs through SAS GLM

```
proc glm data=SASUSER.ANCMULT;
   class VIEW;
   model RATING = VIEW PRE_RATE AGEGRP EXPER ;
run;
```

Dependent Variable: RATING

Source	DF	Sum of Squares	Mean Square	F Value	Pr > F
Model	5	52.93864363	10.58772873	12.77	0.0007
Error	9	7.46135637	0.82903960		
Corrected Total	14	60.40000000			

R-Square	Coeff Var	Root MSE	RATING Mean
0.876468	21.67896	0.910516	4.200000

Source	DF	Type I SS	Mean Square	F Value	Pr > F
VIEW	2	41.20000000	20.60000000	24.85	0.0002
PRE_RATE	1	9.30909091	9.30909091	11.23	0.0085
AGEGRP	1	1.89893201	1.89893201	2.29	0.1645
EXPER	1	0.53062070	0.53062070	0.64	0.4443

Source	DF	Type III SS	Mean Square	F Value	Pr > F
VIEW	2	23.27096810	11.63548405	14.03	0.0017
PRE_RATE	1	0.16418242	0.16418242	0.20	0.6668
AGEGRP	1	1.37132569	1.37132569	1.65	0.2305
EXPER	1	0.53062070	0.53062070	0.64	0.4443

Correlations among CVs are easily obtained through software correlation programs; SMCs (squared multiple correlations) are available to evaluate multicollinearity through most factor analysis programs. SPSS FACTOR provides both bivariate correlations and SMCs. For example, Table 8.21 shows syntax and selected output for an SPSS FACTOR run of the data set with multiple CVs.

We see that, indeed, bivariate correlations are very high, ranging from .77 to .93. And SMCs (Initial Communalities), although not high enough to cause statistical problems, are strong enough to pose questions about the utility of using more than one of the CVs.

8.6.1.2 Choosing Covariates

Experiments typically use a single CV, often a pretest, that is administered before random assignment to treatments. Sometimes, though, circumstances may dictate the availability or potential desirability of multiple CVs. However, a point of diminishing returns in adjustment of the DV is quickly reached when too many CVs are used

TABLE 8.21 Bivariate and squared multiple correlations through SPSS FACTOR (syntax and selected output)

```
FACTOR
 /VARIABLES EXPER PRE_RATE AGEGRP
 /MISSING LISTWISE
 /ANALYSIS EXPER PRE_RATE AGEGRP
 /PRINT INITIAL CORRELATION EXTRACTION
 /CRITERIA MINEIGEN(1) ITERATE(25)
 /EXTRACTION PAF
 /ROTATION NOROTATE
 /METHOD=CORRELATION .
```

Factor Analysis

Correlation Matrix

		EXPER	PRE_RATE	AGEGRP
Correlation	EXPER	1.000	.929	.821
	PRE_RATE	.929	1.000	.769
	AGEGRP	.821	.769	1.000

Communalities

	Initial	Extraction
EXPER	.891	.989
PRE_RATE	.863	.872
AGEGRP	.675	.680

Extraction Method: Principal Axis Factoring.

and correlated with each other. Power is reduced because numerous correlated CVs subtract degrees of freedom from the error term while not removing commensurate sums of squares for error. Preliminary analysis of the CVs improves chances of picking a good set.

Statistically, the goal is to identify a small set of CVs that are uncorrelated with each other but highly correlated with the DV. Conceptually, you want to select CVs that adjust the DV for predictable but unwanted sources of variability. It may be possible to pick the CVs on theoretical grounds or on the basis of knowledge of the literature regarding important sources of variability that should be controlled.

If theory is unavailable or the literature is insufficiently developed to provide a guide to important sources of variability in the DV, statistical considerations assist the selection of CVs. One strategy is to look at correlations among CVs and select one from among each of those sets of potential CVs that are substantially correlated with each other, perhaps by choosing the one with the highest correlation with the DV.

Table 8.22 shows the correlations among CVs and of each CV with the DV, RATING, produced by SPSS CORRELATION and GRAPH.

In Table 8.22, we see that the pretest, PRE_RATE, has the highest correlation with RATING (but not by much) and is the CV of choice among the three. Note also the very high correlations among CVs. A more sophisticated strategy that is useful with a larger set of potential CVs is to use all-subsets or stepwise regression (cf. Tabachnick and Fidell, 2007) to see which (small) set of CVs best predicts the DV, ignoring the IV. SAS REG does all-subsets regression.

Even if *N* is large and power is not a problem, it may still be worthwhile to select and use only a small set of CVs, just for the sake of parsimony.

8.6.2 Test of Homogeneity of Regression

The assumption of homogeneity of regression is that the slopes (regression coefficients or *B* weights) of the regression of the DV on the CV(s) are the same for all cells of a design. Both homogeneity and heterogeneity of regression are illustrated in Figure 8.2. During analysis, slope is computed for every cell of the design and then

TABLE 8.22 Correlations of CVs with each other and with the DV through SPSS CORRELATION and GRAPH

```
CORRELATIONS
 /VARIABLES=PRE_RATE AGEGRP EXPER RATING
 /PRINT=TWOTAIL NOSIG
 /MISSING=PAIRWISE .
GRAPH
 /SCATTERPLOT(MATRIX)=PRE_RATE AGEGRP EXPER RATING
 /MISSING=LISTWISE .
```

Correlations

		PRE_RATE	AGEGRP	EXPER	RATING
PRE_RATE	Pearson Correlation	1	.769**	.929**	.613*
	Sig. (2-tailed)		.001	.000	.015
	N	15	15	15	15
AGEGRP	Pearson Correlation	.769**	1	.821**	.554*
	Sig. (2-tailed)	.001		.000	.032
	N	15	15	15	15
EXPER	Pearson Correlation	.929**	.821**	1	.487
	Sig. (2-tailed)	.000	.000		.066
	N	15	15	15	15
RATING	Pearson Correlation	.613*	.554*	.487	1
	Sig. (2-tailed)	.015	.032	.066	
	N	15	15	15	15

**. Correlation is significant at the 0.01 level (2-tailed).

*. Correlation is significant at the 0.05 level (2-tailed).

TABLE 8.22 Continued

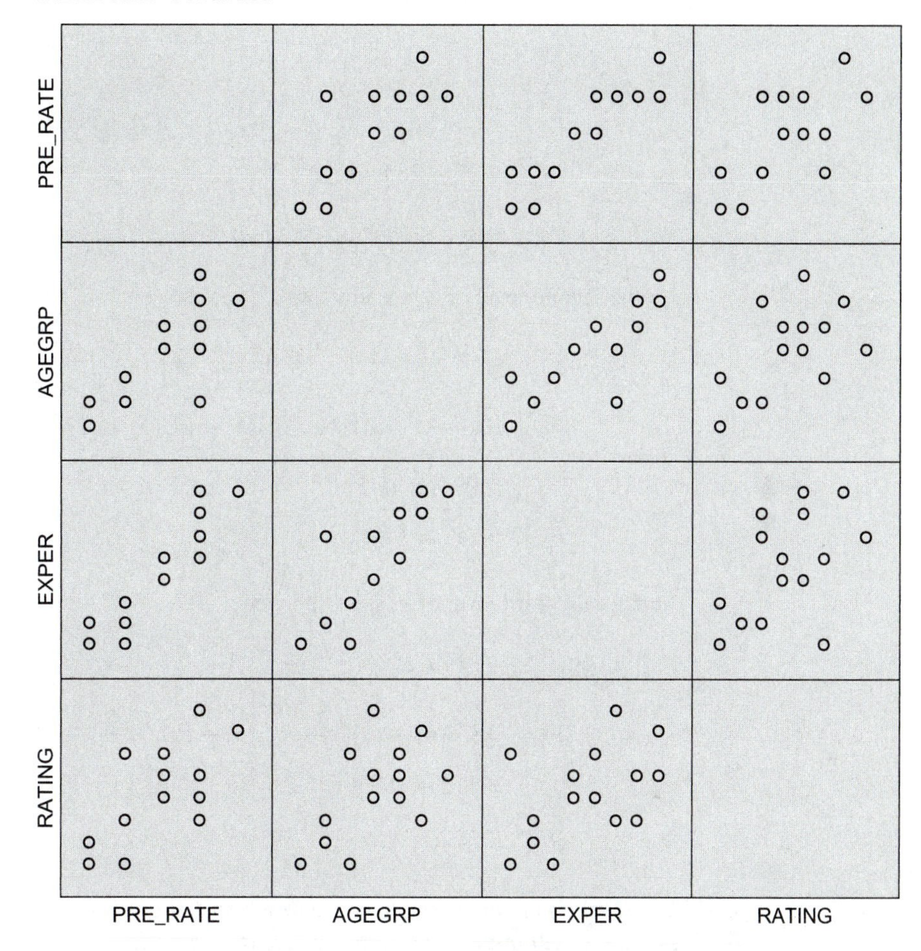

averaged to provide the value used for adjustment. It is assumed that the slopes will differ slightly due to chance, but they are really all estimates of the same population value. If the null hypothesis of equality among slopes is rejected, the analysis of covariance is inappropriate, and an alternative strategy is used, as described in Section 8.6.7.

Hand calculation (with a single CV in a one-way randomized-groups design) requires computing a separate sum of squares for error for the CV for each level of A, a separate sum of products between DV and CV for error for each level of A, and a separate regression coefficient for error for each level of A. These are found as follows:

$$SS_{S/A_{i(x)}} = \sum X_i^2 - \frac{A_{i(x)}^2}{n} \tag{8.19}$$

$$SP_{S/A_i} = \sum X_i Y_i - \frac{A_{i(x)} A_{i(y)}}{n} \tag{8.20}$$

$$B_{S/A_i} = \frac{SP_{S/A_i}}{SS_{S/A_{i(x)}}} \tag{8.21}$$

For the first level of A, parking lot view,

$$\mathrm{SS}_{S/A_{1(x)}} = (2^2 + 1^2 + 4^2 + 4^2 + 1^2) - \frac{12^2}{5} = 38 - 28.8 = 9.2$$

$$\mathrm{SP}_{S/A_1} = (2)(1) + (1)(1) + (4)(3) + (4)(5) + (1)(2) - \frac{(12)(12)}{5}$$

$$= 37 - 28.8 = 8.2$$

$$B_{S/A_1} = \frac{8.2}{9.2} = 0.89$$

For the second level of A, ocean view,

$$\mathrm{SS}_{S/A_{2(x)}} = (2^2 + 4^2 + 4^2 + 4^2 + 3^2) - \frac{17^2}{5} = 61 - 57.8 = 3.2$$

$$\mathrm{SP}_{S/A_2} = (2)(3) + (4)(4) + (4)(3) + (4)(5) + (3)(4) - \frac{(17)(19)}{5}$$

$$= 66 - 64.6 = 1.4$$

$$B_{S/A_2} = \frac{1.4}{3.2} = 0.44$$

And, for the third level of A, ski slope view,

$$\mathrm{SS}_{S/A_{3(x)}} = (3^2 + 4^2 + 2^2 + 5^2 + 3^2) - \frac{17^2}{5} = 63 - 57.8 = 5.2$$

$$\mathrm{SP}_{S/A_3} = (3)(6) + (4)(8) + (2)(6) + (5)(7) + (3)(5) - \frac{(17)(32)}{5}$$

$$= 112 - 108.8 = 3.2$$

$$B_{S/A_3} = \frac{3.2}{5.2} = 0.62$$

These values are then used to calculate a between-groups sum of squares for regression and a within-groups sum of squares for regression:

$$\mathrm{SS}_{\text{between regression}} = \sum\left(\frac{\mathrm{SP}^2_{S/A_i}}{\mathrm{SS}_{S/A_{i(x)}}}\right) - \frac{(\mathrm{SP}_{S/A})^2}{\mathrm{SS}_{S/A_{(x)}}}, \qquad \mathrm{df} = a - 1 \qquad \textbf{(8.22)}$$

and

$$\mathrm{SS}_{\text{within regression}} = \mathrm{SS}_{S/A} - \sum\left(\frac{\mathrm{SP}^2_{S/A_i}}{\mathrm{SS}_{S/A_{i(x)}}}\right), \qquad \mathrm{df} = N - 2 \qquad \textbf{(8.23)}$$

Substituting the values for the data in Table 8.1,

$$\mathrm{SS}_{\text{between regression}} = \left(\frac{(8.2)^2}{9.2} + \frac{(1.4)^2}{3.2} + \frac{(3.2)^2}{5.2}\right) - \frac{(12.8)^2}{17.6} = 9.89 - 9.31 = 0.58$$

$$\mathrm{SS}_{\text{within regression}} = 19.2 - \left(\frac{(8.2)^2}{9.2} + \frac{(1.4)^2}{3.2} + \frac{(3.2)^2}{5.2}\right) = 19.2 - 9.89 = 9.31$$

TABLE 8.23 Source table for testing homogeneity of regression in small-sample ANCOVA

Source	SS	df	MS	F
Between regression	0.58	2	0.29	0.28
Within regression	9.31	9	1.03	
Adjusted *S/A*	9.89	11		

Mean squares and the F ratio are calculated in the usual way, as shown in Table 8.23. These two sources of variance sum to $SS'_{S/A}$, the adjusted error term for the main analysis. Thus, the homogeneity of regression analysis partitions the adjusted error into differences in slopes among groups and residual error within groups.

Homogeneity of regression is present when the obtained F ratio is *smaller* than the tabled value, in this case $F_{.05}(2, 9) = 4.26$. The assumption of homogeneity is met with this small data set.

The most straightforward program for testing homogeneity of regression in randomized-groups designs is SPSS MANOVA. Table 8.24 shows a test for homogeneity of regression through SPSS MANOVA. The inclusion of the IV by CV interaction (PRE_RATE by VIEW) as the last effect in the DESIGN instruction, after the CV and the IV, provides the test for homogeneity of regression. Placing this effect last and requesting METHOD=SEQUENTIAL ensures that the test for the interaction is adjusted for the CV and the IV. The ANALYSIS instruction identifies RATING as the DV.

The test for PRE_RATE BY VIEW shows that there is no violation of homogeneity of regression: $p = .761$.

GLM-based programs test homogeneity of regression by evaluating the CV(s) by IV(s) interaction as the last effect entering a sequential regression model (Section 5.6.5), as demonstrated in Table 8.25.

TABLE 8.24 SPSS MANOVA syntax and selected output for homogeneity of regression

```
MANOVA
 RATING BY VIEW(1,3) WITH PRE_RATE
 /PRINT = SIGNIF(BRIEF)
 /ANALYSIS = RATING
 /METHOD = SEQUENTIAL
 /DESIGN PRE_RATE, VIEW, PRE_RATE BY VIEW.
```

```
* * * * * * A n a l y s i s   o f   V a r i a n c e -- design   1 * * * * * *

 Tests of Significance for RATING using SEQUENTIAL Sums of Squares
 Source of Variation            SS       DF         MS          F  Sig of F

 WITHIN+RESIDUAL              9.31        9       1.03
 PRE_RATE                    22.70        1      22.70      21.95      .001
 VIEW                        27.81        2      13.90      13.44      .002
 PRE_RATE BY VIEW              .58        2        .29        .28      .761

 (Model)                     51.09        5      10.22       9.88      .002
 (Total)                     60.40       14       4.31
```

TABLE 8.25 SAS GLM syntax and selected output testing homogeneity of regression in small-sample ANCOVA

```
proc glm data=SASUSER.SSANCOVA;
   class VIEW;
   model RATING = VIEW PRE_RATE VIEW*PRE_RATE;
run;
```

Dependent Variable: RATING

Source	DF	Sum of Squares	Mean Square	F Value	Pr > F
Model	5	51.09042642	10.21808528	9.88	0.0019
Error	9	9.30957358	1.03439706		
Corrected Total	14	60.40000000			

R-Square	Coeff Var	Root MSE	RATING Mean
0.845868	24.21555	1.017053	4.200000

Source	DF	Type I SS	Mean Square	F Value	Pr > F
VIEW	2	41.20000000	20.60000000	19.91	0.0005
PRE_RATE	1	9.30909091	9.30909091	9.00	0.0150
PRE_RATE*VIEW	2	0.58133551	0.29066776	0.28	0.7614

Source	DF	Type III SS	Mean Square	F Value	Pr > F
VIEW	2	5.27649869	2.63824935	2.55	0.1326
PRE_RATE	1	6.16112489	6.16112489	5.96	0.0373
PRE_RATE*VIEW	2	0.58133551	0.29066776	0.28	0.7614

`Type I SS` and `Type III SS` provide the same test of homogeneity of regression (the interaction between PRE_RATE and VIEW) because that interaction is the last entry in the MODEL equation. The results duplicate those of Tables 8.20 and 8.21.

Only SPSS MANOVA permits a combined test of homogeneity of regression for multiple CVs (Table 8.26). The keyword POOL combines the three CVs in the interaction with the DV, VIEW. The test for homogeneity of regression shows no violation: $F(6, 3) = 1.10, p = .51$.

If this test shows violation of homogeneity of regression, separate tests are done on each CV to help pinpoint the source of the violation. Other programs only test CVs one at a time, so that each is tested in a separate run.

GLM programs automatically test homogeneity of regression for effects of repeated-measures IVs (but not for interaction between randomized-groups and repeated-measures IVs), as shown in Table 8.15. If the assumption is not violated, you may want to eliminate this source of variance, as demonstrated in Section 8.7.2.2.1.

TABLE 8.26 SPSS MANOVA syntax and selected output for homogeneity of regression with multiple CVs

```
MANOVA
 RATING  BY VIEW(1,3)  WITH PRE_RATE AGEGRP EXPER
 /PRINT = SIGNIF(BRIEF)
 /ANALYSIS = RATING
 /METHOD = SEQUENTIAL
 /DESIGN PRE_RATE AGEGRP EXPER VIEW,
      POOL(PRE_RATE AGEGRP EXPER) BY VIEW.
```

```
* * * * * * A n a l y s i s   o f   V a r i a n c e -- design   1 * * * * * *

 Tests of Significance for RATING using SEQUENTIAL Sums of Squares
 Source of Variation            SS        DF        MS          F  Sig of F

 WITHIN+RESIDUAL               2.33        3        .78
 PRE_RATE                     22.70        1      22.70      29.19      .012
 AGEGRP                        1.00        1       1.00       1.28      .340
 EXPER                         5.97        1       5.97       7.67      .070
 VIEW                         23.27        2      11.64      14.96      .028
 POOL(PRE_RATE AGEGRP          5.13        6        .85       1.10      .510
  EXPER) BY VIEW

 (Model)                      58.07       11       5.28       6.79      .071
 (Total)                      60.40       14       4.31

 R-Squared =             .961
 Adjusted R-Squared =    .820

 - - - - - - - - - - - - - - - - - - - - - - - - - - - - - - - - - - - - - - -
```

8.6.3 Effect Size

Effect size in ANCOVA is based on adjusted sums of squares. Either η^2 or partial η^2 are reported, depending on the complexity of the design.

In a one-way design,

$$\eta^2 = \frac{SS'_A}{SS'_T} \tag{8.24}$$

For the small-sample example of Section 8.4,

$$\eta^2 = \frac{27.8}{37.7} = .74$$

Note that this matches the Partial Eta Squared value for VIEW provided by SPSS GLM in Table 8.8. Be sure to include only *adjusted* sums of squares for the IV and for the error when calculating a total from output—the total provided by output may be in error because it includes CVs.

Partial η^2 is less conservative than η^2 in a factorial design (and it matches η^2 in a one-way design):

$$\text{Partial } \eta^2 = \frac{SS'_{\text{effect}}}{SS'_{\text{effect}} + SS'_{\text{error}}} \tag{8.25}$$

where the error term is the one used to test the effect of interest. Smithson's (2003) procedure provides confidence limits around partial η^2 when using his software in SPSS and SAS (Section 5.7.2.1). For the current example, the program provides partial $\eta^2_{\text{view}} = .74$, with confidence limits from .26 to .84. Effect size for the covariate is also substantial—partial $\eta^2_{\text{prerate}} = .67$, with confidence limits from .22 to .81.

The effect size for the IV that is reported has gained increasing importance over the past few years because of its potential use in meta-analyses. ANCOVA poses something of a dilemma because effect size can be computed from either adjusted or unadjusted scores. The issue is which score most closely estimates the impact of the IV on the DV in the population. Because the unadjusted estimate of effect size incorporates differences due to another variable (the CV), it is probably more appropriate to report adjusted effect size in either an experimental or nonexperimental study.

8.6.4 Power

ANCOVA (with a CV) is more powerful than a corresponding ANOVA (without the CV) if the relationship between the DV and CV(s) is large enough to overcome the loss of an error degree of freedom for each CV. An ANOVA without using PRE_RATE as a CV for the small-sample example of Section 8.4 produces Table 8.27. Compared with results in Table 8.6, the unadjusted sums of squares are larger than the adjusted ones, but the *F* ratio is smaller. The loss of 1 df has been offset by subtraction of sum of squares for the CV. The exact probability of *F* with the CV is .0006; without the CV, it is .001.

Power and sample size calculations are identical to those of ANOVA, with the exception that adjusted means and variances (or standard deviations) are used in the equations (Cohen, 1977). Adjusted $MS_{S/A}$ is used for σ^2 in Equation 4.6, and δ refers to the desired minimal difference between adjusted means.

8.6.5 Adjusted Means

ANCOVA adjusts both the means and the error term. Adjustment of the error term is usually more important, but adjustment of the means is sometimes also required. The means are adjusted to the values that are predicted if all cases scored about the same on the CV. In experimental work, random assignment is used in the hope that it equalizes groups before administering treatment; ideally, the groups should have the same mean on the CV before treatment. In any given study, however, random assignment may not create equality. It is common to find adjustment of treatment means in

TABLE 8.27 Source table for one-way ANOVA for the small-sample example of Section 8.4.3

Source	SS	df	MS	*F*
A	41.2	2	20.6	12.88
S/A	19.2	12	1.6	
T	60.4	14		

TABLE 8.28 Adjustment of means based on the CV-DV pattern

	ANOVA Raw Score Means			ANCOVA Adjusted Means		
	a_1	a_2	a_3	a_1	a_2	a_3
CV	Low	Medium	High			
First DV pattern	Low	Medium	High	Medium	Medium	Medium
Second DV pattern	Medium	Medium	Medium	High	Medium	Low
Third DV pattern	High	Medium	Low	Higher	Medium	Lower

nonexperimental work; indeed, this is one reason that ANCOVA is suspect in nonexperimental studies.

To understand the nature of the adjustment, consider the pattern of means in the example of Table 8.28. Rather than equality, the pattern of means for the CV across the three levels of A is low, medium, and high. Suppose the correlation between the CV and DV is relatively strong and positive. The first DV pattern is low, medium, and high, which is the same pattern as the CV obtained before treatment. These results imply that treatment has not changed the DV and significance found in ANOVA should not be found in ANCOVA. The second DV pattern is medium, medium, and medium. In this case, ANOVA would be nonsignificant, but ANCOVA should reach statistical significance, because the treatment at a_1 actually raises scores, whereas the treatment at a_3 actually lowers scores. Finally, both ANOVA and ANCOVA are likely to find significance for the third DV pattern, but the effect of treatment is even stronger than the ANOVA suggests once the pattern in the CV is taken into account. In ANCOVA, it is the adjusted means that are tested for differences and the adjusted means that are interpreted when considering statistically significant effects, doing comparisons, and calculating power. It is useful to compare adjusted and unadjusted means to understand the CV adjustment.

Furthermore, in nonexperimental work with preexisting group differences on the DV, ANCOVA can be quite misleading due to regression to the mean. If there are two groups, each measured during baseline (CV) and then on a DV, ANCOVA lowers the mean for the group with the higher CV and raises the mean for the group with the lower mean. The results, known as Lord's Paradox (Lord, 1967), can be quite misleading, as clearly described by Jamieson (1999).

Each mean in a one-way ANCOVA with a single CV is adjusted by the pooled within-cell regression coefficient, $B_{S/A}$, where

$$B_{S/A} = \frac{\text{SP}_{S/A}}{\text{SS}_{S/A(x)}} \tag{8.26}$$

An adjusted mean is:

$$\overline{Y}'_{A_i} = \overline{Y}_{A_i} - B_{S/A}(\overline{X}_{A_i} - \overline{X}_T) \tag{8.27}$$

The adjusted mean for a level of A is the unadjusted mean for that level minus the difference between the CV mean for that level and the overall CV mean multiplied by the pooled within-cell regression coefficient.

TABLE 8.29 Unadjusted and adjusted means for the small-sample example of Section 8.4.3

Variable	a_1	a_2	a_3	T
X	2.40	3.40	3.40	3.07
Y	2.40	3.80	6.40	4.20
Adjusted Y	2.89	3.56	6.16	4.20

The pooled within-cell regression coefficient for the small sample is

$$B_{S/A} = \frac{12.8}{17.6} = 0.72727$$

from Table 8.2, and the three adjusted means are

$$\bar{Y}'_{A_1} = 2.40 - 0.72727\,(2.40 - 3.07) = 2.89$$
$$\bar{Y}'_{A_2} = 3.80 - 0.72727\,(3.40 - 3.07) = 3.56$$
$$\bar{Y}'_{A_3} = 6.40 - 0.72727\,(3.40 - 3.07) = 6.16$$

Unadjusted means for the CV (X) and DV (Y) for the small-sample example of Section 8.4.3 are in Table 8.29, along with the adjusted DV means.

The means at a_2 and a_3 are pulled down because those two groups scored above the average on the CV to begin with. The a_1 mean is raised because that group scored below the average on the CV to begin with. Because this is an experiment, this suggests that the larger, unadjusted differences among the means are due partially to preexisting differences in rating tendencies between the groups.

Similar to the issue with effect size, there is a dilemma regarding whether to report adjusted or unadjusted means following ANCOVA. The means that are tested are the adjusted ones, so consistency argues in favor of reporting those. In experimental work, the argument is also in favor of the adjusted means, because the unadjusted means are based on group differences on the CV that random assignment should eliminate. A replication of the study with different cases and different random assignment is unlikely to duplicate preexisting group differences on the CV. Therefore, the adjusted means, where those mean differences have been eliminated, are probably the more generalizable. However, in a nonexperimental study, it is probably wise to report both adjusted and unadjusted means, along with the means on the CV (similar to those reported in Table 8.29), to assist the reader in interpreting the results of ANCOVA.

8.6.6 Specific Comparisons

Types of comparisons, and the circumstances under which they are done, are the same as in ANOVA (cf. Sections 4.5.5, 5.6.4, and 6.6.6). The difference is that comparisons are done on the adjusted means and the error term is adjusted beyond that required for the omnibus test, with a separate error term for each comparison.[10]

[10] A simpler method, which uses a single adjusted error term, may be used when $df_{error} \geq 20$ (Keppel, 1991).

The F ratio for a comparison (recalling that $\text{MS}'_{A\text{comp}} = \text{SS}'_{A\text{comp}}$ because $\text{df}_{A\text{comp}} = 1$) is

$$F_{A\text{comp}} = \frac{\text{SS}'_{A\text{comp}}}{\text{MS}'_{\text{error}(A\text{comp})}} \tag{8.28}$$

The numerator, the sum of squares for a comparison, is

$$\text{SS}'_{A\text{comp}} = \frac{n\left(\sum w_j \overline{Y}'_j\right)^2}{\sum w_j^2} \tag{8.29}$$

where $\overline{X}'_j$ is an adjusted mean for the DV and the remaining values are as defined in Equation 4.10.

The adjusted error term for comparisons is

$$\text{MS}'_{\text{error}(A\text{comp})} = \text{MS}'_{S/A}\left(1 + \frac{\text{SS}_{A\text{comp}(x)}}{\text{SS}_{S/A(x)}}\right) \tag{8.30}$$

Notice that $\text{SS}_{A\text{comp}(x)}$ is different for every comparison and is

$$\text{SS}_{A\text{comp}(x)} = \frac{n\left(\sum w_j \overline{X}'_j\right)^2}{\sum w_j^2} \tag{8.31}$$

where $\overline{X}'_j$ is a CV mean, and the remaining terms are as defined in Equation 4.10.

An orthogonal set of comparisons for the small-sample data of Section 8.4.3 compares the parking lot view with the average of the other two views (−2 1 1) and the ocean view with the ski slope view (0 1 −1).

$\text{SS}_{A\text{comp1}}$ for the first comparison, using the adjusted means of Table 8.29, is

$$\text{SS}'_{A\text{comp1}} = \frac{5((-2)(2.89) + (1)(3.56) + (1)(6.16))^2}{(-2)^2 + 1^2 + 1^2} = 12.94$$

The error term for the first comparison requires calculation of $\text{SS}_{A\text{comp}(x)}$ from the CV (X) means of Table 8.29. Thus, $\text{SS}_{A\text{comp1}(x)}$ for the first comparison is

$$\text{SS}_{A\text{comp1}(x)} = \frac{5((-2)(2.4) + (1)(3.4) + (1)(3.4))^2}{(-2)^2 + 1^2 + 1^2} = 3.33$$

and $\text{MS}'_{\text{error}(A\text{comp1})}$ is

$$\text{MS}'_{\text{error}(A\text{comp1})} = 0.90\left(1 + \frac{3.33}{17.60}\right) = 1.07$$

The F ratio for the first comparison—a_1 versus the combination of a_2 and a_3—is

$$F_{A\text{comp1}} = \frac{12.94}{1.07} = 12.09$$

For the second comparison, a_2 is contrasted with a_3. $\text{SS}_{A\text{comp2}}$ for the second comparison, again using the adjusted means of Table 8.29, is

$$\text{SS}'_{A\text{comp2}} = \frac{5((0)(2.89) + (1)(3.56) + (-1)(6.16))^2}{1^2 + (-1)^2} = 16.90$$

The error term for the second comparison requires calculation of $SS_{A\text{comp2}(x)}$ from the means of Table 8.29:

$$SS_{A\text{comp2}(x)} = \frac{5((0)(2.4) + (1)(3.4) + (-1)(3.4))^2}{1^2 + (-1)^2} = 0$$

because the CV means are equal for the last two levels of A, and $MS'_{\text{error}(A\text{comp2})}$ is

$$MS'_{\text{error}(A\text{comp2})} = 0.90\left(1 + \frac{0}{17.6}\right) = 0.90$$

which is the same as the adjusted error term for the omnibus effect of A.

The F ratio for the second comparison, a_2 versus a_3, is

$$F_{A\textbf{comp2}} = \frac{16.90}{0.90} = 18.78$$

Critical F for a planned comparison with 1 and 11 df at $\alpha = .05$ is 4.84, making both of the comparisons statistically significant. Critical F for a post hoc comparison is computed using the Scheffé adjustment of Equation 4.13:

$$F_S = (a-1)F_c = (2)(3.98) = 7.96$$

Both comparisons are statistically significant by either criterion.

Effect sizes for comparisons in ANCOVA use forms of η^2. As usual, effect sizes for comparisons are done with respect to the entire analysis (using SS'_T in the denominator) or the effect they are decomposing. For example, η^2 with respect to the entire analysis for Acomp1 (group 1 versus the other two groups) is

$$\eta^2 = \frac{SS'_{A\text{comp1}}}{SS'_T} = \frac{12.94}{37.70} = .34$$

Thus, 34% of the variance in rating, adjusted for prerating, is associated with the difference between group 1 and the other two groups. And η^2 for the same comparison with respect to the main effect of A, a less conservative measure, is

$$\eta^2 = \frac{SS'_{A\text{comp1}}}{SS'_A} = \frac{12.94}{27.80} = .47$$

This indicates that 47% of the variability attributable to the main effect of A is associated with the difference between group 1 and the other two groups. Smithson's (2003) procedure for finding confidence limits uses df based on the chosen denominator.

Syntax and output for comparisons in ANCOVA are the same as for ANOVA, with inclusion of the CV(s). Table 8.30 shows the orthogonal contrasts just described through SPSS GLM (UNIANOVA). The special contrasts matrix has as many rows as columns, with the first row a series of 1s.

The first contrast (L1) corresponds to the row of 1s in the special contrast matrix and is not of interest. L2 compares the parking lot view with the combination of the other two views, and L3 compares the ski slope and ocean views. The Contrast Estimate is the sum of the weights times the adjusted means—$(-2)(2.89) + (1)(3.56) + (1)(6.16) = 3.94$, which is within rounding error of the L2 value of 3.945 in the table.

TABLE 8.30 Orthogonal comparisons for the small-sample example of Section 8.4.3 through SPSS GLM

```
UNIANOVA
 RATING BY VIEW  WITH PRE_RATE
 /METHOD = SSTYPE(3)
 /INTERCEPT = INCLUDE
 /CRITERIA = ALPHA(.05)
 /CONTRAST(VIEW) = SPECIAL( 1  1  1
                           -2  1  1
                            0  1 -1)
 /DESIGN = PRE_RATE VIEW .
```

Univariate Analysis of Variance

Custom Hypothesis Tests

Contrast Results (K Matrix)

View from room Special Contrast			Dependent Variable RATING
L1	Contrast Estimate		5.909
	Hypothesized Vaue		0
	Difference (Estimate - Hypothesized)		5.909
	Std. Error		2.205
	Sig.		.021
	95% Confidence Interval	Lower Bound	1.055
	for Difference	Upper Bound	10.763
L2	Contrast Estimate		3.945
	Hypothesized Value		0
	Difference (Estimate - Hypothesized)		3.945
	Std. Error		1.133
	Sig.		.005
	95% Confidence Interval	Lower Bound	1.452
	for Difference	Upper Bound	6.439
L3	Contrast Estimate		-2.600
	Hypothesized Value		0
	Difference (Estimate - Hypothesized)		-2.600
	Std. Error		.600
	Sig.		.001
	95% Confidence Interval	Lower Bound	-3.920
	for Difference	Upper Bound	-1.280

A 95% confidence interval is given for the estimate, rather than an F ratio, but this default criterion may be altered. A comparison is statistically significant if the interval does not contain 0. Both comparisons are significant if planned at $\alpha = .05$. A t ratio may be found by dividing the Contrast Estimate by its Std. Error. For L2, $t = 3.945/1.133 = 3.48$. And, as always, $F = t^2 = 12.13$, which is within rounding error of the hand-calculated value of 12.09. The result for the second comparison is identical to that of the hand-calculated value.

SPSS MANOVA and SAS GLM provide either F or t ratios with probabilities, as they do for planned comparisons. Scheffé or other adjustments are needed if the comparisons are post hoc.

Note that in some computer output, the separate adjustments for each ANCOVA comparison show up in the numerator SS_{Acomp} rather than in the error terms, but F ratios match those for the approach in which separate error terms are constructed for each comparison.

8.6.7 Alternatives to ANCOVA

Because of violated assumptions and difficulty in interpreting results of ANCOVA, alternative analytical strategies are sometimes sought. The choice among alternatives depends on the scale of measurement of the CV(s) and the DV, the time that elapses between measurement of the CV and assignment to treatment, the shape of the distributions of the CV and the DV, and the difficulty of interpreting results.

Two alternatives are available when the CV(s) and the DV are measured on the same scale: use of change (difference) scores and conversion of the pretest and posttest scores into a repeated-measures IV. In the first alternative, the difference between the previous CV and DV is computed for each case and used as the DV in randomized-groups ANOVA. In the second alternative, the CV becomes the pretest score and the DV becomes the posttest score, and both compose a repeated-measures IV in a mixed randomized-groups, repeated-measures ANOVA.

The difference among change scores, mixed ANOVA, and ANCOVA is a matter of which research question you want to ask because, actually, the three alternatives yield the same statistical outcome. For example, suppose patient satisfaction with treatment is measured both before and after patients are randomly assigned to one of two new types of treatment. If the research question is, "Does one type of treatment produce a bigger change in patient satisfaction than the other?" then change scores are appropriate. If the research question is, "Do the two treatments produce a different pattern of patient satisfaction in the pretest and posttest scores?" then mixed-design ANOVA is the appropriate analysis. However, if the research question is, "Do the two types of treatment have different mean satisfaction after adjusting for pre-existing group differences in tendency to report satisfaction?" then ANCOVA is the appropriate analysis.

The mixed ANOVA approach preserves information about the means, unlike ANOVA on change scores, in which the means are "lost." This may be an advantage.

A disadvantage of mixed ANOVA is that the effect of the IV of interest is no longer assessed as a main effect but rather as an interaction between the IV and the repeated-measures (pretest, posttest) IV, thus complicating interpretation.

A problem with change scores and mixed ANOVA is ceiling and floor effects (or, more generally, skewness). A change score (or interaction) may be small because the pretest score is very near the end of the scale and no treatment effect can change it very much, or it may be small because the effect of treatment is small. In either case, the result is the same, and the researcher is hard-pressed to decide between them. Furthermore, when the DV is skewed, a change in a mean also produces a change in the shape of the distribution and a potentially misleading significance test (Jamieson and Howk, 1992). If either ANCOVA or ANOVA with change scores is possible, then ANCOVA is usually the better approach when the data are skewed and transformations are not undertaken (Jamieson, 1999).

When CVs are measured on any continuous scale, other alternatives are available: matched randomized blocks (Section 6.5.3) and using the CV as a blocking IV. In the matched randomized-block design, cases are matched into blocks (equated) on the basis of scores on what would have been the CV(s). Each block has as many cases as the number of levels of the IV in a one-way design or number of cells in a larger design (cf. Chapter 6). Cases within each block are randomly assigned to levels or cells of the IV(s) for treatment. In the analytic phase, cases in the same block are treated as if they were the same case in a repeated-measures analysis.

Disadvantages to this approach are the assumption of homogeneity of covariance of a repeated-measures analysis and the loss of df for error without commensurate loss of sums of squares for error if the variables used to block are not highly related to the DV. Implementation of the matched randomized-block design requires the added step of equating cases before randomly assigning them to treatment—a step that may be inconvenient, if not impossible, in some applications. It also requires great faith in the matching procedure.

The last alternative is the use of the CV as a blocking IV. In this instance, cases are measured on potential CV(s) and then grouped according to their scores (e.g., into groups of high, medium, and low consumer satisfaction on the basis of pretest scores). The groups of cases (blocks) become the levels of another full-scale randomized-groups IV that are crossed with the levels of the IV(s) of interest in a factorial design. Interpretation of the main effect of the IV of interest is straightforward, and variation due to the potential CV(s) is removed from the estimate of error variance and assessed as a separate main effect. Furthermore, the assumption of homogeneity of regression shows up as the interaction between the blocking IV and the IV of interest and is available for interpretation.

Blocking has several advantages over ANCOVA and the other alternatives listed here. First, it has none of the special assumptions of ANCOVA or repeated-measures ANOVA. Second, the relationship between the potential CV(s) and the DV need not be linear; curvilinear relationships can be captured in ANOVA when three (or more) levels of a blocking IV are analyzed. Indeed, blocking is more powerful than ANCOVA if the CV-DV relationship is not linear. Blocking, then, is preferable to ANCOVA in many situations, particularly for experimental, rather than nonexperimental, research.

With some difficulty, blocking can also be expanded to multiple CVs. That is, several new IVs (one per CV) are developed through blocking and are crossed in factorial design. However, as the number of IVs increases, the design rapidly becomes very large and cumbersome to implement.

ANCOVA is preferable to blocking for some applications, however. For instance, ANCOVA is more powerful than blocking when the relationship between the DV and the CV is strong and linear. Also, if the assumptions of ANCOVA are met, conversion of a continuous CV to a discrete IV (e.g., small, medium, large) can result in loss of information. Finally, practical limitations may prevent measurement of potential CV(s) sufficiently in advance of treatment to accomplish random assignment of equal numbers of cases to the cells of the design. When blocking is attempted after treatment, sample sizes within cells are likely to be highly discrepant, leading to the problems of unequal *n*.

In some applications, a combination of blocking and ANCOVA may turn out to be best: Some potential CVs are used to create new IVs, whereas others are analyzed as CVs.

A final, but more complex, alternative is multilevel linear modeling (MLM; Tabachnick and Fidell, 2007, Chapter 15). MLM does not assume homogeneity of regression; heterogeneity is dealt with by creating a second level of analysis consisting of groups and specifying that groups may have different slopes (relationships between the DV and CVs), as well as different intercepts (means on the DV).

8.7 COMPLETE EXAMPLES OF ANALYSIS OF COVARIANCE

8.7.1 One-Way Analysis of Covariance with Five Levels and One Covariate

This example of a one-way ANCOVA evaluates the polishing time for Nambeware as a function of five different types of tableware after adjusting for diameter of the tableware as a CV. The data are available on the Internet and are provided by Nambé Mills, Santa Fe, New Mexico. The data set is modified so that IV levels (types of tableware) are coded in a single column, and price has been omitted. The data set contains 59 cases. Data files are NAMBE.*. The goal of analysis is to determine whether polishing time varies with the type of tableware after statistically controlling for its diameter, which presumably is a primary determinant of polishing time.

8.7.1.1 Evaluation of Assumptions

8.7.1.1.1 Sample Sizes, Missing Data, and Normality of Sampling Distributions
Table 8.31 shows SAS Interactive Data Analysis descriptive statistics and histograms for `DIAMETER` and `TIME`, grouped by `TYPE`.

Table 8.31 shows no missing data for either polishing time (the DV) or diameter (the CV). Sample sizes are discrepant, but not terribly so. Standard errors for skewness and kurtosis are found using Equations 2.8 and 2.9. For example, s_{skewness} and s_{kurtosis} for polishing time (TIME) for trays (group 4) with $N = 10$ are

$$s_s = \sqrt{\frac{6}{10}} = 0.775 \qquad \text{and} \qquad s_s = \sqrt{\frac{24}{10}} = 1.549$$

TABLE 8.31 Descriptive statistics for Nambé tableware types (SAS Interactive Data Analysis setup and output)

1. Open SAS Interactive Data Analysis with appropriate data set (here SASUSER.NAMBE).
2. Choose Analyze and then Distribution(Y).
3. Select Y variables: DIAMETER and TIME.
4. Select group variable: TYPE
5. In Output dialog box, select Moments and Histogram/Bar Chart.

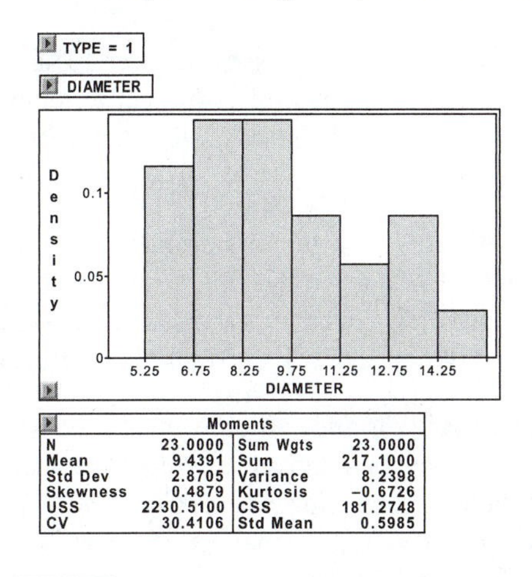

Moments			
N	23.0000	Sum Wgts	23.0000
Mean	9.4391	Sum	217.1000
Std Dev	2.8705	Variance	8.2398
Skewness	0.4879	Kurtosis	−0.6726
USS	2230.5100	CSS	181.2748
CV	30.4106	Std Mean	0.5985

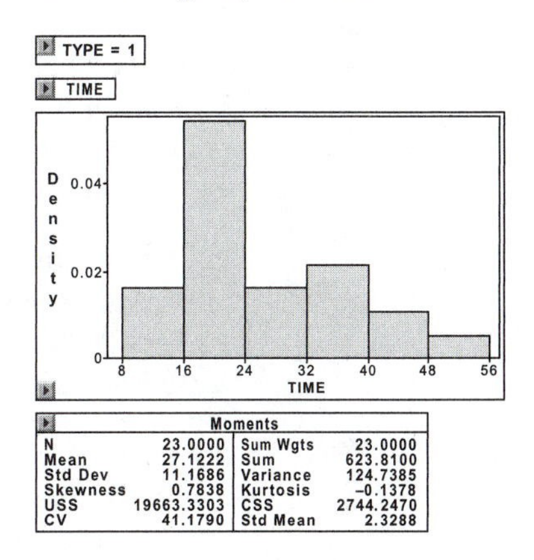

Moments			
N	23.0000	Sum Wgts	23.0000
Mean	27.1222	Sum	623.8100
Std Dev	11.1686	Variance	124.7385
Skewness	0.7838	Kurtosis	−0.1378
USS	19663.3303	CSS	2744.2470
CV	41.1790	Std Mean	2.3288

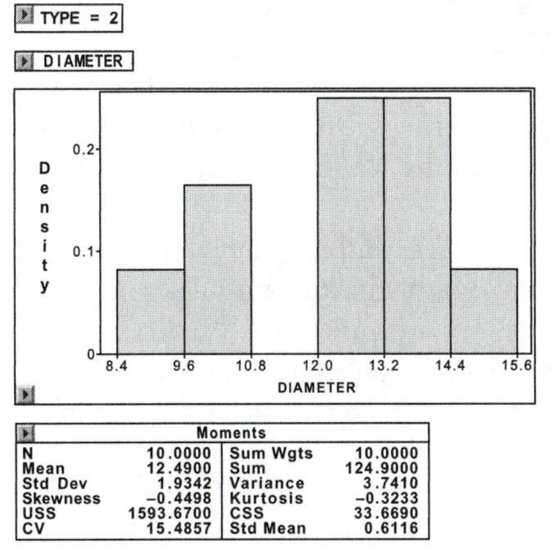

Moments			
N	10.0000	Sum Wgts	10.0000
Mean	12.4900	Sum	124.9000
Std Dev	1.9342	Variance	3.7410
Skewness	−0.4498	Kurtosis	−0.3233
USS	1593.6700	CSS	33.6690
CV	15.4857	Std Mean	0.6116

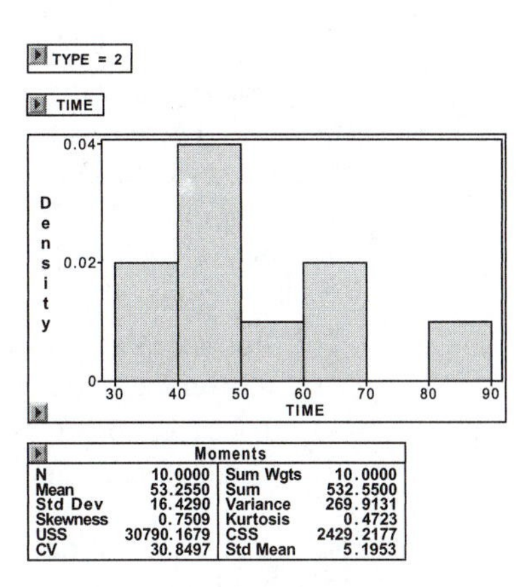

Moments			
N	10.0000	Sum Wgts	10.0000
Mean	53.2550	Sum	532.5500
Std Dev	16.4290	Variance	269.9131
Skewness	0.7509	Kurtosis	0.4723
USS	30790.1679	CSS	2429.2177
CV	30.8497	Std Mean	5.1953

TABLE 8.31 Continued

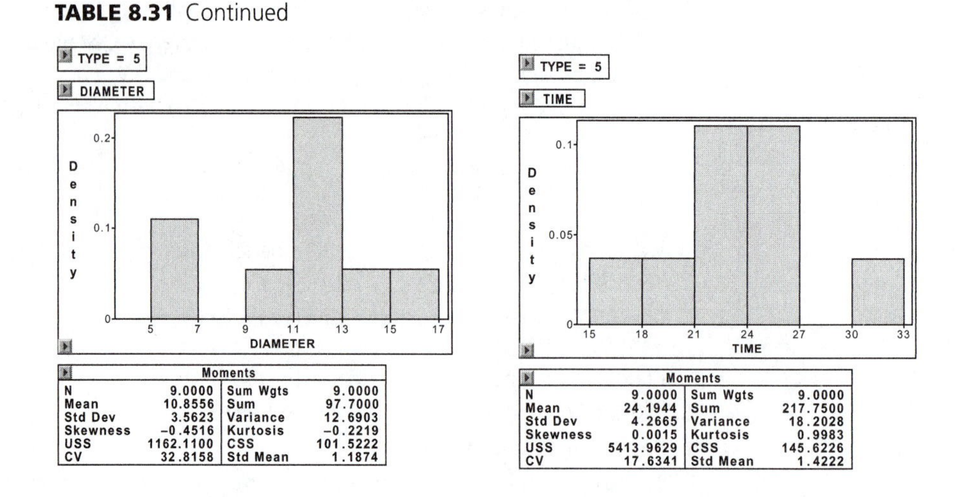

Moments (DIAMETER)

N	9.0000	Sum Wgts	9.0000
Mean	10.8556	Sum	97.7000
Std Dev	3.5623	Variance	12.6903
Skewness	−0.4516	Kurtosis	−0.2219
USS	1162.1100	CSS	101.5222
CV	32.8158	Std Mean	1.1874

Moments (TIME)

N	9.0000	Sum Wgts	9.0000
Mean	24.1944	Sum	217.7500
Std Dev	4.2665	Variance	18.2028
Skewness	0.0015	Kurtosis	0.9983
USS	5413.9629	CSS	145.6226
CV	17.6341	Std Mean	1.4222

Thus, the z value for the highest value of skewness for TIME of trays (TYPE=4, not shown) is 1.67 (1.293/0.775), which is well below a criterion of 2.58 at $\alpha = .01$, appropriate for this small sample. All other tests for skewness and kurtosis produce even smaller z values.

8.7.1.1.2 Outliers Table 8.31 also permits evaluation of univariate outliers. The test for the largest DIAMETER score in the first group (bowls) shows $z = (15.20 - 9.4391)/2.8705 = 2.01$, which is not an outlier at $\alpha = .01$. By this criterion, no cases are univariate outliers on DV or CV (TIME or DIAMETER) in any of the groups.

8.7.1.1.3 Homogeneity of Variance Homogeneity of variance also may be checked in Table 8.31. Sample sizes, though discrepant, do not exceed a ratio of 4:1 for the largest to the smallest group. However, the largest variance for polishing time is 737.7821 in group 4 (trays); the smallest is 18.2028 in group 5 (plates). $F_{max} = 737.7821/18.2028 = 40.53$, far exceeding the criterion of 10. No such problem occurs with the CV (diameter).

Therefore, a variance-stabilizing transformation is considered for TIME. Table 8.32 shows descriptive statistics through SAS MEANS after square root (S_TIME) and logarithmic (L_TIME) transforms. Nambe_X is the data set with the transformed and the original variables.

Table 8.32 shows that F_{max} is still too large with the square root transformation but acceptable using the logarithmic transform: $z = 0.262049/0.0333344 = 8.17$. Therefore, remaining analyses are based on L_TIME rather than on TIME. However, the exceptionally large variance in polishing times among trays and the exceptionally small variance in polishing times among plates are noted in the results.

8.7.1.1.4 Linearity Scatter plots between the DIAMETER and L_TIME produced by SAS Interactive Data Analysis are in Figure 8.5. The plots cause no concern for curvilinearity.

TABLE 8.32 Variances for transformed time variable (SAS MEANS syntax and output)

```
proc means data=SASUSER.NAMBE_X vardef=DF VAR;
  var L_TIME S_TIME ;
  class TYPE ;
run;
```

The MEANS Procedure

TYPE	N Obs	Variable	Label	Variance
1	23	L_TIME	log(TIME)	0.1645774
		S_TIME	sqrt(TIME)	1.0907886
2	10	L_TIME	log(TIME)	0.0923645
		S_TIME	sqrt(TIME)	1.2210720
3	7	L_TIME	log(TIME)	0.1823239
		S_TIME	sqrt(TIME)	1.4506180
4	10	L_TIME	log(TIME)	0.2620459
		S_TIME	sqrt(TIME)	3.2909331
5	9	L_TIME	log(TIME)	0.0333344
		S_TIME	sqrt(TIME)	0.1926938

TABLE 8.33 Test for homogeneity of regression (SAS GLM syntax and selected output)

```
proc glm data=SASUSER.NAMBE_X;
    class TYPE;
    model L_TIME = TYPE DIAMETER TYPE*DIAMETER ;
run;
```

Source	DF	Type III SS	Mean Square	F Value	Pr > F
TYPE	4	0.53415217	0.13353804	1.70	0.1643
DIAMETER	1	1.60815615	1.60815615	20.52	<.0001
DIAMETER*TYPE	4	0.60598720	0.15149680	1.93	0.1198

8.7.1.1.5 Reliability of the Covariate Diameter of a piece of tableware is a straightforward, physical measure, presumably standardized. Thus, there is no reason to question the reliability of the measurement.

8.7.1.1.6 Homogeneity of Regression Homogeneity of regression in either of the GLM programs is done by including the CV by IV interaction in a model statement, as shown in Table 8.33.

The TYPE*DIAMETER effect is the test for homogeneity of regression. The assumption has not been violated because Pr > F = 0.1198 (greater than .05).

FIGURE 8.5 Scatter plots between L_TIME and DIAMETER for five tableware types (setup and output from SAS Interactive Data Analysis)

```
1. Open SAS Interactive Data Analysis with appropriate data set (here SASUSER.NAMBE_X).
2. Choose Analyze and then Scatter Plot(X Y).
3. Select Y (L_TIME) and X (DIAMETER) variables.
4. Select TYPE as the GROUP variable.
```

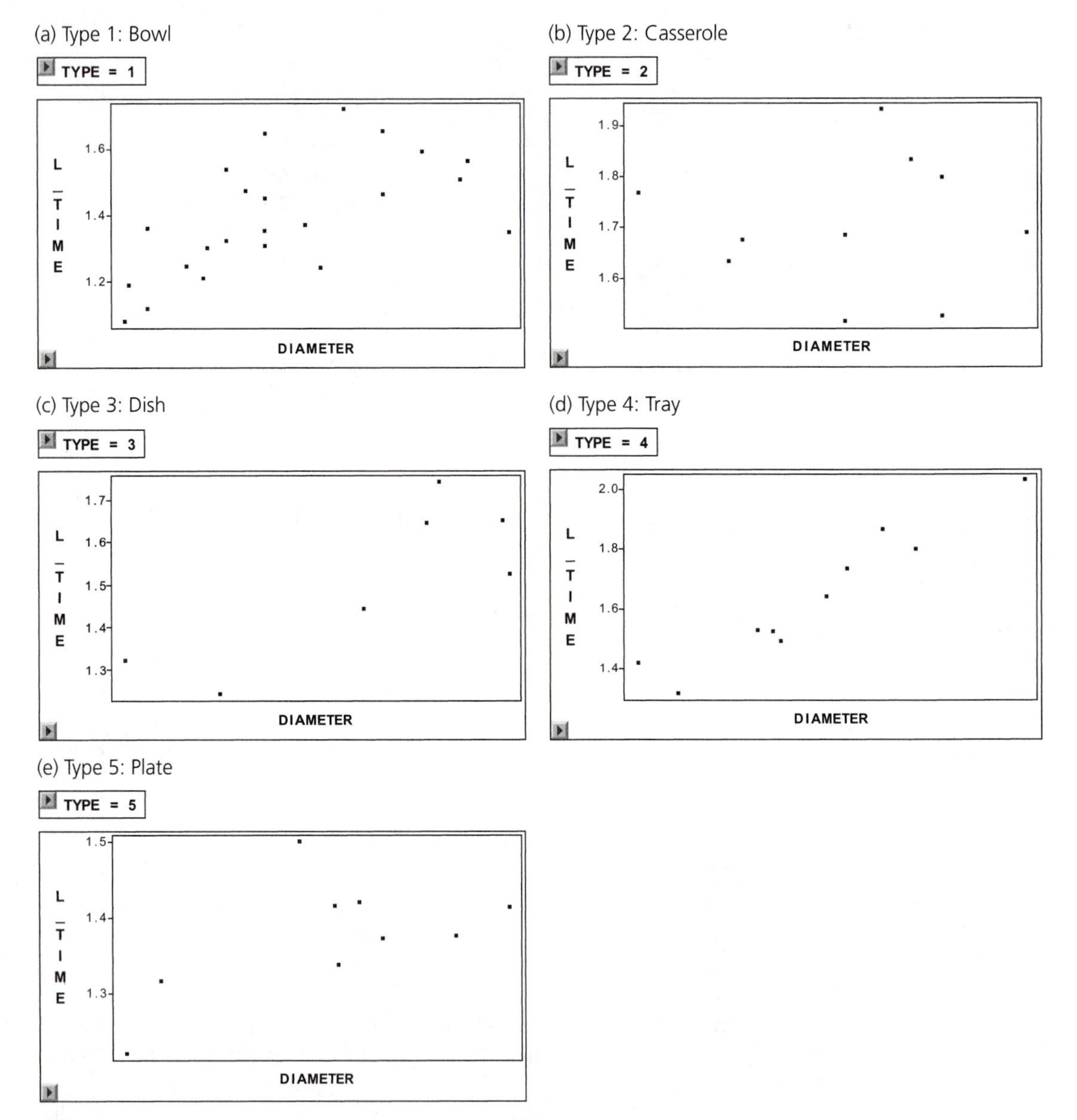

8.7.1.2 One-Way Analysis of Covariance

8.7.1.2.1 Major Analysis The major run for the analysis is similar to that of Table 8.33, but with the CV by IV interaction omitted and requests for means and least squares means (and their standard errors) added. Table 8.34 shows SAS GLM syntax and output.

TABLE 8.34 Analysis of covariance of L_TIME by type of tableware with DIAMETER as a CV (SAS GLM syntax and selected output)

```
proc glm data=SASUSER.NAMBE_X;
    class TYPE;
    model L_TIME = TYPE DIAMETER / solution;
    means TYPE;
    lsmeans TYPE;
run;
```

The GLM Procedure

Dependent Variable: L_TIME   log10(TIME)

Source	DF	Sum of Squares	Mean Square	F Value	Pr > F
Model	5	1.70382624	0.34076525	21.53	<.0001
Error	53	0.83868280	0.01582420		
Corrected Total	58	2.54250904			

R-Square	Coeff Var	Root MSE	L_TIME Mean
0.670136	8.372082	0.125794	1.502545

Source	DF	Type III SS	Mean Square	F Value	Pr > F
TYPE	4	0.46925471	0.11731368	7.41	<.0001
DIAMETER	1	0.70246604	0.70246604	44.39	<.0001

Parameter		Estimate		Standard Error	t Value	Pr > |t|
Intercept		1.018916643	B	0.06822080	14.94	<.0001
TYPE	1	0.068473096	B	0.04995553	1.37	0.1763
TYPE	2	0.276812710	B	0.05836361	4.74	<.0001
TYPE	3	0.199465588	B	0.06414201	3.11	0.0030
TYPE	4	0.150144175	B	0.06023982	2.49	0.0158
TYPE	5	0.000000000	B	.	.	.
DIAMETER		0.033028314		0.00495718	6.66	<.0001

NOTE: The X'X matrix has been found to be singular, and a generalized inverse was used to solve the normal equations. Terms whose estimates are followed by the letter 'B' are not uniquely estimable.

TABLE 8.34 Continued

Level of TYPE	N	L_TIME Mean	L_TIME Std Dev	DIAMETER Mean	DIAMETER Std Dev
1	23	1.39914830	0.17618517	9.4391304	2.87049871
2	10	1.70825299	0.13198871	12.4900000	1.93416649
3	7	1.51186239	0.18544116	8.8857143	2.31835327
4	10	1.64070514	0.22231718	14.2800000	5.72747763
5	9	1.37745734	0.07929224	10.8555556	3.56234161

Least Squares Means

TYPE	L_TIME LSMEAN	Standard Error	Pr > |t|
1	1.44829404	0.02724734	<.0001
2	1.65663365	0.04052708	<.0001
3	1.57928653	0.04861077	<.0001
4	1.52996512	0.04311232	<.0001
5	1.37982094	0.04193293	<.0001

The `Type III SS` table shows a statistically significant effect of tableware type on the logarithm of polishing time after adjusting for differences in diameter: $F(4, 53) = 7.41, p < .001$. There is also a significant effect of diameter after adjusting for differences in tableware types: $F(1, 53) = 44.39, p < .001$.

Adjusted `Least Squares Means` show that casseroles have the longest logarithm of polishing time, on average, and plates have the shortest. However, these observations need to be validated through post hoc analysis. Table 8.35 shows sample sizes, adjusted and unadjusted means, and standard for each type of tableware, in a format suitable for publication.

Effect size for tableware type is found by Equation 8.24, where

$$SS'_T = SS'_{TYPE} + SS'_{ERROR}$$

Therefore,

$$\eta^2 = \frac{0.469}{0.469 + 0.839} = .36$$

The 95% confidence limits using Smithson's (2003) software are from .12 to .49.

TABLE 8.35 Logarithm of polishing time adjusted for diameter for five types of Nambé tableware

Type of Tableware	Sample Size	Unadjusted Mean	Adjusted Mean	Standard Error
Bowl	23	1.399	1.448	0.027
Casserole	10	1.708	1.657	0.041
Dish	7	1.512	1.579	0.049
Tray	10	1.641	1.530	0.043
Plate	9	1.377	1.380	0.042

The effect of the CV, diameter, adjusted for all other omnibus effects is substantial. Smithson's software shows partial $\eta^2 = .46$ with confidence limits from .25 to .59, translating to a pooled within-groups correlation of $(.46)^{1/2} = .68$. The section on parameter estimates shows that the pooled within-groups unstandardized regression coefficient, B, for diameter is 0.033.

8.7.1.2.2 Post Hoc Analyses Differences among types of tableware are most appropriately evaluated through a Tukey test, available in SAS GLM with additional instructions. Table 8.36 shows the additional syntax and output for the Tukey test of tableware types.

Using a criterion $\alpha = .05$, Table 8.36 values of `Pr > | t |` shows that casseroles (TYPE=2) take significantly more time to polish than bowls (TYPE=1) or plates (TYPE=5), and that dishes (TYPE=3) take more time than plates.

Table 8.37 is a checklist for one-way ANCOVA with a single CV. A results section in journal format follows.

TABLE 8.36 Tukey test of types of tableware (SAS GLM syntax and selected output)

```
proc glm data=SASUSER.NAMBE_X;
    class TYPE;
    model L_TIME = TYPE DIAMETER ;
    means TYPE;
    lsmeans TYPE / pdiff=all adjust=tukey;
run;
```

Least Squares Means
Adjustment for Multiple Comparisons: Tukey-Kramer

TYPE	L_TIME LSMEAN	LSMEAN Number
1	1.44829404	1
2	1.65663365	2
3	1.57928653	3
4	1.52996512	4
5	1.37982094	5

Least Squares Means for effect TYPE
Pr > |t| for H0: LSMean(i)=LSMean(j)

Dependent Variable: L_TIME

i/j	1	2	3	4	5
1		0.0010	0.1287	0.5475	0.6488
2	0.0010		0.7520	0.1868	0.0002
3	0.1287	0.7520		0.9484	0.0241
4	0.5475	0.1868	0.9484		0.1078
5	0.6488	0.0002	0.0241	0.1078	

TABLE 8.37 Checklist for one-way randomized-groups ANCOVA with a single CV

1. Issues
   a. Sample sizes and missing data
   b. Normality of sampling distributions
   c. Outliers
   d. Independence of errors
   e. Homogeneity of variance
   f. Linearity
   g. Homogeneity of regression
   h. Reliability of CV
2. Major analyses: Main effect or planned comparisons. If significant:
   a. Effect size(s) and their confidence limits
   b. Parameter estimates (adjusted marginal means and standard deviations or standard errors or confidence intervals)
3. Additional analyses
   a. Evaluation of CV effect
   b. Evaluation of correlation between CV and DV
   c. Post hoc comparisons
   d. Interpretation of departure of homogeneity of variance, homogeneity of covariance, and/or homogeneity of regression

Results

A one-way randomized-groups ANCOVA was performed on polishing time for five types of Nambé tableware—bowls, casseroles, dishes, trays, and plates—with diameter of item as a covariate. Heterogeneity of variance was reduced by a logarithmic transform of polishing time. Raw polishing times were exceptionally variable for trays and exceptionally stable for plates. Remaining assumptions of ANCOVA were met. Sample sizes for the five types of tableware were, in the order listed above, 23, 10, 7, 10, and 9, with a total N = 59. An unweighted means analysis (Type III sums of squares) was used to handle unequal sample sizes and to adjust tests of tableware type and diameter for each other.

Type of tableware predicted logarithm of polishing time after adjusting for diameter, $F(4, 53) = 7.41$, $p < .001$, $\eta^2 = .36$ with 95% confidence limits from .12 to .49. Table 8.34 shows logarithm of adjusted means, standard errors, and sample sizes for each type of tableware. A Tukey test revealed that after adjusting for diameter, casseroles take more time to polish than bowls or plates and dishes take more time than plates, $p < .05$.

Diameter was also related to logarithm of polishing time after adjusting for type of tableware, $F(1, 53) = 44.39$, $p < .001$, $B = 0.033$, partial $\eta^2 = .46$ with confidence limits from .25 to .59, indicating a pooled within-groups correlation of .68.

8.7.2 Mixed Randomized-Groups and Repeated-Measures Analysis of Covariance

This example of a factorial ANCOVA with two CVs uses data collected from the National Transportation Safety Board and available with authorization for free use on the Internet. The data set includes 176 stock automobile crashes in which dummies occupied both the driver and passenger seats on each crash. The injury variable is chest deceleration. The data file also includes information on type of vehicle and its safety features.

The DV for the analysis is chest deceleration, with vehicle weight and model year as CVs. The randomized-groups IV is type of protection (airbag, manual belt, motorized belt, or passive belt). The repeated-measures IV is location of dummy, in either the driver or the passenger seat. Data files are CRASH.*.

The goal is to evaluate chest deceleration as a function of type of safety protection (PROTECT), location of dummy (D_P), and the interaction of the two, adjusted for vehicle weight and year. In addition, the CVs are of interest in their own right, not just as modifiers of the tests of the randomized-groups IV; that is, CVs are considered continuous IVs, which are themselves assessed after adjusting for vehicle protection and location of dummy.

8.7.2.1 Evaluation of Assumptions

8.7.2.1.1 Sample Sizes and Missing Data Table 8.38 shows the sample sizes of available data, produced by SPSS DESCRIPTIVES.

TABLE 8.38 Original sample sizes for complete example of ANCOVA through SPSS DESCRIPTIVES

```
SORT CASES BY PROTECT.
SPLIT FILE BY PROTECT.
DESCRIPTIVES
  VARIABLES=D_CHEST P_CHEST WEIGHT YEAR
  /STATISTICS=MIN MAX .
```

Descriptives

Descriptive Statistics

Type of protection		N	Minimum	Maximum
Driver airbag	Chest deceleration: driver	32	35.00	64.00
	Chest deceleration: passenger	29	31.00	58.00
	Vehicle weight	32	2190.00	4080.00
	Model year	32	87.00	91.00
	Valid N (listwise)	29		
Manual belt	Chest deceleration: driver	97	35.00	97.00
	Chest deceleration: passenger	96	31.00	71.00
	Vehicle weight	98	1590.00	5619.00
	Model year	98	87.00	91.00
	Valid N (listwise)	95		
Motorize	Chest deceleration: driver	20	41.00	64.00
	Chest deceleration: passenger	21	39.00	77.00
	Vehicle weight	22	2200.00	3630.00
	Model year	22	89.00	91.00
	Valid N (listwise)	19		
Passive belt	Chest deceleration: driver	24	35.00	70.00
	Chest deceleration: passenger	22	32.00	82.00
	Vehicle weight	24	2120.00	3240.00
	Model year	24	87.00	91.00
	Valid N (listwise)	22		

TABLE 8.39 Sample sizes used for complete example of ANCOVA

Type of Protection (PROTECT) Randomized Groups	Number of Crashes
Driver airbag	29
Manual belt	95
Motorized belt	19
Passive belt	22

There were no missing data on the CVs, WEIGHT and YEAR; however, several missing values were noted for the DV, chest deceleration. Table 8.38 shows three cases missing passenger values on the DV in the Driver airbag group, one driver and two passenger values missing the DV in the Manual belt group, and so on. Because missing data are scattered over the four protection groups (although a bit heavier in the passenger seat), the decision is made to omit cases with values missing in either location for further analysis. The resulting sample sizes are shown in Table 8.39.

The original sample size of 176 is thus reduced to 165, with a loss of 11 cases due to missing data on the DV. A method 1 adjustment for unequal n will be used (cf. Table 5.44).

8.7.2.1.2 Normality of Sampling Distributions Table 8.40 shows syntax, descriptive statistics, and histograms for each level of PROTECT. The file first is sorted and then split by PROTECT, so that output is organized by groups. The SELECT IF instruction limits the analysis to cases that have valid (nonmissing) data on both passenger and driver chest deceleration.

A glance at the histograms suggests positive skewness in most groups for driver and passenger chest deceleration. Indeed, the skewness for driver chest deceleration in the manual belt group exceeds the $\alpha = .001$ criterion: $z = 0.826/0.247 = 3.34$. Even greater departures from distributional symmetry are noted for passenger chest deceleration in the motorized and passive belt conditions. No departure from normality is evident for the CVs. Therefore, the decision is made to try square root and logarithmic transformations of the DV. Table 8.41 shows the results of the transformation (without histograms).

The square root transform fails to reduce skewness adequately. There is significant skewness in chest deceleration distribution in the motorized group, even after the logarithmic transformation. However, the logarithmic transform will be used and the data checked for outliers that may be contributing to remaining skewness.

8.7.2.1.3 Outliers Table 8.41 shows that the case with the largest value for log of chest deceleration (among passenger dummies) in the motorized belt group is indeed a univariate outlier: $z = (1.89 - 1.6585)/0.0689 = 3.36$. No other univariate outliers are found for logarithm of chest deceleration for either driver or passenger dummies in any of the protection groups. By referring back to Table 8.40, the only univariate outlier noted among the CVs is the largest value of weight in the manual belt condition: $z = (5619 - 2973.716)/726.1261 = 3.64$.

TABLE 8.40 Descriptive statistics and histograms for protection categories through SPSS FREQUENCIES and INTERACTIVE GRAPH (HISTOGRAM)

```
SORT CASES BY PROTECT.
SPLIT FILE BY PROTECT.
SELECT IF (D_CHEST< 100 and P_CHEST< 100).
FREQUENCIES
  VARIABLES=D_CHEST P_CHEST WEIGHT YEAR  /FORMAT=NOTABLE
  /STATISTICS=STDDEV VARIANCE MINIMUM MAXIMUM MEAN SKEWNESS SESKW KURTOS SEKURT
  /ORDER ANALYSIS.
```

Statistics

Type of protection			Chest deceleration: driver	Chest deceleration: passenger	Vehicle weight	Model year
Driver airbag	N	Valid	29	29	29	29
		Missing	0	0	0	0
	Mean		46.7586	42.7586	3097.1379	89.9655
	Std. Deviation		7.71666	6.09247	476.80722	1.11748
	Variance		59.547	37.118	227345.1	1.249
	Skewness		.593	.426	.012	−1.083
	Std. Error of Skewness		.434	.434	.434	.434
	Kurtosis		−.675	.823	−.151	.547
	Std. Error of Kurtosis		.845	.845	.845	.845
	Minimum		35.00	31.00	2190.00	87.00
	Maximum		64.00	58.00	4080.00	91.00
Manual belt	N	Valid	95	95	95	95
		Missing	0	0	0	0
	Mean		53.4421	45.8105	2973.7158	88.3789
	Std. Deviation		10.56159	8.44796	726.12613	1.19555
	Variance		111.547	71.368	527259.2	1.429
	Skewness		.826	.728	.768	.560
	Std. Error of Skewness		.247	.247	.247	.247
	Kurtosis		1.853	.337	1.425	−.541
	Std. Error of Kurtosis		.490	.490	.490	.490
	Minimum		35.00	31.00	1650.00	87.00
	Maximum		97.00	71.00	5619.00	91.00

(continued)

TABLE 8.40 Continued

Type of protection			Chest deceleration: driver	Chest deceleration: passenger	Vehicle weight	Model year
Motorize	N	Valid	19	19	19	19
		Missing	0	0	0	0
	Mean		49.4211	46.1579	2750.5263	90.1053
	Std. Deviation		7.53743	8.59757	383.83827	.87526
	Variance		56.813	73.918	147331.8	.766
	Skewness		.543	2.741	.722	−.220
	Std. Error of Skewness		.524	.524	.524	.524
	Kurtosis		−.995	9.501	.323	−1.711
	Std. Error of Kurtosis		1.014	1.014	1.014	1.014
	Minimum		41.00	39.00	2200.00	89.00
	Maximum		64.00	77.00	3630.00	91.00
Passive belt	N	Valid	22	22	22	22
		Missing	0	0	0	0
	Mean		51.0909	44.2273	2728.1818	88.3636
	Std. Deviation		8.12351	10.03295	343.37688	1.52894
	Variance		65.991	100.660	117907.7	2.338
	Skewness		.544	2.559	−.223	.724
	Std. Error of Skewness		.491	.491	.491	.491
	Kurtosis		.330	9.620	−.951	−1.105
	Std. Error of Kurtosis		.953	.953	.953	.953
	Minimum		35.00	32.00	2120.00	87.00
	Maximum		70.00	82.00	3240.00	91.00

```
IGRAPH /VIEWNAME='Histogram' /X1 = VAR(D_CHEST) TYPE = SCALE /Y = $count
/COORDINATE = VERTICAL  /X1LENGTH=3.0 /YLENGTH=3.0
 /X2LENGTH=3.0 /CHARTLOOK='NONE' /Histogram  SHAPE = HISTOGRAM CURVE = ON X1INTERVAL
AUTO X1START = 0.
EXE.
IGRAPH /VIEWNAME='Histogram' /X1 = VAR(P_CHEST) TYPE = SCALE /Y = $count /COORDINATE =
VERTICAL  /X1LENGTH=3.0 /YLENGTH=3.0
 /X2LENGTH=3.0 /CHARTLOOK='NONE' /Histogram  SHAPE = HISTOGRAM CURVE = ON X1INTERVAL
AUTO X1START = 0.
EXE.
IGRAPH /VIEWNAME='Histogram' /X1 = VAR(WEIGHT) TYPE = SCALE /Y = $count /COORDINATE =
VERTICAL  /X1LENGTH=3.0 /YLENGTH=3.0
 /X2LENGTH=3.0 /CHARTLOOK='NONE' /Histogram  SHAPE = HISTOGRAM CURVE = ON X1INTERVAL
AUTO X1START = 0.
EXE.
IGRAPH /VIEWNAME='Histogram' /X1 = VAR(YEAR) TYPE = SCALE /Y = $count /COORDINATE =
VERTICAL  /X1LENGTH=3.0 /YLENGTH=3.0
 /X2LENGTH=3.0 /CHARTLOOK='NONE' /Histogram  SHAPE = HISTOGRAM CURVE = ON X1INTERVAL
AUTO X1START = 0.
EXE.
```

TABLE 8.40 Continued

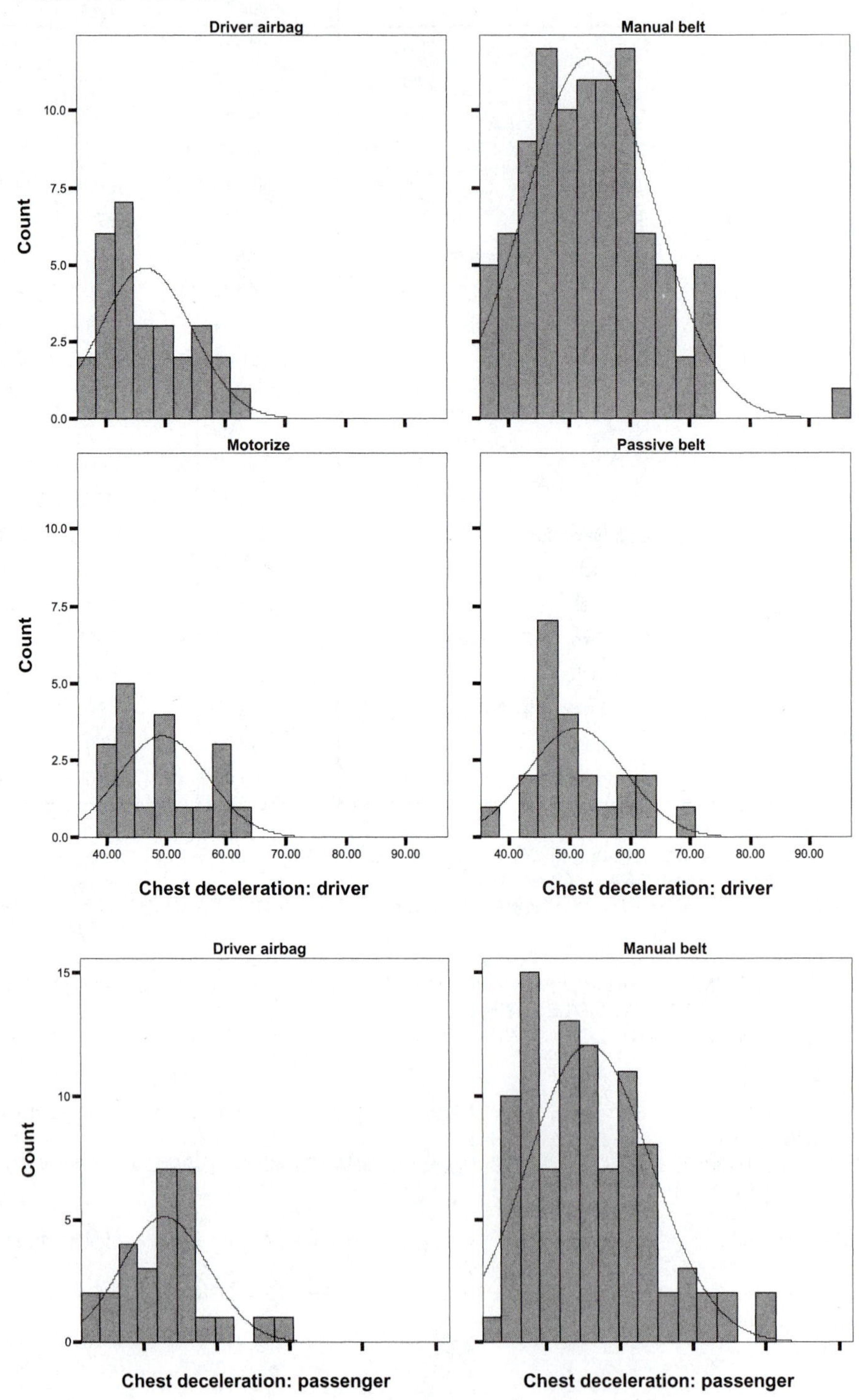

TABLE 8.40 Continued

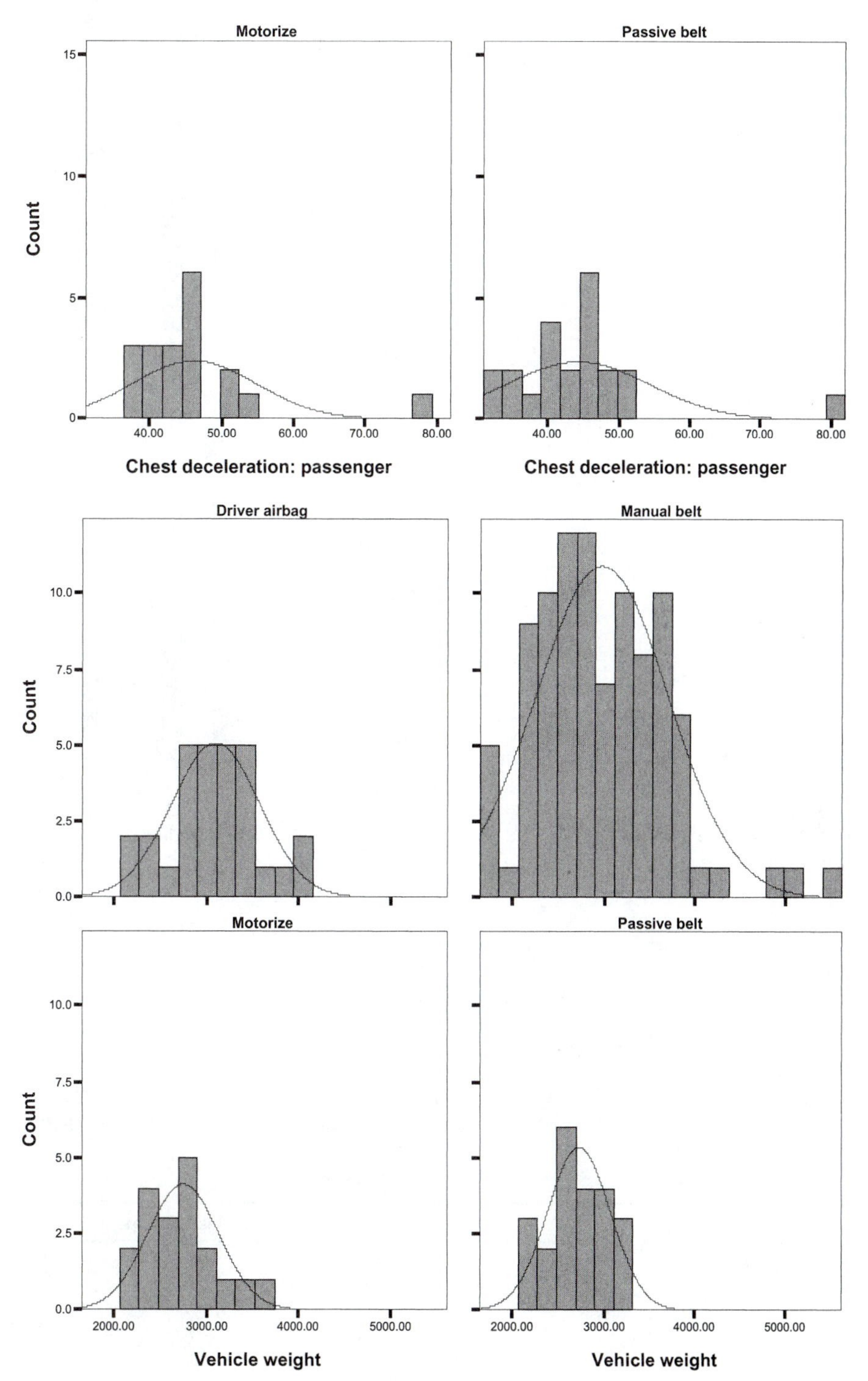

TABLE 8.40 Continued

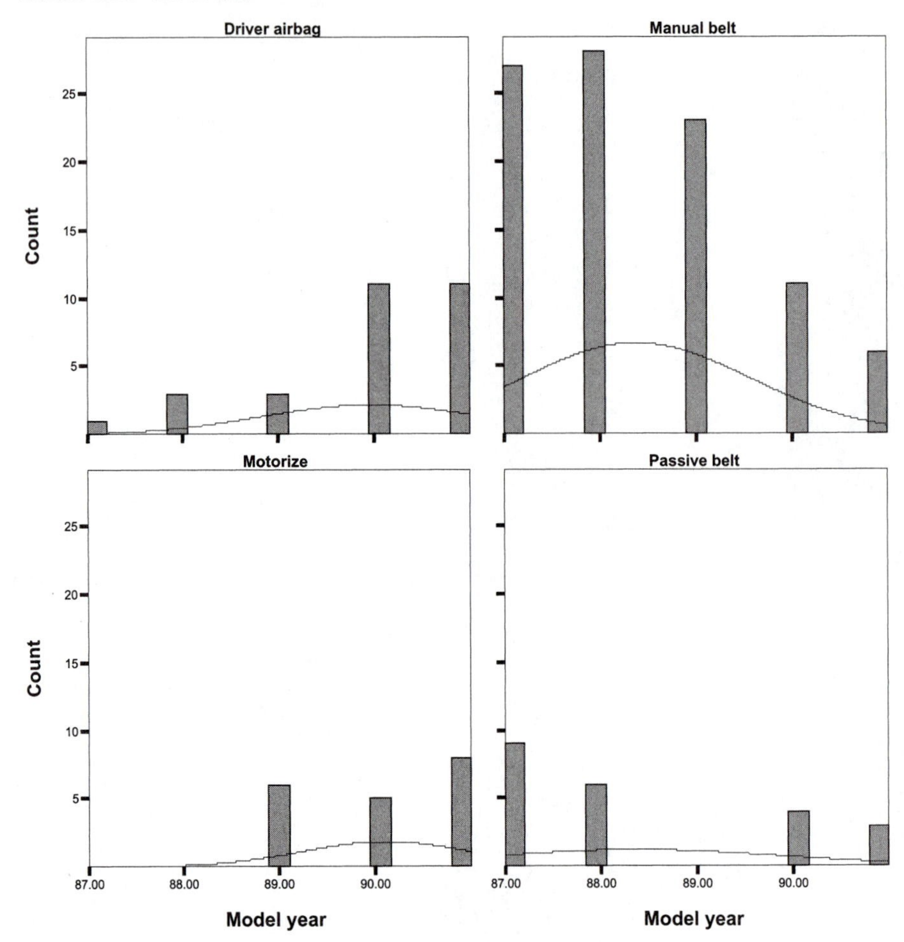

TABLE 8.41 Descriptive statistics for transformed variables in protection categories through SPSS FREQUENCIES

```
SORT CASES BY PROTECT.
SPLIT FILE BY PROTECT.
SELECT IF (D_CHEST< 100 and P_CHEST< 100).
COMPUTE sd_chest = sqrt(D_CHEST) .
COMPUTE sp_chest = sqrt(P_CHEST) .
COMPUTE ld_chest = lg10(D_CHEST) .
COMPUTE lp_chest = lg10(P_CHEST) .
FREQUENCIES
  VARIABLES=sd_chest sp_chest ld_chest lp_chest
  /STATISTICS=STDDEV MINIMUM MAXIMUM SEMEAN MEAN SKEWNESS SESKEW KURTOSIS SEKURT
  /ORDER= ANALYSIS .
```

Statistics

Type of protection			sd_chest	sp_chest	ld_chest	lp_chest
Driver airbag	N	Valid	29	29	29	29
		Missing	0	0	0	0
	Mean		6.8162	6.5231	1.6643	1.6268

(continued)

TABLE 8.41 Continued

Type of protection			sd_chest	sp_chest	ld_chest	lp_chest
Driver airbag (cont.)	Std. Error of Mean		.10316	.08610	.01297	.01147
	Std. Deviation		.55555	.46367	.06986	.06179
	Skewness		.479	.175	.364	−.072
	Std. Error of Skewness		.434	.434	.434	.434
	Kurtosis		−.812	.595	−.903	.496
	Std. Error of Kurtosis		.845	.845	.845	.845
	Minimum		5.92	5.57	1.54	1.49
	Maximum		8.00	7.62	1.81	1.76
Manual belt	N	Valid	95	95	95	95
		Missing	0	0	0	0
	Mean		7.2764	6.7409	1.7198	1.6540
	Std. Error of Mean		.07266	.06277	.00859	.00797
	Std. Deviation		.70821	.61180	.08373	.07773
	Skewness		.462	.509	.153	.303
	Std. Error of Skewness		.247	.247	.247	.247
	Kurtosis		.689	−.096	.050	−.391
	Std. Error of Kurtosis		.490	.490	.490	.490
	Minimum		5.92	5.57	1.54	1.49
	Maximum		9.85	8.43	1.99	1.85
Motorize	N	Valid	19	19	19	19
		Missing	0	0	0	0
	Mean		7.0111	6.7705	1.6893	1.6585
	Std. Error of Mean		.12137	.13308	.01486	.01581
	Std. Deviation		.52905	.58008	.06476	.06891
	Skewness		.462	2.411	.383	2.069
	Std. Error of Skewness		.524	.524	.524	.524
	Kurtosis		−1.111	7.804	−1.211	6.151
	Std. Error of Kurtosis		1.014	1.014	1.014	1.014
	Minimum		6.40	6.24	1.61	1.59
	Maximum		8.00	8.77	1.81	1.89
Passive belt	N	Valid	22	22	22	22
		Missing	0	0	0	0
	Mean		7.1266	6.6161	1.7032	1.6371
	Std. Error of Mean		.11998	.14709	.01457	.01800
	Std. Deviation		.56276	.68990	.06832	.08444
	Skewness		.316	1.994	.068	1.420
	Std. Error of Skewness		.491	.491	.491	.491
	Kurtosis		.312	7.006	.458	4.701
	Std. Error of Kurtosis		.953	.953	.953	.953
	Minimum		5.92	5.66	1.54	1.51
	Maximum		8.37	9.06	1.85	1.91

Table 8.42 shows the syntax for the SPSS DISCRIMINANT analyses to assess multivariate outliers in the set of two CVs. The SELECT IF instruction is used from the prior syntax to ensure that the appropriate cases are analyzed. Mahalanobis distance is based on weight and year. Note that the grouping variable is PROT_NUM, a numeric variable created for type of protection. SPSS DISCRIMINANT cannot accept groups based on string variables (including letters and other symbols, as well as numbers). The /PLOT=CASES instruction produces the casewise results, which show Mahalanobis distances for each case. Note that the file is not split into protection groups for this analysis, accomplished by SPLIT FILE OFF.

TABLE 8.42 SPSS DISCRIMINANT analysis to assess multivariate outliers by protection groups (syntax and selected output)

```
SPLIT FILE OFF.
SELECT IF (D_CHEST< 100 and P_CHEST< 100).
DISCRIMINANT
 /GROUPS=PROT_NUM(1 4)
 /VARIABLES=ld_chest WEIGHT YEAR
 /ANALYSIS ALL
 /PRIORS  EQUAL
 /PLOT=CASES
 /CLASSIFY=NONMISSING POOLED .
```

Discriminant

Casewise Statistics

	Case Number	Actual Group	Predicted Group	P(D>d | G=g) p	Highest Group df	P(G=g | D=d)	Squared Mahalanobis Distance to Centroid
Original	1	1	2	.817	2	.464	.403
	2	1	2	.868	2	.305	.284
	3	1	1	.805	2	.444	.433
	4	1	3	.735	2	.536	.615
	5	1	1	.690	2	.462	.743
	6	1	1	.283	2	.615	2.525
	7	1	3	.758	2	.501	.555
	8	1	2	.370	2	.375	1.987
	9	1	1	.770	2	.450	.524
	10	1	1	.986	2	.368	.028
	11	1	4	.869	2	.290	.281
	12	1	3	.749	2	.523	.577
	13	1	1	.851	2	.436	.322
	14	1	1	.999	2	.381	.002
	15	1	1	.678	2	.482	.776
	16	1	3	.489	2	.625	1.430
	17	1	1	.690	2	.456	.743

TABLE 8.42 Continued

Case Number	Actual Group	Predicted Group	P(D>d | G=g) p	Highest Group df	P(G=g | D=d)	Squared Mahalanobis Distance to Centroid
18	1	1	.906	2	.425	.198
19	1	3	.551	2	.608	1.194
20	1	1	.957	2	.411	.088
21	1	1	.273	2	.521	2.593
22	1	3	.660	2	.494	.831
23	1	1	.997	2	.377	.006
24	1	4	.518	2	.470	1.317
25	1	1	.670	2	.489	.800
26	1	2	.665	2	.488	.817
27	1	3	.756	2	.495	.559
28	1	3	.895	2	.442	.223
29	1	2	.946	2	.427	.111
30	2	4	.475	2	.526	1.490
31	2	4	.869	2	.290	.281
32	2	2	.754	2	.474	.565
33	2	2	.365	2	.528	2.015
34	2	2	.239	2	.391	2.861
35	2	2	.489	2	.511	1.429
36	2	2	.476	2	.364	1.483
37	2	4	.952	2	.412	.098
38	2	2	.459	2	.515	1.559
39	2	2	.002	2	.742	12.324
40	2	2	.097	2	.580	4.656
41	2	2	.000	2	.801	21.747
42	2	2	.627	2	.347	.932
43	2	1	.595	2	.475	1.039

Squared Mahalanobis Distance to Centroid is distributed as an χ^2 variable with df equal to the number of predictors (CVs). Critical χ^2 with 2 df at $\alpha = .001$ is 13.816. Squared Mahalanobis Distance to Centroid for each case from its closest group mean is printed in the last column shown in Table 8.42. A misclassified case will be even farther from its own group centroid. Only the first 43 cases are shown; the only multivariate outlier is case 41 in the manual belt group, with $\chi^2 = 23.312$.

Examination of the data for case 41 revealed that it was the univariate outlier on vehicle weight and was car 27. The outlier on chest deceleration was car 136. The decision was made to omit the two outlying cases from further analysis and to describe their characteristics in the results. Thus, $N = 163$ for subsequent analyses.

8.7.2.1.4 Independence of Errors Independence of errors for the repeated-measures portion of the design (location) is handled by using the appropriate repeated-measures error term for tests of location and the protection by location interaction. Independence of errors is problematic for the randomized-groups IV. It is presumed that cars are not randomly assigned to type of protection. Some stock cars are typically equipped with one (or more) type of protection, whereas other cars are equipped with other types. The hope is that type of belt is not so highly related to other vehicle characteristics as to make generalization beyond the tested vehicles unjustifiable.

8.7.2.1.5 Homogeneity of Variance Table 8.43 shows SPSS simple descriptive statistics for the cases remaining in the analysis. Cars 27 and 136 are the deleted outliers.

TABLE 8.43 SPSS descriptive statistics for transformed variables with outliers deleted

```
SORT CASES BY PROTECT .
SPLIT FILE
  LAYERED BY PROTECT .
SELECT IF (D_CHEST< 100 and P_CHEST< 100 and CAR ~= 27 and CAR ~= 136).
DESCRIPTIVES
  VARIABLES=ld_chest lp_chest WEIGHT YEAR
  /STATISTICS=MEAN STDDEV VARIANCE MIN MAX .
```

Descriptive Statistics

Type of protection		N	Minimum	Maximum	Mean	Std. Deviation	Variance
Driver airbag	ld_chest	29	1.54	1.81	1.6643	.06986	.005
	lp_chest	29	1.49	1.76	1.6268	.06179	.004
	Vehicle weight	29	2190.00	4080.00	3097.1379	476.80722	227345.1
	Model year	29	87.00	91.00	89.9655	1.11748	1.249
	Valid N (listwise)	29					
Manual belt	ld_chest	94	1.54	1.99	1.7200	.08415	.007
	lp_chest	94	1.49	1.85	1.6543	.07808	.006
	Vehicle weight	94	1650.00	5103.00	2945.5745	675.93337	456885.9
	Model year	94	87.00	91.00	88.3936	1.19333	1.424
	Valid N (listwise)	94					
Motorize	ld_chest	18	1.61	1.79	1.6828	.05993	.004
	lp_chest	18	1.59	1.72	1.6458	.04243	.002
	Vehicle weight	18	2200.00	3630.00	2736.1111	389.63828	151818.0
	Model year	18	89.00	91.00	90.1667	.85749	.735
	Valid N (listwise)	18					
Passive belt	ld_chest	22	1.54	1.85	1.7032	.06832	.005
	lp_chest	22	1.51	1.91	1.6371	.08444	.007
	Vehicle weight	22	2120.00	3240.00	2728.1818	343.37688	117907.7
	Model year	22	87.00	91.00	88.3636	1.52894	2.338
	Valid N (listwise)	22					

The ratio between the largest variance (.00708) and the smallest variance (.00359) for log of D_CHEST is far smaller than 10: $F_{max} = 1.97$. Variance ratios for all other variables are similarly acceptable, even though the sample size ratio is a bit greater than 4:1. Therefore, there is no concern about violation of homogeneity of variance within cells.

8.7.2.1.6 Linearity SPSS GRAPH is used to produce scatter plots among the continuous variables: logarithm of chest deceleration (driver and passenger), model year, and vehicle weight. Figures 8.6–8.10 show syntax and output for the five sets of

FIGURE 8.6 Scatter plots between vehicle weight and logarithm of driver chest deceleration for the four safety protection groups (SPSS SCATTERPLOT syntax and output)

```
SORT CASES BY PROTECT .
SPLIT FILE
 LAYERED BY PROTECT .
SELECT IF (D_CHEST< 100 and P_CHEST< 100 and CAR ~= 27 and CAR ~= 136).
IGRAPH /VIEWNAME='Scatterplot' /X1 = VAR(WEIGHT) TYPE = SCALE /Y = VAR(ld_chest)
TYPE = SCALE /COORDINATE = VERTICAL
 /X1LENGTH=3.0 /YLENGTH=3.0 /X2LENGTH=3.0 /CHARTLOOK='NONE' /SCATTER
COINCIDENT = NONE.
EXE.
```

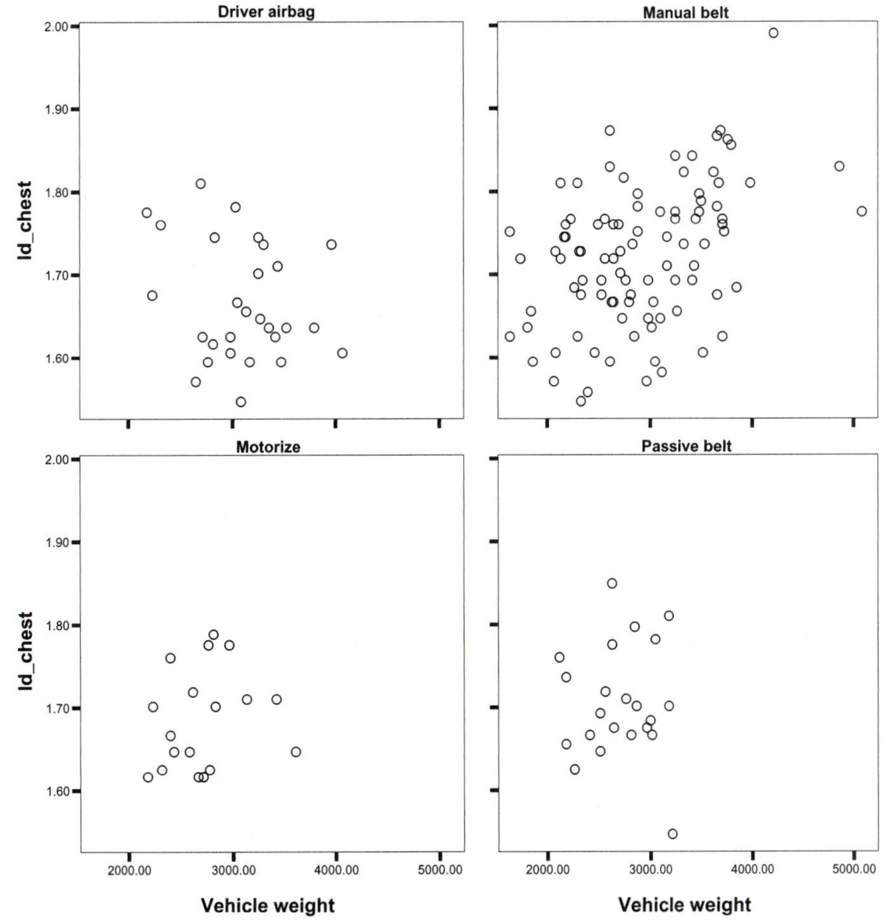

FIGURE 8.7 Scatter plots between vehicle weight and logarithm of passenger chest deceleration for the four safety protection groups (SPSS SCATTERPLOT syntax and output)

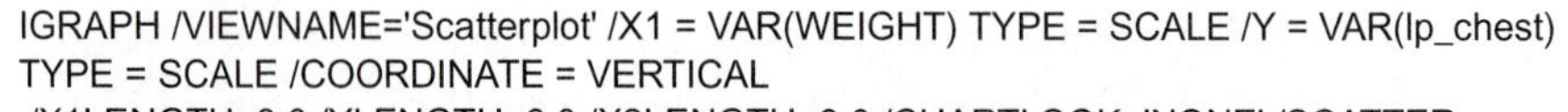

```
IGRAPH /VIEWNAME='Scatterplot' /X1 = VAR(WEIGHT) TYPE = SCALE /Y = VAR(lp_chest)
TYPE = SCALE /COORDINATE = VERTICAL
 /X1LENGTH=3.0 /YLENGTH=3.0 /X2LENGTH=3.0 /CHARTLOOK='NONE' /SCATTER
COINCIDENT = NONE.
EXE.
```

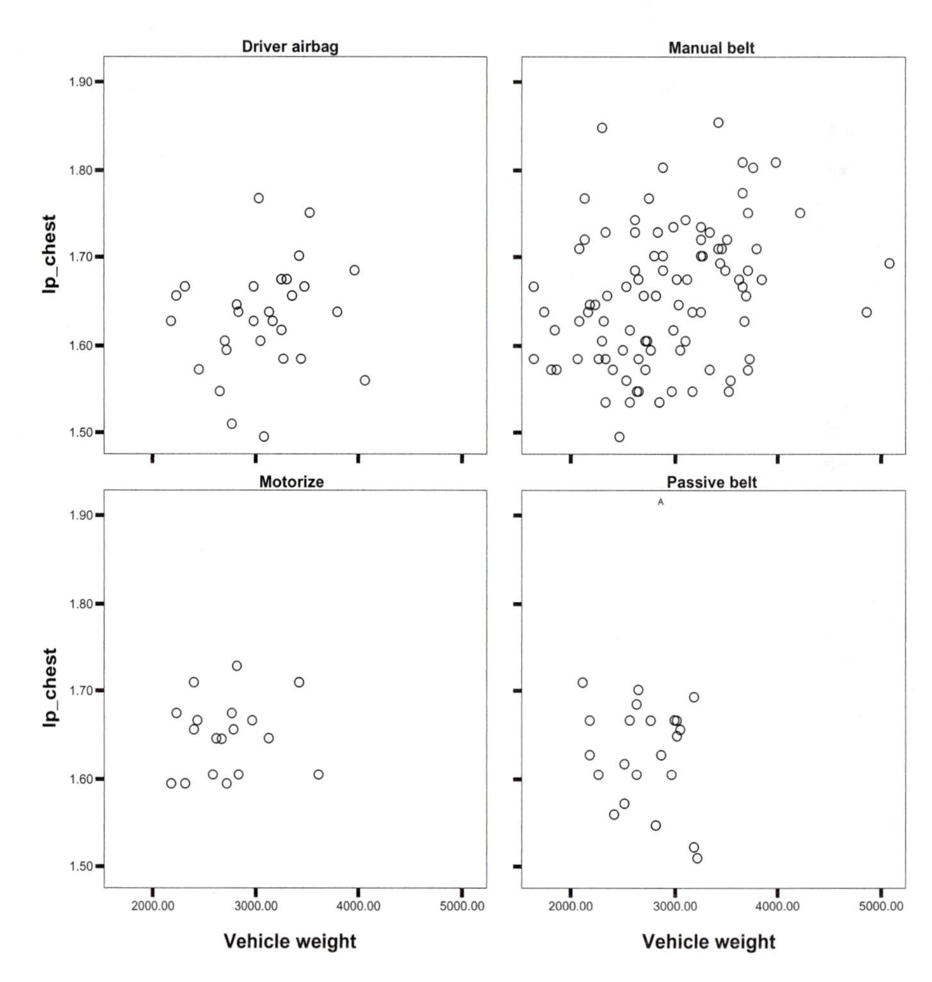

plots: ld_chest by WEIGHT, lp_chest by WEIGHT, ld_chest by YEAR, lp_chest by YEAR, and WEIGHT by YEAR, respectively. Prior filters are left in effect (omitting missing values on the DV and outliers), as well as file splitting by protection.

The most prominent curvilinearity noted is in the relationship between the two CVs, WEIGHT and YEAR, for cases in the passive seatbelt group and possibly in the airbag group. Weights are lower for the earlier and later years than for the middle years. Also, the direction of the relationship appears negative in the motorized belt group. However, because the relationships between the DV and each of the CVs are

FIGURE 8.8 Scatter plots between model year and logarithm of driver chest deceleration for the four safety protection groups (SPSS SCATTERPLOT syntax and output)

```
IGRAPH /VIEWNAME='Scatterplot' /X1 = VAR(YEAR) TYPE = SCALE /Y = VAR(ld_chest)
TYPE = SCALE /COORDINATE = VERTICAL
 /X1LENGTH=3.0 /YLENGTH=3.0 /X2LENGTH=3.0 /CHARTLOOK='NONE' /SCATTER
COINCIDENT = NONE.
EXE.
```

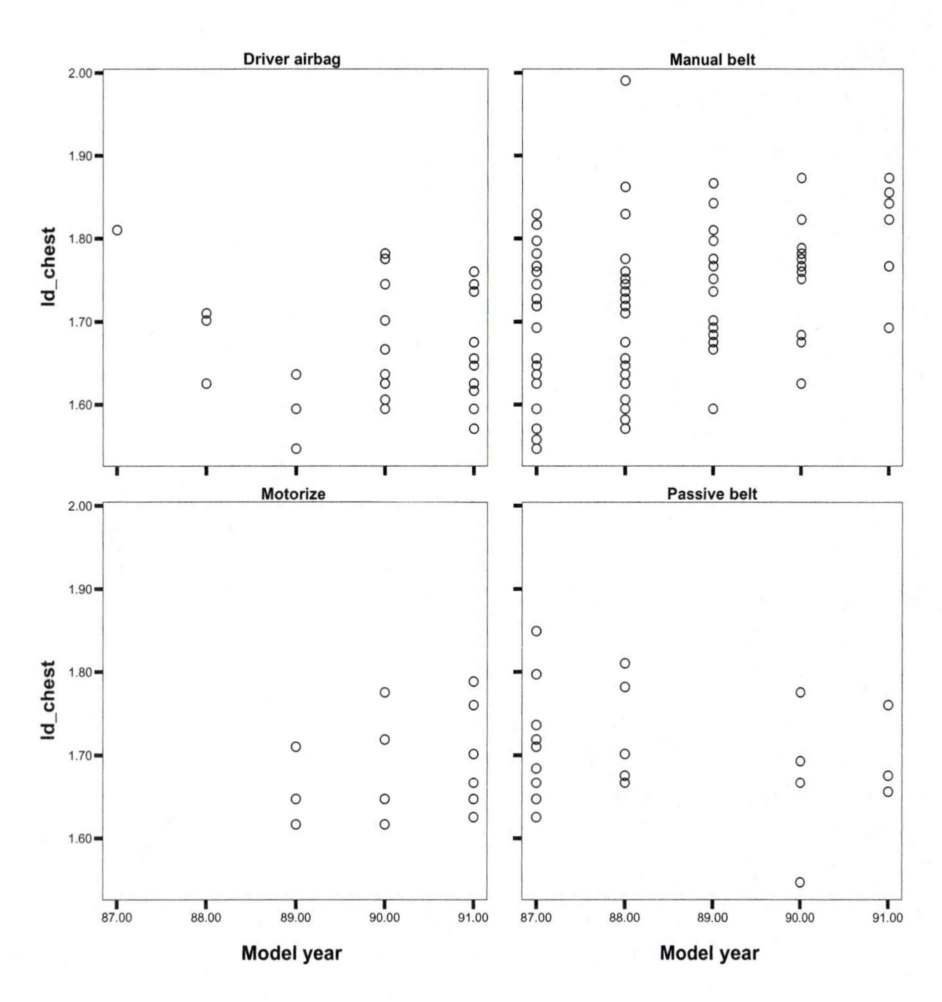

comfortably linear, if not particularly strong, no transformation of the CVs is deemed necessary. The scatter plots suggest that the CVs may not be providing strong adjustment to the DV.

8.7.2.1.7 Reliability of Covariates Measurement of the two CVs, weight of the vehicle and model year, is straightforward. Therefore, there are no concerns about unreliability of measurement.

FIGURE 8.9 Scatter plots between model year and logarithm of passenger chest deceleration for the four safety protection groups (SPSS SCATTERPLOT syntax and output)

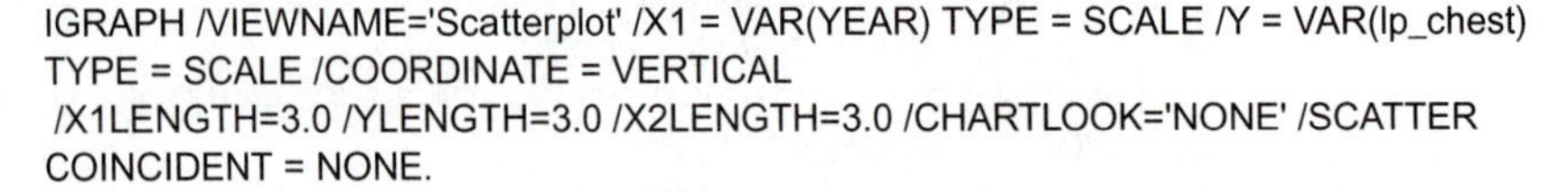

```
IGRAPH /VIEWNAME='Scatterplot' /X1 = VAR(YEAR) TYPE = SCALE /Y = VAR(lp_chest)
TYPE = SCALE /COORDINATE = VERTICAL
 /X1LENGTH=3.0 /YLENGTH=3.0 /X2LENGTH=3.0 /CHARTLOOK='NONE' /SCATTER
COINCIDENT = NONE.
EXE.
```

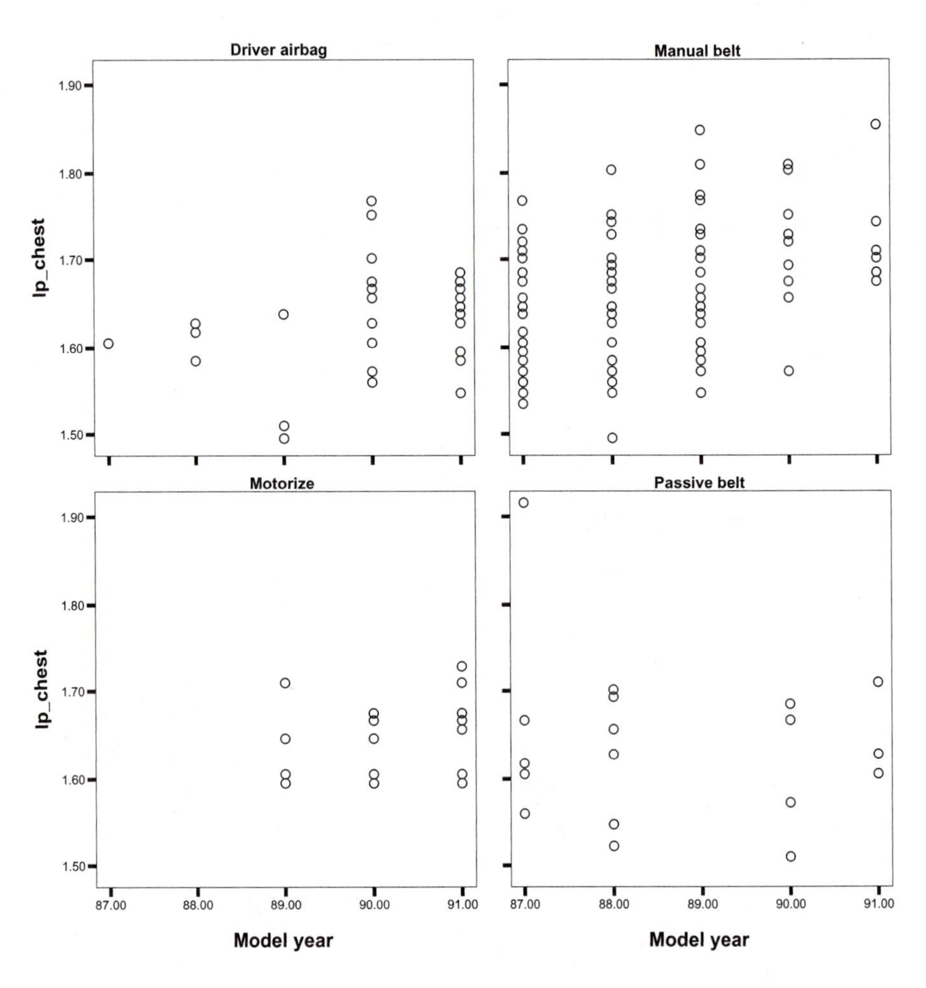

8.7.2.1.8 ***Multicollinearity*** Multicollinearity is only a problem if the two CVs, weight and year, are very highly correlated. No such high correlation is suggested by the scatter plots in Figure 8.10. An SPSS bivariate correlation run (not shown) shows $r = .18$, indicating no threat of multicollinearity in the linear relationships among the CVs.

8.7.2.1.9 ***Homogeneity of Regression*** The check for parallelism of slopes for the CVs is made by including the interactions between each of the IVs and each of the CVs in an ANCOVA run. Table 8.44 shows syntax and selected output for the

FIGURE 8.10
Scatter plots between model year and vehicle weight for the four safety protection groups (SPSS SCATTERPLOT syntax and output)

```
IGRAPH /VIEWNAME='Scatterplot' /X1 = VAR(YEAR) TYPE = SCALE /Y = VAR(WEIGHT)
TYPE = SCALE /COORDINATE = VERTICAL
 /X1LENGTH=3.0 /YLENGTH=3.0 /X2LENGTH=3.0 /CHARTLOOK='NONE' /SCATTER
COINCIDENT = NONE.
EXE.
```

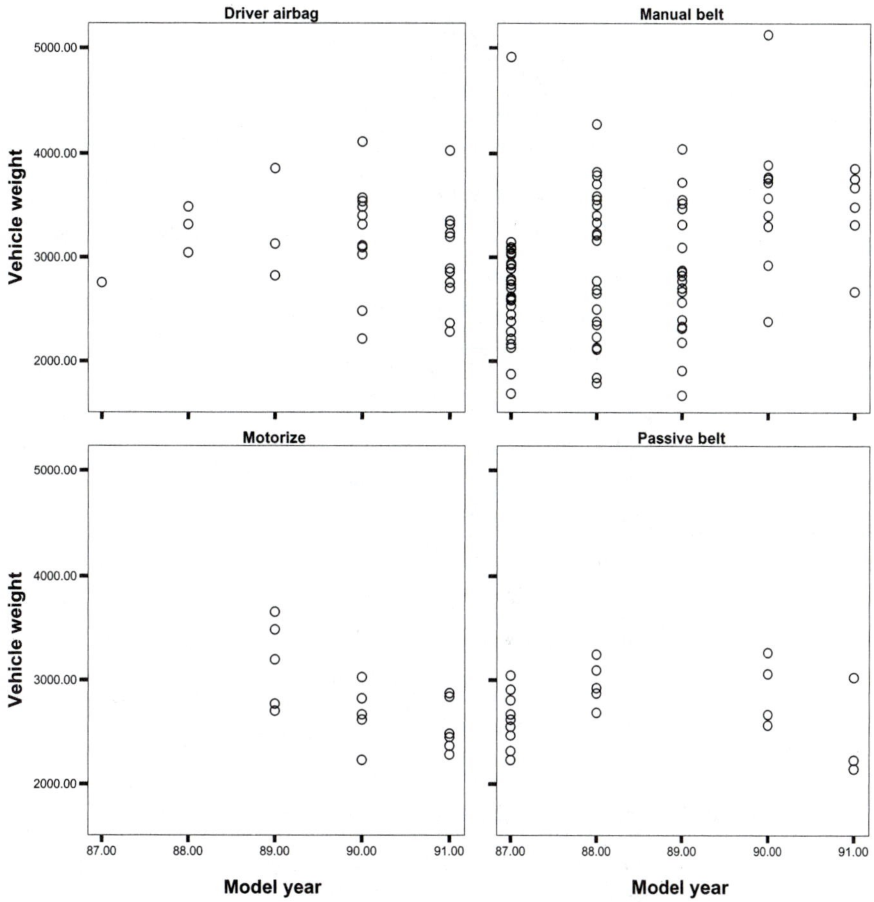

Driver airbag

Manual belt

Motorize

Passive belt

Vehicle weight

5000.00

4000.00

3000.00

2000.00

87.00 88.00 89.00 90.00 91.00

Model year

homogeneity of regression test through SPSS GLM (Repeated Measures). The two interactions relevant to the test are in the /DESIGN instruction.

Homogeneity of regression is assessed through the two interactions, protection by weight and protection by year, in this GLM approach to ANCOVA. The first poses no problem, but the PROTECT*YEAR is statistically significant at $\alpha = .05$. Despite this, the decision is made to retain year as a CV, because the violation is marginal and because an example with two CVs makes a more interesting demonstration.

TABLE 8.44 SPSS GLM analysis to assess homogeneity of regression (syntax and selected output)

```
SPLIT FILE OFF.
GLM
 ld_chest lp_chest BY PROTECT  WITH WEIGHT YEAR
 /WSFACTOR = location 2 Polynomial
 /MEASURE = chest_de
 /METHOD = SSTYPE(3)
 /INTERCEPT = INCLUDE
 /CRITERIA = ALPHA(.05)
 /WSDESIGN = location
 /DESIGN = PROTECT WEIGHT YEAR PROTECT*WEIGHT PROTECT*YEAR.
```

General Linear Model

Tests of Between-Subjects Effects

Measure: chest_de

Transformed Variable: Average

Source	Type III Sum of Squares	df	Mean Square	F	Sig.
Intercept	.009	1	.009	1.156	.284
protect	.068	3	.023	2.996	.033
weight	.009	1	.009	1.171	.281
year	.019	1	.019	2.543	.113
protect * weight	.025	3	.008	1.119	.343
protect * year	.063	3	.021	2.762	.044
Error	1.146	151	.008		

8.7.2.2 Mixed Randomized-Repeated Analysis of Covariance

8.7.2.2.1 Omnibus Analyses Table 8.45 shows SPSS GLM (Repeated Measures) syntax and output for the omnibus ANCOVA of chest deceleration as a function of type of protection and seat location, with vehicle weight and year as CVs. EMMEANS requests adjusted marginal means for both IVs (needed should they turn out to be statistically significant), and a request for printing parameters provides regression coefficients for the CVs.

Multivariate output has been omitted; it does not differ from univariate output when there are only two levels of the repeated-measures IV, and there is no possibility of violation of sphericity (cf. Section 6.6.2). The Tests of Between-Subjects Effects table shows that type of protection affects the logarithm of chest deceleration—$F(3, 157) = 6.293, p = .<.0005$—after adjusting for vehicle weight and model year. Type of protection accounts for about 11% of the variance in the logarithm of chest deceleration, partial $\eta^2 = .107$ with 95% confidence limits (Smithson, 2003) from

TABLE 8.45 Omnibus ANCOVA through SPSS GLM (syntax and selected output)

```
GLM
 ld_chest lp_chest BY PROTECT  WITH WEIGHT YEAR
 /WSFACTOR = location 2 Polynomial
 /MEASURE = chest_de
 /METHOD = SSTYPE(3)
 /EMMEANS = TABLES(PROTECT) WITH(WEIGHT=MEAN YEAR=MEAN)
 /EMMEANS = TABLES(location) WITH(WEIGHT=MEAN YEAR=MEAN)
 /EMMEANS = TABLES(PROTECT*location) WITH(WEIGHT=MEAN YEAR=MEAN)
 /PRINT = DESCRIPTIVE ETASQ OPOWER PARAMETER
 /CRITERIA = ALPHA(.05)
 /WSDESIGN = location
 /DESIGN = WEIGHT YEAR PROTECT .
```

General Linear Model

Within-Subjects Factors

Measure: chest_de

location	Dependent Variable
1	ld_chest
2	lp_chest

Between-Subjects Factors

		Value Label	N
Type of protection	d_airbag	Driver airbag	29
	manual	Manual belt	94
	Motorize		18
	passive	Passive belt	22

Descriptive Statistics

	Type of protection	Mean	Std. Deviation	N
ld_chest	Driver airbag	1.6643	.06986	29
	Manual belt	1.7200	.08415	94
	Motorize	1.6828	.05993	18
	Passive belt	1.7032	.06832	22
	Total	1.7037	.07982	163
lp_chest	Driver airbag	1.6268	.06179	29
	Manual belt	1.6543	.07808	94
	Motorize	1.6458	.04243	18
	Passive belt	1.6371	.08444	22
	Total	1.6462	.07341	163

TABLE 8.45 Continued

Tests of Between-Subjects Effects

Measure: chest_de

Transformed Variable: Average

Source	Type III Sum of Squares	df	Mean Square	F	Sig.	Partial Eta Squared	Noncent. Parameter	Observed Power[a]
Intercept	.019	1	.019	2.425	.121	.015	2.425	.340
weight	.089	1	.089	11.279	.001	.067	11.279	.916
year	.055	1	.055	7.058	.009	.043	7.058	.752
protect	.148	3	.049	6.293	.000	.107	18.879	.963
Error	1.234	157	.008					

a. Computed using alpha = .05

Parameter Estimates

Dependent Variable	Parameter	B	Std. Error	t	Sig.	95% Confidence Interval		Partial Eta Squared	Noncent. Parameter	Observed Power[a]
						Lower Bound	Upper Bound			
ld_chest	Intercept	.916	.436	2.099	.037	.054	1.778	.027	2.099	.550
	weight	3.56E-005	.000	3.460	.001	1.53E-005	5.59E-005	.071	3.460	.930
	year	.008	.005	1.566	.119	−.002	.018	.015	1.566	.343
	[protect=d_airbag]	−.064	.022	−2.872	.005	−.109	−.020	.050	2.872	.814
	[protect=manual]	.009	.018	.502	.616	−.026	.044	.002	.502	.079
	[protect=Motorize]	−.035	.025	−1.382	.169	−.085	.015	.012	1.382	.279
	[protect=passive]	0[b]	.	.	.	.	.	.	.	.
lp_chest	Intercept	.281	.411	.683	.496	−.531	1.092	.003	.683	.104
	weight	2.29E-005	.000	2.370	.019	3.82E-006	4.21E-005	.035	2.370	.654
	year	.015	.005	3.118	.002	.005	.024	.058	3.118	.873
	[protect=d_airbag]	−.042	.021	−1.999	.047	−.084	.000	.025	1.999	.511
	[protect=manual]	.012	.017	.707	.481	−.021	.045	.003	.707	.108
	[protect=Motorize]	−.018	.024	−.753	.452	−.065	.029	.004	.753	.116
	[protect=passive]	0[b]	.	.	.	.	.	.	.	.

a. Computed using alpha = .05
b. This parameter is set to zero because it is redundant.

Tests of Within-Subjects Effects

Measure: chest_de

Source		Type III Sum of Squares	df	Mean Square	F	Sig.	Partial Eta Squared	Noncent. Parameter	Observed Power[a]
location	Sphericity Assumed	.006	1	.006	2.234	.137	.014	2.234	.318
	Greenhouse-Geisser	.006	1.000	.006	2.234	.137	.014	2.234	.318
	Huynh-Feldt	.006	1.000	.006	2.234	.137	.014	2.234	.318
	Lower-bound	.006	1.000	.006	2.234	.137	.014	2.234	.318
location * weight	Sphericity Assumed	.004	1	.004	1.675	.197	.011	1.675	.251
	Greenhouse-Geisser	.004	1.000	.004	1.675	.197	.011	1.675	.251
	Huynh-Feldt	.004	1.000	.004	1.675	.197	.011	1.675	.251
	Lower-bound	.004	1.000	.004	1.675	.197	.011	1.675	.251
location * year	Sphericity Assumed	.005	1	.005	2.081	.151	.013	2.081	.300
	Greenhouse-Geisser	.005	1.000	.005	2.081	.151	.013	2.081	.300
	Huynh-Feldt	.005	1.000	.005	2.081	.151	.013	2.081	.300
	Lower-bound	.005	1.000	.005	2.081	.151	.013	2.081	.300
location * protect	Sphericity Assumed	.004	3	.001	.551	.648	.010	1.653	.162
	Greenhouse-Geisser	.004	3.000	.001	.551	.648	.010	1.653	.162
	Huynh-Feldt	.004	3.000	.001	.551	.648	.010	1.653	.162
	Lower-bound	.004	3.000	.001	.551	.648	.010	1.653	.162
Error(location)	Sphericity Assumed	.388	157	.002					
	Greenhouse-Geisser	.388	157.000	.002					
	Huynh-Feldt	.388	157.000	.002					
	Lower-bound	.388	157.000	.002					

a. Computed using alpha = .05

TABLE 8.45 Continued

Estimated Marginal Means

1. Type of protection

Measure: chest_de

Type of protection	Mean	Std. Error	95% Confidence Interval	
			Lower Bound	Upper Bound
Driver airbag	1.628[a]	.013	1.603	1.653
Manual belt	1.692[a]	.007	1.678	1.705
Motorize	1.655[a]	.016	1.624	1.687
Passive belt	1.681[a]	.014	1.655	1.708

a. Covariates appearing in the model are evaluated at the following values: Vehicle weight = 2920.0675, Model year = 88.8650.

2. location

Measure: chest_de

location	Mean	Std. Error	95% Confidence Interval	
			Lower Bound	Upper Bound
1	1.691[a]	.007	1.677	1.706
2	1.637[a]	.007	1.623	1.650

a. Covariates appearing in the model are evaluated at the following values: Vehicle weight = 2920.0675, Model year = 88.8650.

3. Type of protection * location

Measure: chest_de

Type of protection	location	Mean	Std. Error	95% Confidence Interval	
				Lower Bound	Upper Bound
Driver airbag	1	1.649[a]	.015	1.620	1.679
	2	1.607[a]	.014	1.579	1.634
Manual belt	1	1.723[a]	.008	1.707	1.739
	2	1.661[a]	.008	1.646	1.676
Motorize	1	1.679[a]	.019	1.642	1.716
	2	1.631[a]	.018	1.596	1.666
Passive belt	1	1.714[a]	.016	1.682	1.746
	2	1.649[a]	.015	1.619	1.679

a. Covariates appearing in the model are evaluated at the following values: Vehicle weight = 2920.0675, Model year = 88.8650.

.02 to .19. A look at the marginal means suggests that airbags provide the most protection (leading to the least amount of chest deceleration) and that manual seatbelts provide the least protection. However, this needs to be verified through post hoc comparisons.

There are also statistically significant effects of both CVs. Vehicle weight affects logarithm of chest deceleration after adjusting for all other effects: $F(1, 157) = 11.279$, $p = .001$. The output labeled Parameter Estimates shows the regression coefficients for the two locations: $B = 0.00004$ and partial $\eta^2 = .071$ with 95% confidence limits from .01 to .16 for drivers, and $B = 0.00002$ and partial $\eta^2 = .035$ with confidence limits from .00 to .11 for passengers.[11] Thus, about 3.5% to 7% of the variance over the two locations is accounted for by vehicle weight. The positive coefficients indicate, surprisingly, that the greater the weight of the vehicle, the greater the injury. Model year also affects logarithm of chest deceleration after adjusting for all other effects: $F(1, 157) = 7.058$, $p = .009$. The output labeled Parameter Estimates shows the regression coefficients for the two locations: $B = 0.008$ and partial $\eta^2 = .015$ for drivers with 95% confidence limits from .00 to .07, and $B = 0.015$ and partial $\eta^2 = .058$ with 95% confidence limits from .01 to .14 for passengers. Thus, about 1.5% to 6% of the variance over the two locations is accounted for by model year of the car. The positive coefficients indicate, again surprisingly, that the more recently the vehicle was made, the greater the injury.

The Tests of Within-Subjects Effects output shows no effects on chest deceleration by location of the dummy or by the location by protection interaction. Nor is there a significant interaction between location and either of the CVs, vehicle weight or model year; that is, there is no violation of homogeneity of regression for the repeated-measures IV. Table 8.46 shows an additional SPSS GLM run in which CVs are omitted and only the repeated-measures portion of the output are examined.

The output shows that power for repeated-measures effects was substantially reduced by using the GLM approach in which interaction between the repeated-measures IV and CVs is included in the analysis model. By using the traditional approach, there is a statistically significant and substantial effect of location: $F(1, 159) = 58.568$, $p < .0005$, partial $\eta^2 = .269$ with 95% confidence limits from .16 to .37. The protection by location interaction, however, remains nonsignificant. The location means show that greater average logarithm of chest deceleration is associated with the driver than with the passenger dummy. The lack of a location by protection interaction is especially interesting considering that, with one exception, only the driver's seat was equipped with an airbag.

8.7.2.2.2 Post Hoc Analyses The omnibus analysis suggests two post hoc analyses: comparisons among marginal means for type of protection and evaluation of the bivariate correlations between injury and the CVs.

The SPSS GLM (Repeated Measures) procedure has limited post hoc tests for ANCOVA. Tukey pairwise comparisons are the logical choice for comparisons

[11] Confidence limits are found using Smithson's software. The F value is found as t^2, and each test has 1 df in the numerator and the error df for omnibus between-subjects tests (157) in the denominator.

TABLE 8.46 ANOVA on injury without CVs (syntax and selected output)

```
GLM
  ld_chest lp_chest BY PROTECT
  /WSFACTOR = location 2 Polynomial
  /MEASURE = chest_de
  /METHOD = SSTYPE(3)
  /EMMEANS = TABLES(location)
  /PRINT = DESCRIPTIVE ETASQ OPOWER
  /CRITERIA = ALPHA(.05)
  /WSDESIGN = location
  /DESIGN = PROTECT .
```

General Linear Model

Tests of Within-Subjects Effects

Measure: chest_de

Source		Type III Sum of Squares	df	Mean Square	F	Sig.	Partial Eta Squared	Noncent. Parameter	Observed Power[a]
location	Sphericity Assumed	.146	1	.146	58.568	.000	.269	58.568	1.000
	Greenhouse-Geisser	.146	1.000	.146	58.568	.000	.269	58.568	1.000
	Huynh-Feldt	.146	1.000	.146	58.568	.000	.269	58.568	1.000
	Lower-bound	.146	1.000	.146	58.568	.000	.269	58.568	1.000
location * protect	Sphericity Assumed	.014	3	.005	1.820	.146	.033	5.461	.467
	Greenhouse-Geisser	.014	3.000	.005	1.820	.146	.033	5.461	.467
	Huynh-Feldt	.014	3.000	.005	1.820	.146	.033	5.461	.467
	Lower-bound	.014	3.000	.005	1.820	.146	.033	5.461	.467
Error(location)	Sphericity Assumed	.395	159	.002					
	Greenhouse-Geisser	.395	159.000	.002					
	Huynh-Feldt	.395	159.000	.002					
	Lower-bound	.395	159.000	.002					

a. Computed using alpha = .05

Tests of Between-Subjects Effects

Measure: chest_de

Transformed Variable: Average

Source	Type III Sum of Squares	df	Mean Square	F	Sig.	Partial Eta Squared	Noncent. Parameter	Observed Power[a]
Intercept	608.380	1	608.380	68450.951	.000	.998	68450.951	1.000
protect	.083	3	.028	3.121	.028	.056	9.363	.718
Error	1.413	159	.009					

a. Computed using alpha = .05

Estimated Marginal Means

location

Measure: chest_de

location	Mean	Std. Error	95% Confidence Interval	
			Lower Bound	Upper Bound
1	1.693	.007	1.678	1.707
2	1.641	.007	1.627	1.655

among all levels of protection, but they are a bit inconvenient when sample sizes are unequal. First, we find the critical F_T according to Equation 4.15, using Table A.3 for q_T with four means and $\text{df}_{\text{error}} = 120$ (the largest tabled value short of infinity):

$$F_T = \frac{(q_T)^2}{2} = \frac{(3.68)^2}{2} = 6.77$$

Marginal pairwise comparisons (without Tukey adjustment) are then run through SPSS GLM (as per Section 7.6.1), as shown in Table 8.47.

TABLE 8.47 Pairwise comparisons of levels of protection. SPSS GLM syntax and selected output.

```
GLM
 ld_chest lp_chest BY PROTECT WITH WEIGHT YEAR
 /WSFACTOR = location 2 Polynomial
 /MEASURE = chest_de
 /METHOD = SSTYPE(3)
 /CRITERIA = ALPHA(.05)
 /LMATRIX "air vs man"      protect 1 -1 0 0
 /LMATRIX "air vs. motor"   protect 1  0 -1 0
 /LMATRIX "air vs. pass"    protect 1  0 0 -1
 /LMATRIX "man vs. motor" protect 0 1 -1 0
 /LMATRIX "man vs. pass"   protect 0 1 0 -1
 /LMATRIX "motor vs pass"  protect 0 0 1 -1
 /WSDESIGN = location
 /DESIGN = WEIGHT YEAR PROTECT .
```

General Linear Model

Custom Hypothesis Tests #1

Test Results

Measure: chest_de

Transformed Variable: AVERAGE

Source	Sum of Squares	df	Mean Square	F	Sig.
Contrast	.145	1	.145	18.450	.000
Error	1.234	157	.008		

Custom Hypothesis Tests #2

Test Results

Measure: chest_de

Transformed Variable: AVERAGE

Source	Sum of Squares	df	Mean Square	F	Sig.
Contrast	.016	1	.016	2.000	.159
Error	1.234	157	.008		

TABLE 8.47 Continued

Custom Hypothesis Tests #3

Test Results

Measure: chest_de

Transformed Variable: AVERAGE

Source	Sum of Squares	df	Mean Square	F	Sig.
Contrast	.062	1	.062	7.870	.006
Error	1.234	157	.008		

Custom Hypothesis Tests #4

Test Results

Measure: chest_de

Transformed Variable: AVERAGE

Source	Sum of Squares	df	Mean Square	F	Sig.
Contrast	.032	1	.032	4.129	.044
Error	1.234	157	.008		

Custom Hypothesis Tests #5

Test Results

Measure: chest_de

Transformed Variable: AVERAGE

Source	Sum of Squares	df	Mean Square	F	Sig.
Contrast	.004	1	.004	.475	.492
Error	1.234	157	.008		

Custom Hypothesis Tests #6

Test Results

Measure: chest_de

Transformed Variable: AVERAGE

Source	Sum of Squares	df	Mean Square	F	Sig.
Contrast	.012	1	.012	1.523	.219
Error	1.234	157	.008		

TABLE 8.48 Bivariate correlations among logarithm of chest deceleration, vehicle weight, and model year (SPSS CORRELATIONS syntax and output)

```
CORRELATIONS
 /VARIABLES=WEIGHT YEAR ld_chest lp_chest
 /PRINT=TWOTAIL NOSIG
 /MISSING=LISTWISE .
```

Correlations

Correlations[a]

		Vehicle weight	Model year	ld_chest	lp_chest
Vehicle weight	Pearson Correlation	1	.180*	.262**	.216**
	Sig. (2-tailed)		.021	.001	.006
Model year	Pearson Correlation	.180*	1	.006	.177*
	Sig. (2-tailed)	.021		.943	.024
ld_chest	Pearson Correlation	.262**	.006	1	.573**
	Sig. (2-tailed)	.001	.943		.000
lp_chest	Pearson Correlation	.216**	.177*	.573**	1
	Sig. (2-tailed)	.006	.024	.000	

*. Correlation is significant at the 0.05 level (2-tailed).
**. Correlation is significant at the 0.01 level (2-tailed).
a. Listwise N=163

Two of the comparisons reach statistical significance with $\alpha_{FW} = .05$ and the Tukey adjustment: the difference between airbags and manual belts, with $F(1, 157) = 18.45$, and the difference between airbags and passive belts, with $F(1, 157) = 7.87$. Note that an index table (not shown) identifies the tests using the labels assigned in syntax.

SPSS CORRELATIONS (Table 8.48) provides bivariate correlations among the logarithm of chest deceleration (separately for passenger and driver), vehicle weight, and model year. Cases with missing data and outliers are filtered out, as in previous analyses.

Information about statistical significance is inappropriate for these post hoc analyses and is thus ignored. The output reveals that, indeed, the relationships between injury and the CVs are positive, though small, with correlations ranging from .006 to .262. Thus, weight and year are not particularly impressive CVs in adjusting the DV for testing the effect of protection type. The relationship between the injury to driver and passenger is fairly substantial: $r^2 = (.573)^2 = .33$.

Software is available for finding confidence limits for R^2 (Steiger and Fouladi, 1992), demonstrated in Figure 8.11. The `Confidence Interval` is chosen as the `Option` and `Maximize Accuracy` is chosen as the `Algorithm`. Using the R^2 value of .33, Figure 8.11(a) shows setup values of 163 observations, 2 variables (including the DV and 1 predictor), and probability value of .95. As shown in Figure 8.11(b), the

FIGURE 8.11 Confidence limits around R^2 using Steiger and Fouladi's (1992) software: (a) setup and (b) results

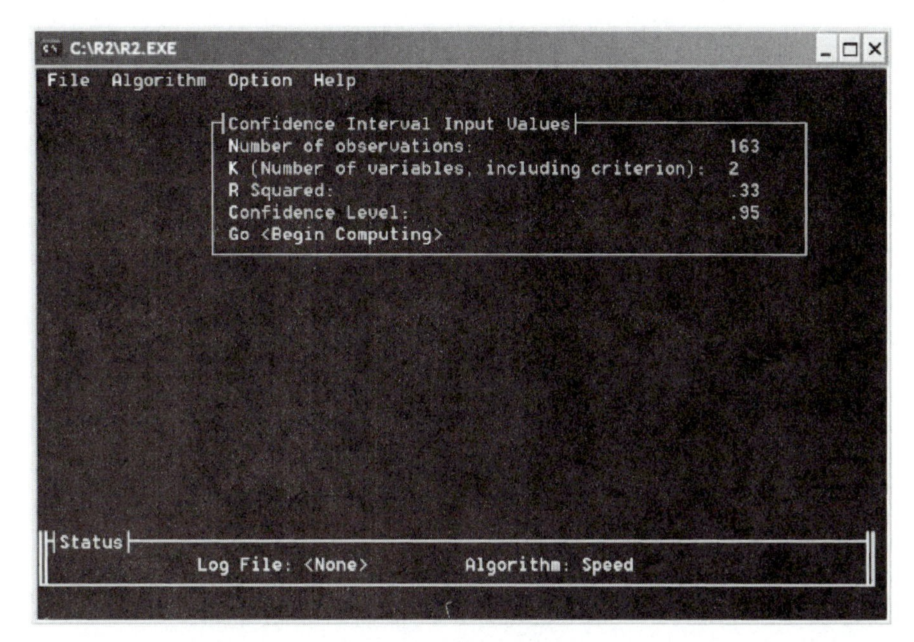

(a)

C:\R2\R2.EXE
File Algorithm Option Help
.95 Confidence Interval Results
Lower Limit: 0.21192
Upper Limit: 0.44610
Lower Bound: 0.23018
Plevel : 0.00000
<Hit any key to continue>
Status
Log File: <None> Algorithm: Speed

(b)

R2 program provides 95% confidence limits from .21 to .45. Remaining correlations are evaluated in the same way (and are reported in the results section that follows).

Table 8.49 shows a checklist for a mixed randomized-groups and repeated-measures ANCOVA. A results section, in journal format, follows.

TABLE 8.49 Checklist for mixed randomized-groups and repeated-measures ANCOVA

1. Issues
   a. Sample sizes and missing data
   b. Normality of sampling distributions
   c. Outliers
   d. Independence of errors
   e. Multicollinearity
   f. Homogeneity of variance
   g. Linearity
   h. Homogeneity of regression
   i. Reliability of CVs
2. Major analyses
   a. Main effects or planned comparisons. If statistically significant:
      (1) Effect size(s) and their confidence limits
      (2) Parameter estimates (adjusted marginal means and standard errors or standard deviations or confidence intervals)
   b. Interactions or planned comparisons. If significant:
      (1) Effect size(s) and their confidence limits
      (2) Parameter estimates (adjusted cell means and standard errors in table or plot)
3. Additional analyses
   a. Evaluation of CV effects
   b. Evaluation of intercorrelations
   c. Post hoc comparisons
   d. Interpretation of departure from homogeneity of variance, homogeneity of covariance, and/or homogeneity of regression, if relevant

Results

A 4 × 2 mixed randomized-groups and repeated-measures ANCOVA was performed on chest deceleration after a vehicle crash. Four types of protection were evaluated: airbags, manual belts, motorized belts, and passive belts. All crashes had dummies in both the driver's and passenger's seats. The repeated-measures independent variable was seat location of the dummy. Covariates were vehicle weight and model year. One hundred seventy-six vehicles were crashed.

There were missing values for either the passenger or dummy chest deceleration for 11 of the crashes. These cases were omitted from analysis, along with two outliers: the vehicle with the heaviest weight (5619 lbs) among cars with manual seatbelts and the vehicle producing the greatest passenger chest deceleration among cars with motorized seatbelts (78 units). Other potential outliers were controlled through logarithmic transformation of both driver and passenger chest deceleration, both of which were generally positively skewed. A sample of 163 crashes remained. Homogeneity of variance; homogeneity of regression; linearity among logarithm of chest deceleration, model year, and vehicle weight; and reliability of weight were acceptable. Unequal sample sizes led to a decision to use Type III sums of squares (unweighted means) for the analyses.

There was a statistically significant effect of type of protection on the logarithm of chest deceleration after adjusting for all other effects, including the covariates, $F(3, 157) = 6.29$, $p < .0005$, partial $\eta^2 = .11$ with 95% confidence limits from .02 to .19. A Tukey test at $\alpha = .05$ showed that two of the six differences among adjusted means were statistically significant. Vehicles with airbags (adjusted mean log of chest deceleration = 1.63, $SE_M = 0.013$) produced less injury than vehicles with manual belts (adjusted mean log of chest deceleration = 1.69, $SE_M = 0.007$) or vehicles with passive belts (adjusted mean log of chest deceleration = 1.68, $SE_M = 0.014$).

Location of dummy was also statistically significant after adjusting for the main effect of protection and the location by protection interaction, $F(1, 159) = 58.57$, $p < .0005$, partial $\eta^2 = .27$ with 95% confidence limits for .16 to .37. Dummies located in the driver's seat showed greater (log of) chest deceleration (mean = 1.69, SE_M = 0.007) than did dummies located in the passenger's seat (mean = 1.64, SE_M = 0.007). There was no statistically significant interaction between location and type of protection.

Vehicle weight predicted (log of) chest deceleration after adjusting for the main effects, interaction, and the other covariate, model year, $F(1, 157) = 11.28$, $p = .001$. Regression coefficients were 0.00004 for driver dummies and 0.00002 for passenger dummies, with $\eta^2 = .07$ with 95% confidence limits from .01 to .16 for the driver seat and $\eta^2 = .035$ with confidence limits from .00 to .11 for the passenger seat. Model year also predicted (log of) chest deceleration after adjusting for all other effects, $F(1, 157) = 7.06$, $p = .009$. Regression coefficients were 0.00781 for driver dummies and 0.0146 for passenger dummies, with $\eta^2 = .015$ with 95% confidence limits from .00 to .07 for the driver seat and $\eta^2 = .06$ and 95% confidence limits from .01 to .14 for the passenger seat.

Raw bivariate correlations between (log of) chest deceleration and the covariates were not particularly high. For vehicle weight, $r^2(161) = .07$ with 95% confidence limits from .01 to .16 for the driver seat and $r^2(161) = .05$ with 95% confidence limits from .01 to .13 for the passenger seat. For model year, $r^2(161) < .01$ with 95% confidence limits from .00 to .01 for the driver seat and $r^2(161) < .03$ with 95% confidence limits from .00 to .10 for the passenger seat. The correlation between injuries in the two locations was more substantial: $r^2(161) = .33$ with 95% confidence limits from .21 to .45.

8.8 COMPARISON OF PROGRAMS

Most programs that do ANOVA also do ANCOVA. About the only exception in the three packages subject to major review in this book is SPSS ANOVA (previously ONEWAY). Thus, most of the packages have more than one program that handles CVs.

Table 8.50 compares the features of SAS GLM, SPSS MANOVA and GLM, and SYSTAT GLM and ANOVA, which are especially relevant to ANCOVA. Other features are reviewed in other chapters.

8.8.1 SPSS Package

The most current program, GLM Univariate, works well for randomized-groups ANCOVA, providing adjusted marginal and cell means and their standard errors and plotting the adjusted means. Tests of individual CVs are available, but a combined test is not.

Marginal means are available for repeated-measures IVs. Syntax is also convenient, with only a single listing of the CV required with repeated-measures IV(s). However, the GLM approach, which adjusts the repeated-measures IV by its interaction with the CV, may be undesirable, and a separate run is required to obtain the traditional repeated-measures portion of the source table. An alternative data setup is required when there are varying CVs for each level of a repeated-measures IV (cf. Section 8.5.2.2.2).

SPSS MANOVA addresses the concerns regarding repeated-measures designs, with its traditional analysis. MANOVA also provides a test of all CVs combined (labeled REGRESSION) and is the only program providing correlations, squared correlations, and standardized regression coefficients between CVs and the predicted DV. However, adjusted means are unavailable for repeated-measures factors.

8.8.2 SAS System

SAS GLM also works well for randomized-groups but not for repeated-measures ANCOVA, and it does not provide either regression coefficients for CVs or adjusted marginal means for repeated-measures IVs. The GLM approach requires a separate run to obtain the traditional repeated-measures portion of the source table, as well as an alternate data setup with varying CVs (cf. Section 8.5.2.2.2). Although a combined test is not given for multiple CVs, the default printing of Type I SS (and an appropriate arrangement of model effects) permits a simple calculation to assess the combined effect (cf. Section 8.6.1).

TABLE 8.50 Comparison of SPSS, SAS, and SYSTAT programs for factorial randomized-groups ANCOVA

Feature	SPSS MANOVA	SPSS GLM	SAS GLM	SYSTAT ANOVA and GLM
Input				
Type of repeated-measures analysis	Traditional	GLM	GLM	GLM
Output				
ANOVA tables include interaction between CV and repeated-measures IV	No	Yes	Yes	Yes
Marginal means adjusted for CV(s) for randomized-groups IVs	PMEANS	EMMEANS	LSMEANS	LS MEANS
Standard errors for adjusted marginal means	No	EMMEANS	STDERR	Yes
Adjusted marginal means for repeated-measures IV	N.A.	Yes	No	No
Cell means adjusted for CV(s) for randomized-groups IVs	PMEANS[a]	EMMEANS	LSMEANS	LS MEANS
Standard errors for adjusted cell means	No	EMMEANS	Data file	LS MEANS
Confidence interval for adjusted cell means	No	Yes	Yes	No
Adjusted cell means for repeated-measures IV(s)	N.A.	EMMEANS	LSMEANS	LS MEANS
Type I as well as Type III sums of squares in a single run by default	No	No	Yes	No
Pooled within-cell regression coefficients	Yes	PARAMETER	SOLUTION	Yes
Standard error and t for regression coefficients	Yes	Yes	Yes	No
Confidence interval for regression coefficients	Yes	Yes	CLPARM	No
Standardized regression coefficients	Yes	No	No	No
Individual tests of multiple CVs	Yes	Yes	Yes	Yes
Combined test of multiple CVs	Yes	No	No	No
Correlations between CVs and predicted DV	Yes	No	No	No
Squared correlations between CVs and predicted DV	Yes	No	No	No
Plots				
Adjusted Means for randomized-groups IVs	Yes	Yes	No	Yes

[a] Only for completely randomized factorial design.

8.8.3 SYSTAT System

Likewise, SYSTAT works better for randomized-groups than for repeated-measures ANCOVA. An advantage over SAS is that adjusted cell means are available for repeated-measures designs; however, adjusted marginal means are not. As for all GLM programs, varying CVs for repeated measures require an alternate data setup. By default, regression coefficients are shown for CVs, but without standard errors.

8.9 PROBLEM SETS

8.1 Use a statistical software program to do an analysis of covariance on the following data set, which is the first problem of Chapter 4 with a CV added. Assume the IV is minutes of presoaking a dirty garment, the DV (Y) is rating of cleanliness, and the CV (X) is the rating of cleanliness before presoaking.

Presoak Time

a_1 5 min		a_2 10 min		a_3 15 min		a_4 20 min	
X	Y	X	Y	X	Y	X	Y
1	2	1	3	5	8	4	6
5	6	8	9	9	11	5	7
5	6	7	9	5	9	7	9
4	6	4	5	6	9	4	6
3	5	5	7	2	6	7	9
2	3	3	5	9	12	8	9

a. Test for homogeneity of variance.
b. Test for homogeneity of regression.
c. Provide a source table.
d. Check results (and homogeneity of regression) by running the data set through at least one statistical software program.
e. Calculate effect size and 95% confidence limits.
f. Write a results section for the analysis.

8.2 Do a computer analysis of variance on the following data set, which adds two CVs to the first problem of Chapter 5. Assume the IV is minutes of presoaking a dirty garment, the DV (Y) is rating of cleanliness, the first CV (X_1) is the rating of cleanliness before presoaking, and the second CV (X_2) is an index of the fabric thread count.

	Presoak Time											
	b_1: 5 min			b_2: 10 min			b_3: 15 min			b_4: 20 min		
	X_1	X_2	Y	X_1	X_2	Y	X_1	X_2	Y	X_1	X_2	Y
	1	4	2	1	2	3	5	2	8	4	7	6
a_1: Cotton	5	5	6	8	7	9	9	3	11	5	3	7
	3	1	6	7	2	9	5	5	9	7	2	9
	4	6	6	4	4	5	6	5	9	4	1	6
a_2: Polyester	3	3	5	5	5	7	2	6	6	7	5	9
	2	4	3	3	8	5	9	1	12	8	4	9

a. Test for homogeneity of variance.

b. Test for homogeneity of regression.

c. Test for absence of univariate and multivariate outliers.

d. Write a results section in journal format. Include evaluation of CVs and the correlation between them, as well as any departure from homogeneity of variance or homogeneity of regression.

8.3 The following mixed randomized-groups, repeated-measures ANCOVA has the same CV for each level of the repeated-measures IV. The randomized-groups IV is keyboard feel on a portable computer: mushy or click. The repeated-measures IV is size of keys: small, medium, or large. The DV is rating of typing speed, with a pretest of typing speed on an IBM desktop computer as a CV.

		Keyboard Size		
	CV	b_1: Small	b_2: Medium	b_3: Large
	7	1	3	6
	8	2	3	5
a_1: Mushy	8	3	4	4
	11	4	3	7
	13	5	5	8
	15	6	7	9
	14	7	7	8
a_2: Click	13	8	7	7
	15	9	8	7
	14	10	6	8

a. Do an analysis of covariance using a GLM computer program, evaluating homogeneity of covariance.

b. Make the adjustment to match results to a traditional analysis.

8.4 Using the results from Problem 8.2, determine the number of cases required to find a significant difference between the two fabrics at $\alpha = .05$ and $\beta = .20$.

8.5 Do an ANOVA (without CV) for the data in Table 8.11, and compare the results with those of Table 8.12.

Extending the Concepts

8.6 The following mixed randomized-groups, repeated-measures ANCOVA has a different CV for each level of the repeated-measures IV. The randomized-groups IV is keyboard feel on a portable computer: mushy or click. The repeated-measures IV is size of keys: small, medium, or large. The DV is rating of typing speed, with a pretest of typing speed on an IBM desktop computer as a CV before each trial. Do an analysis of covariance using a traditional computer program.

Keyboard Size

	b_1: Small		b_2: Medium		b_3: Large	
	CV	DV	CV	DV	CV	DV
	7	1	4	3	3	6
	8	2	5	3	5	5
a_1: Mushy	8	3	4	4	3	4
	11	4	8	3	6	7
	13	5	10	5	9	8
	15	6	11	7	9	9
	14	7	9	7	10	8
a_2: Click	13	8	8	7	9	7
	15	9	12	8	11	7
	14	10	13	6	13	8

9

Latin-Square Designs

9.1 GENERAL PURPOSE AND DESCRIPTION

Latin-square designs are used when the researcher has one IV of overriding interest, as well as other IVs that need to be evaluated in the design because they are expected to have some (often modest) effect on the DV. The other IVs are often considered nuisance variables. Their levels could be crossed with the IV of interest in a fully factorial design, but that is likely to make the design prohibitively large and would probably give the other IVs more "stature" than they deserve. Instead, a Latin-square design is used, in which the main effect of the IV of interest and the main effects of the other IVs are available for analysis, but (most) interactions are not.

In this design, interactions of the IV of interest with the other IVs, and of the other IVs with each other, are confounded with tests of main effects and the error term. This means that Latin-square designs are inappropriate when interactions are anticipated. When choosing this design, a major consideration, then, is expected absence of interaction.

Latin squares are used both with randomized-groups designs and with repeated-measures designs, although their use with repeated measures is more common in many disciplines. In randomized-groups designs, the other IVs are usually nuisance variables that are unavoidable differences in testing conditions (several different pieces of the same equipment, time of day, different operators or researchers, etc.). If not accounted for in the analysis, the differences associated with these IVs inflate the error term but are otherwise not substantive. A Latin-square design and analysis is likely to increase power by assessing the sums of squares for these IVs and subtracting them from the error term.

In repeated-measures designs, the other IVs are usually cases and position (trial) effects. Position effects are unavoidable in a repeated-measures design in which cases are necessarily measured more than once. Latin-square designs are used to remove potential confounding between position effects and treatment effects by giving treatments to cases in different orders. In repeated-measures designs, then, the other IVs may be more substantive (both logically and in terms of sums of squares) than in randomized-groups designs.

The Latin square is an arrangement of treatments. The "square" is a two-dimensional layout of three IVs in which rows represent one IV, columns represent another IV, and the third IV, usually the treatment of greatest interest, appears as entries inside the matrix. The test of the effect of interest is carved out of the interaction of the two nuisance variables. All three IVs have the same number of levels. Table 9.1 shows an example, in which C is the IV of interest.[1]

A defining feature of Latin-square arrangements is that each level of the IV of interest (C) appears once in each row and once in each column. Whatever the effects of the different levels of A and B, they are spread evenly over the various levels of C. Just as in factorial fixed-effects equal-n designs, the test of C is independent of (averaged over) the main effect of A and the main effect of B. However, as previously mentioned, the main effect of C is not independent of the various parts of potential interactions that may be included in the design.

Different Latin squares contain different segments of the interactions from a complete factorial arrangement. In Table 9.2, X indicates the combinations of A, B, and C generated by the Latin square of Table 9.1. Another Latin square would include other combinations of levels from the full factorial. Interaction, as you recall, is a synergistic effect produced by a particular combination of levels of treatments. The particular Latin square generated for a study may or may not include one or more cells with such synergy. This is why the assumption that there is no interaction (additivity) is so important.

TABLE 9.1 Basic Latin-square design with four treatments

	b_1	b_2	b_3	b_4
a_1	c_1	c_2	c_3	c_4
a_2	c_2	c_4	c_1	c_3
a_3	c_3	c_1	c_4	c_2
a_4	c_4	c_3	c_2	c_1

[1] We have diverged from the usual practice of using A for the IV of greatest interest in order to make the relationship between the mixed design and the Latin-square design as clear as possible.

TABLE 9.2 Segments of interaction available in the Latin square of Table 9.1

c_1:

	b_1	b_2	b_3	b_4
a_1	X			
a_2			X	
a_3		X		
a_4				X

c_2:

	b_1	b_2	b_3	b_4
a_1		X		
a_2	X			
a_3				X
a_4			X	

c_3:

	b_1	b_2	b_3	b_4
a_1			X	
a_2				X
a_3	X			
a_4		X		

c_4:

	b_1	b_2	b_3	b_4
				X
		X		
			X	
a_4	X			

Another difference between the full factorial design and the Latin-square design is the number of degrees of freedom generated: The Latin-square design generates far fewer df than does the full factorial. The Latin-square design of Table 9.1, for instance, generates 15 total df, whereas the full factorial design of Table 9.2 generates 63 total df. Degrees of freedom are always a valuable resource to be spent with care, but this is especially true for Latin-square designs.

The same Latin squares are applied to randomized-groups designs and repeated-measures designs, but their purposes are quite different. In a randomized-groups design, a major goal of a Latin-square arrangement is to gain some of the advantages of a full factorial with a substantially reduced number of treatment combinations. A full factorial arrangement of three IVs with four levels each requires 64 treatment combinations, as shown in Table 9.2; the Latin square in Table 9.1 requires only 16 of them. Other advantages include the increased generalizability of results for C over the range of levels of A and B and the increased power that is likely to be achieved by subtracting sums of squares associated with the main effects of A and B from the error term.

A disadvantage compared with a one-way randomized-groups design (with only IV C) is that more groups need to be run. A disadvantage compared with a fully factorial randomized-groups design is lack of information about interactions. And, of course, there is the limitation that all IVs have the same number of levels for the most common applications of Latin-square arrangements: There may be any number of levels, but all IVs have to have the same number of them. The Latin-square design applied to randomized groups is one of a rather large family of incomplete factorial designs. Others are discussed in Chapter 10.

For an example of a randomized-groups Latin-square design, the IV of major interest, C, is four types of assembly procedures potentially affected by two nuisance variables—assembler experience (A) and various temperatures of the room housing the assembly line (B). Sixteen assemblers are blocked into four groups based on experience. The Latin-square arrangement allows the error term to be reduced by variance associated with four levels of assembler experience and four levels of room temperature. However, the design does not permit evaluation of interactions of assembly procedure by experience, assembly procedure by room temperature, experience by room temperature, or assembly procedure by experience by room temperature.

In a repeated-measures design, the purpose of a Latin-square arrangement is to counterbalance position (trial) effects among cases. In Table 9.1, the B variable is position, and the A variable is the case (or cases) to which levels of C are applied.[2] The first case receives c_1, c_2, c_3, and then c_4, and the second case receives c_2, c_4, c_1, and then c_3. The goal is to ensure that the levels of C are not confounded with the trial in which the treatment is applied, B. This feature permits a repeated-measures IV to be experimental, even when all levels of C are applied to the same cases and those cases have memory (for example, experience is carried over from trial to trial). The design also allows evaluation of the effect of position (trial) itself. Sometimes position effects are called *practice effects*, particularly when applied to animate cases (or perhaps computer programs capable of learning).

For an example of a Latin-square repeated-measures design, C is still different assembly procedures, except that the same four assemblers implement all procedures. The A variable represents the four assemblers, and B is the trial in which each procedure is implemented. This application controls for differences that may exist among the four assemblers and for changes in their ability that may occur during the four trials, independent of type of procedure.

9.2 KINDS OF RESEARCH QUESTIONS

The basic question in any Latin-square analysis is whether there are statistically significant mean differences in the DV as a result of different treatments. If there are more than two treatments, comparisons are possible, in addition to or in lieu of the omnibus test.

9.2.1 Effects of the IV(s)

Is there a statistically significant effect of treatment C? Controlling for main effects of A and B and holding all else constant, are changes in the DV that are associated with different levels of C larger than would be expected by chance? For example, is there a mean difference in production efficiency associated with the assembly procedure used after controlling for IVs A and B?

[2] Note that cases here are labeled A rather than the usual S of repeated-measures ANOVA.

In the randomized-groups Latin square, the effects of A and B can also be evaluated. For example, after controlling for assembly procedure and room temperature, is there a difference in production efficiency depending on the assembler's experience? Controlling for assembly procedure and assembler experience, is production efficiency affected by room temperature?

The repeated-measures Latin-square design also permits assessment of A and B, should such be desired. Does production efficiency change over trials? Do assemblers get better as they implement more procedures? Worse? Better and then worse? The test of the main effect of differences among assemblers is not often performed because it is usually assumed that such differences exist.

9.2.2 Interactions among IVs

The analysis of a basic randomized-groups Latin-square design permits no evaluation of interactions, because main effects are confounded with two-way interactions (e.g., the main effect of C is confounded with the $A \times B$ interaction and so on). However, there are expansions of the design that do permit evaluation of some interactions. Section 9.5.3.1 discusses one variant of the Latin-square design that permits tests of interactions among types of treatments. In this variant, c_1, c_2, c_3, and c_4 are not four levels of one treatment, but rather two IVs (C and D) with two levels each in factorial arrangement, so both the main effects and their interaction are available for evaluation. Another expansion involves running the entire Latin-square arrangement at each level of a randomized-groups IV. In this case, the main effect and its interaction with C are available for evaluation. In short, the potential for outfoxing the design is limited only by the ingenuity of the researcher.

9.2.3 Parameter Estimates

What means are expected if a particular treatment is applied to a population? What average production efficiency rating would we expect for a particular assembly procedure? Parameter estimates reported in Latin square, as in any other ANOVA, are means, standard deviations, standard errors, and/or confidence intervals for the mean. The reported values are estimated from the sample; however, the sample size is often quite small, which may limit the generalizability of the findings. If parameters are estimated, means and standard deviations or standard errors (and/or confidence intervals) are reported in a narrative format, shown in a table, or presented in a graph, as discussed in Chapters 4 and 5.

9.2.4 Effect Sizes and Power

As for any other ANOVA, effect size is typically reported as proportion of variance in the DV that is predictable from an IV. For instance, how much of the variability in assembly efficiency is associated with differences in production procedures? Section 9.6.3 demonstrates calculation of effect size in Latin-square designs.

Enhanced power is often a driving force behind the choice of a Latin-square analysis, whether repeated measures or randomized groups. Section 9.6.3 also discusses the increase in power associated with these designs.

9.2.5 Specific Comparisons

If there are more than two levels of an IV, which levels are different from each other? For example, is the assembly procedure in which assembly is completed after all parts are available (say, c_1) more effective than the one in which subassemblies are put together as soon as two parts are available (c_2), or the one where top and bottom are assembled separately before they are put together (c_3), or the one that starts in the middle and proceeds to both ends (c_4)? Is that first procedure more effective than the average of all other procedures? These and similar questions may be answered by applying one of a wide variety of comparison procedures, as discussed in previous chapters and in Section 9.6.4.

As usual, comparisons among specific means can be either planned or post hoc. The advantage of planning, as discussed in Section 4.5.5.4 and elsewhere, is greater power.

9.3 ASSUMPTIONS AND LIMITATIONS

9.3.1 Theoretical Issues

Theoretical issues in a Latin-square design are identical to those in any other analysis of variance. That is, issues of causal inference and generalizability are of similar concern in Latin-square designs.

9.3.2 Practical Issues

The Latin-square arrangement, as an ANOVA model, assumes normality of sampling distributions, homogeneity of variance, independence of errors, and absence of outliers. Additional requirements for the repeated-measures application of Latin-square analysis are sphericity and additivity.

9.3.2.1 Normality of Sampling Distributions

The issue of normality of sampling distributions is associated with sample size, and the criterion is the same as for any other ANOVA: Normality of sampling distributions is tenable if degrees of freedom for error exceed 20. However, because there are fewer df in Latin-square designs, this requirement is not often met. Because df are low, inspection of the distributions of DV scores or of residuals is used to assess normality, as demonstrated in Sections 9.7.1.1.1 and 9.7.2.1.1. Solutions for questionable normality (e.g., transformation) are also the same as for any other ANOVA, except that nonparametric tests are unavailable.

9.3.2.2 Homogeneity of Variance

In many Latin-square applications, there is only a single score in each cell. Therefore, homogeneity of variance is assessed separately on the margins of the three main effects. In other words, variances are compared among levels of *A*, then separately among levels of *B*, and then separately among levels of *C*, using F_{max} and Chapter 3 guidelines for evaluating differences among variances. If there is more than one score per cell, the usual assessment of homogeneity of variance among cells is conducted, as described in Chapter 5.

9.3.2.3 Independence of Errors

The ANOVA model assumes that errors (deviations of individual scores around their cell means) are independent of one another, both within each cell and between cells of a design. This assumption is evaluated in randomized-groups designs by examining the details of the research design rather than by statistical criteria. Independence of errors in repeated-measures designs is dealt with as part of the sphericity assumption, as discussed in Sections 6.3.2.3 and 6.6.2.

9.3.2.4 Absence of Outliers

Univariate outliers can occur in one or more cells (if there are multiple scores in each cell) or in one or more levels of effects in the design. As with homogeneity of variance, outliers are evaluated separately for each level of each IV if there is one score per cell and within cells if there is a sufficient number of scores in each cell.

9.3.2.5 Sphericity

Repeated-measures Latin-square analysis, like other repeated-measures analyses, has the assumption that difference scores among all pairs of levels of the repeated-measures IV have roughly equivalent variances. The assumption is unlikely to be violated for the main effect of interest in a Latin-square design, *C*, because counterbalancing of treatment orders ensures that they are not confounded with time. However, sphericity may still be violated if some levels of treatment are more like each other than like others, because difference scores among pairs of treatments that are similar are probably smaller than difference scores among pairs of treatments that are dissimilar.

The test of position (trials), on the other hand, is likely to be biased by failure of sphericity. This is not a problem, unless position is of research interest. If position is of interest, the research questions are often better answered by a trend analysis (which does not require sphericity) than by the omnibus test. Section 6.6.2 discusses these issues, and Section 9.6.1 demonstrates a test of the assumption.

9.3.2.6 Additivity

The assumption of additivity applies to both repeated-measures Latin square and randomized-groups Latin square, but the issues are slightly different. In repeated-measures Latin square, additivity is part of the sphericity assumption discussed in Section 6.6.2.

The additivity assumption in randomized-groups Latin square is that all interactions, $A \times B$, $B \times C$, $A \times C$, and $A \times B \times C$, are negligible or represent only error variance. If treatments do indeed interact, then tests of main effects are misleading. For example, the main effect of C is completely confounded with the $A \times B$ interaction; an apparently large effect of C could actually be due to a large $A \times B$ interaction. Furthermore, because the $A \times B \times C$ interaction serves as (part of) the error term in a randomized-groups Latin-square analysis, a strong three-way interaction reduces the power of the entire analysis. Section 9.7.1.1.3 shows a plot of residuals against predicted values to assess additivity in a randomized-groups Latin-square analysis with one score per cell. Section 9.5.1 shows how to evaluate this assumption in a randomized-groups design with more than one case per cell.

9.4 FUNDAMENTAL EQUATIONS

Table 9.3 shows a $3 \times 3 \times 3$ Latin-square arrangement[3] randomly chosen by procedures described in Section 9.5.2. This arrangement could represent either a randomized-groups Latin square (where A and B are two nuisance variables) or a repeated-measures Latin square (where B is position and A is case, analogous to S in the designs of Chapters 6 and 7).[4] The IV of primary interest in either design is C. The statistical analysis is the same, whether the design is randomized groups or repeated measures.

TABLE 9.3 Latin-square design for small-sample example

	b_1	b_2	b_3
a_1	c_3	c_1	c_2
a_2	c_2	c_3	c_1
a_3	c_1	c_2	c_3

[3] Some texts refer to this as a 3×3 design, because degrees of freedom are equivalent to a 3×3 full factorial ANOVA. We use the three-way notation to emphasize the existence of three IVs.

[4] Feel free to modify the notation by changing A to S when using the following equations for a repeated-measures Latin-square design.

9.4.1 Allocation of Cases

The Latin square of Table 9.3 provides the arrangement of snow condition (*C*: powder, packed powder, or ice), day of week (*B*: Monday, Wednesday, or Friday), and skiers. The DV is minutes to ski the slope. If this is a randomized-groups study, there are nine different skiers, stratified, say, by skill level (*A*: low, medium, and high). Each skier within each skill level is randomly assigned to a day of the week and, hence, a level of snow condition. If this is a repeated-measures study, there are only three skiers (*A*), each one of whom traverses a run on Monday (with one snow condition), a run on Wednesday (with another snow condition), and a run on Friday (with the third snow condition—and a very weary group of snow makers). Because the orders themselves have been randomized, the three skiers may be assigned sequentially (the first skier is assigned to the first order of snow conditions, and so on).

Note that in either study, there is only one score per cell. Many applications of Latin-square designs are enhanced by running replicates where more than one score is obtained for each of the nine cells, representing combinations of treatment and trials (cf. Section 9.5.4).

9.4.2 Partition of Sources of Variance

The total sum of squared differences is partitioned into four sources:

$$SS_T = SS_A + SS_B + SS_C + SS_{\text{residual}} \tag{9.1}$$

or

$$\sum_j \sum_k \sum_l (Y_{jkl} - \text{GM})^2 = n_j \sum_j (\overline{Y}_A - \text{GM})^2 + n_k \sum_k (\overline{Y}_B - \text{GM})^2 + n_l \sum_l (\overline{Y}_C - \text{GM})^2 + SS_{\text{residual}}$$

The total sum of squared differences between individual scores Y_{jkl} and the grand mean (GM) is partitioned into (1) sum of squared differences between means associated with different levels of A ($\overline{Y}_A$) and the grand mean; (2) sum of squared differences between means associated with different levels of B ($\overline{Y}_B$) and the grand mean; (3) sum of squared differences between means associated with different levels of C ($\overline{Y}_C$) and the grand mean; and (4) SS_{residual}, the remaining sum of squared differences between individual scores and the grand mean. Each n is the number of scores that went into the relevant marginal mean.

The error term (residual) is found by subtracting the three systematic sources of variance from the total:

$$SS_{\text{residual}} = SS_T - SS_A - SS_B - SS_C \tag{9.2}$$

Another way of looking at the partition is that SS_T is first divided into SS_A, SS_B, and $SS_{A\times B}$, and then SS_C is "carved out of" SS_{AB} to produce SS_{residual}:

$$SS_{\text{residual}} = SS_{AB} - SS_C \tag{9.3}$$

This way of conceptualizing the source of variability for *C* and residual is often more helpful later on.

Total degrees of freedom remain the number of scores minus 1, lost when the grand mean is estimated (cf. Equation 3.6). Note, however, that the total number of scores is not *abc* but *ab* or *ac* or *bc* or a^2 because the design is incomplete. Because $a = b = c$, this is most easily expressed as

$$df_T = a^2 - 1 = N - 1 \tag{9.4}$$

Degrees of freedom for *A* are also the number of levels of *A* minus 1, lost when $\overline{Y}_A$ is estimated.

$$df_A = a - 1 \tag{9.5}$$

Degrees of freedom for *B* and *C* are also the number of levels of each minus 1, lost when the means are estimated, so that

$$df_B = df_C = a - 1 \tag{9.6}$$

Finally, degrees of freedom for error are the total degrees of freedom minus degrees of freedom for the three effects, which may be expressed in several different ways:

$$\begin{aligned} df_{\text{residual}} &= (a^2 - 1) - (a - 1) - (b - 1) - (c - 1) \\ &= a^2 - 1 - 3(a - 1) \\ &= a^2 - 3a + 2 \\ &= (a - 1)(a - 2) \end{aligned} \tag{9.7}$$

Figure 9.1 diagrams the partition of sums of squares and degrees of freedom for the Latin-square design. Mean squares are formed by dividing sums of squares by

FIGURE 9.1 Partition of (a) sums of squares and (b) degrees of freedom into systematic sources of differences among marginal means of *A*, *B*, and *C* and remaining (error) differences among cases

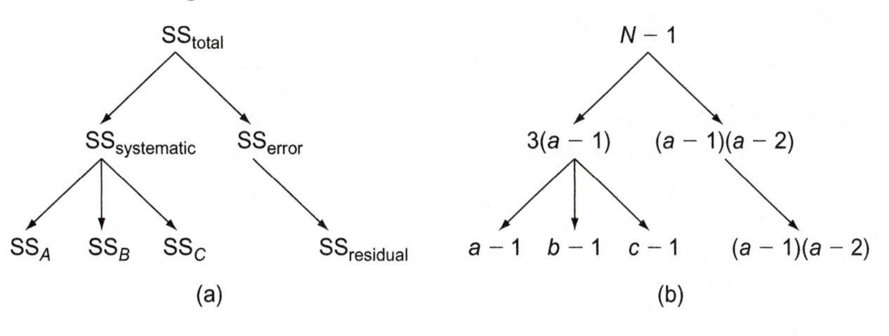

degrees of freedom, as for one-way designs:

$$\mathrm{MS}_A = \frac{\mathrm{SS}_A}{\mathrm{df}_A} \qquad \mathrm{MS}_B = \frac{\mathrm{SS}_B}{\mathrm{df}_B}$$

$$\mathrm{MS}_C = \frac{\mathrm{SS}_C}{\mathrm{df}_C} \qquad \mathrm{MS}_{\text{residual}} = \frac{\mathrm{SS}_{\text{residual}}}{\mathrm{df}_{\text{residual}}} \tag{9.8}$$

F ratios are formed for each of the three sources of variance by dividing the appropriate MS by the single estimate of random variation, $\mathrm{MS}_{\text{residual}}$. F_A evaluates differences among marginal means for levels of *A*, the main effect of *A*. The null hypothesis for this test is $\mu_{a_1} = \mu_{a_2} = \cdots = \mu_{a_j}$, which is the same as discussed in Section 3.2.3.1 for a one-way ANOVA. The obtained F_A is

$$F_A = \frac{\mathrm{MS}_A}{\mathrm{MS}_{\text{residual}}} \qquad \mathrm{df} = (a-1), (a-1)(a-2) \tag{9.9}$$

Obtained F_A is tested against critical *F* with numerator $\mathrm{df} = a - 1$ and denominator $\mathrm{df} = (a-1)(a-2)$. The null hypothesis is rejected if obtained F_A is equal to or larger than critical *F*.

F_B evaluates differences among marginal means for levels of *B*, the main effect of *B*. The null hypothesis for this test is that $\mu_{b_1} = \mu_{b_2} = \cdots = \mu_{b_k}$.

$$F_B = \frac{\mathrm{MS}_B}{\mathrm{MS}_{\text{residual}}} \qquad \mathrm{df} = (a-1), (a-1)(a-2) \tag{9.10}$$

This is also tested against critical *F* with numerator $\mathrm{df} = a - 1$ and denominator $\mathrm{df} = (a-1)(a-2)$. The null hypothesis is rejected if obtained F_B is equal to or exceeds critical *F*.

Similarly, F_C evaluates differences among marginal means for levels of *C*, the main effect of *C*. The null hypothesis for this test is that $\mu_{c_1} = \mu_{c_2} = \cdots = \mu_{c_l}$.

$$F_C = \frac{\mathrm{MS}_C}{\mathrm{MS}_{\text{residual}}} \qquad \mathrm{df} = (a-1), (a-1)(a-2) \tag{9.11}$$

This is also tested against critical *F* with numerator $\mathrm{df} = (a-1)$ and denominator $\mathrm{df} = (a-1)(a-2)$. The null hypothesis is rejected if obtained F_C is equal to or larger than critical *F*.

9.4.3 Traditional ANOVA Approach to 3 × 3 × 3 Latin Square

Table 9.4 shows hypothetical data in the Latin-square arrangement of Table 9.3—where rows represent *A*, columns represent *B*, and cells represent levels of *C*—and then rearranged so that columns represent *C* and cells represent levels of *B*. The first arrangement is convenient for illustrating calculation of sums for *A* and *B*, and the second arrangement is convenient for calculating sums for *C*. *C* is the snow condition with three levels (powder, packed powder, and ice), *A* denotes which of either three (repeated measures) or nine (randomized groups: high, medium, and low skill) skiers

TABLE 9.4 Data for small-sample example in two different Latin-square arrangements

	$A \times B$			
	b_1	b_2	b_3	Sum
a_1	$c_3 = 4$	$c_1 = 9$	$c_2 = 1$	$A_1 = 14$
a_2	$c_2 = 1$	$c_3 = 6$	$c_1 = 7$	$A_2 = 14$
a_3	$c_1 = 8$	$c_2 = 2$	$c_3 = 5$	$A_3 = 15$
Sum	$B_1 = 13$	$B_2 = 17$	$B_3 = 13$	$T = 43$

	$A \times C$			
	c_1	c_2	c_3	Sum
a_1	$b_2 = 9$	$b_3 = 1$	$b_1 = 4$	$A_1 = 14$
a_2	$b_3 = 7$	$b_1 = 1$	$b_2 = 6$	$A_2 = 14$
a_3	$b_1 = 8$	$b_2 = 2$	$b_3 = 5$	$A_3 = 15$
Sum	$C_1 = 24$	$C_2 = 4$	$C_3 = 15$	$T = 43$

traversed the slope, and B is the day of the week (Monday, Wednesday, or Friday). Time to ski the slope, the DV, is measured in minutes.

Table 9.5 shows the computational equations derived from degrees of freedom (which are first expanded if they contain parentheses). Note that $a^2 = N$, the total number of scores in the entire Latin square.

Table 9.6 shows computation of sums of squares for the small-sample example. Note that, as usual, the numbers in the denominators are the number of scores in the

TABLE 9.5 Computational equations derived from degrees of freedom for Latin-square ANOVA

Source	Degrees of Freedom	Expanded df	Sum of Squares Equations
A	$a - 1$		$\frac{\sum A^2}{a} - \frac{T^2}{a^2}$
B	$a - 1 = b - 1$		$\frac{\sum B^2}{a} - \frac{T^2}{a^2}$
C	$a - 1 = c - 1$		$\frac{\sum C^2}{a} - \frac{T^2}{a^2}$
Residual	$a^2 - 1 - (a - 1) - (b - 1) - (c - 1)$	$N - a - b - c + 2 = N - 3a + 2$	$\sum Y^2 - \frac{\sum A^2}{a} - \frac{\sum B^2}{a} - \frac{\sum C^2}{a} + \frac{2T^2}{a^2}$ (or by subtraction: $SS_T - SS_A - SS_B - SS_C$)
T	$a^2 - 1$	$N - 1$	$\sum Y^2 - \frac{T^2}{a^2}$

TABLE 9.6 Calculation of sums of squares for Latin-square ANOVA

$$SS_A = \frac{14^2 + 14^2 + 15^2}{3} - \frac{43^2}{9} = 205.67 - 205.44 = 0.23$$

$$SS_B = \frac{13^2 + 17^2 + 13^2}{3} - \frac{43^2}{9} = 209.00 - 205.44 = 3.56$$

$$SS_C = \frac{24^2 + 4^2 + 15^2}{3} - \frac{43^2}{9} = 272.33 - 205.44 = 66.89$$

$$SS_{residual} = SS_T - SS_A - SS_B - SS_C = 71.56 - 0.23 - 3.56 - 66.89 = 0.88$$

$$SS_T = (4^2 + 9^2 + \cdots + 5^2) - \frac{43^2}{9} = 277.00 - 205.44 = 71.56$$

TABLE 9.7 Source table for 3 × 3 × 3 Latin-square ANOVA

Source	SS	df	MS	F
A	0.23	2	0.12	0.27
B	3.56	2	1.78	4.05
C	66.89	2	33.45	76.02
Residual	0.88	2	0.44	
T	71.56	8		

numerators that are squared. In this example, three scores are summed for each level of A, B, and C before squaring, and the sums of squares are divided by 3. Table 9.7 summarizes the results and shows the mean squares and F values based on Equations 9.1–9.10.

Each obtained F ratio is evaluated separately. With $\alpha = .05$, the same critical value is used to test the main effects of A, B, and C. Based on numerator df = 2 and denominator df = 2, the critical value from Table A.1 is 19.00. Only the obtained F for C exceeds that value. We conclude that there is a difference in time to ski down the slope due to snow condition, but no difference due to day of week or skiers.

9.4.4 Regression Approach to 3 × 3 × 3 Latin Square

Vectors (columns of weighting coefficients) are required for each of the 2 df for the main effect of A, the 2 df for the main effect of B, and the 2 df for the main effect of C. Table 9.8 shows the scores on the DV (Y), codes for the six X vectors for the main effects, and cross products of Y with each of the six X vectors. Arbitrarily, linear and quadratic trend coefficients are used for the first and second vectors for A, B, and C.[5] Therefore, the first X column compares the first and third snow conditions (powder versus ice). The second X column compares the second level of snow with the pooled

[5] Trend coefficients are sensible when the IVs are quantitative or convenient when the researcher just needs an orthogonal set and is not interested in the comparisons, per se.

TABLE 9.8 Vectors for regression approach to 3 × 3 × 3 Latin-square ANOVA

		C		A		B							
	Y	X_1	X_2	X_3	X_4	X_5	X_6	YX_1	YX_2	YX_3	YX_4	YX_5	YX_6
$c_1a_1b_2$	9	−1	1	−1	1	0	−2	−9	9	−9	9	0	−18
$c_1a_2b_3$	7	−1	1	0	−2	1	1	−7	7	0	−14	7	7
$c_1a_3b_1$	8	−1	1	1	1	−1	1	−8	8	8	8	−8	8
$c_2a_1b_3$	1	0	−2	−1	1	1	1	0	−2	−1	1	1	1
$c_2a_2b_1$	1	0	−2	0	−2	−1	1	0	−2	0	−2	−1	1
$c_2a_3b_2$	2	0	−2	1	1	0	−2	0	−4	2	2	0	−4
$c_3a_1b_1$	4	1	1	−1	1	−1	1	4	4	−4	4	−4	4
$c_3a_2b_2$	6	1	1	0	−2	0	−2	6	6	0	−12	0	−12
$c_3a_3b_3$	5	1	1	1	1	1	1	5	5	5	5	5	5
Sum	43	0	0	0	0	0	0	−9	31	1	1	0	−8
Squares summed	277	6	18	6	18	6	18						
N	9												

first and third (packed powder versus powder and ice). The X_5 and X_6 columns code the trend analysis of day of week. These last two columns are difficult to decode because the order of levels of B are scrambled, as they are in the $A \times C$ matrix of Table 9.4. Note that in Table 9.8, C is coded before A and B for ease in decoding. The final six vectors are found by multiplication.

The basic equations for sums of squares and sums of products for the regression approach are the same as illustrated in Table 5.10 for main effects in factorial ANOVA. Table 9.9 shows the equations for finding the sums of squares for the main effect of C, snow condition. Note that this matches the SS_C of Table 9.6 within rounding error. Table 9.10 shows the equations for finding the sums of squares for the main effect of A, skier. Finally, the sum of squares for day of week (B) is calculated in Table 9.11.

TABLE 9.9 Equations for the sum of squares for the main effect of C

$$\text{SS(reg.}X_1) = \frac{[\text{SP}(YX_1)]^2}{\text{SS}(X_1)} = \frac{(-9)^2}{6} = 13.50$$

$$\text{SS(reg.}X_2) = \frac{[\text{SP}(YX_2)]^2}{\text{SS}(X_2)} = \frac{31^2}{18} = 53.38$$

$$\text{SS}_C = \text{SS(reg.}X_1) + \text{SS(reg.}X_2) = 13.50 + 53.38 = 66.88$$

TABLE 9.10 Equations for the sum of squares for the main effect of *A*

$$\text{SS(reg.}X_3) = \frac{[\text{SP}(YX_3)]^2}{\text{SS}(X_3)} = \frac{1^2}{6} = 0.17$$

$$\text{SS(reg.}X_4) = \frac{[\text{SP}(YX_4)]^2}{\text{SS}(X_4)} = \frac{1^2}{18} = 0.06$$

$$\text{SS}_A = \text{SS(reg.}X_3) + \text{SS(reg.}X_4) = 0.17 + 0.06 = 0.23$$

TABLE 9.11 Equations for the sum of squares for the main effect of *B*

$$\text{SS(reg.}X_5) = \frac{[\text{SP}(YX_5)]^2}{\text{SS}(X_5)} = \frac{0^2}{6} = 0.00$$

$$\text{SS(reg.}X_6) = \frac{[\text{SP}(YX_6)]^2}{\text{SS}(X_6)} = \frac{(-8)^2}{18} = 3.56$$

$$\text{SS}_B = \text{SS(reg.}X_5) + \text{SS(reg.}X_6) = 0.00 + 3.56 = 3.56$$

Residual (error) sum of squares is found by subtracting all the sums of squares for effects from the total sum of squares. First, you find SS_Y, identified as SS_T in Table 9.6:

$$\text{SS}_Y = 277 - \frac{43^2}{9} = 71.56$$

and then you obtain $\text{SS}_{\text{residual}}$ by subtraction:

$$\text{SS}_{\text{residual}} = \text{SS}_Y - \text{SS}_A - \text{SS}_B - \text{SS}_C$$
$$= 71.56 - 0.23 - 3.56 - 66.89 = 0.88$$

Each X vector represents a single df. The df for each effect is the number of vectors for that effect (2) in Table 9.8. Total degrees of freedom (df_Y) is $N - 1 = 9 - 1 = 8$, as per Table 9.6. Degrees of freedom for the error term may be found by subtraction or by using the equation $N - v - 1$, where v is the number of X vectors for identified SS. Thus, $\text{df}_{\text{residual}} = 9 - 6 - 1 = 2$.

MS values and F ratios are then calculated as in Table 9.7. Interpretation is identical to that of ANOVA by traditional means.

9.4.5 Computer Analyses of Small-Sample Latin-Square Example

Table 9.12 shows how data are entered for a Latin-square ANOVA. This data format provides the basic source table for either randomized-groups or repeated-measures Latin-square designs. However, it does not permit evaluation of the assumption of sphericity for the repeated-measures design, nor does it provide appropriate error terms for repeated-measures–specific comparisons. The assumption of sphericity is

TABLE 9.12 Data set for small-sample example of Latin-Square analysis

Snow Condition	Day of Week	Case (Skier)	Run Time
3	1	1	4
1	2	1	9
2	3	1	1
2	1	2	1
3	2	2	6
1	3	2	7
1	1	3	8
2	2	3	2
3	3	3	5

tested when either (1) the omnibus effect of position is of research interest, or (2) there is reason to believe that some pairs of treatments are more alike than other pairs of treatments. A data arrangement and analysis that incorporates tests of these assumptions, as well as tests of the treatment and position effects, is described in Section 9.6.1.

Tables 9.13 and 9.14 show computer analyses of the small-sample example of the Latin-square ANOVA of Section 9.4.3 by SPSS GLM and SAS GLM, respectively.

The SPSS GLM (UNIANOVA) instruction lists the DV (TIME), followed by the keyword BY, followed by the list of IVs (SNOW, DAY, and CASE). EMMEANS instructions request marginal means for each IV; ETASQ and OPOWER in the /PRINT instruction request effect size and power statistics. (Note that if you use menus, the choice is General Linear Model > Univariate. . . .) Latin-square analysis is specified in the /DESIGN instruction, where only main effects are listed. This syntax is generated by the menu system when CUSTOM is selected from the MODEL submenu and the three IVs are clicked (highlighted and selected). The remaining syntax is provided by default through the SPSS menu system.

Tests of Between-Subjects Effects shows the source table of Table 9.7 with the error term (Residual in Table 9.7) labeled Error. The Sig. column shows the probability (to three decimal places) that each outcome is a result of chance variation. Thus, the probability that the result with respect to SNOW occurred by chance is .013. The Corrected Model source of variation is the sum of all three main effects plus error. The Total source of variance includes the intercept in its sum; the Corrected Total does not include the intercept and is more useful for calculations of effect size (cf. Section 5.6.2). The column labeled Partial Eta Squared contains effect size measures, described in Section 5.6.2, as partial η^2. R Squared in footnote b is the proportion of total variance that is predictable from the corrected model (the sum of all three effects). Adjusted R Squared adjusts that value for bias due to small sample sizes (cf. Tabachnick and Fidell, 2007, Chapter 5). Observed Power is described in Section 5.6.3. The Noncent. Parameter (noncentrality) is used to calculate power.

TABLE 9.13 Latin-square ANOVA for small-sample example through SPSS GLM (UNIANOVA)

```
UNIANOVA
  TIME  BY SNOW DAY CASE
  /METHOD = SSTYPE(3)
  /INTERCEPT = INCLUDE
  /EMMEANS = TABLES(SNOW)
  /EMMEANS = TABLES(DAY)
  /EMMEANS = TABLES(CASE)
  /PRINT = DESCRIPTIVE ETASQ OPOWER
  /CRITERIA = ALPHA(.05)
  /DESIGN = SNOW DAY CASE .
```

Univariate Analysis of Variance

Tests of Between-Subjects Effects

Dependent Variable: TIME

Source	Type III Sum of Squares	df	Mean Square	F	Sig.	Partial Eta Squared	Noncent. Parameter	Observed Power[a]
Corrected Model	70.667[b]	6	11.778	26.500	.037	.988	159.000	.753
Intercept	205.444	1	205.444	462.250	.002	.996	462.250	1.000
SNOW	66.889	2	33.444	75.250	.013	.987	150.500	.978
DAY	3.556	2	1.778	4.000	.200	.800	8.000	.222
CASE	.222	2	.111	.250	.800	.200	.500	.062
Error	.889	2	.444					
Total	277.000	9						
Corrected Total	71.556	8						

a. Computed using alpha = .05

b. R Squared = .988 (Adjusted R Squared = .950)

Estimated Marginal Means

1. SNOW

Dependent Variable: TIME

			95% Confidence Interval	
SNOW	Mean	Std. Error	Lower Bound	Upper Bound
1.00	8.000	.385	6.344	9.656
2.00	1.333	.385	−.323	2.989
3.00	5.000	.385	3.344	6.656

TABLE 9.13 Continued

2. DAY

Dependent Variable: TIME

DAY	Mean	Std. Error	95% Confidence Interval	
			Lower Bound	Upper Bound
1.00	4.333	.385	2.677	5.989
2.00	5.667	.385	4.011	7.323
3.00	4.333	.385	2.677	5.989

3. CASE

Dependent Variable: TIME

CASE	Mean	Std. Error	95% Confidence Interval	
			Lower Bound	Upper Bound
1.00	4.667	.385	3.011	6.323
2.00	4.667	.385	3.011	6.323
3.00	5.000	.385	3.344	6.656

TABLE 9.14 Latin-square ANOVA for small-sample example through SAS GLM

```
proc    glm data=SASUSER.SS_LATIN;
        class SNOW DAY CASE;
        model TIME = SNOW DAY CASE;
        means SNOW DAY CASE;
run;
```

```
Dependent Variable: TIME

                                          Sum of
  Source                    DF           Squares     Mean Square    F Value    Pr > F

  Model                      6       73.33333333     12.22222222      27.50    0.0355
  Error                      2        0.88888889      0.44444444
  Corrected Total            8       74.22222222

                   R-Square      Coeff Var      Root MSE     TIME Mean

                   0.988024       14.63415      0.666667      4.555556

Source                      DF     Type III SS     Mean Square    F Value    Pr > F

SNOW                         2     66.88888889     33.44444444      75.25    0.0131
DAY                          2      6.22222222      3.11111111       7.00     0.125
CASE                         2      0.22222222      0.11111111       0.25    0.8000

                        Level of             ------------ TIME -----------
                        SNOW          N              Mean                SD

                        1             3       8.00000000         1.00000000
                        2             3       1.33333333         0.57735027
                        3             3       5.00000000         1.00000000
```

TABLE 9.14 Continued

```
Level of              ------------ TIME -----------
DAY          N              Mean            Std Dev

1            3        4.33333333         3.51188458
2            3        5.66666667         3.51188458
3            3        3.66666667         3.05505046

Level of              ------------ TIME -----------
CASE         N              Mean            Std Dev

1            3        4.66666667         4.04145188
2            3        4.66666667         3.21455025
3            3        4.33333333         3.21455025
```

Marginal means appear at the end of the output, with a common standard error for each level of each IV and the 95% confidence interval for the mean. The common standard error is found by dividing the square root of $MS_{residual}$ (labeled Error in the output) by the square root of the number of cases in each marginal mean.

SAS GLM syntax and output appear in Table 9.14. Discrete IVs are identified in the `class` instruction; the `model` instruction shows (in regression format) that the DV is a function of the three IVs. (The intercept is included by default.) Marginal means are requested in the `means` instruction.

SAS GLM shows two source tables. The first is for evaluating the `Model` with all effects combined and includes `Error` (residual) as well as `Total` (identified as `Corrected Total`). This table includes `R-Square`, indicating the proportion of variance predictable from the whole model. Also shown are the square root of the $MS_{residual}$ (`Root MSE`), the grand mean (`TIME Mean`), and the coefficient of variation (`Coeff Var`, the square root of the error mean square divided by the grand mean multiplied by 100).

The second source table shows the three effects, with the last column (`PR > F`) indicating the probability of mean differences of that size or larger resulting from chance variation. Marginal means and standard deviations are then provided. Section 5.6.5 discusses differences between `Type I SS` and `Type III SS` (identical here) for unequal *n*.

9.5 TYPES OF LATIN-SQUARE DESIGNS

The fundamental design in Section 9.4.1 consists of three IVs, each with the same number of levels, and a single score (replication) in each cell. There are advantages for both randomized-groups and repeated-measures designs when each cell contains more than one replication. Two advantages are that the designs generate more degrees of freedom for error, thus greater power, and that other effects become available for analysis.

TABLE 9.15 Allocation of cases in a randomized-groups Latin-square design with replications

	b_1	b_2	b_3
a_1	c_3: s_1	c_1: s_7	c_2: s_{13}
	s_2	s_8	s_{14}
a_2	c_2: s_3	c_3: s_9	c_1: s_{15}
	s_4	s_{10}	s_{16}
a_3	c_1: s_5	c_2: s_{11}	c_3: s_{17}
	s_6	s_{12}	s_{18}

9.5.1 Replicated Randomized-Groups Designs

In Section 9.4.1, each ABC combination is applied to a single case; however, the design is readily extended to replicates where each ABC combination is applied to more than one case. In the small-sample example of Section 9.4.1, for instance, there could be several skiers in each cell, as in Table 9.15 (where there are two skiers in each cell).

When there are replications, the design allows a test of the bits and pieces of interactions that happen to be included in the design. If there is no evidence of interaction, the assumption of additivity is supported, and an error term with more degrees of freedom is available that is likely to increase the power for testing main effects. There are five steps to the analysis:

1. Turn the design into a one-way design with a^2 levels (9 levels, in the example), and perform a one-way randomized-groups ANOVA to get SS_{bg} (between groups) and $MS_{S/ABC}$.
2. Analyze the data as a Latin square to get SS_A, SS_B, and SS_C (and $MS_{residual}$).
3. Compute $SS_{interactions}$ and $df_{interactions}$ where

$$SS_{interactions} = SS_{bg} - SS_A - SS_B - SS_C \tag{9.12}$$

$$df_{interactions} = df_{bg} - df_A - df_B - df_C \tag{9.13}$$

4. Test $MS_{interactions}$ against $MS_{S/ABC}$

$$F_{interactions} = \frac{MS_{interactions}}{MS_{S/ABC}} \qquad df = (a - 1)(a - 2), (N - a^2) \tag{9.14}$$

5. If the test is not significant, breathe a sigh of relief and then use the tests of MS_A, MS_B, and MS_C against $MS_{residual}$ ($= MS_{S/ABC+interactions}$), as obtained in Step 2. If the test is significant, rethink your study because the data violate the additivity assumption of analysis.

Figure 9.2 diagrams the analysis for the randomized-groups Latin-square design with replicates when there are no interactions. $MS_{residual}$ with replicates is usually smaller than that with no replicates because of increased degrees of freedom in the error term. This increases the power of the analysis.

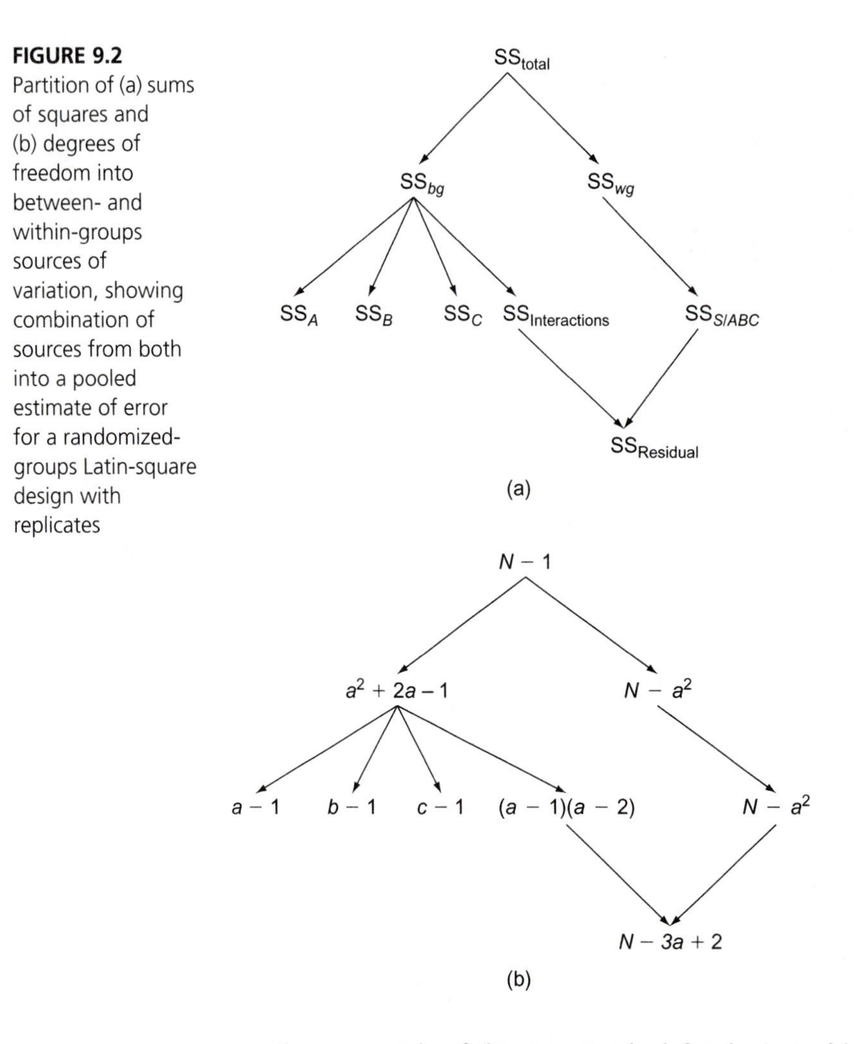

FIGURE 9.2 Partition of (a) sums of squares and (b) degrees of freedom into between- and within-groups sources of variation, showing combination of sources from both into a pooled estimate of error for a randomized-groups Latin-square design with replicates

As an example of the steps required for the test of interaction, consider an expansion of the data in the small-sample example of Table 9.4 to two skiers per cell, as shown in Table 9.16.

The corresponding data set appropriate for computer analysis is in Table 9.17. Notice that a column for group has been added to convert the analysis into a one-way design. Information about group membership can either be inserted into the data set itself (which is often easier) or added by transformation language.

The five steps in the analysis are then performed through SAS GLM:

1. Turn the design into a one-way design with a^2 levels (9 levels, in the example) and perform a one-way randomized-groups ANOVA to get SS_{bg} and $MS_{S/ABC}$, as shown in Table 9.18.
2. Analyze the data as a Latin square to get SS_A, SS_B, and SS_C (and $MS_{residual}$), as shown in Table 9.19.

TABLE 9.16 Data set for a randomized-groups Latin-square design with replications

	b_1	b_2	b_3
a_1	c_3: 4	c_1: 9	c_2: 1
	6	10	4
a_2	c_2: 1	c_3: 6	c_1: 7
	3	5	6
a_3	c_1: 8	c_2: 2	c_3: 5
	11	1	6

TABLE 9.17 Data set for computer analysis of a randomized-groups Latin-square design with replications

A	*B*	*C*	DV	Group
1	1	3	4	1
1	1	3	6	1
1	2	1	9	2
1	2	1	10	2
1	3	2	1	3
1	3	2	4	3
2	1	2	1	4
2	1	2	3	4
2	2	3	6	5
2	2	3	5	5
2	3	1	7	6
2	3	1	6	6
3	1	1	8	7
3	1	1	11	7
3	2	2	2	8
3	2	2	1	8
3	3	3	5	9
3	3	3	6	9

TABLE 9.18 Analysis of interaction in a one-way randomized-groups Latin-square design with replications (SAS GLM syntax and selected output)

```
proc glm data=SASUSER.LATGROUP;
   class GROUP;
   model DV = GROUP;
run;
```

```
Dependent Variable: DV   DV

                                     Sum of
      Source              DF        Squares    Mean Square    F Value    Pr > F

      Model                8    140.1111111     17.5138889      10.17    0.0011
      Error                9     15.5000000      1.7222222
      Corrected Total     17    155.6111111

      Source              DF    Type III SS    Mean Square    F Value    Pr > F

      GROUP                8    140.1111111     17.5138889      10.17    0.0011
```

TABLE 9.19 Analysis of main effects in a one-way randomized-groups Latin-square design with replications (SAS GLM syntax and selected output)

```
proc glm data=SASUSER.LATGROUP;
   class A B C;
   model DV = A B C;
run;
```

```
Dependent Variable: DV   DV
```

Source	DF	Sum of Squares	Mean Square	F Value	Pr > F
Model	6	132.0000000	22.0000000	10.25	0.0006
Error	11	23.6111111	2.1464646		
Corrected Total	17	155.6111111			

Source	DF	Type III SS	Mean Square	F Value	Pr > F
A	2	3.4444444	1.7222222	0.80	0.4729
B	2	1.7777778	0.8888889	0.41	0.6708
C	2	126.7777778	63.3888889	29.53	<.0001

3. Compute $SS_{\text{interactions}}$ and $df_{\text{interactions}}$, using Equations 9.12 and 9.13:

$$SS_{\text{interactions}} = 140.111 - 3.444 - 1.778 - 126.778 = 8.111$$

$$df_{\text{interactions}} = 8 - 2 - 2 - 2 = 2$$

4. Test $MS_{\text{interactions}}$ against $MS_{S/ABC}$, using Equation 9.14:

$$F_{\text{interactions}} = \frac{SS_{\text{interactions}}/df_{\text{interactions}}}{MS_{S/ABC}} = \frac{8.111/2}{1.772} = 2.289$$

5. With 2 and 9 df, the interaction is not significant, so breathe a sigh of relief and then use the tests of MS_A, MS_B, and MS_C against MS_{residual}, shown in Table 9.19 and obtained at Step 2. Note that SS_{residual} (23.611) from Table 9.19 is a sum of $SS_{\text{interactions}}$ (8.111) from Equation 9.11 and $SS_{S/ABC}$ (15.500) from Table 9.18.

If the test of interaction is statistically significant, however, the assumption of additivity is violated, and the study probably needs to be rerun as a complete factorial. One possible rescue for the study as conducted is to demonstrate that C is not confounded with the $A \times B$ interaction (Brown, Michels, and Winer, 1991). Do this by running an analysis in which C is ignored and the data are analyzed as a two-way full factorial $A \times B$ design (cf. Chapter 5). If the $A \times B$ interaction is nonsignificant, the original test of the main effect of C is then defended as unconfounded.

9.5.2 Replicated Repeated-Measures Designs

Replicates in a repeated-measures Latin-square design also increase the power of the analysis. There are at least three ways to replicate a repeated-measures Latin-square design: (1) add cases so that each sequence of *C*s in the same Latin square has more than one case, (2) add a randomized-groups variable with additional cases in the levels of that variable, or (3) use a crossover design. In all three situations, the study becomes a mixed repeated-measures, randomized-groups design, as described in greater detail in Chapter 7.

9.5.2.1 The Replicated Mixed Design

In the replicated mixed design, additional cases are measured for each order of the levels of *C*. The variable *A* becomes a randomized-groups variable with levels that represent the sequence or order in which the levels of *C* are administered. The elements that are part of the Latin square are *A*, *B*, and *C*. The number of levels of *A* is the same as levels of *B* and *C*, because there has to be a level of *A* for each sequence of *C*s. However, there can be any number of cases assigned to each sequence, as shown in Table 9.20, where there are three levels of *A*, *B*, and *C*, but two cases in each sequence.

This mixed-design analysis identifies treatment (*C*) and period (*B*) as repeated-measures effects, with order (*A*) as a randomized-groups effect. The treatment effect (*C*) is "carved out of" the $A \times B$ (period by order) interaction, just as in the design with no replications (Equation 9.3). As in the replicated randomized-groups design of Section 9.5.1, the "leftover" $A \times B$ interaction ($A \times B_{\text{adj}}$) is a test of additivity. There is no interaction of *A* with *C* in this design because there are no combinations of *A* and *C*. Rather, *A* is defined across the levels of *C*.

A data set appropriate for an analysis of this design appears in Table 9.21. There are six skiers randomly assigned to sequence—two receive c_2, c_1, c_3; two receive c_3, c_2, c_1; and two receive c_1, c_3, c_2.

TABLE 9.20 Allocation of cases in a repeated-measures Latin-square design with replications

			Periods: *B*		
			b_1	b_2	b_3
	a_1	s_1 s_2	c_2	c_1	c_3
Order or Sequence: *A*	a_2	s_3 s_4	c_3	c_2	c_1
	a_3	s_5 s_6	c_1	c_3	c_2

TABLE 9.21 Data set for replicated mixed design

			Days (Periods): B			
			Monday b_1	Wednesday b_2	Friday b_3	Sum
	a_1		c_2	c_1	c_3	
		s_1	8	4	5	17
		s_2	9	2	4	15
Order or Sequence: A	a_2		c_3	c_2	c_1	
		s_3	7	7	2	16
		s_4	5	9	4	18
	a_3		c_1	c_3	c_2	
		s_5	3	6	9	18
		s_6	3	5	8	16

There are two ways to organize the data for computer analysis of this design. One way is to enter the data for A and B in a traditional mixed-design format. Information regarding level of C is then made available through transformation language (e.g., if A is 1 and B is 1, then C is 2, etc.). The other way is to enter the data once for A and B and then again for A and C. This organization, which is often easier for a small data set, is shown in Table 9.22.

The easiest way to perform the analysis is to run two mixed-design ANOVAs, one for AB and the other for AC. Most of the sources of variability are obtained directly from the output, and one source is obtained by subtraction. The three steps for this analysis follow.

1. Run a standard A, B mixed-design ANOVA, as shown in Table 9.23.
2. Run a standard A, C mixed-design ANOVA, as shown in Table 9.24.
3. Combine the results into a source table, as shown in Table 9.25.

There are many things to notice about this source table. First, A and S/A from the Tests of Between-Subjects Effects are the same for either analysis. This is because these elements are computed from the sums for the cases, as shown in Table 9.21. These sums are the same whether obtained across the levels of B or across the levels

TABLE 9.22 Data set for computer analysis of a replicated mixed design

Case	A	b_1	b_2	b_3	c_1	c_2	c_3
1	1	8	4	5	4	8	5
2	1	9	2	4	2	9	4
3	2	7	7	2	2	7	7
4	2	5	9	4	4	9	5
5	3	3	6	9	3	9	6
6	3	3	5	8	3	8	5

TABLE 9.23 Analysis of *A*, *B* in a replicated mixed design (SPSS UNIANOVA syntax and output)

```
GLM
  B1 B2 B3 BY A
  /WSFACTOR = b 3 Polynomial
  /METHOD = SSTYPE(3)
  /CRITERIA = ALPHA(.05)
  /WSDESIGN = b
  /DESIGN = A .
```

Tests of Within-Subjects Effects

Measure: MEASURE_1

Source		Type III Sum of Squares	df	Mean Square	F	Sig.
b	Sphericity Assumed	.778	2	.389	.292	.757
	Greenhouse-Geisser	.778	1.046	.743	.292	.635
	Huynh-Feldt	.778	2.000	.389	.292	.757
	Lower-bound	.778	1.000	.778	.292	.627
b * A	Sphericity Assumed	87.222	4	21.806	16.354	.002
	Greenhouse-Geisser	87.222	2.093	41.680	16.354	.022
	Huynh-Feldt	87.222	4.000	21.806	16.354	.002
	Lower-bound	87.222	2.000	43.611	16.354	.024
Error(b)	Sphericity Assumed	8.000	6	1.333		
	Greenhouse-Geisser	8.000	3.139	2.549		
	Huynh-Feldt	8.000	6.000	1.333		
	Lower-bound	8.000	3.000	2.667		

Tests of Between-Subjects Effects

Measure: MEASURE_1

Transformed Variable: Average

Source	Type III Sum of Squares	df	Mean Square	F	Sig.
Intercept	555.556	1	555.556	833.333	.000
A	.444	2	.222	.333	.740
Error	2.000	3	.667		

of C, because the same numbers are involved, just in different orders. Second, the error term for the repeated-measures elements of the design, $B \times S/A$ or $C \times S/A$—called Error(B) or Error(C) in Steps 1 and 2, respectively—are also identical, because the numbers involved in the computations are also the same but for order. Finally, sum of squares (and df) for C is taken out of the $A \times B$ interaction from Step 1 to form $A \times B_{adj}$, as in the design with only one case per sequence. $A \times B_{adj}$ is a test of the assumption of additivity, as in the randomized-groups design with replications. In this example, there are no sequence (carryover) effects (A), no position effects (B), and no violation of the assumption of additivity ($A \times B_{adj}$). There is, however, a statistically significant difference due to treatment (C, snow condition in the example).

TABLE 9.24 Analysis of *A*, *C* in a replicated mixed design (SPSS UNIANOVA syntax and output)

```
GLM
  C1 C2 C3 BY A
  /WSFACTOR = c 3 Polynomial
  /METHOD = SSTYPE(3)
  /CRITERIA = ALPHA(.05)
  /WSDESIGN = c
  /DESIGN = A .
```

Tests of Within-Subjects Effects

Measure: MEASURE_1

Source		Type III Sum of Squares	df	Mean Square	F	Sig.
c	Sphericity Assumed	85.778	2	42.889	32.167	.001
	Greenhouse-Geisser	85.778	1.745	49.144	32.167	.001
	Huynh-Feldt	85.778	2.000	42.889	32.167	.001
	Lower-bound	85.778	1.000	85.778	32.167	.011
c * A	Sphericity Assumed	2.222	4	.556	.417	.792
	Greenhouse-Geisser	2.222	3.491	.637	.417	.772
	Huynh-Feldt	2.222	4.000	.556	.417	.792
	Lower-bound	2.222	2.000	1.111	.417	.692
Error(c)	Sphericity Assumed	8.000	6	1.333		
	Greenhouse-Geisser	8.000	5.236	1.528		
	Huynh-Feldt	8.000	6.000	1.333		
	Lower-bound	8.000	3.000	2.667		

Tests of Between-Subjects Effects

Measure: MEASURE_1

Transformed Variable: Average

Source	Type III Sum of Squares	df	Mean Square	F	Sig.
Intercept	555.556	1	555.556	833.333	.000
A	.444	2	.222	.333	.740
Error	2.000	3	.667		

TABLE 9.25 Source table for analysis of replicated mixed design

Source	SS		df	MS	F
A	0.444	>From either Step 1 or 2	2	0.222	0.333
S/A	2.000	>From either Step 1 or 2	3	0.667	
B	0.778	Step 1	2	0.389	0.292
C	85.778	Step 2	2	42.889	32.167
$A \times B_{adj}$	1.444	$(A \times B - C)$ $87.222 - 85.778 = 1.444$	$(4 - 2) = 2$	0.722	
$B \times S/A$	8.000	From either Step 1 or 2	6	1.333	
Total	98.444	From either Step 1 or 2	17		

9.5.2.2 The Repeated-Measures Design with a Randomized-Groups IV

The second way to add replications to a repeated-measures Latin square is to add a randomized-groups IV with its associated cases. The randomized-groups IV can represent level of treatment (treatment versus control), a blocking variable (such as gender, skill level, or type of disease), or an additional Latin-square arrangement. In this design, the randomized-groups IV can have any number of levels, but the number of cases at each level is the same as, or some multiple of, the number of levels involved in the Latin square.

The allocation of cases for a design of this type is shown in Table 9.26, in which the effects that are part of the Latin square (B, C, and S) each have three levels, but the randomized-groups IV (A) has only two. In this design, A is a regular randomized-groups IV that crosses both the levels of B and the levels of C. That is, all AB and AC combinations are present in the design. The BCS portion of the design at each level of A is identical to that of a repeated-measures Latin square with only one replicate per cell; that entire Latin-square arrangement is duplicated at each level of A.

The small-sample example is readily extended to a design of this type if the randomized-groups variable is defined as, say, gender. As shown in Table 9.27, three male and three female skiers ski on Monday, Wednesday, and Friday under the snow condition prescribed by C.

TABLE 9.26 Allocation of cases in a repeated-measures Latin square with a randomized-groups IV

		Trials		
		b_1	b_2	b_3
a_1	s_1	c_2	c_1	c_3
	s_2	c_3	c_2	c_1
	s_3	c_1	c_3	c_2
a_2	s_4	c_2	c_1	c_3
	s_5	c_3	c_2	c_1
	s_6	c_1	c_3	c_2

TABLE 9.27 Data set for a repeated-measures Latin square with a randomized-groups IV

			Days (Periods): B		
			Monday b_1	Wednesday b_2	Friday b_3
Gender: A	Men a_1	s_1	c_2: 8	c_1: 4	c_3: 5
		s_2	c_3: 4	c_2: 9	c_1: 2
		s_3	c_1: 3	c_3: 6	c_2: 9
	Women a_2	s_4	c_2: 7	c_1: 2	c_3: 7
		s_5	c_3: 5	c_2: 9	c_1: 4
		s_6	c_1: 3	c_3: 5	c_2: 8

The data set is organized the same as in Table 9.22, and the same three steps are used to complete this analysis:

1. Run a standard *A*, *B* mixed-design ANOVA, as shown in Table 9.28, through SPSS GLM.
2. Run a standard *A*, *C* mixed-design ANOVA, as shown in Table 9.29.
3. Combine the results into a source table, as shown in Table 9.30.

TABLE 9.28 Mixed-design analysis of *A*, *B* in a repeated-measures Latin square with a randomized-groups IV (SPSS GLM syntax and selected output)

```
GLM
 B1 B2 B3 BY A
 /WSFACTOR = b 3 Polynomial
 /METHOD = SSTYPE(3)
 /CRITERIA = ALPHA(.05)
 /WSDESIGN = b
 /DESIGN = A .
```

Tests of Within-Subjects Effects

Measure: MEASURE_1

Source		Type III Sum of Squares	df	Mean Square	F	Sig.
b	Sphericity Assumed	2.778	2	1.389	.123	.886
	Greenhouse-Geisser	2.778	1.861	1.493	.123	.873
	Huynh-Feldt	2.778	2.000	1.389	.123	.886
	Lower-bound	2.778	1.000	2.778	.123	.743
b * A	Sphericity Assumed	3.000	2	1.500	.133	.877
	Greenhouse-Geisser	3.000	1.861	1.612	.133	.864
	Huynh-Feldt	3.000	2.000	1.500	.133	.877
	Lower-bound	3.000	1.000	3.000	.133	.734
Error(b)	Sphericity Assumed	90.222	8	11.278		
	Greenhouse-Geisser	90.222	7.444	12.120		
	Huynh-Feldt	90.222	8.000	11.278		
	Lower-bound	90.222	4.000	22.556		

Tests of Between-Subjects Effects

Measure: MEASURE_1

Transformed Variable: Average

Source	Type III Sum of Squares	df	Mean Square	F	Sig.
Intercept	555.556	1	555.556	909.091	.000
A	.000	1	.000	.000	1.000
Error	2.444	4	.611		

TABLE 9.29 Mixed-design analysis of *A*, *C* in a repeated-measures Latin square with a randomized-groups IV (SPSS GLM syntax and selected output)

```
GLM
  C1 C2 C3 BY A
  /WSFACTOR = c 3 Polynomial
  /METHOD = SSTYPE(3)
  /CRITERIA = ALPHA(.05)
  /WSDESIGN = c
  /DESIGN = A .
```

Tests of Within-Subjects Effects

Measure: MEASURE_1

Source		Type III Sum of Squares	df	Mean Square	F	Sig.
c	Sphericity Assumed	85.778	2	42.889	38.600	.000
	Greenhouse-Geisser	85.778	1.629	52.646	38.600	.000
	Huynh-Feldt	85.778	2.000	42.889	38.600	.000
	Lower-bound	85.778	1.000	85.778	38.600	.003
c * A	Sphericity Assumed	1.333	2	.667	.600	.572
	Greenhouse-Geisser	1.333	1.629	.818	.600	.545
	Huynh-Feldt	1.333	2.000	.667	.600	.572
	Lower-bound	1.333	1.000	1.333	.600	.482
Error(c)	Sphericity Assumed	8.889	8	1.111		
	Greenhouse-Geisser	8.889	6.517	1.364		
	Huynh-Feldt	8.889	8.000	1.111		
	Lower-bound	8.889	4.000	2.222		

Tests of Between-Subjects Effects

Measure: MEASURE_1

Transformed Variable: Average

Source	Type III Sum of Squares	df	Mean Square	F	Sig.
Intercept	555.556	1	555.556	909.091	.000
A	.000	1	.000	.000	1.000
Error	2.444	4	.611		

TABLE 9.30 Source table for repeated-measures Latin square with a randomized-groups IV

Source	SS		df	MS	F
A	0.000	>From either Step 1 or 2	1	0.000	0
S/A	2.444	>From either Step 1 or 2	4	0.611	
B	2.778	>From Step 1	2	1.389	1.785
C	85.778	>From Step 2	2	42.889	55.127
$A \times B$	3.000	From Step 1	2	1.500	1.928
$A \times C$	1.333	>From Step 2	2	0.667	0.857
L.S. residual	3.111	$B \times S/A - C - A \times C =$ $90.222 - 85.778 - 1.333$	$8 - 2 - 2 = 4$	0.778	
Total	98.444		17		

The source table has many of the same features as that of the replicated mixed design (Table 9.25), with two exceptions. First, this design generates an $A \times C$ interaction because A is a true randomized-groups IV, and all AC combinations are present in the design. Second, C and $A \times C$ are obtained from $B \times S/A$, rather than from the $A \times B$ interaction, as in Table 9.25. We make a feeble attempt to explain why at the end of this section.

The randomized-groups IV, A, can be a blocking variable (such as gender or age), a manipulated IV (treatment vs. control), or different Latin squares. When it is different Latin squares, the design looks like that in Table 9.31, and it provides a test of potentially different carryover effects if they are of concern in the study. Note that there is a different arrangement of levels of C at a_1 and a_2.

If there is a second randomized-groups IV, D (another blocking variable, another manipulated IV, or another Latin square), then the randomized-groups part of the design has two IVs, the interaction between them, and an error term consisting of cases nested within combinations of levels of both IVs. Thus, the upper part of the source table is the same as any two-way randomized-groups ANOVA: A, D, $A \times D$, and S/AD. The repeated-measures part of the design has the repeated-measures IV of interest, C; the position effect, B; and all interactions with A, D, and $A \times D$. (Myers and Well [1991, pp. 360–363] show a numerical example for this design.)

TABLE 9.31 The randomized-groups IV represents different Latin squares

		Trials		
		b_1	b_2	b_3
a_1	s_1	c_2	c_1	c_3
	s_2	c_3	c_2	c_1
	s_3	c_1	c_3	c_2
a_2	s_4	c_1	c_2	c_3
	s_5	c_3	c_1	c_2
	s_6	c_2	c_3	c_1

The differences between these two designs (replicated mixed design and addition of a randomized-groups IV) are subtle. One difference is the presence of the $A \times C$ interaction. This source is not available in the replicated mixed design, although it is present when A is a genuine randomized-groups IV. In the replicated mixed design, there is no $A \times C$ effect because A does not cross C. For instance, there is no a_1c_1 combination, because A is defined by the order of *all* the levels of C.[6] On the other hand, when A is a regular randomized-groups IV, such as gender, all combinations of A and C are present, as are all combinations of A and B. That is, both interactions are real.

The second difference is what part of the AB analysis the Latin-square segment is derived from. Remember that C (and $A \times C$, if present) are "carved out of" the interaction of the other two elements of the Latin square. But the replicated mixed design and the repeated-measures Latin square with a randomized-groups IV have different elements associated with their Latin squares. In the replicated mixed design, those elements are A, B, and C. In the repeated-measures Latin square with a randomized-groups IV, those elements are B, C, and S. Thus, in the replicated mixed design, C is carved out of the $A \times B$ interaction, whereas in the repeated-measures Latin square, C and $A \times C$ are both carved out of $B \times S$ (nested in A).

How can you tell which elements are part of the Latin square? To answer this, ask yourself which other element has to have the same number of levels as C and B. In the replicated mixed design, there have to be as many levels of A as there are levels of C and B; however, there can be any number of cases at each level of A. In the repeated-measures Latin square with a randomized-groups IV, on the other hand, the Latin square is duplicated at each level of A, so the number of cases has to match the levels of B and C,[7] but there can be any number of levels of A. Once this pattern is apparent to you, the rest of the design can become endlessly complicated, whereas the IVs that are involved in the Latin square itself remain clear.

9.5.2.3 The Crossover Design

The crossover design is very popular, particularly in medical research. At minimum, there are two levels of A, B, and C, with C representing a new pharmaceutical agent tested against a placebo or another pharmaceutical agent. The design is called *crossover* because cases cross over from one level to the other, some going from treatment to control and others from control to treatment. If there are only two or three levels of C, all possible orders of treatment are often included (e.g., c_1, c_2 for some cases and c_2, c_1 for others); if there are numerous levels of C, the order of treatments is most often determined by a Latin square. In drug studies, there is usually a washout period between administrations of the levels of treatment to minimize carryover effects. The test of order (A) is a test of differences in carryover effects, should they be present. The design is popular enough, especially in biomedical research, and complex enough to have generated at least one book (Jones and Kenward, 2003) with an impressive list of references.

[6] Said another way, the Latin square is not complete until you have all the levels of A.

[7] Or be a multiple of those levels.

TABLE 9.32 Minimal crossover design

			B: Period	
			b_1	b_2
A: Order or Sequence	a_1	s_1	c_1	c_2
		s_2	c_1	c_2
		s_3	c_1	c_2
	a_2	s_4	c_2	c_1
		s_5	c_2	c_1
		s_6	c_2	c_1

A minimal crossover design is in Table 9.32. There are two orders (a_1 and a_2), where cases in the a_1 sequence get c_1 and then c_2, while cases in the a_2 sequence get c_2 first and then c_1. If c_1 is placebo and c_2 is a new drug, then cases in the a_1 group get the placebo first, while cases in the a_2 group get the new drug first; during the second period, both groups cross over to the other level of drug. Presumably cases are randomly assigned to either a_1 or a_2. There are also two levels of *B* (period). Notice that if you know both which order a case is in and which period it is, you also know whether a case receives the placebo or the new drug; that is, the $A \times B$ interaction is confounded with *C*. Data for this minimal crossover design are in Table 9.33.

The minimal crossover design can be more or less endlessly tweaked. For example, there can be several levels of *A*, each with different numbers of cases. There can also be multiple periods, although the number of periods must equal the number of levels of treatment. There may also be numerous trials at each period (instead of only one trial in the current example). Section 9.7.2 shows a full analysis of a crossover design with multiple trials in each period, a common complexity added in biostatistical research.

The data set for a minimal crossover design contains a column for order and two columns for the DV scores, one at the first period and one at the second. Each case is in a different row. The analysis for the design appears in Table 9.34.

There is no significant main effect of period (indicating that it does not matter whether the measurement is made first or second) and no significant main effect of

TABLE 9.33 Data set for the minimal crossover design

			B: Period	
			b_1	b_2
A: Order or Sequence	a_1	s_1	c_1: 5	c_2: 9
		s_2	2	6
		s_3	4	5
	a_2	s_4	c_2: 4	c_1: 3
		s_5	7	3
		s_6	5	5

TABLE 9.34 Analysis of a minimal crossover design (SPSS GLM syntax and selected output)

```
GLM
  PERIOD1 PERIOD2 BY ORDER
  /WSFACTOR = period 2 Polynomial
  /METHOD = SSTYPE(3)
  /EMMEANS = TABLES(ORDER*period)
  /PRINT = DESCRIPTIVE
  /CRITERIA = ALPHA(.05)
  /WSDESIGN = period
  /DESIGN = ORDER .
```

Tests of Within-Subjects Effects

Measure: MEASURE_1

Source		Type III Sum of Squares	df	Mean Square	F	Sig.
period	Sphericity Assumed	1.333	1	1.333	.727	.442
	Greenhouse-Geisser	1.333	1.000	1.333	.727	.442
	Huynh-Feldt	1.333	1.000	1.333	.727	.442
	Lower-bound	1.333	1.000	1.333	.727	.442
period * ORDER	Sphericity Assumed	16.333	1	16.333	8.909	.041
	Greenhouse-Geisser	16.333	1.000	16.333	8.909	.041
	Huynh-Feldt	16.333	1.000	16.333	8.909	.041
	Lower-bound	16.333	1.000	16.333	8.909	.041
Error(period)	Sphericity Assumed	7.333	4	1.833		
	Greenhouse-Geisser	7.333	4.000	1.833		
	Huynh-Feldt	7.333	4.000	1.833		
	Lower-bound	7.333	4.000	1.833		

Tests of Between-Subjects Effects

Measure: MEASURE_1

Transformed Variable: Average

Source	Type III Sum of Squares	df	Mean Square	F	Sig.
Intercept	280.333	1	280.333	84.100	.001
ORDER	1.333	1	1.333	.400	.561
Error	13.333	4	3.333		

ORDER (indicating that there are no carryover effects from one drug to the other), but there is a significant interaction of period with ORDER. In this context, the interaction is the test of the difference between the placebo and the new drug.

9.5.3 Less Commonly Encountered Designs

Latin-square designs can become almost endlessly complex. Additional Latin squares can be superimposed on the first, additional IVs (either randomized-groups or repeated-measures) can be added, levels of treatment or position can be randomly selected rather than fixed, and so on. Indeed, Brown, Michels, and Winer (1991) demonstrate 12 Latin square plans. A few are discussed here, with references to texts that provide greater detail.

9.5.3.1 Partition of Main Effects in a Latin-Square Design

If the number of levels is even, the major treatment variable in a Latin-square design, C, can be partitioned into two (or more) sources of variance. For example, four levels of C can be partitioned into two levels of C and two levels of D. In the basic Latin-square design of Table 9.1, for instance, c_1 becomes c_1d_1, c_2 becomes c_1d_2, c_3 becomes c_2d_1, and c_4 becomes c_2d_2. The design permits a test of the $C \times D$ interaction as well as of the main effects of C and D, instead of just a single main effect of C. Myers and Well (1991, pp. 356–357) demonstrate this design numerically for a repeated-measures Latin-square design.

Brown, Michel, and Winer (1991) suggest factorial partition of the A and B variables, as well, in a randomized-groups design. For example, the four levels of A are partitioned into a 2×2 arrangement, and the four levels of B are partitioned into another 2×2 arrangement.

9.5.3.2 Greco-Latin Squares

In another variant, a second Latin square is superimposed on the first to produce a Greco-Latin square, as shown in Table 9.35. There are two separate Latin squares, one for C and one for D. This permits isolation of yet another nuisance or treatment variable in either the randomized-groups or repeated-measures Latin-square design. Latin squares may then be superimposed on Greco-Latin squares, creating hyper-Greco–Latin squares. The number of IVs that may be superimposed, however, is limited by the degrees of freedom. In the example of Table 9.35, for example, there are no df available for the error term unless there are replications, because both C (2 df)

TABLE 9.35 Greco-Latin square

	b_1	b_2	b_3
a_1	c_2, d_1	c_1, d_2	c_3, d_3
a_2	c_3, d_3	c_2, d_1	c_1, d_2
a_3	c_1, d_2	c_3, d_3	c_2, d_1

and D (2 df) are subtracted from the $A \times B$ interaction (4 df). (Montgomery [1984] shows a numerical example of a $4 \times 4 \times 4 \times 4$ Greco-Latin square.)

Although the assumption of additivity extends to Greco-Latin squares, it may be untenable. As more and more IVs are added, the more numerous the potential interactions, and the less reasonable the assumption of no interaction.

9.5.3.3 Youden Square

Incomplete block designs are used when a complete Latin square is impractical, either because there are too many treatments or because it is not possible to expose all cases to all treatments. Several such designs are available; Davies (1963) discusses a wealth of incomplete block designs of all types. The one described here is the Youden square for the balanced incomplete block design. In effect, this is an incomplete Latin square in which one or more of the columns (or rows) is missing, so that it is no longer a square.

In a repeated-measures design, it is sometimes impossible to expose all cases to all treatments due to time or other constraints. Suppose, for example, that there are six treatments, but each case is only available for five of them. Or, in a randomized-groups design with a blocking IV, there may not be enough cases to go around; each block may contain enough cases for $c - 1$ treatments, instead of for all c. For instance, with the same six treatments, there may only be 30 (instead of 36) cases available, so that each of the six blocks has only five cases. This could happen, for example, if there are six treatments to be run, but each batch of material is only large enough to extract five samples. The incomplete blocks design is also applicable to a randomized-groups design when one of the nuisance variables, B, necessarily has one (or more) fewer levels than the other two variables, A and C.

Assignment of cases to treatment is done through the Youden square, in which there are fewer columns (positions) than rows (cases or blocks) of cases or treatments. Table 9.36 is an example of a Youden square with four treatments (C), four cases (A), but only three levels of B; thus, this is a $4 \times 3 \times 4$ design. Each treatment occurs once in each column in a Youden square, and each treatment is deleted once in each row. In a balanced incomplete block design, there is the additional constraint that each pair of treatments occurs together the same number of times as any other pair of treatments. For example, in Table 9.36, each pair of treatments occurs twice.

Davies (1963) provides a variety of Youden-square tables. As with Latin squares, the basic Youden square should be randomized (cf. Section 9.6.2). Total sum of

TABLE 9.36 Basic Youden-square design for four treatments

	b_1	b_2	b_3
a_1	c_1	c_2	c_3
a_2	c_2	c_4	c_1
a_3	c_3	c_1	c_4
a_4	c_4	c_3	c_2

squares and sum of squares for B are calculated in the usual way, but there are adjustments to sums of squares for A and for C. These adjustments are automatically provided when the syntax of Tables 9.13 or 9.14 is used.

9.5.3.4 Digram Balanced Latin Squares: Changeover Designs

When there are more than two levels of the IVs in a repeated-measures Latin square, a changeover analysis permits tests of residual effects of the prior treatment. The desire is to test for the presence of differential carryover when one of the levels of treatment has (relatively) long-lasting impact on the DV and, thus, contaminates responses to the level (or levels) of treatment that follows it. The trick is to analyze for the level of treatment that precedes the current one.

When there is an even number of levels of treatment, a digram balanced Latin square is used, where each treatment follows each other treatment the same number of times. When there is an odd number of levels of treatment, two Latin squares are required (cf. Table 9.31), where one is the mirror image of the other so that each treatment follows each other treatment the same number of times over both Latin squares. In either situation, replicates are highly desirable to generate sufficient df for error and, therefore, power (Cochran and Cox, 1957). A digram balanced Latin-square arrangement with four levels of A is shown in Table 9.37. Note that each level of C follows each other level of C only once.

Table 9.38 contains a data set using the digram-balanced Latin square of Table 9.37 with two replicates (two cases per sequence).

The analysis is most easily accomplished by entering data twice, once in randomized-groups format with cases as an IV (cf. Table 9.12) and then in mixed-design format for AB and AC (cf. Table 9.22). Table 9.39 is the data set of Table 9.38

TABLE 9.37 Digram balanced Latin square for a 4 × 4 × 4 design

	b_1	b_2	b_3	b_4
a_1	c_1	c_2	c_4	c_3
a_2	c_2	c_3	c_1	c_4
a_3	c_3	c_4	c_2	c_1
a_4	c_4	c_1	c_3	c_2

TABLE 9.38 Data set for digram balanced Latin-square design

		b_1	b_2	b_3	b_4
a_1		c_1	c_2	c_4	c_3
	s_1	7	9	7	6
	s_2	6	10	5	8
a_2		c_2	c_3	c_1	c_4
	s_3	5	7	9	11
	s_4	7	6	12	12
a_3		c_3	c_4	c_2	c_1
	s_5	3	6	7	10
	s_6	4	5	5	9
a_4		c_4	c_1	c_3	c_2
	s_7	5	7	8	8
	s_8	5	9	10	8

TABLE 9.39 Data set for computer analysis of a changeover design

Case	A	B	C	Prior	Score
1	1	1	1	0	7
2	1	1	1	0	6
1	1	2	2	1	9
2	1	2	2	1	10
1	1	3	4	2	7
2	1	3	4	2	5
1	1	4	3	4	6
2	1	4	3	4	8
3	2	1	2	0	5
4	2	1	2	0	7
3	2	2	3	2	7
4	2	2	3	2	6
3	2	3	1	3	9
4	2	3	1	3	12
3	2	4	4	1	11
4	2	4	4	1	12
5	3	1	3	1	3
6	3	1	3	1	4
5	3	2	4	3	6
6	3	2	4	3	5
5	3	3	2	4	7
6	3	3	2	4	5
5	3	4	1	2	10
6	3	4	1	2	9
7	4	1	4	0	5
8	4	1	4	0	5
7	4	2	1	4	7
8	4	2	1	4	9
7	4	3	3	1	8
8	4	3	3	1	10
7	4	4	2	3	8
8	4	4	2	3	8

rearranged into randomized-groups format. In addition to the usual columns for A, B, C, and the DV is a column (labeled prior) that codes for the level of C immediately preceding the current one. For treatments at b_1, there are no prior treatments, so prior is coded 0. For treatments at other levels of B, prior reflects the level of C before it. Thus, for the two scores at a_1b_2, prior is 1; for the two scores at a_1b_3, prior is 2; and so forth to the two scores at a_4b_4, in which prior is 3.

The analysis is accomplished in three steps:

1. Run a randomized-groups ANOVA for prior, as shown in Table 9.40, using SAS GLM.
2. Run two mixed-design ANOVAs, one for *AB* and one for *AC*, as shown in Table 9.41.
3. Combine the results into an ANOVA source table, as shown in Table 9.42.

The second step is the routine analysis of a mixed design in which *C*, the effects of the variable of major interest, is carved out of the $A \times B$ interaction. In the changeover design, differences due to the prior treatment are also subtracted from the $A \times B$ interaction. Those sums of squares are computed from a randomized-groups run (at Step 1), where Type I and Type III sums of squares for nonorthogonality are available by default. As you recall from the discussion of unequal n in Chapter 5,

TABLE 9.40 Randomized-groups analysis of prior in a changeover design (SAS GLM syntax and selected output)

```
proc glm data=SASUSER.DIGRAM_R;
  class a b c prior;
  model score = a b c prior / ss3;
run;
```

Dependent Variable: SCORE

Source	DF	Sum of Squares	Mean Square	F Value	Pr > F
Model	13	140.1071429	10.7774725	9.07	<.0001
Error	18	21.3928571	1.1884921		
Corrected Total	31	161.5000000			

R-Square	Coeff Var	Root MSE	SCORE Mean
0.867536	14.78210	1.090180	7.375000

Source	DF	Type I SS	Mean Square	F Value	Pr > F
A	3	25.25000000	8.41666667	7.08	0.0024
B	3	59.25000000	19.75000000	16.62	<.0001
C	3	19.75000000	6.58333333	5.54	0.0071
PRIOR	4	35.85714286	8.96428571	7.54	0.0009

Source	DF	Type III SS	Mean Square	F Value	Pr > F
A	3	9.12229437	3.04076479	2.56	0.0873
B	3	33.94408676	11.31469559	9.52	0.0005
C	3	31.55411255	10.51803752	8.85	0.0008
PRIOR	4	35.85714286	8.96428571	7.54	0.0009

TABLE 9.41 Mixed-design analyses of *AB* and *AC* for a changeover design (SAS GLM syntax and selected output)

```
proc glm data=SASUSER.DIGRAM_F;
   class a;
   model b1 b2 b3 b4 = a ;
   repeated b 4 ;
run;
```

The GLM Procedure

Repeated Measures Analysis of Variance

Tests of Hypotheses for Between Subjects Effects

Source	DF	Type III SS	Mean Square	F Value	Pr > F
A	3	25.25000000	8.41666667	5.39	0.0687
Error	4	6.25000000	1.56250000		

Univariate Tests of Hypotheses for Within Subject Effects

						Adj Pr > F	
Source	DF	Type III SS	Mean Square	F Value	Pr > F	G-G	H-F
b	3	59.25000000	19.75000000	17.24	0.0001	0.0008	0.0001
b*A	9	57.00000000	6.33333333	5.53	0.0039	0.0113	0.0039
Error(b)	12	13.75000000	1.14583333				

```
proc glm data=SASUSER.DIGRAM_F;
   class a;
   model c1 c2 c3 c4 = a ;
   repeated c 4 ;
run;
```

The GLM Procedure

Repeated Measures Analysis of Variance

Tests of Hypotheses for Between Subjects Effects

Source	DF	Type III SS	Mean Square	F Value	Pr > F
A	3	25.25000000	8.41666667	5.39	0.0687
Error	4	6.25000000	1.56250000		

Univariate Tests of Hypotheses for Within Subject Effects

						Adj Pr > F	
Source	DF	Type III SS	Mean Square	F Value	Pr > F	G-G	H-F
c	3	59.25000000	19.75000000	17.24	0.0001	0.0008	0.0001
c*A	9	57.00000000	6.33333333	5.53	0.0039	0.0113	0.0039
Error(c)	12	13.75000000	1.14583333				

TABLE 9.42 Source table for analysis of a changeover design

Source	SS		df	MS	F
A	25.250	>From Step 2 or Step1 Type I SS	3	8.417	5.388
S/A	6.250	>From Step 2	4	1.563	
B	59.250	>From Step 2	3	19.750	17.234
C	19.750	>From Step 2	3	6.583	5.822
Prior	35.857	>From Step 1	4	8.964	7.815
$A \times B$adj	1.393	$A \times B - C -$ Prior $57.000 - 19.750 - 35.857$	$9 - 3 - 4 = 2$	0.697	0.608
$B \times S/A$	13.750	>From Step 2	12	1.146	
Total	161.500	From either Step 1 or 2	31		

when sources of variability in a design are not orthogonal, they have overlapping variability. This variability is removed with Type III because overlapping variability is discarded. (This is why the Type III sums of squares for A, B, and C at Step 1 do not match those of the runs in Step 2.) In the changeover design, prior is not orthogonal to A, B, and C, and, further, it contains differences due to the effects of A, B, and C, as well as differences due to the prior treatment. Differences due to A, B, and C are removed by the Type III adjustment in Step 1, leaving just variability due to prior treatment. These differences are also removed in Type I SS, because `prior` is the last listed effect in the `model`. Sums of squares due to prior treatment are themselves analyzed and subtracted from $A \times B$ to complete the analysis. Sums of squares for A should not be adjusted for other effects. Thus, it is the Type I SS that are used (and that match the output for the other runs).

Another variant of the design permits tests of baseline effects when they are assessed between levels of treatment. Baseline measurements can be taken before the first treatment and/or between treatments. A baseline measurement taken before the first treatment is best handled as a covariate. Jones and Kenward (2003) discuss baseline measures taken between treatments.

9.6 SOME IMPORTANT ISSUES

Issues surrounding basic Latin-square design and analysis are the usual issues of power, effect size, and specific comparisons. Missing data are also a problem, because the balanced nature of the design is lost if some scores are lost. Tests of the assumption of sphericity are required in some repeated-measures applications, but the data set has to be arranged to produce the test. Then there are procedures for randomly generating the Latin squares themselves.

9.6.1 Sphericity in Repeated-Measures Designs

Repeated-measures ANOVA assumes that the variances of difference scores among all pairs of treatments are roughly equivalent (cf. Sections 6.3.2.3 and 6.6.2). The assumption is most often violated when treatments are confounded with time, because DV pairs measured close together in time are more likely to have similar difference scores than DV pairs measured far apart in time. Because treatments in repeated-measures Latin-square designs are counterbalanced over time, a major source of violation of sphericity is eliminated. However, another potential source of violation of sphericity for treatment is variability in the similarity of treatments, as discussed in Sections 6.3.2.3 and 6.6.2. This possible source of violation should be considered in any repeated-measures design, including Latin square. Sphericity is tested if it is possible that the assumption is violated.

Omnibus tests of position effects in repeated-measures Latin squares are expected to violate sphericity because they are necessarily confounded with time. Position effects are not of interest in many applications and are included in the analysis only to reduce error variance. If position effects are of interest, the assumption of sphericity can be circumvented by use of a trend analysis (cf. Sections 6.3.2.3 and 6.6.2), which, in any event, is often more appropriate than the omnibus test of position effects.

Should you wish to test sphericity, SAS and SPSS do it as part of any repeated-measures design. However, the data arrangement of Table 9.12 and the analyses in Section 9.4.5 do not produce the test. Instead, the data are reorganized into the repeated-measures format of Table 9.22, in which all data for each case are entered on a single line. You can enter the data several ways: in position or treatment order, with the other IV created by transformation instructions; in two separate data sets, one in each order; or with both orders in the same data set (as in Table 9.22). In any event, two repeated-measures analyses are then run, one for the position effect and one for the treatment effect, each with its own test of sphericity.

Table 9.43 shows the small-sample data of Table 9.4 reorganized into repeated-measures format in position order. A glance at the $A \times B$ matrix of Table 9.3 should help clarify the data file (recall that A is analogous to S in a repeated-measures Latin-square design).

Table 9.44 shows SPSS syntax and output for two repeated-measures analyses (one of position and one of treatment effects) with tests of sphericity for each. The IF statements create the treatment IV by defining level of treatment in terms of level of position. The remaining GLM instructions request two runs—the first for the effect of day (position) and the second for the effect of snow—in repeated-measures syntax

TABLE 9.43 Small-sample data in repeated-measures format for test of sphericity

CASE	b1=MONDAY	b2=WEDNESDAY	b3=FRIDAY
1	4	9	1
2	1	6	7
3	8	2	5

TABLE 9.44 Tests for sphericity in repeated-measures Latin-square analysis (SPSS syntax and selected output)

```
IF(case eq 1) b1=c2.
IF(case eq 1) b2=c3.
IF(case eq 1) b3=c1.
IF(case eq 2) b1=c3.
IF(case eq 2) b2=c1.
IF(case eq 2) b3=c2.
IF(case eq 3) b1=c1.
IF(case eq 3) b2=c2.
IF(case eq 3) b3=c3.

GLM
 b1 b2 b3
 /WSFACTOR = day 3 Polynomial
 /MEASURE = time
 /METHOD = SSTYPE(3)
 /EMMEANS = TABLES(day)
 /CRITERIA = ALPHA(.05)
 /WSDESIGN
 /DESIGN .

GLM
 c1 c2 c3
 /WSFACTOR = snow 3 Polynomial
 /MEASURE = time
 /METHOD = SSTYPE(3)
 /EMMEANS = TABLES(snow)
 /CRITERIA = ALPHA(.05)
 /WSDESIGN
 /DESIGN .
```

(a) First run

Mauchly's Test of Sphericity[b]

Measure: time

					Epsilon[a]		
Within Subjects Effect	Mauchly's W	Approx. Chi-Square	df	Sig.	Greenhouse-Geisser	Huynh-Feldt	Lower-bound
day	.948	.053	2	.974	.951	1.000	.500

Tests the null hypothesis that the error covariance matrix of the orthonormalized transformed dependent variables is proportional to an identity matrix.

a. May be used to adjust the degrees of freedom for the averaged tests of significance. Corrected tests are displayed in the Tests of Within-Subjects Effects table.

b. Design: Intercept
Within Subjects Design: day

TABLE 9.44 Continued

(b) Second run

Mauchly's Test of Sphericity[b]

Measure: time

Within Subjects Effect	Mauchly's W	Approx. Chi-Square	df	Sig.	Epsilon[a]		
					Greenhouse-Geisser	Huynh-Feldt	Lower-bound
snow	.808	.213	2	.899	.839	1.000	.500

Tests the null hypothesis that the error covariance matrix of the orthonormalized transformed dependent variables is proportional to an identity matrix.

a. May be used to adjust the degrees of freedom for the averaged tests of significance. Corrected tests are displayed in the Tests of Within-Subjects Effects table.

b. Design: Intercept
Within Subjects Design: snow

(see Table 6.34). Case is part of the error term for both runs: Case by day ($A \times B$) is the error term for the effect of position, and case by snow ($A \times C$) is the error term for the effect of treatment. The remaining syntax is produced by the SPSS menu system for General Linear Model, Repeated Measures.

The tests for sphericity appear as Mauchly's Test of Sphericity. The Sig. value of .948 for day indicates no violation of sphericity for the position effect (an unusual result if the data were real). By Mauchly's test, there is also no violation of sphericity for snow.

If sphericity is violated, adjusted df are required (e.g., Huynh-Feldt), or the effect is interpreted from single df comparisons, as discussed in Section 6.6.2. At this point, you have probably realized that a simple repeated-measures Latin-square analysis can also be run either by the procedures of Section 9.4.5 or as two repeated-measures analyses with computation of the appropriate error term[8] and automatic tests for sphericity.

9.6.2 Choosing a Latin Square

Statistical procedures for Latin-square analysis assume that the Latin square has been randomly chosen from a population of Latin squares. The number of squares in the population depends on the number of levels of C. The higher the number of levels, the larger the population of Latin squares that can be generated. A $2 \times 2 \times 2$ arrangement has only two possible Latin squares (c_1 first or c_2 first), a $3 \times 3 \times 3$ arrangement has 12 possible squares, a $4 \times 4 \times 4$ arrangement has 576 possible squares,[9] and so on. Therefore, generating all possible squares and randomly choosing among them rapidly becomes unwieldy. Some helpful shortcuts are available that start with a standard square and then permute rows, columns, and (possibly) levels of A.

[8] Both AB and AC analyses are run. $SS_{residual} = SS_{AB} - SS_C = SS_{AC} - SS_B$; similarly for df.

[9] There are a $(c!)(c-1)!$ possible Latin squares for a $4 \times 4 \times 4$ design.

TABLE 9.45 Simplest 4 × 4 × 4 Latin square

	b_1	b_2	b_3	b_4
a_1	c_1	c_2	c_3	c_4
a_2	c_2	c_3	c_4	c_1
a_3	c_3	c_4	c_1	c_2
a_4	c_4	c_1	c_2	c_3

A standard square is one in which both the first row and the first column contain levels of C in ordinal sequence, for example, c_1, c_2, c_3. The remaining elements can be in a variety of orders, but in the simplest (most standard?) one, levels of C continue in ordinal sequence and wrap around. For example, the simplest $4 \times 4 \times 4$ Latin square is shown in Table 9.45.

This simplest standard square is highly undesirable, because far too often, levels of C with higher numbers follow levels with lower numbers. Therefore, rearrangement of rows and columns is needed. Several texts provide (more or less) detailed procedures for generating Latin squares by starting with a standard square, and then randomly permuting the rows, and then randomly permuting the columns, and then, sometimes, randomly redefining the levels of C. Myers and Well (1991) provide procedures that are easily followed. Brown, Michel, and Winer (1991) show how to generate orthogonal Latin squares, balanced sets of Latin squares, composite squares, and Greco-Latin squares.

The easiest way, by far, to generate random Latin squares is through appropriate software. SAS PLAN, SYSTAT DESIGN, and NCSS Design of Experiments have that capability. Table 9.46 shows SAS PLAN and TABULATE syntax and output for randomly generating a $5 \times 5 \times 5$ Latin square. The `output` instruction requests that 25 runs (cells) be saved to a file named `WORK.g`. The `tabulate` procedure requests that the Latin square itself be printed as output.

In the output, levels of C are designated by numbers inside the square. For example, the first level of B and the first level of A is associated with c_2, the first level of B and the third level of A is associated with c_1, and so on. If this is a repeated-measures Latin square, where case is represented by A and position by B, the first case receives levels of C in c_2, c_5, c_4, c_1, c_3 order.

Table 9.47 shows the output file saved by the instructions in Table 9.46. This file shows the random order of runs, where the first run receives the a_3, b_4, c_5 combination and the last run receives the a_1, b_1, c_2 combination. If a column is added for the DV, the 25 treatment combinations are already appropriately organized as a data file for either a randomized-groups or repeated-measures Latin-square analysis, as shown in Section 9.4.5.

9.6.3 Power, Effect Size, and Relative Efficiency

A basic strategy for increasing power is to decrease the size of the error term in the F ratio. Repeated-measures Latin square analyzes the sum of squares for position effects (B) and subtracts them (and their df) from the error term ($A \times S$) that otherwise would be used for a one-way repeated-measures analysis. Although you can use

TABLE 9.46 Generating a 5 × 5 × 5 Latin-square design through SAS PLAN (syntax and design output)

```
proc plan seed = 37430;
  factors A=5 ordered B=5 ordered / noprint;
  treatments C=5 cyclic;
  output out=g
      A cvals=('a1' 'a2' 'a3' 'a4' 'a5')random
      B cvals=('b1' 'b2' 'b3' 'b4' 'b5')random
      C nvals=(1 2 3 4 5)random;
quit;
proc tabulate;
   class A B;
   var C;
   table A, B*(C*f=6.) /rts=10;
run;
```

	B				
	b1	b2	b3	b4	b5
	C	C	C	C	C
	Sum	Sum	Sum	Sum	Sum
A					
a1	2	5	4	1	3
a2	4	1	3	2	5
a3	1	3	2	5	4
a4	3	2	5	4	1
a5	5	4	1	3	2

TABLE 9.47 Output file produced by SAS PLAN syntax in Table 9.46

WORK.G

3 / 25	Nom A	Nom B	Int C
1	a3	b4	5
2	a3	b2	3
3	a3	b5	4
4	a3	b3	2
5	a3	b1	1
6	a5	b4	3
7	a5	b2	4
8	a5	b5	2
9	a5	b3	1
10	a5	b1	5
11	a4	b4	4
12	a4	b2	2
13	a4	b5	1
14	a4	b3	5
15	a4	b1	3
16	a2	b4	2
17	a2	b2	1
18	a2	b5	5
19	a2	b3	3
20	a2	b1	4
21	a1	b4	1
22	a1	b2	5
23	a1	b5	3
24	a1	b3	4
25	a1	b1	2

a Latin-square arrangement to control position effects in a research study without analyzing them, this is likely to be unwise. If position effects are substantial, removing them from the error term enhances power. If position effects are neutral (neither larger nor smaller than their corresponding df for the error term as a whole), power is unaffected. Power is lower only when position effects are smaller than random error variability, because, in this case, you subtract a very small sum of squares for position for each degree of freedom you lose from the error term. In randomized-groups Latin square, power is enhanced by the inclusion (and subtraction from error variance) of nuisance variables (A and B).

The procedures described in Section 5.6.3 for calculating power and determining the number of cases apply here. The randomized-groups design is treated as a two-way layout, and sample size estimations are typically calculated only for treatment effects. However, estimates can be calculated for other effects as well, and the result used that calls for the largest sample size. Power for a repeated-measures design is calculated using the procedures of Section 6.6.3. Remember that the number of cases must be some multiple of the number of levels of C for a repeated-measures Latin-square design.

Effect size is reported either as η^2 or partial η^2, as described in Section 5.6.2 for randomized-groups and Section 6.6.4 for repeated-measures. Partial η^2 (Equation 5.15) as an estimate of effect size for snow condition (C) in the small-sample example is

$$\text{Partial } \eta^2 = \frac{SS_C}{SS_C + SS_{\text{residual}}} = \frac{66.89}{66.89 + 0.88} = .99$$

Smithson's (2003) software may be used to find effect sizes and their confidence limits, as previously demonstrated.

Relative efficiency may be calculated for an unreplicated Latin-square design relative to either the randomized-groups design or the repeated-measures design (Kirk, 1995). Efficiency relative to the randomized-groups designs takes into account the nuisance variables, A and B, which do not occur in the one-way randomized-groups design. Thus, the error term in the Latin-square design should be reduced relative to a one-way design. However, the Latin-square design also generates fewer degrees of freedom; both differences are taken into account in the following equation:

$$\text{Relative efficiency} = \frac{MS_{S/A}}{MS_{\text{residual}}}\left(\frac{df_{\text{residual}} + 1}{df_{\text{residual}} + 3}\right)\left(\frac{df_{S/A} + 3}{df_{S/A} + 1}\right) \qquad \textbf{(9.15)}$$

The error term for the one-way randomized-groups design ($MS_{S/A}$) is found from a Latin-square data set by analyzing the main effect of C, completely ignoring the existence of A and B, and then performing the following calculation.

$$MS_{S/A} = \frac{SS_T - SS_C}{N - c} \qquad \textbf{(9.16)}$$

with $df_{S/A} = N - c$. For the small-sample example of Table 9.7,

$$MS_{S/A} = \frac{71.56 - 66.89}{9 - 3} = 0.78$$

with $df_{S/A} = 6$, and

$$\text{Relative efficiency} = \frac{0.78}{0.44}\left(\frac{2+1}{2+3}\right)\left(\frac{6+3}{6+1}\right) = 1.37$$

Thus, error variance would be more than 30% larger if the analysis had been done on a one-way randomized-groups design, with the nine cases randomly assigned to the three levels of A.

Efficiency with respect to a one-way repeated-measures design takes into account position effects, B. Recall that the repeated-measures design already takes into account A (cases), usually labeled S in a repeated-measures design. Thus, efficiency relative to a repeated-measures analysis is

$$\text{Relative efficiency} = \frac{MS_{A \times S}}{MS_{\text{residual}}}\left(\frac{df_{\text{residual}}+1}{df_{\text{residual}}+3}\right)\left(\frac{df_{A \times S}+3}{df_{A \times S}+1}\right) \quad \textbf{(9.17)}$$

where MS_{AS} is found by running the analysis as a one-way repeated-measures design (i.e., ignoring B) or by Equation 9.18 if the data from a Latin square are used:

$$MS_{A \times S} = \frac{SS_T - SS_C - SS_A}{(c-1)^2} \quad \textbf{(9.18)}$$

with $df_{A \times S} = (c-1)^2$.

Treating the data from Table 9.7 of Section 9.4.3 as a repeated-measures design,

$$MS_{A \times S} = \frac{71.56 - 66.89 - 0.23}{2(2)} = 1.11$$

with $df_{AS} = 4$, and

$$\text{Relative efficiency} = \frac{1.11}{0.44}\left(\frac{2+1}{2+3}\right)\left(\frac{4+3}{4+1}\right) = 2.12$$

Error variance is more than twice as large when the effect of position is ignored, indicating the advantage of including position effects in the analysis when the Latin-square design is implemented in the study.

9.6.4 Specific Comparisons

Procedures for comparisons differ for randomized-groups and repeated-measures Latin-square designs, just as they do for the ordinary randomized-groups and repeated-measures designs. The difference, as usual, is in the error terms for comparisons. The various forms of effect size also apply to Latin-square analyses.

9.6.4.1 Randomized-Groups Designs

Tests for specific comparisons are straightforward in randomized-groups Latin-square designs. A single error term is used for all comparisons: residual error if there are no replicates or the error term used in Section 9.5.1 if there are replicates. Procedures of Section 4.5.5 are followed for comparisons on marginal means of A, B, or C.

Because these comparisons are usually post hoc, appropriate adjustments are made to control the error rate.

Say, for example, that after performing the small-sample randomized-groups Latin-square analysis, we are interested in the pairwise comparison between packed powder and icy conditions. The weighting coefficients we use for levels of C are 0 −1 1; the corresponding means are 8.00, 1.33, and 5.00. The sample size for each of these means, n, is 3. Following Equation 4.10:

$$\begin{aligned} F &= \frac{n(\Sigma w_j \overline{Y}_j)^2 / \Sigma w_j^2}{\text{MS}_{\text{residual}}} \\ &= \frac{3((0)(8.00) + (-1)(1.33) + (1)(5.00))^2/(0^2 + (-1)^2 + 1^2)}{0.44} \\ &= \frac{40.41/2}{0.44} = 45.92 \end{aligned}$$

(There is a lot of rounding error in this computation; computer output gives an F value of 45.38.)

Following Equation 4.15, Tukey-adjusted critical F for this pairwise comparison is

$$F_T = \frac{q_T^2}{2} = \frac{(8.3)^2}{2} = 34.45$$

where Table A.3 is used to find q_T with three means and 2 df for error. Obtained F is larger than critical F_T, so that this test shows a reliable difference in ski time between packed powder and ice.

Partial η^2 for this effect is formed by dividing the SS for the comparison by the sum of the SS for the comparison and the error term, residual.

$$\text{Partial } \eta^2 = \frac{\text{SS}_{\text{comp}}}{\text{SS}_{\text{comp}} + \text{SS}_{\text{residual}}} = \frac{40.41}{40.41 + 0.88} = .98$$

Such is the power of made-up data.

9.6.4.2 Repeated-Measures Designs

Specific comparisons in a repeated-measures Latin-square analysis are more complicated because each comparison requires a different error term, as in any repeated-measures analysis. In a Latin-square analysis, this is most easily accomplished by subtracting position effects from each score and then treating the adjusted data as a simple one-way repeated-measures design.

The position effect is found by subtracting the grand mean from the mean of B for each level. For the small-sample example, GM = 4.78, and the three position (DAY) totals are 13, 17, and 13, with means of 4.33, 5.67, and 4.33, respectively.

TABLE 9.48 Subtraction of position effects from 3 × 3 × 3 small-sample repeated-measures Latin-square example

	b_1	b_2	b_3
a_1	$c_3 = 4 - (-0.45) = 4.45$	$c_1 = 9 - 0.89 = 8.11$	$c_2 = 1 - (-0.45) = 1.45$
a_2	$c_2 = 1 - (-0.45) = 1.45$	$c_3 = 6 - 0.89 = 5.11$	$c_1 = 7 - (-0.45) = 7.45$
a_3	$c_1 = 8 - (-0.45) = 8.45$	$c_2 = 2 - 0.89 = 1.11$	$c_3 = 5 - (-0.45) = 5.45$

TABLE 9.49 Adjusted scores in treatment by cases matrix for 3 × 3 × 3 small-sample repeated-measures Latin-square example

	c_1	c_2	c_3
a_1	8.11	1.45	4.45
a_2	7.45	1.45	5.11
a_3	8.45	1.11	5.45

Position effects are found by subtracting GM from the mean for each position:

$$b_1: \quad 4.33 - 4.78 = -0.45,$$

$$b_2: \quad 5.67 - 4.78 = 0.89, \text{ and}$$

$$b_3: \quad 4.33 - 4.78 = -0.45.$$

Thus, the effect of the first position, b_1, and the third position, b_3, is to lower scores by .45, whereas the effect of the second position, b_2, is to raise scores by .89. The nine scores in the $A \times B$ Latin-square arrangement of Table 9.4 are adjusted for these position effects, as shown in Table 9.48.

Adjusted scores are then rearranged into treatment (C) order, as in the $A \times C$ arrangement of Table 9.3, as shown in Table 9.49.

This is now treated as the raw data matrix for a standard one-way repeated-measures analysis, with the exception that the degrees of freedom for error, $(a - 1)(c - 1)$, are reduced by 1, lost when position effects are estimated. Section 6.6.6.1.1 shows how to do specific comparisons for a one-way repeated-measures analysis. Effect size for repeated-measures designs is discussed in Section 6.6.4.

9.6.5 Missing Data

Davies (1963) provides a method for estimated missing data in an unreplicated Latin-square analysis that takes into account levels of all three IVs in estimating the missing value. The equation for a missing value, Y^*_{ijk}, is

$$Y^*_{ijk} = \frac{a(A'_i + B'_j + C'_k) - 2T'}{(c-1)(c-2)} \tag{9.19}$$

where c = number of levels of C (or A or B),

A'_i = sum of the known values for that level of A,

B'_j = sum of the known values for that level of B,

C'_k = sum of the known values for that level of C, and

T' = sum of all known values.

Say, for example, that the final value in Table 9.4 is missing; that is, $a_3b_3c_3$ (actually a 5 in the table) is unknown. The known values for the third level of C, 4 and 6, sum to 10 (see the diagonal in Table 9.4). The known values for the third level of B (the third column, 1 and 7) sum to 8, and the known values for the third level of A (the third row, 8 and 2) sum to 10. The sum of all the known values is $4 + 9 + 1 + 1 + 6 + 7 + 8 + 2 = 38$. There are three levels of each IV, so that $a = 3$, and

$$Y^*_{ijk} = \frac{3(10 + 8 + 10) - 2(38)}{(3 - 1)(3 - 2)} = \frac{8}{2} = 4$$

This value is inserted into the data, and a normal analysis follows, except that 1 df is subtracted from the error term to compensate for estimating the missing value.

Things get more complicated if there are several missing values. In a procedure described by Kirk (1995), all except one of the missing values is "guesstimated" (a reasonable guess based on other values in the data set). Then the remaining missing value is estimated using Equation 9.19. The estimated value is entered into the data set; Equation 9.19 is then applied to one of the "guesstimated" values, and so on, until the values stabilize. One degree of freedom for error is deducted for imputing each missing value.

A more complex procedure, described by Kempthorne (1952), circumvents the slight positive bias produced by the procedure just described. Kirk (1995) illustrates this procedure.

9.7 COMPLETE EXAMPLES OF LATIN-SQUARE ANOVA

Section 9.7.1 provides an example of a randomized-groups Latin-square analysis and Section 9.7.2 provides an example of a repeated-measures Latin-square analysis, each with different data sets.

9.7.1 Complete Example of 4 × 4 × 4 Randomized-Groups Latin-Square Analysis

Davies (1963) provides an industrial example in which there are four rubber-covered fabric samples evaluated in four positions of a wear tester. The third factor is the

TABLE 9.50 Data for complete example of 4 × 4 × 4 randomized-groups Latin-square ANOVA

	b_1	b_2	b_3	b_4
a_1	$c_2 = 218$	$c_4 = 236$	$c_1 = 268$	$c_3 = 235$
a_2	$c_4 = 227$	$c_2 = 241$	$c_3 = 229$	$c_1 = 251$
a_3	$c_1 = 274$	$c_3 = 273$	$c_2 = 226$	$c_4 = 234$
a_4	$c_3 = 230$	$c_1 = 270$	$c_4 = 225$	$c_2 = 195$

run number, because all four positions of the wear tester are run at the same time. Data sets are FABRIC.*. The goal is to evaluate differences in loss of weight (greater loss = more wear) in fabric samples (C) after a given number of cycles of abrasion, with adjustments for position (B) and run number (A) being the nuisance variables.

9.7.1.1 Evaluation of Assumptions

The design is fully experimental, in that fabric samples are randomly assigned to positions within runs, and the Latin-square arrangement itself is randomly selected. Table 9.50 shows the Latin-square design and data matrix in $A \times B$ (position by run) format.

Some practical issues are difficult to assess because of the small sample sizes. Nevertheless, descriptive statistics are useful to evaluate distributions for each main effect.

9.7.1.1.1 Sample Sizes, Normality, Independence of Errors, and Homogeneity of Variance Independence of errors is ensured by the nature of the Latin-square design. Although all four positions are run together, dependence due to runs is eliminated from the error term by the nature of the analysis: Error is the residual variance after effects of position and run are subtracted. There are no missing data.

There are no replications within the 16 cells, so that $n = 1$ in this design. With sample sizes this small for tests of each effect, normality of sampling distributions is questionable. However, a standard measure, such as loss of weight in fabric samples, is unlikely to produce peculiar distributions in a population of trials.

Table 9.51 shows the descriptive statistics provided by the major SAS GLM run for the three main effects. Descriptive statistics (means and standard deviations) are requested by the `means` instruction. The ss3 instruction limits the output to Type III sums of squares, which are identical to Type I sums of squares in this balanced design. The `output` instruction is a request for residuals, which are analyzed in Section 9.7.1.1.3. In the second `means` instruction, a Tukey HSD test is requested for later analysis, in the form of a 99% confidence interval.

A glance at the standard deviations, particularly for levels of `FABRIC`, indicates serious violation of homogeneity of variance. With the largest variance

TABLE 9.51 Descriptive statistics from SAS GLM Latin-square analysis (syntax for full analysis and selected output)

```
proc    glm data = SASUSER.FABRIC;
        class CASE POSITION FABRIC;
        model LOSS = CASE POSITION FABRIC / ss3;
        means CASE POSITION FABRIC;
        means FABRIC / TUKEY CLDIFF ALPHA=.01;
        output  OUT=RESID P=PREDICT STUDENT=STDRESID;
run;
```

General Linear Models Procedure

Level of CASE	N	LOSS Mean	LOSS SD
1	4	239.250000	20.8706333
2	4	237.000000	11.1952371
3	4	251.750000	25.3294953
4	4	230.000000	30.8220700

Level of POSITION	N	LOSS Mean	LOSS SD
1	4	237.250000	25.0249875
2	4	255.000000	19.2006944
3	4	237.000000	20.7364414
4	4	228.750000	23.8100119

Level of FABRIC	N	LOSS Mean	LOSS SD
1	4	265.750000	10.1447852
2	4	220.000000	19.2006944
3	4	241.750000	20.9980158
4	4	230.500000	5.3229065

$s^2 = (21.00)^2 = 441.00$ and the smallest variance $(5.32)^2 = 28.30$,

$$F_{max} = \frac{s^2_{largest}}{s^2_{smallest}} = \frac{441.00}{28.30} = 15.58$$

which is well in excess of the recommended ratio of 10 or less. Therefore, a variance-stabilizing transformation of weight loss is considered.

Analyses with log, square root, and inverse transforms of LOSS are evaluated, but none of them produces acceptable homogeneity of variance. Therefore, the decision is made to retain the scores in their original form but to use a more stringent $\alpha = .025$ for the test of fabric. The reliably smaller standard deviation of fabric 4 is also of substantive research interest and is discussed in the section interpreting the major analysis.

9.7.1.1.2 Outliers Outliers cannot be evaluated separately for each cell because there is only one score in each combination of A, B, and C. Therefore, they are evaluated with respect to marginal means. So, for example, the first score in the data matrix, $a_1b_1c_2 = 218$, is evaluated against the first mean of A, the first mean of B, and the second mean of C. By comparing with the second mean of C,

$$z = \frac{218 - 220.00}{19.20} = -0.10$$

Clearly this is not an outlier. A check of the remaining values reveals no outliers in the data set.

9.7.1.1.3 Additivity Additivity cannot be fully assessed in the absence of replications in each cell (cf. Section 9.5.1). Use of the residual error term in the analysis, therefore, contributes to loss of power if the assumption of additivity is violated. The interaction most destructive to interpretability is between position of abrading plates (B) and run (A), because this interaction is confounded with FABRIC. CASE (run number) represents unmeasured variations in temperature, humidity, and the like. Thus, there would be difficulty if some positions were more susceptible to environmental variation than others (e.g., if higher temperatures produced more wear in position 1 and lower temperatures produced more wear in position 2). This seems unlikely enough that threats to power are not considered consequential in this research.

As a partial test of additivity, SAS PLOT is used to find standardized residuals and predicted values and to plot them from the data set created in the syntax of Table 9.51. Figure 9.3 shows the syntax and output plot for this analysis. A linear relationship, either positive or negative, would indicate a nonadditivity problem. The scatter plot shows no indication of a systematic relationship between predicted values and residuals, supporting the assumption of additivity.

9.7.1.2 Analysis of 4 × 4 × 4 Randomized Latin-Square Design

Table 9.52 shows the omnibus tests by SAS GLM for the major analysis (syntax is in Table 9.51).

All three effects are statistically significant, using $\alpha = .05$ for CASE and POSITION and $\alpha = .025$ for FABRIC. CASE is probably not of interest, reflecting only random fluctuation in environment conditions. Indeed, a look at the means in Table 9.51 over the four trials reveals no simple pattern. Differences among the four positions are known and presumably not of special interest in this research. Thus, CASE and POSITION are simply nuisance variables that might obscure differences among fabrics if not considered, but their significant variance attests to the importance of taking them out of the error term.

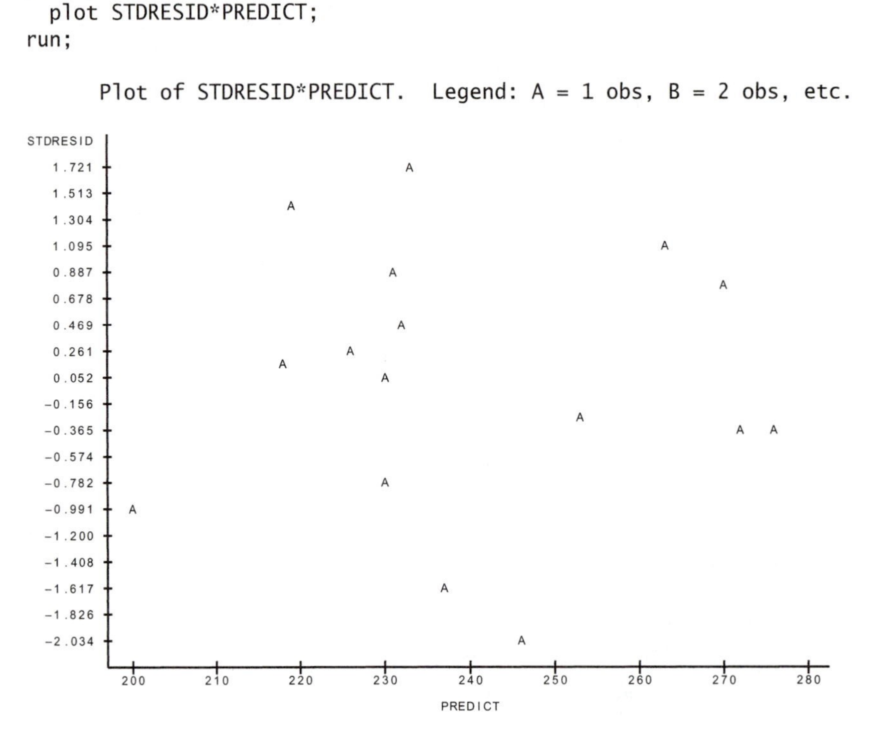

FIGURE 9.3 Standardized residuals as a function of predicted values (syntax and selected output from SAS PLOT)

Differences among the four fabrics are highly statistically significant and considerable. Calculating partial η^2 as per Section 9.6.3,

$$\text{Partial } \eta^2 = \frac{SS_{FABRIC}}{SS_{FABRIC} + SS_{residual}} = \frac{4621.5}{4621.5 + 367.50} = .93$$

Using Smithson's (2003) procedure, confidence limits are found to range from .51 to .95.

A Tukey test is applied to pairwise differences among the four fabrics, as described in Section 4.5.5.5.2. In Table 9.51, the test is requested by the `TUKEY CLDIFF ALPHA=.01` instruction. (`CLDIFF` requests that the results of the Tukey test be presented as confidence intervals; alpha is set to .01 to compensate for heterogeneity of variance, more serious for comparisons than for omnibus tests.) Table 9.53 shows the results of the pairwise comparisons for `FABRIC` only—the results for `CASE` and `POSITION` are also provided (but not shown), but are not of current research interest.

Fabric 1 is reliably different from fabric 2 and fabric 4. There are no reliable differences among the remaining three fabrics: 2, 3, and 4. However, fabric 4 also shows the least variability in wear over the trials and positions (cf. Table 9.51).

TABLE 9.52 Omnibus tests and selected output from SAS GLM Latin-square analysis (syntax is in Table 9.51)

```
                           The GLM Procedure

                        Class Level Information

                         Class         Levels    Values

                         CASE               4    1 2 3 4
                         POSITION           4    1 2 3 4
                         FABRIC             4    1 2 3 4


                      Number of Observations Read          16
                      Number of Observations Used          16


                           The GLM Procedure

Dependent Variable: LOSS

                                         Sum of
    Source                      DF      Squares     Mean Square    F Value    Pr > F

    Model                        9   7076.500000     786.277778      12.84    0.0028
    Error                        6    367.500000      61.250000
    Corrected Total             15   7444.000000


                  R-Square     Coeff Var      Root MSE     LOSS Mean

                  0.950631      3.267740      7.826238      239.5000


    Source                      DF   Type III SS    Mean Square    F Value    Pr > F

    CASE                         3    986.500000     328.833333       5.37    0.0390
    POSITION                     3   1468.500000     489.500000       7.99    0.0162
    FABRIC                       3   4621.500000    1540.500000      25.15    0.0008
```

Means and standard errors for the four fabrics are plotted in Figure 9.4, as produced by SAS GLM and GPLOT. Syntax is generated by the SAS Analyst menu system (syntax for CASE and POSITION plots and for restoring settings is omitted).

Table 9.54 is a checklist for randomized-groups Latin-square ANOVA. An example of a results section, in journal format, follows for the study just described.

TABLE 9.53 Pairwise comparison and selected output from SAS GLM Latin-square analysis (syntax is in Table 9.51)

```
              Tukey's Studentized Range (HSD) Test for LOSS

    NOTE: This test controls the type I experimentwise error rate.

                Alpha                                     0.01
                Error Degrees of Freedom                     6
                Error Mean Square                        61.25
                Critical Value of Studentized Range    7.03327
                Minimum Significant Difference          27.522

   Comparisons significant at the 0.01 level are indicated by ***.

                          Difference       Simultaneous
               FABRIC       Between       99% Confidence
             Comparison       Means          Interval

             1     -3        24.000     -3.522    51.522
             1     -4        35.250      7.728    62.772   ***
             1     -2        45.750     18.228    73.272   ***
             3     -1       -24.000    -51.522     3.522
             3     -4        11.250    -16.272    38.772
             3     -2        21.750     -5.772    49.272
             4     -1       -35.250    -62.772    -7.728   ***
             4     -3       -11.250    -38.772    16.272
             4     -2        10.500    -17.022    38.022
             2     -1       -45.750    -73.272   -18.228   ***
             2     -3       -21.750    -49.272     5.772
             2     -4       -10.500    -38.022    17.022
```

TABLE 9.54 Checklist for randomized-groups Latin-square ANOVA

1. Issues
   a. Independence of errors
   b. Missing data
   c. Normality of sampling distributions
   d. Homogeneity of variance
   e. Outliers
   f. Additivity
2. Major analysis, when statistically significant:
   a. Effect size and confidence interval
   b. Parameter estimates (means and standard deviations or standard errors or confidence intervals for each group)
3. Additional analyses
   a. Post hoc comparisons
   b. Interpretation of departure from homogeneity of variance, if appropriate

FIGURE 9.4
Mean weight loss of four abraded fabrics

```
** Factorial ANOVA ***;
options pageno=1;
proc glm data=SASUSER.FABRIC;
   class CASE POSITION FABRIC;
   model LOSS = CASE POSITION FABRIC / SS3;
** Create output data set for plots **;
   output OUT=work._plotout p=_pred r=_resid student=_stres
      rstudent=_rstres dffits=_dffits h=_h covratio=_covr l95=_l95
      u95=_u95 l95m=_l95m u95m=_u95m;
run; quit;
goptions reset=all device=WIN;
      title;
      footnote;
*** Plots ***;
goptions ftext=SWISS ctext=BLACK htext=1 cells
         gunit=pct htitle=6;
axis1 major=(number=5) label=(a=90 h=4) width=2;
axis2 offset=(10 pct) label=(h=4) width=2;
proc gplot data=work._plotout ;
   where LOSS is not missing and CASE is not missing and POSITION is
      not missing and FABRIC is not missing;
** Main effect plots **;
   goptions reset=symbol;
   symbol1 i=std1mtj v=none color=BLUE height=1 cells width=1;
   plot LOSS * FABRIC = 1 /
      frame hminor=0 vminor=0 vaxis=axis1 haxis=axis2
      cframe=CXF7E1C2 caxis=BLACK  name='MEANS'
      description="Means plot of LOSS by FABRIC";
   run;
```

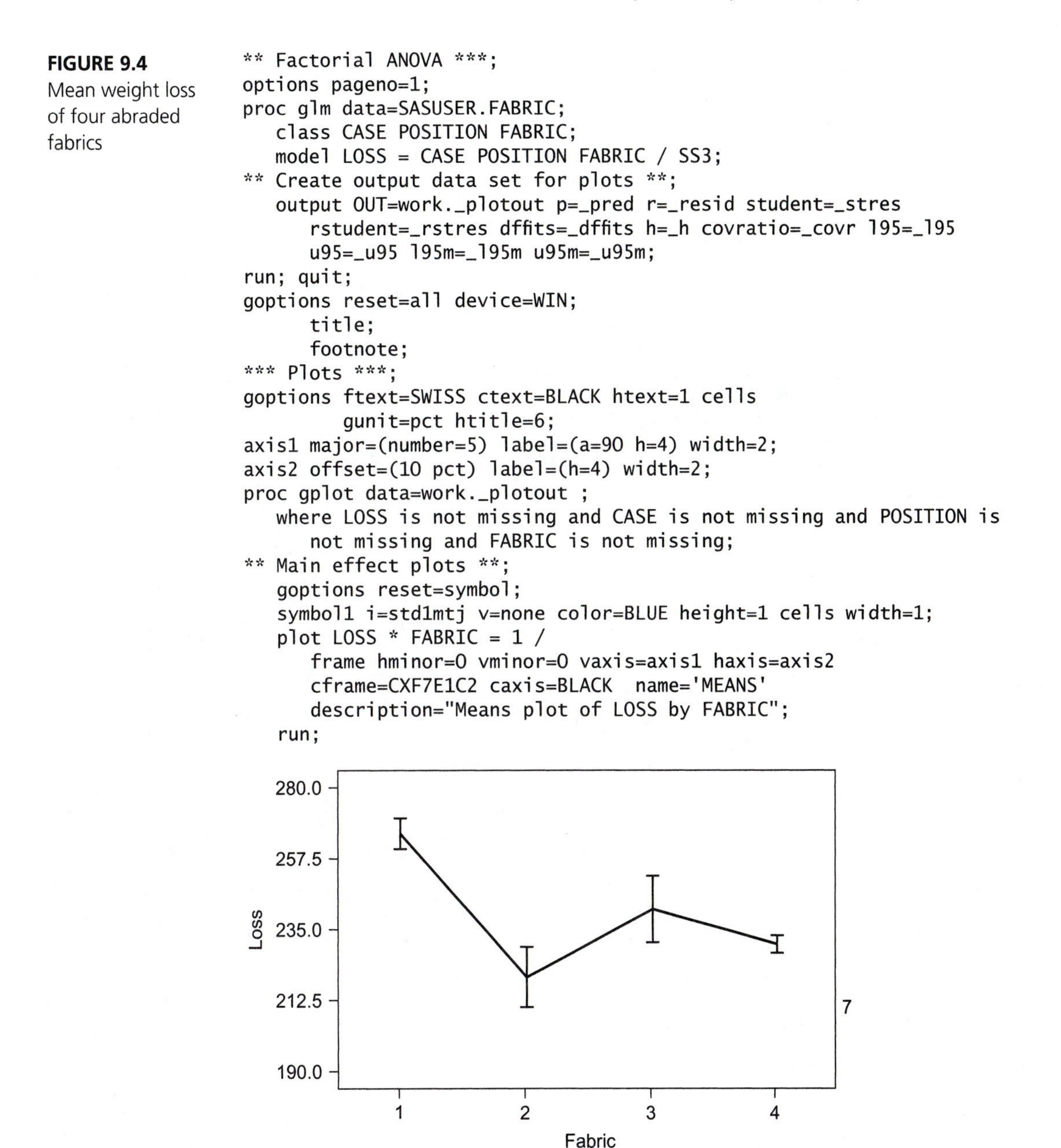

Results

A four-level randomized-groups Latin-square analysis was performed on weight loss (in mg) of four fabrics after abrasion by special-quality emery paper. The test machine consists of plates in four different positions. There were four trials in which the fabrics were assigned to the different plate positions via a randomly chosen Latin-square arrangement.

No outliers were observed; however, homogeneity of variance for the test of differences among fabrics was violated (F_{max} = 15.58). Transformations failed to produce acceptable homogeneity, so Type I error rate for the test of fabrics was set to α = .025, and the untransformed values were used. There was no evidence that main effects were confounded with interactions among the other two independent variables.

All three tests of effects were statistically significant. Weight loss differed by position of plate, $F(3, 6) = 7.99$, $p < .02$. Weight loss also varied over the four trials, $F(3, 6) = 5.37$, $p < .04$. Most important, weight loss differed among the four fabrics, $F(3, 6) = 24.15$, $p < .001$, partial $\eta^2 = .93$ with confidence limits from .51 to .95. Figure 9.4 shows means and standard errors for the amount of wear among the four fabrics.

A Tukey test at α = .01 revealed that fabric 1 was statistically significantly worse than fabric 2 and fabric 4, with no differences among fabrics 2, 3, and 4. Fabric 2 showed the least wear on average; however, fabric 4 had the least variable degree of wear and was not worse than fabric 2. Therefore, fabric 4 is the one of choice in terms of consistent resistance to abrasion.

9.7.2 Complete Example of a Repeated-Measures Crossover Design with Multiple Trials

Hirsch and Johnston (1996) tested whether the presence of a floral scent improved performance. Twenty-one students participated in six trials in which they worked through a set of two paper-and-pencil mazes. Participants wore floral-scented masks on half the trials, either the first three trials or the last three trials (order randomly determined). The DV was length of time to complete each of the six trials. Data sets are SCENT.*.

In terms of the design layout, there is a single randomized-groups effect, order, and three repeated-measures effects: trial, position, and scent. Therefore, this is a 2 × 3 × 2 × 2 design overall. All cases in each level of order have the same sequence. The main effect of interest is the effect of scent (mask): Does it take less time to complete a maze in the presence of a floral scent? Position effects are also (mildly) interesting. Is there a carryover effect from the first set of trials to the second? Does it take less time to complete the second set of mazes? Also, does order make a difference? Does it matter whether the first set of trials is with the scented or unscented mask?

The experimental nature of the design is ensured by randomly assigning participants to orders. Table 9.55 diagrams the layout of the design (but not the analysis), with C = treatment, in which c_1 = unscented mask and c_2 = scented mask; P = first versus second set of trials (position); T = first, second, or third trial within each order; and O = order, where o_1 = unscented trials first and o_2 = scented trials first. Note that the P effect is equivalent to the C by O interaction.

Each case is entered on a separate line in the data set, similar to the arrangement used in Table 9.21 (one case per line). One column is needed to code for order. Then, the six performance times for each case are entered in the following order: unscented trial 1, unscented trial 2, unscented trial 3, scented trial 1, scented trial 2, and scented trial 3. Note that the six DVs could also have been entered by their position instead of their scent. If the DVs are coded by scent, scent is tested as a main effect, and the position effect is tested by the order by scent interaction. If the DVs are coded by position, position is tested as a main effect, and the scent effect is tested by the order by position interaction. Because scent is the more important main effect, DVs are coded by scent.

9.7.2.1 Evaluation of Assumptions

Distributional assumptions are to be tested and outliers screened. However, sphericity is not an issue for the three levels of trials, because they are not of research interest by themselves. The other repeated-measures IVs, scent and position, have only two levels.

TABLE 9.55 Design for complete example of crossover design

			p_1			p_2	
Order	Cases	t_1	t_2	t_3	t_1	t_2	t_3
o_1	$n = 11$	c_1	c_1	c_1	c_2	c_2	c_2
o_2	$n = 10$	c_2	c_2	c_2	c_1	c_1	c_1

9.7.2.1.1 Sample Sizes, Normality, Independence of Errors, and Homogeneity of Variance Independence of errors is ensured by the (presumed) running of each participant individually in the randomized-groups part of the design (order effects) and by the use of appropriate error terms in which individual differences are eliminated in the repeated-measures part of the design.

Sample sizes are unequal between the two orders, with $n_1 = 11$ and $n_2 = 10$. This poses no problem with overlapping variance, as all the sample sizes are equal within each level of order; that is, there are no missing data. However, the inequality in sample sizes makes it slightly more complicated to calculate degrees of freedom for error for evaluating normality of sampling distributions. There are $(n_1 + n_2 - 2) = 19$ df for the test of the order effect, just short of ensuring normality of sampling distributions for the randomized-groups part of the design. The trials effect is tested with $(n_1 + n_2 - 2)(t - 1) = 38$ df. The scent effect and the order by scent (position) effect are also both a bit deficient in df to ensure normality, with df $= (n_1 + n_2 - 2)(o - 1) = 19$. However, none of these insufficiencies is great enough to cause distress, as long as there is adequate homogeneity of variance and there are no outliers.

Homogeneity of variance is most directly evaluated by viewing the 12 cell standard deviations (not those labeled Total), as produced by SPSS GLM (Repeated Measures) in Table 9.56. The /WSFACTOR instruction sets up the repeated-measures factors, scents and trials, and requests that the underlying contrasts for the tests of repeated-measures factors be polynomial (default).

Homogeneity of variance is violated by the exceptionally large variance in the unscented trial 3 data (scented first), with $F_{max} = (29.718)^2/(8.431)^2 = 12.42$. However, as shown in Table 9.57, a square root transformation of the DV (in this case all six of them), reduces F_{max} to an acceptable level of 7.70. (Note that the syntax includes a request for marginal means, /EMMEANS, for the main analysis, as well as requests for effect size and power statistics, ETASQ and OPOWER.)

9.7.2.1.2 Outliers Table 9.57 provides means and standard deviations to permit conversion of raw scores to z scores for testing for outliers in each of the 12 cells. There are no extreme values in the transformed data at $\alpha = .01$ (a reasonable criterion for this sample size).

9.7.2.2 Analysis of Crossover Design

Table 9.57 shows the syntax for SPSS GLM for the major analysis. Note that it is not necessary to do a separate run for the position effect, because it is equivalent to the scent by order interaction. Table 9.58 shows selected output for the omnibus effects' marginal means.

The main effect of greatest interest, scent, is not statistically significant: There is no evidence that performance is affected by the presence of a floral scent: $F(1, 38) = 0.002$, $p = .97$, partial $\eta^2 = .00$ with 95% confidence limits from .00 to .00, using Smithson's (2003) procedure. Scent does, however, interact with trials: $F(2, 38) = 7.27$, $p = .002$, partial $\eta^2 = .28$ with confidence limits from .05 to .45. Thus, the only

TABLE 9.56 Descriptive statistics for crossover example (SPSS GLM syntax and selected output)

```
GLM
 UNSCENT1 UNSCENT2 UNSCENT3 SCENT1 SCENT2 SCENT3 BY ORDER
  /WSFACTOR = scents 2 Polynomial trials 3 Polynomial
  /MEASURE = time
  /METHOD = SSTYPE(3)
  /PRINT = DESCRIPTIVE
  /CRITERIA = ALPHA(.05)
  /WSDESIGN = scents trials scents*trials
  /DESIGN = order .
```

Descriptive Statistics

	Order of treatments	Mean	Std. Deviation	N
unscent1	Unscented first	61.2455	19.89429	11
	Scented first	45.8700	8.71423	10
	Total	53.9238	17.14581	21
unscent2	Unscented first	50.5455	15.60791	11
	Scented first	43.4700	10.28754	10
	Total	47.1762	13.51073	21
unscent3	Unscented first	41.1636	10.42193	11
	Scented first	56.1800	29.71763	10
	Total	48.3143	22.60038	21
scent1	Unscented first	51.3455	18.89933	11
	Scented first	46.3000	10.42678	10
	Total	48.9429	15.30302	21
scent2	Unscented first	54.8818	16.51901	11
	Scented first	56.2500	12.26615	10
	Total	55.5333	14.30508	21
scent3	Unscented first	39.5818	8.43123	11
	Scented first	47.4400	11.88755	10
	Total	43.3238	10.73811	21

table of means reproduced here is the one for scents by trials.[10] The means are plotted in Figure 9.5.

The pattern of means suggests that the floral scent enhances performance on the first and third trials, but hinders it on the second trial. Simple-effects analysis is required to verify these apparent differences. Table 9.59 shows SPSS GLM syntax and

[10] The statistically significant interaction between order and trials is not of research interest.

TABLE 9.57 Descriptive statistics for crossover example with square root transformation (SPSS GLM syntax and selected output)

```
COMPUTE s_u1 = SQRT(UNSCENT1) .
EXECUTE .
COMPUTE s_u2 = SQRT(UNSCENT2) .
EXECUTE .
COMPUTE s_u3 = SQRT(UNSCENT3) .
EXECUTE .
COMPUTE s_s1 = SQRT(SCENT1) .
EXECUTE .
COMPUTE s_s2 = SQRT(SCENT2) .
EXECUTE .
COMPUTE s_s3 = SQRT(SCENT3) .
EXECUTE .
GLM
  s_u1 s_u2 s_u3 s_s1 s_s2 s_s3 BY ORDER
  /WSFACTOR = scents 2 Polynomial trials 3 Polynomial
  /MEASURE = time
  /METHOD = SSTYPE(3)
  /PLOT = PROFILE( trials*scents )
  /EMMEANS = TABLES(scents)
  /EMMEANS = TABLES(trials)
  /EMMEANS = TABLES(ORDER*scents)
  /EMMEANS = TABLES(ORDER*trials)
  /EMMEANS = TABLES(scents*trials)
  /EMMEANS = TABLES(ORDER*scents*trials)
  /PRINT = DESCRIPTIVE ETASQ OPOWER
  /CRITERIA = ALPHA(.05)
  /WSDESIGN = scents trials scents*trials
  /DESIGN = ORDER .
```

Descriptive Statistics

	Order of treatments	Mean	Std. Deviation	N
s_u1	Unscented first	7.7281	1.29413	11
	Scented first	6.7451	.64397	10
	Total	7.2600	1.13006	21
s_u2	Unscented first	7.0284	1.12307	11
	Scented first	6.5530	.76604	10
	Total	6.8020	.97668	21
s_u3	Unscented first	6.3655	.84137	11
	Scented first	7.3011	1.78710	10
	Total	6.8110	1.42139	21
s_s1	Unscented first	7.0558	1.31028	11
	Scented first	6.7645	.77610	10
	Total	6.9171	1.07318	21
s_s2	Unscented first	7.3260	1.15416	11
	Scented first	7.4588	.82760	10
	Total	7.3893	.98938	21
s_s3	Unscented first	6.2576	.68343	11
	Scented first	6.8336	.90805	10
	Total	6.5319	.83156	21

FIGURE 9.5
Square root of mean time to solve mazes over three trials with scented and unscented mask (syntax is in Table 9.57)

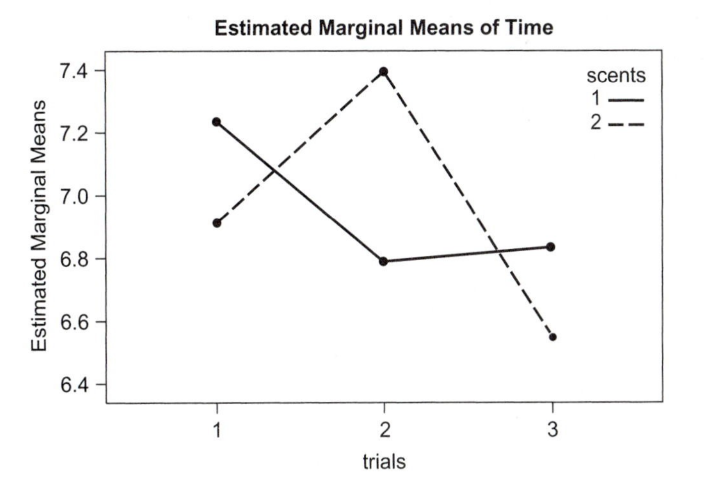

output for a simple-effects analysis of scent at each level of trial. Note that scent and trials are defined as a single one-way repeated-measures IV. The comparison is within the SPECIAL matrix. The first row corresponds to the intercept. The next three rows correspond to the three comparisons of interest: scented versus unscented at trial 1, unscented versus scented at trial 2, and unscented versus scented at trial 3, respectively. The last two comparisons are not of current interest (they correspond to a trend analysis on the main effect of trials) but are included because the comparison matrix must be square—that is, the matrix must have the same number of rows as columns.

The contrasts of interest are labeled tri_scnt L1, tri_scnt L2, and tri_scnt L3 (for the three trials, respectively) in the output. The obtained F ratio for scents at the first trial is 8.63, $p = .008$; at the second trial, $F = 9.54$, $p = .006$; and at the third trial, $F = 1.44$, $p = .246$.

A Bonferroni adjustment is made for these comparisons, in which the number of comparisons (3) affects the critical probability level. To keep familywise error rate at .05, $\alpha_i = .05/3 = .017$. This adjustment does not affect the results; the floral scent enhances performance on the first trial (partial $\eta^2 = .31$ with 95% confidence limits from .02 to .55), hinders it on the second trial (partial $\eta^2 = .33$ with 95% confidence limits from .03 to .57), and does not affect it on the third trial (partial $\eta^2 = .08$ with 95% confidence limits from .00 to .34).

There is also no main effect of order; there is no difference in performance between participants who are exposed to the floral scent first and those who experience no scent first: $F(1, 19) = .002$, $p = .97$, partial $\eta^2 = .00$ with no deviation from 0 in the 95% confidence limits. Nor is the scent by order interaction statistically significant, indicating a lack of position effect: $F(1,38) = 2.29$, $p = .15$, partial $\eta^2 = .06$ with 95% confidence limits from .00 to .23. (Note that the Partial Eta Squared value given by SPSS in Table 9.58 is incorrect, as it is based on 19 df for error instead of 38 df.) That is, there is no evidence that performance on the second set of three trials is any different from performance on the first set of three trials.

Some other effects are noted—a main effect of trials (which occurs within scent conditions), modified by a trials by order interaction—but they are not of current

TABLE 9.58 Selected output from SPSS GLM crossover analysis (syntax is in Table 9.57)

General Linear Model

Tests of Within-Subjects Effects

Measure: time

Source		Type III Sum of Squares	df	Mean Square	F	Sig.	Partial Eta Squared	Noncent. Parameter	Observed Power[a]
scents	Sphericity Assumed	.001	1	.001	.002	.968	.000	.002	.050
	Greenhouse-Geisser	.001	1.000	.001	.002	.968	.000	.002	.050
	Huynh-Feldt	.001	1.000	.001	.002	.968	.000	.002	.050
	Lower-bound	.001	1.000	.001	.002	.968	.000	.002	.050
scents * order	Sphericity Assumed	.772	1	.772	2.291	.147	.108	2.291	.301
	Greenhouse-Geisser	.772	1.000	.772	2.291	.147	.108	2.291	.301
	Huynh-Feldt	.772	1.000	.772	2.291	.147	.108	2.291	.301
	Lower-bound	.772	1.000	.772	2.291	.147	.108	2.291	.301
Error(scents)	Sphericity Assumed	6.400	19	.337					
	Greenhouse-Geisser	6.400	19.000	.337					
	Huynh-Feldt	6.400	19.000	.337					
	Lower-bound	6.400	19.000	.337					
trials	Sphericity Assumed	4.322	2	2.161	5.269	.010	.217	10.539	.804
	Greenhouse-Geisser	4.322	1.870	2.312	5.269	.011	.217	9.852	.783
	Huynh-Feldt	4.322	2.000	2.161	5.269	.010	.217	10.539	.804
	Lower-bound	4.322	1.000	4.322	5.269	.033	.217	5.269	.587
trials * order	Sphericity Assumed	10.535	2	5.267	12.843	.000	.403	25.686	.995
	Greenhouse-Geisser	10.535	1.870	5.635	12.843	.000	.403	24.012	.993
	Huynh-Feldt	10.535	2.000	5.267	12.843	.000	.403	25.686	.995
	Lower-bound	10.535	1.000	10.535	12.843	.002	.403	12.843	.925

Error(trials)	Sphericity Assumed	15.585	38	.410					
	Greenhouse-Geisser	15.585	35.523	.439					
	Huynh-Feldt	15.585	38.000	.410					
	Lower-bound	15.585	19.000	.820					
scents * trials	Sphericity Assumed	5.776	2	2.888	7.274	.002	.277	14.549	.917
	Greenhouse-Geisser	5.776	1.720	3.359	7.274	.004	.277	12.509	.883
	Huynh-Feldt	5.776	1.974	2.926	7.274	.002	.277	14.360	.914
	Lower-bound	5.776	1.000	5.776	7.274	.014	.277	7.274	.725
scents * trials * order	Sphericity Assumed	1.788	2	.894	2.252	.119	.106	4.504	.430
	Greenhouse-Geisser	1.788	1.720	1.040	2.252	.128	.106	3.873	.395
	Huynh-Feldt	1.788	1.974	.906	2.252	.120	.106	4.445	.427
	Lower-bound	1.788	1.000	1.788	2.252	.150	.106	2.252	.297
Error(scents*trials)	Sphericity Assumed	15.087	38	.397					
	Greenhouse-Geisser	15.087	32.674	.462					
	Huynh-Feldt	15.087	37.508	.402					
	Lower-bound	15.087	19.000	.794					

a. Computed using alpha = .05

TABLE 9.58 Continued

Tests of Between-Subjects Effects

Measure: time

Transformed Variable: Average

Source	Type III Sum of Squares	df	Mean Square	F	Sig.	Partial Eta Squared	Noncent. Parameter	Observed Power[a]
Intercept	6074.864	1	6074.864	1264.331	.000	.985	1264.331	1.000
order	.010	1	.010	.002	.965	.000	.002	.050
Error	91.291	19	4.805					

a. Computed using alpha = .05

Estimated Marginal Means

5. scents * trials

Measure: time

scents	trials	Mean	Std. Error	95% Confidence Interval	
				Lower Bound	Upper Bound
1	1	7.237	.227	6.762	7.711
	2	6.791	.212	6.347	7.234
	3	6.833	.300	6.205	7.461
2	1	6.910	.238	6.412	7.409
	2	7.392	.221	6.929	7.855
	3	6.546	.174	6.181	6.910

TABLE 9.59 SPSS GLM syntax and selected output for simple-effects analysis of scents at each trial

```
GLM
  s_u1 s_s1 s_u2 s_s2 s_u3 s_s3 BY ORDER
  /WSFACTOR = tri_scnt 6
   SPECIAL  (1  1  1  1  1  1
             1 –1  0  0  0  0
             0  0  1 –1  0  0
             0  0  0  0  1 –1
            –1  0  1 –1  0  1
             1 –2  1  1 –2  1)
  /MEASURE = time
  /METHOD = SSTYPE(3)
  /PRINT = ETASQ OPOWER
  /CRITERIA = ALPHA(.05)
  /WSDESIGN = tri_scnt
  /DESIGN = ORDER .
```

Tests of Within-Subjects Contrasts

Measure: time

Source	tri_scnt	Type III Sum of Squares	df	Mean Square	F	Sig.	Partial Eta Squared	Noncent. Parameter	Observed Power[a]
tri_scnt	L1	2.233	1	2.233	8.631	.008	.312	8.631	.796
	L2	7.586	1	7.586	9.543	.006	.334	9.543	.834
	L3	1.735	1	1.735	1.436	.246	.070	1.436	.207
	L4	35.014	1	35.014	28.897	.000	.603	28.897	.999
	L5	4.796	1	4.796	.860	.365	.043	.860	.143
tri_scnt * order	L1	2.505	1	2.505	9.684	.006	.338	9.684	.839
	L2	1.937	1	1.937	2.437	.135	.114	2.437	.317
	L3	.677	1	.677	.560	.463	.029	.560	.110
	L4	4.735	1	4.735	3.908	.063	.171	3.908	.467
	L5	21.754	1	21.754	3.900	.063	.170	3.900	.466
Error(tri_scnt)	L1	4.915	19	.259					
	L2	15.103	19	.795					
	L3	22.956	19	1.208					
	L4	23.021	19	1.212					
	L5	105.971	19	5.577					

a. Computed using alpha = .05

TABLE 9.60 Checklist for repeated-measures or crossover Latin-square ANOVA

1. Issues
   a. Missing data
   b. Normality of sampling distributions
   c. Homogeneity of variance
   d. Outliers
   e. Independence of errors/additivity/homogeneity of covariance/sphericity
2. Major analysis, when statistically significant
   a. Effect size and confidence interval
   b. Parameter estimates (means and standard errors or standard deviations or confidence intervals for each group)
3. Additional analyses
   a. Post hoc comparisons
   b. Interpretation of departure from homogeneity of variance, if appropriate

research interest. These findings might be investigated with a strict adjustment for post hoc testing, or follow-up research might be designed to verify these effects were they of interest.

Table 9.60 is a checklist for repeated-measures Latin-square ANOVA. An example of a results section, in journal format, follows for the study just described.

Results

A repeated-measures crossover design was used to investigate the effect of a floral scent on speed of solving mazes. Participants worked through a set of two pencil-and-paper mazes six times, three times while wearing a floral-scented mask and three times while wearing an unscented mask. Participants were randomly assigned to wear the floral mask on either their first three trials (n = 10) or their second three trials (n = 11). The dependent variable was time to complete each maze. Independent variables were the two types of mask, three trials, and two orders of sets of trials. The interaction between type of mask and order indicates the effect of position (first vs. second set of trials).

No outliers were observed; however, there was marked heterogeneity of variance (F_{max} = 12.42). A square root transform of solution times reduced the heterogeneity to an acceptable level.

There was no main effect of type of mask; that is, there was no evidence that the floral scent affected performance ($F < 1$). However, as illustrated in Figure 9.5, there was a statistically significant interaction between type of mask and trials, $F(2, 38) = 7.27$, $p = .002$, partial $\eta^2 = .28$ with 95% confidence limits from .05 to .45. Means, standard errors, and 95% confidence intervals for the six combinations of masks and trials are (*in the last table of output*) in Table 9.48. A simple-effects test with Bonferroni adjustment showed a significant decrease in square root of time to solve mazes while wearing the scented mask (mean = 6.91) as compared with the unscented mask (mean = 7.24) on the first trial, $F(1, 19) = 8.63$, $p = .008$, partial $\eta^2 = .31$ with 95% confidence limits from .02 to .55. The opposite relation occurred on the second trial. There was a statistically significant increase in square root of time to solve the mazes while wearing the scented mask (mean = 7.39) as compared with the unscented mask (mean = 6.79) on the second trial, $F(1, 19) = 9.54$, $p = .006$, partial $\eta^2 = .33$ with 95% confidence limits from .03 to .57. There was no difference between types of mask on the third trial, $F(1,19) = 1.74$, $p > .017$, $\eta^2 = .08$ with 95% confidence limits from .00 to .34.

There was no effect of position (first vs. second set of trials), as tested by the type of mask by order interaction, $F(1,38) = 2.29$, $p = .15$, partial $\eta^2 = .06$ with 95% confidence limits from .00 to .23.

9.8 COMPARISON OF PROGRAMS

SPSS, SAS, and SYSTAT each have one program best suited to Latin-square analysis, whether randomized-groups or repeated-measures. Table 9.61 reviews features of these programs that are especially relevant to randomized-groups Latin-square designs. Other features of the programs that may be applied to the design appear in Chapter 5.

Table 9.62 reviews features of the programs that are especially relevant to repeated-measures Latin-square designs when homogeneity of covariance is of concern. Other features of the programs that may be applied to the design appear in Chapters 6 and 7.

TABLE 9.61 Comparison of SPSS, SAS, and SYSTAT programs for randomized-groups Latin-square ANOVA

Feature	SPSS GLM	SAS GLM	SYSTAT GLM
Input			
Use model instruction to specify main effects only	No	Yes	Yes
Other instructions to specify main effects only	DESIGN	No	No
Request marginal means and standard deviations for each IV	EMMEANS[a]	MEANS	No
Output			
User-specified contrasts	Yes[b]	Yes	Yes
Post-hoc tests (see Sections 4.7 and 5.7 for details)	Yes	Yes	No
Predicted values and residuals added to data set	Yes	No	No
Saved Files			
Predicted values and residuals	Yes	Yes	Yes

[a] Common standard error rather than individual standard deviations.
[b] Confidence intervals, syntax only if other than trend analysis.

TABLE 9.62 Comparison of SPSS, SAS, and SYSTAT programs for repeated-measures Latin-Square ANOVA

Feature	SPSS GLM	SAS GLM	SYSTAT GLM
Input			
Request treatment/position means and standard deviations	EMMEANS[a]	MEAN	No
Output			
Tests and/or adjustments for homogeneity of covariance (see Section 6.7 for details)	Yes	Yes	Yes
User-specified contrasts with repeated-measures format[b]	Yes[c]	No	Yes

[a] Individual standard errors instead of standard deviations.
[b] Requires data matrix of adjusted scores (cf. Section 9.5.6.2).
[c] Confidence intervals, syntax only if other than trend analysis.

9.8.1 SPSS Package

In the SPSS Windows menu, General Linear Model > Univariate . . . analyzes Latin-square designs in the randomized-groups format, and repeated-measures GLM analyzes Latin-square designs in the repeated-measures format. The programs provide all the basics for Latin-square analysis, with the exception of marginal standard deviations for testing homogeneity of variance, which are found through the descriptive statistics program. Predicted values and residuals may be printed or saved to a file. The saved file may then be used to plot values and residuals.

9.8.2 SAS System

SAS GLM is used for all sorts of Latin-square analyses, regardless of the format. Predicted values and residuals are saved to file and may be plotted. The basic Latin-square analysis may be done in repeated-measures format; however, there is no provision for user-specified comparisons or post hoc tests. Performing these tests under the randomized-groups format produces erroneous error terms for a repeated-measures design. SAS PLAN allows you to generate Latin squares and, for a randomized-groups design, a sequence of runs.

9.8.3 SYSTAT System

SYSTAT GLM also analyzes Latin-square designs in either format. User-specified comparisons are available for both formats; however, post hoc tests are unavailable. Predicted values and residuals are saved to file for plotting. Programs for descriptive statistics are required for marginal means and standard deviations. SYSTAT DESIGN generates random Latin-square arrangements and provides a random sequence of runs for randomized-groups setups, as per SAS.

9.9 PROBLEM SETS

9.1 Do a randomized-groups Latin-square analysis on the following data set. Assume that A is assembler experience, B is assembly room temperature, and C is four types of assembly procedures. The DV is quality of finished computer.

	b_1	b_2	b_3	b_4
a_1	$c_1 = 5$	$c_2 = 7$	$c_3 = 8$	$c_4 = 10$
a_2	$c_2 = 3$	$c_4 = 7$	$c_1 = 2$	$c_3 = 5$
a_3	$c_3 = 11$	$c_1 = 8$	$c_4 = 11$	$c_2 = 10$
a_4	$c_4 = 13$	$c_3 = 10$	$c_2 = 9$	$c_1 = 9$

a. Evaluate assumptions of normality of sampling distributions, independence of errors, and homogeneity of variance.
b. Do the analysis using the traditional approach.
c. Do the analysis using the regression approach.
d. Provide a source table.
e. Calculate effect size(s) and find confidence intervals using Smithson's software.
f. Write a results section for the analysis.
g. List an appropriate follow-up procedure.

9.2 Analyze the following data set as a crossover design. Assume that A is sequence, B is position or trial, and C is treatment (four types of aircraft-speed indicators). The cases are pilots. The DV is accuracy of reading in a short duration exposure.

Sequence	b_1	b_2	b_3	b_4
a_1	c_1	c_2	c_3	c_4
a_2	c_3	c_1	c_4	c_2
a_3	c_2	c_4	c_1	c_3
a_4	c_4	c_3	c_2	c_1

Sequence		b_1	b_2	b_3	b_4
a_1	s_1	1	9	3	10
	s_2	3	8	3	10
a_2	s_3	2	1	9	110
	s_4	1	2	7	10
a_3	s_5	8	8	1	4
	s_6	6	6	3	5
a_4	s_7	7	4	9	2
	s_8	7	3	10	3

a. Do the analysis using the mixed-design repeated-measures format.
b. Check your results by running them through at least one statistical software program.
c. Evaluate all appropriate assumptions, using computer programs where necessary.
d. Provide a source table with appropriate error term.
e. Calculate effect size(s) and find confidence intervals using Smithson's software.
f. Write a results section for the analysis.
g. Use one of the statistical software programs to run a trend analysis on the position effect.

9.3 Use a statistical software program to analyze the following data for the following incomplete blocks design. Data are shown in the following Youden square, where C is snow condition (powder, packed powder, ice, and artificial snow), and B and A are defined as in the small-sample example of Section 9.4.3. (*Note:* There now are four cases, but each case skis only three runs.)

	b_1	b_2	b_3
c_1	$a_1 = 9$	$a_2 = 1$	$a_3 = 4$
c_2	$a_2 = 1$	$a_4 = 6$	$a_1 = 7$
c_3	$a_3 = 5$	$a_1 = 8$	$a_4 = 5$
c_4	$a_4 = 4$	$a_3 = 5$	$a_2 = 2$

a. Use the setup of Table 9.12 to enter data for the statistical software program. (Assume position effects are not of interest.)

b. Provide a source table.

c. Calculate effect size(s) and find their confidence limits using software.

d. Write a results section for the analysis.

e. List an appropriate follow-up procedure.

9.4 Find relative efficiency for the following:

a. The randomized-groups design in Problem 9.1

b. The same design assuming it was a repeated-measures Latin square (i.e., *A* representing cases)

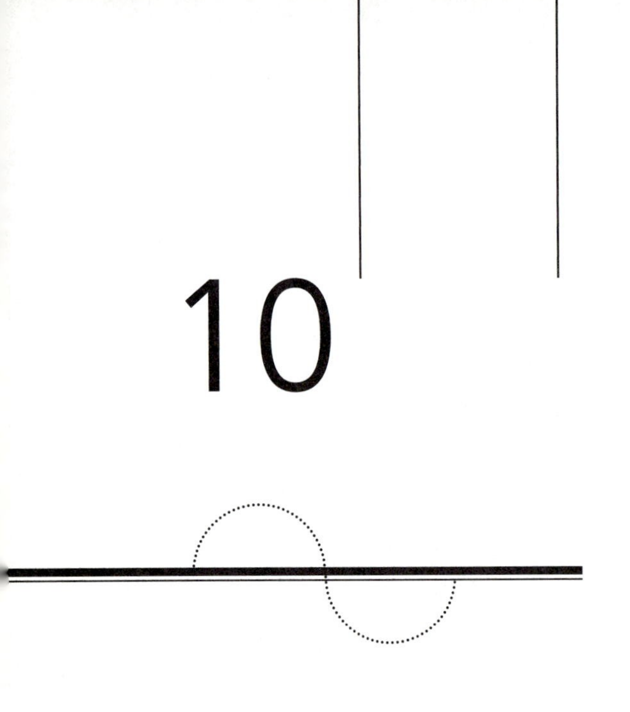

10

Screening and Other Incomplete Designs

10.1 GENERAL PURPOSE AND DESCRIPTION

Screening designs, like Latin-square designs, test numerous main effects that are deliberately confounded with interactions in order to reduce the number of cases required for the study. As *screening* implies, these designs are appropriate for pilot studies in which the researcher has a long list of potentially effective IVs and inadequate information regarding which IVs are important. The results identify important IVs and tell the researcher which to omit in future studies. Another application of this design is when a researcher only needs a quick-and-dirty answer to a research question. A third application is when there are very few cases available (for reasons of expense, rarity, or whatever), and the researcher wants the most information from the fewest cases. Except for human factors research, these designs have not been used extensively in the social sciences.

As one example of this design in use, a researcher is uncertain about the factors that affect the outcome of a new therapy. Is it the duration of therapy (four weeks or eight weeks)? Is it the concentration of sessions (one a week or three a week)? Is it the training level of the therapist (professional or paraprofessional)? Is it the setting in which therapy is delivered (office or clinic)? Is it the privacy of therapy (one-on-one or group)? Five IVs are identified, each with two levels. A complete factorial randomized-groups study of these IVs has $2^5 = 32$ cells, each with a different combination of research conditions. The design requires 64 cases if there are only 2 cases per cell, 96 cases if there are 3 per cell, and so on. However, it is possible to run the study with far fewer cases by giving up analysis of some of the interactions. For example, 32 cases can provide preliminary tests of all five main effects, plus all the two-way interactions. This is accomplished by confounding all the three-way and higher-order interactions with main effects and two-way interactions.

Screening designs are also available in which some or all two-way interactions are confounded with main effects, as they are in Latin-square designs. For some designs, this reduces the number of cases even more (e.g., from 32 to 8, with the current example). Use of these designs, then, is a good strategy if you have a long list of factors that might affect your outcome and you want an efficient way to screen out the unimportant ones. Screening designs let you identify the most important effects to use in future research or to focus efforts at improving the quality of your product.

There are numerous types of screening and other incomplete designs—some available for qualitative IVs and others for when the IVs are quantitative. Two major designs for qualitative IVs are fractional-factorial designs, including the Plackett-Burman designs, and Taguchi procedures. In fractional-factorial designs, some or all two-way interactions can be made available for analysis. The number of interactions that can be tested depends on the number of cases: When more cases are run, more unconfounded interactions can be evaluated. Plackett-Burman designs, a special case of fractional-factorial designs, test no interactions; so, of course, they require the fewest number of cases. Taguchi procedures incorporate many of these same designs but emphasize the stability of the DV under different conditions, the effect size rather than significance testing for evaluation of IVs, and the identification of the best combination of levels of IVs.

The example has both qualitative (e.g., setting for therapy) and quantitative (e.g., duration of therapy) IVs, but each with two levels. If a fractional-factorial design with 32 cases is used, the main effects of all five IVs are tested, in addition to all the two-way interactions. If a Plackett-Burman design is used, only the main effects of the IVs are tested. If a Taguchi approach is taken, the combination of levels of IVs that produces the most stable and best outcome is emphasized.

Other screening designs are used with quantitative IVs or IVs that are blends of two or more factors. Box-Behnken designs, for instance, are used for quantitative IVs that have three levels each and permit tests of linear and quadratic trends of main effects and linear trends of two-way interactions. Central-composite designs are similar to Box-Behnken designs but are used for quantitative IVs with five levels each. Mixture designs are used when conditions are blends of two (or more) IVs (e.g., a class that is 100% lecture versus a class that is 50% lecture and 50% discussion versus a class that is 100% discussion); that is, the levels are proportions. The primary goal with mixture designs is to find the optimal mix (blend) of IVs rather than to see if IVs are statistically significant. Finally, optimal design methods are available that search for the best design when none of the other specific designs is appropriate to your needs.

As with Latin-square designs, major disadvantages of these incomplete designs are forfeiture of tests of at least some interactions and the implicit assumption that confounded interactions are negligible. With so few cases, the stability of the DV and the representativeness of the cases are also important. Remember that screening designs are usually just that—preliminary studies to help you design the "real" study. They typically do not provide final answers.

10.2 KINDS OF RESEARCH QUESTIONS

The major question to be answered by most screening and other incomplete designs is whether main effects of IVs are associated with mean differences on the DV. Some designs also permit evaluation of some interactions among treatments.

10.2.1 Effects of the IV(s)

Are there mean differences between levels of any of the IVs? When the effects of one IV are averaged over the levels of all other IVs and when extraneous variables are held constant by procedures discussed in Sections 1.1 and 1.2.1 and elsewhere, are there mean differences in the DV associated with the levels of an IV? For example, does the average therapeutic outcome depend on the duration of therapy? On the concentration of sessions per week? On the training level of the therapist? On the setting on which therapy is delivered? On the privacy with which therapy is delivered? These questions are usually answered by applying separate *F* tests to each IV, as demonstrated in Section 10.4. Taguchi methods (Section 10.5.4), however, use different procedures.

10.2.2 Effects of Interactions among IV(s)

Some fractional-factorial designs permit questions about some interactions. Holding all else constant, does outcome for one IV depend on the level of another IV? For example, does the effect of concentration of sessions per week depend on the duration of therapy?

Tests of interactions are also *F* tests, conducted separately for each interaction. However, you will rarely be able to test three-way, four-way, or higher-order interactions, no matter how many cases you have. In these designs, higher-order interactions are almost always confounded with main effects or with lower-order interactions and are lost to analysis.

10.2.3 Parameter Estimates

Parameter estimates are sample means, standard deviations, and standard errors or confidence intervals for levels of an IV used to estimate values in a population. Because screening designs are rarely the end in themselves, but instead are used to design future studies, population values are rarely estimated from them. Rather, these statistics provide information about the outcome in different levels of an IV and are used to evaluate the wisdom of including the IV in future studies.

Some fractional factorials provide tests of some interactions, and parameter estimates for interactions are as described in Chapter 5. However, unequal n is often built into the screening designs. As you recall, there are different ways to average over other IVs when sample sizes are not the same for all cells, as discussed in Section 5.6.5.

Some incomplete designs (e.g., mixture designs) use the regression approach to ANOVA and emphasize prediction equations rather than comparisons among means. Parameter estimates for these designs are typically unstandardized regression coefficients (B weights).

10.2.4 Effect Sizes and Power

Effect size represents the proportion of variance in the DV that is predictable from a main effect or interaction. Analyses typically produce F ratios, so that the usual measures of partial η^2 are appropriate, along with the use of Smithson's (2003) software for finding confidence limits around effect size. Taguchi designs (Section 10.5.4) have their own measures of effect size.

Effect size for an IV in a screening design is often used primarily to determine the wisdom of including that IV in subsequent research. However, screening designs typically do not generate a great deal of power. Because few cases are run, power is low and only large effects are sought; thus, a great deal of power is unnecessary.

10.2.5 Specific Comparisons

In many screening designs, because each IV has only two levels, specific comparisons of marginal means are not relevant. The only type of comparison that might be helpful is a simple main effect in a design with a significant interaction, as illustrated in Section 5.6.4.4.1. However, Box-Behnken and central-composite designs, as well as some Taguchi, fractional-factorial, and optimal designs, have more than two levels of at least some IVs (Section 10.5.5), so comparisons may be appropriate. In fact, Box-Behnken and central-composite designs apply to quantitative IVs and have trend analyses and interactions contrasts built into the model. These comparisons are identical to those in Section 5.6.4.

10.3 ASSUMPTIONS AND LIMITATIONS

10.3.1 Theoretical Issues

Issues of causal inference and generalizability are identical to those in any other analysis of variance. However, because of the small number of cases tested in screening designs, the stability of the DV and the representativeness and homogeneity of the cases are especially important. If there is a great deal of error in measuring the DV, you are unlikely to demonstrate statistical significance with such a small sample. Furthermore, if cases are heterogeneous and only a few cases are used to evaluate the IVs, the precise characteristics of the cases assigned to each level may determine the outcome of evaluation of the IV. This is always true, of course, but with larger samples, case differences are expected to cancel each other more effectively than with such small samples. These factors combine to argue against the utility of screening and incomplete designs when the DV is even moderately unreliable or when cases are highly heterogeneous.

10.3.2 Practical Issues

As randomized-groups ANOVA models, these designs assume normality of sampling distributions, homogeneity of variance, independence of errors, and absence of outliers. Additivity is also assumed when interactions are confounded with main effects. For designs in which some interactions are analyzed, unequal sample sizes pose added problems, as discussed in Section 5.6.5.

10.3.2.1 Normality of Sampling Distributions

In routine research studies, normality of sampling distributions is anticipated if there are 20 or more df for error. Screening designs, with their few cases, rarely generate anywhere near 20 df for error. An alternative method for assessing normality is through visual inspection of the normal probability plot of residuals, as demonstrated in the complete example of Section 10.7.2.1. Residuals are deviations of individual scores around their marginal or cell means. All software packages provide normal probability plots of residuals. If the underlying sampling distribution cannot be assumed to be normal, studies that follow results of screening are planned with sufficient df for error.

10.3.2.2 Homogeneity of Variance

Screening designs often have only one or two cases per cell, and many cells are often empty. Therefore, homogeneity of variance is assessed separately on the margins for each main effect. In other words, variances are compared among levels of A, then separately among levels of B, and so on. If interactions are evaluated, homogeneity of variance is assessed among the scores in the cells for that interaction, collapsed over all other IVs. The guidelines of Section 3.4.3 and elsewhere are used to evaluate differences among variances when they can be estimated. Some designs use so few cases that it is impossible to comply with these guidelines. In these situations, the follow-up studies are designed to allow evaluation of homogeneity of variance, as usual.

10.3.2.3 Absence of Outliers

Outliers are also difficult to test in incomplete designs because many cells are empty or have only a single case. However, outliers can be sought on the margins, or they may show up in the normal probability plot of residuals (as illustrated in Section 10.7.2.1) as points that are highly discrepant from all others.

10.3.2.4 Independence of Errors

The ANOVA model assumes that errors (residuals) are independent of one another, both within each cell and between cells of a design. This assumption is evaluated by examining the details of the research design rather than by statistical means (cf. Section 3.4.2).

10.3.2.5 Additivity

Analysis of incomplete designs when main effects are confounded with interactions assumes that all confounded interactions are negligible and represent only error variance. Tests of main effects are erroneous if interactions are present. For example, if the test of the main effect of A is completely confounded with the $B \times C$ interaction, a large main effect of A could actually be due to a large $B \times C$ interaction or a combination of the main effect of A and $B \times C$ interaction. Including only the main effect of A in subsequent research is wasteful, and failure to test the $B \times C$ interaction results in loss of pertinent information. Additivity in screening designs is evaluated by the logic of the choice of IVs and their potential interactions; it is not readily tested in the absence of a full factorial design.

10.4 FUNDAMENTAL EQUATIONS

The fractional-factorial design is chosen to demonstrate fundamental equations because of its popularity and because the other screening designs with two levels of each IV are easily developed from it.

A fractional-factorial design has procedures similar to a full factorial design for allocation of cases to conditions, the partition of variance, and calculation of randomized-groups ANOVA. The difference is that not all combinations of conditions are present, and some interactions are confounded (aliased) with main effects or with other interactions. Sections 10.5.1 and 10.5.2.1 show how to determine which interactions are assessable. Section 10.6.1 shows the formation of screening and other incomplete designs to produce the best pattern of aliasing (confounding) for your application. A 2^5 fractional-factorial design with 16 combinations of treatments is demonstrated in this section. There are five IVs, each with two levels—hence, 2^5. This is a half factorial because only half of the potential 32 combinations of treatments are present; thus, this half-factorial design is called a 2^{5-1} design.

TABLE 10.1 Assignment of cases in a 2^5 half-factorial design

			a_1		a_2	
			b_1	b_2	b_1	b_2
c_1	d_1	e_1	s_1 s_2			s_3 s_4
		e_2		s_5 s_6	s_7 s_8	
	d_2	e_1		s_9 s_{10}	s_{11} s_{12}	
		e_2	s_{13} s_{14}			s_{15} s_{16}
c_2	d_1	e_1		s_{17} s_{18}	s_{19} s_{20}	
		e_2	s_{21} s_{22}			s_{23} s_{24}
	d_2	e_1	s_{25} s_{26}			s_{27} s_{28}
		e_2		s_{29} s_{30}	s_{31} s_{32}	

10.4.1 Allocation of Cases

Table 10.1 shows the layout for a 2^{5-1} half factorial with two cases per cell. The 32 cases are randomly assigned into two replications of 16 cases each, and then the 16 cases within each replication are randomly assigned to runs (cells). A run is a particular combination of levels of IVs assigned to a particular case. The first case in each cell is in the first replication; the second case in the second replication.

This is but one of many half fractional factorials that could be generated, all with a beautiful symmetry. Take a moment to examine the pattern of filled cells in Table 10.1. Half the cells are empty, but each nonempty cell has different cases, making this a randomized-groups design. Because there are two replications (cases) in each cell, there are sufficient degrees of freedom to test all main effects and two-way interactions, but not higher-order interactions. There is stability to the estimates of means on the margins of each IV, because they are each composed of 16 scores. Means for the two-way interactions are each composed of eight scores. Notice that 32 cases are sufficient to assign one case to each cell of a complete factorial, but without df for estimating error. The same 32 cases in the half factorial allow assessment of main effects and two-way interactions for five IVs.

TABLE 10.2 Assignment of cases in a 2^5 quarter-factorial design

			a_1		a_2	
			b_1	b_2	b_1	b_2
c_1	d_1	e_1	s_8			
		e_2				s_5
	d_2	e_1		s_6		
		e_2			s_7	
c_2	d_1	e_1				s_1
		e_2	s_4			
	d_2	e_1			s_3	
		e_2		s_2		

Allocation of cases to a quarter-factorial (2^{5-2}) design is shown in Table 10.2. This is, again, only one of many possible quarter factorials for a design with five IVs. In this design, only 8 of the 32 cells are filled, and there are no replications. Each main effect is assessable but with means estimated from only four scores. No unconfounded interactions are available for analysis. On the other hand, the main effects of five IVs are tested with only eight cases.

10.4.2 Partition of Sources of Variance

As always, partition of sums of squares into main effects and interactions depends on the number of IVs. The difference in fractional-factorial designs is that main effects and interactions are aliased with other effects, as discussed in Section 10.5.1. The following partition of SS for the 2^{5-1} half factorial of Table 10.1 permits tests of all main effects (aliased with four-way interactions) and all two-way interactions (aliased with three-way interactions). This design is sensible if you expect three-way and higher-order interactions to represent only random error.

In this example, the total sum of squares is partitioned into 16 sources:

$$\begin{aligned} SS_T = {} & SS_A + SS_B + SS_C + SS_D + SS_E + SS_{AB} + SS_{AC} + SS_{AD} + SS_{AE} \\ & + SS_{BC} + SS_{BD} + SS_{BE} + SS_{CD} + SS_{CE} + SS_{DE} + SS_{\text{residual}} \end{aligned} \tag{10.1}$$

or, using the deviation approach,

$$\begin{aligned} \sum_j \sum_k \sum_l (\overline{Y}_{ABCDE} - \text{GM})^2 = {} & \left(\frac{1}{2}\right) bcden \sum_j (\overline{Y}_A - \text{GM})^2 + \left(\frac{1}{2}\right) acden \sum_k (\overline{Y}_B - \text{GM})^2 \\ & + \left(\frac{1}{2}\right) abden \sum_l (\overline{Y}_C - \text{GM})^2 + \left(\frac{1}{2}\right) abcen \sum_m (\overline{Y}_D - \text{GM})^2 \\ & + \left(\frac{1}{2}\right) abcdn \sum_o (\overline{Y}_E - \text{GM})^2 \\ & + \ldots \text{interactions calculated as per Equation 5.1} \ldots + SS_{\text{residual}} \end{aligned}$$

The total sum of squared differences between individual scores (Y_{ABCDE}) and the grand mean (GM) is partitioned into a sum of squared differences associated with A, with B, with C, with D, with E; the sum of squared differences associated with all two-way combinations of these five factors; and the residual (the remaining sum of squared differences between individual scores and the grand mean). Each n is the number of scores in each cell.

In this example, the error term (residual) is found by subtracting the 15 known systematic sources of variance from the total:

$$\begin{aligned}\text{SS}_{\text{residual}} = \text{SS}_T - \text{SS}_A - \text{SS}_B - \text{SS}_C - \text{SS}_D - \text{SS}_E - \text{SS}_{AB} - \text{SS}_{AC} - \text{SS}_{AD} \\ - \text{SS}_{AE} - \text{SS}_{BC} - \text{SS}_{BD} - \text{SS}_{BE} - \text{SS}_{CD} - \text{SS}_{CE} - \text{SS}_{DE}\end{aligned} \quad \textbf{(10.2)}$$

Total degrees of freedom remain the number of scores minus 1, lost when the grand mean is estimated (cf. Equation 3.6). However, the total number of scores is not *abcden* but $\frac{1}{2}(abcden)$ because half the cells are empty. In this 2^{5-1} design, with two levels for each of five IVs, the df for each main effect and interaction is 1. Because there are 15 effects, 5 main effects, and 10 two-way interactions,

$$\begin{aligned}\text{df}_{\text{residual}} &= \text{df}_T - \text{df}_A - \text{df}_B - \text{df}_C - \text{df}_D - \text{df}_E - \text{df}_{AB} - \text{df}_{AC} - \text{df}_{AD} - \text{df}_{AE} \\ &\quad - \text{df}_{BC} - \text{df}_{BD} - \text{df}_{BE} - \text{df}_{CD} - \text{df}_{CE} - \text{df}_{DE} \\ &= \left(\tfrac{1}{2}(abcden) - 1\right) - 15\end{aligned} \quad \textbf{(10.3)}$$

Figure 10.1 diagrams the partition of sums of squares and degrees of freedom for a 2^{5-1} half fractional-factorial design. As in any ANOVA, mean squares for all main effects and interactions are produced by dividing sums of squares by degrees of freedom:

$$\begin{aligned}\text{MS}_A &= \frac{\text{SS}_A}{\text{df}_A} & \text{MS}_B &= \frac{\text{SS}_B}{\text{df}_B} \\ \text{MS}_C &= \frac{\text{SS}_C}{\text{df}_C} \quad \cdots & \text{MS}_{\text{residual}} &= \frac{\text{SS}_{\text{residual}}}{\text{df}_{\text{residual}}}\end{aligned} \quad \textbf{(10.4)}$$

F ratios are formed for all main effects and two-way interactions as in Section 5.4.2, with $\text{MS}_{\text{residual}}$ substituted for the randomized-groups error term. For this example, each F ratio is tested against critical F from Table A.1, with numerator $\text{df} = 1$ and denominator $\text{df} = (\frac{1}{2}(abcden) - 1) - 15$. The null hypothesis for each effect is rejected if the obtained F is equal to or exceeds the tabled F.

10.4.3 Regression Approach to a 2^5 Half-Factorial ANOVA

The regression approach and the concept of orthogonality "come into their own" with screening and incomplete designs. Basically, the effects that can be tested unambiguously are those that are in orthogonal columns. Orthogonal columns are first set up to test main effects; then, as usual, columns are cross multiplied to produce interactions. However, because the design is not fully factorial, some of the codes in

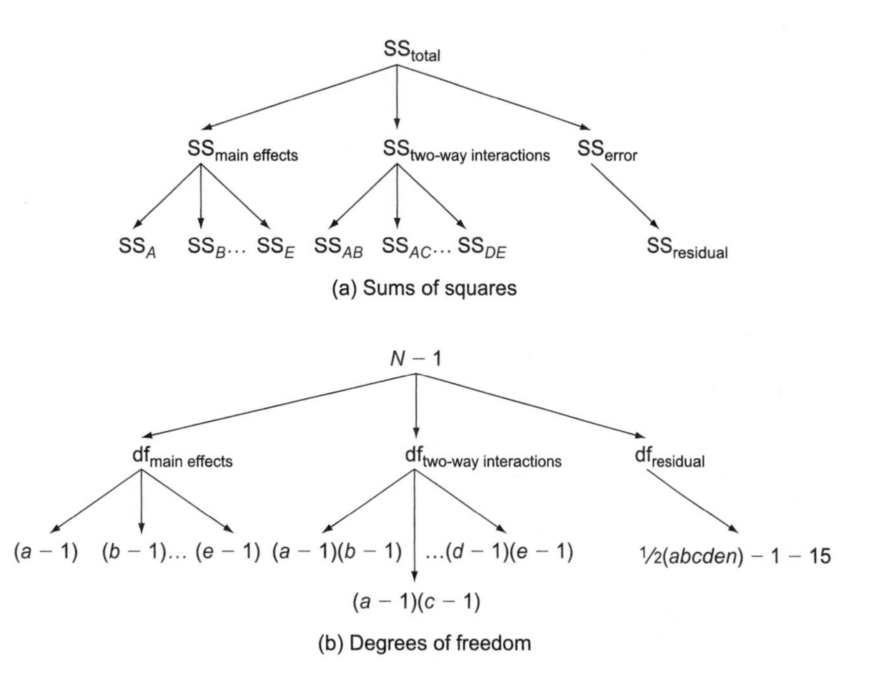

FIGURE 10.1 Partition of (a) sums of squares and (b) degrees of freedom into systematic sources of differences among marginal means of A, B, C, D, and E; differences due to two-way interactions $A \times B$, $A \times C$, $A \times D$, $A \times E$, $B \times C$, $B \times D$, $B \times E$, $C \times D$, $C \times E$, and $D \times E$; and remaining (error) differences among cases

the columns for interactions are the same as codes in other columns for either main effects or other interactions. The columns with the same codes are aliased. Because the regression approach is clear for these designs (and is the approach taken by most of the software), only data analysis through the regression approach is illustrated.

A sequence of runs is developed either by hand or through software to determine the combinations of levels of IVs presented to cases. Each case is randomly assigned to a run, and that run determines the combination of levels of IVs assigned to the case. In Table 10.1, for instance, the first case (s_1) is assigned to the run with the a_1, b_1, c_1, d_1, e_1 combination; the ninth case (s_9) is assigned to the run with the a_1, b_2, c_1, d_2, e_1 combination; and so on.

The half-factorial design for this fictitious example has five IVs (each with two levels), and each combination has two replications (two cases per cell). The goal of the research is to determine which factors contribute to the joy of dancing. The DV is the rating of enjoyment a dancer gives after performing a belly dancing routine.[1] The first IV (A) is whether the dancer views a film of a professional belly dancer before performing his or her own routine. The second IV (B) is whether the dancer is wearing a professional or a practice costume. The third IV (C) is whether the routine is being videotaped. The fourth IV (D) is whether the dancer is performing with live musicians or to taped music. The fifth IV (E) is whether there are other dancers in the audience. Two dancers are assigned to each of the 16 cells in the design, as in Table 10.1.

Table 10.3 shows the sequence of runs and the combination of levels of IVs associated with each run. The runs are developed by software in a format appropriate for the regression approach to screening designs, which is easily used to analyze

[1] Dancers are extensively trained to produce reliable ratings of enjoyment.

TABLE 10.3 Small-sample hypothetical data set for 2^5 half-fractional–factorial ANOVA

		A	B	C	D	E	AB	AC	AD	AE	BC	BD	BE	CD	CE	DE
Run	Y	X_1	X_2	X_3	X_4	X_5	X_6	X_7	X_8	X_9	X_{10}	X_{11}	X_{12}	X_{13}	X_{14}	X_{15}
2	8	1	−1	−1	−1	−1	−1	−1	−1	−1	1	1	1	1	1	1
10	8	1	−1	−1	1	1	−1	−1	1	1	1	−1	−1	−1	−1	1
6	7	1	−1	1	−1	1	−1	1	−1	1	−1	1	−1	−1	1	−1
14	9	1	−1	1	1	−1	−1	1	1	−1	−1	−1	1	1	−1	−1
4	7	1	1	−1	−1	1	1	−1	−1	1	−1	−1	1	1	−1	−1
12	7	1	1	−1	1	−1	1	−1	1	−1	−1	1	−1	−1	1	−1
8	8	1	1	1	−1	−1	1	1	−1	−1	1	−1	−1	−1	−1	1
16	6	1	1	1	1	1	1	1	1	1	1	1	1	1	1	1
1	3	−1	−1	−1	−1	1	1	1	1	−1	1	1	−1	1	−1	−1
9	1	−1	−1	−1	1	−1	1	1	−1	1	1	−1	1	−1	1	−1
5	2	−1	−1	1	−1	−1	1	−1	1	1	−1	1	1	−1	−1	1
13	2	−1	−1	1	1	1	1	−1	−1	−1	−1	−1	−1	1	1	1
3	2	−1	1	−1	−1	−1	−1	1	1	1	−1	−1	−1	1	1	1
11	3	−1	1	−1	1	1	−1	1	−1	−1	−1	1	1	−1	−1	1
7	4	−1	1	1	−1	1	−1	−1	1	−1	1	−1	1	−1	1	−1
15	3	−1	1	1	1	−1	−1	−1	−1	1	1	1	−1	1	−1	−1
2	7	1	−1	−1	−1	−1	−1	−1	−1	−1	1	1	1	1	1	1
10	7	1	−1	−1	1	1	−1	−1	1	1	1	−1	−1	−1	−1	1
6	8	1	−1	1	−1	1	−1	1	−1	1	−1	1	−1	−1	1	−1
14	8	1	−1	1	1	−1	−1	1	1	−1	−1	−1	1	1	−1	−1
4	8	1	1	−1	−1	1	1	−1	−1	1	−1	−1	1	1	−1	−1
12	7	1	1	−1	1	−1	1	−1	1	−1	−1	1	−1	−1	1	−1
8	8	1	1	1	−1	−1	1	1	−1	−1	1	−1	−1	−1	−1	1
16	6	1	1	1	1	1	1	1	1	1	1	1	1	1	1	1
1	4	−1	−1	−1	−1	1	1	1	1	−1	1	1	−1	1	−1	−1
9	3	−1	−1	−1	1	−1	1	1	−1	1	1	−1	1	−1	1	−1
5	2	−1	−1	1	−1	−1	1	−1	1	1	−1	1	1	−1	−1	1
13	1	−1	−1	1	1	1	1	−1	−1	−1	−1	−1	−1	1	1	1
3	3	−1	1	−1	−1	−1	−1	1	1	1	−1	−1	−1	1	1	1
11	4	−1	1	−1	1	1	−1	1	−1	−1	−1	1	1	−1	−1	1
7	3	−1	1	1	−1	1	−1	−1	1	−1	1	−1	1	−1	1	−1
15	4	−1	1	1	1	−1	−1	−1	−1	1	1	1	−1	1	−1	−1
Sum	163	0	0	0	0	0	0	0	0	0	0	0	0	0	0	0
Squares Summed	1031	32	32	32	32	32	32	32	32	32	32	32	32	32	32	32

Run	YX_1	YX_2	YX_3	YX_4	YX_5	YX_6	YX_7	YX_8	YX_9	YX_{10}	YX_{11}	YX_{12}	YX_{13}	YX_{14}	YX_{15}
2	8	−8	−8	−8	−8	−8	−8	−8	−8	8	8	8	8	8	8
10	8	−8	−8	8	8	−8	−8	8	8	8	−8	−8	−8	−8	8
6	7	−7	7	−7	7	−7	7	−7	7	−7	7	−7	−7	7	−7
14	9	−9	9	9	−9	−9	9	9	−9	−9	−9	9	9	−9	−9
4	7	7	−7	−7	7	7	−7	−7	7	−7	−7	7	7	−7	−7
12	7	7	−7	7	−7	7	−7	7	−7	−7	7	−7	−7	7	−7
8	8	8	8	−8	−8	8	8	−8	−8	8	−8	−8	−8	−8	8
16	6	6	6	6	6	6	6	6	6	6	6	6	6	6	6
1	−3	−3	−3	−3	3	3	3	3	−3	3	3	−3	3	−3	−3
9	−1	−1	−1	1	−1	1	1	−1	1	1	−1	1	−1	1	−1
5	−2	−2	2	−2	−2	2	−2	2	2	−2	2	2	−2	−2	2
13	−2	−2	2	2	2	2	−2	−2	−2	−2	−2	−2	2	2	2
3	−2	2	−2	−2	−2	−2	2	2	2	−2	−2	−2	2	2	2
11	−3	3	−3	3	3	−3	3	−3	−3	−3	3	3	−3	−3	3
7	−4	4	4	−4	4	−4	−4	4	−4	4	−4	4	−4	4	−4
15	−3	3	3	3	−3	−3	−3	−3	3	3	3	−3	3	−3	−3
2	7	−7	−7	−7	−7	−7	−7	−7	−7	7	7	7	7	7	7
10	7	−7	−7	7	7	−7	−7	7	7	7	−7	−7	−7	−7	7
6	8	−8	8	−8	8	−8	8	−8	8	−8	8	−8	−8	8	−8
14	8	−8	8	8	−8	−8	8	8	−8	−8	−8	8	8	−8	−8
4	8	8	−8	−8	8	8	−8	−8	8	−8	−8	8	8	−8	−8
12	7	7	−7	7	−7	7	−7	7	−7	−7	7	−7	−7	7	−7
8	8	8	8	−8	−8	8	8	−8	−8	8	−8	−8	−8	−8	8
16	6	6	6	6	6	6	6	6	6	6	6	6	6	6	6
1	−4	−4	−4	−4	4	4	4	4	−4	4	4	−4	4	−4	−4
9	−3	−3	−3	3	−3	3	3	−3	3	3	−3	3	−3	3	−3
5	−2	−2	2	−2	−2	2	−2	2	2	−2	2	2	−2	−2	2
13	−1	−1	1	1	1	1	−1	−1	−1	−1	−1	−1	1	1	1
3	−3	3	−3	−3	−3	−3	3	3	3	−3	−3	−3	3	3	3
11	−4	4	−4	4	4	−4	4	−4	−4	−4	4	4	−4	−4	4
7	−3	3	3	−3	3	−3	−3	3	−3	3	−3	3	−3	3	−3
15	−4	4	4	4	−4	−4	−4	−4	4	4	4	−4	4	−4	−4
Sum	75	3	−1	−5	−1	−13	3	−1	−9	3	−1	−1	−1	−13	−9
Squares Summed															

the results. There are many things to notice about Table 10.3. The first column gives the numbers for the runs. Because there are two replications, the sequence of runs is repeated twice—one block of 16 runs above the second block of 16 runs. The second column is for the DV score (enjoyment, rated on a 10-point scale). Columns 3–7 code levels for IVs *A–E*, respectively, where 1 or −1 designates the level. Each row of these five columns is a different run because each row represents a different combination of levels of the five IVs. Consider the first row. It is run 2 with a DV score of 8. This case received the a_1, b_2, c_2, d_2, e_2 combination; that is, the dancer randomly assigned to the second run saw a tape of a professional dancer before performing her own routine, but danced in a practice costume, was not videotaped, danced to taped music, and had no other dancers in the audience. Going back to Table 10.1, this dancer was either s_{29} or s_{30}. In Table 10.3, columns 8–17 for the two-way interactions are cross products of columns 3–7. The remaining columns cross the DV values in column 2 with the codes of columns 3–17 to produce the sums of products. Preliminary sums of squares are at the bottom of columns 3–17 and sums of products are at the bottom of columns 18–32.

This half factorial is only one of a very large number of sets of runs that accomplish the same purpose. For instance, assignment of IVs to columns is arbitrary. A different assignment results in a different set of runs. The five columns that code for IVs (columns 3–7) meet the criteria for orthogonality: Their coefficients sum to 0, as do the pairwise cross products of the columns. Thus, the five main effects and their 10 two-way interactions are mutually orthogonal. Any other set of orthogonal coefficients chosen for the columns coding the main effects has the same properties but results in a different set of runs.

With a bit of effort, the pattern of aliasing is also available from Table 10.3. A three-way interaction in full factorial ANOVA is the product of the three columns that code for the main effects. For example, the *ABC* interaction is the product of the column for *A* times the column for *B* times the column for *C*. If *A*, *B*, and *C* are multiplied in Table 10.3, the result for the first (and second) block of 16 runs is 1, 1, −1, −1, −1, −1, 1, 1, −1, −1, 1, 1, 1, 1, −1, −1. However, this is exactly the same set of coefficients as for the *DE* interaction. Thus, in this reduced design, *ABC* is aliased with *DE*. With some patience, the other sets of confounded effects in the design are found.

The equations for preliminary sums of squares and sums of products for the regression approach are the same as illustrated in Table 5.10 for main effects in factorial ANOVA. Table 10.4 shows the equations for finding the sums of squares for the main effects *A–E*. Table 10.5 shows the equations for the two-way interactions.

You have to find the sum of squares total (SS_T) in order to compute the sum of squares for residual. Residual sum of squares is found by applying Equation 10.2:

$$SS_T = 1031 - \frac{(163)^2}{32} = 200.72$$

$$\begin{aligned} SS_{residual} = {} & 200.72 - 175.78 - 0.28 - 0.03 - 0.78 - 0.03 - 5.28 - 0.28 - 0.03 \\ & - 2.53 - 0.28 - 0.03 - 0.03 - 0.03 - 5.28 - 2.53 = 7.52 \end{aligned}$$

and residual df is found by applying Equation 10.3:

$$df_{residual} = \tfrac{1}{2}(2)(2)(2)(2)(2)(2) - 1 - 15 = 16$$

TABLE 10.4 Equations for the sum of squares for the main effects

$$SS(reg.X_1) = SS_A = \frac{[SP(YX_1)]^2}{SS(X_1)} = \frac{75^2}{32} = 175.78$$

$$SS(reg.X_2) = SS_B = \frac{[SP(YX_2)]^2}{SS(X_2)} = \frac{3^2}{32} = 0.28$$

$$SS(reg.X_3) = SS_C = \frac{[SP(YX_3)]^2}{SS(X_3)} = \frac{(-1)^2}{32} = 0.03$$

$$SS(reg.X_4) = SS_D = \frac{[SP(YX_4)]^2}{SS(X_4)} = \frac{(-5)^2}{32} = 0.78$$

$$SS(reg.X_5) = SS_E = \frac{[SP(YX_5)]^2}{SS(X_5)} = \frac{(-1)^2}{32} = 0.03$$

TABLE 10.5 Equations for the sum of squares for the two-way interactions

$$SS(reg.X_6) = SS_{AB} = \frac{[SP(YX_6)]^2}{SS(X_6)} = \frac{(-13)^2}{32} = 5.28$$

$$SS(reg.X_7) = SS_{AC} = \frac{[SP(YX_7)]^2}{SS(X_7)} = \frac{3^2}{32} = 0.28$$

$$SS(reg.X_8) = SS_{AD} = \frac{[SP(YX_8)]^2}{SS(X_8)} = \frac{(-1)^2}{32} = 0.03$$

$$SS(reg.X_9) = SS_{AE} = \frac{[SP(YX_9)]^2}{SS(X_9)} = \frac{(-9)^2}{32} = 2.53$$

$$SS(reg.X_{10}) = SS_{BC} = \frac{[SP(YX_{10})]^2}{SS(X_{10})} = \frac{3^2}{32} = 0.28$$

$$SS(reg.X_{11}) = SS_{BD} = \frac{[SP(YX_{11})]^2}{SS(X_{11})} = \frac{(-1)^2}{32} = 0.03$$

$$SS(reg.X_{12}) = SS_{BE} = \frac{[SP(YX_{12})]^2}{SS(X_{12})} = \frac{(-1)^2}{32} = 0.03$$

$$SS(reg.X_{13}) = SS_{CD} = \frac{[SP(YX_{13})]^2}{SS(X_{13})} = \frac{(-1)^2}{32} = 0.03$$

$$SS(reg.X_{14}) = SS_{CE} = \frac{[SP(YX_{14})]^2}{SS(X_{14})} = \frac{(-13)^2}{32} = 5.28$$

$$SS(reg.X_{15}) = SS_{DE} = \frac{[SP(YX_{15})]^2}{SS(X_{15})} = \frac{(-9)^2}{32} = 2.53$$

Table 10.6 summarizes the results of the calculations and shows the mean squares and F values. Each F ratio is evaluated separately, as usual. At $\alpha = .05$, the critical value from Table A.1 is 4.49, based on numerator df = 1 and denominator df = 16 for all effects. The main effect of A (viewing a film of a professional dancer

TABLE 10.6 Source table for 2^5 fractional-factorial ANOVA

Source	SS	df	MS	F
A	175.78	1	175.78	374.00
B	0.28	1	0.28	0.60
C	0.03	1	0.03	0.06
D	0.78	1	0.78	1.66
E	0.03	1	0.03	0.06
AB	5.28	1	5.28	11.23
AC	0.28	1	0.28	0.60
AD	0.03	1	0.03	0.06
AE	2.53	1	2.53	5.38
BC	0.28	1	0.28	0.60
BD	0.03	1	0.03	0.06
BE	0.03	1	0.03	0.06
CD	0.03	1	0.03	0.06
CE	5.28	1	5.28	11.23
DE	2.53	1	2.53	5.38
Residual	7.52	16	0.47	
T	200.75	31		

prior to the dance) and the interactions $A \times B$ (viewing the film or not by wearing a professional costume when dancing or not), $A \times E$ (viewing the film or not by dancing with other dancers in the audience or not), $C \times E$ (being videotaped and dancing with other dancers in the audience or not), and $D \times E$ (dancing with live musicians or not by dancing for audience of other dancers) are all significant. Thus, all the IVs participate in at least one statistically significant effect—not an entirely happy result for a screening run. (See Section 5.6.1 for interpretation of interactions.)

Note that there are 15 F tests for the small data set—an invitation to inflated familywise Type I error. A prudent researcher would consider a Bonferroni-type adjustment to α_{FW}, $(.05/15) = .003$, if this analysis is to be used for final decision making rather than just screening. By this criterion, only the main effect of A is statistically significant. The remaining borderline effects would probably be considered for further research, however.

10.4.4 Computer Analyses of the Small-Sample 2^{5-1} Half-Factorial ANOVA

Data are entered in the format of the first seven columns of Table 10.3: run number (optional), DV, and the codes (1 or −1) for the five IVs. The columns may be in any order. Tables 10.7 and 10.8 show computer analyses of the small-sample example of the 2^{5-1} half-factorial ANOVA of Section 10.4.3 by SPSS GLM and SAS GLM, respectively.

The SPSS GLM syntax in Table 10.7 is the same as for the Latin-square design of Chapter 9, except that two-way interactions are requested in addition to main effects in the TABLES instructions and in the DESIGN instruction. ETASQ and OPOWER request effect sizes and observed power statistics.

The output is the same as for Latin square but longer due to inclusion of all two-way interactions. The *F* ratios in the source table (Tests of Between-Subject Effects) differ slightly from those of Table 10.6, due to rounding error in the hand calculations. At the end of the output are marginal means for main effects and cell means for the two-way interactions (only two of the tables are reproduced here). The standard error for each marginal mean is the same—the square root of $MS_{residual}$ divided by the square root of the 16 cases in each marginal mean. The standard error for each cell mean is the same—the square root of $MS_{residual}$ divided by the square root of the 8 cases in each two-way cell. Other output is as described in Section 9.4.4 for the Latin-square design.

TABLE 10.7 Fractional-factorial ANOVA for small-sample example through SPSS GLM

```
UNIANOVA
  rating BY prof_dan costume video music audience
  /METHOD = SSTYPE(3)
  /INTERCEPT = INCLUDE
  /EMMEANS = TABLES(prof_dan) /EMMEANS = TABLES(costume)
  /EMMEANS = TABLES(video) /EMMEANS = TABLES(music)
  /EMMEANS = TABLES(audience) /EMMEANS = TABLES(costume*prof_dan)
  /EMMEANS = TABLES(prof_dan*video) /EMMEANS = TABLES(music*prof_dan)
  /EMMEANS = TABLES(audience*prof_dan) /EMMEANS = TABLES(costume*video)
  /EMMEANS = TABLES(costume*music) /EMMEANS = TABLES(audience*costume)
  /EMMEANS = TABLES(music*video) /EMMEANS = TABLES(audience*video)
  /EMMEANS = TABLES(audience*music)
  /PRINT = ETASQ OPOWER
  /CRITERIA = ALPHA(.05)
  /DESIGN = prof_dan costume video music audience costume*prof_dan prof_dan
 *video music*prof_dan audience*prof_dan costume*video costume*music audience
 *costume music*video audience*video audience*music .
```

Univariate Analysis of Variance

Between-Subjects Factors

		Value Label	N
Professional dancer film	−1.00	No	16
	1.00	Yes	16
Professional costume	−1.00	No	16
	1.00	Yes	16
Videotaping routine	−1.00	No	16
	1.00	Yes	16
Live musicians	−1.00	No	16
	1.00	Yes	16
Audience of other dancers	−1.00	No	16
	1.00	Yes	16

TABLE 10.7 Continued

Tests of Between-Subjects Effects

Dependent Variable: rating

Source	Type III Sum of Squares	df	Mean Square	F	Sig.	Partial Eta Squared	Noncent. Parameter	Observed Power[a]
Corrected Model	193.219[b]	15	12.881	27.480	.000	.963	412.200	1.000
Intercept	830.281	1	830.281	1771.267	.000	.991	1771.267	1.000
prof_dan	175.781	1	175.781	375.000	.000	.959	375.000	1.000
costume	.281	1	.281	.600	.450	.036	.600	.113
video	.031	1	.031	.067	.800	.004	.067	.057
music	.781	1	.781	1.667	.215	.094	1.667	.229
audience	.031	1	.031	.067	.800	.004	.067	.057
prof_dan * costume	5.281	1	5.281	11.267	.004	.413	11.267	.883
prof_dan * video	.281	1	.281	.600	.450	.036	.600	.113
prof_dan * music	.031	1	.031	.067	.800	.004	.067	.057
prof_dan * audience	2.531	1	2.531	5.400	.034	.252	5.400	.588
costume * video	.281	1	.281	.600	.450	.036	.600	.113
costume * music	.031	1	.031	.067	.800	.004	.067	.057
costume * audience	.031	1	.031	.067	.800	.004	.067	.057
video * music	.031	1	.031	.067	.800	.004	.067	.057
video * audience	5.281	1	5.281	11.267	.004	.413	11.267	.883
music * audience	2.531	1	2.531	5.400	.034	.252	5.400	.588
Error	7.500	16	.469					
Total	1031.000	32						
Corrected Total	200.719	31						

a. Computed using alpha = .05

b. R Squared = .963 (Adjusted R Squared = .928)

Estimated Marginal Means

1. Professional dancer film

Dependent Variable: rating

Professional dancer film	Mean	Std. Error	95% Confidence Interval	
			Lower Bound	Upper Bound
No	2.750	.171	2.387	3.113
Yes	7.438	.171	7.075	7.800

6. Professional costume * Professional dancer film

Dependent Variable: rating

Professional costume	Professional dancer film	Mean	Std. Error	95% Confidence Interval	
				Lower Bound	Upper Bound
No	No	2.250	.242	1.737	2.763
	Yes	7.750	.242	7.237	8.263
Yes	No	3.250	.242	2.737	3.763
	Yes	7.125	.242	6.612	7.638

SAS GLM requires a MODEL instruction, as shown in Table 10.8, which includes all the effects to be assessed: 5 main effects and 10 interactions. All 15 effects are also in the request for means. Output is limited by requesting only Type III SS (ss3), which for this analysis is identical to that of Type I SS. (Section 5.6.5 discusses these types of sums of squares.)

Output follows that of the Latin-square example of Section 9.4.4. Cell means and standard deviations follow those for marginal values.

TABLE 10.8 Fractional-factorial ANOVA for small-sample example through SAS GLM (syntax and selected output)

```
proc glm data=SASUSER.FRACFAC2
     class PROF_DAN COSTUME VIDEO MUSIC AUDIENCE;
     model RATING = PROF_DAN COSTUME VIDEO MUSIC AUDIENCE
                PROF_DAN*COSTUME PROF_DAN*VIDEO PROF_DAN*MUSIC
                PROF_DAN*AUDIENCE COSTUME*VIDEO COSTUME*MUSIC
                COSTUME*AUDIENCE VIDEO*MUSIC VIDEO*AUDIENCE
                MUSIC*AUDIENCE /ss3;
     means PROF_DAN COSTUME VIDEO MUSIC AUDIENCE
             PROF_DAN*COSTUME PROF_DAN*VIDEO PROF_DAN*MUSIC
             PROF_DAN*AUDIENCE COSTUME*VIDEO COSTUME*MUSIC
             COSTUME*AUDIENCE VIDEO*MUSIC VIDEO*AUDIENCE
             MUSIC*AUDIENCE;
run;
```

The GLM Procedure

Dependent Variable: RATING

Source	DF	Sum of Squares	Mean Square	F Value	Pr > F
Model	15	193.2187500	12.8812500	27.48	<.0001
Error	16	7.5000000	0.4687500		
Corrected Total	31	200.7187500			

R-Square	Coeff Var	Root MSE	RATING Mean
0.962634	13.44104	0.684653	5.093750

Source	DF	Type III SS	Mean Square	F Value	Pr > F
PROF_DAN	1	175.7812500	175.7812500	375.00	<.0001
COSTUME	1	0.2812500	0.2812500	0.60	0.4499
VIDEO	1	0.0312500	0.0312500	0.07	0.7995
MUSIC	1	0.7812500	0.7812500	1.67	0.2150
AUDIENCE	1	0.0312500	0.0312500	0.07	0.7995
PROF_DAN*COSTUME	1	5.2812500	5.2812500	11.27	0.0040
PROF_DAN*VIDEO	1	0.2812500	0.2812500	0.60	0.4499
PROF_DAN*MUSIC	1	0.0312500	0.0312500	0.07	0.7995
PROF_DAN*AUDIENCE	1	2.5312500	2.5312500	5.40	0.0336
COSTUME*VIDEO	1	0.2812500	0.2812500	0.60	0.4499
COSTUME*MUSIC	1	0.0312500	0.0312500	0.07	0.7995
COSTUME*AUDIENCE	1	0.0312500	0.0312500	0.07	0.7995
VIDEO*MUSIC	1	0.0312500	0.0312500	0.07	0.7995
VIDEO*AUDIENCE	1	5.2812500	5.2812500	11.27	0.0040
MUSIC*AUDIENCE	1	2.5312500	2.5312500	5.40	0.0336

TABLE 10.8 Continued

Level of PROF_DAN	N	RATING Mean	RATING Std Dev
-1	16	2.75000000	1.00000000
1	16	7.43750000	0.81394103

Level of PROF_DAN	Level of COSTUME	N	RATING Mean	RATING Std Dev
-1	-1	8	2.25000000	1.03509834
-1	1	8	3.25000000	0.70710678
1	-1	8	7.75000000	0.70710678
1	1	8	7.12500000	0.83452296

10.5 TYPES OF SCREENING AND OTHER INCOMPLETE DESIGNS

There are several types of screening and other incomplete designs for randomized-groups IVs with two or more levels. Before discussing these designs, however, it is useful to define *resolution.*

10.5.1 Resolution of Incomplete Designs

Incomplete designs differ in degree of aliasing and in the number of interactions that can be tested. The more interactions that can be tested, the higher the resolution. With low-resolution designs, only main effects are tested; with higher resolution, some or all of the two-way interactions can also be tested. Three-way and higher-order interactions are aliased with main effects and with two-way interactions in all of these designs, and so they are unavailable for analysis. Table 10.9 describes designs in terms of their resolution, the kinds of aliasing that occur, and the effects that can be tested.

The 2^5 half-factorial design with replications of Table 10.1 (also called 2^{5-1}) is resolution V. The quarter-factorial design of Table 10.2 with no replications is resolution III. Latin-square, Plackett-Burman, and most Taguchi designs are also resolution III. The 2^{4-1} half-fractional factorial of Section 10.5.2.1 is resolution IV. Some resolution III designs are turned into resolution IV designs by a fold-over technique, in which the number of runs is doubled and all the coefficients are reversed (1 changed to −1 and vice versa).

10.5.2 Fractional-Factorial Designs

The fractional-factorial design of Table 10.1 is a half factorial because half the combinations of a full factorial are used while the other half of the cells are empty. Other

TABLE 10.9 Resolution in various screening designs

Resolution	Description	When to Use
III	Main effects are independent of each other but are aliased with interactions.	You are willing to assume that all interactions are negligible and will base future research only on screened main effects.
IV	Main effects are independent of each other and independent of two-way interactions, but some two-way interactions are aliased with others.	You are willing to assume that all third- and higher-order interactions are negligible. You are unable to determine ahead of time which two-way interactions are negligible and which are worth testing.
V	Main effects and two-way interactions are all independent of each other.	You are willing to assume that all third- and higher-order interactions are negligible. You want to test all main effects and two-way interactions.

fractions of a full factorial are also possible, as shown in Table 10.2. This section discusses other fractions in designs with two levels of each IV and in designs with three levels of each IV.

10.5.2.1 Two-Level Fractional-Factorial Designs

The number of runs in a fractional-factorial design in which all IVs have 2 levels (sometimes called a Box-Hunter design) is always 2 to some power. Recall that a run is the combination of levels of IVs assigned to a case. Two raised to a power is the number of combinations in the study. For example, a full factorial with five IVs has $2^5 = 32$ combinations. A half factorial has 1 less than the power of 2 in the full factorial. So a half-fractional factorial with five IVs has $2^{5-1} = 2^4 = 16$ runs (as in Table 10.1). A quarter factorial has 2 fewer than the exponent of 2 in the full factorial. Therefore, a quarter factorial with five IVs has $2^{5-2} = 2^3 = 8$ runs (as in Table 10.2), and so on.

In general, the number of runs (not including replicates) in a fractional factorial is

$$r = 2^{k-q} \tag{10.5}$$

The number of runs (r) in a two-level fractional-factorial design is 2 to the power of $(k - q)$, where k is the number of IVs and q is some number less than k. In a half factorial, $q = 1$; in a quarter factorial, $q = 2$; and so on.

The smaller the number of runs relative to the full factorial, the lower the resolution and the fewer the effects that can be analyzed. A 2^5 quarter factorial permits analyses only of main effects (which are aliased with three-way interactions); the two-way interactions are all aliased with each other and with bits and pieces of higher-order interactions. Other parts of higher-order interactions are folded into the residual error term. As shown in Section 10.4, a 2^{5-1} half factorial with two replications permits analysis of main effects (aliased with four-way interactions) and all two-way interactions (aliased with three-way interactions). Typically, fractional-factorial designs are not used to analyze interactions higher than two-way.

The number of replicates affects both the power of the study and the effects that are assessable. With one replicate per run, there is one case for each combination of levels of IVs. With two replicates per run, there are two cases for each combination of levels of IVs, and so on. When there is one replication and sample size is small, analyses of some effects are sacrificed to provide degrees of freedom for error. Obviously, omitted effects should be those that are not of research concern and that are likely to be negligible.

How do you generate a fractional-factorial design by hand,[2] and how do you know which effects are aliased? The first step is to determine the number of IVs (the value of k) and whether tests of only main effects or tests of main effects plus two-way interactions are needed. This determines the resolution of the design, the fraction required, and thus the value of q. The second step is to use Equation 10.5 to determine the number of runs required. This also determines the minimum number of cases needed if there is only one replication. The third step is to generate a set of orthogonal coefficients for each IV for the required number of runs, as shown, for example, in Table 10.3. The last step is to cross multiply the coefficients in the IV columns for main effects to produce various interactions. This reveals the pattern of aliasing, because columns with the same codes are aliased. Only in fully factorial designs are all the columns for main effects and interactions different from each other (orthogonal). Once the columns for main effects are coded, the cross products of columns for various interactions are easily found using a spreadsheet program, such as Quattro or Excel or one of the data-entry programs of the statistical packages.

Schmidt and Launsby (1994) present a more formal approach (with jargon) to identifying the pattern of aliasing. Suppose in a 2^{4-1} design, for example, D (the last main effect) is aliased with ABC (expected to be negligible).[3] The expression, $D = ABC$, is called the *half fraction design generator*, and ABC is called the *defining word*. The *defining relation* is $I = ABCD$. The rest of the aliasing pattern is identified by multiplying both sides of the defining relation by the effect in question. To identify the factor aliased with AB,

$$\begin{aligned} AB(I) &= AB(ABCD) \\ &= A^2B^2CD = IICD = CD \end{aligned}$$

where I is the identity column with all entries equal to 1, because any column multiplied by itself equals I (e.g., $A^2 = I$). Therefore, the AB interaction is confounded with the CD interaction. Testing one of these interactions lets you know that there is a significant two-way effect of either $A \times B$ or $C \times D$ (or both), but you cannot tell which. Similarly, AC is aliased with BD and so on.

Other main effects (besides D) are also aliased. For example,

$$\begin{aligned} A(I) &= A(ABCD) \\ &= A^2BCD = I(BCD) = BCD \end{aligned}$$

This design is of resolution IV, because it permits tests of main effects (aliased with three-way interactions) and of three of the six two-way interactions (each aliased with one of the other three). Three-way interactions are aliased with main effects. The four-way interaction is I, a column of ones, and is equivalent to the intercept in a regression model.

[2] The easiest way, by far, to generate a fractional-factorial design is through software, as illustrated later in this section.

[3] In your own study, you can determine what is aliased with what by your assignment of IVs to the letters A, B, C, and D.

Schmidt and Launsby (1994, pp. 483–486) demonstrate aliasing patterns in a 2^{7-3} design. Other tables of aliasing patterns are provided by Davies (1963), Montgomery (1984, pp. 338–339), and Woodward, Bonett, and Brecht (1990, pp. 246–247). Further discussion of these issues in terms of resolution is in Section 10.6.1.

SAS FACTEX generates a fractional-factorial design when you provide the number of runs, the number of factors, the resolution (Section 10.6.1), and, optionally, a blocking variable to represent replications, if any. (SAS FACTEX labels replicates as blocks.) Table 10.10 shows syntax and printed output for the 2^{4-1} design ($2^3 = 8$ runs). A data file, SASUSER.FRACFAC, is requested. Each run of this syntax produces a different output table because of the `randomize` instruction.

TABLE 10.10 Generating a 2^{4-1} fractional-factorial design through SAS FACTEX (syntax and output)

```
proc factex ;
  factors
     A B C D;
      size design = 8;
      model resolution = 4;
run;
examine aliasing;
quit;
proc factex ;
   factors A B C D;
      size design = 8;
      model resolution = 4;
run;
output out = SASUSER.FRACFAC
   randomize novalran;
quit;
run;
data SASUSER.FRACFAC; set SASUSER.FRACFAC end=adxeof;
   Y = .;
run;
data SASUSER.FRACFAC; set SASUSER.FRACFAC; _adxran = ranuni(0);
proc sort data=SASUSER.FRACFAC
     out=SASUSER.FRACFAC(drop=_adxran);
     by  _adxran;
run;
proc print data=SASUSER.FRACFAC;
run;

Aliasing Structure

 A
 B
 C
 D
 A*B = C*D
 A*C = B*D
 A*D = B*C
```

TABLE 10.11 Output data set (SASUSER.FRACFAC) from SAS FACTEX run (syntax is in Table 10.10)

OBS	A	B	C	D	Y
1	1	−1	1	−1	12
2	−1	−1	−1	−1	5
3	−1	−1	1	1	6
4	−1	1	−1	1	3
5	−1	1	1	−1	4
6	1	1	−1	−1	11
7	1	−1	−1	1	9
8	1	1	1	1	8

The output shows the pattern of aliasing: All main effects are unconfounded with each other and with two-way interactions (although they are confounded with three-way interactions), and each two-way interaction is aliased with another. The set of runs for one replicate (block) is output in spreadsheet format in Table 10.11. You may copy the design for as many replicates as you want. Hypothetical values of the DV, Y, have been added by hand in Table 10.11.

Other programs that help you generate two-level fractional-factorial designs are SYSTAT DESIGN and NCSS Design of Experiments.

The analysis proceeds as in Section 10.4, where only those effects that are interpretable are included in the model. This means that in a resolution III design (cf. Section 10.5.1), you analyze only main effects. In a resolution IV design, you analyze the main effects and the two-way interactions of interest by choosing one from among each set that are aliased. For example, if AB and CD are aliased, you analyze the AB interaction if you believe the CD interaction is negligible. Otherwise, the result may be due to a CD interaction or to both AB and CD interactions. In a resolution V design, all main effects and two-way interactions are available for analysis. In resolution IV and V designs, you can always pool unimportant two-way interactions into the residual term to gain degrees of freedom and power if you expect those interactions to be negligible.

Table 10.12 shows the analysis only of main effects through SAS GLM, using the data in Table 10.11. Two-way interactions are pooled into residual, the error term, because they are expected to be negligible. The request for `lsmeans` is for marginal means for all main effects.

Only the main effect of A is statistically significant, and only the marginal means for A are shown; remaining marginal means are deleted. Further research is designed to explore A, perhaps through an experiment with more levels of A.

10.5.2.2 Three-Level Fractional-Factorial Designs

Fractional-factorial designs in which IVs have three levels are more complicated because interactions, as well as main effects, have more than 1 df, and there is much

TABLE 10.12 Analysis of 2^{4-1} fractional-factorial design through SAS GLM (syntax and selected output)

```
proc glm Data=SASUSER.FRACFAC;
     class A B C D;
     model Y = A B C D / ss3;
     lsmeans A B C D/ ;
run;
```

The GLM Procedure
Class Level Information

Class	Levels	Values
A	2	−1 1
B	2	−1 1
C	2	−1 1
D	2	−1 1

Number of Observations Read	8
Number of Observations Used	8

The GLM Procedure

Dependent Variable: Y

Source	DF	Sum of Squares	Mean Square	F Value	Pr > F
Model	4	70.00000000	17.50000000	9.55	0.0470
Error	3	5.50000000	1.83333333		
Corrected Total	7	75.50000000			

R-Square	Coeff Var	Root MSE	Y Mean
0.927152	18.67595	1.354006	7.250000

Source	DF	Type III SS	Mean Square	F Value	Pr > F
A	1	60.50000000	60.50000000	33.00	0.0105
B	1	4.50000000	4.50000000	2.45	0.2152
C	1	0.50000000	0.50000000	0.27	0.6376
D	1	4.50000000	4.50000000	2.45	0.2152

Least Squares Means

A	Y LSMEAN
−1	4.5000000
1	10.0000000

more opportunity for confounding. In a 1/3 fractional factorial (3^{k-1}), for instance, each effect has three aliases, instead of the two in the two-level half-fractional factorial; in a 1/9 fractional factorial (3^{k-2}), each effect has nine aliases; and so on. In addition, more than five IVs are necessary for reasonable tests of two-way interactions, and even then they are often difficult to interpret (Kirk, 1995). For these reasons, the design is not discussed further here; if you are interested, Montgomery (1984) discusses 3^{k-q} designs, and Kirk (1995) provides a general discussion of 3^{k-q} designs and a complete worked-out example of the 3^{4-1} design. When IVs have three levels, the Latin-square design of Chapter 9 and the Box-Behnken and Taguchi designs described in this chapter are probably more practical than the fractional factorial.

10.5.3 Plackett-Burman Designs

Plackett-Burman designs are of resolution III (cf. Section 10.5.1) and maximize the number of main effects that can be analyzed from as few runs as possible. They are used to screen a very large number of IVs and are more flexible than resolution III fractional-factorial designs in the number of runs that can be used.

The most common Plackett-Burman designs have numerous IVs, each with two levels. The number of runs (cases) required is some multiple of four, instead of the number computed using Equation 10.5 for the two-level fractional-factorial design. The design allows you to study up to k IVs in $k + 1$ runs, as long as $k + 1$ is a multiple of four. For example, you can study 11 IVs in 12 runs. Schmidt and Launsby (1994) show how to generate Plackett-Burman designs by hand.

Table 10.13 shows a Plackett-Burman analysis of 11 IVs in 12 runs with two replicates through SAS ADX, including setup and a generated data set with responses filled in. Later in this chapter, Table 10.16 shows setup and analysis of the 11 main effects.

There are two blocks of runs representing the two replications, 1–12 and 13–24. The first column (`RUN`) shows the original order of runs before randomization. Columns for IVs, `X1`–`X11`, use −1 and 1 to designate the two levels of each IV; there are 12 observations at each of the two levels. The columns are orthogonal, but only if computed over the entire 24 runs. Hypothetical data are added to the last column (`Y1`) of the data set; in practice, data are added after they are collected. SAS ADX also analyzes the data through the `Fit...` menu.

Table 10.14 shows SAS ADX output. Only main effects are analyzed—no interactions. The table shows that mean differences on `X5` are statistically significant at $\alpha = .05$. Further research examines this IV, and perhaps `X2`, with $p = .0904$. These variables might be examined in factorial arrangement with analysis of the interaction.

SAS ADX also provides various diagnostic and descriptive plots. Mean values may be requested through the `Explore... > Box Plot` menu for the design. Alternatively, the data of Table 10.13 may be exported to a standard SAS data set and a main-effects model run through GLM.

SAS ADX permits up to 48 IVs for Plackett-Burman designs.[4] Plackett-Burman designs are generated through NCSS Design of Experiments by specifying the number of runs ($k + 1$) and the number of replicates in the Screening Designs menu. No output is printed, but a data set is generated, which looks similar to Table 10.13. SYSTAT DESIGN also generates Plackett-Burman designs and extends to designs where IVs have three or more levels. A request for 9, 27, or 81 runs provides designs with three levels, each of 4, 13, or 40 IVs, respectively. A request for 25 or 125 runs provides designs with five levels, each of 6 or 31 IVs, respectively.

[4] Are we serious about 48 factors?

TABLE 10.13 Plackett-Burman design for 11 factors in 12 runs (SAS ADX setup and data set)

1. In ADX (Solutions, Design of Experiments), select File to Create New Design . . ., then Two-level . . .
2. Choose Select Design.
3. Fill in the Two-Level Design Specifications window as follows:
   a. Number of factors = 11.
   b. Choose Plackett-Burman designs.
   c. Click on design with 11 factors, 12 runs, 3 Main Effects only
4. Choose Customize . . .
   a. Click on Replicate runs.
   b. Fill in REPS column with 2 for each run.
5. Choose Edit Response.
6. Click on Design on Toolbar and choose Randomize Design.
7. Choose a seed number and return to spreadsheet.

Save the file, run the experiment, and then fill in the response values (within Edit Responses).

ADX: Two-level Design

plackett-burman

	RUN	X1	X2	X3	X4	X5	X6	X7	X8	X9	X10	X11	Y1
1	11	1	1	1	-1	1	1	-1	1	-1	-1	-1	6
2	3	1	1	-1	1	-1	-1	-1	1	1	1	-1	5
3	2	1	-1	1	-1	-1	-1	1	1	1	-1	1	4
4	18	-1	-1	-1	1	1	1	-1	1	1	-1	1	2
5	6	-1	1	1	-1	1	-1	-1	-1	1	1	1	8
6	5	-1	1	1	-1	1	-1	-1	-1	1	1	1	7
7	10	1	1	-1	1	1	-1	1	-1	-1	-1	1	5
8	23	-1	-1	-1	-1	-1	-1	-1	-1	-1	-1	-1	9
9	4	1	1	-1	1	-1	-1	-1	1	1	1	-1	9
10	15	-1	-1	1	1	1	-1	1	1	-1	1	-1	1
11	20	1	-1	-1	-1	1	1	1	-1	1	1	-1	9
12	9	1	1	-1	1	1	-1	1	-1	-1	-1	1	2
13	13	-1	1	1	1	-1	1	1	-1	1	-1	-1	8
14	8	1	-1	1	1	-1	1	-1	-1	-1	1	1	6
15	19	1	-1	-1	-1	1	1	1	-1	1	1	-1	3
16	21	-1	1	-1	-1	-1	1	1	1	-1	1	1	5
17	7	1	-1	1	1	-1	1	-1	-1	-1	1	1	6
18	12	1	1	1	-1	1	1	-1	1	-1	-1	-1	6
19	22	-1	1	-1	-1	-1	1	1	1	-1	1	1	8
20	1	1	-1	1	-1	-1	-1	1	1	1	-1	1	8
21	14	-1	1	1	1	-1	1	1	-1	1	-1	-1	9
22	17	-1	-1	-1	1	1	1	-1	1	1	-1	1	3
23	24	-1	-1	-1	-1	-1	-1	-1	-1	-1	-1	-1	4
24	16	-1	-1	1	1	1	-1	1	1	-1	1	-1	3

Define Variables...
Design Details...
Customization Details...
Edit Responses...
Explore...
Fit...
Optimize...
Experiment Notes...
Report...

TABLE 10.14 Plackett-Burman design for 11 factors in 12 runs (SAS ADX setup and output)

Click on Fit . . . to get significance tests.

Effect	Estimate	Std Error	t Ratio	P Value
X1	0.16667	0.90523	0.18411	0.8570
X2	1.6667	0.90523	1.8411	0.0904
X3	0.66667	0.90523	0.73646	0.4756
X4	-1.5	0.90523	-1.657	0.1234
X5	-2.1667	0.90523	-2.3935	0.0339
X6	0.5	0.90523	0.55234	0.5909
X7	-0.5	0.90523	-0.55234	0.5909
X8	-1.3333	0.90523	-1.4729	0.1665
X9	1.1667	0.90523	1.2888	0.2218
X10	0.33333	0.90523	0.36823	0.7191
X11	-0.66667	0.90523	-0.73646	0.4756

10.5.4 Taguchi Designs

Taguchi designs were developed to facilitate quality control. Like Plackett-Burman designs, they provide tests of the maximum number of IVs with a minimum number of runs. However, the emphasis is not so much on the performance of individual IVs as on effect size and the combination of levels of IVs that produces the optimum (largest or smallest) value of the DV. Taguchi designs can be used to identify the optimal combination of conditions for producing anything from a high-quality paint job to the most favorable therapeutic outcome.

Taguchi's approach differs from the classic ANOVA approach in at least two respects. The first is that two (or more) responses are usually taken for each run under different environmental (noise) conditions. These two (or more) responses form the levels of the *outer array*. The individual responses are the same as responses to any of the other screening designs. However, the two (or more) responses from different environmental conditions are combined into a single DV that reflects both the central tendency and the variability among the responses. The combined DVs are then used to form the *inner array*. That is, each combination of levels of IVs (run) has a response in the inner array that is a combination of responses in the outer array.

Taguchi methods emphasize that the optimum combination of levels of IVs is the one where responses are not only at their highest (or lowest) level but also are stable. Thus, the levels of the outer array are combined into a single DV that reflects both central tendency and variability, η (eta; details to follow).

A second difference is in the statistics used for analysis. The DV is a signal-to-noise ratio statistic, η, that takes into account both the central tendency and variability of responses in the outer array. The combination of levels of IVs is sought to produce maximum η. When the importance of individual IVs is evaluated, a degree-of-contribution measure, ρ (%), often replaces the F ratio. The statistical approach is discussed in detail by Taguchi (1993) and by Schmidt and Launsby (1994).[5]

The designs are named by the number of runs (treatment combinations) formed by the inner array of IVs, where L_4 is a design with 4 runs, L_{12} a design with 12 runs, and so on. Taguchi designs are typically forms of other resolution III designs, such as the fractional factorial, Plackett-Burman, Latin square, and so on (Schmidt and Launsby, 1994). Although most are resolution III screening designs (cf. Section 10.5.1), some also permit evaluation of some two-way interactions. Table 10.15 summarizes Taguchi designs in terms of the number of IVs and their levels. For example, an L_{18} design has 18 runs with 8 IVs: 1 two-level IV and 7 three-level IVs.

If a traditional ANOVA approach (with F ratios) is used to evaluate individual IVs and all effects are evaluated, there are often no degrees of freedom left for error. Either more than one replicate is run in order to have some degrees of freedom for error, or a preliminary analysis determines which effects are unimportant and these are pooled in subsequent analyses to provide an estimate of error.

L_8, L_{12}, L_{16}, and L_{32} are Plackett-Burman designs and are coded as described in Section 10.5.3. L_4, L_8, L_{16}, and L_{32} are two-level fractional factorials that are coded as described in Section 10.5.2.1. L_{27} is a three-level fractional-factorial design.

TABLE 10.15 Taguchi designs

Design	Number of Runs	Number of IVs	Number of Levels
L_4	4	3	2 each
L_8	8	7	2 each
L_9	9	4	3 each
L_{12}	12	11	2 each
L_{16}	16	15	2 each
L'_{16}	16	5	4 each
L_{18}	18	1 7	2 3 each
L_{25}	25	6	5 each
L_{27}	27	13	3 each
L'_{27}	27	22	3 each
L_{32}	32	31	2 each
L'_{32}	32	1 9	3 4 each
L''_{32}	32	1 9	2 4 each
L_{36}	36	11 12	2 each 3 each
L'_{36}	36	3 3	2 each 3 each
L_{50}	50	1 11	2 5 each
L_{54}	54	1 25	2 3 each
L_{64}	64	63	2

[5] These sources also discuss a loss function, which translates differences between ideal and actual responses into a financial measure.

Coding of levels of IVs for most remaining designs is complex and best left to SYSTAT DESIGN, SAS ADX, or NCSS. SAS ADX offers a variety of Taguchi designs in its Mixed Level design option. However, designs with only two levels are set up as two-level fractional factorials. The SAS on-disk documentation is helpful, with plenty of details and demonstrations of Taguchi design creation and analysis.

Taguchi (1993) provides tables for coding many of these designs and also shows patterns of aliasing for them. These tables are useful if you are interested in testing some interactions in lieu of some of the main effects. For example, if the main effect of C, expected to be negligible, is aliased with the interesting $A \times B$ interaction, the statistical result of the column labeled C is interpreted as the $A \times B$ interaction. SYSTAT DESIGN also shows aliasing patterns for some Taguchi designs.

Table 10.16 shows a Taguchi L_{12} design generated through SYSTAT DESIGN. Letter labels are requested, as is randomization of runs, with one replicate per cell. Entries in the columns indicate the levels of each IV for that particular run. The pattern of aliasing for Taguchi designs is also available in SYSTAT DESIGN if you request `PRINT=LONG`, although only full aliasing (complete confounding) is shown here. Effects that are only partially confounded are not identified—that output (not shown) revealed that none of the interactions is fully aliased in the L_{12} design of Table 10.16.

TABLE 10.16 Taguchi L_{12} design through SYSTAT DESIGN

```
SAVE 'C:\DATA\BOOK.ANV\SCREEN\TAGL12.SYD'
DESIGN
 TAGUCHI / REPS=1 TYPE=L12 LETTERS RAND

  Taguchi Design L12:  12 Runs; 11 Factors, Each With 2 Levels
```

	Factor										
Run	A	B	C	D	E	F	G	H	I	J	K
1	1	1	2	2	2	1	1	1	2	2	2
2	2	1	2	1	2	2	2	1	1	1	2
3	2	1	1	2	2	2	1	2	2	1	1
4	1	2	1	2	2	1	2	2	1	1	2
5	2	2	2	1	1	1	1	2	2	1	2
6	1	2	2	2	1	2	2	1	2	1	1
7	2	2	1	1	2	1	2	1	2	2	1
8	2	2	1	2	1	2	1	1	1	2	2
9	2	1	2	2	1	1	2	2	1	2	1
10	1	1	1	1	1	1	1	1	1	1	1
11	1	1	1	1	1	2	2	2	2	2	2
12	1	2	2	1	2	2	1	2	1	2	1

```
The design matrix has been saved.
```

Table 10.17 shows the output data set with IV entries generated by SYSTAT DESIGN with two hypothetical response values added to the data set; these two responses form the outer array. The two responses, `RESP1` and `RESP2`, represent responses taken by running the experiment under two different environmental (noise) conditions. Each case provides values for both responses, which are then combined over the outer array into a single S:N (signal-to-noise) ratio (`ETA` in the data set) to serve as the DV.

The final column, ETA, is the S:N ratio, converted to a dB (decibel) scale and calculated as follows:

$$\eta = -10 \log \left(\frac{1}{p} \sum_{1}^{p} Y^2 \right) \tag{10.6}$$

> The S:N ratio in dB units (η) for a case is found by squaring each p response for the case, summing all of them, dividing by the number of responses, taking the log to the base 10, and multiplying the result by -10.

For example, the case that received the first run has responses in the outer array of $Y_1 = 5$ and $Y_2 = 7$. The S:N ratio for the responses for that case is

$$\eta = -10 \log \left(\frac{1}{2}(5^2 + 7^2) \right) = -15.682 \text{ dB}$$

The combination of levels of IVs in the inner array with the largest η (closest to 0, in this example, because all of the η values are negative) is the most desirable; in this example, it is the case that received the 12th run where $\eta = -3.979$ dB.

Equation 10.6 is appropriate when the DV is scaled so that smaller values are better. For the example, the case that received the first run has values in the outer array of 5 and 7, whereas the case that received the last run has values of 1 and 2, which is the best of the outcomes if smaller values are better. If larger DV values in the outer array are better, however, Equation 10.6 is modified by using the reciprocal of Y^2 $(1/Y^2)$ in place of Y^2 in the equation:

$$\eta = -10 \log \left(\frac{1}{p} \sum_{1}^{p} \frac{1}{Y^2} \right)$$

For the example, the case that received the 12th run, with scores of 1 and 2, now has the *least* favorable outcome and an η value of 2.04. The case that received the 10th run has both the highest DV values (7 and 7) and the most stable DV values and produces an η value of 16.90. This value is higher than the value of 16.635, achieved by the case that received the 6th run, with DV scores in the outer array of 6 and 8. With this scaling, the highest value of η is also the most favorable outcome, but now all the η values are positive (above 0).[6]

Individual IVs may be analyzed with traditional ANOVA using η as the DV. However, without replicates, there are no degrees of freedom for error, unless some effects are pooled to serve as error. Alternatively, measures of effect size are calculated and interpreted instead of F ratios. In either event, a preliminary ANOVA is used to determine which effects are unimportant.

[6] Other variations of the S:N equation are available when the goal is for Y to achieve a target value or when the response standard deviation is related to its mean (Taguchi, 1993).

TABLE 10.17 Data set and hypothetical response values for Taguchi L_{12} design through SYSTAT DESIGN

RUN	A	B	C	D	E	F	G	H	I	J	K	RESP1	RESP2	ETA
1	1	1	2	2	2	1	1	1	2	2	2	5	7	-15.682
2	2	1	2	1	2	2	2	1	1	1	2	4	5	-13.118
3	2	1	1	2	2	2	1	2	2	1	1	3	6	-13.522
4	1	2	1	2	2	1	2	2	1	1	2	2	3	-8.129
5	2	2	2	1	1	1	1	2	2	1	2	1	3	-6.99
6	1	2	2	2	1	2	2	1	2	1	1	6	8	-16.99
7	2	2	1	1	2	1	2	1	2	2	1	4	7	-15.119
8	2	2	1	2	1	2	1	1	1	2	2	5	4	-13.118
9	2	1	2	2	1	1	2	2	1	2	1	2	3	-8.129
10	1	1	1	1	1	1	1	1	1	1	1	7	7	-16.902
11	1	1	1	1	1	2	2	2	2	2	2	3	4	-10.969
12	1	2	2	1	2	2	1	2	1	2	1	1	2	-3.979

TABLE 10.18 Preliminary analysis of Taguchi L_{12} design through SYSTAT GLM (syntax and selected output)

```
GLM
CATEGORY A B C D E F G H I J K / EFFECT
MODEL ETA = CONSTANT + A+B+C+D+E+F+G+H+I+J+K
ESTIMATE

Effects coding used for categorical variables in model.
```

Analysis of Variance

Source	Sum of Squares	df	Mean Square	F ratio	P
A	0.587	1	0.587	.	.
B	16.326	1	16.326	.	.
C	13.805	1	13.805	.	.
D	6.011	1	6.011	.	.
E	1.050	1	1.050	.	.
F	0.046	1	0.046	.	.
G	0.426	1	0.426	.	.
H	128.125	1	128.125	.	.
I	21.060	1	21.060	.	.
J	6.242	1	6.242	.	.
K	3.669	1	3.669	.	.
Error	0.0	0	.		

Table 10.18 shows SYSTAT GLM syntax and selected output for the preliminary analysis to determine which effects to pool for error in this single-replicate design.

F ratios are not calculated, because there are 0 df for error. Taguchi provides no clear guidelines for pooling small effects into error. One strategy is to begin with a few very small effects, in this case *A*, *F*, and *G*, all of which have mean square values much smaller than the others. SYSTAT GLM syntax is modified to eliminate these as IVs during the next analysis, as shown in Table 10.19.

By traditional ANOVA methods, all effects except *E* are statistically significant at $\alpha = .05$. However, *E* is retained in the model because its *F* ratio exceeds 2, a criterion recommended by Schmidt and Launsby (1994). It should be clear that one's choice of which terms to combine into error after the preliminary analysis has a direct impact on the significance of effects in the final analysis.

The Taguchi approach advocates replacing the *F* ratio with ρ (%) as a criterion for interpretation, calculated as:

$$\rho_{\text{effect}} = \frac{\text{SS}_{\text{effect}} - \text{df}_{\text{effect}}(\text{MS}_{\text{error}})}{\text{SS}_{\text{total}}}(100) \tag{10.7}$$

This measure of effect size (converted to a percentage) falls between η^2 and T^2 (Section 4.5.1) in size. This effect size, because it contains MS_{error}, is also sensitive to the choices of IVs to pool into error.[7] For the main effect of *B* in Table 10.19,

$$\rho_B = \frac{16.326 - (1)(0.353)}{197.348}(100) = 8.09$$

[7] Note, however, that ρ may be calculated even when there are no df for error, as in the output in Table 10.18. The second term in the numerator of Equation 10.7 is simply 0.

TABLE 10.19 Analysis of Taguchi L_{12} design through SYSTAT GLM (syntax and selected output)

```
GLM
CATEGORY B C D E H I J K / EFFECT
MODEL ETA = CONSTANT + B+C+D+E+H+I+J+K
ESTIMATE

Effects coding used for categorical variables in model.
Dep Var: ETA    N: 12    Multiple R: 0.997    Squared multiple
R: 0.995
```

Analysis of Variance

Source	Sum of Squares	df	Mean Square	F ratio	P
B	16.326	1	16.326	46.221	0.007
C	13.805	1	13.805	39.083	0.008
D	6.011	1	6.011	17.017	0.026
E	1.050	1	1.050	2.972	0.183
H	128.125	1	128.125	362.728	0.000
I	21.060	1	21.060	59.620	0.005
J	6.242	1	6.242	17.673	0.025
K	3.669	1	3.669	10.386	0.048
Error	1.060	3	0.353		

As a form of a correlation, R^2 effect size software (Steiger and Fouladi, 1993)—demonstrated in Figure 8.11—may be used to find confidence intervals around $\rho/100$, a form of r^2. Here, the number of observations is 12, the number of variables is 2 (B and ETA), and the R squared is .08. The 95% confidence limits for this example turn out to be from .00 to .50.

Table 10.20 shows the ANOVA table, including both traditional and Taguchi information.

H is the strongest IV by far, accounting for about twice as much variance as all the other IVs combined. A Taguchi-based strategy also examines the combination of levels of IVs with the largest S:N ratio. As shown in Table 10.17, the best combination of factors occurs in the final run, where $\eta = -3.979$. Ignoring the IVs with nonsignificant F ratios and/or trivial ρ, the best combination of factors includes lower levels of *D*, *I*, and *K* and higher levels of *B*, *C*, *E*, *H*, and *J*. Follow-up research could include intermediate, lower, or higher levels of those IVs and possibly some interactions among them. Levels of *H*, in particular, could be extended to higher levels, if possible. Further experimentation also might be specially designed to assess interactions between *H* and other statistically significant effects.

There is far more to the Taguchi approach than is dealt with here. For example, Taguchi (1993, pp. 75–76) provides the details, including statistics, to estimate the optimum combination of levels and the improvement in quality expected from it.

10.5.5 Response-Surface Methodology

Response-surface methodology is used for designs with quantitative IVs. As in Taguchi designs, the goal is not just to identify statistically significant IVs but also to

TABLE 10.20 Source table for Taguchi L_{12} design

Source	SS	df	MS	*F*	ρ (%)
(*A*)	0.587	1	0.587		
B	16.326	1	16.326	46.22	8.1
C	13.805	1	13.805	39.08	6.8
D	6.011	1	6.011	17.02	2.9
E	1.050	1	1.050	2.97	0.4
(*F*)	0.046	1	0.046		
(*G*)	0.426	1	0.426		
H	128.125	1	128.125	362.73	64.7
I	21.060	1	21.060	59.62	10.5
J	6.242	1	6.242	17.67	3.0
K	3.669	1	3.669	10.39	1.7
Error	1.060	3	0.353		
T	197.348	12			

Note: Parentheses indicate effects to be pooled into error.

reveal the combination of levels of IVs that produces the most desirable response. The three most popular designs are Box-Behnken designs for three-level IVs; central-composite designs for five-level IVs; and mixture designs, used when the combination of IVs forms a constant quantity (e.g., 100%). Taguchi methods may also be applied to response-surface designs, as illustrated by Myers and Montgomery (1995, Chapter 10).

Response-surface methodology is typically applied after the list of potential IVs has been reduced through one or more of the screening designs described earlier. The researcher then tries additional levels of the surviving IVs to determine when the response approaches some optimum level. When the region of optimum response is identified, further analyses are conducted to build a mathematical model to find the combination of levels that produces the optimum response (Myers and Montgomery, 1995). Model adequacy is assessed at each phase of the study by examining residuals, as in any regression model (cf. Tabachnick and Fidell, 2007, Chapter 5).

Many of the designs and techniques in response-surface methodology are beyond the scope of this book. Some, for example, include multiple-response measures. Only the more straightforward popular designs and analyses—and those most easily implemented through software—are outlined here. A good resource for a more detailed, complete treatment of response surface methodology is Myers and Montgomery (1995).

Response-surface methodology takes the regression approach to analysis. Early stages in the research focus on IVs with two levels or linear trends of quantitative IVs; later phases often include some (but not all) interactions among IVs and usually explore quadratic trends of IVs as well. Higher-order interactions and trends are rarely investigated.

10.5.5.1 Box-Behnken Design

Box-Behnken designs are used when there are numerous quantitative IVs, each with three levels. For this reason, the design is infrequently employed in the social sciences. An example, however, might involve therapeutic outcome tested against length of therapy (2, 4, or 6 weeks), concentration of therapy (1, 2, or 3 sessions per week), number of other patients included in the sessions (0, 2, or 4), age range of client (20–29, 30–39, 40–49), severity of illness (1 = mild, 2 = moderate, 3 = severe), and so on.

Linear and quadratic trends of main effects and interaction contrasts of linear trends of all two-way interactions are analyzed. For the example, does outcome improve as the duration of therapy increases (linear trend)? Or, is outcome better at four weeks than at either two or six weeks (quadratic trend)? Does the linear trend of duration of therapy depend on the linear trend of, say, age range of clients so that, for instance, older patients require longer therapies (linear by linear interaction contrast)?

Figure 10.2 diagrams the layout for a three-way Box-Behnken design. A Box-Behnken design is especially amenable to a regression approach, because each effect that is tested has a single degree of freedom.

Table 10.21 shows a hypothetical data set with four IVs and codes for the 14 effects in this Box-Behnken design. The 14 effects consist of four linear and four quadratic tests of main effects and six linear by linear interaction contrasts. Codes for linear effects are available through software (e.g., SYSTAT DESIGN, which has several Box-Behnken design options; SAS ADX, which has options for three to seven IVs; or Schmidt and Launsby, 1994, who show codes for three, four, and five IVs). Squaring codes for linear trend generates codes for quadratic trend. Note that this is

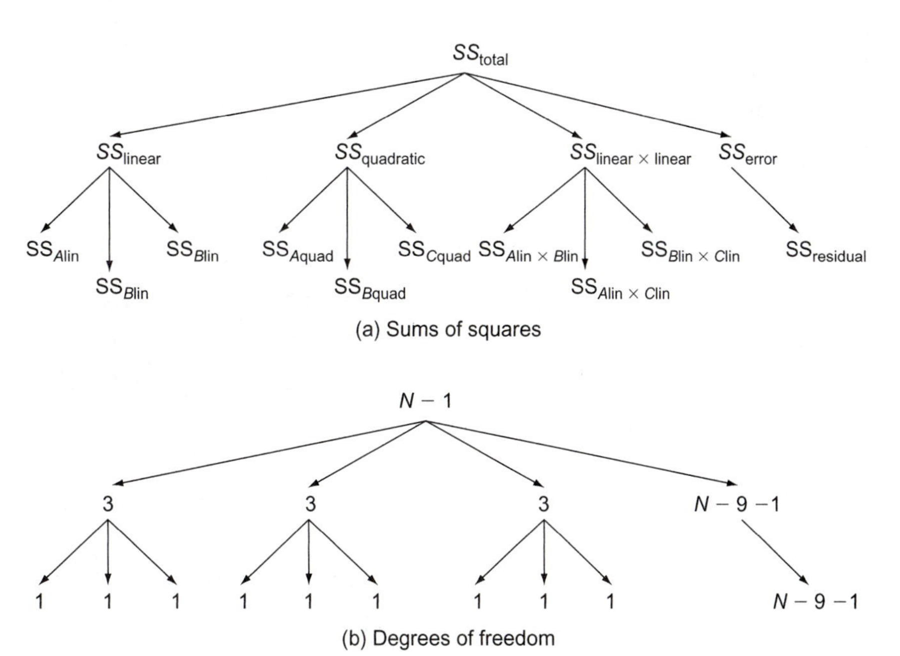

FIGURE 10.2 Partition of (a) sums of squares and (b) degrees of freedom into systematic sources of linear trends of *A*, *B*, and *C*; quadratic trends of *A*, *B*, and *C*; linear by linear interaction contrasts among *A*, *B*, and *C*; and remaining (error) differences among cases

TABLE 10.21 Example of Box-Behnken design with four IVs

Run	Y	A_{LIN}	B_{LIN}	C_{LIN}	D_{LIN}	A_{QUAD}	B_{QUAD}	C_{QUAD}	D_{QUAD}	$A_{LIN}*B_{LIN}$	$A_{LIN}*C_{LIN}$	$A_{LIN}*D_{LIN}$	$B_{LIN}*C_{LIN}$	$B_{LIN}*D_{LIN}$	$C_{LIN}*D_{LIN}$
1	9	−1	−1	0	0	1	1	0	0	1	0	0	0	0	0
2	6	1	−1	0	0	1	1	0	0	−1	0	0	0	0	0
3	1	−1	1	0	0	1	1	0	0	−1	0	0	0	0	0
4	6	1	1	0	0	1	1	0	0	1	0	0	0	0	0
5	3	0	0	−1	−1	0	0	1	1	0	0	0	0	0	1
6	3	0	0	1	−1	0	0	1	1	0	0	0	0	0	−1
7	3	0	0	−1	1	0	0	1	1	0	0	0	0	0	−1
8	2	0	0	1	1	0	0	1	1	0	0	0	0	0	1
9	4	0	0	0	0	0	0	0	0	0	0	0	0	0	0
10	5	−1	0	0	−1	1	0	0	1	0	0	1	0	0	0
11	7	1	0	0	−1	1	0	0	1	0	0	−1	0	0	0
12	5	−1	0	0	1	1	0	0	1	0	0	−1	0	0	0
13	7	1	0	0	1	1	0	0	1	0	0	1	0	0	0
14	8	0	−1	−1	0	0	1	1	0	0	0	0	1	0	0
15	3	0	1	−1	0	0	1	1	0	0	0	0	−1	0	0
16	7	0	−1	1	0	0	1	1	0	0	0	0	−1	0	0
17	2	0	1	1	0	0	1	1	0	0	0	0	1	0	0
18	6	0	0	0	0	0	0	0	0	0	0	0	0	0	0
19	6	−1	0	−1	0	1	0	1	0	0	1	0	0	0	0
20	8	1	0	−1	0	1	0	1	0	0	−1	0	0	0	0
21	5	−1	0	1	0	1	0	1	0	0	−1	0	0	0	0
22	8	1	0	1	0	1	0	1	0	0	1	0	0	0	0
23	6	0	−1	0	−1	0	1	0	1	0	0	0	0	1	0
24	3	0	1	0	−1	0	1	0	1	0	0	0	0	−1	0
25	7	0	−1	0	1	0	1	0	1	0	0	0	0	−1	0
26	3	0	1	0	1	0	1	0	1	0	0	0	0	1	0
27	8	0	0	0	0	0	0	0	0	0	0	0	0	0	0

a type of coding that uses only 0, 1, and -1 as coefficients (cf. Section 3.5.5, effects coding), which is different from the contrast coding used elsewhere in this book. Codes for the linear by linear interaction contrasts are, as usual, the products of the linear codes for the pairs of component IVs.

Table 10.22 shows the setup for generating the linear codes of Table 10.21 and the data file output through SAS ADX. The response scores are added to the file by hand after data are collected. (A blocking factor may be selected to represent replications. The number of blocks depends on the number of IVs; a blocked design is available in ADX but only with four or five factors.)

The data set is evaluated through regression, with columns for quadratic effects and interaction contrasts of linear trend added if the data are analyzed by hand. However, the normal regression approach described in Section 10.4.3 and elsewhere is inadequate, because not all the effects are orthogonal. In particular, the codes for the quadratic effects are correlated with other effects, so the cross products of columns do not sum to 0; that is, they are not orthogonal. Therefore, matrix algebra is required for hand calculation of the regression approach, a grisly procedure with 14 predictors.

TABLE 10.22 Box-Behnken design generated through SAS ADX (setup and data file output)

1. In ADX (Solutions > Analysis > Design of Experiments), select File to Create New Design . . . , then Response Surface. . . .
2. Choose Select Design.
3. Fill in the Response Surface Design Specifications window as follows:
    a. Number of factors = 4.
    b. Click on Box-Behnken design with 4 factors, 27 runs, 3 center points, 1 block.
    c. Exit window and choose to use the selected design when prompted.
4. Choose Edit Response.
5. Click on Design on Toolbar and choose Randomize Design.
6. Choose a seed number and return to spreadsheet.

Save the file, run the experiment, and then fill in the response values (within Edit Responses).

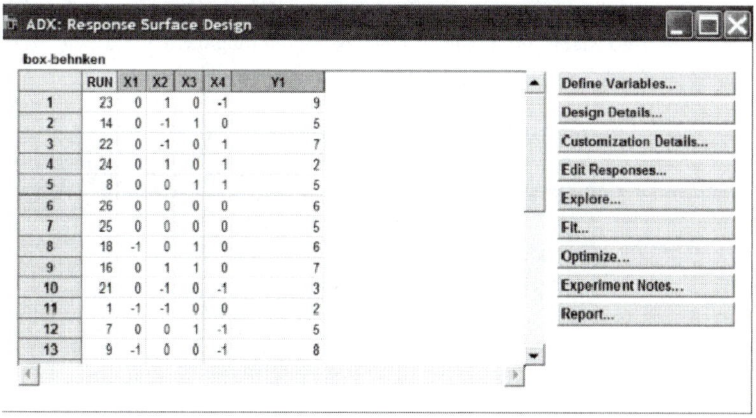

ADX: Response Surface Design

box-behnken

	RUN	X1	X2	X3	X4	Y1
1	23	0	1	0	-1	9
2	14	0	-1	1	0	5
3	22	0	-1	0	1	7
4	24	0	1	0	1	2
5	8	0	0	1	1	5
6	26	0	0	0	0	6
7	25	0	0	0	0	5
8	18	-1	0	1	0	6
9	16	0	1	1	0	7
10	21	0	-1	0	-1	3
11	1	-1	-1	0	0	2
12	7	0	0	1	-1	5
13	9	-1	0	0	-1	8

Define Variables...
Design Details...
Customization Details...
Edit Responses...
Explore...
Fit...
Optimize...
Experiment Notes...
Report...

However, analysis is available through SAS ADX and several GLM programs, as well as NCSS. Table 10.23 shows setup and output for the analysis through SAS ADX. Quadratic effects are identified in the output as factor multiplied by itself (e.g., X1*X1), and interactions are identified as cross-multiplied factors (e.g., X1*X2).

There are two statistically significant effects at $\alpha = .05$: X4 has a quadratic effect, and there is an interaction between linear components of X2 and X4. Neither X1 nor X3 appear in any significant effects. SAS ADX also has a module to help identify regions of optimal response, as well as exploratory and descriptive plots. Constituent main effects—here, X2 and X4—are added to a model to optimize it. The output prediction plot shows that response is greatest when X2 is at its highest value and X4 is either at its highest or lowest value.

The design can also be analyzed through NCSS Response Surface Analysis, which includes modules that do not require specification of quadratic components

TABLE 10.23 Box-Behnken design analyzed through SAS ADX (setup and selected output)

Click on the Fit . . . option.

Effect	Estimate	Std Error	t Ratio	P Value
X1	0.5	0.59463	0.84086	0.4169
X2	0.58333	0.59463	0.981	0.3460
X3	0.25	0.59463	0.42043	0.6816
X4	-0.33333	0.59463	-0.56057	0.5854
X1*X1	0.95833	0.89195	1.0744	0.3038
X1*X2	0.5	1.0299	0.48547	0.6361
X1*X3	-0.25	1.0299	-0.24273	0.8123
X1*X4	-0.75	1.0299	-0.7282	0.4805
X2*X2	-0.16667	0.89195	-0.18686	0.8549
X2*X3	0.5	1.0299	0.48547	0.6361
X2*X4	2.75	1.0299	2.6701	0.0204
X3*X3	0.33333	0.89195	0.37371	0.7151
X3*X4	-0.5	1.0299	-0.48547	0.6361
X4*X4	1.9583	0.89195	2.1956	0.0485

Add X2 and X4 to the Fit table by highlighting them.

Click on Optimize. . . .

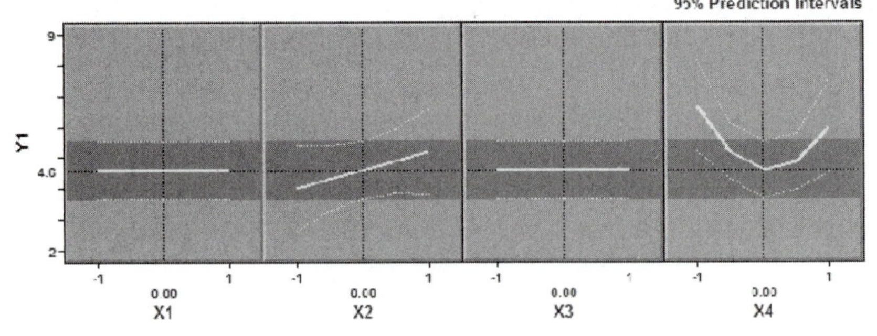

and interaction contrasts. This program also offers plots to help identify regions of optimal response.

Qualitative IVs can be analyzed through Box-Behnken designs, as well, if codes can be applied to levels in a way that makes research sense. For example, if the levels of an IV represent two types of treatment and a control, codes of 1 and −1 are applied to the treatments and 0 to the control. The linear trend compares the treatments with each other, whereas the quadratic effect evaluates the average of the treatments versus the control. The interaction contrasts are between treatments, with the control groups omitted.

10.5.5.2 Central-Composite Designs

Each IV has five levels in central-composite designs, which is the prototype of the response surface design. The IVs are considered continuous, with discrete coding used for convenience during experimentation. Central-composite designs, like Box-Behnken designs, test linear and quadratic effects of each IV and the linear by linear components of the two-way interactions. Thus, the partition of sums of squares and degrees of freedom follows Figure 10.2. Variations among central composite designs and their properties are discussed in detail by Montgomery (1984).

Standard central-composite analysis is a sequential procedure that begins with a test of the combination of linear effects of all IVs. This first test asks whether a model with only linear components is related to the DV. A test of the combination of quadratic effects of IVs and linear interaction contrasts follows. The second test asks whether the prediction of the DV is improved by addition of quadratic and interaction-contrast components. Follow-up tests determine which individual effects are statistically significant.

Response surfaces are examined through pairwise plots of IVs, in which contours are shown for values of the DV. A circle inside such a plot indicates that highest (or lowest) values of the DV fall at the levels of the two IVs aligned with the circle. Figure 10.3

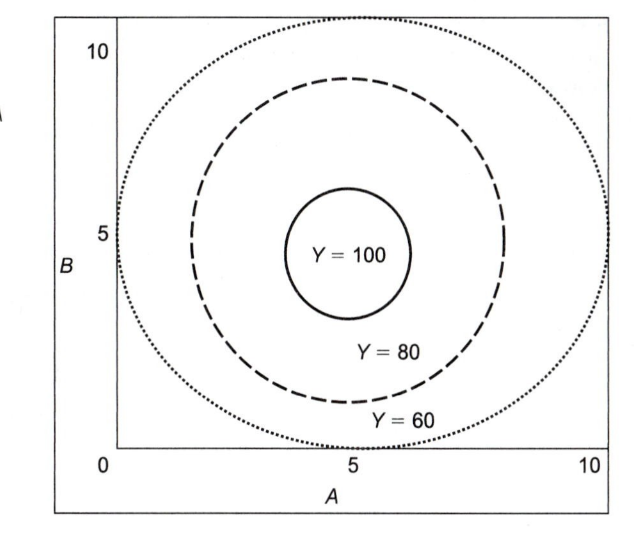

FIGURE 10.3 Response-surface contour plot of Y as a function of A and B

shows such a contour plot, in which high values of the DV, Y, are desirable, and the two axes represent levels of A and B. In this plot, the optimum response occurs in the region of $A = B = 5$.

Sometimes the point of optimum response appears to lie outside the range of the levels chosen for some of the IVs. The plots are examined to estimate the direction in which the optimum region lies to help design follow-up experiments. SAS ADX and NCSS Response Surface Analysis provide printed estimates of optimum areas for each IV or show directions in which optimum points are likely to be found. A central-composite design generated and analyzed through SAS ADX is demonstrated in Section 10.7.

There are some alternatives available when analyzing data generated by a central-composite design. For example, if you are uninterested in the interaction contrasts, you can do a full trend analysis (to fourth order) of the main effects.

10.5.5.3 Mixture Models

Mixture designs are used when cases are formed by mixtures or blends of levels of quantitative IVs. These designs require that sum of the levels of the IVs equal some fixed total for every case or run (e.g., 100%, or "a glassful of beverage"). If, for example, the two IVs are percentage of rum (A) and percentage of cola (B) in a drink, levels of each might be 0%, 50%, and 100%. Three runs (blends or cases) cover all possible mixtures for two IVs with three levels: all rum and no cola, all cola and no rum, and 50% of each. (We leave the DV to the imagination of the reader.) Table 10.24 shows the levels in this mixture design. Two additional runs are required if there are five levels (adding 25% and 75%), as shown in Table 10.25.

Proportions are chosen to cover the response surface of all potential mixtures of these components to facilitate prediction of the optimal blend. Obviously, not all blends can be investigated, particularly as the IVs are considered continuous. Several strategies have been proposed to determine mixtures likely to cover the response surface under different research circumstances (e.g., reduce the number of IVs in each blend). Cornell (1990) provides guidelines for generating many mixture designs, or those designs can be generated through software such as SYSTAT DESIGN, SAS OPTEX, or ADX.

TABLE 10.24 Diagram of a 3 × 3 mixture design

Run	A: Rum	B: Cola
1	1.0	0
2	0	1.0
3	0.5	0.5

TABLE 10.25 Diagram of a 5 × 5 mixture design

Run	A: Rum	B: Cola
1	1.0	0
2	0	1.0
3	0.5	0.5
4	0.25	0.75
5	0.75	0.25

10.5.5.3.1 Lattice Designs Lattice designs (labeled *simplex-lattice* by Cornell, 1990) are the most popular of the mixture designs. The number of IVs is flexible, but all IVs have the same number of levels. As shown in Table 10.24, two IVs (A and B) each can have three levels (1.0, 0.5, and 0), or, as in Table 10.25, each can have five levels (1.0, 0.75, 0.50, 0.25, and 0), or any other number of levels. The number of IVs and their levels determine the minimum number of runs, without replication.

Lattice designs are neither factorial nor orthogonal. For example, the 3×3 design of Table 10.24 would have nine cells in a full factorial design, one of them representing 0% of both A and B, another representing 100% of both A and B, and so on. Clearly, not all combinations are even feasible. Furthermore, the correlation between A and B is not 0, because once the level of A is chosen, the level of B is fixed since A and B sum to 100%. Consequently, there are not enough degrees of freedom to include all IVs in the analysis. On the other hand, the quantitative nature of the IVs makes trend analysis appealing.

The most popular analysis, based on Scheffé and reviewed by Cornell (1990) and Myers and Montgomery (1995), is a multiple-regression approach in which one of the IVs is dropped (and thus acts as the intercept), and interactions are formed for all the IVs, including the dropped one. The original IVs, coded by their proportions in the mixture (as in Tables 10.25 and 10.26), form the linear portion of the model; two-way interactions form the quadratic portion of the model; and so on.

Table 10.26 shows a data set from Cornell (1990), elaborated by Myers and Montgomery (1995). The three IVs are components of a yarn mixture: $A =$ polyethylene, $B =$ polystyrene, and $C =$ polypropylene. The two types of mixtures of

TABLE 10.26 Lattice design from Cornell (1990)

A	B	C	ELONG
1	0	0	11.0
1	0	0	12.4
.5	.5	0	15.0
.5	.5	0	14.8
.5	.5	0	16.1
0	1	0	8.8
0	1	0	10.0
0	.5	.5	10.0
0	.5	.5	9.7
0	.5	.5	11.8
0	0	1	16.8
0	0	1	16.0
.5	0	.5	17.7
.5	0	.5	16.4
.5	0	.5	16.6

interest in this lattice design with three components are pure blends (all one component) and binary blends (.50-.50 mixes of two of the three components). There are no three-component blends. Two replications of each pure blend and three replications of each binary blend are assessed. The DV is elongation of the fabric when a certain force is applied (ELONG).

Because of the nonorthogonality of *A*, *B*, and *C* in this lattice design and because trends are interesting with these quantitative IVs, a sequential regression analysis (cf. Section 5.6.5) is run in which the first step includes the linear portions of *A*, *B*, and *C*, and the second step includes their interactions. Recall that sequential regression provides an analysis for nonorthogonal data because variables entered first take all their own variability, plus any overlapping variability shared with variables not yet entered. Variables entered second are assigned only the remaining variability they share with the DV. Thus, each portion of overlapping variability is analyzed only once.

Table 10.27 shows syntax and output from SPSS COMPUTE and REGRESSION for this analysis. The interactions are created through COMPUTE instructions before REGRESSION is run.[8] The first three lines of the REGRESSION syntax are produced by default, except for the request for CHANGE statistics, which asks whether the interactions add to prediction of elongation beyond that afforded by the linear portion. The NOORIGIN instruction makes the CONSTANT equivalent to *A*, so the main effect

TABLE 10.27 Analysis of lattice design through SPSS REGRESSION (syntax and selected output)

```
COMPUTE ab = a*b.
COMPUTE ac = a*c.
COMPUTE bc = b*c.
REGRESSION
  /MISSING LISTWISE
  /STATISTICS COEFF OUTS R ANOVA CHANGE
  /CRITERIA=PIN(.05) POUT(.10)
  /NOORIGIN
  /DEPENDENT elong
  /METHOD=ENTER b c  /METHOD=ENTER ab ac bc.
```

Regression

Variables Entered/Removed[b]

Model	Variables Entered	Variables Removed	Method
1	c, b[a]	.	Enter
2	bc, ac, ab[a]	.	Enter

a. All requested variables entered.
b. Dependent Variable: elong

[8] Because linear codes are used for *A*, *B*, and *C* (0, .5, and 1.0), the cross products of the terms also represent a quadratic trend. In this lattice design, these terms represent the tendency for a .50-.50 mix to be stronger (or weaker) than the pure blends (e.g., to have quadratic trend).

TABLE 10.27 Continued

Model Summary

Model	R	R Square	Adjusted R Square	Std. Error of the Estimate	Change Statistics				
					R Square Change	F Change	df1	df2	Sig. F Change
1	.654[a]	.427	.332	2.53684	.427	4.477	2	12	.035
2	.975[b]	.951	.924	.85375	.524	32.317	3	9	.000

a. Predictors: (Constant), c, b
b. Predictors: (Constant), c, b, bc, ac, ab

ANOVA[c]

Model		Sum of Squares	df	Mean Square	F	Sig.
1	Regression	57.629	2	28.815	4.477	.035[a]
	Residual	77.227	12	6.436		
	Total	134.856	14			
2	Regression	128.296	5	25.659	35.203	.000[b]
	Residual	6.560	9	.729		
	Total	134.856	14			

a. Predictors: (Constant), c, b
b. Predictors: (Constant), c, b, bc, ac, ab
c. Dependent Variable: elong

Coefficients[a]

Model		Unstandardized Coefficients		Standardized Coefficients	t	Sig.
		B	Std. Error	Beta		
1	(Constant)	14.995	1.410		10.632	.000
	b	−5.164	2.163	−.602	−2.387	.034
	c	.800	2.163	.093	.370	.718
2	(Constant)	11.700	.604		19.381	.000
	b	−2.300	.854	−.268	−2.694	.025
	c	4.700	.854	.548	5.505	.000
	ab	19.000	2.608	.634	7.285	.000
	ac	11.400	2.608	.380	4.371	.002
	bc	−9.600	2.608	−.320	−3.681	.005

a. Dependent Variable: elong

of A is left out. The two METHOD instructions list the IVs to be entered on the two regression steps. The first ENTER specifies entry of B and C; A is omitted because it serves as the intercept, or baseline, value. This step analyzes the linear portion of the model, because coding for each effect is quantitative: 0, .5, and 1. The second step includes the three interaction terms, AB, AC, and BC. The final line of syntax requests predicted scores, residuals, and standardized residuals.

The Variables Entered/Removed table reminds us that variables enter into the regression equation in two steps: first the linear components and then the interactions. In the output, Model 1 refers to an equation with A (the intercept; called Constant by SPSS Regression), B, and C. Model 2 refers to the complete equation with both linear components and interaction. The Model Summary table shows significant prediction of the response with only the linear effects of each component—$F(2, 12) = 4.48, p = .035$—and enhancement of that prediction when interaction components are added—$F(3, 9) = 32.32, p < .0005$. This result indicates that combinations of components have effects on elongation above and beyond the linear effect of each one separately. When all components are included, the model predicts 95% (R Square = .951) of the variance in elongation, with $F(5, 9) = 35.203, p < .0005$.

With a bit of maneuvering, the Coefficients table gives us the regression equation for predicting elongation. Component A is represented in the equation by the intercept. In the full model (Model 2), this value is 11.7, which is the mean of the two data points for a pure mixture of A. Thus, A is acting as a baseline, supplemented by coefficients for B and C. The coefficients for B and C are adjusted by adding the constant to them. The corrected coefficient for B, then, is the constant plus the printed value—$11.7 + (-2.3) = 9.4$—and the corrected coefficient for C is $11.7 + 4.7 = 16.4$. (These coefficients now equal the means of B and C, respectively.) The prediction equation with corrected coefficients is then

$$Y' = (11.7)A + (9.4)B + (16.4)C + (19.0)AB + (11.4)AC + (-9.6)BC$$

All components are statistically significant at $\alpha = .05$, but $\alpha = .01$ provides better control of familywise error. At the more stringent α level, the linear contribution of B is problematic. Component C (polypropylene), with the highest coefficient, makes the largest contribution to yarn elongation, producing the strongest yarn. However, the positive coefficients of the AB and AC interactions suggest that these combinations produce yarn of greater strength than expected from simply adding their linear effects, whereas the negative coefficient of the BC interaction indicates that this combination produces a *weaker* yarn than expected from their linear components.

10.5.5.3.2 Other Mixture Designs The lattice design is the most flexible mixture design, with the number of IVs and the number of levels independent (although each IV must have the same number of levels). In the three remaining mixture designs mentioned here—centroid, axial, and screening—the number of IVs determines the number of levels.

Centroid designs (called *simplex-centroid* by Cornell, 1990) are those in which each of the mixtures has equal proportions of each included component. So, for example, a design with five IVs has one mixture composed of one-third each of, say,

A, *B*, and *C*, and no *D* or *E*; and then other one-third mixtures of all other combination of three IVs from the five. Other mixtures are composed one-half each of, say, *C* and *D*, with no *A*, *B*, or *E*, and one-half mixtures of all other pairwise combinations of two IVs, and so on. Pure mixtures of each component are also included in the design.

Axial designs have some proportion of each component in almost all mixtures. These designs allow an in-depth analysis of the response surface (on-disk SYSTAT manual).

Screening designs are useful when there is a large number of potential components, and the goal is to rule out some of them out. The design involves running mixtures with proportions of numerous components; however, during analysis, some components are effectively reduced to 0 by combining terms in statistical tests of the model (Cornell, 1990). The design is similar to the centroid design, and differs from the axial design, in that most mixtures contain proportions of a few components, rather than emphasizing complete mixtures (SPSS, 1999e).

These designs, and variations on them (including the lattice design), are comprehensively covered by Cornell (1990). Myers and Montgomery (1995) offer a discussion of these and other mixture designs (with several detailed analyses.)

10.5.6 Optimal Designs

The appropriate design for your experiment may not fit the specifications of any of the designs in the preceding sections. For example, you may have both qualitative and quantitative IVs; or not all combinations of factor levels are feasible; or you simply cannot manage the number of runs specified in the most appropriate design. Optimal design methods let you specify your model just as you want it, along with any restrictions that you impose.

There may be a number of combinations of IV levels possible, but the optimal design is chosen on the basis of efficiency.[9] You want to choose the most efficient design that meets your needs. SAS OPTEX, SYSTAT DESIGN Wizard, and NCSS search for and identify optimal designs.

Finding an optimal design begins with a definition of the IVs of interest and their levels (the candidate set of points). Then, combinations of levels to be excluded, if any, are specified. You may also specify combinations of levels that must be included. The optimal design is chosen from among the remaining combinations of levels.

Consider an example in which there are a total of eight IVs—six IVs with two levels each (*A–F*) and two IVs with five levels each (*G–H*). These might be alternate items on a questionnaire, with rating of clarity of the whole questionnaire as a DV. There are two possible ways of asking questions *A–F*. There are five possible ways of asking questions *G* and *H*, but some combinations of *G* and *H* do not make sense when asked together. Say, for example, that six combinations of levels of *G* and *H* cannot occur: g_1 and h_5, h_1 and g_5, g_1 and h_4, h_1 and g_4, g_2 and h_5, h_2 and g_5. A second

[9] There are a number of efficiency criteria that, under most circumstances, give similar results. The most popular is D-optimality, available in the three programs that produce experimental designs.

hitch might be that only about 20 potential respondents are available to pilot the questionnaire—say only 18 combinations of all questions can be run.

SAS uses a set of procedures to search for optimal designs. The candidate set is defined through PROC PLAN, which produces a data set. Then the optimization procedure is applied to this data set according to your criteria. Table 10.28 shows the SAS procedures for finding the optimal design for the example. SAS PROC PLAN identifies the eight IVs (`factors`) and the name of the output data file, CAN. This is a full eight-way factorial design. Then a DATA step modifies the data set, identifying combinations to exclude (^). PROC OPTEX is then run on the modified data set, in which all eight IVs are identified as `class` (classification) IVs (quantitative), and a main effects model is specified (without a response variable). The `generate` instruction limits the design to 18 runs. Finally, the optimal design is saved as a data set called OPTIMAL and printed out.

TABLE 10.28 Optimal design for eight factors with restrictions on combinations and number of runs (SAS PLAN and OPTEX syntax and selected output)

```
proc plan;
   factors A=2 B=2 C=2 D=2 E=2 F=2 G=5 H=5;
   output out=CAN
     A B C D E F nvals = (1 to 2)
             G H nvals = (1 to 5);
data CAN; set CAN;
    if (^((G=1) & (H=5)));
    if (^((G=5) & (H=1)));
    if (^((G=1) & (H=4)));
    if (^((G=4) & (H=1)));
    if (^((G=2) & (H=5)));
    if (^((G=5) & (H=2)));
proc print data=CAN;
run;
proc optex data=CAN;
     class A B C D E F G H;
     model A B C D E F G H;
     generate n=18;
run;
     output out=OPTIMAL;
proc print data=OPTIMAL;
run;
```

The PLAN Procedure

Factor	Select	Levels	Order
A	2	2	Random
B	2	2	Random
C	2	2	Random
D	2	2	Random
E	2	2	Random
F	2	2	Random
G	5	5	Random
H	5	5	Random

TABLE 10.28 Continued

The OPTEX Procedure

Design Number	D-Efficiency	A-Efficiency	G-Efficiency	Average Prediction Standard Error
1	46.3151	30.6023	69.0687	1.0061
2	46.0897	28.9577	66.5476	1.0137
3	45.9952	29.3663	70.7333	1.0148
4	45.9293	28.9389	61.9689	1.0187
5	45.7735	28.3197	62.8887	1.0199
6	45.6116	29.1300	64.3602	1.0304
7	45.1925	26.1608	59.7607	1.0481
8	44.9295	25.9535	62.5567	1.0532
9	44.4299	28.0258	62.1805	1.0669
10	44.3151	28.0040	62.1337	1.0518

OBS	A	B	C	D	E	F	G	H
1	2	2	2	2	1	1	2	4
2	2	2	1	2	2	2	4	5
3	2	2	1	1	2	1	5	4
4	2	1	2	2	1	1	3	1
5	2	1	2	1	2	2	1	3
6	2	1	2	1	1	1	5	5
7	2	1	1	1	1	2	3	2
8	2	1	1	1	1	1	2	3
9	1	2	2	2	1	2	2	2
10	1	2	2	1	2	1	4	2
11	1	2	2	1	1	1	3	3
12	1	2	1	1	1	2	1	1
13	1	1	2	2	1	2	5	3
14	1	1	2	1	2	2	2	1
15	1	1	2	1	1	2	3	4
16	1	1	1	2	2	1	3	5
17	1	1	1	2	2	1	1	2
18	1	1	1	2	1	1	4	4

The output first shows the design and indicates that runs may be done in any order. Then, all 1216[10] runs are printed (not shown here). SAS randomly selects 10 designs with 18 runs and orders them in terms of D-efficiency. This means that some nonselected designs could have greater efficiency, and every run with this syntax produces a different output. Two other efficiency criteria are shown in this table, along with an estimate of prediction error for each. Then, the optimum set among the 10 designs (the one with the highest value of D-efficiency—design 1, in this example) is printed out. In this design, the questionnaire for the first respondent has the second

[10] A full factorial combination ($2 \times 2 \times 2 \times 2 \times 2 \times 2 \times 5 \times 5$) produces 1600 runs; 384 of these are the combinations that are excluded.

alternative for the first four questions (*A*–*D*), the first alternative for questions *E* and *F*, the second alternative for question *G*, and the fourth alternative for question *H*. The second respondent has a questionnaire composed of the second alternatives for questions *A*, *B*, *D*, *E*, and *F*; the first alternative for question *C*; the fourth alternative for question *G*; the fifth alternative for question *H*; and so on through the 18 respondents. At least four of the questionnaires should be duplicated and given to additional respondents to provide some estimate of error variance in the analysis phase (Hintze, 2001). Data are then collected and analyzed through an ANOVA program.

10.6 SOME IMPORTANT ISSUES

The two prominent issues in screening and other incomplete designs are using software to generate the designs and choosing the design most appropriate for your needs.

10.6.1 Generating Screening and Other Incomplete Designs

Incomplete designs are tedious and difficult to generate by hand. Although Section 10.4.3 discusses development of a 2^5 half factorial by hand, all these designs are much more easily developed through software. Several software packages are available. All of them provide output files in which the levels of the IVs for each run are already coded into columns; DV scores for each run are added as data are collected.

SYSTAT has design generation built into its statistics packages. Another package that is particularly easy to use is NCSS, also reviewed in this chapter.

SAS QC and ADX are modules separate from the statistics package. SAS QC has two programs, FACTEX and OPTEX, which are accessed through syntax only. ADX is a menu-driven system that does not generate syntax. The programs vary widely in the types of designs that are generated and the way they are generated. Table 10.29 compares features of SAS FACTEX, SYSTAT DESIGN, and NCSS.

SAS has four programs, FACTEX and OPTEX (both in the QC package), PLAN (in the statistics package), and ADX. PLAN is best used for designs with repeated measures, such as the Latin-square designs of Chapter 9, and for creating factorial designs. OPTEX searches for the optimal experimental design. FACTEX (accessible through syntax) generates two-level and response-surface designs. Tables 10.9 and 10.10 show a fractional-factorial design generated through FACTEX. Other two-level designs accessible through FACTEX are Taguchi and Plackett-Burman designs. The FACTEX response-surface procedure generates Box-Behnken and central-composite designs. OPTEX (also accessible through syntax) generates mixture and optimal designs. Table 10.28 shows an optimal design generated through OPTEX and PLAN.

SAS ADX, which overlaps FACTEX and OPTEX, generates and analyzes a wide variety of designs through a menu system: two-level, including fractional-factorial and

TABLE 10.29 Comparison of SAS, SYSTAT, and NCSS for the design of screening and other incomplete research designs

Feature	SAS FACTEX	SYSTAT DESIGN	NCSS
Fractional-factorial Designs	Two-Level	Box-Hunter	Fractional-Factorial
Input			
Specify number of factors	Yes	Yes	Yes
Specify number of runs	Yes	Yes	Yes
Specify number of replicates	No	Yes	No
Specify blocking variable	Yes	No[a]	Yes
Request simulated response variable	No	No	Yes
Specify coding of factor levels	Yes	No	Yes
Specify fraction of a full factorial	Yes	No	No
Specify resolution	Yes	No	No
Specify effects to be estimated	Yes	No	No
Specify generator	No	No	No
Specify folding	Yes	No	No
Printed Output			
Design matrix	Yes	Yes	No
Design generator	No	Yes	Yes
Resolution	No	Yes	No
Defining relation	No	No	Yes
Pattern of aliasing	Yes	No	Yes
Special analysis module with plots	No	No	Yes[e]
Plackett-Burman Designs	ADX Two-Level	Plackett	Screening
Input			
Maximum number of levels	2	7	2
Specify number of runs	Yes	No	Yes
Specify number of factors	Yes	Yes	Yes
Specify blocking variable	Yes	No[a]	No
Specify number of replicates	Yes	Yes	No
Request simulated response variable	Yes	No	Yes
Specify coding of factor levels	Yes	No	Yes
Special analysis module with plots[c]	Yes	No	No
Printed output			
Design matrix	Yes	Yes	No

(continued)

TABLE 10.29 Continued

Feature	SAS FACTEX	SYSTAT DESIGN	NCSS
Taguchi Designs	ADX Mixed Level	Taguchi	Taguchi
Input			
Specify Taguchi type (see Table 10.15)	Yes	Yes	Yes
Maximum number of levels	3	5	5
Specify number of factors	Yes	No	Yes
Specify number of runs	Yes	No	No
Specify number of replicates	Yes	Yes	No
Distinguish between inner and outer arrays	Yes	No	No
Specify coding of factor levels	Yes	No	No
Online control for nested models	No	Yes	No
Calculate a loss function	No	Yes	No
Printed output			
Design matrix	Yes	Yes	No
Pattern of aliasing	No	Yes	No
Box-Behnken Designs	ADX Response Surface	Box-Behnken	Response Surface
Input			
Maximum number of factors	7	16	6
Specify blocking variable	Yes	Yes	Yes
Specify simulated response variable	Yes	No	Yes
Specify coding of levels	Yes	No	Yes
Specify random number seed	Yes	No	No
Printed output			
Design matrix	Yes	Yes	No
Special analysis module with plots[c]	Yes	No	Yes
Central-Composite Designs	ADX Response surface	Central Composite	Response surface
Input			
Maximum number of factors	8	Unlimited	6
Specify blocking variable	Yes	Yes	Yes

TABLE 10.29 Continued

Feature	SAS FACTEX	SYSTAT DESIGN	NCSS
Specify simulated response variable	Yes	No	Yes
Specify coding of levels	Yes	No	Yes
Specify replications	Yes	No	No
Specify number of center points	Yes	Yes	No
Specify value of M to construct cube	No	Yes	No
Specify value of α for placement of axial points	Yes	Yes	No
Specify incomplete blocks	No	Yes	No
Printed output			
Design matrix	Yes	Yes	No
Complete defining relation	No	Yes	No
Design generators	No	Yes	No
Special analysis module with plots[c]	Yes	No	Yes
Mixture Models	OPTEX	Mixture	Not available separately[d]
Input			
Maximum number of factors	10	Unlimited	N.A.
Specify number of factor levels	No[b]	Yes	N.A.
Specify number of runs	Yes	No	N.A.
Specify blocking variable	Yes	No	N.A.
Specify number of replications	Yes	No	N.A.
Lattice design	Yes	Yes	N.A.
Centroid design	No	Yes	N.A.
Axial design	Yes	Yes	N.A.
Screening design	Yes	Yes	N.A.
Constrained mixture designs	Yes	Yes	N.A.
Printed output			
Design matrix	No	Yes	N.A.
Special analysis module with plots[c]	No	No	N.A.
Optimal Designs	PLAN and OPTEX	Optimal	D-optimal
Input			
Maximum number of factors	Unlimited	Unlimited	Unlimited
Specify optimality criterion	Yes	Yes	No

(*continued*)

TABLE 10.29 Continued

Feature	SAS FACTEX	SYSTAT DESIGN	NCSS
Specify optimization method	Yes	Yes	No
Specify candidate design point (data set)	Yes	Yes	Yes
Specify number of solutions	Yes	Yes	No
Specify number of runs	Yes	Yes	Yes
Specify number of points to exchange at each iteration	Yes	Yes	Yes
Specify maximum number of iterations	Yes	Yes	Yes
Specify convergence tolerance	Yes	Yes	No
Specify which solutions are to be displayed	Yes	Yes	No
Specify whether design region is continuous or discrete for each IV	CLASS	Yes	Yes
Specify number of duplicates (replicates)	No	No	Yes
Specify whether intercept is to be included	Yes	Yes	Yes
Specify simulated response variable	No	No	Yes
Indicate combinations of levels to be forced	Yes	No	Yes
Indicate combinations of levels to be excluded	Yes	No	Yes
Specify random number seed	Yes	Yes	No
Printed output			
Efficiencies of designs	Yes	Yes	Yes
Model summary	Yes	No	Yes
Statistics regarding individual df effects	No	No	Yes
Aliasing structure	Yes	No	No

[a] Codes may be added to data set to turn replicates into levels of a blocking factor.
[b] Depends on number of factors and number of runs.
[c] See Table 10.37 for features of analysis modules.
[d] May be generated through D-optimal module.
[e] The Analysis of Two-Level Designs module may be used if the number of runs is a power of 2.

Plackett-Burman designs (illustrated in Table 10.15); response-surface, including Box-Behnken and central-composite designs (illustrated in Section 10.7); mixture; mixed-level, including factorial and Taguchi designs; optimal designs; split plot; and general factorial designs. (Special analytic procedures in ADX are reviewed later in Table 10.37.)

Beginning with Version 9, SYSTAT DESIGN has two modes: classic and wizard. The classic mode generates and can be accessed through syntax; the wizard can not. Both modes generate full factorial designs, as well as fractional-factorial (Box-Hunter), Taguchi, Plackett-Burman, Box-Behnken, and mixture designs. Wizard mode generates central-composite and optimal designs as well. SYSTAT DESIGN also generates Latin-square designs (Chapter 9). Patterns of aliasing are available for Box-Hunter and Taguchi designs. Randomization of runs may be specified for all designs. Designs are analyzed through SYSTAT ANOVA, GLM, or REGRESSION. Table 10.16 is an example of a Taguchi experiment planned through SYSTAT DESIGN.

NCSS Design of Experiments generates fractional-factorial, screening, Taguchi, and response-surface designs (Box-Behnken and central-composite), as well as, in the most recent version, D-optimal designs. Fractional-factorial designs are limited to two levels. Screening designs in NCSS include Plackett-Burman designs, as well as designs in which the number of runs is a power of 2. All the procedures have the option to simulate responses, so you can ensure that the design provides the tests you really want. Generated files are fully modifiable, so that any number of responses may be added and codes for IVs altered. Indeed, the method for excluding combinations of levels of IVs in optimal designs requires deleting unwanted rows from a fully factorial candidate set of runs. Special analysis modules with plots are available for two-level designs without interactions and for response-surface designs (cf. Table 10.37). D-optimal designs may be generated with replicates. The D-optimal module also permits design of mixture experiments. Most of the modules permit generation of a simulated response for each run, which is useful in planning phases to ensure that the analysis will provide desired information.

10.6.2 Choosing among Screening and Other Incomplete Designs

The first issue in choosing a design is whether one of the incomplete designs in this chapter or in Chapter 9 is more appropriate. All of the designs in this chapter apply to randomized-groups IVs; Latin-square designs are the only readily available incomplete designs that apply to either randomized-groups or repeated-measures IVs. Latin square is useful for randomized groups when there are three IVs (with two of them usually "nuisance" IVs) and all IVs have the same number of levels. Otherwise, the designs of this chapter are likely to be more useful.

In this chapter, choice of designs is made on the basis of the type of IVs (qualitative, quantitative, or a mix of qualitative and quantitative) and the number of levels of each IV. Table 10.30 summarizes these choices.

10.6.2.1 Qualitative IVs

Four types of incomplete designs are available when IVs are qualitative: fractional-factorial, Plackett-Burman, Taguchi, and optimal designs. Fractional-factorial designs are useful when all IVs have two levels or when all IVs have three levels (although use with three levels is rare and rather complicated). With replication, you may be able to test at least some two-way interactions. Similarly, some Taguchi designs allow tests of

TABLE 10.30 Choosing a screening or other incomplete design

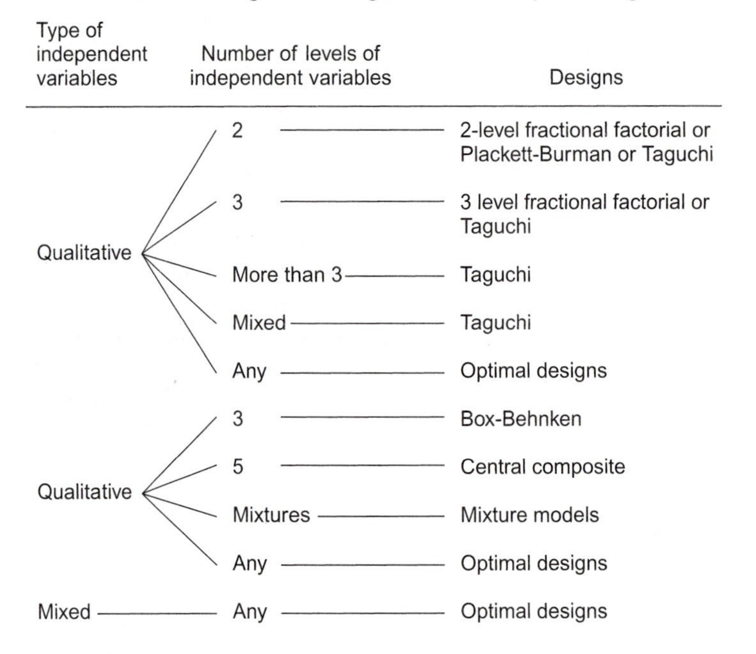

Type of independent variables	Number of levels of independent variables	Designs
Qualitative	2	2-level fractional factorial or Plackett-Burman or Taguchi
	3	3 level fractional factorial or Taguchi
	More than 3	Taguchi
	Mixed	Taguchi
	Any	Optimal designs
Qualitative	3	Box-Behnken
	5	Central composite
	Mixtures	Mixture models
	Any	Optimal designs
Mixed	Any	Optimal designs

two-way interactions in lieu of tests of main effects.[11] Plackett-Burman designs are useful for screening a very large number of IVs, each with two levels, in the smallest possible number of runs; however, they are resolution III designs in which no interactions are tested. Optimal design methods are available when none of these designs is appropriate for your purposes.

For two-level IVs, then, choose a Plackett-Burman design if you want the fewest number of runs and can do without testing any interactions; choose a Taguchi design if you find one that meets your particular needs (see Table 10.15); otherwise, choose a fractional-factorial design. For IVs with three levels or studies in which IVs have different numbers of levels, your best bet is one of the Taguchi designs. Or, when IVs have three levels, you might choose a Box-Behnken design (see below) if you can find a coding scheme that makes sense for your research.

10.6.2.2 Quantitative IVs

Box-Behnken (three levels of each IV), central-composite (five levels of each IV), or mixture designs are used with quantitative IVs. In their standard form, both Box-Behnken and central-composite designs test linear and quadratic trends of main effects and linear interaction contrasts. Therefore, the Box-Behnken design is probably adequate for most purposes; the central-composite design requires more runs to

[11] Taguchi analyses (S:N ratios, calculation of ρ) may be used with any screening or other incomplete design, even with quantitative IVs (Section 10.5.3).

accommodate the extra two levels of IVs but does not provide additional information. Optimal design methods are available if your design does not meet these criteria.

Mixture designs are used when the response is to a mixture of levels of IVs and the various blends add up to a fixed total. The degree of trend is limited by the number of levels of the IVs and, therefore, the degrees of freedom available for evaluating each IV. Thus, some designs permit tests of trends higher than quadratic, although such trends are often difficult to interpret. Taguchi analyses are frequently useful with mixture designs to identify the most effective blend(s).

10.6.2.3 Mixed Qualitative and Quantitative

None of the specific designs accommodates a mix of qualitative and quantitative IVs, unless the particular characteristics of the qualitative IVs happen to make a trend analysis meaningful. However, any such mixed designs can be handled through optimal design methods.

10.7 COMPLETE EXAMPLE OF A CENTRAL-COMPOSITE DESIGN

The example illustrated here is an experiment conducted by one of us one hot August afternoon on a Pentium 166 MHz computer running main-effects ANOVA on SPSS 7.5 for Windows. The goal was to determine which factors, from among the number of IVs, the number of covariates, or sample size, had an impact on time to complete an analysis.[12] The desire was to determine which factor or factors in the design influenced the processing time of the computer and whether there were quadratic effects of any of the factors or interactions among them. Each factor had five levels, in keeping with the requirements of a central-composite design. The first factor was number of IVs (each with two levels) included in the model (IVNO: 1 to 5); the second factor was number of covariates included in the model (COVAR: 0 to 4); and the third factor was sample size (SIZE: 50, 150, 250, 350, 450). The DV was processing time for GLM for main effects only; all other settings were left at default for the program. Each of the 20 runs included a different combination of the three IVs.

The data set used was from the sequential logistic regression example of Tabachnick and Fidell (2007, Section 10.7.3). The 5 two-level IVs were created from demographic variables representing marital status, working status, presence of children, race, and age group. The four continuous covariates were attitudes toward marital status, attitudes toward role of women, attitudes toward housework, and socioeconomic level. When the number of IVs (or the number of covariates) was less

[12] This experiment was impossible to conduct at a later date on a Pentium II 450 MHz computer running SPSS 9.0. Run time was simply too fast to record without special timing equipment.

than five (or four), in a particular run, the variables were randomly selected from among the sets of five (or four). To create the factor of sample size, samples of appropriate size were randomly selected for each run from the 465 cases available. The DV for all runs was educational attainment. Data files are COMPUTER.*.

10.7.1 Generating the Design

SAS ADX was used to create the pattern of runs for the three-factor central-composite design (choose Solutions > Analysis > Design of Experiments from the SAS menus). Table 10.31 shows the setup for designing a three-factor central-composite experiment and the resulting runs where the DV (elapsed time in seconds to perform the analysis) from the experimental runs have been added. Variable names were changed in the data window from the X1, X2, X3 notation to IVNO, COVAR, and SIZE, respectively, while Y1, the DV, was changed to TIME. Coding involves specifying the range of the three middle values when highest and lowest values are requested—that is, 2 and 4 for IVNO, 1 and 3 for COVAR, 150 and 350 for SIZE, replacing the -1 and 1 that are the defaults. Highest and lowest values are generated by setting alpha to 2 in the Axial Scaling menu.

The levels of the IVs are specified with unequal frequency. Among the 20 runs, there is a single run with the highest level and a single run with the lowest level of each IV. The middle level of each IV is the most commonly specified. The second and fourth levels of each IV are less frequently specified than the middle level but are more frequent than the extreme levels, as is typical in central-composite designs.

10.7.2 Assumptions and Limitations

10.7.2.1 Normality of Sampling Distributions, Outliers, and Independence of Errors

Degrees of freedom are inadequate to assume normality. With 20 runs (cases) and 9 effects to be tested (linear and quadratic effects of all three IVs and linear by linear interaction contrasts of the 3 two-way interactions), df for error $= N - k - 1 = 10$. SAS ADX provides several plots: outliers, influential runs, and residuals. This assessment of normality does not take into account sampling distributions for cells; instead, it considers all data as a single group. Table 10.32 shows the setup for SAS ADX analysis, a normal probability plot of residuals, and a plot of outliers, with a criterion set to $\alpha = .001$.

SAS ADX flags runs that are outliers on the residuals plot, if there are any. None are marked here. In the normal plot of residuals, the data points deviate from the major diagonal (representing normality) at both the lower and upper levels. This suggests that the distribution is not skewed but rather has heavier-than-normal tails. This should pose no problem for the current screening analysis. Independence of errors is ensured by the random assignment of levels to runs.

TABLE 10.31 Generating a three-factor central-composite design through SAS ADX (setup and data file)

1. In ADX, select File to Create New Design . . . , then Response Surface.
2. Choose Select Design.
3. Fill in the Response Surface Design window as follows:
   a. Number of factors = 3.
   b. Choose Central Composite: Uniform Precision.
   c. Click on Axial Scaling and specify alpha = 2.
   d. Close out of the window, agreeing to use the selected design.
4. Fill in the Define Variables window as follows:
   a. Under Factor tab, fill in variable names and low level and high level: Low and high levels are 2 and 4 for IVNO; low and high levels are 1 and 3 for COVAR; low and high levels are 150 and 350 for SIZE.
   b. Under Response tab, fill in TIME as name.
5. Choose Edit Response.
6. Click on Design on the toolbar and choose Randomize Design.
7. Choose a seed number and return to the spreadsheet.

Save the file, run the experiment, and then fill in the response values (within Edit Response)

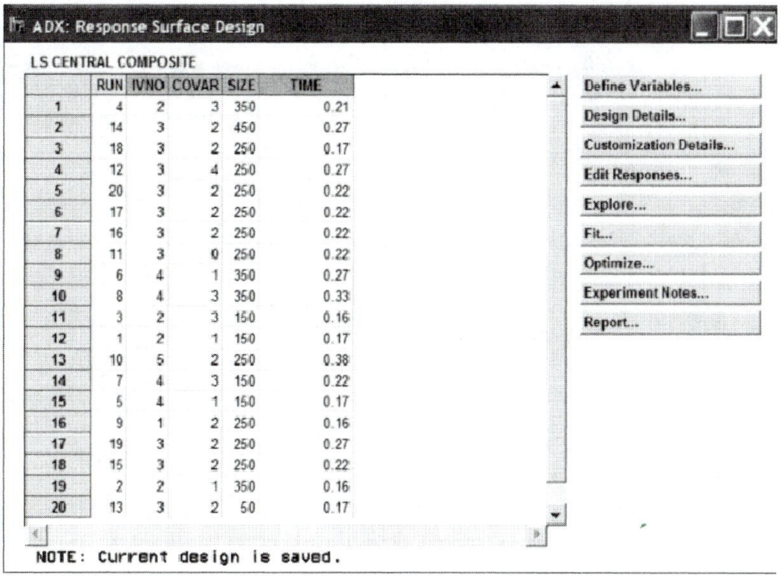

ADX: Response Surface Design

LS CENTRAL COMPOSITE

	RUN	IVNO	COVAR	SIZE	TIME
1	4	2	3	350	0.21
2	14	3	2	450	0.27
3	18	3	2	250	0.17
4	12	3	4	250	0.27
5	20	3	2	250	0.22
6	17	3	2	250	0.22
7	16	3	2	250	0.22
8	11	3	0	250	0.22
9	6	4	1	350	0.27
10	8	4	3	350	0.33
11	3	2	3	150	0.16
12	1	2	1	150	0.17
13	10	5	2	250	0.38
14	7	4	3	150	0.22
15	5	4	1	150	0.17
16	9	1	2	250	0.16
17	19	3	2	250	0.27
18	15	3	2	250	0.22
19	2	2	1	350	0.16
20	13	3	2	50	0.17

Define Variables...
Design Details...
Customization Details...
Edit Responses...
Explore...
Fit...
Optimize...
Experiment Notes...
Report...

NOTE: Current design is saved.

10.7.2.2 Homogeneity of Variance

There are insufficient data to evaluate homogeneity of variance for the main effects because there is only one trial of the extreme values for each IV. Also, some combinations of IVs (e.g., IVNO = 1, COVAR = 2, SIZE = 250) occur only once, so that homogeneity of variance cannot be assessed for the interaction terms. Follow-up

TABLE 10.32 Analyzing a three-factor central composite design through SAS ADX (setup with outliers and residuals output)

1. Open the Response Surface Design window showing the spreadsheet.
2. Click on Fit but don't peek at the results yet.
3. From the main toolbar, choose Model and then Check Assumptions. . . .
   a. Within Residuals tab, change Y axis to Normal score for Residuals.
   b. Within the Outliers tab, change Outlier probability to .001 and Y Axis to _COOKD.

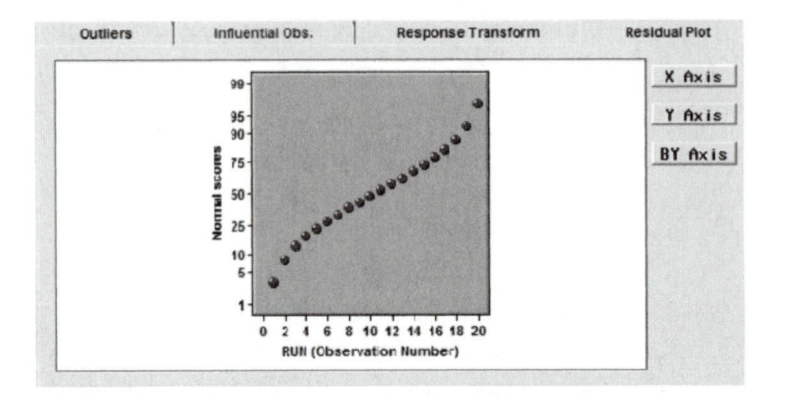

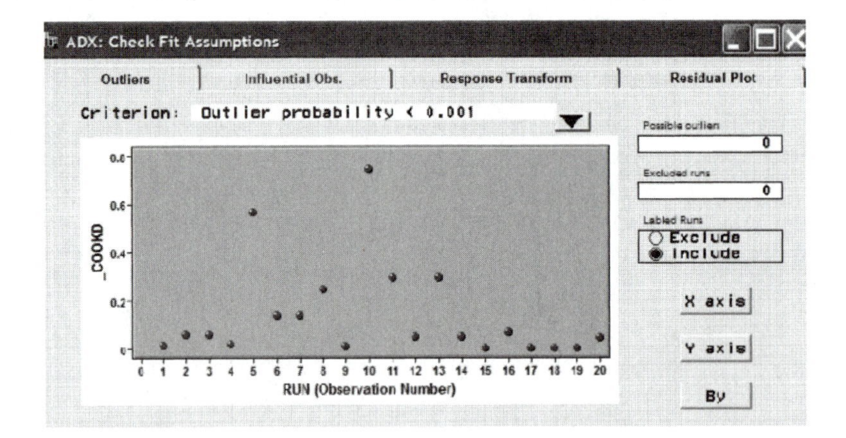

research with a full factorial design would resolve the issue. The effect of heterogeneity of variance is to raise the Type I error rate beyond the nominal level. This is not fatal if the purpose of the experiment is to screen effects for planning further research. On the other hand, if the results of this experiment are themselves to be interpreted, a prudent course is to set $\alpha = .025$ to compensate for possible violation of homogeneity of variance.

TABLE 10.33 Tests of individual effects by SAS ADX (setup is in Table 10.32)

Effect	Estimate	Std Error	t Ratio	P Value
IVNO	0.045625	0.0080356	5.6779	0.0002
COVAR	0.015625	0.0080356	1.9445	0.0805
SIZE	0.028125	0.0080356	3.5001	0.0057
IVNO*IVNO	0.01	0.0064102	1.56	0.1498
IVNO*COVAR	0.00875	0.011364	0.76997	0.4591
IVNO*SIZE	0.02125	0.011364	1.8699	0.0910
COVAR*COVAR	0.00375	0.0064102	0.58501	0.5715
COVAR*SIZE	0.00875	0.011364	0.76997	0.4591
SIZE*SIZE	-0.0025	0.0064102	-0.39001	0.7047

10.7.3 Three-Factor Central-Composite Design

Table 10.33 shows the output of the preliminary analysis through SAS ADX, as produced by the setup of Table 10.32. This part of the output, visible when `Fit...` is chosen in the main spreadsheet menu, indicates which of the effects affects computer processing time. Only two of the main effects are statistically significant predictors: linear effects of the number of IVs ($p = .0002$) and the sample size ($p = .0057$). None of the quadratic effects (e.g., `IVNO*IVNO`) or two-way interactions (`IVNO*COVAR`) adds to the predictability of the model. This is strictly an additive model in which the number of IVs in a design adds to the sample size to lengthen processing time.

Details can be accessed for this basic model, as shown in Table 10.34. Two models are printed out: a full model and one that includes only the two significant main effects. Lack of fit tests the adequacy of the model and is evaluated in both analyses. A significant result would suggest that a higher-order model is appropriate (e.g., a model that includes cubic terms or quadratic by quadratic interaction terms). Note that a conventional source table is available under the `Model Terms` tab.

The "pure error" in both models is provided by the replicated observations from at least some of the cells and is what permits a test of lack of fit. The overall model, including all terms in the model, is statistically significant: $F(9, 10) = 6.23, p = .009$, with an impressive 84% of the variance (71% adjusted R^2) in processing time predictable from the combination of nine effects. Even the simplified model, which includes only the main effects of IVNO and SIZE, has $R^2 = .67$ (.63 adjusted).

Table 10.35 shows regression coefficients, standard errors, and t-ratios for all effects, as well as effect sizes and their confidence limits, in a form suitable for publication. Effect sizes and their confidence limits are found using Smithson's (2003) software, using $t^2 = F$ with 1 and 10 df.

Response-surface methodology provides a contour plot to show optimum combinations of factors to produce the desired outcome. One of these plots that includes

TABLE 10.34 Model details for SAS ADX analysis (setup and output)

Click on Model on the main toolbar, then Fit Details

(a) Master model

ADX: Fit Details for TIME

Model Type: ○ Predictive Model ◉ Master Model

Overall ANOVA | Model Terms | Parameters | Canonical Analysis | Ridge Analysis

Source	DF	SS	MS	F	Pr > F
Model	9	0.0579	0.0064	6.2323	.0042400
(Linear)	3	0.0499	0.0166	16.0899	.0003730
(Quadratic)	3	0.0032	0.0011	1.0462	.4141006
(Cross Product)	3	0.0048	0.0016	1.5608	.2594242
Error	10	0.0103	0.0010		
(Lack of fit)	5	0.0053	0.0011	1.0662	.4727981
(Pure Error)	5	0.0050	0.0010		
Total	19	0.0683			

Response Mean	R-Square	Adjusted R-Square	Root MSE	C.V.
0.224	0.848693	0.712516	0.032142	14.34922

Include block effect in predictive model

(b) Predictive model

ADX: Fit Details for TIME

Model Type: ◉ Predictive Model ○ Master Model

Overall ANOVA | Model Terms | Parameters | Canonical Analysis | Ridge Analysis

Source	DF	SS	MS	F	Pr > F
Model	2	0.0460	0.0230	17.5056	.0001000
Error	17	0.0223	0.0013		
(Lack of fit)	6	0.0108	0.0018	1.7130	.2078119
(Pure Error)	11	0.0115	0.0010		
Total	19	0.0683			

Response Mean	R-Square	Adjusted R-Square	Root MSE	C.V.
0.224	0.673147	0.634694	0.036233	16.17522

Include block effect in predictive model

IVNO and SIZE is shown in Figure 10.4. Because this is a simple main-effects model, the plot shows that time is minimized when there are fewer IVs and smaller size, with number of covariates at midlevel. Note that standard errors are minimized in the center of the plot, because there are more replications of levels of IVs that are in the middle than at the upper and lower extremes.

Note that the combination of smallest sample (50) and fewest number of IVs (1) does not occur in this central-composite design (it is the nature of the central-composite

TABLE 10.35 Parameter estimates and effect sizes for all effects in central-composite analysis

Effect	B Coefficient	Standard Error	t-Ratio (df = 10)	Partial η^2	95% Confidence Limits for Partial η^2 Lower	Upper
IVNO: Number of IVs	0.046	0.008	5.68	.77	.34	.86
COVAR: Number of covariates	0.017	0.008	1.94	.27	0	.58
SIZE: Sample size	0.031	0.008	3.50	.55	.07	.75
Quadratic IVNO	0.010	0.006	1.56	.20	0	.52
Quadratic COVAR	0.001	0.006	0.59	.03	0	.35
Quadratic SIZE	−0.007	0.006	−0.39	.01	0	.30
IVNO by COVAR interaction	0.009	0.011	0.77	.06	0	.38
IVNO by SIZE interaction	0.021	0.011	1.87	.26	0	.57
COVAR by SIZE interaction	0.009	0.011	0.77	.01	0	.30

FIGURE 10.4 Contour plot of time as a function of number of IVs and sample size (setup and output)

1. From the Effect window of Table 10.33, Choose Contour Plots.
2. From the View menu on the main toolbar, choose Change Plot. . . .
3. Select Simple Plot, then Next, and choose IVNO as the F1 variable and SIZE as the F2 variable.
4. When the plot appears, jazz it up by selecting options with right-click on contours and overall plot.

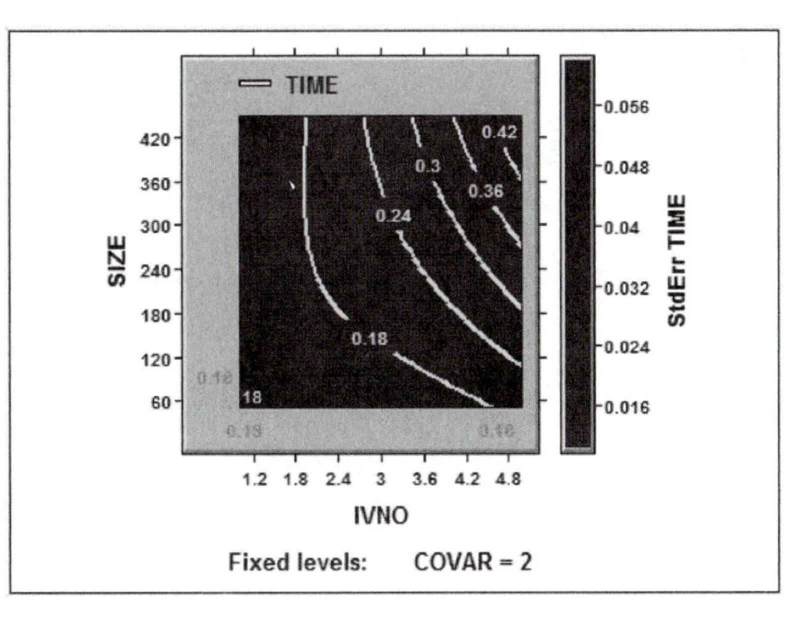

TABLE 10.36 Checklist for central-composite design

1. Generate the design
2. Issues
   a. Normality of sampling distributions and outliers
   b. Independence of errors
   c. Homogeneity of variance
3. Major analysis
   a. Statistical significance of levels of model (linear, quadratic, interaction). If significant:
      (1) Significance of factors
      (2) Parameter estimates
      (3) Effect size and confidence limits
   b. Design further research
      (1) Optimum values
      (2) Contour plots

design that most of the combinations of levels are in the central range). Without quadratic effects or interactions, the only further research that might be of interest would be that which explores further minimization and/or maximization of processing time: larger samples and greater numbers of IVs to explore further maximization of processing time and/or smaller sample sizes to explore further minimization of processing time, if research using smaller samples would be powerful enough to warrant exploration.

Table 10.36 shows a checklist for a central-composite design. A results section, in journal format, follows for the study just described.

Results

A central-composite screening design was used to analyze factors that increase processing time for a general factorial main-effects ANOVA through SPSS for Windows 7.5, running on a Pentium 166. The three independent variables were number of two-level IVs in the ANOVA (1-5), the number of covariates (0-4), and the sample size for the run (50, 150, 250, 350, 450). There were 20 runs, with varying combinations of number of factors, number of covariates, and sample size. The design was generated through SAS ADX, with random order of runs. Residuals analysis showed acceptable normality and no outliers, but homogeneity of variance could not be assessed because of small cell sizes. Therefore, the criterion for statistical reliability was set to .025 to compensate for possible violation.

The overall analysis through SAS ADX, including linear and quadratic trends of the three main effects and linear by linear interaction contrasts, was statistically significant, $F(9, 10) = 6.23$, $p = .004$, $R^2 = .85$. However, this analysis showed that only the linear trends of main effects were statistically significant, $F(3, 10) = 16.09$, $p = .003$. Parameter estimates (unstandardized coefficients, B) for all effects with standard errors, t-ratios, effect sizes, and their 95% confidence limits are in Table 10.35. Among the nine effects tested, only two were statistically significant: number of IVs, $t(10) = 5.68$, and sample size, $t(10) = 3.50$, $p < .025$.

Contour plots show that the minimum processing time is spent with fewer number of IVs and smaller sample sizes.

10.8 COMPARISONS OF PROGRAMS

SPSS, SAS, and SYSTAT all have ANOVA programs to analyze incomplete designs. SAS and NCSS also have modules to analyze some specific designs.

Screening designs are diverse enough that various features of the general ANOVA programs could be applicable to any one of them. Consult the Comparison of Programs sections in Chapters 5–9 for features of the SAS, SYSTAT, and SPSS GLM programs. Table 10.37 compares features of analysis modules within NCSS and SAS ADX, both programs to generate designs.

TABLE 10.37 Comparison of NCSS and SAS FACTEX analysis modules

Feature	SAS ADX	NCSS
Fractional-factorial, Plackett-Burman, and Taguchi Designs	Two-level	Two-level
Input		
Specify terms in model	Yes	No
Specify pooling of terms	No	Yes
Specify significance level for tests of effects	No	Yes
Specify size of aliasing to be reported	No	Yes
Specify number of decimals in means and mean squares	No	Yes
Specify plot characteristics	Yes	Yes
Save residuals to data set	No	Yes
Select ANOVA or regression-type output or both	Yes	No
Printed Output		
ANOVA source table	Yes	Yes
Regression source table	Yes	No
Estimated marginal and cell means	Yes	Yes
Mean differences for each main effect	No	Yes
Sorted means and mean differences	No	Yes
Lack of fit tests (residual and pure error)	Yes	No
Parameter estimates and tests	Yes	No
Overall R^2 and adjusted R^2	Yes	No
Plots		
Normal probability plot of residuals	Yes	Yes
Plot of residuals (raw or standardized) vs. predicted (or observed) score	Yes	No
Store residuals in data file	No	Yes
Normal probability plot of effects	No	Yes
Interaction plots of means (profiles)	Yes	No
Residuals vs. observation number	Yes	No
Cube plot	Yes	No
Box-Behnken and Central Composite Designs	Response surface	Response surface
Input		
Specify terms in model	Yes	No
Specify covariates (continuous factors)	No	Yes
Specify minimums, maximums, and number of decimals	No	Yes
Specify maximum exponents for terms in model	No	Yes
Specify whether minimum or maximum is sought	No	Yes
Specify characteristics of linear terms	No	Yes

TABLE 10.37 Continued

Feature	SAS ADX	NCSS
Specify characteristics of cubic terms	No	Yes
Request search for most parsimonious hierarchical model	No	Yes
Specify minimum R^2 to avoid removing term	No	Yes
Specify precision accuracy	No	Yes
Specify plot characteristics	Yes	Yes
Specify maximum number of evaluations	No	Yes
Save residuals and predicted values to data set	No	Yes
Specify sum of squares type	Yes	No
Specify significance level for tests of effects	No	No
Request main effect comparisons	No	No
Specify multiple responses to optimize	No	No
Printed output		
Count, mean, minimum, maximum for each variable	No	Yes
Estimated marginal means	Yes	Yes
Estimated cell means	Yes	No
Sequential ANOVA table with trend components	Yes	Yes
Sequential ANOVA table with trend components using pure error	Yes	Yes
ANOVA table with IVs (main effect and interactions combined)	No	Yes
ANOVA table with IVs using pure error	No	Yes
ANOVA table with tests of each effect	Yes	No
Regression source table	Yes	No
Parameter estimates for all effects	Yes	Yes
Parameter estimates using pure error	No	Yes
Incremental, term, and last R^2s	No	Yes
Overall R^2 and adjusted R^2	Yes	No
Optimum value of each IV (main effect and interaction combined)	No	Yes
Function at optimum	No	Yes
Scores at minimum and maximum ridges	Yes	No
Canonical analysis of response surface	Yes	No
Predicted value at stationary point	Yes	No
Residuals and predicted values	No	Yes
Effect size	No	Yes
Observed power	No	No
Lack of fit tests	Yes	Yes

(continued)

TABLE 10.37 Continued

Feature	SAS ADX	NCSS
Plots		
Normal probability plot of residuals	No	Yes
Plot of residuals (raw or standardized) vs. predicted (or observed) score	Yes	No
Contour plots	Yes	Yes
Surface plots	Yes	No
Interaction plots of means (profiles)	Yes	No
Residuals vs. observation number	Yes	No
Cube plot	Yes	No
Save residuals and predicted values to data file	No	Yes
Mixture Models	Mixture	Not available separately
Input		
Specify terms in model	Yes	N.A.
Specify blocks	Yes	N.A.
Printed output		
ANOVA source table	Yes	N.A.
Regression source table	Yes	N.A.
Parameter estimates	Yes	N.A.
Scores at minimum and maximum of ridges	Yes	N.A.
Plots		
Normal probability plot of residuals	Yes	N.A.
Plot of residuals (raw or standardized) vs. predicted (or observed) score	Yes	N.A.
Contour plots	Yes	N.A.
Surface plots	Yes	N.A.
Interaction plots of means (profiles)	Yes	N.A.
Residuals vs. observation number	Yes	N.A.

10.8.1 SAS ADX

Designs generated through the ADX menu system may be analyzed through that same menu system after responses are added for the runs. Any of the two-level designs can be analyzed through the menu system. The ADX menu system analyzes IVs with more than two levels only in response surface (Box-Behnken, central-composite). Any data set may be saved to an SAS file and analyzed through GLM.

Options for modifying designs are limited in SAS ADX; however, the program provides output of both a standard ANOVA table and a regression table, accompanied by parameter estimates. Standard residuals and profile plots are available; however, residuals cannot be added to the data file. The program plots residuals as a function of observation number, which can be of interest if there is reason to suspect a sequential change in response over runs (cases).

10.8.2 NCSS Design of Experiments

NCSS provides a full range of analyses for two-level and response-surface (Box-Behnken and central-composite) designs. Mixture designs, and fractional-factorial and Taguchi designs that have IVs with more than two levels, are analyzed through regular regression and ANOVA modules.

There is detailed information about aliasing in two-level designs. Effects cannot be selected specifically during analysis, but they can be pooled and added to the error term if preliminary analysis shows that they are negligible and the design provides no degrees of freedom for error. This is the only program that provides tables of means and mean differences for each two-level effect and combination of two-level effects; one table is sorted by size of effect. For response-surface designs, covariates can be specified. There are many choices for analyzing these designs, including a stepwise (called hierarchical) analysis; ANOVA tables are presented in several formats. Contour plots are available for response-surface analysis. For all the designs, the only available plot of residuals is a normal probability plot; however, predicted scores for response-surface analysis and residuals for any analysis can be added to the data file for plotting in any format within another NCSS module.

10.9 PROBLEM SETS

10.1 IVs in the following 2^{3-1} half-factorial design are type of cement (Portland or other), reinforcement (yes or no), and additive (yes or no). In each case, the first listed level is coded 1, and the second level is coded −1. There are two replicates (blocks). The DV is strength of the dried cement.

Run	Cement Type	Reinforcement	Additive	Strength
1	1	1	1	9
2	1	−1	−1	3
3	1	−1	−1	4
4	−1	−1	1	6
5	−1	−1	1	7
6	1	1	1	8
7	−1	1	−1	7
8	−1	1	−1	8

a. Set up the full coding for the regression approach.

b. Show the aliasing pattern for this design.

c. Do the ANOVA using the regression approach.

d. Calculate effect sizes(s) and their confidence limits.

e. Write a results section for this analysis.

f. Use one of the statistical software programs to check your results.

10.2 Generate a Plackett-Burman design with seven factors and eight runs.

10.3 The following L_8 Taguchi design has seven IVs with two levels and a response, which is replicated.

Run	A	B	C	D	E	F	G	Response 1	Response 2
1	1	1	1	1	1	1	1	3	5
2	1	1	1	2	2	2	2	4	5
3	1	2	2	1	1	2	2	5	7
4	2	1	2	2	1	2	1	9	8
5	2	1	2	1	2	1	2	8	6
6	1	2	2	2	2	1	1	6	4
7	2	2	1	1	2	2	1	8	9
8	2	2	1	2	1	1	2	7	9

a. Find the S:N ratio, assuming that small values of Y are desirable.

b. Use a statistical software program to do a preliminary ANOVA on the main effects to determine ignorable effects.

c. Rerun the ANOVA to find F ratios after removing ignorable effects.

d. Calculate p (%) for all significant effects.

e. Summarize the results in a source table.

10.4 IVs in the following Box-Behnken design are length of ski vacation (2, 4, and 6 days), size of hotel room in square meters (50, 100, 150), and cost of lift tickets ($45, $50, $55). The DV is the rating of the ski trip.

Days	Room Size	Tickets	Rating
−1	0	1	2
0	−1	−1	6
1	0	−1	8
1	0	1	9
0	0	0	5
−1	1	0	3
−1	0	−1	1
1	1	0	10
0	−1	1	4
1	−1	0	7
−1	−1	0	1
0	1	−1	7
0	0	0	6
0	1	1	6
0	0	0	4

a. Code the quadratic trends of the main effects and the linear by linear trends of the interactions.

b. Do the analysis using the regression approach.

c. Write a results section.

d. Check your work through a statistical software program.

10.5 The following data are from a mixture design to investigate optimum proportions of cinnamon and sugar for cinnamon toast. The DV is a rating of taste on scale of 1 to 10.

Run	Sugar	Cinnamon	Rating
1.00	0.33	0.67	4.00
2.00	1.00	0.00	2.00
3.00	0.67	0.33	10.00
4.00	0.50	0.50	7.00
5.00	0.17	0.83	3.00
6.00	0.00	1.00	1.00
7.00	0.83	0.17	6.00
8.00	0.50	0.50	6.00
9.00	0.00	1.00	2.00
10.00	0.67	0.33	9.00
11.00	0.83	0.17	7.00
12.00	0.33	0.67	5.00
13.00	1.00	0.00	2.00
14.00	0.17	0.83	5.00

a. Calculate the codes for the sugar by cinnamon interaction. (*Hint:* See Table 10.26 syntax.)

b. Run a sequential regression to assess the effects of proportions of sugar and cinnamon on rating of taste, being sure to include the interaction as an effect.

c. Indicate which effects are statistically significant.

d. Show the prediction equation, using the corrected coefficients.

e. Write a results section for this analysis.

f. Using software, find an optimal design in which there are three qualitative IVs, with $a = 2$, $b = 2$, and $c = 3$. The design should include 12 runs and permit tests of all three main effects and the A by B interaction.

g. Show the runs that are chosen.

h. What is the design's D-efficiency?

i. Show a source table with df for each effect (including error and total), assuming there are no replicates.

11

Analysis of Variance with Random Effects

11.1 GENERAL PURPOSE AND DESCRIPTION

Random-effects ANOVA is used when levels of an IV are randomly chosen from a population of possible levels. Suppose, for example, a researcher defines the range of reasonable doses of a new drug as 0 to 500 mg, and then randomly chooses five levels of them, say, 15, 89, 236, 251, and 402 mg. Any set of five doses between 0 and 500 mg could be chosen for the levels, including, say, 1, 2, 3, 4, and 5 mg. Because the five doses are randomly selected, dosage is a random-effects IV.

Often, random effects are part of a mixed design that has both fixed and random IVs.[1] For instance, if time of administering the doses is added as a second IV and its levels are purposefully chosen (say, 8 a.m., 4 p.m., and 12 p.m.), then time is a fixed IV, and the design has one fixed effect and one random effect. On the other hand, if the times for administering the doses are also randomly chosen, there are two random-effects IVs.

[1] This is the second use of the term *mixed* to characterize factorial designs. The other use is to label designs with a combination of randomized-groups and repeated-measures IVs.

The differences between random and fixed IVs are in how the levels are selected and the generalizability of the findings. Levels of fixed-effects IVs are purposefully chosen to cover the range of levels of interest; the researcher is often unwilling to leave the choice to chance, because a randomly selected range could be quite narrow and some levels could be very similar. If the fixed IV is quantitative, levels are usually also chosen with equal spacing between them. The findings generalize only to the range of levels included in the study. Levels of random effects are randomly chosen and, if quantitative, are not equally spaced. The findings, however, generalize to the entire range of values from which the levels are chosen. With the exception of cases, all the IVs discussed in Chapters 2–10 are fixed.

Cases (or subjects) are almost always considered random. When treatments are applied to a set of cases, the researcher is rarely interested in those particular cases, but rather in a larger population. In many experiments, the population of cases comprises the cases potentially available for the study; for example, all the employees of a firm could be considered a population. If numerous cases are chosen and randomly assigned to various levels of treatment, we generalize the results to all the employees of that firm. If cases are first randomly selected from a larger population (say, the firm is randomly selected from among many), and then employees are randomly selected and randomly assigned to levels of treatment, the generalization has a much wider scope.[2]

The goal, then, of a random-effects IV is to widen the scope of generalization. However, there is a penalty to pay. Random-effects ANOVA is often less powerful than fixed-effects ANOVA because different, and frequently larger, error terms are developed. You are permitted a wider scope of generalization only as long as the effect of treatment is large enough to withstand the reduction in power.

Random-effects IVs appear in a multitude of applications, with both randomized-groups and repeated-measures IVs; with one-way, full, and incomplete factorials; and with covariates. One frequent application of random-effects IVs is in studies that use generalizability theory (Chronbach, 1972) for the analysis. The goal of these studies is to determine how reliable (dependable) scores for cases are when used in different situations or at different times. Each score is thought to have both a "true" component (the part that is reliably measuring what you want measured) and error. Generalizability theory uses ANOVA to partition the error component into numerous IVs that represent different features of the testing situation (test items, types of tests, scoring procedures, raters, testing environments, time frames, and so on) to see which are important. The levels of the error IVs are considered to be randomly selected. The theory has been applied in a wide variety of research areas and summarized by Brennan (1992), Shavelson and Webb (1991), and Marcoulides (1999), among others.

However, the most common use of a random-effects IV is with a nested design, as was illustrated in Table 1.3(c), where there is both a fixed- and a random-effects IV.

[2] As discussed in Chapter 1, statistical generalization is much more limited than logical generalization, which often extends beyond the population that has been randomly sampled.

For example, suppose you have three training programs and 15 groups of students, already formed. You randomly assign five groups to each of the three training programs. Training is a fixed-effect IV. Groups are nested within (do not cross) training programs and form a random-effect IV. Cases (students) are nested within groups and form another random-effect IV.

Nested designs can also be useful for increasing the scope of generalization. Consider, for example, a memory experiment in which the researcher is interested in the effects of the emotional content of words (negative or positive) on the number of words recalled. The researcher could put together one list of negative and one list of positive words, but there is the danger that recall depends on the particular lists used and not on their emotional content. One solution is to generate several lists of negative words and several lists of positive words and nest them within levels of emotional content. If an effect is found for emotional content but not for list, the researcher has evidence that the emotional content, and not the particular list, affects memory.

Nested designs are like the incomplete factorial designs of Chapter 10 in that not all combinations of levels are present. They are like mixed randomized-groups, repeated-measures designs, in that cases nest within randomized-groups IVs but cross repeated-measures IVs. There can be additional randomized-groups and repeated-measures IVs, any of which may be fixed or random.

Many random-effects ANOVA designs have largely been supplanted by the more complex and flexible features of multilevel modeling. Both nested and repeated-measures designs are appropriately analyzed in multilevel modeling; it is occasions that are nested within participants in repeated-measures designs. So, for example, reading achievement may be measured among children in classrooms that have been assigned different texts, with children nested within classrooms.

A major advantage of multilevel modeling is that there may be predictors at each level of a hierarchical design and the relationship between these predictors and the DV may vary among units at a higher level of the design. For example, hours of homework may be added as a predictor at the child level of analysis, and teacher experience may be added as a predictor at the class level of analysis. Groups may differ not only in their average score on the DV but also in their slopes relating predictors to the DV. That is, different classrooms may have different relationships between hours of homework and reading achievement. Furthermore, there may be relationships between slopes and means among groups and between interactions across the hierarchy. For example, teacher experience may be related to the amount of time children spend on homework.

Although multilevel modeling offers a much richer strategy than does random-effects ANOVA, it is beyond the scope of this book. Random-effects ANOVA is appropriate, however, for the simpler designs discussed in this chapter. Several books on multilevel modeling are available to help sort out this fascinating topic, including those by Bryk and Raudenbush (2001), Kreft and DeLeeuw (1998), Heck and Thomas (2000), and Hox (2002). Tabachnick and Fidell (2007) also devote a chapter to multilevel modeling.

11.2 KINDS OF RESEARCH QUESTIONS

11.2.1 Effect of the IV(s)

Questions about the effect of a random IV are the same as for a fixed IV. Are there mean differences among the levels of treatment, holding all else constant? For example, is there a mean difference in productivity among groups of employees who have different types of training? Do different dosages lead to different degrees of symptom alleviation? These questions are answered by applying an F test to each IV, as demonstrated in Section 11.4. However, the F tests differ in their error terms, depending on whether IVs are random or fixed, as discussed in Section 11.6.1.

11.2.2 Effects of Interactions among IVs

Questions about interactions are also the same for random and fixed IVs. Holding all else constant, does the pattern of means in the DV over the levels of one IV depend on the level of another IV? For example, does the efficacy of different dosages depend on time of administration of the drug? Interactions are tested by applying F tests, but these tests also use different error terms, depending on whether IVs are random or fixed, as discussed in Section 11.6.1.

An important limitation to nested designs is that nested factors do not interact with the IVs in which they are nested. For instance, in the example with type of training applied to different groups, groups nest within level of training; there is no interaction of groups with type of training to test.

11.2.3 Specific Comparisons

Specific comparisons also follow the same format as fixed-effects ANOVA but differ in the error terms used to test them. For a quantitative, random-effects IV, such as dose, for instance, trends are typically the comparisons of interest. Section 11.6.3 demonstrates trend analysis when spacing among levels of the IV is unequal, as happens when levels are randomly chosen.

11.2.4 Parameter Estimates

Parameter estimates for random IVs are the same as for fixed IVs. Sample means, standard deviations, standard errors, and confidence intervals provide estimates of the central tendency and variability in a population at each level of a statistically significant IV. For example, if mean differences in symptom alleviation are found at different dosages, what is the expected amount of symptom alleviation for each dose, and what are the standard deviations or standard errors or confidence intervals for each dose? Parameter estimates are calculated as for fixed-effects ANOVA, with the same ambiguity when there are unequal sample sizes in factorial designs (cf. Section 5.6.5).

11.2.5 Effect Sizes

Effect size indicates the proportion of variance in a DV that is associated with a statistically significant main effect or interaction. How much of the variance in symptom alleviation is associated with dosage? How much of the variance in productivity is associated with differences in training? The η^2 and partial η^2 of Sections 4.5.1 and 5.6.2 apply to nested designs. As always, the appropriate error term for the effect is used for partial η^2. The $\hat{\omega}^2$ and partial η^2 of earlier chapters, however, do *not* apply to designs in which there are random effects. Alternative measures are in Section 11.6.5.

11.2.6 Power

What is the probability of finding an IV significant, if indeed, the IV affects the DV? What is the probability of finding an effect of dosage if size of dose does affect symptoms? Power is an important issue in random-effects ANOVA because increased generalizability is often gained at the expense of power. Tests of random IVs are typically less powerful than tests of the same IVs when fixed because of differences in the required error terms. For an IV in a factorial random-effects design, the error term is an interaction (or some other combination of terms) that typically has fewer degrees of freedom than the error term in fixed-effects analysis.

In nested designs, the problem is that the nested IV serves as the error term for the effect of treatment. For example, if five groups of students are assigned to each of three levels of training, the error term for training is based on the variance among means for the groups and on the number of groups within each level, not on scores for the total number of students at each level of training. The reduction in power due to smaller degrees of freedom may be partially offset by lower variability among group means than among individual scores—but, on balance, only partially.

With either design, the problem of reduced power is sometimes dealt with by pooling terms from nonsignificant effects to increase the degrees of freedom. For instance, in the nested design, researchers test the effects of groups first, and then, if there is no significant difference among groups, they pool error terms or use the error term based on cases. However, it is much safer to plan the study with sufficient numbers of groups to ensure necessary power than to rely on finding no significant differences among groups. Section 11.6.1.2 discusses pooling of error terms. Power and sample size estimation for random-effects ANOVA are available in the GANOVA program (Woodward, Bonett, and Brecht, 1990).

11.3 ASSUMPTIONS AND LIMITATIONS

Assumptions of random-effects ANOVA are the same as fixed-effects ANOVA, with the added assumptions that levels of the random IVs are randomly chosen from a population of levels and that the population of levels is normally distributed with independent errors. It is assumed that scores within levels of the random IV have equal variances (homogeneity of variance) and that in a factorial design, all pairs of levels of the random IV(s) have homogeneity of covariance, even if they are not repeated measures. Beyond that, assumptions and limitations depend on whether the IV is a repeated-measures IV or a randomized-groups IV.

11.3.1 Theoretical Issues

As with any ANOVA, causal inference depends on the manner in which cases are assigned to the levels of the IV(s), manipulation of the levels of the IV(s), and controls used in the research. Statistical tests may also be used in nonexperimental studies, but causal inference is not attempted.

It is sometimes difficult to decide whether an IV is fixed or random. If an IV is random, the levels of the IV are randomly selected from a larger population of levels to which the researcher wants to generalize. An extension of random selection is that a different set of levels would probably be selected in any replication of the study. Some guidelines for deciding whether an effect is random or fixed are as follows:

1. Cases are always considered a random factor. Although they are usually randomly assigned to levels of treatment, they often are not randomly selected from a population. However, cases would almost always be a different group in a replication, as is true for other random effects.
2. Groups in nested designs are always considered random factors because, at the very least, they are randomly assigned to levels of treatment.
3. With other IVs, the decision is based on how the particular levels are chosen. Suppose, for example, you have a collection of vacation slides categorized as either landscapes or close-ups, and you want to study how well they are remembered. If you just select your favorite five slides in each category, the IV is fixed. However, you probably want to generalize your results to memory for landscapes and close-ups in general, not to the particular ones used in your experiment. Therefore, you randomly select some number of slides—say, 5—from each category in your collection or, better yet, from several collections. Type of slide (landscape vs. close-up) is a fixed factor; the five slides in each category are a random factor.
4. A trial factor in a repeated-measures study is typically fixed. However, suppose there is a very large number of observations—say, a movement response

recorded every half-second during a night of sleep. A random trial factor is formed if you randomly select some manageable number of half-second time periods for each case (say, 10 of them) and include only those 10 movement responses in the analysis.

5. A blocking factor can be either fixed or random, depending on how the blocks are formed (Lee, 1975). Blocks form a fixed factor if the ranges defining the levels are chosen in advance and are constant through replications. For example, if you decide to block on operator experience by choosing ranges such as 0–5 years, 6–10 years, 11–15 years, the IV is a fixed factor. On the other hand, if you simply record the years of experience of available operators and then count off equal numbers of operators into least, middle, and most experienced, the IV is a random factor, because replication of the study would typically produce different cutoffs for the three levels of experience.
6. Qualitative IVs are typically fixed. The levels differ in kind, and it is likely that the ones chosen for study are the ones of primary interest (or availability). If, however, the number of "kinds" is very large, and some are randomly selected for study, the levels form a random-effects IV.
7. Quantitative IVs can easily be either fixed or random. The researcher could be interested only in specified levels or in randomly selected levels from a range.

Lee (1975) recommends treating a factor as fixed when in doubt. However, be sure that cases and groups nested within treatments are treated as random, and do not generalize the results of analyses of fixed factors outside the range of levels included in the study.

11.3.2 Practical Issues

Practical issues are the same as for other ANOVA designs, and there are additional assumptions that depend on whether an IV is randomized groups or repeated measures, as discussed in what follows. Violations of assumptions of normality and homogeneity of variance are more consequential in designs with random effects.

11.3.2.1 Randomized-Groups Designs

In a completely randomized-groups design, the assumptions and limitations of Section 5.3.2 apply to all IVs, whether random or fixed effects: normality of sampling distributions, homogeneity of variance, independence of errors, and absence of outliers. If a Latin-square or some other incomplete design is used, there is also the assumption of additivity when only one score is in each cell of the design (cf. Sections 9.3.2.6 and 10.3.2.5). If there are covariates, the additional assumptions and limitations of Section 8.3.2 apply: multicollinearity, linearity, homogeneity of regression, and reliability of covariates.

In factorial designs, the additional assumption for random-effects IVs is that covariances between all pairs of levels are equal. For example, if five dosages are levels of one random-effects IV and three times of administration are levels of another random-effects IV, the assumption of homogeneity of covariance is that the relationships among all pairs of levels of dosage are the same (collapsed over levels of time) and that the relationships among all pairs of levels of time are the same (collapsed over levels of dosage). Evaluation of homogeneity of covariance is discussed in Section 11.6.4.

11.3.2.2 Repeated-Measures and Mixed (Randomized-Repeated) Designs

The assumptions and limitations of repeated measures apply to random-effects, repeated-measures IVs, whether in a fully repeated-measures design (Section 6.3) or a mixed randomized-repeated design (Section 7.3). The additional assumption, you recall, is of sphericity (Sections 6.3.2.3 and 6.6.2).

11.4 FUNDAMENTAL EQUATIONS

There are two basic types of random-effects ANOVA: nested designs and designs in which one or more factors have levels that are randomly chosen but factors are not nested within other factors. The simplest of the latter designs is a one-way ANOVA with randomly chosen levels of the single IV. However, analysis of this one-way, random-effects design is identical to that of a one-way fixed-effects design, whether randomized groups or repeated measures. It is only when additional IVs are added to the design that the analysis changes, because with additional IVs there is overlap between tests of main effects and interactions.[3] Therefore, the design demonstrated here is the simplest nested design, with a single fixed IV and with a random IV nested within the fixed IV.

The fixed IV (A) is length of ski training with two levels: a half-day class or a full-day class. The random effect IV (B) is group, with two groups of three skiers in each level of training time. B is a random-effect IV because the study is intended to draw conclusions about groups of skiers in general, not just the groups of skiers in the study. The DV is time to ski a standard run. Groups form the error term for length of training, and skiers within groups form the error term for groups. Although both length of training and groups can be tested, only length of training is of interest. The traditional ANOVA approach is shown first, followed by the regression approach for nested designs.

11.4.1 Allocation of Cases

Three skiers are randomly assigned to each of four groups, and then two groups are randomly assigned to each of the two treatments. Table 11.1 shows the assignment of

[3] There is similar overlap, at least conceptually, in unequal-n factorial designs.

TABLE 11.1 Assignment of cases in a simple nested design

Treatment A	a_1		a_2	
Groups B	b_1	b_2	b_3	b_4
	s_1	s_4	s_7	s_{10}
Cases S	s_2	s_5	s_8	s_{11}
	s_3	s_6	s_9	s_{12}

the 12 skiers to the two groups of B at each of the two levels of A. Cases are designated S, as usual. The design, of course, is far too small for an actual experiment.

In practice, this design is often used with intact groups (such as classrooms), which are then randomly assigned to levels of treatment. Intact groups are typically formed by convenience or by self-selection, rarely by random selection and assignment. The test of treatment is legitimate because the error term for treatment is based on groups (where random assignment has occurred[4]) rather than on individual cases (where random assignment has not occurred). However, the test of groups uses individual cases for the error term and is problematic when random assignment to groups has not occurred.

Nesting is also present when cases are first randomly assigned to levels of A and then randomly split into groups (B) within each level of treatment. This second step is sometimes necessary because treatment is something that must be applied to groups, not to individuals (e.g., when studying social interaction in a group). Otherwise, the more efficient one-way randomized-groups design is used, in which each case within a treatment level is run individually or all levels of treatment are run at one time.

Allocation of cases is the same as for fixed-effects designs in studies in which the levels of one or more IVs are randomly selected but there is no nesting.

11.4.2 Partition of Sources of Variance

The total sum of squares is divided into three sources: differences due to levels of treatment (A), differences due to groups (B) nested within levels of A (B/A), and differences due to cases nested within levels of B nested within levels of A ($S/B/A$).

$$\text{SS}_T = \text{SS}_A + \text{SS}_{B/A} + \text{SS}_{S/B/A} \tag{11.1}$$

or

$$\sum_i \sum_j \sum_k (Y_{ijk} - \text{GM})^2 = n_j \sum_j (\overline{Y}_A - \text{GM})^2$$
$$+ \left[n_{jk} \sum_j \sum_k (\overline{Y}_B - \text{GM})^2 - n_j \sum_j (\overline{Y}_A - \text{GM})^2 \right]$$
$$+ \sum_i \sum_j \sum_k (Y_{ijk} - \overline{Y}_B)^2$$

[4] The assumption of independence of errors is also violated if individual cases are used to form the error term when cases within groups are exposed to a level of treatment together.

> The total sum of squared differences between individual scores (Y_{ijk}) and the grand mean (GM) is partitioned into (1) sum of squared differences between means associated with different levels of A ($\overline{Y}_A$) and the grand mean; (2) sum of squared differences between means associated with levels B ($\overline{Y}_B$), which also contain differences due to levels of A, and the grand mean minus the sum of squared differences between means associated with different levels of A ($\overline{Y}_A$) and the grand mean, which results in the sum of squared differences associated with different levels of B; and (3) sum of squared differences between individual scores (Y_{ijk}) and the means associated with levels of B ($\overline{Y}_B$).

Because A and B do not interact, the "real" sum of squares for B is found by subtracting sum of squares for A from the regularly computed sum of squares for B that, in this case, contains differences due to both levels of A and levels of B. The notation for groups is B/A (B nested within A), and the notation for cases nested within B is $S/B/A$.

Total degrees of freedom is the number of scores minus 1 (or $abs - 1$), lost when the grand mean is estimated (cf. Equation 3.6). Note that the number of levels of B is now defined as *the number at each level of A*; for the example, there are two levels of B at each level of A. There are as many levels of B altogether as number of levels of B within each level of A times the number of levels of A.

Degrees of freedom for A are the number of levels of A minus 1, lost because the $\overline{Y}_{a_j}$ average to the grand mean; thus, $a - 1$ of the means can take on any values whatever, but the last mean is fixed.

$$\text{df}_A = a - 1 \tag{11.2}$$

Degrees of freedom for B are derived from the notational scheme in the usual way:

$$\text{df}_{B/A} = a(b - 1) \tag{11.3}$$

Degrees for cases within levels of B and A are also derived from the notation, as follows:

$$\text{df}_{S/B/A} = ab(n - 1) \tag{11.4}$$

To verify the equality of degrees of freedom,

$$N - 1 = abn - 1 = a - 1 + ab - a + abn - ab$$

Figure 11.1 shows the partitions of sums of squares and degrees of freedom. Mean squares are formed by dividing sums of squares by degrees of freedom, as for all ANOVA designs:

$$\text{MS}_A = \frac{\text{SS}_A}{\text{df}_A} \qquad \text{MS}_{B/A} = \frac{\text{SS}_{B/A}}{\text{df}_{B/A}} \qquad \text{MS}_{S/B/A} = \frac{\text{SS}_{S/B/A}}{\text{df}_{S/B/A}} \tag{11.5}$$

F ratios are formed for each of the two sources of variance to be evaluated: A and B/A. F_A evaluates the main effect of A. The null hypothesis for this test is the same as for a one-way fixed-effects ANOVA, expressed as $\mu_{a_1} = \mu_{a_2} = \cdots = \mu_{a_j}$. However,

FIGURE 11.1 Partitions of (a) total sum of squares into sum of squares due to differences among treatments, differences among groups within treatments, and differences among cases within groups and (b) total degrees of freedom into degrees of freedom associated with treatments, degrees of freedom associated with groups within treatments, and degrees of freedom associated with cases within the groups.

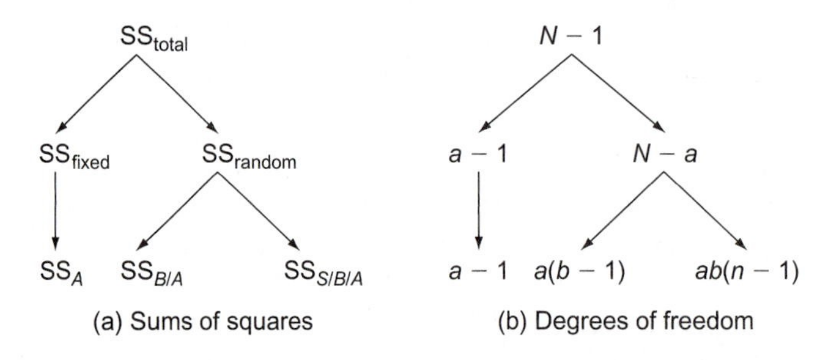

the appropriate error term is the variation among groups (B) within levels of A. Obtained F_A is computed as

$$F_A = \frac{\text{MS}_A}{\text{MS}_{B/A}} \qquad \text{df} = (a-1), a(b-1) \tag{11.6}$$

Obtained F_A is tested against critical F from Table A.1, with numerator df $= a - 1$ and denominator df $= a(b - 1)$. The null hypothesis is rejected if obtained F_A is equal to or exceeds critical F.

$F_{B/A}$ evaluates differences among marginal means for levels of B (groups) nested within levels of A, called the main effect of B nested within A. The null hypothesis is that group means are the same. This test is not often of research interest[5] but is sometimes used to decide whether to pool error terms (cf. Section 11.6.1).

$$F_{B/A} = \frac{\text{MS}_{B/A}}{\text{MS}_{S/B/A}} \qquad \text{df} = a(b-1), ab(n-1) \tag{11.7}$$

Obtained $F_{B/A}$ is tested against critical F from Table A.1, with numerator df $= a(b - 1)$ and denominator df $= ab(n - 1)$. The null hypothesis of no mean difference among groups is rejected if obtained $F_{B/A}$ is equal to or larger than critical F.

11.4.3 Traditional ANOVA Approach (One Treatment Factor and One Level of Nesting)

Table 11.2 shows an unrealistically small hypothetical data set to evaluate speed of skiing as a function of length of class. Treatment A, length of class, has two levels (a half-day class or a full-day class) and is a fixed-effect IV. B/A is a second,

[5] More complex analyses are useful if differences among groups are of research interest (cf. Section 11.6.2.)

TABLE 11.2 Hypothetical data for nested ANOVA

Class Length A	a_1: Half day		a_2: Full day	
Groups B	b_1	b_2	b_3	b_4
Cases S	7	7	5	4
	6	6	3	2
	10	8	5	5
Sums for B	23	21	13	11
Sums for A	44		24	

$\Sigma Y^2 = 7^2 + 6^2 + \cdots + 5^2 = 438$
$T = 68$
$a = 2, b = 2, n = 3$

TABLE 11.3 Computational equations for sums of squares for nested ANOVA

$$SS_A = \frac{\sum A^2}{bn} - \frac{T^2}{abn} = \frac{44^2 + 24^2}{6} - \frac{68^2}{12} = 33.33$$

$$SS_{B/A} = \frac{\sum B^2}{n} - \frac{A^2}{bn} = \frac{23^2 + 21^2 + 13^2 + 11^2}{3} - \frac{44^2 + 24^2}{6} = 1.33$$

$$SS_{S/B/A} = \sum Y^2 - \frac{B^2}{n} = 7^2 + 6^2 + 10^2 + 7^2 + \cdots + 2^2 + 5^2 - \frac{23^2 + 21^2 + 13^2 + 11^2}{3} = 18.00$$

$$SS_T = \sum Y^2 - \frac{T^2}{abn} = 7^2 + 6^2 + 10^2 + 7^2 + \cdots + 2^2 + 5^2 - \frac{68^2}{12} = 52.66$$

random-effects IV, in which two different groups of skiers ($b = 2$) are randomly assigned to each length of class; there are four groups altogether. Each group has three skiers, so $n = 3$ for each level of B. There are a total of 12 skiers. The DV is time to ski a common run.

Table 11.3 shows computation equations for sums of squares for A, B/A, $S/B/A$, and T (total).

Divisors as well as summations sometimes differ from those of the factorial fixed-effects ANOVA of Chapter 5. The divisor for B is n rather than an, because B does not cross A. That is, there are only n scores in each level of B, not an scores as there would be in a fully crossed $A \times B$ design. Table 11.4 summarizes the results and shows mean squares and F values based on Equations 11.1–11.7.

TABLE 11.4 Source table for nested ANOVA

Source	SS	df	MS	F
A	33.33	1	33.33	49.75
B/A	1.33	2	0.67	0.30
$S/B/A$	18.00	8	2.25	
T	52.66	11		

The error term for treatment A is B/A, and the error term for B/A is $S/B/A$. Therefore, the degrees of freedom for evaluating F_A are 1 $(a - 1 = 2 - 1)$ and 2 $[a(b - 1) = 2(1)]$, and the degrees of freedom for evaluating $F_{B/A}$ are 2 and 8 $[ab(n - 1) = 12 - 4]$. With $\alpha = .05$, the critical value for F_A is 18.51 and the critical value for $F_{B/A}$ is 4.46. Therefore, there is a statistically significant effect of length of ski class but no significant differences among groups. Looking at the means, we conclude that skiers trained for a full day take less time to ski a run than do skiers trained for only half a day.

11.4.4 Regression Approach (One Treatment Factor and One Level of Nesting)

The regression approach can be taken with nested designs and other varieties of random-effects designs, as long as the correct error term is chosen for each effect (cf. Sections 11.5 and 11.6.1) and there is equal n.[6] The only exception is for quantitative IVs that have unequal spacing between levels. When levels are unequally spaced, special coefficients for trend analysis must be derived, because the coefficients that are readily available are for equally spaced levels. Keppel (1982, Appendix C-3) describes procedures for generating an appropriate set of trend coefficients by hand.

Table 11.5 shows the regression analysis of the nested design of data in Table 11.2. The Y column contains the DV scores. The X_1 column contains coefficients to separate the two levels of A. Columns X_2 and X_3 contain coefficients for separating the four levels of B, two nested in each level of A. Notice that nesting of B is represented the same as nesting of cases in the mixed (randomized-groups, repeated-measures) design of Table 7.7 (columns X_9–X_{32}). Columns X_4–X_{11} separate cases, nested in levels of B nested in levels of A. Because there are three cases at each level of B, two columns are required for each AB combination. As in Table 7.7, the coding used to separate cases is completely arbitrary. Columns X_{12}–X_{22} are the cross of Y with columns X_1–X_{11}, respectively. This table is generated by a spreadsheet program, where the facilities of the spreadsheet are used to compute SS(reg.) for columns X_1–X_{11}.

Values for ANOVA are readily computed by summing appropriate SS(reg.) for columns X_1–X_{11}, as shown in Table 11.6. Within rounding error, these values agree with the values in Table 11.4.

11.4.5 Computer Analyses of Small-Sample Nested Example

Table 11.7 shows how data are organized for the small-sample nested example. Data may be converted from a spreadsheet program or entered through facilities provided by the statistical program. Columns can be in any order.

[6] As usual, the coefficients are not orthogonal if the design is not balanced, and matrix algebra is required to find sums of squares adjusted for overlap.

TABLE 11.5 Regression approach to nested design ANOVA

				A	$B/a1$	$B/a2$	$S1/b1/a1$	$S2/b1/a1$	$S1/b2/a1$	$S2/b2/a1$	$S1/b3/a2$	$S2/b3/a2$
			Y	X_1	X_2	X_3	X_4	X_5	X_6	X_7	X_8	X_9
a_1	b_1	s_1	7	1	1	0	2	0	0	0	0	0
		s_2	6	1	1	0	−1	1	0	0	0	0
		s_3	10	1	1	0	−1	−1	0	0	0	0
	b_2	s_4	7	1	−1	0	0	0	2	0	0	0
		s_5	6	1	−1	0	0	0	−1	1	0	0
		s_6	8	1	−1	0	0	0	−1	−1	0	0
a_2	b_3	s_7	5	−1	0	1	0	0	0	0	2	0
		s_8	3	−1	0	1	0	0	0	0	−1	1
		s_9	5	−1	0	1	0	0	0	0	−1	−1
	b_4	s_{10}	4	−1	0	−1	0	0	0	0	0	0
		s_{11}	2	−1	0	−1	0	0	0	0	0	0
		s_{12}	5	−1	0	−1	0	0	0	0	0	0
sum			68	0	0	0	0	0	0	0	0	0
sum sq			438	12	6	6	6	2	6	2	6	2
SS			52.67	12	6	6	6	2	6	2	6	2
SP				20	2	2	−2	−4	0	−2	2	−2
ss(reg)				33.33	0.67	0.67	0.67	8	0	2	0.67	2

			$S1/b4/a2$	$S2/b4/a2$	YX_1	YX_2	YX_3	YX_4	YX_5	YX_6	YX_7	YX_8	YX_9	YX_{10}	YX_{11}
			X_{10}	X_{11}	X_{12}	X_{13}	X_{14}	X_{15}	X_{16}	X_{17}	X_{18}	X_{19}	X_{20}	X_{21}	X_{22}
a_1	b_1	s_1	0	0	7	7	0	14	0	0	0	0	0	0	0
		s_2	0	0	6	6	0	−6	6	0	0	0	0	0	0
		s_3	0	0	10	10	0	−10	−10	0	0	0	0	0	0
	b_2	s_4	0	0	7	−7	0	0	0	14	0	0	0	0	0
		s_5	0	0	6	−6	0	0	0	−6	6	0	0	0	0
		s_6	0	0	8	−8	0	0	0	−8	−8	0	0	0	0
a_2	b_3	s_7	0	0	−5	0	5	0	0	0	0	10	0	0	0
		s_8	0	0	−3	0	3	0	0	0	0	−3	3	0	0
		s_9	0	0	−5	0	5	0	0	0	0	−5	−5	0	0
	b_4	s_{10}	2	0	−4	0	−4	0	0	0	0	0	0	8	0
		s_{11}	−1	1	−2	0	−2	0	0	0	0	0	0	−2	2
		s_{12}	−1	−1	−5	0	−5	0	0	0	0	0	0	−5	−5
sum			0	0	20	2	2	−2	−4	0	−2	2	−2	1	−3
sum sq			6	2											
SS			6	2											
SP			1	−3											
ss(reg)			0.17	4.5											

TABLE 11.6 Sums of squares for the regression approach to nested-design ANOVA

Source	Columns		SS	df (number of columns)
A	SS(reg. $X1$)	33.33	33.33	1
B/A	SS(reg. X_2) + SS(reg. X_3)	0.67 + 0.67	1.34	2
$S/B/A$	SS(reg. X_4) + SS(reg. X_5) + · · · + SS(reg. X_{11})	0.67 + 8 + · · · + 4.5	18.00	8
Total	SS_Y	52.67	52.67	11

TABLE 11.7 Data set for small-sample example of nested design

Group	Time	Training
1	7	1
1	6	1
1	10	1
2	7	1
2	6	1
2	8	1
3	5	2
3	3	2
3	5	2
4	4	2
4	2	2
4	5	2

Tables 11.8 and 11.9 show computer analyses of the small-sample example of the nested design with one fixed and one random IV by SPSS GLM (UNIANOVA) and SAS GLM.

SPSS GLM (UNIANOVA) requires two additional lines of syntax to specify the design, as shown in Table 11.8. First, the nested IV, group, is declared a RANDOM variable. Second, the /DESIGN instruction shows both TRAINING and GROUP(TRAINING) to indicate that groups are nested within training. Remaining options are produced by default through the Windows menu system.

The segment labeled Tests of Between-Subjects Effects shows the tests of TRAINING and GROUP(TRAINING), along with the appropriate Error terms for each, identified in footnotes. Note that the label for $S/B/A$ is Error. The segment labeled Expected Mean Squares shows the basis for choice of error terms, as described in Section 11.6.1.

Table 11.9 shows that SAS GLM also requires two special types of syntax for a nested design. The `model` instruction includes `GROUP(TRAINING)` to indicate the

TABLE 11.8 Nested ANOVA for small-sample example through SPSS GLM (UNIANOVA)

```
UNIANOVA
 TIME BY GROUP TRAINING
 /RANDOM = GROUP
 /METHOD = SSTYPE(3)
 /INTERCEPT = INCLUDE
 /CRITERIA = ALPHA(.05)
 /DESIGN = TRAINING GROUP(TRAINING).
```

Tests of Between-Subjects Effects

Dependent Variable: TIME

Source		Type III Sum of Squares	df	Mean Square	F	Sig.
Intercept	Hypothesis	385.333	1	385.333	578.000	.002
	Error	1.333	2	.667[a]		
TRAINING	Hypothesis	33.333	1	33.333	50.000	.019
	Error	1.333	2	.667[a]		
GROUP(TRAINING)	Hypothesis	1.333	2	.667	.296	.751
	Error	18.000	8	2.250[b]		

a. MS(GROUP(TRAINING))

b. MS(Error)

Expected Mean Squares[a,b]

Source	Variance Component		
	Var(GROUP (TRAINING))	Var(Error)	Quadratic Term
Intercept	3.000	1.000	Intercept, TRAINING
TRAINING	3.000	1.000	TRAINING
GROUP(TRAINING)	3.000	1.000	
Error	.000	1.000	

a. For each source, the expected mean square equals the sum of the coefficients in the cells times the variance components, plus a quadratic term involving effects in the Quadratic Term cell.

b. Expected Mean Squares are based on the Type III Sums of Squares.

nesting arrangement, and the `random` instruction shows that the nested effect is a random IV and is to be tested (`test`) using the appropriate error term.

The initial segments of output are the same as results of fixed-effects ANOVA and use inappropriate error terms. The portion of output labeled `Type III Expected Mean Square` shows formation of error terms, as explained in Section 11.6.1. The final table of output shows tests of `TRAINING` and `GROUP(TRAINING)`, using appropriate error terms that match.

TABLE 11.9 Nested ANOVA for small-sample example through SAS GLM

```
proc glm Data=SASUSER.SSNESTED;
     class GROUP TRAINING;
     model TIME = TRAINING  GROUP(TRAINING)  ;
     random GROUP(TRAINING) / test ;
run;
```

The GLM Procedure

Class Level Information

Class	Levels	Values
GROUP	4	1 2 3 4
TRAINING	2	1 2

Number of Observations Read	15
Number of Observations Used	12

The GLM Procedure

Dependent Variable: TIME

Source	DF	Sum of Squares	Mean Square	F Value	Pr > F
Model	3	34.66666667	11.55555556	5.14	0.0286
Error	8	18.00000000	2.25000000		
Corrected Total	11	52.66666667			

R-Square	Coeff Var	Root MSE	TIME Mean
0.658228	26.47059	1.500000	5.666667

Source	DF	Type I SS	Mean Square	F Value	Pr > F
TRAINING	1	33.33333333	33.33333333	14.81	0.0049
GROUP(TRAINING)	2	1.33333333	0.66666667	0.30	0.7514

Source	DF	Type III SS	Mean Square	F Value	Pr > F
TRAINING	1	33.33333333	33.33333333	14.81	0.0049
GROUP(TRAINING)	2	1.33333333	0.66666667	0.30	0.7514

Source	Type III Expected Mean Square
TRAINING	Var(Error) + 3 Var(GROUP(TRAINING)) + Q(TRAINING)
GROUP(TRAINING)	Var(Error) + 3 Var(GROUP(TRAINING))

Tests of Hypotheses for Mixed Model Analysis of Variance

Dependent Variable: TIME

Source	DF	Type III SS	Mean Square	F Value	Pr > F
TRAINING	1	33.333333	33.333333	50.00	0.0194
Error	2	1.333333	0.666667		

Error: MS(GROUP(TRAINING))

Source	DF	Type III SS	Mean Square	F Value	Pr > F
GROUP(TRAINING)	2	1.333333	0.666667	0.30	0.7514

11.5 TYPES OF DESIGNS WITH RANDOM EFFECTS

Types of designs with random-effects IVs are numerous: one-way designs (either randomized groups or repeated measures) with a random-effects IV; factorial designs with all random-effects IVs (which are, in turn, any combination of randomized groups and repeated measures); mixed random fixed-effects designs (with any combination of randomized-groups and repeated-measures IVs); and nested designs (with any degree of nesting) under either randomized-groups or repeated-measures fixed-effects IVs. Nesting under a repeated-measures fixed-effects IV, for instance, might be something like multiple lists of words given to all cases, with Latin-square counterbalancing. The designs may incorporate other complications, such as incomplete crossing of factors (Chapter 10) and covariates (Chapter 8).[7]

11.5.1 Nested Designs

The simple example of Section 11.4 has one randomized-groups IV nested within a fixed-effects IV. Additional levels of nesting also are possible. Figure 11.2 shows two levels of nesting: B/A and $C/B/A$.

In this example, groups could be nested within ski resorts, which, in turn, are nested within training. That is, four resorts (Bear Mountain, Snow Summit, Mammoth Mountain, and Heavenly Valley) are randomly selected from a list of resorts in California that have both half-day and full-day classes. Two (Bear Mountain and Snow

FIGURE 11.2 Partitions of (a) total sum of squares into sum of squares due to differences among treatments, differences among groups within treatments, differences among subgroups within groups, and differences among cases within subgroups within groups within treatments and (b) total degrees of freedom into degrees of freedom associated with treatments, degrees of freedom associated with groups within treatments, differences among subgroups within groups within treatments, and degrees of freedom associated with cases within subgroups within groups within treatments

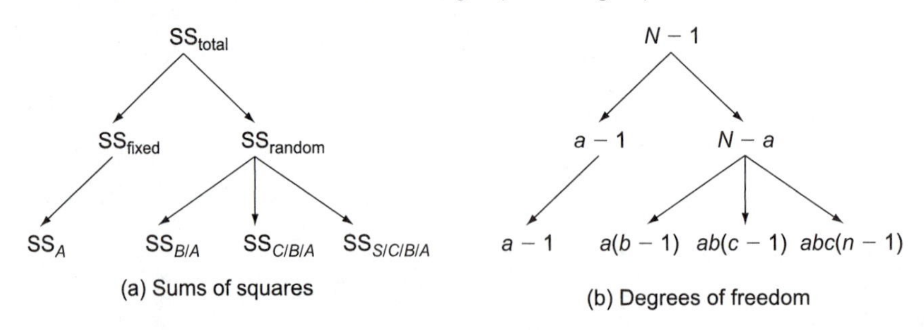

[7] We recommend a simulated data set and trial analyses before embarking on data collection for highly complex designs to avoid ending up with a data set you do not know how to analyze.

TABLE 11.10 Hypothetical data for doubly nested design

Class Length A	a_1: Half Day				a_2: Full Day			
Resort B	b_1: Bear Mountain		b_2: Snow Summit		b_3: Mammoth Mountain		b_4: Heavenly Valley	
Groups C	c_1	c_2	c_3	c_4	c_5	c_6	c_7	c_8
Cases S	7	7	8	7	5	4	4	1
	6	6	5	10	3	2	6	2
	10	8	9	7	5	5	4	3

Summit) are randomly assigned to half-day classes, and the other two (Mammoth Mountain and Heavenly Valley) are randomly assigned to full-day classes. Resorts and groups are random-effects IVs. Length of training is a fixed-effect IV. Table 11.10 shows a hypothetical data set for this design.

Table 11.11 shows SPSS GLM (UNIANOVA) syntax and selected output for the analysis of this data set. The double nesting is indicated in the /DESIGN instruction, with RESORT(TRAINING) indicating that resorts are nested within types of training and with GROUP(RESORT(TRAINING)) indicating that groups are nested within resorts, which, in turn, are nested within types of training. Both resorts and groups are identified as RANDOM.

The Tests of Between-Subjects Effects table shows that the mean difference between the two levels of TRAINING is statistically significant, even with only 2 df associated with RESORT(TRAINING) as its error term. The Estimated Marginal Means table shows that groups trained for a full day ski faster than those trained for half a day. (Remaining tables of means are not reproduced here.) Neither of the random (nested) effects is statistically significant.

Nested designs can have additional IVs that cross treatments, rather than being nested within them. For example, we might add gender to the preceding design, so that there are three women and three men within each of the eight groups, as shown in Table 11.12.

Degrees of freedom for A (training) are $a - 1 = 1$. Degrees of freedom for B (resort) are $a(b - 1) = 2(2 - 1) = 2$. Degrees of freedom for C (groups) are $ab(c - 1) = 2(2)(2 - 1) = 4$, and degrees of freedom for D (gender) are $d - 1 = 1$. Degrees of freedom for the interaction of gender with training are $(d - 1)(a - 1) = 1$. Gender is also crossed with resort and with groups. The degrees of freedom for the gender by resort within training interaction are $(d - 1)a(b - 1) = 1(2)(1) = 2$; and finally degrees of freedom for the interaction of gender by groups within resort within training are $(d - 1)ab(c - 1) = 1(2)(2)(1) = 4$.

Table 11.13 shows the analysis through SAS GLM. Note the difference in specifying double nesting between SPSS and SAS. Double parentheses are not used in SAS; instead, the double nesting is indicated by including all higher level IVs within a single set of parentheses. (Note that multiple terms within a set of parentheses are separated by * in output.)

TABLE 11.11 Doubly nested ANOVA through SPSS GLM (UNIANOVA) (syntax and selected output)

```
UNIANOVA
 TIME BY TRAINING GROUP RESORT
  /RANDOM = GROUP RESORT
  /METHOD = SSTYPE(3)
  /INTERCEPT = INCLUDE
  /EMMEANS = TABLES(RESORT)
  /EMMEANS = TABLES(GROUP)
  /EMMEANS = TABLES(TRAINING)
  /CRITERIA = ALPHA(.05)
  /DESIGN = TRAINING RESORT(TRAINING) GROUP(RESORT(TRAINING)).
```

Univariate Analysis of Variance

Between-Subjects Factors

		Value Label	N
TRAINING	1	Half day classes	12
	2	Full day classes	12
GROUP	1		3
	2		3
	3		3
	4		3
	5		3
	6		3
	7		3
	8		3
RESORT	1	Bear Mountain	6
	2	Snow Summit	6
	3	Mammoth Mountain	6
	4	Heavenly Valley	6

Tests of Between-Subjects Effects

Dependent Variable: TIME

Source		Type III Sum of Squares	df	Mean Square	F	Sig.
Intercept	Hypothesis	748.167	1	748.167	897.800	.001
	Error	1.667	2	.833[a]		
TRAINING	Hypothesis	88.167	1	88.167	105.800	.009
	Error	1.667	2	.833[a]		
RESORT(TRAINING)	Hypothesis	1.667	2	.833	.263	.781
	Error	12.667	4	3.167[b]		
GROUP(RESORT (TRAINING))	Hypothesis	12.667	4	3.167	1.357	.292
	Error	37.333	16	2.333[c]		

a. MS(RESORT(TRAINING))

b. MS(GROUP(RESORT(TRAINING)))

c. MS(Error)

TABLE 11.11 Continued

Expected Mean Squares[a,b]

Source	Variance Component			
	Var(RESORT (TRAINING))	Var(GROUP (RESORT (TRAINING)))	Var(Error)	Quadratic Term
Intercept	6.000	3.000	1.000	Intercept, TRAINING
TRAINING	6.000	3.000	1.000	TRAINING
RESORT(TRAINING)	6.000	3.000	1.000	
GROUP(RESORT (TRAINING))	.000	3.000	1.000	
Error	.000	.000	1.000	

a. For each source, the expected mean square equals the sum of the coefficients in the cells times the variance components, plus a quadratic term involving effects in the Quadratic Term cell.

b. Expected Mean Squares are based on the Type III Sums of Squares.

Estimated Marginal Means

3. TRAINING

Dependent Variable: TIME

TRAINING	Mean	Std. Error	95% Confidence Interval	
			Lower Bound	Upper Bound
Half day classes	7.500[a]	.441	6.565	8.435
Full day classes	3.667[a]	.441	2.732	4.601

a. Based on modified population marginal mean.

TABLE 11.12 Hypothetical data for doubly nested design with additional crossed IV

Class Length A		a_1: Half Day				a_2: Full Day			
Resort B		b_1: Bear Mountain		b_2: Snow Summit		b_3: Mammoth Mountain		b_4: Heavenly Valley	
Groups C		c_1	c_2	c_3	c_4	c_5	c_6	c_7	c_8
Gender D	d_1: Women	7	7	8	7	5	4	4	1
		6	6	5	10	3	2	6	2
		10	8	9	7	5	5	4	3
	d_2: Men	6	9	7	9	5	3	4	2
		9	5	7	8	4	6	5	2
		8	7	6	9	4	5	6	1

$a = 2, b = 2, c = 2, d = 2, n = 3$

TABLE 11.13 Doubly nested ANOVA with additional crossed IV (SAS GLM syntax and selected output)

```
proc glm Data=SASUSER.DBLNEST;
    class TRAINING GROUP RESORT GENDER;
    model TIME = TRAINING GROUP(RESORT TRAINING) RESORT(TRAINING)
                 GENDER TRAINING*GENDER
                 GENDER*GROUP(RESORT TRAINING) GENDER*RESORT(TRAINING) /ss3 ;
    random GROUP(RESORT TRAINING) RESORT(TRAINING)
           GENDER*GROUP(RESORT TRAINING) GENDER*RESORT(TRAINING) /TEST ;
    lsmeans TRAINING GROUP(RESORT TRAINING) RESORT(TRAINING)
           GENDER*TRAINING GENDER*RESORT(TRAINING) GENDER*GROUP(RESORT TRAINING);
run;
```

```
Source                 Type III Expected Mean Square

TRAINING               Var(Error) + 6 Var(RESOR*GENDER(TRAINI)) + 3 Var(GROU*GEND(TRAI*RESO)) +
                       12 Var(RESORT(TRAINING)) + 6 Var(GROUP(TRAINI*RESORT)) +
                       Q(TRAINING,TRAINING*GENDER)
GROUP(TRAINI*RESORT)   Var(Error) + 3 Var(GROU*GEND(TRAI*RESO)) + 6 Var(GROUP(TRAINI*RESORT))
RESORT(TRAINING)       Var(Error) + 6 Var(RESOR*GENDER(TRAINI)) + 3 Var(GROU*GEND(TRAI*RESO)) +
                       12 Var(RESORT(TRAINING)) + 6 Var(GROUP(TRAINI*RESORT))

GENDER                 Var(Error) + 6 Var(RESOR*GENDER(TRAINI)) + 3 Var(GROU*GEND(TRAI*RESO)) +
                       Q(GENDER,TRAINING*GENDER)
TRAINING*GENDER        Var(Error) + 6 Var(RESOR*GENDER(TRAINI)) + 3 Var(GROU*GEND(TRAI*RESO)) +
                       Q(TRAINING*GENDER)
GROU*GEND(TRAI*RESO)   Var(Error) + 3 Var(GROU*GEND(TRAI*RESO))

RESOR*GENDER(TRAINI)   Var(Error) + 6 Var(RESOR*GENDER(TRAINI)) + 3 Var(GROU*GEND(TRAI*RESO))

                  Tests of Hypotheses for Mixed Model Analysis of Variance

Dependent Variable: TIME

        Source                 DF    Type III SS    Mean Square    F Value    Pr > F

    *   TRAINING                1     165.020833     165.020833      57.82    0.0169
        Error                   2       5.708333       2.854167

    Error: MS(RESORT(TRAINING))
```

```
* This test assumes one or more other fixed effects are zero.

Source                        DF    Type III SS    Mean Square    F Value    Pr > F

GROUP(TRAINI*RESORT)           4      33.750000       8.437500      13.97    0.0128
RESOR*GENDER(TRAINI)           2       0.375000       0.187500       0.31    0.7494
Error                          4       2.416667       0.604167

Error: MS(GROU*GEND(TRAI*RESO))

Source                        DF    Type III SS    Mean Square    F Value    Pr > F

RESORT(TRAINING)               2       5.708333       2.854167       0.36    0.7228
Error                     3.5927      28.816638       8.020833

Error: MS(GROUP(TRAINI*RESORT)) - MS(GROU*GEND(TRAI*RESO))
+ MS(RESOR*GENDER(TRAINI))

  Source                    DF    Type III SS    Mean Square    F Value    Pr > F

* GENDER                     1       0.187500       0.187500       1.00    0.4226
  TRAINING*GENDER            1       0.187500       0.187500       1.00    0.4226
  Error                      2       0.375000       0.187500

Error: MS(RESOR*GENDER(TRAINI))

* This test assumes one or more other fixed effects are zero.

  Source                    DF    Type III SS    Mean Square    F Value    Pr > F

  GROU*GEND(TRAI*RESO)       4       2.416667       0.604167       0.33    0.8585
  Error: MS(Error)          32      59.333333       1.854167

                        Least Squares Means

                      TRAINING    TIME LSMEAN

                      1            7.50000000
                      2            3.79166667

              GROUP    TRAINING    RESORT    TIME LSMEAN

              1        1           1          7.66666667
              2        1           1          7.00000000
              5        1           3          7.00000000
              6        1           3          8.33333333
              3        2           2          4.33333333
              4        2           2          4.16666667
              7        2           4          4.83333333
              8        2           4          1.83333333
```

Only the training and group (within resorts and within training) are statistically significant. This result is interpreted from the `Least Squares Means` tables. The means for `TRAINING` show that skiers with full-day classes ski faster than those who have only been trained for half a day. Group (within resort and training type) differences are not of particularly substantive interest; however, a look at the means shows that group 8 (within resort 4 with full-day training) did especially well. Other `Least Squares Means` tables (for nonsignificant interactions) are not shown here.

11.5.2 One-Way Random-Effects Design

The IV in a one-way random-effects design can be either randomized groups (different cases in each level of the IV) or repeated measures (the same cases in each level of the IV). In either case, the analysis is the same as for a one-way fixed-effects design. This is because it is interaction that creeps into the test of a main effect with a random IV, and a one-way design has no interactions.[8]

11.5.3 Factorial Random-Effects Designs

This section discusses factorial designs where all IVs are random. Such designs can be any combination of randomized groups and repeated measures (see below). The difference between random and fixed-effect ANOVA is in the error terms. Error terms for main effects are two-way interactions, and error terms for interactions are higher-order interactions. Only the highest-order interaction uses the error term based on within-cell variability. Calculation of sums of squares, degrees of freedom, and mean squares are otherwise exactly the same as for fixed-effects analysis.

11.5.3.1 Factorial Random-Effects ANOVA with Randomized Groups

The most straightforward factorial random-effects design is one where all IVs are randomized groups (different cases in all combinations of levels). Understanding the rationale for use of interaction as the error term is based on understanding the deviation approach to ANOVA. In an ordinary fixed-effects analysis, the levels of main effects chosen are the only ones of interest, and they cover the range of interest. When you look at one main effect, say, A, you collapse over levels of all other effects, say, B. The deviations due to B sum to 0 over the levels of B—some levels of B raise scores and some lower scores, but when all are considered, they sum to 0. Because the test of A is conducted over the levels of B, it is a pure test of the main effect of A; whatever the effects of B, they cancel each other when examining A. With random effects, however, the levels of B are randomly chosen, and there is no guarantee that the deviations produced by these particular levels sum to 0. Instead, the sums of

[8] Cases are a random factor in any design, including repeated-measures designs. Technically, cases should be declared a random factor in a one-way repeated-measures design (with both cases and the IV declared between-subjects factors) and in a Latin-square repeated-measures design (where time is a third between-subjects factor; cf. Section 9.4.5). Because no interactions are computed in these designs, however, the software develops the appropriate error term.

squares for the main effect of A are often contaminated with differences associated with the particular levels of B randomly selected for the study. Because the apparent effects of A depend somewhat on choice of B, it is interaction of those particular Bs with levels of A that contaminates the look at A.

Suppose, for example, that the small-sample example of Section 5.4.3 is redefined so that the three professions (administrators, belly dancers, and politicians) are randomly chosen from a compendium of all possible professions and that vacation lengths are randomly chosen from a set of all possible vacation lengths from 1 day to 35 days (from which we happened to select 7 days, 12 days, and 31 days). For these particular levels of both IVs, there may well be special interaction. Any special interaction of these particular levels of profession with these particular levels of vacation length contaminates both the test of the main effect of profession and the test of the main effect of vacation length. Had we randomly chosen other levels of either variable, their special interactions would similarly confound tests of the two main effects. The contamination is removed by using interaction as the error term, so that F for an IV is a pure test of the main effect. Lindeman (1974) and Maxwell and Delaney (1990) offer more detailed discussions of the issue, along with illustrative examples.

The partitions of sums of squares and degrees of freedom are the same as in Figure 5.1, and all computations of sums of squares, degrees of freedom, and mean squares for this random-effects example are the same as in Table 5.7. Only the F ratios for the main effects change, reflecting the random nature of the choice of IV levels.

$$F_A = \frac{\text{MS}_A}{\text{MS}_{AB}} \qquad \text{df} = (a-1), (a-1)(b-1) \qquad \textbf{(11.8)}$$

$$F_B = \frac{\text{MS}_B}{\text{MS}_{AB}} \qquad \text{df} = (b-1), (a-1)(b-1) \qquad \textbf{(11.9)}$$

Table 11.14 shows the sources of variance, error terms, and F tests for the two-way random-effects ANOVA applied to the data of Table 5.3.

Tests of main effects are evaluated with 2 and 4 df; at $\alpha = .05$, the critical value is 6.94. The test for interaction remains the same as in Section 5.4.3. The consequence of interpreting these IVs as random, rather than fixed, effects is to greatly reduce power for testing both main effects. In most randomized-groups designs, power is reduced because the degrees of freedom for interaction are markedly smaller than degrees of freedom for S/AB. This example also happens to have a very strong interaction, which further reduces power.

TABLE 11.14 Source table for two-way randomized-groups, random-effects ANOVA

Source	SS	df	MS	Error Term	F
A	33.85	2	16.93	AB	0.65
B	32.30	2	16.15	AB	0.62
AB	104.15	4	26.04	S/AB	23.45
S/AB	20.00	18	1.11		
T	190.30	26			

TABLE 11.15 Hypothetical data set for three-way randomized-groups, random-effects ANOVA

Age	Dosage	Severity	Relief	Age	Dosage	Severity	Relief
1	1	1	3	3	1	1	6
1	1	1	4	3	1	1	8
1	1	2	1	3	1	2	5
1	1	2	2	3	1	2	6
1	2	1	5	3	2	1	9
1	2	1	4	3	2	1	8
1	2	2	2	3	2	2	7
1	2	2	3	3	2	2	10
2	1	1	8	4	1	1	10
2	1	1	9	4	1	1	11
2	1	2	8	4	1	2	8
2	1	2	7	4	1	2	7
2	2	1	9	4	2	1	12
2	2	1	7	4	2	1	13
2	2	2	5	4	2	2	10
2	2	2	6	4	2	2	11

Although with higher-order factorials, partition of sums of squares and degrees of freedom remains the same, error terms are yet more complicated. Take, for example, a hypothetical drug study with two randomly chosen levels of reasonable drug dosages, two randomly chosen levels of symptom severity, and four randomly chosen ages (chosen from the population of ages of patients). The DV is amount of symptom relief. Table 11.15 shows the data set, and Table 11.16 shows the analysis through SPSS GLM.

As shown in the footnotes for Table 11.16, error terms are very complicated for some effects, and there is sometimes disagreement among statistical packages regarding the appropriate error term for a given effect. Furthermore, because of the pattern of addition and subtraction, the values of error terms can become negative. Error terms and difficulties associated with them are discussed in Section 11.6.1.

Because the mean squares for interactions are substantial, all tests, except the three-way interaction, are penalized in this analysis. The tests of the two-way interactions are not very powerful, with the MS for the three-way interaction, with its 3 df, serving as error term.

11.5.3.2 Factorial Random-Effects ANOVA with Repeated Measures

Repeated measures in a factorial random-effects ANOVA are most easily handled by considering cases as just another random factor; although the data set is rearranged,

TABLE 11.16 Three-way randomized-groups, random-effects ANOVA through SPSS GLM (syntax and selected output)

```
UNIANOVA
 RELIEF  BY AGE DOSAGE SEVERITY
 /RANDOM = AGE DOSAGE SEVERITY
 /METHOD = SSTYPE(3)
 /INTERCEPT = INCLUDE
 /CRITERIA = ALPHA(.05)
 /DESIGN = AGE DOSAGE SEVERITY AGE*DOSAGE AGE*SEVERITY DOSAGE*SEVERITY
  AGE*DOSAGE*SEVERITY .
```

Tests of Between-Subjects Effects

Dependent Variable: RELIEF

Source		Type III Sum of Squares	df	Mean Square	F	Sig.
Intercept	Hypothesis	1568.000	1	1568.000	15.680	.016
	Error	412.713	4.127	100.000[a]		
AGE	Hypothesis	214.750	3	71.583	11.767	.035
	Error	18.526	3.045	6.083[b]		
DOSAGE	Hypothesis	10.125	1	10.125	1.976	.283
	Error	11.431	2.230	5.125[c]		
SEVERITY	Hypothesis	24.500	1	24.500	73.500	.540
	Error	.056	.168	.333[d]		
AGE * DOSAGE	Hypothesis	17.625	3	5.875	6.714	.076
	Error	2.625	3	.875[e]		
AGE * SEVERITY	Hypothesis	3.250	3	1.083	1.238	.432
	Error	2.625	3	.875[e]		
DOSAGE * SEVERITY	Hypothesis	.125	1	.125	.143	.731
	Error	2.625	3	.875[e]		
AGE * DOSAGE * SEVERITY	Hypothesis	2.625	3	.875	.933	.447
	Error	15.000	16	.938[f]		

a. MS(AGE) + 1.000 MS(DOSAGE) + 1.000 MS(SEVERITY) – 1.000 MS(AGE * DOSAGE) – 1.000 MS(AGE * SEVERITY) – 1.000 MS(DOSAGE * SEVERITY) + 1.000 MS(AGE * DOSAGE * SEVERITY)

b. MS(AGE * DOSAGE) + MS(AGE * SEVERITY) – MS(AGE * DOSAGE * SEVERITY)

c. MS(AGE * DOSAGE) + MS(DOSAGE * SEVERITY) – MS(AGE * DOSAGE * SEVERITY)

d. MS(AGE * SEVERITY) + MS(DOSAGE * SEVERITY) – MS(AGE * DOSAGE * SEVERITY)

e. MS(AGE * DOSAGE * SEVERITY)

f. MS(Error)

the partitions of sums of squares and degrees of freedom do not change from fixed-effects ANOVA with repeated measures.

Consider a design similar to the earlier drug study, with two levels of severity and two of dosage, both with levels randomly selected. Instead of four levels of age, however, there are four patients and only one score for each combination of patient, severity, and dosage. Table 11.17 shows a data set for a "three-way" random-effects ANOVA, where cases form the third "way."

TABLE 11.17 Hypothetical data set for two-way repeated-measures random-effects ANOVA

Patient	Severity	Dosage	Relief	Patient	Severity	Dosage	Relief
1	1	1	1	3	1	1	3
1	1	2	6	3	1	2	5
1	2	1	4	3	2	1	5
1	2	2	6	3	2	2	3
2	1	1	7	4	1	1	5
2	1	2	8	4	1	2	8
2	2	1	6	4	2	1	6
2	2	2	3	4	2	2	4

Table 11.18 shows the syntax and selected output for the analysis through SPSS GLM (UNIANOVA). The two substantive IVs are SEVERITY and DOSAGE. Table 11.18 shows that neither of these main effects produces a mean difference in RELIEF of symptoms. However, there is a statistically significant interaction between DOSAGE and SEVERITY. The main effect of PATIENT represents individual differences and is ordinarily not of interest; however, it could be tested in this random-effects model, unlike a corresponding fixed-effects model, because there is an unambiguous error term.

Table 11.19 compares the error terms in a fixed-effects and a two-way random-effects repeated-measures design. The error terms for the main effects of dosage and severity follow the same pattern as in a randomized-groups design, both with complicated adjustments to degrees of freedom. The error term for the dosage by severity interaction is the same as in a repeated-measures fixed-effects design (cf. Section 6.4.5).

Note that sphericity is not tested when the problem is set up this way. When repeated-measures IVs have only two levels, there is no opportunity to violate the assumption. However, when there are more than two levels of a repeated-measures IV, the assumption could be violated. To test the assumption, a separate run is required in which data are set up with one line of data per patient, as in the usual repeated-measures analysis. Or a planned trend analysis could be substituted for the omnibus test of the effect.

11.5.4 Mixed Fixed-Random Designs

This section discusses designs in which some effects are fixed and some random. These designs also do not differ from fixed-effects designs in partition of sources of variability, computation of sums of squares, degrees of freedom, mean squares, and the error term used to test the highest-order interaction. They do differ in which error terms are used to test some of the main effects and lower-order interactions. Mixed fixed-random designs can include randomized-groups IVs, repeated-measures IVs, and a mix of randomized-groups and repeated-measures IVs, with any number of IVs

TABLE 11.18 Two-way repeated-measures, random-effects ANOVA through SPSS GLM (UNIANOVA)

```
UNIANOVA
 RELIEF BY PATIENT SEVERITY DOSAGE
 /RANDOM = PATIENT SEVERITY DOSAGE
 /METHOD = SSTYPE(3)
 /INTERCEPT = INCLUDE
 /CRITERIA = ALPHA(.05)
 /DESIGN = PATIENT SEVERITY DOSAGE PATIENT*SEVERITY PATIENT*DOSAGE SEVERITY*DOSAGE
 PATIENT*SEVERITY*DOSAGE .
```

Tests of Between-Subjects Effects

Dependent Variable: RELIEF

Source		Type III Sum of Squares	df	Mean Square	F	Sig.
Intercept	Hypothesis	400.000	1	.	.	.
	Error	.	.[a]	.		
PATIENT	Hypothesis	12.500	3	4.167	.568	.657
	Error	42.025	5.731	7.333[b]		
SEVERITY	Hypothesis	2.250	1	2.250	.115	.777
	Error	28.809	1.471	19.583[c]		
DOSAGE	Hypothesis	2.250	1	2.250	.115	.777
	Error	28.809	1.471	19.583[d]		
PATIENT * SEVERITY	Hypothesis	11.250	3	3.750	22.500	.015
	Error	.500	3	.167[e]		
PATIENT * DOSAGE	Hypothesis	11.250	3	3.750	22.500	.015
	Error	.500	3	.167[e]		
SEVERITY * DOSAGE	Hypothesis	16.000	1	16.000	96.000	.002
	Error	.500	3	.167[e]		
PATIENT * SEVERITY * DOSAGE	Hypothesis	.500	3	.167	.	.
	Error	.000	0	.[f]		

a. Cannot compute the error degrees of freedom using Satterthwaite's method.
b. MS(PATIENT * SEVERITY) + MS(PATIENT * DOSAGE) – MS(PATIENT * SEVERITY * DOSAGE)
c. MS(PATIENT * SEVERITY) + MS(SEVERITY * DOSAGE) – MS(PATIENT * SEVERITY * DOSAGE)
d. MS(PATIENT * DOSAGE) + MS(SEVERITY * DOSAGE) – MS(PATIENT * SEVERITY * DOSAGE)
e. MS(PATIENT * SEVERITY * DOSAGE)
f. MS(Error)

TABLE 11.19 Comparison of error terms in fixed-effects and two-way random-effects repeated-measures designs

Source	Fixed-Effects Error	Random-Effects Error
Dosage	Dosage × Patient	Dosage × Patient + Dosage × Severity – Dosage × Severity × Patient
Severity	Severity × Patient	Severity × Patient + Dosage × Severity – Dosage × Severity × Patient
Dosage × Severity	Dosage × Severity × Patient	Dosage × Severity × Patient

of any particular type. Recall that there are two uses of the term *mixed*: (1) fixed and random effects and (2) randomized-groups and repeated-measures IVs.

There are two forms of the randomized-groups mixed fixed-random design: the restricted and unrestricted models. This chapter, like most textbooks, illustrates the restricted model; however, most software programs are based on the unrestricted model. The two models differ in their choice of error term for the random factor: The restricted model uses $\text{MS}_{S/AB}$ as the error term for dosage, whereas the unrestricted model uses more complicated error terms, as indicated in the footnotes to the accompanying tables in what follows. The restricted model is recommended because it is likely to provide a more powerful test of the random factor and because of the rationale described below.

This section discusses the following: a fully randomized-groups design with some fixed and some random-effects IVs; a fully repeated-measures design with some fixed and some random-effects IVs; a "mixed-mixed" design, in which there are both randomized-groups and repeated-measures IVs, some fixed effects and some random effects.

11.5.4.1 Mixed Fixed-Random-Effects ANOVA with Randomized Groups

The simplest of these designs is a two-way randomized-groups factorial, with one IV fixed and the other random. Suppose that in the example of Section 5.3.3, levels of profession are randomly selected but vacation lengths are fixed at one, two, and three weeks. The F ratios, reflecting the error terms for main effects, are

$$F_A = \frac{\text{MS}_A}{\text{MS}_{AB}} \qquad \text{df} = (a-1), (a-1)(b-1) \tag{11.10}$$

$$F_B = \frac{\text{MS}_B}{\text{MS}_{S/AB}} \qquad \text{df} = (b-1), ab(n-1) \tag{11.11}$$

The random effect factor, B (profession), uses the same error term as in a fixed-effect design. It is the *fixed* factor, A (vacation length), that uses a different error term—the $A \times B$ interaction. Thus, the design can be analyzed through any ANOVA software. If there is no random-effects capability, you simply recalculate F_A using MS_{AB} as the error term and assess statistical significance using Table A.1 with numerator df $= b - 1$ and denominator df $= (a-1)(b-1)$.

The analysis is run through SAS in Table 5.16. The F ratios for profession and the profession by length interaction are interpreted as for the fixed-effects design. To conform to the present example, the F ratio for vacation length is recalculated as

$$F_B = \frac{16.148}{26.037} = 0.62 \qquad \text{df} = 2, 4$$

Because of the reduction in degrees of freedom, the factor has suffered a reduction in power. Table 11.20 summarizes the sources of variance, error terms, and F tests for the two-way randomized-groups, fixed-random ANOVA.

A three-way randomized-groups, fixed-random design can have either one or two random IVs. Consider, for example, that the levels of age and severity are fixed in the data of Table 11.15 and the dosage levels are randomly selected. Table 11.21 shows the analysis through SPSS GLM.

TABLE 11.20 Source table for two-way randomized-groups, fixed-random-effects ANOVA

Source	SS	df	MS	Error Term	F
A (random)	33.85	2	16.93	S/AB	15.23
B (fixed)	32.30	2	16.15	AB	0.62
AB	104.15	4	26.04	S/AB	23.45
S/AB	20.00	18	1.11		
T	190.30	26			

TABLE 11.21 Three-way randomized-groups, fixed-fixed-random-effects ANOVA through SPSS GLM

```
UNIANOVA
  RELIEF  BY AGE SEVERITY DOSAGE
  /RANDOM = DOSAGE
  /METHOD = SSTYPE(3)
  /INTERCEPT = INCLUDE
  /CRITERIA = ALPHA(.05)
  /DESIGN = AGE SEVERITY DOSAGE AGE*SEVERITY AGE*DOSAGE SEVERITY*DOSAGE
  AGE*SEVERITY*DOSAGE .
```

Tests of Between-Subjects Effects

Dependent Variable: relief

Source		Type III Sum of Squares	df	Mean Square	F	Sig.
Intercept	Hypothesis	1568.000	1	1568.000	154.864	.051
	Error	10.125	1	10.125[a]		
age	Hypothesis	214.750	3	71.583	12.184	.035
	Error	17.625	3	5.875[b]		
severity	Hypothesis	24.500	1	24.500	196.000	.045
	Error	.125	1	.125[c]		
dosage	Hypothesis	10.125	1	10.125	1.976	.283
	Error	11.431	2.230	5.125[d]		
age * severity	Hypothesis	3.250	3	1.083	1.238	.432
	Error	2.625	3	.875[e]		
age * dosage	Hypothesis	17.625	3	5.875	6.714	.076
	Error	2.625	3	.875[e]		
severity * dosage	Hypothesis	.125	1	.125	.143	.731
	Error	2.625	3	.875[e]		
age * severity * dosage	Hypothesis	2.625	3	.875	.933	.447
	Error	15.000	16	.938[f]		

a. MS(dosage)
b. MS(age * dosage)
c. MS(severity * dosage)
d. MS(age * dosage) + MS(severity * dosage) − MS(age * severity * dosage)
e. MS(age * severity * dosage)
f. MS(Error)

TABLE 11.22 Error terms in three-way randomized-groups, fixed-fixed-random-effects design.

Source	Error Term
Age (fixed)	Age × Dosage
Severity (fixed)	Severity × Dosage
Dosage (random)	*S/ABC*
Age × Severity	Age × Severity × Dosage
Severity × Dosage	*S/ABC*
Age × Dosage	*S/ABC*
Age × Severity × Dosage	*S/ABC*

Table 11.22 shows the error terms for this three-way mixed design. Most effects suffer loss of power by using error terms based on interactions. SAS GLM adds a little "fudge factor" to the derived error term: 22E − 17 × *S/ABC.*

Compare these results with those of the fully random-effects model of Table 11.16. The error terms are the same for the three-way interaction and for the main effect of the random effect (dosage), but they differ for the remaining main effects and interactions. The error terms are smaller in the model when only dosage is a random effect, which is enough, in this example, to change the effect of severity from insignificant ($p = .540$) to significant ($p = .045$).

11.5.4.2 Mixed Fixed-Random-Effects ANOVA with Repeated Measures

Another variation is the fully repeated-measures design with some IVs fixed and some random. For example, the two-way repeated-measures design of Section 6.4.2.3 could easily have both a fixed effect (month) and a random effect (novel) if the types of novel were randomly chosen from a population of all possible types.

The trick with the analysis is to organize the data as if all IVs, including cases, are randomized-groups IVs. Then both cases and the actual random-effects IV(s) are defined as random. The IVs are treated as if they were randomized-groups. Table 11.23 shows the data setup for a "three-way" mixed fixed-random-effects ANOVA. Table 11.24 shows the analysis through SPSS GLM (UNIANOVA). In Table 11.25, the error terms are summarized and compared with those in a two-way fixed effects analysis.

The effects of interest in this analysis are MONTH, NOVEL, and their interaction. These effects are reported in Table 11.24 in the section labeled Tests of Between-Subjects Effects, because the data are organized as randomized groups. Power is lost for both tests of main effects, as compared with that of the fixed-effects analysis of Section 6.4.2.3, because of the change in error terms. The test of the MONTH*NOVEL interaction is the same as in the fixed-effects analysis. Again, setting up the data and analysis this way provides no test of sphericity.

TABLE 11.23 Hypothetical two-way repeated-measures data set in one-line-per-cell format

Case	Novel	Month	Books	Case	Novel	Month	Books
1	1	1	1	3	2	1	4
1	1	2	3	3	2	2	5
1	1	3	6	3	2	3	3
1	2	1	5	4	1	1	5
1	2	2	4	4	1	2	5
1	2	3	1	4	1	3	7
2	1	1	1	4	2	1	3
2	1	2	4	4	2	2	2
2	1	3	8	4	2	3	0
2	2	1	8	5	1	1	2
2	2	2	8	5	1	2	4
2	2	3	4	5	1	3	5
3	1	1	3	5	2	1	5
3	1	2	3	5	2	2	6
3	1	3	6	5	2	3	3

TABLE 11.24 Two-way repeated-measures, fixed-random ANOVA through SPSS GLM (UNIANOVA) (syntax and selected output)

```
UNIANOVA
 BOOKS  BY MONTH NOVEL CASE
 /RANDOM = NOVEL CASE
 /METHOD = SSTYPE(3)
 /INTERCEPT = INCLUDE
 /CRITERIA = ALPHA(.05)
 /DESIGN = MONTH NOVEL CASE MONTH*NOVEL MONTH*CASE NOVEL*CASE
 MONTH*NOVEL*CASE .
```

Univariate Analysis of Variance

Between-Subjects Factors

		N
month	1	10
	2	10
	3	10
novel	1	15
	2	15
case	1	6
	2	6
	3	6
	4	6
	5	6

TABLE 11.24 Continued

Tests of Between-Subjects Effects

Dependent Variable: books

Source		Type III Sum of Squares	df	Mean Square	F	Sig.
Intercept	Hypothesis	512.533	1	.	.	.
	Error	.	.[a]	.		
month	Hypothesis	2.867	2	1.433	.045	.957
	Error	60.236	1.911	31.517[b]		
novel	Hypothesis	.133	1	.133	.003	.957
	Error	113.699	2.887	39.383[c]		
case	Hypothesis	16.467	4	4.117	.537	.722
	Error	25.325	3.303	7.667[d]		
month * novel	Hypothesis	64.467	2	32.233	26.135	.000
	Error	9.867	8	1.233[e]		
month * case	Hypothesis	4.133	8	.517	.419	.880
	Error	9.867	8	1.233[e]		
novel * case	Hypothesis	33.533	4	8.383	6.797	.011
	Error	9.867	8	1.233[e]		
month * novel * case	Hypothesis	9.867	8	1.233	.	.
	Error	.000	0	.[f]		

a. Cannot compute the error degrees of freedom using Satterthwaite's method.
b. MS(month * novel) + MS(month * case) - MS(month * novel * case)
c. 1.000 MS(month * novel) + 1.000 MS(novel * case) - 1.000 MS(month * novel * case)
d. MS(month * case) + 1.000 MS(novel * case) - 1.000 MS(month * novel * case)
e. MS(month * novel * case)
f. MS(Error)

TABLE 11.25 Comparison of error terms in two-way repeated-measures designs, with both effects fixed, both random, and mixed

Source	Mixed and Completely Random-Effects Error Terms	Fixed-Effects Error Terms
Month	Month × Novel + Month × Case − Month × Novel × Case	Month × Case
Novel	Month × Novel + Novel × Case − Month × Novel × Case	Novel × Case
Month × Novel	Month × Novel × Case	Month × Novel × Case

11.5.4.3 Mixed Fixed-Random-Effects ANOVA with Randomized Groups and Repeated Measures

Finally, there are mixed-mixed designs, which have both random and fixed effects and both randomized groups and repeated measures. Using the example of Section 7.4.3, let us assume that types of novels are randomized groups and are randomly chosen,

TABLE 11.26 Hypothetical data set for mixed-mixed design

CASE	NOVEL	MONTH	BOOKS	CASE	NOVEL	MONTH	BOOKS
1	1	1	1	8	2	3	2
1	1	2	3	9	2	1	4
1	1	3	6	9	2	2	2
2	1	1	1	9	2	3	0
2	1	2	4	10	2	1	4
2	1	3	8	10	2	2	5
3	1	1	3	10	2	3	3
3	1	2	3	11	3	1	4
3	1	3	6	11	3	2	2
4	1	1	5	11	3	3	0
4	1	2	5	12	3	1	2
4	1	3	7	12	3	2	6
5	1	1	2	12	3	3	1
5	1	2	4	13	3	1	3
5	1	3	5	13	3	2	3
6	2	1	3	13	3	3	3
6	2	2	1	14	3	1	6
6	2	3	0	14	3	2	2
7	2	1	4	14	3	3	1
7	2	2	4	15	3	1	3
7	2	3	2	15	3	2	3
8	2	1	5	15	3	3	2
8	2	2	3				

while the three months are repeated measures and fixed effects. The first task is to reorganize the data into one-line-per-cell format, so that cases are also set up as a random randomized-groups IV, as shown in Table 11.26.

Table 11.27 shows the analysis through SPSS GLM (UNIANOVA). The syntax uses nesting specification to indicate that cases nest within type of novel (the randomized-groups IV) but cross levels of month (the repeated-measures IV).

The effects of interest are NOVEL, MONTH, and their interaction. The error terms for NOVEL, which is tested against CASE(NOVEL) + MONTH*NOVEL–MONTH*CASE(NOVEL), and MONTH, which is tested against MONTH*NOVEL, are both considerably larger than those in the fixed-effects analysis of Section 7.4.3. However, the test of the MONTH*NOVEL interaction uses the same error term, MONTH*CASE(NOVEL), as in the fixed-effects analysis and is unaffected by the

TABLE 11.27 Analysis of mixed-mixed design: Novel is a randomized-group, random-effect IV, and month is a repeated-measures, fixed-effects IV (SPSS GLM syntax and selected output)

```
UNIANOVA
 BOOKS BY MONTH CASE NOVEL
 /RANDOM = CASE NOVEL
 /METHOD = SSTYPE(3)
 /INTERCEPT = INCLUDE
 /CRITERIA = ALPHA(.05)
 /DESIGN = MONTH NOVEL CASE(NOVEL) MONTH*NOVEL MONTH*CASE(NOVEL).
```

Tests of Between-Subjects Effects

Dependent Variable: BOOKS

Source		Type III Sum of Squares	df	Mean Square	F	Sig.
Intercept	Hypothesis	473.689	1	473.689	46.039	.021
	Error	20.578	2	10.289[a]		
MONTH	Hypothesis	.711	2	.356	.020	.980
	Error	71.422	4	17.856[b]		
NOVEL	Hypothesis	20.578	2	10.289	.556	.610
	Error	79.011	4.270	18.506[c]		
CASE(NOVEL)	Hypothesis	26.400	12	2.200	1.419	.224
	Error	37.200	24	1.550[d]		
MONTH * NOVEL	Hypothesis	71.422	4	17.856	11.520	.000
	Error	37.200	24	1.550[d]		
MONTH * CASE(NOVEL)	Hypothesis	37.200	24	1.550	.	.
	Error	.000	0	.[e]		

a. MS(NOVEL)
b. MS(MONTH * NOVEL)
c. 1.000 MS(CASE(NOVEL)) + 1.000 MS(MONTH * NOVEL) − 1.000 MS(MONTH * CASE(NOVEL))
d. MS(MONTH * CASE(NOVEL))
e. MS(Error)

presence of a random-effects IV. Recall that the test of the highest-order interaction is always the same as for a fixed-effects analysis. Again, no test of sphericity is available with this arrangement of data.

11.6 SOME IMPORTANT ISSUES

A major issue in random-effects ANOVA is the choice of error terms. Indeed, this is the only difference between random and fixed-effects ANOVA. Alternative strategies for analysis of random effects—particularly hierarchical designs—have recently been implemented in software that uses iterative algorithms to circumvent the issue of choice of error terms.

Trend analysis, as discussed in Section 11.6.3, of a random IV usually requires adjustment for unequal spacing of levels. Homogeneity of covariance, Section 11.6.4, is an issue in factorial random-effects designs—that is, designs in which random effects cross each other or cross fixed effects.

11.6.1 Error Terms in Random-Effects ANOVA

Error terms are what distinguish between analyses of random and fixed-effects ANOVA. The reason is that an interaction between a random factor (B) and a fixed factor (A) intrudes on the test of the main effect of the fixed factor, because the matrix of cell means from any particular study represents all possible levels of A but not all possible levels of B. This means you can "trust" the marginal means of B, because they are averaged over *all* levels of A, but you can't trust the marginal means of A, because they are averaged over only some of the possible levels of B. Suppose there are no real (population) main effects of A or B. In other words, averaging over all possible levels of B produces about the same means for each level of A, but averaging over only the particular randomly selected levels of B results in different means for each level of A. This happens when the particularly selected levels of B do not happen to have the average response when combined with the levels of A—that is, these particular levels of B interact with A.

Although one fondly hopes that such interaction is absent, the prudent course is to assume the interaction is present and include it in tests of the main effects (and of lower-order interactions, if there are more than two factors). Little is lost if there really is no interaction, because little is added to the numerator and denominator of the F ratio—only a little more error, resulting in a little less power.

Of course, this applies to all main effects in a fully random factorial design. Interaction intrudes on main effects, and higher-order interactions intrude on lower-order interactions. Therefore, the expected mean square (EMS; a term we have avoided so far but cannot any longer) in the numerator of an F ratio includes not only treatment plus error but also interactions between treatment and random effects.

Recall from Section 3.2.3.4 that in a fixed-effects ANOVA, the test of treatment is based on the assumption that the components of variance in the F ratio are

$$F = \frac{\text{treatment} + \text{error}}{\text{error}}$$

A second, random, IV in the design changes the numerator to include the potential interaction between treatment and the second IV. Therefore, for the F to be equal to 1 when the null hypothesis is true, the interaction must be added to the denominator as well. Thus, the components of variance in the test for treatment become

$$F = \frac{\text{treatment} + \text{interaction} + \text{error}}{\text{interaction} + \text{error}}$$

So, let us consider expected mean squares. They are based on population parameters and are represented by Greek letters—σ^2 or θ^2 (theta squared)—which are used instead of MS to represent expected population variance. The symbol used for population error variance (the real stuff that we can only estimate) is σ_ε^2. This source of variability appears in every expected mean square. In the simplest factorial design—a two-way randomized-groups fixed layout—this is estimated by $MS_{S/AB}$ and is all that is needed to serve as the error term for A, B, and the $A \times B$ interaction in that design. However . . .

11.6.1.1 Expected Mean Squares

Population variances for random effects are denoted σ^2; population variances for fixed effects are denoted θ^2. So, for a fixed two-way factorial design, σ^2 (error) is in the denominator, and both θ^2 (effect) and σ^2 (error) are in the numerator. There is an additional hitch: The population variance for treatment (θ^2) is multiplied by the number of scores in the marginal mean (i.e., the number of levels of the other factor times the number of scores in each cell), because it affects the calculated mean square. Thus, the expected MS (EMS) for A in a two-way fixed-effects ANOVA is $(b)(n)(\theta_A^2)$. Table 11.28 shows the expected mean squares and error terms for three two-way randomized-groups factorial designs: both IVs fixed, A fixed and B random, and both IVs random. The appropriate error term for an effect is the one that has all the components of the effect *except* the effect itself. Thus, in the design with both IVs fixed, S/AB has the only component that is in A in addition to $(b)(n)(\theta_A^2)$; that is, σ_ε^2.

Table 11.28 shows that the test of the $A \times B$ interaction is the same in all designs and the error term for the test of the main effects depends on whether there are any random IVs—not whether the IV being tested is fixed or random. Note that the last column of Table 11.28 matches that of the error terms in Table 11.14. Table 11.29 compares error terms in a variety of two-way designs.

Nested designs have their own expected mean squares. Table 11.30 shows EMS and error terms for the simple nested design of Section 11.4, where A is the fixed treatment IV, B is the random IV nested with A, and cases are nested within B.

The EMS for A includes B/A (the random effect), as well as $S/B/A$; therefore, B/A, which also includes $S/B/A$, is used as the error term. These error terms are shown in the SPSS GLM output Tests of Between-Subject Effects of Table 11.8,

TABLE 11.28 Expected mean squares and error terms for sources of variance in three randomized-groups two-way factorial designs

	Both IVs Fixed		*A* Fixed, *B* Random		Both IVs Random	
Source	*EMS*	*Error Term*	*EMS*	*Error Term*	*EMS*	*Error Term*
A	$\sigma_\varepsilon^2 + (b)(n)(\theta_A^2)$	S/AB	$\sigma_\varepsilon^2 + (n)(\sigma_{AB}^2) + (b)(n)(\theta_A^2)$	$A \times B$	$\sigma_\varepsilon^2 + (n)(\sigma_{AB}^2) + (b)(n)(\sigma_A^2)$	$A \times B$
B	$\sigma_\varepsilon^2 + (a)(n)(\theta_B^2)$	S/AB	$\sigma_\varepsilon^2 + (a)(n)(\sigma_B^2)$	S/AB	$\sigma_\varepsilon^2 + (n)(\sigma_{AB}^2) + (a)(n)(\sigma_B^2)$	$A \times B$
$A \times B$	$\sigma_\varepsilon^2 + (n)(\theta_{AB}^2)$	S/AB	$\sigma_\varepsilon^2 + (n)(\sigma_{AB}^2)$	S/AB	$\sigma_\varepsilon^2 + (n)(\sigma_{AB}^2)$	S/AB
S/AB	σ_ε^2		σ_ε^2		σ_ε^2	

TABLE 11.29 Comparision of error terms in various combinations of randomized-groups, repeated-measures, fixed and random two-way designs

	Randomized Groups			**Repeated Measures**			**Mixed**
Source	Both Fixed	Both Random	*A* Fixed *B* Random	Both Fixed	Both Random	*A* Fixed *B* Random	*A* RG[a] and Random *B* RM[b] and Fixed
A	S/AB	AB	AB	AS	$AS + AB - ABS$	$AS + AB - ABS$	$S/A + AB - B \times S/A$
B	S/AB	AB	S/AB	BS	$BS + AB - ABS$	$BS + AB - ABS$	AB
$A \times B$	S/AB	S/AB	S/AB	ABS	ABS	ABS	$B \times S/A$

[a] RG = randomized groups
[b] RM = repeated measures

TABLE 11.30 Expected mean squares and error terms for sources of variance in a simple nested design

Source	EMS	Error Term
A	$\sigma_\varepsilon^2 + (n)(\sigma_{B/A}^2) + (b)(n)(\theta_A^2)$	B/A
B/A	$\sigma_\varepsilon^2 + (n)(\sigma_{B/A}^2)$	$S/A/B$
$S/B/A$	σ_ε^2	

where the error term identified in footnote a is MS(GROUP(TRAINING)), which corresponds to $MS_{B/A}$. In the portion of the table labeled Expected Mean Squares, the value of 3 corresponds to the three cases in each group (n); the quadratic term includes the number of scores in each level of TRAINING, although not so specified in output. SAS GLM output (cf. Table 11.9) similarly shows expected mean squares for each effect, with Q indicating the quadratic term. Q corresponds to $(b)(n)$ in the EMS for A in Table 11.30. For the small-sample example of Section 11.4.3, the number of levels of B is 2, and the number of scores in each level of B/A is 3; thus, the number of scores in each level of A is $(b)(n) = (2)(3) = 6 = Q$.

So far, EMS and choice of error terms are fairly straightforward: You figure out the interactions that could be included in main effects and add their variances (multiplied by number of scores in the relevant cell) to the EMS that already includes the variance for the effect being tested plus population error, σ_ε^2. Then you look for a term that includes all the components, except the effect to be tested, and use that as the error term. Sometimes, however, that term does not exist. Consider, for example, the three-way randomized-groups designs in which all factors are random. Table 11.31 shows the EMS for that design.

Error terms for the interactions are easily found. Each two-way interaction includes population error, its own effect, and the three-way interaction. Because the three-way interaction includes its own effect and population error, it serves as an error term for all the two-way interactions. The three-way interaction needs only population error as its error term—S/ABC does the job.

TABLE 11.31 Expected mean squares and error terms for sources of variance in two randomized-groups three-way factorial designs

	All IVs Fixed		All IVs Random	
Source	*EMS*	*Error Term*	*EMS*	*Error Term*
A	$\sigma_\varepsilon^2 + (b)(c)(n)(\theta_A^2)$	S/ABC	$\sigma_\varepsilon^2 + (n)(\sigma_{ABC}^2) + (n)(c)(\sigma_{AB}^2) + (n)(b)(\sigma_{AC}^2) + (b)(c)(n)(\sigma_A^2)$	$\text{MS}_{AB} + \text{MS}_{AC} - \text{MS}_{ABC}$
B	$\sigma_\varepsilon^2 + (a)(c)(n)(\theta_B^2)$	S/ABC	$\sigma_\varepsilon^2 + (n)(\sigma_{ABC}^2) + (n)(c)(\sigma_{AB}^2) + (n)(a)(\sigma_{BC}^2) + (a)(c)(n)(\sigma_B^2)$	$\text{MS}_{AB} + \text{MS}_{BC} - \text{MS}_{ABC}$
C	$\sigma_\varepsilon^2 + (a)(b)(n)(\theta_C^2)$	S/ABC	$\sigma_\varepsilon^2 + (n)(\sigma_{ABC}^2) + (n)(b)(\sigma_{AC}^2) + (n)(a)(\sigma_{BC}^2) + (a)(b)(n)(\sigma_C^2)$	$\text{MS}_{AC} + \text{MS}_{BC} - \text{MS}_{ABC}$
$A \times B$	$\sigma_\varepsilon^2 + (c)(n)(\theta_{AB}^2)$	S/ABC	$\sigma_\varepsilon^2 + (n)(\sigma_{ABC}^2) + (c)(n)(\sigma_{AB}^2)$	$A \times B \times C$
$A \times C$	$\sigma_\varepsilon^2 + (b)(n)(\theta_{AC}^2)$	S/ABC	$\sigma_\varepsilon^2 + (n)(\sigma_{ABC}^2) + (b)(n)(\sigma_{AC}^2)$	$A \times B \times C$
$B \times C$	$\sigma_\varepsilon^2 + (a)(n)(\theta_{BC}^2)$	S/ABC	$\sigma_\varepsilon^2 + (n)(\sigma_{ABC}^2) + (a)(n)(\sigma_{BC}^2)$	$A \times B \times C$
$A \times B \times C$	$\sigma_\varepsilon^2 + (n)(\theta_{ABC}^2)$	S/ABC	$\sigma_\varepsilon^2 + (n)(\sigma_{ABC}^2)$	S/ABC
S/ABC	σ_ε^2		σ_ε^2	

Main effects, however, each have 2 two-way interactions in their EMS, along with the three-way interaction (population error) and its own effect. There are no terms to serve as error that have the three-way interaction (population error) and 2 two-way interactions. Therefore, the strategy is to combine effects to end up with the correct components. For example, for the error term for A, EMS_{AB} has one of the interaction components needed, and EMS_{AC} has the other; both have the three-way component and population error. So, adding them together produces an error term with all the needed components, but it also produces an extra three-way interaction component and an extra population error. These are dealt with by subtracting out EMS_{ABC}, which only has the extra components, so that $\text{EMS}_{A\text{error}}$ can be formed as $\text{MS}_{AB} + \text{MS}_{AC} - \text{MS}_{ABC}$. This error term and the terms for B and C appear in Table 11.31. F ratios based on this term are called *quasi-F* (also called *pseudo-F*) *ratios,* or F'. For example, the main effect of A is tested by

$$F'_A = \frac{\text{MS}_A}{\text{MS}_{AB} + \text{MS}_{AC} - \text{MS}_{ABC}} \tag{11.12}$$

Adjusted degrees of freedom for the error term are then calculated using Equation 11.13, so that standard F tables (Appendix A.1) may be used to evaluate F' (Satterthwaite, 1946).

$$\text{df}_{A_{\text{adf}}} = \frac{(\text{MS}_{AB} + \text{MS}_{AC} - \text{MS}_{ABC})^2}{(\text{MS}_{AB})^2/\text{df}_{AB} + (\text{MS}_{AC})^2/\text{df}_{AC} + (\text{MS}_{ABC})^2/\text{df}_{ABC}} \tag{11.13}$$

SAS GLM output for the three-way random-effects ANOVA (Table 11.16) shows these error terms and their use in evaluating main effects. This strategy can be applied to any design in which there is no straightforward error term for testing an effect—for example, the mixed-mixed design of Table 11.27. One problem with this strategy, however, is that the error MS can be less than 0 if the MS that are added are less than the MS that are subtracted. Thus, other strategies for calculating quasi-F ratios are available (Myers, 1979). There also are alternatives to ANOVA in dealing with random-effects designs, as discussed in Section 11.6.2.

11.6.1.2 Pooled Error Terms

Loss of power in random-effects designs stems from loss of degrees of freedom in error terms. Some researchers recommend pooling error terms to overcome this loss of power under some circumstances. For example, Brown, Michel, and Winer (1991) recommend pooling MS_{AB} with $MS_{S/AB}$ if there is no statistically significant interaction or if prior research indicates that no interaction is to be expected.[9] To consider pooling, preliminary tests are performed on higher-order effects before testing lower-order effects of interest. The procedure is as follows: If an interaction is apparently random variability, it can be pooled with another mean square that has the same expected value except for interaction. Pooling is done by adding the SS for the two effects, adding their df, and then forming a new MS by dividing the new SS by the new df.

Consider, for example, the three-way analysis of Table 11.16. All the two-way effects use the three-way interaction as the error term, but it is not statistically significant ($p = .477$). Thus, MS_{ADS} (AGE by DOSAGE by SEVERITY) may be pooled with MS_{error} in Table 11.18. The pooled sums of squares and degrees of freedom are

$$SS_{pooled} = 15.000 + 2.625 = 17.625$$

$$df_{pooled} = 16 + 3 = 19$$

The new MS to use as error for the tests of two-way interactions is

$$MS_{pooled} = \frac{17.625}{19} = 0.928$$

The test of AGE by DOSAGE now has $F = 5.8750/0.928 = 6.33$, which is statistically significant at $\alpha = .05$ with df $= 2$ and 19. Although the F ratios change little, the increased df_{error} are sufficient to find a significant $A \times D$ interaction. However, AGE by SEVERITY and DOSAGE by SEVERITY still are not statistically significant.

Nested models may also use pooled error terms if the designated error terms are not statistically significant. For our earlier example, neither of the random error terms—RESORT(TRAINING) or GROUP(RESORT(TRAINING))—is statistically significant in the doubly nested ANOVA of Table 11.13. Therefore, the EMS for each term reduces to σ^2, just as it is for $MS_{S/C/B/A}$, labeled `MS(error)` in Table 11.13. All three effects may be pooled to provide a stronger test of TRAINING, although more

[9] There is a strong interaction in the two-way random-effects example of Table 11.14, so pooling is not feasible.

power is unnecessary in this case. As a demonstration:

$$SS_{pooled} = 1.667 + 12.667 + 37.333 = 51.667$$

$$df_{pooled} = 2 + 4 + 16 = 22$$

The new MS error for the test of TRAINING is

$$MS_{pooled} = \frac{51.67}{22} = 2.35$$

$F_{training} = 88.167/2.35 = 37.52$, which is actually less than with the nonpooled error term. However, because df_{error} increases to 22, the p value becomes less than .001.

One problem with pooling is that Type I error increases once preliminary tests are done; the test of interest can be considered post hoc. This is why some statisticians recommend that pooling never be done. A compromise position to deal with increased Type I error rate is to apply a more conservative α level when using pooled error terms, perhaps .025 instead of .05. However, this may obviate the gain in power from pooling. Another recommendation is to pool terms only when the preliminary test of significance has $\alpha > .25$ (Myers and Well, 1991).

11.6.2 Alternative Strategies to ANOVA

Alternatives to the ANOVA strategy in this chapter are based on maximum-likelihood procedures (Eliason, 1993), sometimes referred to as *variance components analysis*. Effects in the ANOVA design are considered variance components in the maximum-likelihood procedures, which are iterative procedures designed to deal with such problems as unbalanced designs (including unequal sample sizes) and negative variance and df estimates. They also handle repeated measures without some of the restrictive assumptions of the ANOVA models, such as sphericity. Disadvantages of maximum-likelihood strategies are that they require large samples, and, sometimes, they produce negative variance estimates. SAS VARCOMP, SPSS GLM (VARCOMP), SAS MIXED, and SPSS MIXED all may be used for maximum-likelihood analysis of random-effects models. The latter two (MIXED) programs are the more recent and flexible programs.

Tables 11.32 and 11.33 show analysis of the three-way random-effects randomized-groups factorial data of Table 11.15 through SPSS MIXED and SAS MIXED, respectively. Both analyses specify restricted maximum likelihood as the (default) estimation procedure. Table 11.32 shows syntax and selected output for SPSS MIXED. All three IVs and all interactions are declared RANDOM, and REML is chosen as the (default) /METHOD. TESTCOV is requested to print the test of covariance parameters (effects). Remaining instructions are produced through the SPSS menu system.

The parameter estimate for age is 8.189, the component for dosage is 0.266, and so on. Wald Z is calculated by dividing each parameter estimate by its standard error. Setting $\alpha = .05$, the Sig. column shows no statistically significant predictors of relief. This is a more conservative outcome than noted in the traditional analysis of Table 11.16, but then maximum-likelihood methods are for large samples.

TABLE 11.32 Three-way random-effects maximum-likelihood analysis (SPSS MIXED syntax and selected output)

```
MIXED
  relief  BY age dosage severity
  /CRITERIA = CIN(95) MXITER(100) MXSTEP(5) SCORING(1) SINGULAR(0.000000000001)
   HCONVERGE(0, ABSOLUTE) LCONVERGE(0, ABSOLUTE)   PCONVERGE(0.000001, ABSOLUTE)
  /FIXED = | SSTYPE(3)
  /METHOD = REML
  /PRINT = TESTCOV
  /RANDOM age dosage severity age*dosage age*severity dosage*severity age*dosage*
   severity  | COVTYPE(VC) .
```

Estimates of Covariance Parameters[b]

Parameter		Estimate	Std. Error	Wald Z	Sig.	95% Confidence Interval	
						Lower Bound	Upper Bound
Residual		.887500	.280652	3.162	.002	.477524	1.649461
age	Variance	8.189063	7.331426	1.117	.264	1.416376	47.346706
dosage	Variance	.265625	.943816	.281	.778	.000251	281.026945
severity	Variance	1.463542	2.166220	.676	.499	.080449	26.624848
age * dosage	Variance	1.246875	1.201280	1.038	.299	.188689	8.239451
age * severity	Variance	.048958	.231999	.211	.833	4.53152E-006	528.943902
dosage * severity	Variance	.000000[a]	.000000	.	.	.	.
age * dosage * severity	Variance	.000000[a]	.000000	.	.	.	.

a. This covariance parameter is redundant. The test statistic and confidence interval cannot be computed.
b. Dependent Variable: relief.

Table 11.33 shows the same analysis through SAS MIXED. The options chosen are for the `reml` method (default), `covtest`, which requests Wald z tests of the parameter estimates. Unlike SAS GLM, this program assumes that effects listed in the `model` instruction are fixed; random effects are specified in the `random` instruction.

The `Covariance Parameter Estimates` are the same as the SPSS estimates. Again, no effects are statistically significant at $\alpha = .05$.

In addition to restricted maximum-likelihood estimation, SAS and SPSS MIXED provide unrestricted maximum-likelihood estimation, a procedure similar to maximum-likelihood estimation that also takes into account degrees of freedom for fixed effects when estimating variance components.

Readers interested in further information about maximum-likelihood methods for models with random effects and nesting are referred to Rao and Kleffe (1988), Searle, Casella, and McColloch (1992), Kreft and DeLeeuw (1998), and Bryk and Raudenbush (2001). The latter two sources are devoted to nested (hierarchical) models, also known as *multilevel modeling,* and introduce random-coefficients models, an elaboration of random-effects models in which the higher levels of nesting (e.g., groups within treatments) may differ not only in level (means) but also in slope (relationships between DVs and CVs).

TABLE 11.33 Three-way random-effects maximum likelihood analysis (SAS MIXED syntax and selected output)

```
proc mixed data=SASUSER.FAC3WAY method=reml covtest;
   class AGE DOSAGE SEVERITY;
   model RELIEF =;
   random AGE|DOSAGE|SEVERITY;
run;
```

The Mixed Procedure

Covariance Parameter Estimates

Cov Parm	Estimate	Standard Error	Z Value	Pr Z
AGE	8.1888	7.3309	1.12	0.1320
DOSAGE	0.2656	0.9438	0.28	0.3892
AGE*DOSAGE	1.2469	1.2012	1.04	0.1496
SEVERITY	1.4635	2.1662	0.68	0.2496
AGE*SEVERITY	0.04896	0.2320	0.21	0.4164
DOSAGE*SEVERITY	0	.	.	.
AGE*DOSAGE*SEVERITY	0	.	.	.
Residual	0.8875	0.2807	3.16	0.0008

For example, consider an affirmative action program versus a no-program control as the fixed treatment IV. Organizations, differing in size, are randomly assigned to treatment or control conditions. The DV is minority salary, measured at the individual employee level; employees, of course, are nested within organizations. Average salaries may differ between organizations, as well as between treatments, and this difference is handled in nested ANOVA, because organizations within treatments serve as the error term. In addition, however, the relationship between salary and treatment may differ among organizations of different sizes. That is, the relationship between salary and treatment may depend on organization size. This is similar to the classic ANCOVA problem of heterogeneity of regression (cf. Section 8.3.2.6), but here organization size (the covariate) and salary (the DV) are measured at different levels of the hierarchical design (at the organization level vs. the employee level). All employees within a single organization have the same size score, leading to nonindependence of errors if classic ANCOVA is used and nesting is ignored. A nested ANCOVA, then, is plagued with heterogeneity of regression because of the different slopes between treatments. Random-coefficients models, however, are designed to take into account both the hierarchical levels of analysis and the differing slopes associated with heterogeneity of regression. Among the programs described in this section, SAS MIXED and SPSS MIXED handle the random-coefficients models.

11.6.3 Trend Analysis with Unequal Spacing

Trend analysis is usually interesting whenever there is a quantitative IV with more than two levels. Trend analysis is especially interesting when the levels of the quantitative IV are randomly selected from the range of levels of interest, because the results generalize to the entire range. Random choice of levels, however, rarely provides levels that are equally spaced, making the trend analysis a bit more demanding, as discussed in Section 4.5.3.6.

Suppose, for example, we are interested in a trend analysis on age in the data of Table 11.15. Suppose, in addition, that the levels of age randomly selected from a range of interest, say, 15–75, turned out to be 16, 29, 33, and 56. Table 11.34 shows

TABLE 11.34 Trend analysis with unequal spacing of levels in a three-way random-effects design (SPSS GLM syntax and selected output)

```
UNIANOVA
 RELIEF  BY AGE DOSAGE SEVERITY
 /RANDOM = AGE DOSAGE SEVERITY
 /CONTRAST (AGE)=Polynomial (16 29 33 56)
 /METHOD = SSTYPE(3)
 /INTERCEPT = INCLUDE
 /EMMEANS = TABLES(AGE)
 /CRITERIA = ALPHA(.05)
 /DESIGN = AGE DOSAGE SEVERITY AGE*DOSAGE AGE*SEVERITY
  DOSAGE*SEVERITY AGE*DOSAGE*SEVERITY .
```

Univariate Analysis of Variance

Custom Hypothesis Tests

Contrast Results (K Matrix)

age Polynomial Contrast[a]			Dependent Variable relief
Linear	Contrast Estimate		4.894
	Hypothesized Value		0
	Difference (Estimate - Hypothesized)		4.894
	Std. Error		.342
	Sig.		.000
	95% Confidence Interval for Difference	Lower Bound	4.168
		Upper Bound	5.620
Quadratic	Contrast Estimate		-1.578
	Hypothesized Value		0
	Difference (Estimate - Hypothesized)		-1.578
	Std. Error		.342
	Sig.		.000
	95% Confidence Interval for Difference	Lower Bound	-2.304
		Upper Bound	-.853
Cubic	Contrast Estimate		.633
	Hypothesized Value		0
	Difference (Estimate - Hypothesized)		.633
	Std. Error		.342
	Sig.		.083
	95% Confidence Interval for Difference	Lower Bound	-.093
		Upper Bound	1.359

a. Metric = 16.000, 29.000, 33.000, 56.000

TABLE 11.34 Continued

Estimated Marginal Means

age

Dependent Variable: relief

age	Mean	Std. Error	95% Confidence Interval	
			Lower Bound	Upper Bound
1.00	3.000	.342	2.274	3.726
2.00	7.375	.342	6.649	8.101
3.00	7.375	.342	6.649	8.101
4.00	10.250	.342	9.524	10.976

the trend analysis through SPSS GLM (UNIANOVA). Unequal spacing for the trend analysis of age is specified in the /CONTRAST instruction by the spacing of ages within parentheses. Least squares means are requested for AGE to help interpret the trend analysis.

The Custom Hypothesis Tests output shows the results of the trend analysis. The 95% confidence intervals for the linear and quadratic effects do not include 0, indicating that both effects are statistically significant at $\alpha = .05$. The cubic trend is not statistically significant. Figure 11.3 plots the means, showing the direction of the trends.

The rate of growth is steeper between 16 and about 30 than between 30 and 56. The last portion of the line is speculative, drawn by eye, to continue the decrease in rate of growth with increasing age.

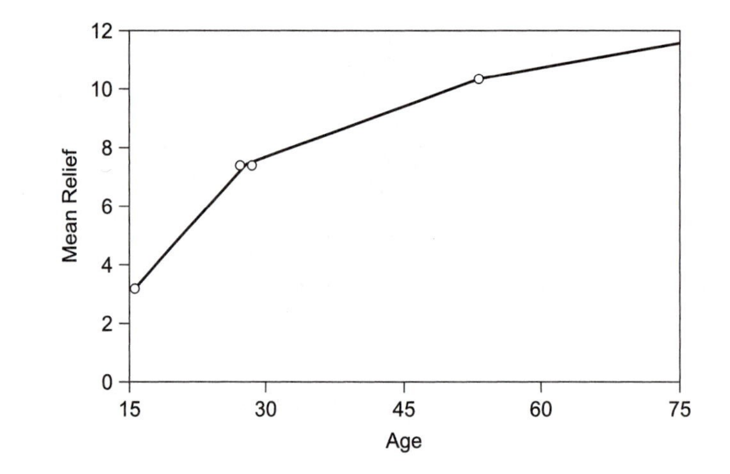

FIGURE 11.3
Mean relief as a function of age

11.6.4 Homogeneity of Covariance

All repeated-measures models assume homogeneity of covariance/sphericity (cf. Section 6.3.2.3 and 6.6.2). The assumption is also required for a random-effects IV in a factorial ANOVA. That is, it's assumed that the relationships between all pairs of levels of any IV are the same at all levels of a random IV.

There is no formal test for homogeneity of covariance in random-effects designs akin to Mauchly's sphericity test for repeated measures. However, some idea of the violation of the assumption may be gleaned through viewing the variance-covariance matrices formed around cell means (Lindeman, 1974). For the data of Table 5.3, with the random-effects ANOVA in Table 11.14, cell means are shown in Table 11.35.

Because both A and B are random, two variance-covariance matrices are formed. The diagonal elements in the first matrix (A) are formed by finding the variance of cell means for A over the three levels of B. The off-diagonal elements are formed by finding the covariances of all pairs of cell means for A over levels of B.[10] The reverse is done for the variance-covariance matrix for B—variances of cell means are found over levels of A. Table 11.36 shows the variance-covariance matrices for A and B.

Both variances and covariances are far from homogeneous in these two matrices. However, the estimates are highly tentative, because of the small samples used to generate them (three samples for each parameter estimate). This possible violation of assumptions of homogeneity of variance and homogeneity of covariance puts the ANOVA results in doubt. A variance-stabilizing transformation might be considered, or a more conservative α level for testing effects—perhaps .025 or even .01. Another option is to use SAS MIXED to specify an appropriate variance-covariance matrix.

TABLE 11.35 Cell means for data in two-way random-effects ANOVA

	b_1: 7 days	b_2: 12 days	b_3: 31 days
a_1: Administrators	0.33	5.67	6.33
a_2: Belly Dancers	6.00	6.00	8.33
a_3: Politicians	6.33	9.00	2.67

TABLE 11.36 Variance-covariance matrices for means in Table 11.35

(a) Variance-covariance matrix for A

	a_1	a_2	a_3
a_1	10.83	2.59	−1.98
a_2	2.59	1.81	−3.88
a_3	−1.98	−3.88	10.10

(b) Variance-covariance matrix for B

	b_1	b_2	b_3
b_2	11.38	3.81	−2.08
b_2	3.81	3.37	−4.75
b_3	−2.08	−4.75	8.24

[10] Chapter 2 shows how to calculate variances and covariances.

11.6.5 Effect Size

The $\hat{\omega}^2$ and partial $\hat{\omega}^2$ of Chapter 5 do not apply to ANOVA designs with random effects.

An alternative measure, $\hat{\rho}$ (rho, the intraclass correlation), has been developed as an estimate of population effect size (Vaughan and Corballis, 1969).[11] For a one-way random-effects ANOVA,

$$\hat{\rho} = \frac{\text{MS}_A - \text{MS}_{S/A}}{\text{MS}_A + (n-1)\text{MS}_{S/A}} \tag{11.14}$$

Note that although the test for significance is the same for one-way fixed- and one-way random-effects designs, the effect size estimate is not.

Partial $\hat{\rho}$ is available for statistically reliable effects in a two-way design with both IVs random (Kirk, 1995):

$$\text{Partial } \hat{\rho}_A = \frac{(\text{MS}_A - \text{MS}_{AB})/bn}{\text{MS}_{S/AB} + (\text{MS}_A - \text{MS}_{AB})/bn} \tag{11.15}$$

$$\text{Partial } \hat{\rho}_B = \frac{(\text{MS}_B - \text{MS}_{AB})/an}{\text{MS}_{S/AB} + (\text{MS}_B - \text{MS}_{AB})/an} \tag{11.16}$$

and

$$\text{Partial } \hat{\rho}_{AB} = \frac{(\text{MS}_{AB} - \text{MS}_{S/AB})/n}{\text{MS}_{S/AB} + (\text{MS}_{AB} - \text{MS}_{S/AB})/n} \tag{11.17}$$

Any values that turn out to be negative are changed to 0. Applying this to the interaction (the only reliable effect) in the two-way random-effects design of Section 11.5.3.1 (cf. Table 11.16):

$$\text{Partial } \hat{\rho}_{AB} = \frac{(26.04 - 1.11)/3}{1.11 + (26.04 - 1.11)/3} = .88$$

For a mixed design with A fixed and B random, partial $\hat{\omega}^2$ is as defined in Equation 5.16, partial $\hat{\rho}^2_B$ is as defined in Equation 11.15, and partial $\hat{\rho}_{AB}$ is as defined in Equation 11.16.

Nested models and those with repeated measures are most easily handled by relying on η^2 and partial η^2 of Section 5.6.2, using the applicable error term for partial η^2. Smithson's (2003) software for finding confidence limits for partial η^2 applies to nested models, as well as to fixed-effects models.

11.7 COMPLETE EXAMPLE OF RANDOM-EFFECTS ANOVA

The example involves testing the fat content of dried eggs. Data are published in Bliss (1967) and are available on the Internet. The original article describing the experiment (Mitchell, 1950) was not available to us.

[11] We use the conventional notation, although *correlation* is a misnomer. This is actually a variance/effect size (actually a squared correlation).

Four samples, each drawn from a well-mixed can of dried eggs, were sent to six testing laboratories, deliberately chosen. Samples were arbitrarily labeled either G or H. Two technicians in each laboratory used the same chemical technique to provide two tests of samples labeled G and two tests of samples labeled H. Therefore, laboratory is a fixed IV with six levels; technicians within laboratories is a second, nested, random IV with two levels; and label of sample (G or H) within technician within laboratory is another two-level, nested, random IV. Therefore, this is a doubly nested $6 \times 2 \times 2$ fixed-random-random design with two replicates (i.e., two testings of each sample by each technician). The DV is fat content determined by acid hydrolysis. The metric is percentage of fat minus 41.70% (the latter value is presumably the nominal fat content). The layout of the design is similar to that of Table 11.10, with six levels of A, two levels of B, and two levels of C.

The major effect of interest is differences among laboratories, although differences between technicians within laboratories would likely be of interest to the laboratories. Differences among samples are also potentially interesting because they were randomly selected and arbitrarily labeled from the well-mixed can and, therefore, should differ only by random error. The analysis produces tests of all three main effects (but no interactions, because they are precluded by the nesting arrangement).

11.7.1 Evaluation of Assumptions

The usual distributional assumptions for fixed-effects randomized-groups designs apply to this doubly nested design. Because there are no interactions, cells are not defined for this nested design. Therefore, assumptions are evaluated separately on the margin of each of the three main effects: laboratory, technician, and sample.

The design does include repeated measures for the technicians. Each technician provides two tests of the sample labeled G and two tests of the sample labeled H. Individual differences due to cases are dealt with by including technicians as an IV (nested within laboratories). Despite the use of repeated measures, sphericity is not an issue because each of the repeated-measures factors (label of sample and replicates) has only two levels.

11.7.1.1 Sample Sizes, Normality, and Independence of Errors

Independence of errors for the test of laboratories is ensured by sending random samples to the six laboratories. Nonindependence of errors within the four tests produced by each technician is accounted for by treating technician as an IV, so that individual differences among technicians do not bias any of the other tests.

Table 11.37 shows syntax and descriptive statistics from SAS MEANS analyses. Data must be sorted by a variable before separate statistics are given for each level of that variable. The 24 levels of sample (with only two scores each) are not shown in the table.

This is an equal-n design, with 8 scores in each level of laboratory, 4 scores for each technician, and 24 scores for each label of sample. Normality of sampling

TABLE 11.37 Descriptive statistics for egg-testing data through SAS MEANS

```
proc sort data=SASUSER.EGGS;
  by LAB;
run;
proc means vardef=DF
    N MIN MAX MEAN STD VAR SKEWNESS KURTOSIS;
    var FATCONT;
    by LAB;
run;
proc sort;
   by TECH;
run;
proc means vardef=DF
    N MIN MAX MEAN STD VAR SKEWNESS KURTOSIS;
    var FATCONT;
    by TECH;
run;
```

```
----------------------------------- LAB=1 -----------------------------------

                              The MEANS Procedure

                         Analysis Variable : FATCONT

N       Minimum      Maximum         Mean      Std Dev     Variance      Skewness
---------------------------------------------------------------------------------
8     0.2400000    0.8000000    0.5800000    0.1969046    0.0387714    -0.8799415
---------------------------------------------------------------------------------

                         Analysis Variable : FATCONT

                                   Kurtosis
                                ------------
                                  -0.2925716
                                ------------

----------------------------------- LAB=2 -----------------------------------

                         Analysis Variable : FATCONT

N       Minimum      Maximum         Mean      Std Dev     Variance      Skewness
---------------------------------------------------------------------------------
8     0.1800000    0.4300000    0.3400000    0.0824621    0.0068000    -1.0190572
---------------------------------------------------------------------------------

                         Analysis Variable : FATCONT

                                   Kurtosis
                                ------------
                                   0.7497281
                                ------------

----------------------------------- LAB=3 -----------------------------------

                         Analysis Variable : FATCONT

N       Minimum      Maximum         Mean      Std Dev     Variance      Skewness
---------------------------------------------------------------------------------
8     0.2700000    0.5400000    0.4075000    0.0799553    0.0063929    -0.0656791
---------------------------------------------------------------------------------
```

TABLE 11.37 Continued

```
                     Analysis Variable : FATCONT
                            Kurtosis
                          ------------
                           0.8137434
                          ------------
----------------------------------- LAB=4 -----------------------------------
                     Analysis Variable : FATCONT
N      Minimum      Maximum         Mean      Std Dev     Variance      Skewness
--------------------------------------------------------------------------------
8    0.1800000    0.5300000    0.3762500    0.1082243    0.0117125    -0.5109438
--------------------------------------------------------------------------------
                     Analysis Variable : FATCONT
                            Kurtosis
                          ------------
                           0.4917539
                          ------------
----------------------------------- LAB=5 -----------------------------------
                     Analysis Variable : FATCONT
N      Minimum      Maximum         Mean      Std Dev     Variance      Skewness
--------------------------------------------------------------------------------
8    0.2000000    0.4200000    0.3537500    0.0690626    0.0047696    -1.8081903
--------------------------------------------------------------------------------
                     Analysis Variable : FATCONT
                            Kurtosis
                          ------------
                           3.9948441
                          ------------
----------------------------------- LAB=6 -----------------------------------
                     Analysis Variable : FATCONT
N      Minimum      Maximum         Mean      Std Dev     Variance      Skewness
--------------------------------------------------------------------------------
8    0.0600000    0.4300000    0.2675000    0.1200893    0.0144214    -0.4125213
--------------------------------------------------------------------------------
                     Analysis Variable : FATCONT
                            Kurtosis
                          ------------
                          -0.2608303
                          ------------
---------------------------------- TECH=1 ----------------------------------
                       The MEANS Procedure
                     Analysis Variable : FATCONT
N      Minimum      Maximum         Mean      Std Dev     Variance      Skewness
--------------------------------------------------------------------------------
4    0.2400000    0.6200000    0.4375000    0.1774589    0.0314917    -0.1346073
--------------------------------------------------------------------------------
```

TABLE 11.37 Continued

```
                          Analysis Variable : FATCONT

                                  Kurtosis
                                ------------
                                  -3.8153829
                                ------------

-------------------------------- TECH=2 --------------------------------

                          Analysis Variable : FATCONT

N     Minimum      Maximum        Mean      Std Dev     Variance      Skewness
-------------------------------------------------------------------------------
4   0.6500000    0.8000000   0.7225000   0.0694622    0.0048250     0.1200936
-------------------------------------------------------------------------------

                          Analysis Variable : FATCONT

                                  Kurtosis
                                ------------
                                  -3.6286612
                                ------------

-------------------------------- TECH=3 --------------------------------

                          Analysis Variable : FATCONT

N     Minimum      Maximum        Mean      Std Dev     Variance      Skewness
-------------------------------------------------------------------------------
4   0.3000000    0.4300000   0.3650000   0.0602771    0.0036333  1.962811E-15
-------------------------------------------------------------------------------

                          Analysis Variable : FATCONT

                                  Kurtosis
                                ------------
                                  -3.7274640
                                ------------

-------------------------------- TECH=4 --------------------------------

                            The MEANS Procedure

                          Analysis Variable : FATCONT

N     Minimum      Maximum        Mean      Std Dev     Variance      Skewness
-------------------------------------------------------------------------------
4   0.1800000    0.4000000   0.3150000   0.1027943    0.0105667    -0.8838208
-------------------------------------------------------------------------------

                          Analysis Variable : FATCONT

                                  Kurtosis
                                ------------
                                  -1.0693111
                                ------------

-------------------------------- TECH=5 --------------------------------

                          Analysis Variable : FATCONT

N     Minimum      Maximum        Mean      Std Dev     Variance      Skewness
-------------------------------------------------------------------------------
4   0.2700000    0.4600000   0.3700000   0.0778888    0.0060667    -0.3809319
-------------------------------------------------------------------------------
```

TABLE 11.37 Continued

Analysis Variable : FATCONT

Kurtosis
1.5000000

------------------------------------ TECH=6 ------------------------------------

Analysis Variable : FATCONT

N	Minimum	Maximum	Mean	Std Dev	Variance	Skewness
4	0.3700000	0.5400000	0.4450000	0.0714143	0.0051000	0.7687812

Analysis Variable : FATCONT

Kurtosis
1.0434448

------------------------------------ TECH=7 ------------------------------------

The MEANS Procedure

Analysis Variable : FATCONT

N	Minimum	Maximum	Mean	Std Dev	Variance	Skewness
4	0.1800000	0.5300000	0.3750000	0.1571623	0.0247000	−0.5152103

Analysis Variable : FATCONT

Kurtosis
−1.9514170

------------------------------------ TECH=8 ------------------------------------

Analysis Variable : FATCONT

N	Minimum	Maximum	Mean	Std Dev	Variance	Skewness
4	0.3100000	0.4300000	0.3775000	0.0512348	0.0026250	−0.7528372

Analysis Variable : FATCONT

Kurtosis
0.3428571

------------------------------------ TECH=9 ------------------------------------

Analysis Variable : FATCONT

N	Minimum	Maximum	Mean	Std Dev	Variance	Skewness
4	0.3300000	0.3900000	0.3600000	0.0258199	0.000666667	6.054399E−15

TABLE 11.37 Continued

```
                    Analysis Variable : FATCONT

                            Kurtosis
                          ------------
                           -1.2000000
                          ------------

---------------------------------- TECH=10 ----------------------------------

                       The MEANS Procedure

                    Analysis Variable : FATCONT

N      Minimum      Maximum         Mean      Std Dev     Variance     Skewness
-------------------------------------------------------------------------------
4    0.2000000    0.4200000    0.3475000    0.1017759    0.0103583   -1.6327103
-------------------------------------------------------------------------------

                    Analysis Variable : FATCONT

                            Kurtosis
                          ------------
                           2.5381952
                          ------------

---------------------------------- TECH=11 ----------------------------------

                    Analysis Variable : FATCONT

N      Minimum      Maximum         Mean      Std Dev     Variance     Skewness
-------------------------------------------------------------------------------
4    0.2800000    0.4300000    0.3600000    0.0616441    0.0038000   -0.4781263
-------------------------------------------------------------------------------

                    Analysis Variable : FATCONT

                            Kurtosis
                          ------------
                           1.5000000
                          ------------

---------------------------------- TECH=12 ----------------------------------

                    Analysis Variable : FATCONT

N      Minimum      Maximum         Mean      Std Dev     Variance     Skewness
-------------------------------------------------------------------------------
4    0.0600000    0.2600000    0.1750000    0.0838650    0.0070333   -1.0070352
-------------------------------------------------------------------------------

                    Analysis Variable : FATCONT

                            Kurtosis
                          ------------
                           1.8293390
                          ------------
```

distributions is safely assumed for the test of the main effect of samples because $df_{error} = N -$ (number of samples) $= 48 - 24 = 24$ (see Section 11.5.1). However, error degrees of freedom for tests of laboratory and technician are insufficient to assume normality of sampling distributions. Therefore, skewness and kurtosis are examined within each level of laboratory and technician. SAS provides values for these, but not standard errors to use to form tests of them. Standard errors and z tests for skewness and kurtosis are easy to compute, however, using Equations 2.8–2.11.

The standard error for skewness for each of the six levels of laboratory is

$$s_s = \sqrt{\frac{6}{8}} = 0.866$$

and the standard error for kurtosis is

$$s_k = \sqrt{\frac{24}{8}} = 1.732$$

For the first level of laboratory, then, $z = -0.880/0.866 = 1.02$ for skewness and $z = -0.293/1.732 = -0.169$ for kurtosis. At $\alpha = .01$ (an appropriate level for evaluating small samples), critical $z = \pm 2.58$. Therefore, neither skewness nor kurtosis exceeds acceptable limits for normality. The only level of laboratory that approaches the limits is laboratory 5, with $z = 2.09$ for skewness and $z = 2.31$ for kurtosis.

Standard errors for the 12 technicians are 1.225 for skewness and 2.449 for kurtosis, by using Equations 2.8 and 2.9. Using these errors to form z tests (Equations 2.10 and 2.11), there is no concern about departure from normality of sampling distributions for any of the technicians' distributions.

11.7.1.2 Homogeneity of Variance and Outliers

Homogeneity of variance for the test of laboratories is evaluated by calculating the ratio of highest to lowest variance among the six levels: $F_{max} = 0.0388/.0048 = 8.08$. With $F_{max} < 10$ and equal sample sizes in each level, there is no concern about violation of homogeneity of variance.

The largest variance among technicians is 0.0315 and the smallest is 0.0007. This yields $F_{max} = 45$, which seemingly violates the assumption of homogeneity of variance. However, the test of F_{max} is not statistically significant, due to the large number of variances and small number of df for each variance; the critical value for F_{max} with 3 df (four cases in each variance) and 12 variances is 124 at $\alpha = .05$ (Pearson and Hartley, 1958).

A look at maximum and minimum scores for technician 1 reveals that neither is an outlier: $z = (0.24 - 0.4375)/0.1775 = -1.11$ for the minimum score and $z = (0.62 - 0.4375)/0.1775 = 1.03$ for the maximum score. Therefore, there are no outlying scores for the first technician. Evaluation of minimum scores for the other technicians reveals that there are no outliers to cloud the test of technicians within laboratories.

Tests for outliers are also applied to each of the six laboratories. For example, in the first laboratory, $z = (0.24 - 0.58)/0.1969 = -1.73$ for the minimum score and $z = (0.80 - 0.58)/0.1969 = 1.12$ for the maximum score, indicating no outliers at $\alpha = .01$. Tests for the remaining laboratories produce similar results.

Therefore, the only problem identified is the impossibility of testing homogeneity of variance and of evaluating outliers among label of samples within technicians because there are only two replicates. Therefore, the decision is made to set $\alpha = .01$ for the test of label of sample.

11.7.2 ANOVA for Doubly Nested Design

Table 11.38 shows SAS GLM syntax and output for a traditional analysis of the fat content of eggs in this doubly nested design.

Only the test of technicians within laboratories is statistically significant at $\alpha = .05$: $F(6, 12) = 3.0954, p = .0453$. Neither laboratory nor label of samples within technicians within laboratories displays statistical significance. Using Smithson's (2003) software, effect size (partial η^2) for technicians within laboratories is .61, with 95% confidence limits from .00 to .69. For the effect of laboratory, partial $\eta^2 = .64$, with 95% confidence limits from .00 to .71. For the effect of samples, partial $\eta^2 = .48$, with confidence limits from .00 to .50.

Means and standard deviations for the 12 technicians are in the runs producing Table 11.37. With four samples per technician, standard error is calculated as `Std Dev`/$\sqrt{4}$. Means and standard errors are in Table 11.39 in a form appropriate for a journal report.

TABLE 11.38 Traditional ANOVA on fat content of eggs in doubly nested design (SAS GLM syntax and selected output)

```
proc glm data=SASUSER.EGGS;
    class LAB TECH SAMPLE;
    model FATCONT = LAB TECH(LAB) SAMPLE(TECH LAB)   ;
    random TECH(LAB) SAMPLE(TECH LAB) /TEST ;
run;
```

```
                          The GLM Procedure
                      Class Level Information

    Class        Levels   Values

    LAB               6   1 2 3 4 5 6
    TECH             12   1 2 3 4 5 6 7 8 9 10 11 12
    SAMPLE           24   1 2 3 4 5 6 7 8 9 10 11 12 13 14 15 16 17 18 19 20 21 22 23 24

                      Number of Observations Read         48
                      Number of Observations Used         48

                          The GLM Procedure

Dependent Variable: FATCONT

                                         Sum of
     Source                    DF       Squares    Mean Square    F Value    Pr > F

     Model                     23    0.85040000     0.03697391       5.14    <.0001
     Error                     24    0.17270000     0.00719583
     Corrected Total           47    1.02310000

              R-Square     Coeff Var      Root MSE    FATCONT Mean

              0.831199      21.89116      0.084828        0.387500
```

TABLE 11.38 Continued

```
     Source                 Type III Expected Mean Square

     LAB                    Var(Error) + 2 Var(SAMPLE(LAB*TECH)) + 4 Var(TECH(LAB)) + Q(LAB)
     TECH(LAB)              Var(Error) + 2 Var(SAMPLE(LAB*TECH)) + 4 Var(TECH(LAB))
     SAMPLE(LAB*TECH)       Var(Error) + 2 Var(SAMPLE(LAB*TECH))

                Tests of Hypotheses for Mixed Model Analysis of Variance

Dependent Variable: FATCONT

     Source                            DF    Type III SS    Mean Square    F Value    Pr > F

     LAB                                5       0.443025       0.088605       2.15    0.1895
     Error: MS(TECH(LAB))               6       0.247475       0.041246

     Source                            DF    Type III SS    Mean Square    F Value    Pr > F

     TECH(LAB)                          6       0.247475       0.041246       3.10    0.0453
     Error                             12       0.159900       0.013325

     Error: MS(SAMPLE(LAB*TECH))

     Source                            DF    Type III SS    Mean Square    F Value    Pr > F

     SAMPLE(LAB*TECH)                  12       0.159900       0.013325       1.85    0.0962
     Error: MS(Error)                  24       0.172700       0.007196
```

TABLE 11.39 Means and standard errors for twelve technicians in six laboratories

Lab	Technician	Mean	Standard Error
1	1	0.4375	0.0887
	2	0.7225	0.0347
2	3	0.3650	0.0301
	4	0.3150	0.0514
3	5	0.3700	0.0389
	6	0.4450	0.0357
4	7	0.3750	0.0786
	8	0.3775	0.0256
5	9	0.3600	0.0129
	10	0.3475	0.0509
6	11	0.3600	0.0308
	12	0.1750	0.0419

TABLE 11.40 Checklist for doubly nested analysis of variance

1. Issues
   a. Independence of errors
   b. Unequal sample sizes and missing data
   c. Normality of sampling distributions
   d. Outliers
   e. Homogeneity of variance
2. Major analyses (planned comparisons or omnibus *F*s). If statistically significant:
   a. Effect size and confidence limits
   b. Parameter estimates (means and standard deviations or standard errors or confidence intervals for each group)
3. Additional analyses
   a. Post hoc comparisons
   b. Interpretation of departure from homogeneity of variance, if appropriate

Tukey tests are used to compare all pairs of technicians. SAS GLM does not generate Tukey tests for this design. Therefore, the tests are done by hand using $MS_{SAMPLE(LAB*TECH)} = 0.013325$, from the final source table of Table 11.38. The easiest way to do this uses Equation 4.16, modified to incorporate the current error term.

$$\bar{d}_T = q_T\sqrt{\frac{MS_{SAMPLE(LAB*TECH)}}{n}} = 5.62\sqrt{\frac{0.013325}{4}} = 0.324$$

with number of means = 12, $df_{SAMPLE(LAB*TECH)} = 12$, and $\alpha = .05$.

Technician 2 reports the higher fat content—significantly higher than most of the other technicians, but not significantly higher than the fat content reported by technicians 1 and 6. None of the other differences among technicians is statistically significant.

Table 11.40 is a checklist of items to consider for doubly nested ANOVA. An example of a results section in journal format follows.

Results

A 6 × 2 × 2 doubly nested fixed-random-random ANOVA was performed on fat content (percentage of fat minus 41.70%) of samples from a single well-mixed can of dried eggs. Six laboratories, deliberately chosen, were provided with four samples each, two arbitrarily labeled G and two arbitrarily labeled H. Within each laboratory, two technicians were given two of each sample (G and H) to test. Thus, technicians were nested within laboratories, and samples (G and H) were nested within technicians within laboratories. There were two replicates for each sample.

All assumptions evaluated were met. However, assumptions about normality and homogeneity of variance could not be evaluated for the test of sample labels because of the small sample size ($n = 2$). Therefore, α was set to .01 for that test, with $\alpha = .05$ for the tests of laboratory and technicians within laboratories. There were no outlying cases within levels of laboratory or within measurements of technicians. Outliers could not be evaluated among samples.

No differences were found among laboratories, nor among samples within technicians within laboratories. However, technicians within laboratories were found to differ, $F(6, 12) = 3.10$, $p = .0453$, partial $\eta^2 = .61$, with confidence limits from .00 to .69. For the effect of laboratory, partial $\eta^2 = .64$, with 95% confidence limits from .00 to .71. For the effect of samples, partial $\eta^2 = .48$, with confidence limits from .00 to .50.

A post hoc Tukey test showed that differences among technicians were largely due to one technician whose reports of fat content (mean = 0.72) were higher than those of all except two of the other technicians; means for the other 11 technicians ranged from 0.18 to 0.45, as shown in Table 11.38. The remaining technicians did not differ from one another.

11.8 COMPARISON OF PROGRAMS

SPSS and SAS have a variety of procedures for handling random-effects designs. Although SYSTAT is not designed for random-effects analysis, it may be used to analyze some designs with some hand calculation.

Table 11.41 shows features of programs that use the traditional ANOVA approach. Only those features relevant to random effects are noted in the table. Other features appear in the Comparison of Programs sections of other chapters. Table 11.42 shows features of programs that use variance components or maximum-likelihood approaches.

11.8.1 SAS System

SAS has a comprehensive set of programs for analyzing random-effects designs. PROC GLM makes available all of its many features to traditional random and mixed fixed-random-effects ANOVA, as well as nested designs. About the only shortcoming is the lack of ability to specify trend analysis with unequal spacing to randomized-groups effects.

Maximum-likelihood methods are available through VARCOMP and MIXED. The latter is the more recent program and subsumes the former, except that VARCOMP has one method (and not a very popular one) that is unavailable in MIXED.

The newer program is extensive and meant to be at least as comprehensive as SAS GLM, with highly developed facilities for both repeated-measures and random-effects designs. An alternative covariance structure may be specified when the

TABLE 11.41 Comparison of SPSS, SYSTAT, and SAS programs for random-effects ANOVA

Feature	SPSS GLM (UNIANOVA)	SAS GLM	SYSTAT GLM
Input			
Special specification for trend (polynomial) analysis with unequal spacing	Polynomial	No	Yes
Indicate random and/or fixed effects	Yes	Yes	Fixed only
Indicate nesting arrangements	Yes	Yes	Yes
Random interactions automatically identified	Yes	No	N.A.
Output			
ANOVA table	Yes	Yes	Yes
Expected mean squares	Yes	Yes	No
Error terms shown for tests of effects	Yes	Yes	No

TABLE 11.42 Comparison of SPSS and SAS programs for alternative approaches to random-effects designs

Feature	SPSS GLM (VARCOMP)	SPSS MIXED	SAS VARCOMP	SAS MIXED
Input				
Indicate random and/or fixed effects	Yes	Yes	Yes	Yes
Random interactions automatically identified	Yes	Yes	Yes	Yes
Indicate nesting arrangements	Yes	Yes	Yes	Yes
Estimation methods				
Maximum likelihood	ML	Yes	ML	ML
Restricted maximum likelihood	REML	Yes	REML	REML
Minimum norm unbiased with 0 weight for random effects	MINQUE(0)	No	MIVQUE0	MIVQUE0
Minimum norm unbiased with unit weight for random effects	MINQUE(1)	No	No	No
ANOVA with Type I SS	SSTYPE(1)	SSTYPE(1)	TYPE1	HTYPE=1[a]
ANOVA with Type III SS	SSTYPE(3)	SSTYPE(3)	No	HTYPE=3[a]
Option to exclude intercept	EXCLUDE	Yes	No	NOINT
Specify covariates	No	Yes	Yes	Yes[b]
Specify hypothesis tests and/or model	No	Yes	No	COVTEST
Specify contrasts	No	No	No	CONTRAST, ESTIMATE
Specify convergence criterion	CONVERGE	Yes	No	Various
Specify maximum number of iterations	ITERATE	Yes	MAXITER	MAXITER
Specify maximum number of likelihood evaluations	No	No	No	MAXFUNC
Specify epsilon value for tolerance	EPS	No	EPSILON	SINGULAR
Request iteration information	HISTORY	Yes	Default	ITDETAILS
Specify α level for tests or confidence limits	No	Yes	No	ALPHA
Specify initial estimates of parameters	No	No	No	PARMS
Sampling based Bayesian analysis	No	No	No	PRIOR
Optional covariance structure for repeated-measures and random factors	No	TYPE	No	TYPE
Specify denominator df for fixed effects	No	No	No	DDF
Specify method for computing denominator df for fixed effects	No	No	No	DDFM
Specify random coefficients model for multilevel modeling	No	Yes	No	Yes

TABLE 11.42 Continued

Feature	SPSS GLM (VARCOMP)	SPSS MIXED	SAS VARCOMP	SAS MIXED
Output				
Estimates of variance components	Yes	Yes	Yes	Yes
Asymptotic variance-covariance matrix	Yes	Yes	Yes	ASYCOV
Confidence limits for parameter estimates	No	Yes	No	CL
Standard errors and Wald tests for parameter estimates	No	Yes	No	COVTEST
Means for fixed effects	No	Yes	No	LSMEANS
Multiple comparisons for fixed effects	No	No	No	Several
Correlation matrix of fixed effect parameter estimates	No	Yes	No	CORRB
Covariance matrix of fixed effect parameter estimates	No	Yes	No	COVB
Multivariate tests for repeated-measures effects	No	No	No	Yes
Residuals analysis	No	Yes	No	Yes
Imputes missing values with repeated measures	No	No	No	Yes
Save to data file				
Variance components estimates	VAREST	No	No[c]	MAKE
Asymptotic variance-covariance matrix	COVB	No	No[c]	MAKE
Asymptotic correlation matrix	CORB	No	No[c]	MAKE

[a] For fixed effects.
[b] Variables not included in the CLASS instruction are treated as continuous covariates.
[c] Any printed output may be saved to data sets through PROC OUTPUT.

assumption of homogeneity of covariance is not met in either repeated measures or random-effects analysis. The program provides many GLM-type features that are unavailable in other variance components programs, such as least squares means, contrasts, and post hoc tests of differences among means. The program has capability for random coefficients and multilevel modeling (cf. Section 11.6.2). Anything that can be printed out can also be saved to a data file through the MAKE procedure. Indeed, Table 11.42 highlights only some of the more prominent features of PROC MIXED; many others are documented in the on-disk manual.

11.8.2 SPSS Package

SPSS GLM can analyze all types of random effects. Traditional ANOVA is done through the UNIANOVA module, and maximum-likelihood procedures are implemented in the VARCOMP module. UNIANOVA implements all the features of fixed-effects analysis for random effects—covariates, comparisons among means, and the like. VARCOMP provides only the basics for a random-effects design, with no information about fixed effects in a mixed random-fixed design.

SPSS has recently added MIXED MODELS, similar to SAS MIXED, with highly developed facilities for both repeated-measures and random-effects designs. There is flexibility to specify an alternative covariance structure when the assumption of homogeneity of covariance is not met in either repeated-measures or random-effects analysis. The program provides least squares means and has capability for random coefficients and multilevel modeling (cf. Section 11.6.2). Table 11.42 highlights only some of the more prominent features of SPSS MIXED; many others are documented in the on-disk manual.

11.8.3 SYSTAT System

SYSTAT GLM analyzes nested effects but does not currently have facilities for random-effects ANOVA. However, random-effects designs may be analyzed using the traditional approach and then recalculating F ratios using the appropriate error term. Other random-effects designs may be similarly analyzed if you are willing to compute expected mean squares and degrees of freedom and then apply them to F ratios by hand. Maximum-likelihood estimation is available in the nonlinear regression module, but its use requires skills beyond the scope of this book.

11.9 PROBLEM SETS

11.1 For the data in Problem 5.1, assume that presoak time is a random IV with levels = 7, 16, 39, and 45 minutes.

a. Test all assumptions for a mixed fixed-random ANOVA and do the analysis.

b. Provide a source table.

c. Calculate effect size(s).

d. Do a trend analysis and apply an appropriate post hoc adjustment.

e. Write a results section for the analyses.

f. Use one of the statistical software programs to do a variance components (maximum-likelihood) analysis of this data set.

11.2 Do an analysis of variance on the following data set, in which organization is nested within treatments and the DV is minority employee salary.

Affirmative Action Treatment

Org1	Org2	Org3
34437	31734	23605
37641	28947	30621
30046	25695	20736
32755	21747	23596

No Treatment Control

Org4	Org5	Org6
22189	31037	16241
15874	21550	16962
18630	25390	21673
20891	25517	22972

a. Test all relevant assumptions.
b. Provide a source table with pooled error, if appropriate.
c. Calculate effect size(s) and confidence limits.
d. Write a results section for the analysis.

Extending the Concepts

11.3 Show the table of expected mean squares for a doubly nested design.

APPENDIX A

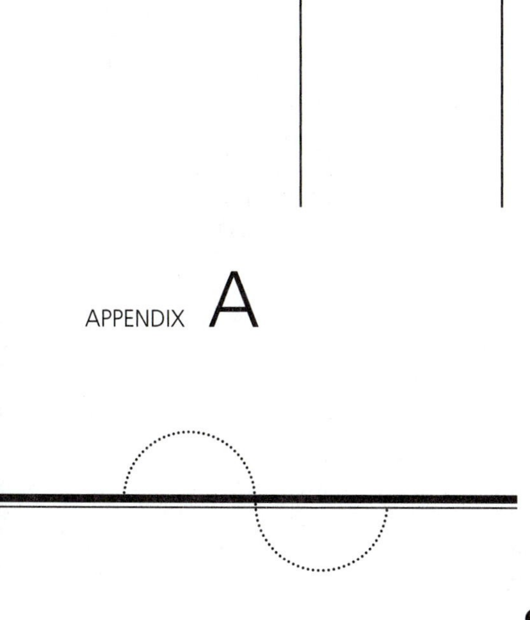

Statistical Tables

TABLE A.1 Critical Values of F Distribution

df_2 \ df_1		1	2	3	4	5	6	8	12	24	∞
1	0.1%	405284	500000	540379	562500	576405	585937	598144	610667	623497	636619
	0.5%	16211	20000	21615	22500	23056	23437	23925	24426	24940	25465
	1%	4052	4999	5403	5625	5764	5859	5981	6106	6234	6366
	2.5%	647.79	799.50	864.16	899.58	921.85	937.11	956.66	976.71	997.25	1018.30
	5%	161.45	199.50	215.71	224.58	230.16	233.99	238.88	243.91	249.05	254.32
	10%	39.86	49.50	53.59	55.83	57.24	58.20	59.44	60.70	62.00	63.33
	20%	9.47	12.00	13.06	13.73	14.01	14.26	14.59	14.90	15.24	15.58
	25%	5.83	7.50	8.20	8.58	8.82	8.98	9.19	9.41	9.63	9.85
2	0.1	998.5	999.0	999.2	999.2	999.3	999.3	999.4	999.4	999.5	999.5
	0.5	198.50	199.00	199.17	199.25	199.30	199.33	199.37	199.42	199.46	199.51
	1	98.49	99.00	99.17	99.25	99.30	99.33	99.36	99.42	99.46	99.50
	2.5	38.51	39.00	39.17	39.25	39.30	39.33	39.37	39.42	39.46	39.50
	5	18.51	19.00	19.16	19.25	19.30	19.33	19.37	19.41	19.45	19.50
	10	8.53	9.00	9.16	9.24	9.29	9.33	9.37	9.41	9.45	9.49
	20	3.56	4.00	4.16	4.24	4.28	4.32	4.36	4.40	4.44	4.48
	25	2.56	3.00	3.15	3.23	3.28	3.31	3.35	3.39	3.44	3.48
3	0.1	167.5	148.5	141.1	137.1	134.6	132.8	130.6	128.3	125.9	123.5
	0.5	55.55	49.80	47.47	46.20	45.39	44.84	44.13	43.39	42.62	41.83
	1	34.12	30.81	29.46	28.71	28.24	27.91	27.49	27.05	26.60	26.12
	2.5	17.44	16.04	15.44	15.10	14.89	14.74	14.54	14.34	14.12	13.90
	5	10.13	9.55	9.28	9.12	9.01	8.94	8.84	8.74	8.64	8.53
	10	5.54	5.46	5.39	5.34	5.31	5.28	5.25	5.22	5.18	5.13
	20	2.68	2.89	2.94	2.96	2.97	2.97	2.98	2.98	2.98	2.98
	25	2.02	2.28	2.36	2.39	2.41	2.42	2.44	2.45	2.46	2.47
4	0.1	74.14	61.25	56.18	53.44	51.71	50.53	49.00	47.41	45.77	44.05
	0.5	31.33	26.28	24.26	23.16	22.46	21.98	21.35	20.71	20.03	19.33
	1	21.20	18.00	16.69	15.98	15.52	15.21	14.80	14.37	13.93	13.46
	2.5	12.22	10.65	9.98	9.60	9.36	9.20	8.98	8.75	8.51	8.26
	5	7.71	6.94	6.59	6.39	6.26	6.16	6.04	5.91	5.77	5.63
	10	4.54	4.32	4.19	4.11	4.05	4.01	3.95	3.90	3.83	3.76
	20	2.35	2.47	2.48	2.48	2.48	2.47	2.47	2.46	2.44	2.43
	25	1.81	2.00	2.05	2.05	2.07	2.08	2.08	2.08	2.08	2.08
5	0.1	47.04	36.61	33.20	31.09	29.75	28.84	27.64	26.42	25.14	23.78
	0.5	22.79	18.31	16.53	15.56	14.94	14.51	13.96	13.38	12.78	12.14
	1	16.26	13.27	12.06	11.39	10.97	10.67	10.29	9.89	9.47	9.02
	2.5	10.01	8.43	7.76	7.39	7.15	6.98	6.76	6.52	6.28	6.02
	5	6.61	5.79	5.41	5.19	5.05	4.95	4.82	4.68	4.53	4.36
	10	4.06	3.78	3.62	3.52	3.45	3.40	3.34	3.27	3.19	3.10
	20	2.18	2.26	2.25	2.24	2.23	2.22	2.20	2.18	2.16	2.13
	25	1.70	1.85	1.89	1.89	1.89	1.89	1.89	1.89	1.88	1.87

TABLE A.1 Continued

df_2 \ df_1		1	2	3	4	5	6	8	12	24	∞
6	0.1%	35.51	27.00	23.70	21.90	20.81	20.03	19.03	17.99	16.89	15.75
	0.5%	18.64	14.54	12.92	12.03	11.46	11.07	10.57	10.03	9.47	8.88
	1%	13.74	10.92	9.78	9.15	8.75	8.47	8.10	7.72	7.31	6.88
	2.5%	8.81	7.26	6.60	6.23	5.99	5.82	5.60	5.37	5.12	4.85
	5%	5.99	5.14	4.76	4.53	4.39	4.28	4.15	4.00	3.84	3.67
	10%	3.78	3.46	3.29	3.18	3.11	3.05	2.98	2.90	2.82	2.72
	20%	2.07	2.13	2.11	2.09	2.08	2.06	2.04	2.02	1.99	1.95
	25%	1.62	1.76	1.78	1.79	1.79	1.78	1.78	1.77	1.75	1.74
7	0.1	29.22	21.69	18.77	17.19	16.21	15.52	14.63	13.71	12.73	11.69
	0.5	16.24	12.40	10.88	10.05	9.52	9.16	8.68	8.18	7.65	7.08
	1	12.25	9.55	8.45	7.85	7.46	7.19	6.84	6.47	6.07	5.65
	2.5	8.07	6.54	5.89	5.52	5.29	5.12	4.90	4.67	4.42	4.14
	5	5.59	4.74	4.35	4.12	3.97	3.87	3.73	3.57	3.41	3.23
	10	3.59	3.26	3.07	2.96	2.88	2.83	2.75	2.67	2.58	2.47
	20	2.00	2.04	2.02	1.99	1.97	1.96	1.93	1.91	1.87	1.83
	25	1.57	1.70	1.72	1.72	1.71	1.71	1.70	1.68	1.67	1.65
8	0.1	25.42	18.49	15.83	14.39	13.49	12.86	12.04	11.19	10.30	9.34
	0.5	14.69	11.04	9.60	8.81	8.30	7.95	7.50	7.01	6.50	5.95
	1	11.26	8.65	7.59	7.01	6.63	6.37	6.03	5.67	5.28	4.86
	2.5	7.57	6.06	5.42	5.05	4.82	4.65	4.43	4.20	3.95	3.67
	5	5.32	4.46	4.07	3.84	3.69	3.58	3.44	3.28	3.12	2.93
	10	3.46	3.11	2.92	2.81	2.73	2.67	2.59	2.50	2.40	2.29
	20	1.95	1.98	1.95	1.92	1.90	1.88	1.86	1.83	1.79	1.74
	25	1.53	1.66	1.67	1.55	1.66	1.65	1.64	1.62	1.60	1.58
9	0.1	22.86	16.39	13.90	12.56	11.71	11.13	10.37	9.57	8.72	7.81
	0.5	13.61	10.11	8.72	7.96	7.47	7.13	6.69	6.23	5.73	5.19
	1	10.56	8.02	6.99	6.42	6.06	5.80	5.47	5.11	4.73	4.31
	2.5	7.21	5.71	5.08	4.72	4.48	4.32	4.10	3.87	3.61	3.33
	5	5.12	4.26	3.86	3.63	3.48	3.37	3.23	3.07	2.90	2.71
	10	3.36	3.01	2.81	2.69	2.61	2.55	2.47	2.38	2.28	2.16
	20	1.91	1.94	1.90	1.87	1.85	1.83	1.80	1.76	1.72	1.67
	25	1.51	1.62	1.63	1.63	1.62	1.61	1.60	1.58	1.56	1.53
10	0.1	21.04	14.91	12.55	11.28	10.48	9.92	9.20	8.45	7.64	6.76
	0.5	12.83	9.43	8.08	7.34	6.87	6.54	6.12	5.66	5.17	4.64
	1	10.04	7.56	6.55	5.99	5.64	5.39	5.06	4.71	4.33	3.91
	2.5	6.94	5.46	4.83	4.47	4.24	4.07	3.85	3.62	3.37	3.08
	5	4.96	4.10	3.71	3.48	3.33	3.22	3.07	2.91	2.74	2.54
	10	3.28	2.92	2.73	2.61	2.52	2.46	2.38	2.28	2.18	2.06
	20	1.88	1.90	1.86	1.83	1.80	1.78	1.75	1.72	1.66	1.62
	25	1.49	1.60	1.60	1.60	1.59	1.58	1.56	1.54	1.52	1.48

(continued)

TABLE A.1 Continued

df_2	df_1	1	2	3	4	5	6	8	12	24	∞
11	0.1%	19.69	13.81	11.56	10.35	9.58	9.05	8.35	7.63	6.85	6.00
	0.5%	12.23	8.91	7.60	6.88	6.42	6.10	5.68	5.24	4.76	4.23
	1%	9.65	7.20	6.22	5.67	5.32	5.07	4.74	4.40	4.02	3.60
	2.5%	6.72	5.26	4.63	4.28	4.04	3.88	3.66	3.43	3.17	2.88
	5%	4.84	3.98	3.59	3.36	3.20	3.09	2.95	2.79	2.61	2.40
	10%	3.23	2.86	2.66	2.54	2.45	2.39	2.30	2.21	2.10	1.97
	20%	1.86	1.87	1.83	1.80	1.77	1.75	1.72	1.68	1.63	1.57
	25%	1.46	1.58	1.58	1.58	1.56	1.55	1.54	1.51	1.49	1.45
12	0.1	18.64	12.97	10.80	9.63	8.89	8.38	7.71	7.00	6.25	5.42
	0.5	11.75	8.51	7.23	6.52	6.07	5.76	5.35	4.91	4.43	3.90
	1	9.33	6.93	5.95	5.41	5.06	4.82	4.50	4.16	3.78	3.36
	2.5	6.55	5.10	4.47	4.12	3.89	3.73	3.51	3.28	3.02	2.72
	5	4.75	3.88	3.49	3.26	3.11	3.00	2.85	2.69	2.50	2.30
	10	3.18	2.81	2.61	2.48	2.39	2.33	2.24	2.15	2.04	1.90
	20	1.84	1.85	1.80	1.77	1.74	1.72	1.69	1.63	1.60	1.54
	25	1.46	1.56	1.56	1.55	1.54	1.53	1.51	1.49	1.46	1.42
13	0.1	17.81	12.31	10.21	9.07	8.35	7.86	7.21	6.52	5.78	4.97
	0.5	11.37	8.19	6.93	6.23	5.79	5.48	5.08	4.64	4.17	3.65
	1	9.07	6.70	5.74	5.20	4.86	4.62	4.30	3.96	3.59	3.16
	2.5	6.41	4.97	4.35	4.00	3.77	3.60	3.39	3.15	2.89	2.60
	5	4.67	3.80	3.41	3.18	3.02	2.92	2.77	2.60	2.42	2.21
	10	3.14	2.76	2.56	2.43	2.35	2.28	2.20	2.10	1.98	1.85
	20	1.82	1.83	1.78	1.75	1.72	1.69	1.66	1.62	1.57	1.51
	25	1.45	1.55	1.55	1.53	1.52	1.51	1.49	1.47	1.44	1.40
14	0.1	17.14	11.78	9.73	8.62	7.92	7.43	6.80	6.13	5.41	4.60
	0.5	11.06	7.92	6.68	6.00	5.56	5.26	4.86	4.43	3.96	3.44
	1	8.86	6.51	5.56	5.03	4.69	4.46	4.14	3.80	3.43	3.00
	2.5	6.30	4.86	4.24	3.89	3.66	3.50	3.29	3.05	2.79	2.49
	5	4.60	3.74	3.34	3.11	2.96	2.85	2.70	2.53	2.35	2.13
	10	3.10	2.73	2.52	2.39	2.31	2.24	2.15	2.05	1.94	1.80
	20	1.81	1.81	1.76	1.73	1.70	1.67	1.64	1.60	1.55	1.48
	25	1.44	1.53	1.53	1.52	1.51	1.50	1.48	1.45	1.42	1.38
15	0.1	16.59	11.34	9.34	8.25	7.57	7.09	6.47	5.81	5.10	4.31
	0.5	10.80	7.70	6.48	5.80	5.37	5.07	4.67	4.25	3.79	3.26
	1	8.68	6.36	5.42	4.89	4.56	4.32	4.00	3.67	3.29	2.87
	2.5	6.20	4.77	4.15	3.80	3.58	3.41	3.20	2.96	2.70	2.40
	5	4.54	3.68	3.29	3.06	2.90	2.79	2.64	2.48	2.29	2.07
	10	3.07	2.70	2.49	2.36	2.27	2.21	2.12	2.02	1.90	1.76
	20	1.80	1.79	1.75	1.71	1.68	1.66	1.62	1.58	1.53	1.46
	25	1.43	1.52	1.52	1.51	1.49	1.48	1.46	1.44	1.41	1.36

TABLE A.1 Continued

df_2 \ df_1		1	2	3	4	5	6	8	12	24	∞
16	0.1%	16.12	10.97	9.00	7.94	7.27	6.81	6.19	5.55	4.85	4.06
	0.5%	10.58	7.51	6.30	5.64	5.21	4.91	4.52	4.10	3.64	3.11
	1%	8.53	6.23	5.29	4.77	4.44	4.20	3.89	3.55	3.18	2.75
	2.5%	6.12	4.69	4.08	3.73	3.50	3.34	3.12	2.89	2.63	2.32
	5%	4.49	3.63	3.24	3.01	2.85	2.74	2.59	2.42	2.24	2.01
	10%	3.05	2.67	2.46	2.33	2.24	2.18	2.09	1.99	1.87	1.72
	20%	1.79	1.78	1.74	1.70	1.67	1.64	1.61	1.56	1.51	1.43
	25%	1.42	1.51	1.51	1.50	1.48	1.47	1.45	1.43	1.39	1.34
17	0.1	15.72	10.66	8.73	7.68	7.02	6.56	5.96	5.32	4.63	3.85
	0.5	10.38	7.35	6.16	5.50	5.07	4.78	4.39	3.97	3.51	2.98
	1	8.40	6.11	5.18	4.67	4.34	4.10	3.79	3.45	3.08	2.65
	2.5	6.04	4.62	4.01	3.66	3.44	3.28	3.06	2.82	2.56	2.25
	5	4.45	3.59	3.20	2.96	2.81	2.70	2.55	2.38	2.19	1.96
	10	3.03	2.64	2.44	2.31	2.22	2.15	2.06	1.96	1.84	1.69
	20	1.78	1.77	1.72	1.68	1.65	1.63	1.59	1.55	1.49	1.42
	25	1.42	1.51	1.51	1.49	1.47	1.46	1.44	1.41	1.38	1.33
18	0.1	15.38	10.39	8.49	7.46	6.81	6.35	5.76	5.13	4.45	3.67
	0.5	10.22	7.21	6.03	5.37	4.96	4.66	4.28	3.86	3.40	2.87
	1	8.28	6.01	5.09	4.58	4.25	4.01	3.71	3.37	3.00	2.57
	2.5	5.98	4.56	3.95	3.61	3.38	3.22	3.01	2.77	2.50	2.19
	5	4.41	3.55	3.16	2.93	2.77	2.66	2.51	2.34	2.15	1.92
	10	3.01	2.62	2.42	2.29	2.20	2.13	2.04	1.93	1.81	1.66
	20	1.77	1.76	1.71	1.67	1.64	1.62	1.58	1.53	1.48	1.40
	25	1.41	1.50	1.49	1.48	1.46	1.45	1.43	1.40	1.37	1.32
19	0.1	15.08	10.16	8.28	7.26	6.61	6.18	5.59	4.97	4.29	3.52
	0.5	10.07	7.09	5.92	5.27	4.85	4.56	4.18	3.76	3.31	2.78
	1	8.18	5.93	5.01	4.50	4.17	3.94	3.63	3.30	2.92	2.49
	2.5	5.92	4.51	3.90	3.56	3.33	3.17	2.96	2.72	2.45	2.13
	5	4.38	3.52	3.13	2.90	2.74	2.63	2.48	2.31	2.11	1.88
	10	2.99	2.61	2.40	2.27	2.18	2.11	2.02	1.91	1.79	1.63
	20	1.76	1.75	1.70	1.66	1.63	1.61	1.57	1.52	1.46	1.39
	25	1.41	1.50	1.49	1.48	1.46	1.44	1.42	1.40	1.36	1.31
20	0.1	14.82	9.95	8.10	7.10	6.46	6.02	5.44	4.82	4.15	3.38
	0.5	9.94	6.99	5.82	5.17	4.76	4.47	4.09	3.68	3.22	2.69
	1	8.10	5.85	4.94	4.43	4.10	3.87	3.56	3.23	2.86	2.42
	2.5	5.87	4.46	3.86	3.51	3.29	3.13	2.91	2.68	2.41	2.09
	5	4.35	3.49	3.10	2.87	2.71	2.60	2.45	2.28	2.08	1.84
	10	2.97	2.59	2.38	2.25	2.16	2.09	2.00	1.89	1.77	1.61
	20	1.76	1.75	1.70	1.65	1.62	1.60	1.56	1.51	1.45	1.37
	25	1.40	1.49	1.48	1.47	1.45	1.44	1.42	1.39	1.35	1.29

(continued)

TABLE A.1 Continued

df_2 \ df_1		1	2	3	4	5	6	8	12	24	∞
21	0.1%	14.59	9.77	7.94	6.95	6.32	5.88	5.31	4.70	4.03	3.26
	0.5%	9.83	6.89	5.73	5.09	4.68	4.39	4.01	3.60	3.15	2.61
	1%	8.02	5.78	4.87	4.37	4.04	3.81	3.51	3.17	2.80	2.36
	2.5%	5.83	4.42	3.82	3.48	3.25	3.09	2.87	2.64	2.37	2.04
	5%	4.32	3.47	3.07	2.84	2.68	2.57	2.42	2.25	2.05	1.81
	10%	2.96	2.57	2.36	2.23	2.14	2.08	1.98	1.88	1.75	1.59
	20%	1.75	1.74	1.69	1.65	1.61	1.59	1.55	1.50	1.44	1.36
	25%	1.40	1.49	1.48	1.46	1.44	1.43	1.41	1.38	1.34	1.29
22	0.1	14.38	9.61	7.80	6.81	6.19	5.76	5.19	4.58	3.92	3.15
	0.5	9.73	6.81	5.65	5.02	4.61	4.32	3.94	3.54	3.08	2.55
	1	7.94	5.72	4.82	4.31	3.99	3.76	3.45	3.12	2.75	2.31
	2.5	5.79	4.38	3.78	3.44	3.22	3.05	2.84	2.60	2.33	2.00
	5	4.30	3.44	3.05	2.82	2.66	2.55	2.40	2.23	2.03	1.78
	10	2.95	2.56	2.35	2.22	2.13	2.06	1.97	1.86	1.73	1.57
	20	1.75	1.73	1.68	1.64	1.61	1.58	1.54	1.49	1.43	1.35
	25	1.40	1.48	1.47	1.46	1.44	1.42	1.40	1.37	1.33	1.28
23	0.1	14.19	9.47	7.67	6.69	6.08	5.65	5.09	4.48	3.82	3.05
	0.5	9.63	6.73	5.58	4.95	4.54	4.26	3.88	3.47	3.02	2.48
	1	7.88	5.66	4.76	4.26	3.94	3.71	3.41	3.07	2.70	2.26
	2.5	5.75	4.35	3.75	3.41	3.18	3.02	2.81	2.57	2.30	1.97
	5	4.28	3.42	3.03	2.80	2.64	2.53	2.38	2.20	2.00	1.76
	10	2.94	2.55	2.34	2.21	2.11	2.05	1.95	1.84	1.72	1.55
	20	1.74	1.73	1.68	1.63	1.60	1.57	1.53	1.49	1.42	1.34
	25	1.39	1.47	1.47	1.45	1.43	1.41	1.40	1.37	1.33	1.27
24	0.1	14.03	9.34	7.55	6.59	5.98	5.55	4.99	4.39	3.74	2.97
	0.5	9.55	6.66	5.52	4.89	4.49	4.20	3.83	3.42	2.97	2.43
	1	7.82	5.61	4.72	4.22	3.90	3.67	3.36	3.03	2.66	2.21
	2.5	5.72	4.32	3.72	3.38	3.15	2.99	2.78	2.54	2.27	1.94
	5	4.26	3.40	3.01	2.78	2.62	2.51	2.36	2.18	1.98	1.73
	10	2.93	2.54	2.33	2.19	2.10	2.04	1.94	1.83	1.70	1.53
	20	1.74	1.72	1.67	1.63	1.59	1.57	1.53	1.48	1.42	1.33
	25	1.39	1.47	1.46	1.44	1.43	1.41	1.39	1.36	1.32	1.26
25	0.1	13.88	9.22	7.45	6.49	5.88	5.46	4.91	4.31	3.66	2.89
	0.5	9.48	6.60	5.46	4.84	4.43	4.15	3.78	3.37	2.92	2.38
	1	7.77	5.57	4.68	4.18	3.86	3.63	3.32	2.99	2.62	2.17
	2.5	5.69	4.29	3.69	3.35	3.13	2.97	2.75	2.51	2.24	1.91
	5	4.24	3.38	2.99	2.76	2.60	2.49	2.34	2.16	1.96	1.71
	10	2.92	2.53	2.32	2.18	2.09	2.02	1.93	1.82	1.69	1.52
	20	1.73	1.72	1.66	1.62	1.59	1.56	1.52	1.47	1.41	1.32
	25	1.39	1.47	1.46	1.44	1.42	1.41	1.39	1.36	1.32	1.25

TABLE A.1 Continued

df_2 \ df_1		1	2	3	4	5	6	8	12	24	∞
26	0.1%	13.74	9.12	7.36	6.41	5.80	5.38	4.83	4.24	3.59	2.82
	0.5%	9.41	6.54	5.41	4.79	4.38	4.10	3.73	3.33	2.87	2.33
	1%	7.72	5.53	4.64	4.14	3.82	3.59	3.29	2.96	2.58	2.13
	2.5%	5.66	4.27	3.67	3.33	3.10	2.94	2.73	2.49	2.22	1.88
	5%	4.22	3.37	2.98	2.74	2.59	2.47	2.32	2.15	1.95	1.69
	10%	2.91	2.52	2.31	2.17	2.08	2.01	1.92	1.81	1.68	1.50
	20%	1.73	1.71	1.66	1.62	1.58	1.56	1.52	1.47	1.40	1.31
	25%	1.38	1.46	1.45	1.44	1.42	1.41	1.38	1.35	1.31	1.25
27	0.1	13.61	9.02	7.27	6.33	5.73	5.31	4.76	4.17	3.52	2.75
	0.5	9.34	6.49	5.36	4.74	4.34	4.06	3.69	3.28	2.83	2.29
	1	7.68	5.49	4.60	4.11	3.78	3.56	3.26	2.93	2.55	2.10
	2.5	5.63	4.24	3.65	3.31	3.08	2.92	2.71	2.47	2.19	1.85
	5	4.21	3.35	2.96	2.73	2.57	2.46	2.30	2.13	1.93	1.67
	10	2.90	2.51	2.30	2.17	2.07	2.00	1.91	1.80	1.67	1.49
	20	1.73	1.71	1.66	1.61	1.58	1.55	1.51	1.46	1.40	1.30
	25	1.38	1.46	1.45	1.43	1.42	1.40	1.38	1.35	1.31	1.24
28	0.1	13.50	8.93	7.19	6.25	5.66	5.24	4.69	4.11	3.46	2.70
	0.5	9.28	6.44	5.32	4.70	4.30	4.02	3.65	3.25	2.79	2.25
	1	7.64	5.45	4.57	4.07	3.75	3.53	3.23	2.90	2.52	2.06
	2.5	5.61	4.22	3.63	3.29	2.06	2.90	2.69	2.45	2.17	1.83
	5	4.20	3.34	2.95	2.71	2.56	2.44	2.29	2.12	1.91	1.65
	10	2.89	2.50	2.29	2.16	2.06	2.00	1.90	1.79	1.66	1.48
	20	1.72	1.71	1.65	1.61	1.57	1.55	1.51	1.46	1.39	1.30
	25	1.38	1.46	1.45	1.43	1.41	1.40	1.38	1.34	1.30	1.24
29	0.1	13.39	8.85	7.12	6.19	5.59	5.18	4.64	4.05	3.41	2.64
	0.5	9.23	6.40	5.28	4.66	4.26	3.98	3.61	3.21	2.76	2.21
	1	7.60	5.42	4.54	4.04	3.73	3.50	3.20	2.87	2.49	2.03
	2.5	5.59	4.20	3.61	3.27	3.04	2.88	2.67	2.43	2.15	1.81
	5	4.18	3.33	2.93	2.70	2.54	2.43	2.28	2.10	1.90	1.64
	10	2.89	2.50	2.28	2.15	2.06	1.99	1.89	1.78	1.65	1.47
	20	1.72	1.70	1.65	1.60	1.57	1.54	1.50	1.45	1.39	1.29
	25	1.38	1.45	1.45	1.43	1.41	1.40	1.37	1.34	1.30	1.23
30	0.1	13.29	8.77	7.05	6.12	5.63	5.12	4.58	4.00	3.36	2.59
	0.5	9.18	6.35	5.24	4.62	4.23	3.95	3.58	3.18	2.73	2.18
	1	7.56	5.39	4.51	4.02	3.70	3.47	3.17	2.84	2.47	2.01
	2.5	5.57	4.18	3.59	3.25	3.03	2.87	2.65	2.41	2.14	1.79
	5	4.17	3.32	2.92	2.69	2.53	2.42	2.27	2.09	1.89	1.62
	10	2.88	2.49	2.28	2.14	2.05	1.98	1.88	1.77	1.64	1.46
	20	1.72	1.70	1.64	1.60	1.57	1.54	1.50	1.45	1.38	1.28
	25	1.38	1.45	1.44	1.42	1.41	1.39	1.37	1.34	1.29	1.23

(continued)

TABLE A.1 Continued

df_2 \ df_1		1	2	3	4	5	6	8	12	24	∞
40	0.1%	12.61	8.25	6.60	5.70	5.13	4.73	4.21	3.64	3.01	2.23
	0.5%	8.83	6.07	4.98	4.37	3.99	3.71	3.35	2.95	2.50	1.93
	1%	7.31	5.18	4.31	3.83	3.51	3.29	2.99	2.66	2.29	1.80
	2.5%	5.42	4.05	3.46	3.13	2.90	2.74	2.53	2.29	2.01	1.64
	5%	4.08	3.23	2.84	2.61	2.45	2.34	2.18	2.00	1.79	1.51
	10%	2.84	2.44	2.23	2.09	2.00	1.93	1.83	1.71	1.57	1.38
	20%	1.70	1.68	1.62	1.57	1.54	1.51	1.47	1.41	1.34	1.24
	25%	1.36	1.44	1.42	1.41	1.39	1.37	1.35	1.31	1.27	1.19
60	0.1	11.97	7.76	6.17	5.31	4.76	4.37	3.87	3.31	2.69	1.90
	0.5	8.49	5.80	4.73	4.14	3.76	3.49	3.13	2.74	2.29	1.69
	1	7.08	4.98	4.13	3.65	3.34	3.12	2.82	2.50	2.12	1.60
	2.5	5.29	3.93	3.34	3.01	2.79	2.63	2.41	2.17	1.88	1.48
	5	4.00	3.15	2.76	2.52	2.37	2.25	2.10	1.92	1.70	1.39
	10	2.79	2.39	2.18	2.04	1.95	1.87	1.77	1.66	1.51	1.29
	20	1.68	1.65	1.59	1.55	1.51	1.48	1.44	1.38	1.31	1.18
	25	1.35	1.42	1.41	1.39	1.37	1.35	1.32	1.29	1.24	1.15
120	0.1	11.38	7.31	5.79	4.95	4.42	4.04	3.55	3.02	2.40	1.56
	0.5	8.18	5.54	4.50	3.92	3.55	3.28	2.93	2.54	2.09	1.43
	1	6.85	4.79	3.95	3.48	3.17	2.96	2.66	2.34	1.95	1.38
	2.5	5.15	3.80	3.23	2.89	2.67	2.52	2.30	2.05	1.76	1.31
	5	3.92	3.07	2.68	2.45	2.29	2.17	2.02	1.83	1.61	1.25
	10	2.75	2.35	2.13	1.99	1.90	1.82	1.72	1.60	1.45	1.19
	20	1.66	1.63	1.57	1.52	1.48	1.45	1.41	1.35	1.27	1.12
	25	1.34	1.40	1.39	1.37	1.35	1.33	1.30	1.26	1.21	1.10
∞	0.1	10.83	6.91	5.42	4.62	4.10	3.74	3.27	2.74	2.13	1.00
	0.5	7.88	5.30	4.28	3.72	3.35	3.09	2.74	2.36	1.90	1.00
	1	6.64	4.60	3.78	3.32	3.02	2.80	2.51	2.18	1.79	1.00
	2.5	5.02	3.69	3.12	2.79	2.57	2.41	2.19	1.94	1.64	1.00
	5	3.84	2.99	2.60	2.37	2.21	2.09	1.94	1.75	1.52	1.00
	10	2.71	2.30	2.08	1.94	1.85	1.77	1.67	1.55	1.35	1.00
	20	1.64	1.61	1.55	1.50	1.46	1.43	1.38	1.32	1.23	1.00
	25	1.32	1.39	1.37	1.35	1.33	1.37	1.28	1.24	1.18	1.00

Adapted from Table 18 in *Biometrika Tables for Statisticians,* vol. 1, 3rd ed., edited by E. S. Pearson and H. O. Hartley (New York: Cambridge University Press, 1958). Reproduced by permission of the Biometrika trustees.

TABLE A.2 Critical Values of χ^2 Distribution

df	0.250	0.100	0.050	0.025	0.010	0.005	0.001
1	1.32330	2.70554	3.84146	5.02389	6.63490	7.87944	10.828
2	2.77259	4.60517	5.99147	7.37776	9.21034	10.5966	13.816
3	4.10835	6.25139	7.81473	9.34840	11.3449	12.8381	16.266
4	5.38527	7.77944	9.48773	1.1433	13.2767	14.8602	18.467
5	6.62568	9.23635	11.0705	12.8325	15.0863	16.7496	20.515
6	7.84080	10.6446	12.5916	14.4494	16.8119	18.5476	22.456
7	9.03715	12.0170	14.0671	16.0128	18.4753	20.2777	24.322
8	10.2188	13.3616	15.5073	17.5346	20.0902	21.9550	26.125
9	11.3887	14.6837	16.9190	19.0228	21.6660	23.5893	27.877
10	12.5489	15.9871	18.3070	20.4831	23.2093	25.1882	29.586
11	13.7007	17.2750	19.6751	21.9200	24.7250	26.7569	31.264
12	14.8454	18.5494	21.0261	23.3367	26.2170	28.2995	32.909
13	15.9839	19.8119	22.3621	24.7356	27.6883	29.8194	34.528
14	17.1770	21.0642	23.6848	26.1190	29.1413	31.3193	36.123
15	18.2451	22.3072	24.9958	27.4884	30.5779	32.8013	37.697
16	19.3688	23.5418	26.2962	28.8454	31.9999	34.2672	39.252
17	20.4887	24.7690	27.5871	30.1910	33.4087	35.7185	40.790
18	21.6049	25.9894	28.8693	31.5264	34.8053	37.1564	42.312
19	22.7178	27.2036	30.1436	32.8523	36.1908	38.5822	43.820
20	23.8277	28.4120	31.4104	34.1696	37.5662	39.9968	45.315
21	24.9348	29.6151	32.6705	35.4789	38.9321	41.4010	46.797
22	26.0393	30.8133	33.9244	36.7807	40.2894	42.7956	48.268
23	27.1413	32.0069	35.1725	38.0757	41.6384	44.1813	49.728
24	28.2412	33.1963	36.4151	39.3641	42.9798	45.5585	51.179
25	29.3389	34.3816	37.6525	40.6465	44.3141	46.9278	52.620
26	30.4345	35.5631	38.8852	41.9232	45.6417	48.2899	54.052
27	31.5284	36.7412	40.1133	43.1944	46.9630	49.6449	55.476
28	32.6205	37.9159	41.3372	44.4607	48.2782	50.9933	56.892
29	33.7109	39.0875	42.5569	45.7222	49.5879	52.3356	58.302
30	34.7998	40.2560	43.7729	46.9792	50.8922	53.6720	59.703
40	45.6160	51.8050	65.7585	59.3417	63.6907	66.7659	73.402
50	56.3336	63.1671	67.5048	71.4202	76.1539	79.4900	86.661
60	66.9814	74.3970	79.0819	83.2976	88.3794	91.9517	99.607
70	77.5766	85.5271	90.5312	95.0231	100.425	104.215	112.317
80	88.1303	96.5782	101.879	106.629	112.329	116.321	124.839
90	98.6499	107.565	113.145	118.136	124.116	128.299	137.208
100	109.141	118.498	124.342	129.561	135.807	140.169	149.449

Adapted from Table 8 in *Biometrika Tables for Statisticians,* vol. 1, 3rd ed., edited by E. S. Pearson and H. O. Hartley (New York: Cambridge University Press, 1958). Reproduced by permission of the Biometrika trustees.

TABLE A.3 Critical Values of Studentized Range Statistic Distribution

df for $s_{\overline{X}}$	$1 - a$	Number of Ordered Means													
		2	3	4	5	6	7	8	9	10	11	12	13	14	15
1	.95	18.0	27.0	32.8	37.1	40.4	43.1	45.4	47.4	49.1	50.6	52.0	53.2	54.3	55.4
	.99	90.0	135	164	186	202	216	227	237	246	253	260	266	272	277
2	.95	6.09	8.3	9.8	10.9	11.7	12.4	13.0	13.5	14.0	14.4	14.7	15.1	15.4	15.7
	.99	14.0	19.0	22.3	24.7	26.6	28.2	29.5	30.7	31.7	32.6	33.4	34.1	34.8	35.4
3	.95	4.50	5.91	6.82	7.50	8.04	8.48	8.85	9.18	9.46	9.72	9.95	10.2	10.4	10.5
	.99	8.26	10.6	12.2	13.3	14.2	15.0	15.6	16.2	16.7	17.1	17.5	17.9	18.2	18.5
4	.95	3.93	5.04	5.76	6.29	6.71	7.05	7.35	7.60	7.83	8.03	8.21	8.37	8.52	8.66
	.99	6.51	8.12	9.17	9.96	10.6	11.1	11.5	11.9	12.3	12.6	12.8	13.1	13.3	13.5
5	.95	3.64	4.60	5.22	5.67	6.03	6.33	6.58	6.80	6.99	7.17	7.32	7.47	7.60	7.72
	.99	5.70	6.97	7.80	8.42	8.91	9.32	9.67	9.97	10.2	10.5	10.7	10.9	11.1	11.2
6	.95	3.46	4.34	4.90	5.31	5.63	5.89	6.12	6.32	6.49	6.65	6.79	6.92	7.03	7.14
	.99	5.24	6.33	7.03	7.56	7.97	8.32	8.61	8.87	9.10	9.30	9.49	9.65	9.81	9.95
7	.95	3.34	4.16	4.69	5.06	5.36	5.61	5.82	6.00	6.16	6.30	6.43	6.55	6.66	6.76
	.99	4.95	5.92	6.54	7.01	7.37	7.68	7.94	8.17	8.37	8.55	8.71	8.86	9.00	9.12
8	.95	3.26	4.04	4.53	4.89	5.17	5.40	5.60	5.77	5.92	6.05	6.18	6.29	6.39	6.48
	.99	4.74	5.63	6.20	6.63	6.96	7.24	7.47	7.68	7.87	8.03	8.18	8.31	8.44	8.55
9	.95	3.20	3.95	4.42	4.76	5.02	5.24	5.43	5.60	5.74	5.87	5.98	6.09	6.19	6.28
	.99	4.60	5.43	5.96	6.35	6.66	6.91	7.13	7.32	7.49	7.65	7.78	7.91	8.03	8.13
10	.95	3.15	3.88	4.33	4.65	4.91	5.12	5.30	5.46	5.60	5.72	5.83	5.93	6.03	6.11
	.99	4.48	5.27	5.77	6.14	6.43	6.67	6.87	7.05	7.21	7.36	7.48	7.60	7.71	7.81
11	.95	3.11	3.82	4.26	4.57	4.82	5.03	5.20	5.35	5.49	5.61	5.71	5.81	5.90	5.99
	.99	4.39	5.14	5.62	5.97	6.25	6.48	6.67	6.84	6.99	7.13	7.26	7.36	7.46	7.56
12	.95	3.08	3.77	4.20	4.51	4.75	4.95	5.12	5.27	5.40	5.51	5.62	5.71	5.80	5.88
	.99	4.32	5.04	5.50	5.84	6.10	6.32	6.51	6.67	6.81	6.94	7.06	7.17	7.26	7.36
13	.95	3.06	3.73	4.15	4.45	4.69	4.88	5.05	5.19	5.32	5.43	5.53	5.63	5.71	5.79
	.99	4.26	4.96	5.40	5.73	5.98	6.19	6.37	6.53	6.67	6.79	6.90	7.01	7.10	7.19

14	.95	3.03	3.70	4.11	4.41	4.64	4.83	4.99	5.13	5.25	5.36	5.46	5.55	6.64	5.72
	.99	4.21	4.89	5.32	5.63	5.88	6.08	6.26	6.41	6.54	6.66	6.77	6.87	6.96	7.05
16	.95	3.00	3.65	4.05	4.33	4.56	4.74	4.90	5.03	5.15	5.26	5.35	5.44	5.52	5.59
	.99	4.13	4.78	5.19	5.49	5.72	5.92	6.08	6.22	6.35	6.46	6.56	6.66	6.74	6.82
18	.95	2.97	3.61	4.00	4.28	4.49	4.67	4.82	4.96	5.07	5.17	5.27	5.35	5.43	5.50
	.99	4.07	4.70	5.09	5.38	5.60	5.79	5.94	6.08	6.20	6.31	6.41	6.50	6.58	6.65
20	.95	2.95	3.58	3.96	4.23	4.45	4.62	4.77	4.90	5.01	5.11	5.20	5.28	5.36	5.43
	.99	4.02	4.64	5.02	5.29	5.51	5.69	5.84	5.97	6.09	6.19	6.29	6.37	6.45	6.52
24	.95	2.92	3.53	3.90	4.17	4.37	4.54	4.68	4.81	4.92	5.01	5.10	5.18	5.25	5.32
	.99	3.96	4.54	4.91	5.17	5.37	5.54	5.69	5.81	5.92	6.02	6.11	6.19	6.26	6.33
30	.95	2.89	3.49	3.84	4.10	4.30	4.46	4.60	4.72	4.83	4.92	5.00	5.08	5.15	5.21
	.99	3.89	4.45	4.80	5.05	5.24	5.40	5.54	5.56	5.76	5.85	5.93	6.01	6.08	6.14
40	.95	2.86	3.44	3.79	4.04	4.23	4.39	4.52	4.63	4.74	4.82	4.91	4.98	5.05	5.11
	.99	3.82	4.37	4.70	4.93	5.11	5.27	5.39	5.50	5.60	5.69	5.77	5.84	5.90	5.96
60	.95	2.83	3.40	3.74	3.98	4.16	4.31	4.44	4.55	4.65	4.73	4.81	4.88	4.94	5.00
	.99	3.76	4.28	4.60	4.82	4.99	5.13	5.25	5.36	5.45	5.53	5.60	5.67	5.73	5.79
120	.95	2.80	3.36	3.69	3.92	4.10	4.24	4.36	4.48	4.56	4.64	4.72	4.78	4.84	4.90
	.99	3.70	4.20	4.50	4.71	4.87	5.01	5.12	5.21	5.30	5.38	5.44	5.51	5.56	5.61
∞	.95	2.77	3.31	3.63	3.86	4.03	4.17	4.29	4.39	4.47	4.55	4.62	4.68	4.74	4.80
	.99	3.64	4.12	4.40	4.60	4.76	4.88	4.99	5.08	5.16	5.23	5.29	5.35	5.40	5.45

Adapted from Table 29 in *Biometrika Tables for Statisticians,* vol. 1, 3rd ed., edited by E. S. Pearson and H. O. Hartley (New York: Cambridge University Press, 1966). Reproduced by permission of the Biometrika trustees.

TABLE A.4 Critical Values of Dunnett's *d*-Statistic in Comparing Treatment Means with a Control (1-sided test)

df for MS_{error}	$1 - a$	Number of Means (Including Control)								
		2	3	4	5	6	7	8	9	10
6	.95	1.94	2.34	2.56	2.71	2.83	2.92	3.00	3.07	3.12
	.975	2.45	2.86	3.18	3.41	3.60	3.75	3.88	4.00	4.11
	.99	3.14	3.61	3.88	4.07	4.21	4.33	4.43	4.51	4.59
	.995	3.71	4.22	4.60	4.88	5.11	5.30	5.47	5.61	5.74
7	.95	1.89	2.27	2.48	2.62	2.73	2.82	2.89	2.95	3.01
	.975	2.36	2.75	3.04	3.24	3.41	3.54	3.66	3.76	3.86
	.99	3.00	3.42	3.66	3.83	3.96	4.07	4.15	4.23	4.30
	.995	3.50	3.95	4.28	4.52	4.17	4.87	5.01	5.13	5.24
8	.95	1.86	2.22	2.42	2.55	2.66	2.74	2.81	2.87	2.92
	.975	2.31	2.67	2.94	3.13	3.28	3.40	3.51	3.60	3.68
	.99	2.90	3.29	3.51	3.67	3.79	3.88	3.96	4.03	4.09
	.995	3.36	3.77	4.06	4.27	4.44	4.58	4.70	4.81	4.90
9	.95	1.83	2.18	2.37	2.50	2.60	2.68	2.75	2.81	2.86
	.975	2.26	2.61	2.86	3.04	3.18	3.29	3.39	3.48	3.55
	.99	2.82	3.19	3.40	3.55	3.66	3.75	3.82	3.89	3.94
	.995	3.25	3.63	3.90	4.09	4.24	4.37	4.48	4.57	4.65
10	.95	1.81	2.15	2.34	2.47	2.56	2.64	2.70	2.76	2.81
	.975	2.23	2.57	2.81	2.97	3.11	3.21	3.31	3.39	3.46
	.99	2.76	3.11	3.31	3.45	3.56	3.64	3.71	3.78	3.83
	.995	3.17	3.53	3.78	3.95	4.10	4.21	4.31	4.40	4.47
11	.95	1.80	2.13	2.31	2.44	2.53	2.60	2.67	2.72	2.77
	.975	2.20	2.53	2.76	2.92	3.05	3.15	3.24	3.31	3.38
	.99	2.72	3.06	3.25	3.38	3.48	3.56	3.63	3.69	3.74
	.995	3.11	3.45	3.68	3.85	3.98	4.09	4.18	4.26	4.33
12	.95	1.78	2.11	2.29	2.41	2.50	2.58	2.64	2.69	2.74
	.975	2.18	2.50	2.72	2.88	3.00	3.10	3.18	3.25	3.32
	.99	2.68	3.01	3.19	3.32	3.42	3.50	3.56	3.62	3.67
	.995	3.05	3.39	3.61	3.76	3.89	3.99	4.08	4.15	4.22
13	.95	1.77	2.09	2.27	2.39	2.48	2.55	2.61	2.66	2.71
	.975	2.16	2.48	2.69	2.84	2.96	3.06	3.14	3.21	3.27
	.99	2.65	2.97	3.15	3.27	3.37	3.44	3.51	3.56	3.61
	.995	3.01	3.33	3.54	3.69	3.81	3.91	3.99	4.06	4.13
14	.95	1.76	2.08	2.25	2.37	2.46	2.53	2.59	2.64	2.69
	.975	2.14	2.46	2.67	2.81	2.93	3.02	3.10	3.17	3.23
	.99	2.62	2.94	3.11	3.23	3.32	3.40	3.46	3.51	3.45
	.995	2.98	3.29	3.49	3.64	3.75	3.84	3.92	3.99	4.05
16	.95	1.75	2.06	2.23	2.34	2.43	2.50	2.56	2.61	2.65
	.975	2.12	2.42	2.63	2.77	2.88	2.96	3.04	3.10	3.16
	.99	2.58	2.88	3.05	3.17	3.26	3.33	3.39	3.44	3.48
	.995	2.92	3.22	3.41	3.55	3.65	3.74	3.82	3.88	3.93

TABLE A.4 Continued

df for MS_{error}	$1 - a$	Number of Means (Including Control) 2	3	4	5	6	7	8	9	10
18	.95	1.73	2.04	2.21	2.32	2.41	2.48	2.53	2.58	2.62
	.975	2.10	2.40	2.59	2.73	2.84	2.92	2.99	3.05	3.11
	.99	2.55	2.84	3.01	3.12	3.21	3.27	3.33	3.38	3.42
	.995	2.88	3.17	3.35	3.48	3.58	3.67	3.74	3.80	3.85
20	.95	1.72	2.03	2.19	2.30	2.39	2.46	2.51	2.56	2.60
	.975	2.09	2.38	2.57	2.70	2.81	2.89	2.96	3.02	3.07
	.99	2.53	2.81	2.97	3.08	3.17	3.23	3.29	3.34	3.38
	.995	2.85	3.13	3.31	3.43	3.53	3.61	3.67	3.73	3.78
24	.95	1.71	2.01	2.17	2.28	2.36	2.43	2.48	2.53	2.57
	.975	2.06	2.35	2.53	2.66	2.76	2.84	2.91	2.96	3.01
	.99	2.49	2.77	2.92	3.03	3.11	3.17	3.22	3.27	3.31
	.995	2.80	3.07	3.24	3.36	3.45	3.52	3.58	3.64	3.69
30	.95	1.70	1.99	2.15	2.25	2.33	2.40	2.45	2.50	2.54
	.975	2.04	2.32	2.50	2.62	2.72	2.79	2.86	2.91	2.96
	.99	2.46	2.72	2.87	2.97	3.05	3.11	3.16	3.21	3.24
	.995	2.75	3.01	3.17	3.28	3.37	3.44	3.50	3.55	3.59
40	.95	1.68	1.97	2.13	2.23	2.31	2.37	2.42	2.47	2.51
	.975	2.02	2.29	2.47	2.58	2.67	2.75	2.81	2.86	2.90
	.99	2.42	2.68	2.82	2.92	2.99	3.05	3.10	3.14	3.18
	.995	2.70	2.95	3.10	3.21	3.29	3.36	3.41	3.46	3.50
60	.95	1.67	1.95	2.10	2.21	2.28	2.23	2.39	2.44	2.48
	.975	2.00	2.27	2.43	2.55	2.63	2.70	2.76	2.81	2.85
	.99	2.39	2.64	2.78	2.87	2.94	3.00	3.04	3.08	3.12
	.995	2.66	2.90	3.04	3.14	3.22	3.28	3.33	3.38	3.42
120	.95	1.66	1.93	2.08	2.18	2.26	2.32	2.37	2.41	2.45
	.975	1.98	2.24	2.40	2.51	2.59	2.66	2.71	2.76	2.80
	.99	2.36	2.60	2.73	2.82	2.89	2.94	2.99	3.03	3.06
	.995	2.62	2.84	2.98	3.08	3.15	3.21	3.25	3.30	3.33
∞	.95	1.64	1.92	2.06	2.16	2.23	2.29	2.34	2.38	2.42
	.975	1.96	2.21	2.37	2.47	2.55	2.62	2.67	2.71	2.75
	.99	2.33	2.56	2.68	2.77	2.84	2.89	2.93	2.97	3.00
	.995	2.58	2.79	2.92	3.01	3.08	3.14	3.18	3.22	3.25

Adapted from C. W. Dunnett, "A Multiple Comparison Procedure for Comparing Several Treatments with a Control," *Journal of the American Statistical Association,* 1955, 50, 1096–1121. Reprinted with permission from the *Journal of the American Statistical Association.*

TABLE A.5 Coefficients of Orthogonal Polynomials

a	Polynomial	*X* = 1	2	3	4	5	6	7	8	9	10	$\sum w_i^2$
3	Linear	−1	0	1								2
	Quadratic	1	−2	1								6
4	Linear	−3	−1	1	3							20
	Quadratic	1	−1	−1	1							4
	Cubic	−1	3	−3	1							20
5	Linear	−2	−1	0	1	2						10
	Quadratic	2	−1	−2	−1	2						14
	Cubic	−1	2	0	−2	1						10
	Quartic	1	−4	6	−4	1						70
6	Linear	−5	−3	−1	1	3	5					70
	Quadratic	5	−1	−4	−4	−1	5					84
	Cubic	−5	7	4	−4	−7	5					180
	Quartic	1	−3	2	2	−3	1					28
7	Linear	−3	−2	−1	0	1	2	3				28
	Quadratic	5	0	−3	−4	−3	0	5				84
	Cubic	−1	1	1	0	−1	−1	1				6
	Quartic	3	−7	1	6	1	−7	3				154
8	Linear	−7	−5	−3	−1	1	3	5	7			168
	Quadratic	7	1	−3	−5	−5	−3	1	7			168
	Cubic	−7	5	7	3	−3	−7	−5	7			264
	Quartic	7	−13	−3	9	9	−3	−13	7			616
	Quintic	−7	23	−17	−15	15	17	−23	7			2184
9	Linear	−4	−3	−2	−1	0	1	2	3	4		60
	Quadratic	28	7	−8	−17	−20	−17	−8	7	28		2772
	Cubic	−14	7	13	9	0	−9	−13	−7	14		990
	Quartic	14	−21	−11	9	18	9	−11	−21	14		2002
	Quintic	−4	11	−4	−9	0	9	4	−11	4		468
10	Linear	−9	−7	−5	−3	−1	1	3	5	7	9	330
	Quadratic	6	2	−1	−3	−4	−4	−3	−1	2	6	132
	Cubic	−42	14	35	31	12	−12	−31	−35	−14	42	8580
	Quartic	18	−22	−17	3	18	18	3	−17	−22	18	2860
	Quintic	−6	14	−1	−11	−6	6	11	1	−14	6	780

Adapted from Table 47 in *Biometrika Tables for Statisticians,* vol. 1, 3rd ed., edited by E. S. Pearson and H. O. Hartley (New York: Cambridge University Press, 1966). Reproduced by permission of the Biometrika trustees.

TABLE A.6 Critical Values of F_{max} Distribution

df for s_X^{2j}	$1-a$	a = Number of Variances								
		2	3	4	5	6	7	8	9	10
4	.95	9.60	15.5	20.6	25.2	29.5	33.6	37.5	41.4	44.6
	.99	23.2	37.	49.	59.	69.	79.	89.	97.	106.
5	.95	7.15	10.8	13.7	16.3	18.7	20.8	22.9	24.7	26.5
	.99	14.9	22.	28.	33.	38.	42.	46.	50.	54.
6	.95	5.82	8.38	10.4	12.1	13.7	15.0	16.3	17.5	18.6
	.99	11.1	15.5	19.1	22.	25.	27.	30.	32.	34.
7	.95	4.99	6.94	8.44	9.70	10.8	11.8	12.7	13.5	14.3
	.99	8.89	12.1	14.5	16.5	18.4	20.	22.	23.	24.
8	.95	4.43	6.00	7.18	8.12	9.03	9.78	10.5	11.1	11.7
	.99	7.50	9.9	11.7	13.2	14.5	15.8	16.9	17.9	18.9
9	.95	4.03	5.34	6.31	7.11	7.80	8.41	8.95	9.45	9.91
	.99	6.54	8.5	9.9	11.1	12.1	13.1	13.9	14.7	15.3
10	.95	3.72	4.85	5.67	6.34	6.92	7.42	7.87	8.28	8.66
	.99	5.85	7.4	8.6	9.6	10.4	11.1	11.8	12.4	12.9
12	.95	3.28	4.16	4.79	5.30	5.72	6.09	6.42	6.72	7.00
	.99	4.91	6.1	6.9	7.6	8.2	8.7	9.1	9.5	9.9
15	.95	2.86	3.54	4.01	4.37	4.68	4.95	5.19	5.40	5.59
	.99	4.07	4.9	5.5	6.0	6.4	6.7	7.1	7.3	7.5
20	.95	2.46	2.95	3.29	3.54	3.76	3.94	4.10	4.24	4.37
	.99	3.32	3.8	4.3	4.6	4.9	5.1	5.3	5.5	5.6
30	.95	2.07	2.40	2.61	2.78	2.91	3.02	3.12	3.21	3.29
	.99	2.63	3.0	3.3	3.4	3.6	3.7	3.8	3.9	4.0
60	.95	1.67	1.85	1.96	2.04	2.11	2.17	2.22	2.26	2.30
	.99	1.96	2.2	2.3	2.4	2.4	2.5	2.5	2.6	2.6
∞	.95	1.00	1.00	1.00	1.00	1.00	1.00	1.00	1.00	1.00
	.99	1.00	1.00	1.00	1.00	1.00	1.00	1.00	1.00	1.00

Adapted from Table 31 in *Biometrika Tables for Statisticians,* vol. 1, 3rd ed., edited by E. S. Pearson and H. O. Hartley (New York: Cambridge University Press, 1966). Reproduced by permission of the Biometrika trustees.

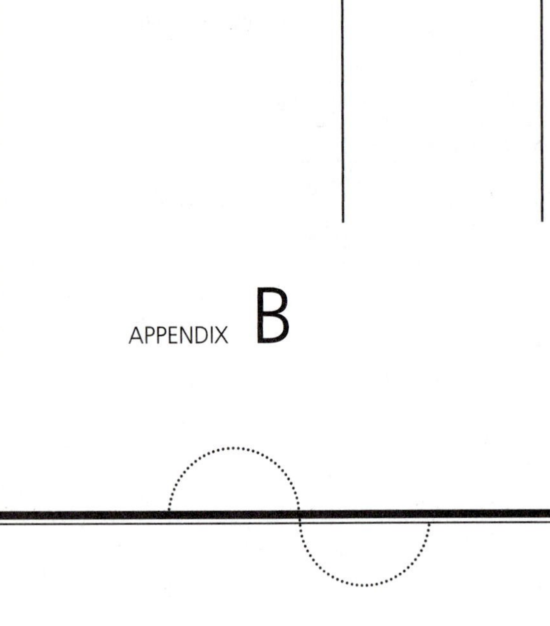

APPENDIX B

Research Designs for Complete Examples

B.1 FACETS IN FLY'S EYES

The data on which the complete example of a one-way randomized-groups ANOVA are based were collected by Hersh (1924). The purpose of the experiment was to determine the influences of temperature on flies that are heterozygous for genes of the bar series. Ten probable virgins of one stock (stocks are described in detail by Hersh) were mated with ten males of another stock in an 8-dram vial containing banana agar inoculated with yeast, and then the bottle was placed at the experimental temperatures. The hatched-out females were counted for facet number and then preserved. The crosses were made in either direction.

The method of counting is described by Hersh as follows:

> The flies were counted upon blackened paraffin in a Syracuse glass. The best slope for placing the flies was easily secured by chipping away at the paraffin. The flies were placed upon the inclined slope of the paraffin and counted under a compound microscope with a Leitz 4 ocular and a Leitz 3 objective. In flies of high facet count, the dorsal and ventral areas of the eye were counted separately, and the results added to give the total facet count of the individual. (p. 56)

Temperatures were chosen between 15° C and 31° C, with an interval of 2° C. Cool and warm rooms in the vivarium building provided the 15° C and 25° C temperatures. Incubators in those rooms were converted to produce the 17° C and 27° C temperatures. The laboratory incubator provided the 29° C condition. The remaining conditions were maintained in aquaria fitted with temperature control. Because previous research has shown that facet numbers are "not a good measure of the values of the temperatures," facet counts were tabulated on a "factorial system" described elsewhere (Zeleny, 1920).

B.2 BONDING STRENGTH OF PERIODONTAL DRESSINGS

The data on which the complete example of a randomized-groups factorial ANOVA is based were collected by the periodontal department at the University of Southern California Dental School in 1983 (Farnoush, Joseph, and Tabachnick, 1985; Sachs, Joseph, Farnoush, and Tabachnick, 1985).

The purpose of the study was to measure the tensile shear bond strengths of five commercially available non-eugenol periodontal dressings, three of which had not been previously investigated. The five periodontal dressing materials were Coe Pak (Coe Laboratories, Chicago), PerioCare (Pulpdent Corp., Brookline, Massachusetts), Peripac (De Trey Freres, S.A., Zurich, Switzerland), Perio Putty, and Zone (both from Cadco Inc., Los Angeles). Perio Putty is a rigid material that hardens with a snap set. The other materials all set with a lesser degree of hardness and rigidity.

In all trials, a clean, dry, smoothly polished glass slab 15 cm × 7.8 cm × 1.9 cm was used for adherence of materials. Standardized, machined brass cylinders with an outside diameter of 9 mm were filled with periodontal dressing, which was mixed according to the manufacturer's directions and standardized weight ratios to ensure uniformity of mix. Peripac was supplied ready-to-use with no mixing required. The materials were pressed into contact with the glass surface for 15 s and then were released and allowed to set for either 3 h or 24 h, during which time they were stored in a humidor at 37.8° C. Each trial was repeated at least 10 times.

For the tensile test, the glass slab with the cylinders adhered to it was placed in a jig on an Instron Universal Testing Machine (Instron Corp, Canton, Massachusetts), with the cylinder attached to a wire loop engaged by a torque-free tension chain to allow application of tensile force. For the shear test, the glass slab was placed in a jig with the specimens at right angles to a piston (in the same machine), which generated a compressive force. Thus, a tensile or shear force was applied to the specimens, with the machine set at a crosshead speed of 1 mm/m. Bond failure was characterized by visual inspection as either cohesive or adhesive.

B.3 REACTION TIME TO IDENTIFY FIGURES

Data for this study were collected by Damos (1989). Twenty right-handed males were required to respond "same" or "different" on a keypad as quickly as possible to a stimulus (the letter F or G) and its mirror image. Each trial consisted of 30 presentations of the stimulus and 30 presentations of its mirror image, with five presentations in each of six orientations: upright (0), 60°, 120°, 180°, 240°, and 300° rotation. In all, the experiment required 10 sessions over two consecutive weeks. Each session had four blocks of nine trials distributed over morning and afternoon periods. Thus, each subject made 21,600 responses during the study. Half the subjects were given stimulus F in the first week and G in the second week; the others were given the letter stimuli in the reverse order. Order of presentation of all stimuli was random on all trials.

The two DVs were average correct reaction time and error rate. Responses to 60° of absolute rotation were averaged with those of 300°, and responses to 120° were averaged with those of 240°. Averages were calculated separately for standard and mirror-image trials.

Data selected for analysis in Section 6.7 were averages from the final (afternoon) period of the last session (day 10), so that all subjects were well practiced with both F and G stimuli. Previous research had shown no reliable differences in response due to letter, so no distinction between F and G stimuli was made in the current analysis. Correct reaction time was the DV selected for analysis. Thus, each of the 20 subjects provided eight reaction time averages: standard image at four orientations and mirror image at four orientations.

B.4 AUTO POLLUTION FILTER NOISE

Data for the mixed randomized-repeated–measures design were available on the Data and Storage Library (DASL) Web page, with free use authorization given. The reference given is Lewin and Shakun (1976, p. 313):

> The data are from a statement by Texaco, Inc., to the Air and Water Pollution Subcommittee of the Senate Public Works Committee on June 26, 1973. Mr. John McKinley, President of Texaco, cited an automobile filter developed by Associated Octel Company as effective in reducing pollution. However, questions had been raised about the effects of filters on vehicle performance, fuel consumption, exhaust gas back pressure, and silencing. On the last question, he referred to the data included here as evidence that the silencing properties of the Octel filter were at least equal to those of standard silencers.

B.5 WEAR TESTING OF FABRIC SAMPLES

The data for the randomized-group Latin-square analysis are from Davies (1963), who describes the experiment "relating to the testing of rubber-covered fabric in the Martindale wear tester carried out as a 4 × 4 Latin square":

> The machine consists of four rectangular brass plates on each of which is fastened an abrading surface consisting of special-quality emery paper. Four weighted bushes, into which test samples of fabric are fixed, rest on the emery surfaces, and a mechanical device moves the bushes over the surface of the emery, thus abrading the test specimens. The loss in weight after a given number of cycles is used as a criterion of resistance to abrasion. There are slight differences between the four positions on the machine, and it is known from past experience that if a run is repeated under apparently identical conditions and using apparently identical samples, slightly different results are obtained, ascribable partly to variations in the emery paper and partly to variations in temperature, humidity, etc. (pp. 163–164)

B.6 ODORS AND PERFORMANCE

The data for the crossover Latin-square analysis were provided by Hirsch and Johnston (1996), of the Smell and Taste Treatment and Research Foundation. The goal of the research was to assess the effects of odors on performance of cognitive tasks. They recruited "22 volunteers, 12 men and 10 women ranging in age from 15 to 65 years (mean 36, median 34) from the shopping mall at Water Tower Place in Chicago." The volunteers were paid a nominal sum for their participation. Participants rated themselves as normally achieving and were given Amoore's pyridine odor threshold test "to establish they were normosmic."

The instruments used were two trail-making (maze) tests, "modified from the trail making subtest of the Halsted-Reitan Neuropsychological Test Battery which is used to detect neurological problems." Hirsch and Johnston described the following procedure:

> Prior to testing, subjects accustomed themselves to the masks by wearing unscented masks for one minute. They then underwent testing in randomized, single-blinded fashion. Each subject performed the trials twice: once wearing an unscented mask and once wearing a floral-scented mask. The scented masks had one drop of a mixed floral odorant applied, resulting in a suprathreshold level of scent.
>
> The order of presentation of scented versus unscented masks was random, but the order of maze presentation was constant. Each subject completed the set of two mazes a total of three times sequentially with each mask. The time required to carry out each of the trials was measured. (pp. 119–120)

B.7 PROCESSING TIME FOR ANOVA

Data for the screening experiment were collected on a hot August afternoon in the San Fernando Valley by the first author, who was enjoying her then new Pentium 166 in an air-conditioned office. The purpose was to determine which factors affect the computation speed of the SPSS 7.5 GLM procedure: number of two-level IVs, number of covariates, and sample size using a data set described in Tabachnick and Fidell (2007, Chapter 9). Five levels of each factor were chosen to demonstrate a central composite design.

Five dichotomous variables were chosen (or created) from the data set to serve as IVs. In runs with fewer than five IVs, the particular variables were chosen in a haphazard fashion. Four continuous variables were also chosen from the data set to serve as covariates. Again, in runs with fewer than four covariates, the variables were chosen haphazardly, except for runs with no covariates at all. Samples were chosen randomly through SPSS 7.5 for each of the 20 runs. Each run then was executed through SPSS 7.5, and the elapsed time in seconds was recorded from the "Notes" portion of the output.

B.8 NAMBEWARE POLISHING TIMES

Data for the randomized-groups ANCOVA example were available from the DASL Web page, with free use authorization. The reference given is Nambé Mills, Santa Fe, New Mexico. The following descriptions are from the DASL Web page:

> Nambé Mills manufactures a line of tableware made from sand casting a special alloy of several metals. After casting, the pieces go through a series of shaping, grinding, buffing, and polishing steps. In 1989 the company began a program to rationalize its production schedule of some 100 items in its tableware line.
>
> The average polishing times for 59 products are related in general to the sizes of the products as indicated by product diameter (or equivalent). However, different types of products have special problems in polishing. For example, casseroles are deeper than ordinary plates and on that account would require more polishing time than a plate of equal diameter. Also it is possible that the increase in polishing time per additional inch of diameter could be different for casseroles than for ordinary plates.

B.9 CHEST DECELERATION INJURIES IN AUTOMOBILE CRASHES

Data for the factorial randomized-repeated ANCOVA example were available from the DASL Web page, with free use authorization. Reference is the National Transportation Safety Board. The following description is provided:

> Stock automobiles containing dummies in the driver and front passenger seats crashed into a wall at 35 miles per hour. National Transportation Safety Board officials collected information how the crash affected the dummies. The injury variables describe the extent of head injuries, chest deceleration, and left and right femur load. The data file also contains information on the type and safety features of each crashed car.

B.10 FAT CONTENT OF EGGS

The following description is from the abstract provided on the Web page accompanying the data set. Data were taken from Bliss (1967), based on an experiment by Mitchell (1950).

A single can of dried eggs was stirred well. Samples were drawn, and a pair of samples (claimed to be of two "types") was sent to each of six commercial laboratories to be analyzed for fat content. Each laboratory assigned two technicians, who each analyzed both "types." Because the data were all drawn from a single well-mixed can, the null hypothesis for ANOVA that the mean fat content of each sample is equal is true. The experiment is thus really a study of the laboratories.

References

Bliss, C.I. (1967). *Statistics in Biology: Statistical Methods for Research in the Natural Sciences.* New York: McGraw-Hill.

Box, G.E.P., and Cox, D.R. (1964). An analysis of transformations. *Journal of the Royal Statistical Society 26*(Series B): 211–243.

Bradley, D.R. (1988). *Datasim.* Lewiston, ME: Desktop Press. (Available by writing to Drake Bradley, 90 Bardwell St., Lewiston, ME 04240.)

Bradley, J.V. (1982). The insidious L-shaped distribution. *Bulletin of the Psychonomic Society 20*(2): 85–88.

Brennan, R.L. (1992). NCME instructional module: Generalizability theory. *Educational Measurement: Issues and Practice 11*(4): 27–34.

Brown, D.R., Michels, K.M., and Winer, B.J. (1991). *Statistical Principles in Experimental Design,* 3rd ed. New York: McGraw-Hill.

Bryk, A.S., and Raudenbush, S.W. (2001). *Hierarchical Linear Models: Applications and Data Analysis Methods,* 2nd ed. Newbury Park, CA: Sage.

Cleveland, W.S. (1994). *The Elements of Graphing Data,* 2nd ed. Murray Hills, NJ: AT&T Bell Laboratories.

Cochran, W.G., and Cox, G.M. (1957). *Experimental Designs,* 2nd ed. New York: McGraw-Hill.

Cohen, J. (1965). Some statistical issues in psychological research. In B.B. Wolman (ed.), *Handbook of Clinical Psychology* (pp. 95–121). New York: McGraw-Hill.

Cohen, J. (1977). *Statistical Power Analysis for the Behavioral Sciences.* New York: Academic Press.

Cohen, J. (1988). *Statistical Power Analysis for the Behavioral Sciences,* 2nd ed. Mahwah, NJ: Lawrence Erlbaum Associates.

Cohen, J. (1994). The earth is round ($p < .05$). *American Psychologist 49*: 997–1003.

Conover, W.J. (1980). *Practical Nonparametric Statistics,* 2nd ed. New York: John Wiley & Sons.

Cook, T.D., and Campbell, D.T. (1979). *Quasi-Experimentation: Design and Analysis Issues for Field Settings.* Chicago: Rand-McNally.

Cornell, J.A. (1990). *Experiments with Mixtures: Designs, Models, and the Analysis of Mixture Data,* 2nd ed. New York: John Wiley & Sons.

Cronbach, L.J. (1972). *The Dependability of Behavioral Measurements: Theory of Generalizability for Scores and Profiles.* New York: John Wiley & Sons.

Dallal, G.E. (1986). PC-SIZE: A program for sample size determinations. *American Statistician 40*: 52. (Available by writing Gerard E. Dallal, 53 Beltran St., Malden, MA 02148.)

Damos, D.L. (1989). *Transfer of Mental Rotation Skill* (technical report). Los Angeles: University of Southern California, Department of Human Factors.

Davies, O.L., ed. (1963). *Design and Analysis of Industrial Experiments,* 2nd ed. New York: Hafner.

Dixon, W.J., ed. (1992). *BMDP Statistical Software Manual.* Berkeley: University of California Press.

Dunnett, C.W. (1980). Pairwise multiple comparisons in the unequal variance case. *Journal of the American Statistical Association 70:* 796–800.

Egan, W.J., and Morgan, S.L. (1998). Outlier detection in multivariate analytical chemical data. *Analytical Chemistry 70*: 2372–2379.

Eliason, S.R. (1993). *Maximum Likelihood Estimation: Logic and Practice.* Newbury Park, CA: Sage.

Emerson, J.D. (1991). Introduction to transformation. In D.C. Hoaglin, F. Mosteller, and J.W. Tukey (eds.), *Fundamentals of Exploratory Analysis of Variance.* New York: John Wiley & Sons.

Farnoush, A., Joseph, C.E., and Tabachnick, B. (1985, March). *Clinical Evaluation of Ceramic Bone Implants in Treatment of Advanced Periodontal Defects*. Paper presented at the International Association of Dental Research, Las Vegas.

Fisher, R.A. (1935). *The Design of Experiments*. Edinburgh and London: Oliver & Boyd.

Greenhouse, S.W., and Geisser, W. (1959). On methods in the analysis of profile data. *Psychometrika 24*: 95–112.

Hadi, A.S., and Simonoff, J.W. (1993). Procedures for the identification of multiple outliers in linear models. *Journal of the American Statistical Association 88*: 1264–1272.

Harlow, L.L., Mulaik, S.A., and Steiger, J.H. (eds.). (1997). *What If There Were No Significance Tests?* Mahweh, NJ: Lawrence Erlbaum Associates.

Heck, R., and Thomas, S. (2000). *An Introduction to Multilevel Modeling Techniques.* Mahwah, NJ: Lawrence Erlbaum Associates.

Hersh, A. (1924). The effect of temperature upon the heterozygotes in the bar series of drosophila. *Journal of Experimental Zoology 39*: 55–71.

Himmelfarb, S. (1975). What do you do when the control group doesn't fit the factorial design? *Psychological Bulletin 82*: 363–368.

Hintze, J. (2001). *NCSS and PASS. Number Cruncher Statistical Systems.* [Computer software]. Kaysville, UT: Number Cruncher Statistical System.

Hirsch, A.R., and Johnston, L.H. (1996). Odors and learning. *Journal of Neurological and Orthopaedic Medicine and Surgery 17*: 119–126.

Hoaglin, D.C., Mosteller, F., and Tukey, J.W. (1991). *Fundamentals of Exploratory Analysis of Variance.* New York: John Wiley & Sons.

Hox, J.J. (2002). *Multilevel Analysis.* Mahwah, NJ: Lawrence Erlbaum Associates.

Huynh, H., and Feldt, L.S. (1976). Estimation of the Box correction for degrees of freedom from sample data in randomized block and split-plot designs. *Journal of the American Statistical Association 65*: 1582–1589.

Jamieson, J. (1999). Dealing with baseline differences: Two principles and two dilemmas. *International Journal of Psychophysiology 31:* 155–161.

Jamieson, J., and Howk, S. (1992). The law of initial values: A four factor theory. *International Journal of Psychophysiology 12*: 53–61.

Jones, B., and Kenward, M.G. (2003). *Design and Analysis of Cross-Over Trials,* 2nd ed. London: Chapman and Hall.

Kempthorne, O. (1952). *The Design and Analysis of Experiments.* New York: John Wiley & Sons.

Keppel, G. (1982). *Design and Analysis: A Researcher's Handbook,* 2nd ed. Englewood Cliffs, NJ: Prentice-Hall.

Keppel, G. (1991). *Design and Analysis: A Researcher's Handbook,* 3rd ed. Englewood Cliffs, NJ: Prentice-Hall.

Keppel, G., Saufley, W.H., Jr., and Tokunaga, H. (1992). *Introduction to Design and Analysis: A Student's Handbook,* 2nd ed. New York: W.H. Freeman.

Keren, G., and Lewis, C. (1979). Partial omega squared for ANOVA. *Educational and Psychological Measurement 39*: 119–128.

Kirk, R.E. (1995). *Experimental Design,* 3rd ed. Pacific Grove, CA: Brooks/Cole.

Kreft, I.G.G., and DeLeeuw, J. (1998). *Introducing Multilevel Modeling*. Thousand Oaks, CA: Sage.

Lee, W. (1975). *Experimental Design and Analysis*. San Francisco: W.H. Freeman.

Levene, H. (1960). Robust tests for equality of variance. In I. Olkin (ed.), *Contributions to Probability and Statistics.* Palo Alto, CA: Stanford University Press.

Lewin, A.Y., and Shakun, M.F. (1976). *Policy Sciences: Methodology and Cases.* New York: Pergammon Press.

Lindeman, H.R. (1974). *Analysis of Variance in Complex Experimental Designs.* San Francisco: W.H. Freeman.

Lord, F.E. (1967). A paradox in the interpretation of group comparisons. *Psychological Bulletin 68*: 304–305.

Marcoulides, G.A. (1999). Generalizability theory: Picking up where the Rasch IRT model leaves off. In S.E. Embretson and S.L. Hershberger (eds.), *The New Rules of Measurement: What Every Psychologist and Educator Should Know* (pp. 129–152). Mahwah, NJ: Lawrence Erlbaum Associates.

Mauchly, J.W. (1940). Significance test for sphericity of a normal *n*-variate distribution. *Annals of Mathematical Statistics 11*: 204–209.

Maxwell, S.E., and Delaney, H.D. (1990). *Designing Experiments and Analyzing Data: A Model Comparison Perspective.* Belmont, CA: Wadsworth.

Medical Research Working Party on Misonidazole in Gliomas. (1983). A study of the effect of misonidazole in conjunction with radiotherapy for the treatment of grades 3 and 4 astrocytomas. *British Journal of Radiology 56*: 673–682.

Minitab, Inc. (2003). *MINITAB Release 14.* [Computer software]. State College, PA: Minitab.

Mitchell, L.C. (1950). Report on fat by acid hydrolysis in eggs. *Journal of the Association Office of Agricultural Chemists 33*: 699–703.

Montgomery, D.C. (1984). *Design and Analysis of Experiments,* 2nd ed. New York: John Wiley & Sons.

Mulaik, S.A., Raju, N.S., and Harshman, R.A. (1997). There is a time and a place for significance testing. In L.L. Harlow, S.A. Mulaik, and J.H. Steiger (eds.), *What If There Were No Significance Tests?* Mahweh, NJ: Lawrence Erlbaum Associates.

Myers, J.L. (1979). *Fundamentals of Experimental Design,* 3rd ed. Boston: Allyn and Bacon.

Myers, J.L., and Well, A.D. (1991). *Research Design and Statistical Analysis.* New York: HarperCollins.

Myers, R.H., and Montgomery, D.C. (1995). *Response Surface Methodology: Process and Product Optimization Using Designed Experiments.* New York: John Wiley & Sons.

Overall, J.E., and Spiegel, D.K. (1969). Concerning least squares analysis of experimental data. *Psychological Bulletin 72*: 311–322.

Pearson, E.S., and Hartley, H.O. (1958). *Biometrika Tables for Statisticians.* New York: Cambridge University Press.

Rao, C.R., and Kleffe, J. (1988). *Estimation of Variance Components and Applications.* Amsterdam: North-Holland.

Rea, L.M., and Parker, R.A. (1997). *Designing and Conducting Survey Research: A Comprehensive Guide.* San Francisco: Jossey-Bass.

Rossi, J.S. (1990). Statistical power of psychological research: What have we gained in 20 years? *Journal of Consulting and Clinical Pscyhology 58*: 646–656.

Rousseeuw, P.J., and van Zomeren, B.C. (1990). Unmasking multivariate outliers and leverage points. *Journal of the American Statistical Association 85*: 633–639.

Sachs, H., Joseph, C.E., Farnoush, A., and Tabachnick, B. (1985, March). Evaluation of periodontal dressings in vitro. Presented to the International Association of Dental Research, Las Vegas.

Satterthwaite, F.E. (1946). An approximate distribution of estimates of variance components. *Biometrics Bulletin 2*: 110–114.

Scheffé, H.A. (1953). A method of judging all contrasts in the analysis of variance. *Biometrika 40*: 87–104.

Schmidt, S.R., and Launsby, R.G. (1994). *Understanding Industrial Designed Experiments,* 4th ed. Colorado Springs: Air Academy Press.

Searle, S.R., Casella, G., and McColloch, C.E. (1992). *Variance Components.* New York: John Wiley & Sons.

Sedlmeier, P., and Gigerenzer, G. (1989). Do studies of statistical power have an effect on the power of studies? *Psychological Bulletin 105*: 309–316.

Shavelson, R.J., and Webb, N.M. (1991). *Generalizability theory: A Primer.* Newbury Park, CA: Sage.

Siegel, S., and Castellan, N.J., Jr. (1988). *Nonparametric Statistics for the Behavioral Sciences,* 2nd ed. New York: McGraw-Hill.

Smithson, M.J. (2003). *Confidence intervals.* Belmont, CA: Sage.

Spatz, C. (1997). *Basic Statistics: Tales of Distribution,* 6th ed. Pacific Grove, CA: Brooks/Cole.

Statistical Solutions, Inc. (1997). *SOLAS for Missing Data Analysis 1.0.* Cork, Ireland: Statistical Solutions.

Steiger, J.H., and Fouladi, R.T. (1992). R2: A computer program for interval estimation, power calculation, and hypothesis testing for the squared multiple correlation. *Behavior Research Methods, Instruments, and Computers 4*: 581–582.

Steiger, J.H., and Fouladi, R.T. (1997). When confidence intervals should be used instead of statistical significance tests, and vice versa. In L.L. Harlow, S.A. Mulaik, and J.H. Steiger (eds.), *What If There Were No Significance Tests?* Mahweh, NJ: Lawrence Erlbaum Associates.

Tabachnick, B.G., and Fidell, L.S. (2006). *Using Multivariate Statistics,* 5th ed. Boston: Allyn & Bacon.

Taguchi, G. (1993). *Taguchi Methods: Design of Experiments.* Dearborn, MI: ASI Press.

Tufte, E.R. (1983). *The Visual Display of Quantitative Information.* Cheshire, CT: Graphics Press.

Tukey, J.W. (1949). One degree-of-freedom for nonadditivity. *Biometrics, 5*: 232–242.

Tukey, J.W. (1977). *Exploratory Data Analysis.* Reading, MA: Addison-Wesley.

Vaughan, G.M, and Corballis, M.C. (1969). Beyond tests of significance: Estimating strength of effects in selected ANOVA designs. *Psychological Bulletin, 72*: 204–213.

Wang, Y. (1971). Probabilities of the type I errors of the Welch tests for the Behrens-Fisher problem. *Journal of the American Statistical Association 66*: 605–608.

Welch, B.L. (1947). The generalization of student's problem when several different population variances are involved. *Biometrika 34*: 28–35.

Winer, B.J. (1971). *Statistical Principles in Experimental Design,* 2nd ed. New York: McGraw-Hill.

Woodward, J.A., Bonett, D.G., and Brecht, M.-L. (1990). *Introduction to Linear Models and Experimental Design.* San Diego: Harcourt Brace Jovanovich.

Zeleny, C. (1920). The tabulation of factorial values. *American Nature 54*: 358–362.

Index